Table of Atomic Numbers and Element Symbols

The elements are listed in alphabetical order. If you want a table of element symbols in the order that the elements appear on the Periodic Table of Elements, see page 93.

Element	Atomic Number	Symbol
Actinium	89	Ac
Aluminum	13	Al
Americium	95	Am
Antimony	51	Sb
Argon	18	Ar
Arsenic	33	As
Astatine	85	At
Barium	56	Ba
Berkelium	97	Bk
Beryllium	4	Be
Bismuth	83	Bi
Bohrium	107	Bh
Boron	5	B
Bromine	35	Br
Cadmium	48	Cd
Calcium	20	Ca
Californium	98	Cf
Carbon	6	C
Cerium	58	Ce
Cesium	55	Cs
Chlorine	17	Cl
Chromium	24	Cr
Cobalt	27	Co
Copper	29	Cu
Curium	96	Cm
Dubnium	105	Db
Dysprosium	66	Dy
Einsteinium	99	Es
Erbium	68	Er
Europium	63	Eu
Fermium	100	Fm
Fluorine	9	F
Francium	87	Fr
Gadolinium	64	Gd
Gallium	31	Ga
Germanium	32	Ge
Gold	79	Au
Hafnium	72	Hf
Hassium	108	Hs
Helium	2	He
Holmium	67	Ho
Hydrogen	1	H
Indium	49	In
Iodine	53	I
Iridium	77	Ir
Iron	26	Fe
Krypton	36	Kr
Lanthanum	57	La
Lawrencium	103	Lr
Lead	82	Pb
Lithium	3	Li
Lutetium	71	Lu
Magnesium	12	Mg
Manganese	25	Mn
Meitnerium	109	Mt
Mendelevium	101	Md
Mercury	80	Hg
Molybdenum	42	Mo
Neodymium	60	Nd
Neon	10	Ne
Neptunium	93	Np
Nickel	28	Ni
Niobium	41	Nb
Nitrogen	7	N
Nobelium	102	No
Osmium	76	Os
Oxygen	8	O
Palladium	46	Pd
Phosphorus	15	P
Platinum	78	Pt
Plutonium	94	Pu
Polonium	84	Po
Potassium	19	K
Praseodymium	59	Pr
Promethium	61	Pm
Protactinium	91	Pa
Radium	88	Ra
Radon	86	Rn
Rhenium	75	Re
Rhodium	45	Rh
Rubidium	37	Rb
Ruthenium	44	Ru
Rutherfordium	104	Rf
Samarium	62	Sm
Scandium	21	Sc
Seaborgium	106	Sg
Selenium	34	Se
Silicon	14	Si
Silver	47	Ag
Sodium	11	Na
Strontium	38	Sr
Sulfur	16	S
Tantalum	73	Ta
Technetium	43	Tc
Tellurium	52	Te
Terbium	65	Tb
Thallium	81	Tl
Thorium	90	Th
Thulium	69	Tm
Tin	50	Sn
Titanium	22	Ti
Tungsten	74	W
Ununbium	112	Uub
Ununnilium	110	Uun
Unununium	111	Uuu
Uranium	92	U
Vanadium	23	V
Xenon	54	Xe
Ytterbium	70	Yb
Yttrium	39	Y
Zinc	30	Zn
Zirconium	40	Zr

Exploring Creation

with

Chemistry
2nd Edition

by Dr. Jay L. Wile

This work is dedicated to my beautiful daughter, Dawn.

Even though you hate chemistry, I love you more than you'll ever know.

Exploring Creation With Chemistry, 2nd Edition

Published by
Apologia Educational Ministries, Inc.
1106 Meridian Plaza, Suite 220
Anderson, IN 46016
www.apologia.com

Manufactured in the United States of America
Seventh Printing 2009

ISBN: 978-1-932012-26-2

Printed by Courier, Inc., Kendallville, IN

 Cover design by Kim Williams

Any photo or illustration not credited in the text was done by the author.

Need Help?

Apologia Educational Ministries, Inc. Curriculum Support

If you have any questions while using Apologia curriculum,
feel free to contact us in any of the following ways:

By Mail: Dr. Jay L. Wile
Apologia Educational Ministries, Inc.
1106 Meridian Plaza, Suite 220
Anderson, IN 46016

By E-MAIL: help@apologia.com

On The Web: http://www.apologia.com

By FAX: (765) 608 - 3290

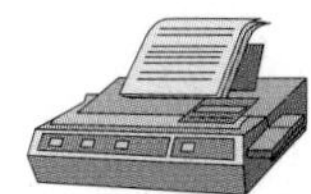

By Phone: (765) 608 - 3280

Illustrations from the MasterClips collection and the Microsoft Clip Art Gallery

STUDENT NOTES

Exploring Creation With Chemistry, 2nd Edition

You are about to embark upon an amazing journey! In this text, you will be introduced to the fascinating subject of chemistry. You will learn about the matter that makes up God's creation and how it changes. Although this course will be hard work, you will learn some truly incredible things. As a result, I hope that you develop an even deeper appreciation for the wonderful creation that God has given us!

Pedagogy of the Text

This text contains 16 modules. Each module should take you about 2 weeks to complete, as long as you devote 45 minutes to an hour of every school day to studying chemistry. At this pace, you will complete the course in 32 weeks. Since most people have school years which are longer than 32 weeks, there is some built-in "flex time." You should not rush through a module just to make sure that you complete it in 2 weeks. Set that as a goal, but be flexible. Some of the modules might come harder to you than others. On those modules, take more time on the subject matter.

To help you guide your study, there are several student exercises which you should complete:

- The "On Your Own" problems should be solved as you read the text. The act of working out these problems will cement in your mind the concepts you are trying to learn. Complete solutions to these problems appear at the end of the module. Once you have solved an "On Your Own" problem, turn to the end of the module and check your work. If you did not get the correct answer, study the solution to learn why.

- The review questions are conceptual in nature and should be answered after you have completed the entire module. They will help you recall the important concepts from the reading.

- The practice problems should also be solved after the module has been completed, allowing you to review the important quantitative skills from the module.

Your teacher/parent has the solutions to the review questions and practice problems.

Any information that you must memorize is centered in the text and put in boldface type. In addition, all definitions presented in the text need to be memorized. Words that appear in boldface type (centered or not) in the text are important terms that you should know. Finally, if any student exercise requires the use of a formula or skill, you must have that memorized for the test.

Learning Aids

Extra material is available to aid you in your studies. For example, Apologia Educational Ministries, Inc. has produced a multimedia companion CD that accompanies this course. It contains videos of experiments that you would probably not be able to perform yourself. These experiments demonstrate concepts that are discussed in the course. In addition, it contains animated white board solutions of many of the example problems in the course, along with an audio explanation that is different than the explanation given in this book. Thus, if you are having trouble understanding how I worked a certain example problem, you might find more explanation on the multimedia CD. The following graphic in the book:

indicates that there is a video or animation on the CD that further explains a concept or example problem.

Finally, the CD contains audio pronunciations of the technical words used in this book. Even though the book gives pronunciation guides for most of the technical words used, nothing beats actually hearing someone say the word! As you read through the book, you will see words that have pronunciation guides in parentheses. If you would like to hear one of those words pronounced for you, you will find it on the multimedia companion CD.

In addition to the multimedia companion CD, there is a special website for this course that you can visit. The website contains links to web-based materials related to the course. These links are arranged by module, so if you are having trouble with a particular subject in the course, you can go to the website and look at the links for that module. Most likely, you will find help there. In addition, there are answers to many of the frequently-asked questions regarding the material. For example, many people ask us for examples of how to properly record experiments in your laboratory notebook. Those examples can be found at the website. Finally, if you are enjoying a particular module in the course and would like to learn more about it, there are links which will lead you to advanced material related to that module.

To visit the website, go to the following address:

http://www.apologia.com/bookextras

When you get to the address, you will be asked for a password. Type the following into the password box:

Godisthegreatestchemist

Be sure that you do not put spaces between any of the letters and that the first letter is capitalized. When you click on the button labeled "Submit," you will be sent to the course website. You must use Internet Explorer 5.1 or higher to view this website.

There are also several items at the end of the book that you will find useful in your studies. There is a glossary that defines many of the terms used in the course and an index that will tell you where topics can be found in the course. In addition, there are three appendices in the course. Appendix A lists tables and formulas that are found throughout the reading. Appendix B contains extra problems for each module of the course. If you are having difficulty with a certain type of problem, you can get more practice by solving the related problems in Appendix B. Your parent/teacher has the worked-out solutions for those problems. Appendix C contains a complete list of all of the supplies you need to perform the experiments in this course.

Experiments

The experiments in this course are designed to be done as you are reading the text. I recommend that you keep a notebook of these experiments. This notebook serves two purposes. First, as you write about the experiment in the notebook, you will be forced to think through all of the concepts that were explored in the experiment. This will help you cement them into your mind. Second, certain colleges might actually ask for some evidence that you did, indeed, have a laboratory component to your chemistry course. The notebook will not only provide such evidence but will also show the college administrator the quality of your chemistry instruction. I recommend that you perform the experiments in the following way:

- When you get to an experiment, read through it in its entirety. This will allow you to gain a quick understanding of what you are to do.

- Once you have read the experiment, start a new page in your laboratory notebook. The first page should be used to write down all of the data taken during the experiment. What do I mean by "data"? Any observations or measurements you make during the experiment are considered data. Thus, if you see a solution begin to bubble when two chemicals are mixed, that is part of the experiment's data and should be written down. If you measure the mass of something during the experiment, that is part of the experiment's data and should be written down. In addition, any calculation that you are asked to do as a part of the experiment should be done on this page.

- When you have finished the experiment and any necessary calculations, write a brief report in your notebook, right after the page where the data and calculations were written. The report should be a brief discussion of what was done and what was learned. You should not write a step-by-step procedure. Instead, write a brief summary that will allow someone who has never read the text to understand what you did and what you learned.

PLEASE OBSERVE COMMON SENSE SAFETY PRECAUTIONS! The experiments in this course are no more dangerous than most normal, household activity. Remember, however, that the vast majority of accidents do happen in the home. Chemicals used in the experiments should never be ingested; hot beakers and flames should be regarded with care; and all chemistry experiments should be performed while wearing eye protection such as safety glasses or goggles.

Although the chemicals used in the experiments are common substances that you probably have around the home, there are specialized materials (such as glassware) that make the experiments much easier to perform. You can purchase a laboratory kit that contains these materials by contacting Apologia Educational Ministries, Inc. (888-524-4724, www.apologia.com) or Nature's Workshop Plus (888-393-5663, www.natureswokshopplus.com). The current cost of the kit is $55.00 (post paid), but that price is subject to change. A list of its contents is given on the next page:

Laboratory Equipment Set Available for *Exploring Creation With Chemistry, 2nd Edition*

Item	*# of Experiments***
Safety goggles*	30
Mass scale (0-500 grams)*	8
Blue and red litmus paper*	1
Thermometer (-10 - 110 °C)*	6
6 test tubes	6
Test tube cleaning brush	4
Funnel	2
Alcohol burner with stand	6
250 mL beaker	11
100 mL beaker	11
Watch glass	2
2 eyedroppers	4
Stirring rod	7
50 mL glass graduated cylinder	8
50 mL plastic graduated cylinder	1
Round pieces of filter paper	2

* The first four items on this list are necessary for completing all of the experiments in the course. If you already have items that meet these specifications, or if you do not wish to do *every* experiment in the course, you need not order anything. The remaining items make it *easier* for you to do the experiments, but there are common household substitutes you can use for each of them. These substitutes are detailed in the experiment itself.

**The "# of Exps" column tells you how many experiments require the item or its household substitute.

If you are interested in performing more than the 31 experiments contained in this course, there is a second kit that you can purchase. You can also purchase this kit through Nature's Workshop Plus. You must buy both the kit and the laboratory manual that accompanies it. This kit contains a total of 17 experiments, 13 of which are referenced in this course. The following graphic will alert you to an opportunity to perform a MicroChem experiment:

Next to the graphic, you will find instructions indicating which experiment from the MicroChem kit you can do at that time. Please note that these experiments are *optional.* You will have a solid high school experience in chemistry without doing these experiments. However, they do enhance this course if you choose to do them. If you look up "MicroChem experiments" in the index, you will find the page numbers that contain references to these experiments.

Exploring Creation With Chemistry

Table of Contents

MODULE #7: Atomic Structure **201**

MODULE #8: Molecular Structure **247**

MODULE #1: Measurement and Units

Introduction

What is chemistry? That's a very good question. Chemistry is, quite simply, the study of **matter**. Of course, this definition doesn't do us much good unless we know what matter is. So, in order to understand what chemistry is, we first need to define matter. A good working definition for matter is:

Matter - Anything that has mass and takes up space

If you have a problem with the word "mass," don't worry about it. We will discuss this concept in a little while. For right now, you can replace the word "mass" with the word "weight." As we will see later, this isn't quite right, but it will be okay for now.

If matter is defined in this way, almost *everything* around us is matter. Your family car has a lot of mass. That's why it's so heavy. It also takes up a lot of space sitting in the driveway or in the garage. Thus, your car must be made of matter. The food you eat isn't as heavy as a car, but it still has some mass. It also takes up space. So food must be made up of matter as well. Indeed, almost everything you see around you is made up of matter because nearly everything has mass and takes up space. There is one thing, however, that has no mass and takes up no space. It's all around you right now. Can you think of what it might be? What very common thing that is surrounding you right now has no mass and takes up no space?

You might think that the answer is "air." Unfortunately, that's not the right answer. Perform the following experiments to see what I mean:

EXPERIMENT 1.1

Air Has Mass

Supplies:

- A meterstick (A yardstick will work as well; a 12-inch ruler is not long enough.)
- Two 8-inch or larger balloons
- Two pieces of string long enough to tie the balloons to the meterstick
- Tape
- Safety goggles

1. Without blowing them up, tie the balloons to the strings. Be sure to make the knots loose so that you can untie one of the balloons later in the experiment.
2. Tie the other end of each string to each end of the meterstick. Try to attach the strings as close to the ends of the meterstick as possible.
3. Once the strings have been tied to the meterstick, tape them down so that they cannot move.
4. Go into your bathroom and pull back the shower curtain so that a large portion of the curtain rod is bare. Balance the meterstick (with the balloons attached) on the bare part of the shower curtain rod. You should be able to balance it very well. If you don't have a shower curtain rod or you are having trouble using yours, you can use any surface that is adequate for delicate balancing.

5. Once you have the meterstick balanced, stand back and look at it. The meterstick balances right now because the total mass on one side of the meterstick equals the total mass on the other side of the meterstick. In order to knock it off balance, you would need to move the meterstick or add more mass to one side. You will do the latter.
6. Have someone else hold the meterstick so that it does not move. In order for this experiment to work properly, the meterstick must stay stationary.
7. While the meterstick is held stationary, remove one of the balloons from its string (do not untie the string from the meterstick), and blow up the balloon.
8. Tie the balloon closed so that the air does not escape, then reattach it to its string.
9. Have the person holding the meterstick let go. If the meterstick was not moved while you were blowing up the balloon, it will tilt toward the side with the inflated balloon as soon as the person lets it go. This is because you added air to the balloon. Since air has mass, it knocks the meterstick off balance. Thus, air does have mass!
10. Clean up your mess.

EXPERIMENT 1.2

Air Takes Up Space

Supplies:

- A tall glass
- A paper towel
- A sink full of water
- Safety goggles

1. Fill your sink with water until the water level is high enough to submerge the entire glass.
2. Make sure the inside of the glass is dry.
3. Wad up the paper towel and shove it down into the bottom of the glass.
4. Turn the glass upside down and be sure that the paper towel does not fall out of the glass.
5. Submerge the glass upside down in the water, being careful not to tip the glass at any time.
6. Wait a few seconds and remove the glass, still being careful not to tilt it.
7. Pull the paper towel out of the glass. You will find that the paper towel is completely dry. Even though the glass was submerged in water, the paper towel never got wet. Why? When you tipped the glass upside down, there was air inside the glass. When you submerged it in the water, the air could not escape the glass. Thus, the glass was still full of air. Since air takes up space, there was no room for water to enter the glass, so the paper towel stayed dry.
8. Repeat the experiment, but this time be sure to tip the glass while it is underwater. You will see large bubbles rise to the surface of the water, and when you pull the glass out, you will find that it has water in it and that the paper towel is wet. This is because you allowed the air trapped inside the glass to escape when you tilted the glass. Once the air escaped, there was room for the water to come into the glass.
9. Clean up your mess.

Now that you see that air does have mass and does take up space, have you figured out the correct answer to my original question? What very common thing that is surrounding you right now has no mass and takes up no space? The answer is *light.* As far as scientists can tell, light does not have any mass and takes up no space. Thus, light is not considered matter. Instead, it is pure energy. Everything else that you see around you, however, is considered matter. Chemistry, then, is the study of nearly everything! As you can imagine, studying nearly everything can be a very daunting task. However, chemists have found that even though there are many forms of matter, they all behave according to a few fundamental laws. If we can clearly understand these laws, then we can clearly understand the nature of the matter that exists in God's creation.

Before we start trying to understand these laws, however, we must first step back and ask a more fundamental question. *How* do we study matter? Well, the first thing we have to be able to do in order to study matter is to measure it. If I want to study an object, I first must learn things like how big it is, how heavy it is, and how old it is. In order to learn these things, I have to make some measurements. The rest of this module explains how scientists measure things and what those measurements mean.

Units of Measurement

Let's suppose I'm making curtains for a friend's windows. I ask him to measure the window and give me the dimensions so that I can make the curtains the right size. My friend tells me that his windows are 50 by 60, so that's how big I make the curtains. When I go over to his house, it turns out that my curtains are more than twice as big as his windows! My friend tells me that he's certain he measured the windows correctly, and I tell my friend that I'm certain I measured the curtains correctly. How can this be? The answer is quite simple. My friend measured the windows in *centimeters*. I, on the other hand, measured the curtains in *inches*. Our problem was not caused by one of us measuring incorrectly. Instead, our problem was the result of measuring with different **units.**

When we are making measurements, the units we use are just as important as the numbers that we get. If my friend had told me that his windows were 50 centimeters by 60 centimeters, then there would have been no problem. I would have known exactly how big to make the curtains. Since he failed to do this, the numbers that he gave me (50 by 60) were essentially useless. Please note that a failure to indicate the units involved in measurements can lead to serious problems. For example, the Mars Climate Orbiter, a NASA (National Aeronautics and Space Administration) spacecraft built for the exploration of Mars, vanished during an attempt to put the craft into orbit around the planet. In an investigation that followed, NASA determined that a units mix-up had caused the disaster. One team of engineers had used metric units in its designs, while another team had used English units. The teams did not indicate the units they were using, and as a result, the designs were incompatible.

In the end, then, scientists should never simply report numbers. They must always include units with those numbers so that everyone knows exactly what those numbers mean. That will be the rule in this chemistry course. If you answer a question or a problem and do not list units with the numbers, your answer will be considered incorrect. In science, numbers mean nothing unless there are units attached to them.

Window illustration by Megan Whitaker

FIGURE 1.1
Two Consequences of Not Using Units Properly

Mars Climate Orbiter image Courtesy of NASA/JPL/Caltech

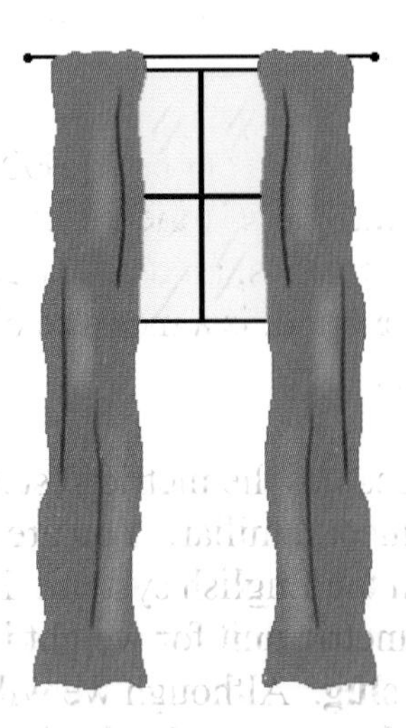

These curtains are too long for this window because the window was measured in centimeters, but the curtains were made assuming the measurements were in inches.

The Mars Climate Orbiter did not successfully make it into orbit because two of the engineering teams involved used different units in their designs.

Since scientists use units in all of their measurements, it is convenient to define a standard set of units that will be used by everyone. This system of standard units is called the **metric system**. If you do not fully understand the metric system, don't worry. By the end of this module, you will be an expert at using it. If you do fully understand the metric system, you can probably skip ahead to the section labeled "Converting Between Units."

The Metric System

There are many different things that we need to measure when studying nature. First, we must determine how much matter exists in the object that we want to study. We know that there is a lot more matter in a car than there is in a feather, since a car is heavier. In order to study an object precisely, however, we need to know *exactly* how much matter is in the object. To accomplish this, we measure the object's **mass.** In the metric system, the unit for mass is the **gram**. If an object has a mass of 10 grams, we know that it has 10 times the matter that is in an object with a mass of 1 gram. To give you an idea of the size of a gram, the average mass of a housefly is just about 1 gram. Based on this fact, we can say that a gram is a rather small unit. Most of the things that we will measure will have masses of 10 to 10,000 grams. For example, this book has a mass of about 2,300 grams.

Now that we know what the metric unit for mass is, we need to know a little bit more about the concept itself. I said in the beginning that we could think of mass as weight. That's not exactly true. Mass and weight are two different things. Mass measures how much matter exists in an object. Weight, on the other hand, measures how hard gravity pulls on that object.

For example, if I were to get on my bathroom scale and weigh myself, I would find that I weigh 170 pounds. However, if I were to take that scale to the moon and weigh myself, I would find that I weighed only 28 pounds there. Does that mean I'm thinner on the moon than I am at home? Of course not. It means that on the moon, gravity is not as strong as it is in my house.

On the other hand, if I were to measure my mass at home, I would find it to be 77,000 grams. If I were to measure my mass on the moon, it would still be 77,000 grams. That's the difference between mass and weight. Since weight is a measure of how hard gravity pulls, an object weighs different amounts depending on where that object is. Mass, on the other hand, is a measure of how much matter is in an object and does not depend on where that object is.

Unfortunately, there are many other unit systems in use today besides the metric system. In fact, the metric system is probably not the system with which you are most familiar. You are probably most familiar with the English system. The unit of pounds comes from the English system. However, pounds are not a measure of mass; they are a measure of weight. The metric unit for weight is called the **Newton**. The English unit for mass is (believe it or not) called the **slug**. Although we will not use the slug often, it is important to understand what it means, especially when you study physics.

There is more to measurement than just grams, however. We might also want to measure how big an object is. For this, we must use the metric system's unit for distance, which is the **meter**. You are probably familiar with a yardstick. Well, a meter is just slightly longer than a yardstick. The English unit for distance is the **foot**. What about inches, yards, and miles? We'll talk about those a little later.

We also need to be able to measure how much space an object occupies. This measurement is commonly called "volume" and is measured in the metric system with the unit called the **liter**. The main unit for measuring volume in the English system is the **gallon**. To give you an idea of the size of a liter, it takes just under four liters to make a gallon.

Finally, we have to be able to measure the passage of time. When studying matter, we will see that it has the ability to change. The shape, size, and chemical properties of certain substances change over time, so it is important to be able to measure time so that we can determine how quickly the changes take place. In both the English and metric systems, time is measured in **seconds.**

Since it is very important for you to be able to recognize which units correspond to which measurements, Table 1.1 summarizes what you have just read. The letters in parentheses are the commonly used abbreviations for the units listed.

TABLE 1.1
Physical Quantities and Their Base Units

Physical Quantity	Base Metric Unit	Base English Unit
Mass	gram (g)	slug (sl)
Distance	meter (m)	foot (ft)
Volume	liter (L)	gallon (gal)
Time	second (s)	second (s)

Manipulating Units

Now, let's suppose I asked you to measure the width of your home's kitchen using the English system. What unit would you use? Most likely, you would express your measurement in feet. However, suppose instead I asked you to measure the length of a sewing needle. Would you still use the foot as your measurement unit? Probably not. Since you know the English system already, you would probably recognize that inches are also a unit for distance and, since a sewing needle is relatively small, you would use inches instead of feet. In the same way, if you were asked to measure the distance between two cities, you would probably express your measurement in terms of miles, not feet. This is why I used the term "Base English Unit" in Table 1.1. Even though the English system's normal unit for distance is the foot, there are alternative units for length if you are trying to measure very short or very long distances. The same holds true for all English units. Volume, for example, can be measured in cups, pints, and ounces.

This concept exists in the metric system as well. There are alternative units for measuring small things as well as alternative units for measuring big things. These alternative units are called "prefix units" and, as you will soon see, prefix units are much easier to use and understand than the alternative English units! The reason that prefix units are easy to use and understand is that they always have the same relationship to the base unit, regardless of what physical quantity you are interested in measuring. You will see how this works in a minute.

In order to use a prefix unit in the metric system, you simply add a prefix to the base unit. For example, in the metric system, the prefix "centi" means one hundredth, or 0.01. So, if I wanted to measure the length of a sewing needle in the metric system, I would probably express my measurement with the centimeter unit. Since a centimeter is one hundredth of a meter, it can be used to measure relatively small things. On the other hand, the prefix "kilo" means 1,000. So, if I want to measure the distance between two states, I would probably use the kilometer. Since each kilometer is 1,000 times longer than the meter, it can be used to measure long things.

Now, the beauty of the metric system is that these prefixes mean the same thing *regardless of the physical quantity that you want to measure!* So, if I were measuring something with a very large mass (such as a car), I would probably use the kilogram unit. One kilogram is the same as 1,000 grams. In the same way, if I were measuring something that had a large volume, I might use the kiloliter, which would be 1,000 liters.

Compare this incredibly logical system of units to the chaotic English system. If you want to measure something short, you use the inch unit, which is equal to one twelfth of a foot. On the other hand, if you want to measure something with small volume, you might use the quart unit, which is equal to one fourth of a gallon. In the English system, every alternative unit has a different relationship to the base unit, and you must remember all of those crazy numbers. You have to remember that there are 12 inches in a foot, 3 feet in a yard, and 5,280 feet in a mile, while at the same time remembering that for volume there are 8 ounces in a cup, 2 cups in a pint, 2 pints in a quart, and 4 quarts in a gallon.

In the metric system, all you have to remember is what the prefix means. Since the "centi" prefix means one hundredth, then you know that 1 centimeter is one hundredth of a meter, 1 centiliter is one hundredth of a liter, and 1 centigram is one hundredth a gram. Since the "kilo" prefix means

1,000, you know that there are 1,000 meters in a kilometer, 1,000 grams in a kilogram, and 1,000 liters in a kiloliter. Doesn't that make a lot more sense?

Another advantage to the metric system is that there are many, many more prefix units than there are alternative units in the English system. Table 1.2 summarizes the most commonly used prefixes and their numerical meanings. The prefixes in boldface type are the ones that we will use over and over again. You will be expected to have those three prefixes and their meanings memorized before you take the test for this module. Once again, the commonly used abbreviations for these prefixes are listed in parentheses.

TABLE 1.2
Common Prefixes Used in the Metric System

PREFIX	NUMERICAL MEANING
micro (μ)	0.000001
milli (m)	**0.001**
centi (c)	**0.01**
deci (d)	0.1
deca (D)	10
hecta (H)	100
kilo (k)	**1,000**
Mega (M)	1,000,000

Remember that each of these prefixes, when added to a base unit, makes an alternative unit for measurement. So, if you wanted to measure the length of something small, the only unit you could use in the English system would be the inch. However, if you used the metric system, you would have all sorts of options for which unit to use. If you wanted to measure the length of someone's foot, you could use the decimeter. Since the decimeter is one tenth of a meter, it measures things that are only slightly smaller than a meter. On the other hand, if you wanted to measure the length of a sewing needle, you could use the centimeter, because a sewing needle is significantly smaller than a meter. Or, if you want to measure the thickness of a piece of paper, you might use the millimeter, since it is one thousandth of a meter, which is a *really* small unit.

So you see that the metric system is more logical and versatile than the English system. That is, in part, why scientists use it as their main system of units. The other reason that scientists use the metric system is that most countries in the world use it. With the exception of the United States, almost every other country in the world uses the metric system as its standard system of units. Since scientists in the United States frequently work with scientists from other countries around the world, it is necessary that American scientists use and understand the metric system. Throughout all of the modules of this chemistry course, the English system of measurement will only be presented for illustration purposes. Since scientists must thoroughly understand the metric system, it will be the main system of units that we will use.

Converting Between Units

Now that we understand what prefix units are and how they are used in the metric system, we must become familiar with converting between units within the metric system. In other words, if you measure the length of an object in centimeters, you should also be able to convert your answer to any other distance unit. For example, if I measure the length of a sewing needle in centimeters, I should be able to convert that length to millimeters, decimeters, meters, etc. Accomplishing this task is relatively simple as long as we remember a trick we can use when multiplying fractions. Suppose I asked you to complete the following problem:

$$\frac{7}{64} \times \frac{64}{13} =$$

There are two ways to figure out the answer. The first way would be to multiply the numerators and the denominators together and, once you had accomplished that, simplify the fraction. If you did it that way, it would look something like this:

$$\frac{7}{64} \times \frac{64}{13} = \frac{448}{832} = \frac{7}{13}$$

You could get the answer much more quickly, however, if you remember that when multiplying fractions, common factors in the numerator and the denominator cancel each other out. Thus, the 64 in the numerator cancels with the 64 in the denominator, and the only factors left are the 7 in the numerator and the 13 in the denominator. In this way, you reach the final answer in one less step:

$$\frac{7}{\cancel{64}} \times \frac{\cancel{64}}{13} = \frac{7}{13}$$

We will use the same idea in converting between units. Suppose I measure the length of a pencil to be 15.1 centimeters, but suppose the person who wants to know the length of the pencil would like me to tell him the answer in meters. How would I convert between centimeters and meters? First, I would need to know the relationship between centimeters and meters. According to Table 1.2, "centi" means 0.01. So, 1 centimeter is the same thing as 0.01 meters. In mathematical form, we would say:

$$1 \text{ centimeter} = 0.01 \text{ meter}$$

Now that we know how centimeters and meters relate to one another, we can convert from one to another. First, we write down the measurement that we know:

$$15.1 \text{ centimeters}$$

We then realize that any number can be expressed as a fraction by putting it over the number one. So we can rewrite our measurement as:

$$\frac{15.1 \text{ centimeters}}{1}$$

Now we can take that measurement and convert it into meters by multiplying it with the relationship we determined above. We have to do it the right way, however, so that the units work out properly. Here's how we do it:

$$\frac{15.1 \cancel{\text{centimeters}}}{1} \times \frac{0.01 \text{ meters}}{1 \cancel{\text{centimeters}}} = 0.151 \text{ meters}$$

So, 15.1 centimeters is the same as 0.151 meters. There are two reasons this conversion method, called the **factor-label method**, works. First, since 0.01 meters is the same as 1 centimeter, multiplying our measurement by 0.01 meters over 1 centimeter is the same as multiplying by one. Since nothing changes when we multiply by one, we haven't altered the value of our measurement at all. Second, by putting the 1 centimeters in the denominator of the second fraction, we allow the centimeters unit to cancel (just like the 64 canceled in the previous discussion). Once the centimeters unit has canceled, the only thing left is meters, so we know that our measurement is now in meters.

This is how we will do all of our unit conversions. We will first write the measurement we know in fraction form by putting it over one. We will then find the relationship between the unit we have and the unit to which we want to convert. We will then use that relationship to make a fraction that, when multiplied by our first fraction, cancels out the unit we have and replaces it with the unit we want to have. We will see many examples of this method, so don't worry if you are a little confused right now.

It may seem odd to you that words can be treated exactly the same as numbers. Measuring units, however, have just that property. Whenever a measurement is used in any mathematical equation, the units for that measurement must be included in the equation. Those units are then treated the same way numbers are treated. I will come back to this point in an upcoming section of this module.

We will be using the factor-label method for many other types of problems throughout this course, so it is very, very important for you to learn it. Also, since we will be using it so often, we should start abbreviating things so that they will be easier to write down. We will use the abbreviations for the base units that have been listed in Table 1.1 along with the prefix abbreviations listed in Table 1.2. Thus, kilograms will be abbreviated as "kg," while milliliters will be abbreviated as "mL."

Since the factor-label method is so important in our studies of chemistry, let's see how it works in another example:

EXAMPLE 1.1

A student measures the mass of a rock to be 14,351 grams. What is the rock's mass in kilograms?

First, we use the definition of "kilo" to determine the relationship between grams and kilograms:

$$1 \text{ kg} = 1{,}000 \text{ g}$$

Then we put our measurement in fraction form:

$$\frac{14{,}351 \text{ g}}{1}$$

Then we multiply our measurement by a fraction that contains the relationship noted above, making sure to put the 1,000 g in the denominator so that the unit of grams will cancel out:

$$\frac{14{,}351\,\cancel{\text{g}}}{1} \times \frac{1\text{ kg}}{1{,}000\,\cancel{\text{g}}} = 14.351\text{ kg}$$

So 14,351 grams is the same as 14.351 kilograms.

Because we will use it over and over again, you must master this powerful technique. Also, you will see towards the end of this module that the factor-label method can become extremely complex; therefore, it is very important that you take the time now to perform the following "On Your Own" problems. Once you have solved the problems on your own, check your answers using the solutions provided at the end of the module.

ON YOUR OWN

1.1 A student measures the mass of a book as 12,321 g. What is the book's mass in kg?

1.2 If a glass contains 0.121 L of milk, what is the volume of milk in mL?

1.3 On a professional basketball court, the distance from the three-point line to the basket is 640.08 cm. What is this distance in meters?

Converting Between Unit Systems

As you may have guessed, the factor-label method can also be used to convert *between systems* of units as well as within systems of units. Thus, if a measurement is done in the English system, the factor-label method can be used to convert that measurement to the metric system, or vice versa. In order to be able to do this, however, we must learn the relationships between metric and English units. Although these relationships are important, we will not use them very often, so you need not memorize them.

TABLE 1.3
Relationships Between English and Metric Units

Measurement	English/Metric Relationship
Distance	1 inch = 2.54 cm
Mass	1 slug = 14.59 kg
Volume	1 gallon = 3.78 L

We can use this information in the factor-label method the same way we used the information in Table 1.2. Study the following example to see what I mean.

EXAMPLE 1.2

The length of a tabletop is measured to be 37.8 inches. How many cm is that?

To solve this problem, we first put the measurement in its fraction form:

$$\frac{37.8 \text{ in}}{1}$$

We then multiply this fraction by the conversion relationship so that the inches unit cancels:

$$\frac{37.8 \cancel{\text{in}}}{1} \times \frac{2.54 \text{ cm}}{1 \cancel{\text{in}}} = 96.012 \text{ cm}$$

So a measurement of 37.8 inches is equivalent to <u>96.012 cm</u>.

Give yourself a little more practice with the factor-label method by answering the following problems:

ON YOUR OWN

1.4 How many slugs are there in 123.5 kg?

1.5 If an object occupies 3.2 gallons of space, how many liters of space does it occupy?

<u>More Complex Unit Conversions</u>

Now that we have seen some simple applications of the factor-label method, let's look at more complex problems. For example, suppose I measure the volume of a liquid to be 4,523 centiliters but would like to convert this measurement into kiloliters. This is a more complicated problem because we do not have a direct relationship between cL and kL. In all of the previous examples, we knew a relationship between the unit we had and the unit to which we wanted to convert. In this problem, however, no such relationship exists.

We do, however, have an *indirect* relationship between the two units. Although we don't know how many cL are in a kL, we do know how many cL are in a L and how many L are in a kL. We can use these two relationships in a two-step conversion. First, we can convert centiliters into liters:

$$\frac{4{,}523 \cancel{\text{cL}}}{1} \times \frac{0.01 \text{ L}}{1 \cancel{\text{cL}}} = 45.23\text{L}$$

Then we can convert liters into kiloliters:

$$\frac{45.23\ \cancel{\text{L}}}{1} \times \frac{1\ \text{kL}}{1{,}000\ \cancel{\text{L}}} = 0.04523\ \text{kL}$$

We are forced to do this two-step process because we do not have a direct relationship between two prefix units in the metric system. However, we can always convert between two prefix units if we first convert to the base unit. In order to speed up this kind of conversion, we can take these two steps and combine them into one line:

$$\frac{4{,}523\ \cancel{\text{cL}}}{1} \times \frac{0.01\ \cancel{\text{L}}}{1\ \cancel{\text{cL}}} \times \frac{1\ \text{kL}}{1{,}000\ \cancel{\text{L}}} = 0.04523\ \text{kL}$$

You will be seeing mathematical equations like this one as we move through the subject of chemistry, so it is important for you to be able to understand what's going on in it. The first fraction in the equation above represents the measurement that we were given. Since we have no relationship between the unit we were given and the unit to which we will convert, we first convert the given unit to the base unit. This is accomplished with the second fraction in the equation. When the first fraction is multiplied by the second fraction, the "cL" unit cancels and is replaced by the "L" unit. The third fraction then cancels the "L" unit and replaces it with the "kL" unit, which is the unit we wanted. This, then, gives us our final answer. Try to follow this reasoning in Example 1.3.

EXAMPLE 1.3

The mass of an object is measured to be 0.030 kg. What is the object's mass in mg?

We have no direct relationship between milligrams and kilograms, but we do have an indirect relationship between the two. First, we know that

$$1\ \text{kg} = 1{,}000\ \text{g}$$

We also know that

$$1\ \text{mg} = 0.001\ \text{g}$$

So, we need to take our original measurement in fraction form and multiply it with both of these relationships. We must do it in such a way as to cancel out the kg and replace it with g, and then cancel out the g and replace it with mg:

$$\frac{0.030\ \cancel{\text{kg}}}{1} \times \frac{1{,}000\ \cancel{\text{g}}}{1\ \cancel{\text{kg}}} \times \frac{1\ \text{mg}}{0.001\ \cancel{\text{g}}} = 30{,}000\ \text{mg}$$

The object's mass is 0.030 kg, which is the same as <u>30,000 mg</u>.

Once again, the factor-label method is one of the most important tools you can learn for the study of chemistry (and physics, for that matter). Thus, you must become a veritable expert at it. Try your hand at a few more "On Your Own" problems so that you can get some more practice.

ON YOUR OWN

1.6 A balloon is blown up so that its volume is 1,500 mL. What is its volume in kL?

1.7 If the length of a race car track is 2.0 km, how many cm is that?

1.8 How many Mg are there in 10,000,000 mg?

Derived Units

I mentioned previously that units can be used in mathematical expressions in the same way that numbers can be used. Just as there are rules for adding, subtracting, multiplying, and dividing numbers, there are also rules governing those operations when using units. You will have to become very adept at using units in mathematical expressions, so I want to discuss those rules now.

Adding and Subtracting Units: When adding and subtracting units, the most important thing to remember is that the units you are adding or subtracting must be identical: You cannot add grams and liters. The result would not make sense physically. Since gram is a mass unit and liter is a volume unit, there is no way you can add or subtract the two. You also cannot add or subtract kilograms and grams. Even though both units measure mass, you are not allowed to add or subtract them unless they are *identical.* Thus, if I did want to add or subtract them, I would have to convert the kilograms into grams. Alternatively, I could convert the grams into kilograms. It doesn't matter which way I go, as long as the units I add or subtract are identical.

Once I have identical units, I can add and subtract them using the rules of algebra. Since 2x + 3x = 5x, we know that 2 cm + 3 cm = 5 cm. In the same way, 3.1 g - 2.7 g = 0.4 g. So, when adding or subtracting units, you add or subtract the numbers they are associated with and then simply carry the unit along in the answer.

Multiplying and Dividing Units: When multiplying and dividing units, it doesn't matter whether or not the units are identical. Unlike addition and subtraction, you are allowed to multiply or divide any unit by any other unit. In algebra,

$$3x \cdot 4y = 12xy$$

When multiplying units,

$$3\,\text{kg} \cdot 4\,\text{mL} = 12\ \text{kg} \cdot \text{mL}$$

Similarly, in algebra,

$$6x \div 2y = 3\frac{x}{y}$$

When dividing units,

$$6\,\text{g} \div 2\,\text{mL} = 3\ \frac{\text{g}}{\text{mL}}$$

So, when multiplying or dividing units, you multiply or divide the numbers and then do exactly the same thing to the units.

Let's use the rules we've just learned to explore a few other things about units. First, let's see what happens when we multiply measurements that have the same units. Suppose I wanted to measure the surface area of a rectangular table. From geometry, we know that the area of a rectangle is the length times the width. So, let's suppose we measure the length of a table to be 1.1 meters and the width to be 2.0 meters. Its area would be

$$1.1 \cdot 2.0 = 2.2$$

What would the units be? In algebra, we would say that

$$1.1\text{x} \cdot 2.0\text{x} = 2.2\text{x}^2$$

Thus,

$$1.1\text{ m} \cdot 2.0\text{ m} = 2.2\text{ m}^2$$

This tells us that m^2 is a unit for area.

Let's take this one step further and suppose we measure the length, width, and height of a small box to be 1.2 cm, 3.1 cm, and 1.4 cm, respectively. What would the volume of the box be? From geometry, we know that volume is length times width times height, so the volume would be:

$$1.2\text{ cm} \cdot 3.1\text{ cm} \cdot 1.4\text{ cm} = 5.208\text{ cm}^3$$

Thus, cm^3 (usually called "cubic centimeters," or "cc's") is a unit for volume. If you've ever listened to doctors or nurses talking about how much liquid to put in a syringe when administering a shot, they usually use "cc's" as the unit. When a doctor tells a nurse, "Give the patient 4 cc's of penicillin," he or she is telling the nurse to inject a 4 cm^3 volume of penicillin into the patient.

Wait a minute. Wasn't the metric unit for volume the liter? Well, yes, but another metric unit for volume is the cm^3. Additionally, m^3 (cubic meters) and km^3 (cubic kilometers) are also possible units for volume. This is a very important point. Often, several different units exist for the same measurement. The units you use will depend, to a large extent, on what information you are given in the first place. We'll see more about this fact later.

Units like cm^3 are called **derived units** because they are derived from the basic units that make up the metric system. It turns out that many of the units you will use in chemistry are derived units. I'll talk about one very important physical quantity with derived units in an upcoming section of this module, but first I want to make sure that you understand exactly how to use derived units in mathematical equations.

Let's suppose I would like to take the volume that we previously determined for the box and convert it from cubic centimeters to cubic meters. You might think the conversion would look something like this:

$$\frac{5.208\ \text{cm}^3}{1} \times \frac{0.01\ \text{m}}{1\ \text{cm}}$$

Unfortunately, even though this conversion might look correct, there is a major problem with it. Remember what the factor-label method is designed to accomplish. In the end, the old units are supposed to cancel out, leaving the new units in their place. The way this conversion is set up, however, the old units *do not cancel!* When I take these two fractions and multiply them together, the cm in the denominator does not cancel out the cm^3 in the numerator. When multiplying fractions, the numerator and denominator must be *identical* in order for them both to cancel. Thus, the cm in the denominator above must be replaced with a cm^3.

How is this done? Actually it's quite simple. Just take the second fraction and raise it to the third power:

$$\frac{5.208\ \text{cm}^3}{1} \times \left(\frac{0.01\ \text{m}}{1\ \text{cm}}\right)^3$$

That way, the m becomes m^3, the cm becomes cm^3, the 0.01 becomes 0.000001, and the 1 stays as 1:

$$\frac{5.208\ \cancel{\text{cm}^3}}{1} \times \frac{0.000001\ \text{m}^3}{1\ \cancel{\text{cm}^3}} = 0.000005208\ \text{m}^3$$

Now, since both the numerator and denominator have a unit of cm^3, that unit cancels and is replaced with the m^3. So, a volume of 5.208 cm^3 is equivalent to a volume of 0.000005208 m^3.

Since cubic meters, cubic centimeters and the like are measurements of volume, you might have already guessed that there must be a relationship between these units and the other volume units we discussed earlier. In fact, 1 cm^3 is the same thing as 1 mL. This is a very important relationship, and it is something you will have to know before you can finish this module. So, commit it to memory now:

1 cubic centimeter is the same as 1 milliliter.

Let's combine this fact with the mathematics we just learned and perform a very complicated unit conversion. If you can understand this next example and successfully complete the "On Your Own" problems that follow it, then you have mastered the art of unit conversion. If things are still a bit shaky for you, don't worry. There are plenty of practice problems at the end of this module to help you shore up your unit conversion skills!

EXAMPLE 1.4

The length, width, and height of a small box are measured to be 1.1 in, 3.2 in, and 4.6 in, respectively. What is the box's volume in liters?

In order to solve this problem, we first use the geometric equation for the volume of a box:

$$V = l \cdot w \cdot h$$

Inserting our numbers:

$$V = 1.1 \text{ in} \cdot 3.2 \text{ in} \cdot 4.6 \text{ in} = 16.192 \text{ in}^3$$

Now that we know the volume, we just have to convert from in^3 to L. This is a little more difficult than it sounds, however. Since we have no direct relationship between in^3 and L, we must go through a series of conversions to get to the desired unit. First, we can convert our unit from the English system to the metric system by using the relationship

$$1 \text{ in} = 2.54 \text{ cm}$$

To do this, however, we will have to cube the fraction we multiply by so that we have cm^3 and in^3!

$$\frac{16.192 \text{ in}^3}{1} \times \left(\frac{2.54 \text{ cm}}{1 \text{ in}}\right)^3 = \frac{16.192 \cancel{\text{in}^3}}{1} \times \frac{16.387 \text{ cm}^3}{1 \cancel{\text{in}^3}} = 265.338 \text{ cm}^3$$

Now that we have the metric volume unit, we can use the fact that a cm^3 is the same as a mL:

$$265.338 \text{ cm}^3 = 265.338 \text{ mL}$$

Now we can convert from mL to L:

$$\frac{265.338 \cancel{\text{mL}}}{1} \times \frac{0.001 \text{ L}}{1 \cancel{\text{mL}}} = 0.265338 \text{ L}$$

The volume, then, is 0.265338 L.

ON YOUR OWN

1.9 A braggart tells you that he walks 100,000 cm each day. He expects you to be impressed with such a big number. Should you be impressed? Convert the distance measurement to miles in order to determine whether or not to be impressed. (HINT: Earlier in this module, you were told how many inches are in a foot and how many feet are in a mile. You must use those numbers to solve this problem.)

1.10 How many cm^3 are in 0.0045 kL?

1.11 The area of a room is 16 m^2. What is the area of the room in mm^2?

Making Measurements

Now that we've learned so much about measurement units, we need to spend a little time learning how to make measurements. After all, being able to manipulate units in mathematical equations isn't going to help us unless we can make measurements with those units to begin with. In order to learn how to make measurements properly, let's start with something simple: a ruler.

Suppose I wanted to measure the length of a small ribbon with an English ruler. I would make my measurement something like this:

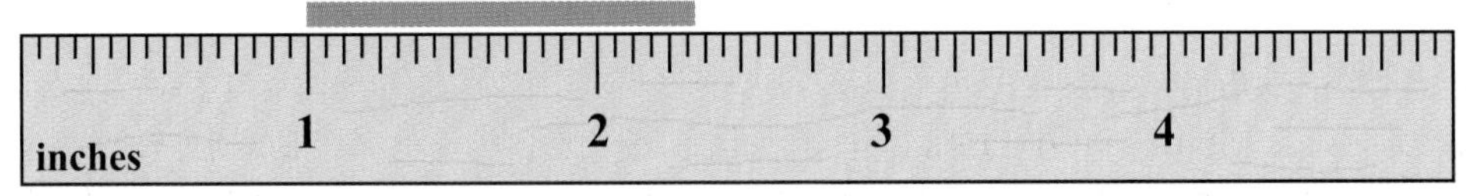

Illustration by Megan Whitaker

First, notice that I did not start my measurement at the beginning of the ruler. Instead, I lined up the ribbon with the first inch mark. The reason I did this is that it is slightly more accurate. It is very difficult to line up the edge of a ruler with the edge of the object you are measuring. This is especially true when the ruler is old and the edges are worn. So, the first rule for measuring with a ruler is to start at "1," not "0."

Now, how would you read this measurement? Well, first you need to see what the scale on the ruler is. If you count the number of dashes between 1 inch and 2 inches, you will find that there are 15 of them. That means that every dash is worth one sixteenth of an inch. This is because 15 dashes break up the area between 1 inch and 2 inches into 16 equal regions.

Now that we know the scale is marked off in sixteenths of an inch, we can see that the ribbon is a little bigger than 1 and 5/16 of an inch. Is that the best we can do? Of course not! Because the edge of the ribbon falls between 5/16 (10/32) and 6/16 (12/32) of an inch, we can estimate that it is approximately 11/32. Thus, the proper length of the ribbon is 1 and 11/32 inches. Generally, chemists do not like fractions in their final measurements, so we will convert 11/32 into its decimal form to get a measurement of 1.34375 inches. Later on we will see that this measurement has far too many digits in it, but for right now we will assume that it is okay.

Let's take that same ribbon and now measure it with a metric ruler:

Illustration by Megan Whitaker

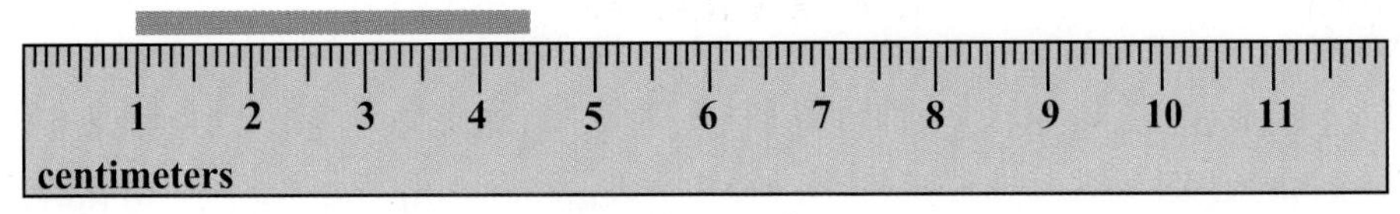

Now what measurement do we get? Well, there are 9 small dashes between each cm mark; therefore, the scale of this ruler is one tenth of a cm, or 0.1 cm. This is typical of metric rulers. They are almost always marked off in tenths since the prefixes in the metric system are all multiples of 10. If you think about it, 1 mm = 0.1 cm, so you could also say that each small dash is 1 mm. Clearly, the ribbon is between 3.4 cm and 3.5 cm long. Using our method of approximating between the dashes, we would say that the ribbon is 3.41 cm long.

Whenever you are using a measuring device that has a scale on it, be sure to use it the way we have used the rulers here. First, determine what the dashes on the scale mean. Then, try to estimate in between the dashes if the object you are measuring does not exactly line up with a dash. That gives you as accurate a measurement as possible. You should always strive to read the scale to *the next decimal place* if possible. In the metric ruler, for example, the scale is marked off in 0.1 cm, so you should read the ruler to 0.01 cm, as I discussed above.

One physical quantity that chemists measure quite a lot is volume. Since chemists spend a great deal of time mixing liquids, volume is often a very important factor and must be measured. When chemists measure volume, one of the most useful tools is the **graduated cylinder**. This device looks a lot like a glass rain gauge. It is a hollow glass cylinder with markings on it. These markings, called graduations, measure the volume of liquid that is poured into the cylinder.

In the last experiment you will perform in this module, you will use a graduated cylinder (or a suitable substitute) to measure volume; thus, you need to be aware of how to do this. When liquid is poured into a cylinder, the liquid tends to creep up the edges of the cylinder. This is because there are attractive forces between the liquid and the cylinder. Thus, liquid poured into a graduated cylinder does not have a flat surface. Instead, it looks something like this:

Illustration by Megan Whitaker

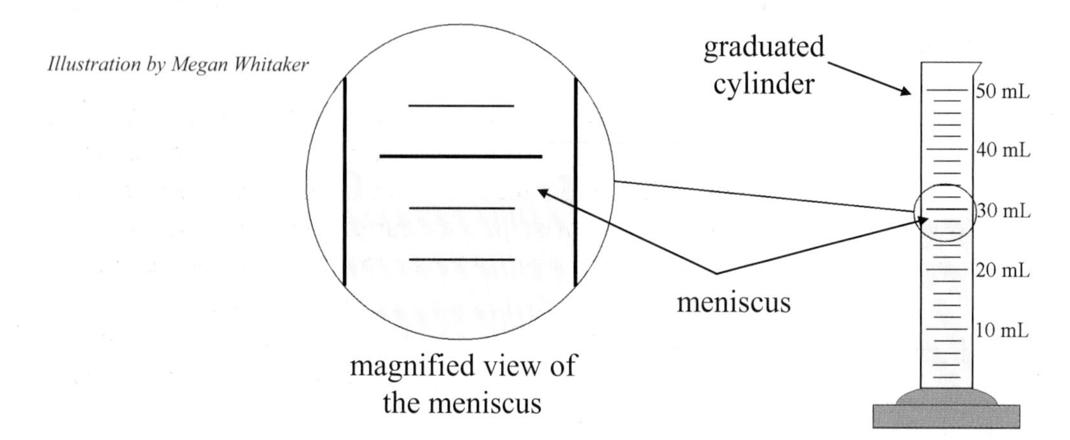

The curved surface of the yellow liquid is called the **meniscus** (muh nis' kus). In order to determine the volume of the liquid that is in any graduated cylinder, you must read the level of the liquid from the *bottom of the meniscus*. Looking at the graduated cylinder pictured above, there are 4 dashes in between each marking of 10 mL. Thus, the dashes split the distance between the 10 mL marks into 5 divisions. This means that each dash must be worth 2 mL. If you look at the bottom of the meniscus in the drawing above, you will see that it is between 28 and 30 mL. Is the volume 29 mL then? No, not quite.

In the two examples you saw before, it was hard to guess how far between the dashes the object's edge was because the dashes were very close together. In this example, the dashes are farther apart, so we can be a bit more precise in our final answer. In order for the volume to be 29 mL, the bottom of the meniscus would have to be exactly halfway between 28 and 30. Clearly, the meniscus is much closer to 28 than to 30. So the volume is really between 28 and 29, and probably a little closer to 28. I would estimate that the proper reading is 28.3 mL. It could be as low as 28.1 or as high as 28.5, so 28.3 is a good compromise. Remember, you need to try to read the scale to the next decimal place. Since the scale is marked off in increments of 2 mL, you must try to read the answer to 0.1 mL. Experiments 1.3 and 1.4 will give you some practice at this kind of estimation.

Accuracy, Precision, and Significant Figures

Now that we've learned a little bit about making measurements, we need to talk about when measurements are good and when they are not. In chemistry, we can describe a measurement in two ways. We can talk about its **accuracy** or we can talk about its **precision**. Believe it or not, these two words mean two entirely different things to a chemist:

Accuracy - An indication of how close a measurement is to the true value

Precision - An indication of the scale on the measuring device that was used

In other words, the more correct a measurement is, the more accurate it is. On the other hand, the smaller the scale on the measuring instrument, the more precise the measurement.

For example, let's go back to the idea of measuring a ribbon. Suppose we used a ruler with a scale marked off in 0.1 cm and got a length of 3.45 cm. What does that number mean? It means that the length of the ribbon, as far as we could tell, was somewhere between 3.445 and 3.454 cm long. Since both of those numbers round to 3.45, any length within that range is consistent with our measurement. We could not determine the length of the ribbon any better than that because our ruler was not precise enough to do any better.

On the other hand, suppose we found a ruler whose scale was marked off in 0.01 cm. With that ruler, we could get a measurement of 3.448 cm, because estimating between the dashes gives us one more decimal place. Thus, since the ruler is marked off in hundredths, we can get a measurement out to the thousandths place. Now this measurement is still consistent with our previous one, because it rounds up to 3.45 cm. That extra digit in the thousandths place, however, "nails down" the length better. It would be impossible to obtain so precise a measurement from the ruler we used in the example above, so the new ruler provided us with a way of being more precise. Thus, the precision of your measurement depends completely on the measuring device you use. The smaller the difference in the dashes on the scale, the more precise your answer will be.

However, suppose you used the second ruler improperly. Maybe you read the scale incorrectly or didn't line the ribbon up to the ruler very well and ended up getting a measurement of 3.118 cm. Even though this measurement is more *precise* than the one we made with the first ruler, it is significantly less accurate because it is way off of the correct value of 3.448 cm. Thus, the accuracy of your measurement depends on how carefully and correctly you used the measuring device. In other words, a measurement's precision depends upon the instrument, whereas a measurement's accuracy depends upon the person doing the measurement.

Since a measurement's precision depends on the instrument used, the only way to improve your precision is to get a better instrument. However, there are other ways to improve your accuracy. First, you can make sure you understand the proper methods of using each instrument at your disposal. Second, you can practice making measurements, which will help you in being careful.

The most practical way to help your accuracy in measurement, however, is to make your measurement several times and average the results. This tends to average out all of the little differences between measurements. An even better way of assuring accuracy is to have several *different* people make the measurements and average all of their answers together. The more individual measurements you can make, the more accurate the average of them will be.

Let me illustrate this whole idea of accuracy and precision in another way. Suppose you were throwing darts on a dart board. Here are three possible outcomes and the way that I would characterize them:

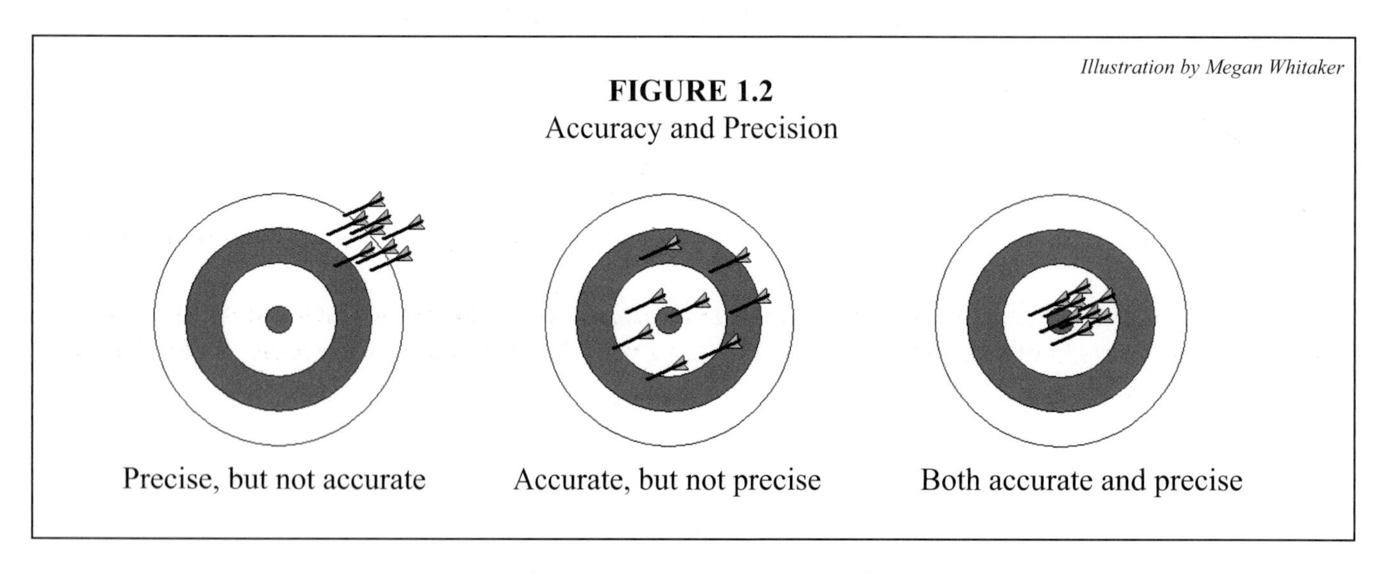

FIGURE 1.2
Accuracy and Precision

The first target on the left has all of the darts clumped together but way off the bull's eye of the target. If I made several measurements with a precise device, but I used the device wrongly every time, or if the device had a flaw, I would get a bunch of measurements very close to one another, but they would be far from the true value (the bull's eye). Thus, like the result on the dart board, I would have a lot of precision, but not much accuracy.

In the middle target, the darts are surrounding the bull's eye, but they are far from one another. This is what would happen if I used a measuring device that was not very precise, but I used it correctly. Because the estimation between the marks on the scale would be harder to do if the marks were far from each other, I would get a lot of different measurements from estimating between the dashes on the scale. However, if I used the device properly, and if the device was good, the average measurement would be accurate. Thus, like the result on the dart board, I would have accuracy, but not much precision.

In the target on the right side of the figure, I have the darts in a tight clump right around the bull's eye. This, then, is an illustration of measurements that are both accurate and precise. They are accurate because they average out to the correct value (the bull's eye). They are precise because they are very close to one another, indicating a precise measuring device was used.

Since accuracy and precision are very important, we need to know how to evaluate the accuracy and precision of our measurements. The way to determine the accuracy of a measurement is to compare it to the correct value. If you have no idea what the correct value is, determining your measurement's accuracy is a little more difficult. It is not impossible, but it is so complex that we will not discuss it in this course.

Determining the precision of a measurement, however, is quite easy. In order to determine the precision of a measurement, we merely need to look at its **significant figures**. If two instruments measure the same thing, the one which gives you a significant figure in the smallest decimal place is the more precise instrument. Of course, this fact does us no good if we do not know what a significant figure is:

A digit within a number is considered to be a significant figure if:

i. It is non-zero OR

ii. It is a zero that is between two significant figures OR

iii. It is a zero at the end of the number and to the right of the decimal point

Counting significant figures will be very important in our ability to understand measurements, so you must have a firm grasp of this concept. Read through Example 1.5 and follow the logic. After that, try the "On Your Own" problems that follow.

EXAMPLE 1.5

Count the significant figures in each of the following numbers:

a. 3.234 **b. 6.016** **c. 105.340** **d. 0.00450010**

(a) Since every digit is non-zero in this number, every digit is a significant figure. Thus, there are <u>four</u> significant figures in this number.

(b) The three non-zero digits are all significant figures and the zero is also significant because it is between two significant figures. Thus, there are <u>four</u> significant figures in this number.

(c) The four non-zero digits are all significant figures. The zero between the 1 and the 5 is significant because it is between two significant figures. The last zero is also significant because it is at the end of the number and to the right of the decimal point. Thus, there are <u>six</u> significant figures in the number.

(d) The first three zeros are not significant because they are not between two significant figures and they are not at the end of the number. The three non-zero digits are all significant figures as are the zeros between the 5 and the 1. The zero at the end is also a significant figure. Thus, there are <u>six</u> significant figures in the number.

ON YOUR OWN

1.12 How many significant figures are in the following measurements?

a. 3.0220 cm b. 0.0060 m c. 1.00450 L d. 61.054 kg

In the end, then, the precision of our instrument determines the number of significant figures we can report in a measurement. In our graduated cylinder example, we decided that we could reasonably approximate our measurement to somewhere between 28 and 29 mL. If we can do that, then we can report our answer to the nearest tenth of a mL, giving us a three-significant-figure answer of 28.3 mL. If, instead, our graduated cylinder had been marked off in tenths of a mL, we could probably approximate the measurement to somewhere between 28.2 and 28.3 mL, allowing us to have a four-significant-figure answer like 28.26 mL.

In the example which used an English ruler, I said that our final answer, 1.34375 inches, had too many digits in it. Now hopefully you can see why. According to this number, our ruler was precise enough to measure distance of one hundred thousandth of an inch! That's far too much precision. Most English rulers are precise to, at best, 0.01 inches. Thus, the proper length of the ribbon is 1.34 in.

Now, suppose we had another ribbon to measure :

Illustration by Megan Whitaker

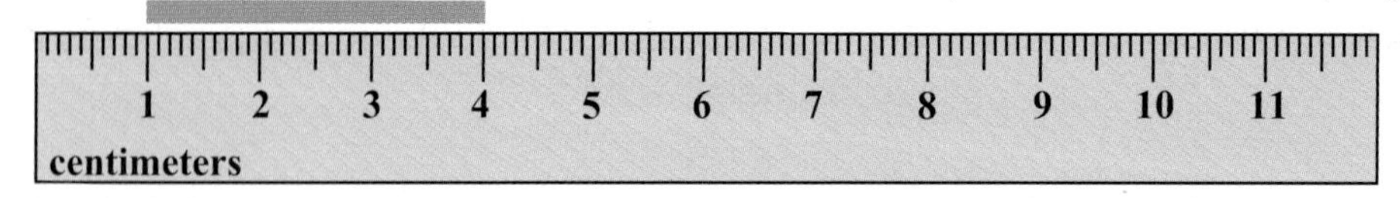

How would we report its measurement? Would we say that this ribbon is 3 cm long? Actually, that measurement is not quite right. The ribbon does seem to end right on the 4 cm line, so there is no need to do any approximations here. Why, then, is 3 cm a wrong answer for the length of the ribbon?

The problem with reporting the length of the ribbon as 3 cm is that you are not being as precise as you can be. Since the ruler's scale is marked off in 0.1 cm, you can safely report your answers to the hundredths of a cm, because if the ribbon's edge had fallen between two dashes, you could have approximated as we did above. Thus, the precision of the ruler is to the hundredths place. Therefore, if the object's edge falls right on one of the dashes, do not throw away your precision. Report this length as 3.00 cm. This tells someone who reads the measurement that the ribbon's length was measured to a precision in the hundredths of a cm.

If you report the length as 3 cm, that means the ribbon could be as short as 2.5 cm or as long as 3.4 cm. Both of those measurements round to 3. But the ruler you used was much more precise. It determined the length of the ribbon to be 3.00 cm. Thus, the ribbon is somewhere between 2.995 and 3.004 cm long. Keeping all of the significant figures that you can is a very important part of doing chemistry experiments. You will get some practice at this in Experiments 1.3 and 1.4.

Reporting the precision of a measurement is just as important as reporting the number itself. Why? Well, let's suppose that the NO-WEIGHT COOKIE CO. just produced a diet cookie that they claim has only 5 Calories per cookie. In order to confirm this claim, researchers did several careful experiments and found that, in fact, there were 5.4 Calories per cookie. Does this result mean that the NO-WEIGHT COOKIE CO. lied about the number of Calories in its cookies? No. When the company reported that there are 5 Calories per cookie, the precision of their claim indicated that there could be anywhere from 4.5 to 5.4 Calories per cookie. The researchers' finding was more precise than the company's claim, but, nevertheless, the company's claim was accurate.

Scientific Notation

Since reporting the precision of a measurement is so important, we need to be able to develop a notation system that allows us to do this no matter what number is involved. As numbers get very large, it becomes more difficult to report their precision properly. For example, suppose you measured the distance between two cities as 100.0 km. According to our rules of precision, reporting 100.0 km as the distance means that your measuring device was marked off in units of 1 km, and you estimated between the marks to come up with 100.0 km. However, suppose your measurement wasn't that precise. Suppose the instrument you used could only determine the distance to within 10 km? How could you write down a distance of 100 km and indicate that the precision was only to within 10 km?

The answer to this question lies in the technique of **scientific notation**. In scientific notation, we write numbers so that no matter how large or how small they are, they always have a decimal point in them. The way we do this is to remember that a number can be represented in many, many different ways. The number 4, for example, could be written as "2 x 2" or "4 x 1" or simply "4." Each one of these are appropriate representations of the number "4." In scientific notation, we always represent a number as a something times a power of 10. For example, "50" could be written in scientific notation as "5 x 10." The number "150" could be written as "1.5 x 100."

Do you see why this helps us in writing down the precision of our original measurement? Instead of writing the distance as 100 km, I could write it as 1.0 x 100. How does this help? Well, according to our rules of significant figures, the zero in 1.0 is significant, because it is at the end of the number and to the right of the decimal. Thus, by writing down our measurement this way, we indicate that the zero was actually measured and that the measurement is precise to within 10 km. There is no way to do that with normal, decimal notation, because neither of the zeros in 100 are significant. Scientific notation, then, gives us a way to *make* zeros significant if they need to be. If our measurement of 100 km was precise to within 1 km, we could indicate that by reporting the measurement as 1.00 x 100 km. Since both zeros in 1.00 are significant, this tells us that both zeros were measured, so our precision is within 1 km.

Now, since numbers that we deal with in chemistry can be very big or very small, we use one piece of mathematical shorthand in scientific notation. Recall from algebra that "100" is the same as "10^2." We will use this shorthand to make the numbers easier to write down. In the end, then, scientific notation always has a number with a decimal point right after the first digit times a 10 raised to some power.

One other advantage of using scientific notation is that you can use it to simplify the job of recording very large or very small numbers. For example, there are roughly 20,000,000,000,000,000,000,000 particles in each breath of air that you take. Numbers like that are very common in chemistry. In scientific notation, the number would look like 2×10^{22}. That's much easier to write down!

How did I know that I needed to raise the 10 to the 22nd power? I saw that in order to get the decimal point right after the two, I would have to move it to the left 22 digits. Moving the decimal 22 digits is equivalent to multiplying by 10^{22}. So, when putting a large number into scientific notation, all you need to do is count the number of spaces the decimal point needs to move, and raise the 10 to that power.

Chemistry also deals with very small numbers. For example, one of the things we will discuss in great detail in several upcoming modules is a particle called a proton. The proton has a mass of about 0.00000000000000000000000000167 kg. Once again, this number is a real pain to write down. We will use scientific notation to make our job a little easier. In scientific notation, the proton's mass is 1.67×10^{-27} kg. Why raise the 10 to the -27th power? When numbers are raised to the negative power, they are also inverted. So, when you multiply a number by 10 raised to a negative power, you end up shifting the decimal place the other way! In order to get the decimal point to be right after the "1," I had to move it 27 places. Since I moved it to the right 27 places, I multiply it by 10^{-27}. See how this works by following the two examples below, and then make sure you understand this technique by completing the "On Your Own" problems that follow.

EXAMPLE 1.6

Convert the following numbers into scientific notation:

a. 20,300 b. 3,151,367 c. 234,000 d. 0.000002340 e. 0.000875

(a) The decimal place must be moved to the left by 4 digits to get it next to the "2." Since we are dealing with a big number, we have to multiply by a 10 raised to a positive power. Thus, the answer is $\underline{2.03 \times 10^{4}}$. Since the last two zeros are not significant as the number is written, we must drop them in our answer, because all zeros are significant in 2.0300×10^{4}.

(b) The decimal place must be moved to the left 6 places and the number is big, so the answer is $\underline{3.151367 \times 10^{6}}$.

(c) The decimal place must be moved to the left 5 places, and since it is a big number, the answer is $\underline{2.34 \times 10^{5}}$. Once again, the last three zeros were dropped, because as written, they are not significant.

(d) The decimal must be moved 6 places to the right. Since this is a small number, we are dealing with a negative exponent, so the answer is $\underline{2.340 \times 10^{-6}}$. In this case, the final zero cannot be dropped because, based on our rules of significant figures, a zero at the end of a number to the right of the decimal is significant.

(e) The decimal point must be moved 4 places to the right and since it is a small number, the answer is $\underline{8.75 \times 10^{-4}}$.

Convert the following numbers from scientific notation back into decimal form.

(a) 3.45×10^{-5} (b) 2.3410×10^{7} (c) 1.89×10^{-9} (d) 3.0×10

(a) Since the 10 is raised to a negative power, the decimal point must be moved in order to make it small. The power of -5 tells us that we move it 5 spaces, so the answer is <u>0.0000345</u>.

(b) Since the power of ten is positive, we must move the decimal point to make the number bigger. The power of 7 tells us we must move it 7 places, so the answer is <u>23,410,000</u>. Note that we cannot indicate that the zero after the 1 is significant in this notation. It is clearly significant in the original number, so it is impossible to properly represent the precision of this number in decimal form.

(c) We must move the decimal point 9 places and make the number smaller, so the answer is 0.00000000189.

(d) Since the exponent is not listed, we assume it's "1." That means that we move the decimal point 1 place so that the number gets bigger, so the answer is 30. Once again, there is no way to indicate that the zero is significant, as it is in the original number.

ON YOUR OWN

1.13 Convert the following numbers into scientific notation.

(a) 26,789,000 (b) 123 (c) 0.00009870 (d) 0.980

1.14 Convert the following numbers from scientific notation to decimal form.

(a) 3.456×10^{14} (b) 1.2341×10^{3} (c) 3.45×10^{-5} (d) 3.1×10^{-1}

Using Significant Figures in Mathematical Problems

Now that we have the ability to write down any measurement with its proper precision, there is only one more topic on significant figures that we need to discuss. We need to know how to use our concepts of significant figures when we work mathematical problems. Suppose I had two measurements and I wanted to add them together. Since each measurement has its own precision, the final answer would also have a certain precision. How do I know what the precision of my answer is?

For example, suppose I measured the total length of a knife to be 25.46 cm. Later on, someone else measured the length of the knife handle with a less precise ruler and got 7.8 cm. If I wanted to determine the length of the knife's blade, I could either go and measure it, or I could say that the blade's length was the total length of the knife minus the length of the handle, or 25.46 cm - 7.8 cm. If I do the subtraction, I get 17.66 cm. It turns out, however, that this answer is too precise. Since the knife handle was measured with a less precise ruler, when I use its measurement in a calculation, the answer is also less precise. In the end, the proper answer would be 17.7 cm.

In order to add, subtract, multiply, or divide measurements, we have to learn two rules about using significant figures in mathematical equations. You will be using these rules over and over again throughout this course, so you will be expected to know them:

Adding and Subtracting with Significant Figures: **When adding and subtracting measurements, round your answer so that it has the same precision as the *least precise* measurement in the equation.**

Multiplying and Dividing with Significant Figures:* When multiplying and dividing measurements, round the answer so that it has the *same number of significant figures as the measurement with the fewest significant figures.

Here's how these rules work:

EXAMPLE 1.7

A student measures the mass of a jar that is filled with sand and finds it to be 546.2075 kg. The jar has a note on it which says, "When empty, this jar has a mass of 87.61 kg." What is the mass of the sand that is in the jar?

Since 546.2075 kg is the mass of both the jar and the sand, and since 87.61 kg is the mass of the jar alone, the mass of the sand must be the difference between the two:

$$546.2075\text{ kg} - 87.61\text{ kg} = 458.5975\text{ kg}$$

However, since the precision of the jar's mass only goes out to the hundredths place, that's the best we can do in our final answer. Thus, the mass of the sand is 458.60 kg. Note that this number has more significant figures than 87.61. That doesn't matter, however, because in addition and subtraction, you *do not* count significant figures; you only look at precision.

A person runs 3.012 miles in 0.430 hours. What is the person's average speed?

We can get the person's average speed by taking the distance traveled divided by the time:

$$\text{Speed} = 3.012\text{ miles} \div 0.430\text{ hours} = 7.004651163\ \frac{\text{miles}}{\text{hours}}$$

The 3.012 miles has 4 significant figures while 0.430 hours has 3. Thus, our final answer must have 3 significant figures, making it $7.00\ \frac{\text{miles}}{\text{hour}}$.

Now that we have learned these rules, you will be expected to use them *in all further mathematical operations!* Whether you are working an "On Your Own" problem, a practice problem, a test problem, or an experiment, you will be expected to use these rules. In the examples and answers for all previous problems, these rules have not been followed, but they will from now on. By the time you finish the next couple of modules, keeping track of significant figures and precision should be second nature to you.

There is one point that I must make about significant figures before you get some practice using the rules. When making unit conversions, you might be tempted to round everything to one significant figure, because of the conversion relationships. For example, when you convert 121 g into kg, you use the following equation:

$$\frac{121\ \cancel{g}}{1} \times \frac{1\ kg}{1{,}000\ \cancel{g}}$$

Note that the "1 kg," the "1" in the denominator of the first fraction, and "1,000 g" all look like they have only one significant figure. Thus, you might be tempted to round your answer to one significant figure. However, that would not be right. The reason is simple: these three numbers all come from definitions. They are actually infinitely precise. The "1 kg" is really "1.000... kg," and the "1,000 g" is really "1,000.000... g." This is because *exactly* 1 kilogram is defined to be *exactly* 1,000 g. In the same way, the "1" on the bottom of the first fraction is really "1.000…," because it is an integer. In the end, then, the *only* number that has a limited number of significant figures is the 121 g (it is a measurement), so the answer is 0.121 kg.

In general, then, the prefixes used in the metric system as well as the integers that we use in fractions are infinitely precise and thus have an infinite number of significant figures. As a result, we ignore them when determining the significant figures in a problem. This is a very important rule:

The definitions of the prefixes in the metric system and the integers we use in fractions are not considered when determining the significant figures in the answer.

Get some practice making measurements, using them in mathematical equations, and keeping track of significant figures by performing the following experiment:

EXPERIMENT 1.3

Comparing Conversions to Measurements

Note: A sample set of calculations is available in the solutions and tests guide. It is with the solutions to the practice problems.

Supplies:

- Book (not oversized)
- Metric/English ruler or rulers
- Safety goggles

1. Lay the book on a table and measure its length in inches. Read the ruler as I showed you in the measurement section above, estimating any answer that falls in between the markings on the scale. Once you do that, convert the fraction to a decimal (as we did in the measurement section above) and round it to the hundredths place, because that's the precision of an English ruler.
2. Measure the width of the book in the same way.
3. Now that you have the length and width measured, multiply them together to get the surface area of the book. Since you are multiplying inches by inches, your area unit should be in^2. Remember to count the significant figures in each of the measurements and round your final answer so that it has the same number of significant figures as the measurement with the least number of significant figures.
4. Now take the length measurement and use the relationship given in Table 1.3 to convert it into cm. Do the same thing to the width measurement, making sure to keep the proper number of significant figures. Note that the relationship between inches and centimeters is *exact*, so the "2.54 cm" should not be taken into account when considering the significant figures, because 1 inch is exactly 2.54 cm.
5. Now use your metric ruler to measure the length and width of the book in centimeters. Once again, do it just like I showed you in the measurements section above. If the scale of the ruler is marked

off in 0.1 cm, then your length and width measurements should be written to the hundredths of a centimeter. Compare these answers to the length and width you calculated by converting from inches. They should be nearly the same. If they are different by only a few percent, there is no problem. However, if they differ by more than a few percent, recheck your measurements and your conversions.

6. Finally, multiply the length and width measurements you took with the metric ruler to calculate the surface area of the book in cm^2. Use the relationship between inches and centimeters to convert your answer into in^2. Remember, since you are using a derived unit, the conversion is more complicated. You might want to review Example 1.4.
7. Now compare the converted value for the surface area to the one you calculated in step (3) using your English measurements. Once again, they should be equal or close to equal. If not, you have either measured wrongly or made a mistake in your conversion.
8. Clean up your mess.

Density

Before we leave this module, I would like to introduce one very important quantity that is measured quite frequently in chemistry: **density**. The definition of density is as follows:

Density - An object's mass divided by the volume that the object occupies

This definition for density can be mathematically represented as:

$$\rho = \frac{m}{V} \tag{1.1}$$

In the equation above, the "ρ" is the Greek letter "rho" and is typically the symbol used to represent density. The letter "m" stands for the mass of the object, and the letter "V" stands for the object's volume. From this equation, we can determine the units that describe density. Since mass is measured in grams and the most common volume unit in chemistry is mL, the equation says that the units for density will be the unit one gets when grams are divided by mL, or $\frac{\text{grams}}{\text{mL}}$. In words, we would call this unit "grams per mL." Since mL and cm^3 are equivalent, we often see density expressed in grams per cm^3. These, of course, are not the only units for density. Any mass unit divided by any volume unit is a possible unit for density, but these two are the most common in chemistry.

Density is a very important quantity in chemistry. It gives you an idea of how tightly packed the matter is in an object. For example, suppose you had a ball made out of plastic and another ball of precisely the same size made out of lead. Which ball would be heavier? The lead ball would be much heavier than the plastic ball, because there would be a lot more matter in the lead ball than in the plastic ball. This is because matter is packed very tightly together in lead, but matter is packed pretty loosely in plastic. Thus, for the same size ball, the lead ball is much heavier.

Density gives us a way to determine exactly how heavy the lead ball is compared to the plastic ball. Lead has a density of 11.4 grams per mL, whereas most plastic has a density of less than 1 gram per mL. Let's suppose we had a sample of plastic with a density of 0.54 grams per mL. Since the density of lead is 21 times (11.4 ÷ 0.54) greater than the density of the sample of plastic, we know that the matter inside of lead is packed 21 times tighter than the matter inside of the sample of plastic.

It also turns out that every substance on earth has its own characteristic density. This is nice because it provides a way for us to identify a substance if we don't know exactly what it is. For example, years ago people used density to determine whether or not something that looked golden was really made out of pure gold. They would measure the substance's mass, measure its volume and then divide the two. If the result turned out to be 19.3 grams per mL (the density of pure gold), then they knew it was pure gold because no other substance on earth has that density. If the density was not 19.3 grams per mL, then they knew it was not pure gold.

Before showing you some examples of how to use density in problems, I want you to perform another experiment. This experiment will teach you something very important about the physical meaning of density.

EXPERIMENT 1.4
The Density of Liquids

Note: A sample set of calculations is available in the solutions and tests guide. It is with the solutions to the practice problems.

Supplies:

- Water
- Vegetable oil
- Something that measures the volume of a liquid, preferably in mL or cm^3. A graduated cylinder would be ideal, but measuring cups will work as well.
- Maple syrup (Natural syrup does not work as well as something like Mrs. Butterworth's®.)
- A large glass
- A mass scale, preferably one that reads in grams. (The scale should not go much over 500 grams, or it will be very difficult for you to read the mass of the objects in this experiment.)
- Safety goggles

1. First, measure the mass of the graduated cylinder, or whatever you have that measures the volume of a liquid. Be sure to write it down with the correct precision. If you are using a standard mass scale from a grocery store, its scale is probably marked off in units of 10 grams. Thus, you should be able to report the mass to a precision of 1 g.
2. Next, measure out 50.0 mL (1/4 cup if you are using measuring cups) of syrup. Now put the graduated cylinder (with the syrup in it) back on the scale and measure the total mass. Subtract the mass of the graduated cylinder from this number (using our rules for significant figures) to get the mass of the table syrup by itself. This method of measuring mass is called the **difference method**. Chemists often call it "measuring the mass by difference."
3. Now that you have the mass of the table syrup, and you know that its volume was 50.0 mL (because that's what you measured out), divide the mass by the volume to get the density. Be sure to follow our significant figure rules when you do this! Finally, pour the syrup into the tall glass. Repeat this procedure for both the water and the vegetable oil.
4. Once you have measured the density of all three substances, look at the tall glass. You should see that the table syrup is all at the bottom of the glass, the water forms a layer above that, and the vegetable oil is all in one layer on top!
5. Clean up your mess.

What explains the layering you saw in the glass? Look at the densities you calculated from your measurements. Which substance has the largest density? The syrup. Where was the syrup? It was on the bottom of the glass. In the same way, the vegetable oil has the smallest density and was at the top of the glass, while the water's density is in between the two, so it stays in between the syrup and the vegetable oil. If we think about what density means, this should make sense. Density tells us how tightly packed the matter is inside a substance. Syrup's matter is very tightly packed, so it falls through the more loosely packed water and vegetable oil until it reaches the bottom of the glass. Water falls through the more loosely packed matter in vegetable oil but then is stopped by the more tightly packed matter in syrup. The vegetable oil could not fall through either water or syrup, so it stayed at the top of the glass.

(The multimedia CD has a video demonstration of the fact that even gases have density.)

Now that you have a better understanding of what density means, follow the example below and then do the "On Your Own" problems after it.

EXAMPLE 1.8

A gold miner has just found a nugget of pure gold. He measures its dimensions and then calculates its volume to be 0.125 L. Knowing that the density of gold is 19.3 grams per mL, calculate the mass of the miner's nugget.

Our equation for density is:

$$\rho = \frac{m}{V}$$

The problem is, the equation is not much good to us the way it is currently written. Since we would like to calculate the mass of the nugget, we need an equation that starts "m=." So we use algebra to rearrange the equation so that it reads:

$$m = \rho \cdot V$$

Now we can simply stick in our numbers for volume and density, and we will get the mass, right? There's only one small problem. If we use the numbers as they are given in the problem, we would not get the correct answer. Why? If we did that, it would look like:

$$m = 19.3\frac{g}{mL} \times 0.125 \text{ L}$$

According to our rules for multiplying numbers with units, we multiply the numbers, and then we do the same with the units. In the end, since we are calculating mass, we know that our answer should end up with the unit of grams. If we multiply these two units together, however, we don't get grams. In order to get grams, the mL in the denominator would have to cancel with the L in the volume measurement. That doesn't work. Instead, we need to change one of the units. It doesn't matter which, but I will choose to change the volume measurement into mL.

$$\frac{0.125\ \cancel{L}}{1} \times \frac{1\text{ mL}}{0.001\ \cancel{L}} = 125\text{ mL}$$

Now that I have the proper units, I can put everything into the equation:

$$m = 19.3\frac{\text{g}}{\cancel{\text{mL}}} \times 125\ \cancel{\text{mL}} = 2{,}412.5\text{ g}$$

We're still not quite done, however. According to our rules of significant figures, we can only have 3 significant figures in our final answer, because there are 3 significant figures in each of our numbers. Thus, the answer is 2,410 g. In scientific notation, the answer is 2.41×10^3 g. Either expression is correct, as they both have 3 significant figures.

This problem illustrates two important things about studying chemistry. First, many chemistry problems require a great deal of algebra in order to solve them. Thus, you must be pretty familiar with algebra to be successful at chemistry. Second, you must always watch your units. Be sure that units cancel when they should and do not cancel when they shouldn't. The best way to do this is to write down all of your units in every equation. That way, you can see if they work out once you do the math. Try your hand at the following "On Your Own" problems to see how well you grasp this. You will have to use algebra, watch your units, and keep track of significant figures.

ON YOUR OWN

1.15 The density of silver is 10.5 grams per cm^3. If a jeweler makes a silver bracelet out of 0.081 kg of silver, what is the bracelet's volume in mL?

1.16 A gold miner tries to sell some gold that he found in a nearby river. The person who is thinking about purchasing the gold measures the mass and volume of one nugget. The mass is 0.319 kg and the volume is 0.065 liters. Is this nugget really made out of gold? (Remember that the density of gold is 19.3 grams per mL)

Now that you have completed reading this module and have done the "On Your Own" problems, it is time for you to shore up your new skills and knowledge with the practice problems and review questions at the end of this module. As you go through them, check your answers with the solutions provided, and be sure you understand any mistakes you made. If you need more practice problems, you can find them in Appendix B of this book. Once you are confident in your abilities, take the test. You should try to score at least 70% on the test. If you do not, you should probably spend time reviewing this module before you proceed to the next one.

ANSWERS TO THE "ON YOUR OWN" PROBLEMS

1.1 $\frac{12{,}321\ \cancel{g}}{1} \times \frac{1\ kg}{1{,}000\ \cancel{g}} = \underline{12.321\ kg}$

1.2 $\frac{0.121\ \cancel{L}}{1} \times \frac{1\ mL}{0.001\ \cancel{L}} = \underline{121\ mL}$

1.3 $\frac{640.08\ \cancel{cm}}{1} \times \frac{0.01\ m}{1\ \cancel{cm}} = \underline{6.4008\ m}$

1.4 $\frac{123.5\ \cancel{kg}}{1} \times \frac{1\ sl}{14.59\ \cancel{kg}} = \underline{8.465\ sl}$

1.5 $\frac{3.2\ \cancel{gal}}{1} \times \frac{3.78\ L}{1\ \cancel{gal}} = \underline{12.096\ L}$

1.6 $\frac{1{,}500\ \cancel{mL}}{1} \times \frac{0.001\ \cancel{L}}{1\ \cancel{mL}} \times \frac{1\ kL}{1{,}000\ \cancel{L}} = \underline{0.0015\ kL}$

1.7 $\frac{2.0\ \cancel{km}}{1} \times \frac{1{,}000\ \cancel{m}}{1\ \cancel{km}} \times \frac{1\ cm}{0.01\ \cancel{m}} = \underline{200{,}000\ cm}$

1.8 $\frac{10{,}000{,}000\ \cancel{mg}}{1} \times \frac{0.001\ \cancel{g}}{1\ \cancel{mg}} \times \frac{1\ Mg}{1{,}000{,}000\ \cancel{g}} = \underline{0.01\ Mg}$

1.9 $\frac{100{,}000\ \cancel{cm}}{1} \times \frac{1\ \cancel{in}}{2.54\ \cancel{cm}} \times \frac{1\ \cancel{ft}}{12\ \cancel{in}} \times \frac{1\ mile}{5280\ \cancel{ft}} = \underline{0.621371\ miles}$

Since that's not even 1 mile, you should not be impressed by the braggart.

1.10 $\frac{0.0045\ \cancel{kL}}{1} \times \frac{1{,}000\ \cancel{L}}{1\ \cancel{kL}} \times \frac{1\ mL}{0.001\ \cancel{L}} = 4{,}500\ mL = \underline{4{,}500\ cm^3}$ (mL and cm^3 equivalent)

1.11 The relationship between m and mm is easy:

$$1\text{ mm} = 0.001\text{ m}$$

Thus, to set up the conversion, we start with:

$$\frac{16\text{ m}^2}{1} \times \frac{1\text{ mm}}{0.001\text{ m}}$$

This expression, however, *does not cancel* m^2. There is a m^2 on the top of the first fraction and only a m on the bottom of the second fraction. In order to cancel m^2 (which we must do to get the answer), we have to square the conversion fraction:

$$\frac{16\text{ m}^2}{1} \times \left(\frac{1\text{ mm}}{0.001\text{ m}}\right)^2$$

Then we get:

$$\frac{16\text{ }\cancel{\text{m}}^2}{1} \times \frac{1\text{ mm}^2}{0.000001\text{ }\cancel{\text{m}}^2} = \underline{16{,}000{,}000\text{ mm}^2}$$

1.12 (a) All three non-zero digits are significant figures as are both zeros. One zero is between two significant figures while the other is at the end of the number to the right of the decimal. Thus, there are five significant figures.

(b) The first three zeros are not significant because they are not between two significant figures. The six is a significant figure, as is the last zero because it is at the end of the number to the right of the decimal. Thus, there are two significant figures.

(c) All digits are significant figures here. The first two zeros are between significant figures and the last one is at the end of the number to the right of the decimal. Thus, there are six significant figures.

(d) All digits are significant figures. The zero is between two significant figures. Thus, there are five significant figures.

1.13 (a) 2.6789×10^7 (b) 1.23×10^2 (c) 9.870×10^{-5} (d) 9.80×10^{-1}

1.14 (a) 345,600,000,000,000 (b) 1,234.1 (c) 0.0000345 (d) 0.31

1.15 Since we are trying to calculate volume, we must first use algebra to rearrange the equation so that it begins "V=." If you do not understand how this is done, go back to your algebra book and review these skills. You cannot be successful in this chemistry course without a command of algebra! The rearranged equation looks like:

$$V = \frac{m}{\rho}$$

We can't put our numbers into the equation yet, however, because the units would not work out if we did. In the end, we need an answer with a volume unit, so the mass units must cancel. Right now, however, density is in grams per cm^3 while mass is in kg. So we first have to change kg into grams:

$$\frac{0.081\ \cancel{\text{kg}}}{1} \times \frac{1{,}000\ \text{g}}{1\ \cancel{\text{kg}}} = 81\ \text{g}$$

Now we can take 81 grams and divide by 10.5 grams per cm^3. Remember from math that dividing is the same as inverting and multiplying, so:

$$V = m \times \frac{1}{\rho} = 81\ \cancel{\text{g}} \times \frac{1\ \text{cm}^3}{10.5\ \cancel{\text{g}}} = 7.7\ \text{cm}^3 = \underline{7.7\ \text{mL}}$$

The answer was rounded to 7.7 because there are only 2 significant figures in the mass measurement, and the volume unit was changed to mL since cm^3 and mL are the same thing.

1.16 Since we only know gold's density in the units of grams per mL, our mass has to be in grams and our volume has to be in mL:

$$\frac{0.319\ \cancel{\text{kg}}}{1} \times \frac{1{,}000\ \text{g}}{1\ \cancel{\text{kg}}} = 319\ \text{g}$$

$$\frac{0.065\ \cancel{\text{L}}}{1} \times \frac{1\ \text{mL}}{0.001\ \cancel{\text{L}}} = 65\ \text{mL}$$

Now we can take these numbers and divide them to get the density:

$$\rho = \frac{319\ \text{g}}{65\ \text{mL}} = 4.9\ \frac{\text{g}}{\text{mL}}$$

The answer can have only 2 significant figures, since 65 has only 2 significant figures. Since the density did not work out to be 19.3 grams per mL, then we know that <u>this is not gold</u>. In fact, this density is equal to the density of iron pyrite, also known as "fool's gold" because it looks a lot like gold but has little value.

REVIEW QUESTIONS FOR MODULE #1

1. Which of the following contains no matter?
   a. A rock
   b. A balloon full of air
   c. A balloon full of helium
   d. A lightning bolt

2. List the base metric units used to measure length, mass, time, and volume.

3. In the metric system, what does the prefix "centi" mean?

4. Which has more liquid: a glass holding 0.5 kL or a glass holding 120 mL?

5. How long is the bar in the picture below?

Illustration by Megan Whitaker

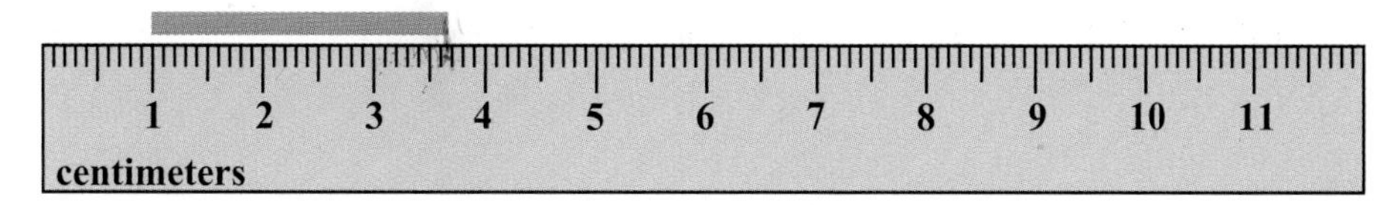

6. Two students measure the mass of a 502.1 gram object. The first student measures the mass to be 496.8123 grams. The second measures the mass to be 501 grams. Which student was more precise? Which student was more accurate?

7. How many significant figures are in the following numbers?
   a. 0.0120350
   b. 10.020
   c. 12
   d. 3.40×10^3

8. A student measures the mass of an object as 2.32 grams and its volume as 34.56 mL. The student then calculates the density to be 0.067129629. There are two things wrong with the student's value for density. What are they?

9. Why does ice float on top of water?

10. Lead has a density of 11.4 grams per mL, whereas gold has a density of 19.3 grams per cc. If I were to make two identical statues, one out of gold and the other out of lead, which would be heavier?

PRACTICE PROBLEMS FOR MODULE #1

Be sure to use the proper number of significant figures in ALL of your answers!

1. Convert 1.2 mL into L.

2. Convert 34.50 km into m.

3. Convert 0.045 km into cm.

4. If an object has a volume of 34.6 mL, how many kL of space does it occupy?

5. A box is measured to be 2.3 m by 4.2 m by 3.5 m. What is its volume in cubic centimeters?

6. A nurse injects 34.5 cc of medicine into a patient. How many liters is that?

7. Convert the following decimal numbers into scientific notation:
   a. 123.45
   b. 0.0003040
   c. 6,100,000
   d. 0.1234

8. Convert the following numbers back into decimal:
   a. 6.54×10^3
   b. 3.450×10^{-3}
   c. 3.56×10^7
   d. 4.050×10^{-7}

9. Lead has a density of 11.4 grams per mL. If I make a statue out of 3.45 L of lead, what is the statue's mass?

10. Gold has a density of 19.3 grams per cc. If a gold nugget has a mass of 45.6 kg, what is its volume?

MODULE #2: Energy, Heat, and Temperature

Introduction

Have you ever sat down and watched a fire? The flames dancing along the wood, the crackling sounds, and the soft light make fire a pleasant source of extra heat. But have you ever wondered what exactly is going on while the flames consume the wood? What happens to the wood during a fire, and what is the nature of the ash that the wood leaves behind? Where does the heat come from? What makes the flames?

The answers to these questions lie in a study of how matter is able to change. During the course of a fire, the matter that makes up the wood undergoes substantial change. Some of the matter in the wood changes into water vapor, some of it changes into a gas (carbon dioxide), and most of the rest of it changes into what we call "ash."

The details of this change are very interesting and will be discussed in several upcoming modules. For right now, however, we want to concentrate on understanding the nature of the flames. Flames are a form of **energy**. It turns out that whenever matter undergoes a substantial change, that change is accompanied by either a release or an intake of energy. As we go through this module, we will learn more about the nature of this energy, how to measure it, and how it relates to the study of chemistry.

Energy and Heat

Before we can get started, we have to make sure we understand what the term "energy" really means. Although you have probably used this word several times in normal conversation, you probably don't have a good idea of its definition:

Energy - The ability to do work

This definition probably doesn't mean a lot to us until we also define work:

Work - The force applied to an object times the distance that the object travels parallel to that force

Let's think about these two definitions for a moment. First, let's concentrate on **work**. Our definition of work tells us that in order for work to occur, there must be motion. After all, the object must travel a distance parallel to the force that is applied, or the work will be the force times zero, which is zero. Thus, as strange as it might sound, when you push against a wall with all of your might, *you are doing no work*. Only if the wall actually moved would you, in fact, be doing work. You might exhaust yourself trying, but unless you got the wall to move, you would not actually be doing any work, because in order to do work, the object to which you applied a force must move some distance parallel to that force.

Now that we understand the definition of work, we can see that the definition of energy makes a lot more sense. Energy is, essentially, the ability to cause motion. I mentioned earlier that the flames in a fire are a form of energy. Based on our definition of energy, then, the flames in a fire must be able to cause motion. What motion can the flames cause?

For one thing, you can see smoke rising up out of the fire. The energy in the flames is causing that motion. Also, if you were to put a pot of water on the flames, it would eventually start to boil. The energy in the flames causes the motion of the bubbles that result from boiling. Finally, think about the fact that you can feel heat coming from a fire. The only way that the heat can get to you is by traveling; thus, the fire's very act of warming you involves motion.

Now that we're on the subject of heat, we had better come up with a good definition of it:

Heat - Energy that is transferred as a consequence of temperature differences

One thing we see from this definition is that heat is really energy. This should make sense if you think about it. Energy is, essentially, the ability to create motion. If we used the heat from a fire to boil water, the heat would create motion; thus, heat must be energy. What we see from the definition, however, is that heat is a special form of energy: energy that is on its way from a hot body to a cold body. Heat always travels from hot to cold. If there is no difference in temperature, there is no heat.

Let's go back to discussing the fire. We know that the fire is sending out heat, because we can feel it. The question is, where did that heat come from? Since heat is energy that is being transferred from a hot body to a cold body, there must be energy in the fire that is being transferred to the things surrounding the fire. Where did that energy first come from?

It turns out that the energy you feel coming from the fire was actually there long before the fire started. The wood that made up the fire and the oxygen in the air surrounding the fire contained all of the energy that the fire transfers to you in the form of heat. So when a pile of wood is just sitting in a fireplace, it has an enormous amount of energy in it. That energy has the potential of heating an entire room for a long period of time. Why doesn't the wood by itself feel hot, then? If there is energy contained in a pile of wood, why don't we feel heat coming from it? Why do you first have to light the wood and make a fire before you feel heat? Because the energy is trapped, or stored, inside the wood and cannot be transferred to you until the wood is burned. When the wood burns, the energy that has been stored there is released and transferred to you in the form of heat. Only then will it feel hot!

This discussion leads us to one of the most important concepts in all of science: **The First Law of Thermodynamics** (ther' moh dye nam' iks). Before we discuss this very important law, however, we have to understand exactly what is meant by the term "scientific law." As you will see from even a brief look at history, science is fallible. Things that scientists once thought were absolutely true have later been shown to be quite wrong. That's the nature of science. As human beings begin to investigate God's wonderful creation, we are bound to make mistakes in our attempts to understand it. As time goes on, however, many of those mistakes will hopefully be corrected, and we will come to a better understanding of the nature of creation.

Keeping this in mind, we cannot put as much faith in scientific laws as we can in the laws of God. God says, "For by grace you have been saved through faith; and that not of yourselves, it is the gift of God; not as a result of works, so that no one may boast" (Ephesians 2:8,9; NASB). This might be called the law of salvation through faith. This law, since it comes from God, can always be relied upon. God will always save you if you have faith in Christ. Scientific laws, on the other hand, are not nearly as reliable. Since they are created by human beings, and since human beings are fallible, scientific laws are also fallible.

The Nature of a Scientific Law

The way scientists develop laws is through experimentation and observation. After experimenting on or observing some facet of nature, scientists formulate a **hypothesis** to explain their observations. A hypothesis is no more than an educated guess that attempts to explain some aspect of the world around us. For example, when early scientists observed rotting meat, they always saw maggots crawling around on it. This led them to form the hypothesis that maggots are created from rotting meat.

Once a hypothesis has been formulated, scientists test it with more rigorous experiments. For example, after forming the hypothesis that maggots are created from rotting meat, early scientists did experiments to make sure that the maggots were not coming from something else. They would put rotting meat on a shelf high in the air to make sure that no maggot could crawl up to it. Even when the rotting meat was put high in the air, maggots still appeared on it. To early scientists, such experiments confirmed their hypothesis. Despite the fact that there was no way for maggots to crawl up to the rotting meat, they did indeed appear on it. Many similar experiments convinced early scientists that their original hypothesis was correct.

Once a hypothesis is confirmed by more rigorous experimentation, it is considered a **theory**. Early scientists called their theory "the theory of spontaneous generation." As the centuries passed, many more experiments were done to test the theory. Those experiments seemed to support the idea that life, such as maggots, can be created from non-life, such as rotting meat. After centuries of such experiments, the theory of spontaneous generation was considered a **scientific law**.

So we see that a scientific law is really nothing more than an educated guess that has been confirmed over and over again by experimentation. The problem with putting too much faith in a scientific law is that the experiments which established it might be flawed, making the scientific law itself flawed.

Such was the case with the law of spontaneous generation. All of the experiments done to confirm this law were flawed. Francesco Redi, an Italian physician, showed that if you were careful to completely isolate the rotting meat from the outside world, no maggots would appear. Those who believed in the law of spontaneous generation, however, were slow to give up faith in it. They believed that although maggots did not appear spontaneously on rotting meat, microscopic organisms did. The great French scientist Louis Pasteur performed careful experiments that finally overturned the law of spontaneous generation. His work showed that even microscopic organisms could not arise from non-life, but came to the meat by dust particles that blew in the wind.

The point of this rather long story is to illustrate to you that when you read about scientific laws, you must be careful not to put too much faith in them. Science is an endeavor created by flawed human beings, so science itself is flawed. Its laws are a good guide for trying to understand the nature of God's creation, but they are not absolute truth. Remember that.

The First Law of Thermodynamics

Now that we understand the limitations of science and its laws, let's learn what is probably the most important law of science:

The First Law of Thermodynamics - Energy cannot be created or destroyed. It can only change form.

What this law tells us is that God only made so much energy for the entire universe. This energy is used to do all of the work that has to be done in order for the universe to exist. Since humans are not as smart or powerful as God, there is no way that we can make energy. We cannot destroy it either.

"Wait a minute," you might say. "Aren't we creating energy when we burn something? Before we start the fire, the wood is not hot, but afterward, it is. Isn't that creating energy?" No. Remember what I said earlier: The energy was already in the wood. It was stored there, unable to move to you and warm you up. The act of starting the fire merely *released* the stored energy that was *already there*. Once the stored energy was released, it became heat and was able to warm you up.

In terms of the First Law of Thermodynamics, then, when you start a fire, you are changing the *form* of the energy that is already in the wood and the surrounding air. There are basically two forms of energy:

Potential energy - Energy that is stored

Kinetic energy - Energy that is in motion

In our fire example, then, the wood had a lot of *potential* energy. The energy was there, but was being stored. When the fire got started, that potential energy was changed, or converted, into *kinetic energy*. So in a fire, energy is not created; it is simply changed from potential to kinetic energy.

In chemistry, the difference between potential energy and kinetic energy is very straightforward. Heat is considered kinetic energy. This should make sense to you. After all, heat is defined as energy that is being transferred as a result of temperature differences. In order for something to be transferred, it must be moved. Since kinetic energy is energy in motion, it becomes pretty obvious that heat is kinetic energy. On the other hand, all substances have energy stored in them; thus, all substances have potential energy. These are not the only types of potential and kinetic energy in chemistry, however. In later modules, we will introduce more types of potential and kinetic energy. For right now, however, these ideas will suffice.

 (The multimedia CD has a video demonstration of The First Law of Thermodynamics.)

Units for Measuring Heat and Energy

Now that we know the definitions of heat and energy, the forms of energy, and a law of chemistry that governs energy, it is time to learn how we measure energy. In order to do that, we must first understand the units that govern energy measurement. In the metric system, energy is measured in a unit called the **Joule**. This unit is named after James Prescott Joule, a brilliant scientist who worked during the mid-1800s. Joule, a devout Christian, did pioneering research on energy, and we honor him by using his name as the metric unit for energy.

To give you an idea of how much energy is in one Joule, lifting a 102-gram object straight up (at a constant speed) for 1.0 meter takes 1.0 Joule of energy. Now remember how much each of these metric units is worth. One gram is approximately the mass of an ordinary housefly. So a 102-gram object has the same mass as about 102 flies (not all that much). One meter is just a little longer than a yardstick. Thus, 1.0 Joule is the energy needed to lift about 102 flies up in the air (at a constant speed) just a bit higher than a yard.

How do we measure energy? That's actually a tricky question. When in potential form, energy is very difficult to measure. There *are* ways of doing it, but they go beyond the scope of this course. On the other hand, it is relatively easy to measure energy in its kinetic form. Thus, we will spend a considerable amount of time on the measurement of heat.

In order to measure heat, we must first measure temperature. It is important to understand that temperature and heat are not the same, but they are related to each other. That's why we can use temperature to measure heat. We all know that in order to measure temperature we use a thermometer. But how does a thermometer work?

A thermometer makes use of the fact that substances tend to expand as they get warmer. Thus, a thermometer consists of a substance (usually mercury or alcohol) that is contained in a column of glass. As the substance gets warmer, it begins to expand. Since the only way it can expand is in the upward direction, it begins to move up the glass column. The height that it moves up the column is directly related to the temperature.

In the end, then, we measure temperature *indirectly*. Instead of finding some way to measure the temperature directly, we find it easier to measure the expansion of a substance and relate it back to temperature. This turns out to be quite common in science. Most measurements we do are, in fact, measurements of something else that we then can use to relate back to the physical quantity in which we are interested.

Now that we know *how* temperature is measured, we need to know what units are used in the measurement. The temperature unit that you are probably most familiar with is **Fahrenheit** (abbreviated as ^{o}F). Although this is a very common temperature unit, it is not used by chemists. Instead, chemists use one of two temperature units: **Celsius** (sel' see us, abbreviated ^{o}C) or **Kelvin** (kel' vuhn, abbreviated K). First we need to see how these units are defined, then we will see how they relate to Fahrenheit and why chemists use them.

Remember, when we measure temperature, we are really measuring how much a certain substance within the thermometer is expanding. Thus, we must find a way to relate that measurement to a unit which means temperature. We do this in the following manner:

1. We immerse a thermometer in a mixture of ice and water.
2. We make a mark where we see the liquid in the thermometer, and we assign a value to that mark. In the Celsius temperature scale, we call it exactly 0 ^{o}C. In the Fahrenheit scale, we give that mark a value of exactly 32 ^{o}F.
3. We then immerse the thermometer in a pot of boiling water.
4. We make a mark where we see the liquid in the thermometer, and we assign it a value of exactly 100 ^{o}Celsius or exactly 212 ^{o}Fahrenheit.

5. We divide the distance between the two marks into equal divisions, and then we have a temperature scale.

This method for defining a temperature scale is illustrated in Figure 2.1:

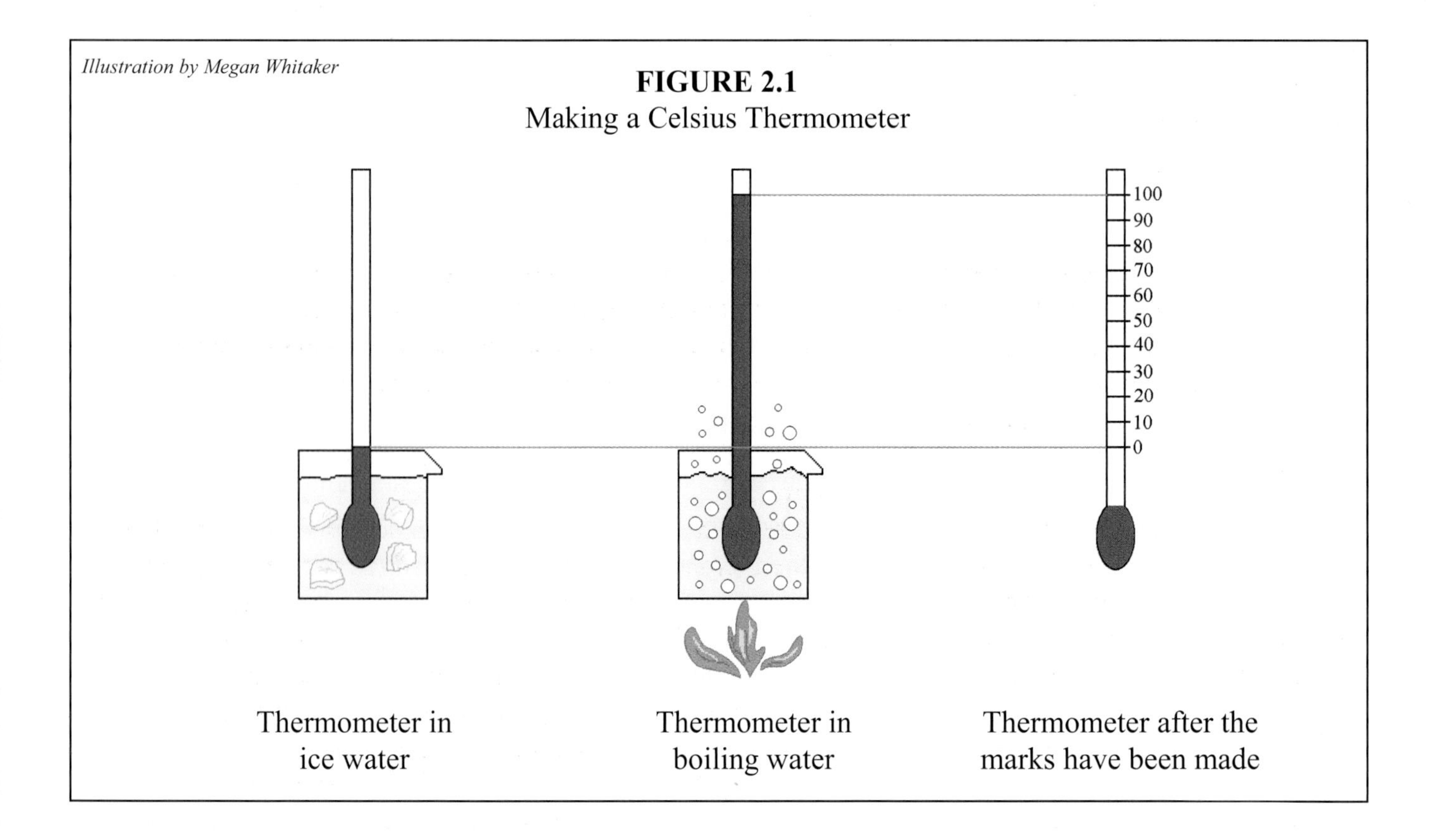

FIGURE 2.1
Making a Celsius Thermometer

This process of using certain physical measurements to define the scale of a measuring device is called **calibration.** This particular calibration makes use of a surprising fact in chemistry:

If ice and water are thoroughly mixed, the temperature of the mixture will stay the same (0.0 °C or 32.0 °F), regardless of the amount of ice or water present.

This might sound like a rather surprising statement, but it is, nevertheless, true. Even though you might think that a little water with a lot of ice is colder than a lot of water with a little ice, they are, in fact, the same temperature! Equally surprising:

Boiling water is always at the same temperature (100.0 °C or 212.0 °F at standard atmospheric pressure) whether it is boiling rapidly or hardly boiling at all.

You will learn in a later module *why* these statements are true. For right now, however, you will perform the following experiment that demonstrates their truth.

EXPERIMENT 2.1
Calibrating Your Thermometer

Supplies:

- A thermometer that you intend to use in future experiments (preferably marked in Celsius)
- A 250 mL beaker (A small glass container will do, but make sure it is safe for boiling water.)
- Water (preferably distilled water, which can be purchased at any supermarket)
- Ice (preferably crushed)
- An alcohol burner, or something like a hot plate or stove, to heat the water
- Safety goggles

1. Fill the beaker three-quarters full with ice (crushed if possible).
2. Add enough water so that the ice begins to float. Take your thermometer and *carefully* use it to stir the ice and water for five minutes. This will allow the ice to fully mix with the water.
3. After five minutes, read your thermometer (without removing it from the ice water) and write down the temperature reading. Remember that according to the rules we developed for reading a scale, you should be able to estimate between the dashes marked on the thermometer. Since most thermometers are marked off in increments of one degree, you should be able to write down your measurement to the nearest tenth of a degree.
4. Your thermometer, if it is Celsius, should read 0.0 degrees. If it is Fahrenheit, it should read 32.0 degrees. Now it is very likely that your thermometer will not, in fact, read the correct temperature. This is because most thermometers are not very accurately marked when they are made; thus, your thermometer may be as much as two degrees off. This is okay, as long as you write the reading down so that you can refer back to it often, because you will use it to correct any inaccuracy in your thermometer. For example, let's say your thermometer reads 1.8 degrees Celsius. This tells you that your thermometer reads 1.8 degrees too hot. In the future, when you measure temperature, you will correct for this inaccuracy by subtracting 1.8 from your measurements. Whatever you have to add or subtract to your measurement so that it turns into the correct value (0.0 degrees Celsius or 32.0 degrees Fahrenheit), you will do the same thing to all of your future measurements. This is another example of *calibration.*
5. Now that you have a calibrated thermometer, it is time to demonstrate the fact that any mixture of ice and water is always at the same temperature once it is thoroughly mixed. Slowly begin to heat up the container, while constantly stirring carefully with your thermometer. Keep your flame, stove, or hot plate on low. If you add heat too quickly, you will not be able to stir quickly enough to evenly distribute the heat throughout the ice water.
6. While heat is being added to the ice water, read the temperature every minute. Write down both the time of the reading and the temperature, being sure to use your calibration to correct your thermometer. In order for this experiment to work properly, you must keep stirring the ice water mixture at all times except for the instances in which you read the thermometer. In addition, *be sure that your thermometer never touches the edges or bottom of the glass container.* Keep the thermometer in the ice water, but do not allow it to come in contact with the container!
7. As you watch your experiment, you should see that even though the ice starts melting, the temperature of the ice water does not change!
8. Continue to heat the container and make your temperature readings every minute until all of the ice has melted. Note the time at which all of the ice is gone.
9. After that, continue for at least five more minutes. If you are using a beaker (or a glass container that is safe for boiling), you can continue this experiment until the water begins boiling vigorously. If you do this, it may take a while. To speed things up, you could increase the heat source to add

heat more quickly. Continue to make temperature measurements until you have taken several measurements while the water is boiling vigorously.

10. Take your temperature measurements and use graph paper to graph all of the results you obtained up to five minutes after all of the ice melted. Put the time (in minutes) on the x-axis and the temperature on the y-axis. Your y-axis should have a range of -1.0 °C to 5 °C, and your x -axis should start at 0 minutes and end five minutes after all of the ice melted. If any of your temperature measurements within this time frame were greater than 5 °C, then the maximum on your y-axis should be equal to the maximum temperature. Your graph should look something like this:

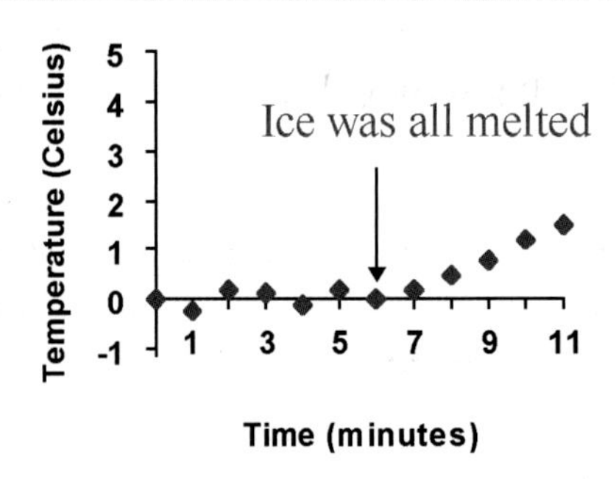

11. You should notice a couple of things in this graph. First of all, notice how the first several data points bounce up and down right around a temperature of 0.0 °C. This is called **experimental scatter** and is simply a reflection of the fact that all measurements have some inaccuracy in them. Any single measurement may not be all that accurate, but the more measurements you have, the more accurate your final result will be. Since all of the data points scatter around 0.0 degrees, you can assume that the ice water stayed at 0.0 °C during the entire time ice was present. Second, notice that after 7 minutes, the temperature of the water begins a slow but steady increase. This represents the point at which all of the ice was melted. Only after the ice was melted did the temperature of the water increase. Thus, no matter how much ice was present, the ice water mixture was always at a temperature of 0.0 °C! Even though the ice water was being heated, it would not change its temperature until all of the ice had melted. The reason for this surprising result is actually pretty simple:. It takes energy for ice to melt. It also takes energy to increase temperature. When a mixture of ice and water is heated, all of the energy it receives is used to melt the ice, and none of it can be used to raise the temperature. Once all of the ice is melted, then the incoming energy can be used to increase the temperature of the water.

12. If you continued the experiment through boiling, make another graph that includes that data as well. The graph should look something like this:

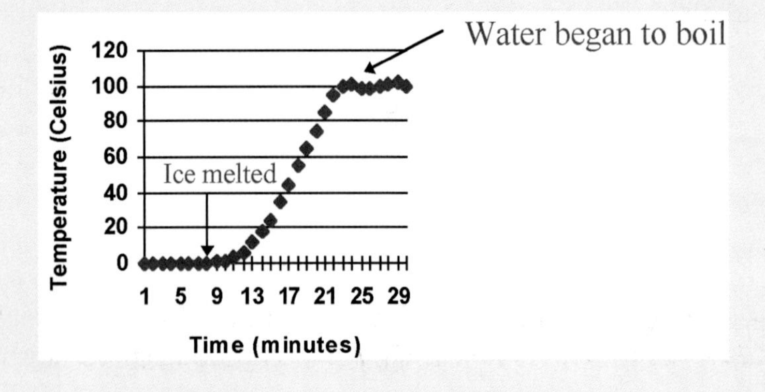

13. Notice how the data levels off at 100.0 °C. Once again, this is because when the water begins boiling, all of the energy that is being transferred to the water is being used to create the motion of boiling. As long as the water continued to boil, it would always stay at the same temperature. Graphs like the one pictured above are often called **heating curves**.
14. If your graph does not level off at precisely 100.0 °C, don't worry about it. We will learn in a later module that the boiling temperature of water depends a bit on the atmospheric pressure. If the atmospheric pressure is a little higher than normal on the day you perform this experiment, the boiling temperature will be slightly higher than 100.0 °C. If the atmospheric pressure is a little lower than normal, then your boiling temperature will be slightly under 100.0 °C. One other thing to consider is that the atmospheric pressure depends on your elevation. If you live somewhere like Denver, Colorado, your graph will probably level off well under 100.0 °C.

Now that we know how the Celsius temperature unit is defined, we can learn how it relates to the Fahrenheit unit. It should make sense to you that Fahrenheit and Celsius relate to one another since they both measure the same thing: temperature. It turns out that they are related by a very simple equation:

$$^{\circ}C = \frac{5}{9}(^{\circ}F - 32) \tag{2.1}$$

In this equation, °C represents the temperature in Celsius and °F stands for the temperature in degrees Fahrenheit. So if you must use a Fahrenheit thermometer in your experiments, you can use this equation to convert your measurements into Celsius. Please note one very important thing about this equation: The 5, 9, and 32 are all exact. Thus, you need not worry about their significant figures. They have infinite precision and an infinite number of significant figures. Thus, the only significant figures you must worry about are those of the original measurement. The following example problems will help show you how this is done:

EXAMPLE 2.1

A student uses a Fahrenheit thermometer to do a chemistry experiment but then must convert his or her answer to Celsius. If the temperature reading was 50.0 °F, what is the temperature in Celsius?

To solve this one, we simply use Equation (2.1):

$$^{\circ}C = \frac{5}{9}(50.0 - 32)$$

$$^{\circ}C = 10.0$$

There are three significant figures in the original measurement. Since the other numbers in this equation are exact, the answer must have three significant figures. Thus, the answer is 10.0 °C.

We usually say that room temperature is about 25 °C. What is this temperature in Fahrenheit?

In order to solve this one, we must first rearrange Equation (2.1) using algebra. Once we do this, we get the following equation:

$$^{\circ}F = \frac{9}{5}(^{\circ}C) + 32$$

$$^{\circ}F = \frac{9}{5}(25) + 32$$

$$^{\circ}F = 77$$

The presence of only two significant figures in the original number means only two significant figures in the end; therefore, the temperature is 77 °F.

Try the following problems to see whether or not you fully understand this type of conversion:

ON YOUR OWN

2.1 Normal body temperature is 98.6 °F. What is this temperature in Celsius?

2.2 If water boils at 100.0 °C, what is the boiling temperature of water in Fahrenheit?

What about the other unit I mentioned earlier? The Kelvin temperature unit is a special unit that we will use quite a bit in later modules. It is special because you can never reach a temperature of 0 Kelvin or lower. This fact makes the Kelvin temperature scale different from most others. After all, anything colder than ice water has a negative temperature in Celsius units. This means that temperatures less than zero are quite common in the Celsius scale. Although not quite as common, it is also possible to reach temperatures less than zero in the Fahrenheit scale as well. It is impossible, however, for anything in nature to reach 0 Kelvin or below. We will see why this is the case in later modules. Since you can never get to or go below 0 Kelvin, the Kelvin temperature scale is often called an **absolute temperature scale**.

Once you have your temperature in units of degrees Celsius, converting it to Kelvin is a snap. All you do is add 273.15 to your measurement. In mathematical terms, we would say:

$$K = {}^{\circ}C + 273.15 \tag{2.2}$$

where K is the temperature in units of Kelvin and °C is the temperature in units of Celsius. In this equation, the 273.15 is *not* exact. Thus, its precision plays a role. Note that since this equation involves adding, we use the rules of addition and subtraction when determining the significant figures involved. Those rules are *different* than the ones for multiplication and division (see Module #1), so you need to be aware of that. The following two example problems will show you how to use this equation:

EXAMPLE 2.2

What is the boiling temperature of water in Kelvin?

This conversion is a snap. We just realize that water boils at 100.0 °C. If we put that temperature into Equation (2.2), we get:

$$K = 100.0 + 273.15 = 373.15$$

The original temperature goes out to the tenths place, while 273.15 goes out to the hundredths place. The rules for significant figures in adding tell us that the answer must have the same precision as the least precise number in the equation. Thus, our final answer is 373.2 K.

The lowest temperature that has ever been recorded in the United States of America is -80.0 °F. What is this temperature in Kelvin?

Since the only way we can get to Kelvin is by adding 273.15 to the temperature in Celsius, we must first convert °F into °C. After that, we can convert to Kelvin. First, then,

$$^{\circ}C = \frac{5}{9}(-80.0 - 32)$$

$$^{\circ}C = -62.2$$

Now that we have the answer in °C, we can easily convert to Kelvin:

$$K = -62.2 + 273.15 = 210.95$$

Thus, our final answer is 211.0 K. Once again, our rules for adding tell us that the precision must be kept to the tenths place, as that is the same precision as the least precise number in the equation. That's why our final answer goes out to the tenths place. So you see, even very, very cold temperatures in the Celsius and Fahrenheit temperature scales are still rather large numbers in the Kelvin temperature scale!

Now cement your knowledge of temperature conversions with this problem:

ON YOUR OWN

2.3 What is the Fahrenheit equivalent of 0.00 Kelvin?

The Calorie Unit

Now that we know how temperature is measured and what units it is measured in, we can begin our discussion of measuring heat. The first thing we need to know is that chemists sometimes use a unit other than the Joule for measuring heat. This unit is the *calorie* and is defined as follows:

Calorie (cal) - The amount of heat necessary to warm one gram of water one degree Celsius

You might wonder why chemists sometimes use the calorie unit to measure heat. After all, since heat is energy, it could be measured in Joules. That's true, but chemists didn't always know that heat was just another form of energy. Thus, they would measure heat with the calorie unit and energy with a different unit. When James Joule demonstrated that heat was just another form of energy, it was a major breakthrough in our understanding of energy. It was such a profound accomplishment that we honor him by naming the metric energy unit after him. However, since old habits die hard, chemists sometimes use the calorie to measure heat, even though the standard metric unit is the Joule.

Since Joules and calories really measure the same thing, there must be a relationship between them:

1 calorie = 4.184 Joules

Since it takes a little more than 4 Joules to make a calorie, you can see that the calorie is a bigger energy unit than the Joule. This relationship can be used to convert between calories and Joules using the factor-label method that we employed in Module #1. Note that the 4.184 is not exact; thus, its significant figures must be taken into account when using it in a conversion.

Most people have heard of calories before. If you look at the nutritional information on any food-product label, you will see a listing of how many Calories are contained in one serving of the product. Those who are watching their weight are usually quite conscious of this number. Do the Calories that you see on a food label have anything to do with the calories we are discussing here? Yes.

When you eat food, one of the most important things that the food can give you is energy. Most of the things that you do (walking, running, breathing, even sitting down and reading a book) involve work. Since work requires energy, your body needs a steady supply of it. Your body gets the energy it needs from the food you eat. The Calories listed on the label of a soup can, for example, measure how much energy a serving of that soup provides to your body. Your body can then use that energy for any work it needs to do.

For some strange reason, however, the food Calorie (notice I use a capital "C" here) is not the same as a chemistry calorie (written with a small "c"). Food generally contains an enormous amount of energy; therefore, the calorie as we defined it earlier is simply too small a unit to use when measuring the energy content of food. Instead, 1 food Calorie is equal to 1,000 chemistry calories.

1 food calorie (Cal) = 1,000 chemistry calories (cal)

Based on our discussion of the metric system in the previous module, we should call the food Calorie a "kilocalorie." Unfortunately, food scientists do not like to use that term, so they drop the "kilo" prefix and just use a capital "C" instead. It is a little confusing, but we will just have to live with it.

Think for a moment about what all of this means: We defined a calorie as the amount of energy required to heat one gram of water one degree Celsius. We now know that a food Calorie is, in fact, 1,000 of those calories. Thus, a food Calorie can heat up 1,000 grams of water 1.0 degree Celsius. Now consider the fact that in an normal day, the average human uses 2,500 food Calories to do all its work. That's enough energy to heat 2.5 *million* grams of water 1.0 degree Celsius. That's a lot of energy! The amazing thing is that God has designed your body to use all of that energy without overheating!

Measuring Heat

Although chemists have traditionally measured heat in calories, most newer chemistry textbooks use the standard metric unit, the Joule. Since you will probably have to use Joules in your college chemistry course, it is probably a good idea for you to start using them now. Thus, most of the problems in this course will use Joules rather than calories. However, I will have you use calories occasionally, just to keep them in your mind.

As I mentioned at the beginning of the module, when heat is given to an object, it can cause that object's temperature to increase. Thus, if I measure the temperature of an object and notice that it begins to rise, I can assume that the object is absorbing some heat. Alternatively, if I notice that the temperature is decreasing, that tells me the object is releasing (losing) heat. The temperature increase or decrease I measure should be, in some way, proportional to the amount of heat the object absorbs or releases.

It turns out that the temperature change in an object is related to the amount of heat it absorbs or releases by the following equation:

$$q = m \cdot c \cdot \Delta T \qquad (2.3)$$

In this equation, "q" stands for the amount of heat absorbed or released, "m" represents the mass of the object, and "c" is the **specific heat** of the object, which will be discussed momentarily. The last symbol in the equation, "ΔT" (pronounced "Delta T"), stands for the change in temperature and is defined as the difference between the final temperature and the initial temperature:

$$\Delta T = T_{final} - T_{initial} \qquad (2.4)$$

The "Δ" symbol is frequently used in chemistry. It is the Greek capital letter "Delta," and it means "change in." So whenever you see a "Δ" in front of a variable, it means you are looking for a change in that variable.

The variable "c" that I used in Equation (2.3) above is a very important quantity in the study of energy. It is called **specific heat capacity** or, usually, specific heat. The physical meaning of specific heat is a very important concept to learn. The definition is as follows:

Specific heat - The amount of heat necessary to raise the temperature of 1 gram of a substance by 1 degree Celsius

In other words, specific heat tells how easy it is to heat up a substance. For example, if I put a metal plate on the burner of a stove and I put a wooden plate on another burner of the stove, which one do you think would get hot faster? As long as the wood did not catch on fire, the metal plate would get much hotter much faster. This is because the metal plate has a lower specific heat than the wood. Since metals have low specific heats, a small amount of heat changes the temperature significantly. On the other hand, wood has a relatively large specific heat, so it takes more heat to produce the same temperature change in wood.

Like density, each substance has a unique specific heat. Table 2.1 lists some common substances and their specific heats. You needn't memorize any of these numbers; they are simply provided for illustration. When doing problems, however, you may need to refer back to this table.

TABLE 2.1
Specific Heats of Common Substances

Substance	Specific heat $\left(\frac{J}{g \cdot {}^{o}C}\right)$
Copper	0.3851
Iron	0.4521
Glass	0.8372
Aluminum	0.9000

In order to determine whether or not you understand the concept of specific heat, answer the following "On Your Own" questions:

ON YOUR OWN

2.4 Which substance takes the least energy to heat up, copper or glass? (refer to Table 2.1)

2.5 A copper kettle and an iron kettle are placed side by side on a fire. After 3.0 minutes on the fire, the iron kettle reaches a temperature of 100.0 °C. Will the copper kettle's temperature be higher or lower than 100.0 °C? (refer to Table 2.1)

Now that we know what specific heat is, we need to learn its units. Looking back at Equation (2.3), you should recognize the units that we will use for all quantities in that equation except the specific heat. The units for heat (q) will either be calories or Joules, the units for mass (m) will be something like grams, kilograms, milligrams, etc., and the units for temperature will either be °F, °C, or K. What will the units for specific heat be? That can be determined from the equation.

Let's suppose you wanted to measure heat with the units of Joules. In the end, then, all of the units on the right-hand side of Equation (2.3) would have to work out to a final unit of Joules. If you measured temperature in °C and mass in grams, those units would have to somehow cancel and leave the unit of Joules in their place so that heat (q) would have the proper unit. How does this happen? It happens by referring to the units for specific heat.

The only way for °C to cancel out is for that same unit to be in the denominator of the unit for specific heat. That way, when specific heat is multiplied by the change in temperature, °C would cancel. In the same way, in order for grams to cancel, they would also have to appear in the denominator of the unit for specific heat. On the other hand, since the final answer must be in Joules, then Joules must appear in the numerator of the unit for specific heat. In the end, then, the unit for specific heat in this case must be $\frac{\text{J}}{\text{g}\cdot{}^{\text{o}}\text{C}}$.

Of course, if you decided to measure heat in calories instead of Joules, that would change the unit for specific heat. It would then become $\frac{\text{cal}}{\text{g}\cdot{}^{\text{o}}\text{C}}$. If you wanted to measure the mass in kg, the specific heat unit would be $\frac{\text{cal}}{\text{kg}\cdot{}^{\text{o}}\text{C}}$. In general, then, the unit for specific heat will have an energy unit (calories or Joules) in the numerator, a mass unit (grams, kg, mg, etc.) in the denominator, and a temperature unit (°C, °F, or K) in the denominator. See if you can follow this type of reasoning in these two examples:

EXAMPLE 2.3

In order to cook an omelet, a 5.0-kg iron skillet must be heated from room temperature (25.0 °C) to 150.0 °C. How many Joules of heat must be used to accomplish this?

In order to solve this, we use Equation (2.3) and Table 2.1. The table gives us the specific heat of iron in $\frac{\text{J}}{\text{g}\cdot{}^{\text{o}}\text{C}}$. Thus, we must use temperature in °C, mass in grams, and we will get our answer in Joules. So first we have to convert 5.0 kg into grams:

$$\frac{5.0\,\cancel{\text{kg}}}{1} \times \frac{1{,}000\,\text{g}}{1\,\cancel{\text{kg}}} = 5.0 \times 10^3\ \text{g}$$

Since there are two significant figures in our mass measurement, we can only have two significant figures after the conversion, so our mass is $5.0\text{x}10^3$ grams. Note that scientific notation is the only way we can express the mass properly, since the zero must be significant. Had we used 5,000 grams, there would only be one significant figure. Now that we have everything in the proper units, we can use Equation (2.3):

$$q = (5.0\text{x}10^3\ \text{g}) \cdot \left(0.4521\,\frac{\text{J}}{\text{g}\ {}^{\circ}\text{C}}\right) \cdot (150.0\ {}^{\circ}\text{C} - 25.0\ {}^{\circ}\text{C})$$

$$q = (5.0\text{x}10^3\ \text{g}) \cdot \left(0.4521\ \frac{\text{J}}{\text{g}\cdot{}^\circ\text{C}}\right) \cdot (125.0\ {}^\circ\text{C})$$

$$\underline{q = 280{,}000\text{J}}$$

Notice that since we used all of the same units that we found in our specific heat, everything except Joules cancels out.

Also, you need to pay attention to the significant figures here. Notice that in order to calculate ΔT, I had to subtract 25.0 °C from 150.0 °C. When I did that, I had to follow the rules of addition and subtraction. Both temperatures went out to the tenths place, so the answer can go out to the tenths place as well. Thus, the value for ΔT has four significant figures. In the end, however, I had to multiply that number by 5.0×10^3 g, which has only two significant figures. Thus, the final answer can have only two significant figures. I could have expressed the answer in scientific notation (2.8×10^5 J) as well. Either expression would be correct.

If a 305.4-gram mass of copper that is initially at 75 °F is given 1,234.0 Joules of energy, what will its final temperature be?

In order to solve this one, we must solve for ΔT. Since we already know the initial temperature, solving for ΔT will allow us to calculate the final temperature. So, the first thing we have to do is use algebra to rearrange the equation:

$$\Delta T = \frac{q}{m \cdot c}$$

Once we do that, we now have to worry about units. The only unit we have specific heat in is $\frac{\text{J}}{\text{g}\cdot{}^\text{o}\text{C}}$.

Thus, we need to convert our temperature into °C:

$${}^\circ\text{C} = \frac{5}{9}({}^\circ\text{F} - 32)$$

$${}^\circ\text{C} = \frac{5}{9}(75 - 32) = 24$$

Since our temperature in °F only had two significant figures, our final answer can have only two, so it was rounded to 24. Now that we have all of our numbers in the correct units, we can use the first equation to calculate ΔT:

$$\Delta T = \frac{1234.0\ \text{J}}{305.4\ \text{g} \cdot 0.3851 \frac{\text{J}}{\text{g}\cdot{}^\circ\text{C}}}$$

$$\Delta T = 10.49\ {}^\circ\text{C}$$

Since all of the numbers in the equation have at least four significant figures, ΔT can also have four significant figures. Now that we have ΔT, we can use its definition to solve for the final temperature:

$$\Delta T = T_{final} - T_{initial}$$

$$T_{final} = \Delta T + T_{initial}$$

$$T_{final} = 10.49\ ^{\circ}C + 24\ ^{\circ}C = 34\ ^{\circ}C$$

Notice that in this case, we are adding, so we look at decimal place. The value for ΔT goes out to the hundredths place, but the value for $T_{initial}$ goes out to the ones place. Thus, the answer can go out only to the ones place. The chunk of copper, then, reaches a final temperature of 34 °C.

Now see if you really understand this by doing the following problems:

ON YOUR OWN

2.6 A 502 g iron rod is heated to an unknown initial temperature. If, while cooling down to a temperature of 22 °C, the rod releases 597.5 Joules of energy, what was its initial temperature? (HINT: When an object loses energy, its "q" is negative. We will discuss this fact in detail shortly.)

2.7 If a 15.6 g object requires 836.8 J of heat to raise its temperature by 21 °C, what is the specific heat of the object?

Calorimetry

Even though I mentioned that you needn't memorize any of the specific heats listed in Table 2.1, there is one specific heat that you do need to memorize: the specific heat of water. This is an easy one to memorize, because it relates back to the definition of a calorie. Remember that we defined a calorie as the amount of heat necessary to raise one gram of water 1.0 degree Celsius. If that is the definition of a calorie, what does that tell us about the specific heat of water?

The specific heat of water is $\mathbf{1.000 \frac{cal}{g \cdot {}^{o}C}}$ **or** $\mathbf{4.184 \frac{J}{g \cdot {}^{o}C}}$

These two numbers are very important to chemists, so you really need to memorize them.

Now that we understand the equation that relates temperature changes to heat measurements, we can finally see how heat is actually measured by chemists. The experimental process that chemists use to do this is called **calorimetry** (kal uh rim' uh tree) and is illustrated by Figure 2.2:

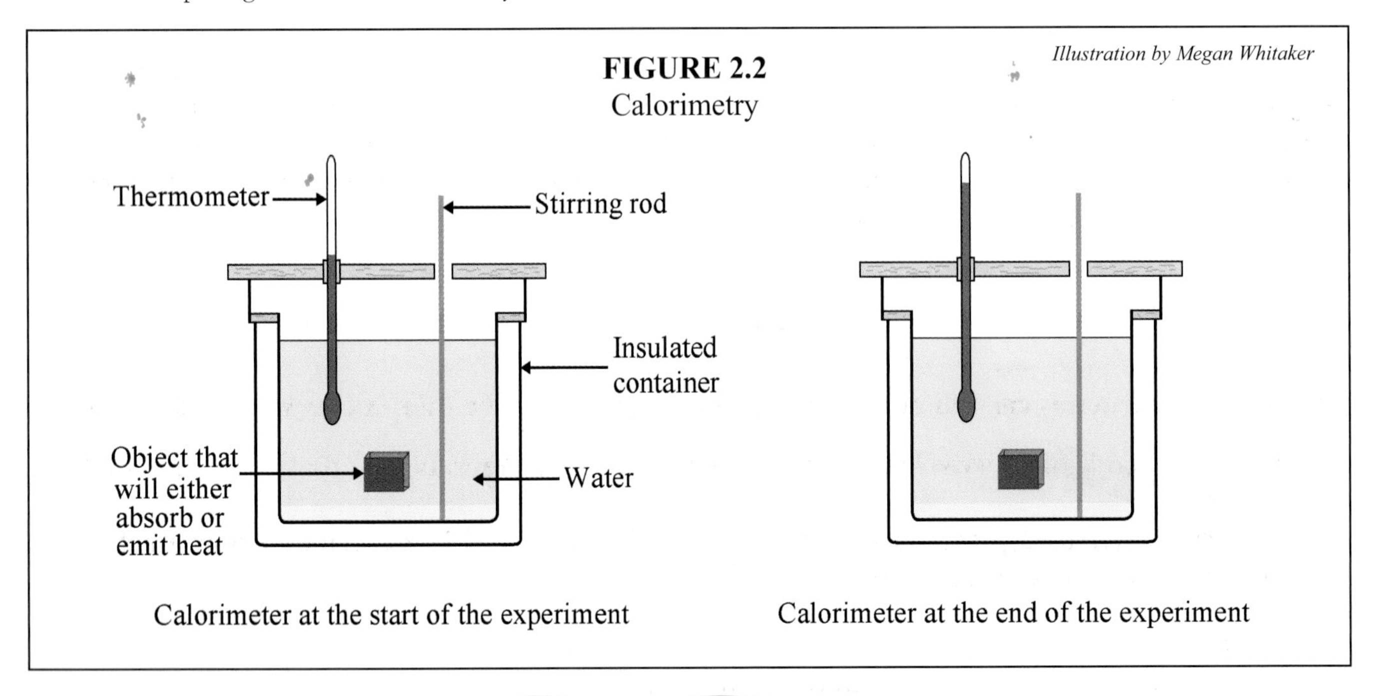

FIGURE 2.2
Calorimetry

In calorimetry, chemists use an apparatus known as a **calorimeter** (kal uh rim' uh ter). A calorimeter consists of an insulated container that holds a known mass of water. The temperature of the water is constantly measured by a thermometer, and there is usually a rod that stirs the water to keep the heat distributed evenly. When a hot object is placed in the calorimeter, it begins to emit heat. That heat is absorbed by the water and the calorimeter, causing the temperature measured by the thermometer to increase. Because heat is being transferred from the object to the water and the calorimeter, the water and calorimeter will begin to warm up and the object will begin to cool down. At some point, they will reach the same temperature, and heat will no longer be transferred from the object to the water and calorimeter. Thus, the temperature measured by the thermometer will stop increasing. This tells the chemist that the experiment is finished.

The difference in temperature between the end of the experiment and the beginning of the experiment can then be used as the "ΔT" in Equation (2.3). Since we know the specific heat of water, the specific heat of the calorimeter, and the mass of each, we can use Equation (2.3) to calculate the total amount of heat that the object emitted. If you think about it, we can write a simple equation to relate the heat gained or lost by the object, the water, and the calorimeter.

$$-q_{object} = q_{water} + q_{calorimeter} \qquad (2.5)$$

You can think of this equation as the **calorimetry equation**. It tells us that any heat gained by the calorimeter and water must be lost by the object placed in the calorimeter. This makes sense in light of the First Law of Thermodynamics. Energy cannot be created or destroyed. Thus, if the water and calorimeter *gain* energy, that energy must come from somewhere. It comes from the hot object.

Why is there a negative sign in front of "q_{object}"? Think about it. The object is *releasing* heat. That means it is *losing* energy. If it is losing energy, its "q" will be negative. However, the water and the calorimeter are *gaining heat*. That means they will each have a *positive* "q." In order to make the two sides equal, then, we must put a negative sign in front of "q_{object}". You will see how this works in a moment.

In the end, then, if we want to measure the amount of heat released by a substance, we can surround that substance with water and a calorimeter, and measure the change in temperature of the water and the calorimeter. Equations (2.5), (2.4), and (2.3) will then tell us exactly how much heat was released. We can also measure the specific heat of a substance with calorimetry. The following two examples should help you see how this is done:

EXAMPLE 2.4

A chemical reaction is carried out in a 5.0-gram calorimeter. The calorimeter has a specific heat of 1.95 $\frac{J}{g \cdot {}^{o}C}$ and is filled with 150.0 grams of water. If the temperature of the water was 24.0 °C before the reaction took place, and it rose to 32.0 °C by the end of the experiment, how much heat was released by the reaction?

Equation (2.5) tells us that:

$$-q_{object} = q_{water} + q_{calorimeter}$$

Equation (2.3) gives us a way to calculate "q_{water}" and "$q_{calorimeter}$". After all, Equation (2.3) tells us that to calculate the heat gained or lost by a substance, we need to know the mass, specific heat, and change in temperature of the substance. We know all of those things for both the calorimeter and the water. First, let's calculate ΔT with Equation (2.4). Since the water and calorimeter start out in contact with each other and remain in contact with each other the entire time, they are at the same temperature throughout the course of the experiment. Thus, they start out at the same temperature (24.0 °C), and they end up at the same temperature (32.0 °C). This means they have the same ΔT:

$$\Delta T = T_{final} - T_{initial} = 32.0\ {}^{o}C - 24.0\ {}^{o}C = 8.0\ {}^{o}C$$

Notice that since we are subtracting here, we must look at decimal place to determine what digits to keep. Since both temperatures go out to the tenths place, our answer can go out to the tenths place. Now that we know ΔT for both the water and the calorimeter, we can calculate the heat they gained. Let's start with the calorimeter. We know its mass and specific heat, and we just determined its ΔT:

$$q_{calorimeter} = m \cdot c \cdot \Delta T$$

$$q_{calorimeter} = (5.0\ \cancel{g}) \cdot \left(1.95\ \frac{J}{\cancel{g} \cdot {}^{o}\cancel{C}}\right) \cdot (8.0\ {}^{o}\cancel{C})$$

$$q_{calorimeter} = 78\ J$$

We can now calculate the heat gained by the water using the mass of the water, the specific heat of the water (which you should memorize), and the same ΔT:

$$q_{water} = m \cdot c \cdot \Delta T$$

$$q_{water} = (150.0\,\cancel{g}) \cdot \left(4.184 \frac{J}{\cancel{g} \cdot {}^{o}\cancel{C}}\right) \cdot (8.0\ {}^{o}\cancel{C})$$

$$q_{water} = 5.0 \times 10^{3}\ J$$

Now that we know the "q" for both the water and the calorimeter, we can use Equation (2.5):

$$-q_{object} = q_{water} + q_{calorimeter} = 5.0 \times 10^{3}\ J + 78\ J = 5{,}078\ J$$

$$q_{object} = \underline{-5{,}100\ J}$$

Notice the significant figures here. The "q" for the water has its last significant figure in the hundreds place. Don't let the scientific notation fool you. The zero is to the right of the decimal, but that is still the hundreds place, because $0.1 \times 10^{3} = 100$. Thus, its last significant figure is in the hundreds place. The "q" for the calorimeter has its last significant figure in the ones place. Based on the rules of addition and subtraction, then, the final answer can have its last significant figure in the hundreds place.

What does the negative sign mean? It means that the reaction *lost* energy. That makes sense, because the water and calorimeter warmed up, which means they gained energy. Since they gained energy, the chemical reaction must have lost energy, and that's why it has a negative "q."

Although calorimetry is often used to determine the heat released by a chemical reaction (as was done in the example above), it can also be used to determine the specific heat of an object. In fact, the specific heats listed in Table 2.1 were measured using calorimetry. To see how this is done, study the following example.

EXAMPLE 2.5

A 28.5-gram block of an unknown metal is heated to a temperature of 151.7 °C. It is then dropped into the same calorimeter that was used in the previous example (m = 5.0 g, c = $1.95 \frac{J}{g \cdot {}^{o}C}$, mass of water = 150.0 grams). If the initial temperature of the water was 22.0 °C and its final temperature was 24.2 °C, what is the specific heat of the metal and what is its identity?

Once again, we can use Equation (2.5), but we must first calculate all of the "q"s that we can. We can first calculate the ΔT of the water and calorimeter:

$$\Delta T = T_{final} - T_{initial} = 24.2\ {}^{o}C - 22.0\ {}^{o}C = 2.2\ {}^{o}C$$

We can now calculate "q" for both the water and calorimeter, because we have all the information we need for Equation (2.3). Let's start with the calorimeter:

$$q_{calorimeter} = m \cdot c \cdot \Delta T$$

$$q_{calorimeter} = (5.0\,\cancel{g}) \cdot \left(1.95 \frac{J}{\cancel{g} \cdot {}^{o}\cancel{C}}\right) \cdot (2.2\ {}^{o}\cancel{C})$$

$$q_{calorimeter} = 21\ J$$

Now we can move on to the water:

$$q_{water} = m \cdot c \cdot \Delta T$$

$$q_{water} = (150.0\,\cancel{g}) \cdot \left(4.184 \frac{J}{\cancel{g} \cdot {}^{o}\cancel{C}}\right) \cdot (2.2\ {}^{o}\cancel{C})$$

$$q_{water} = 1{,}400\ J$$

Now that we have calculated the "q"s of the water and calorimeter, we can use Equation (2.5) to determine the "q" of the metal. Since the metal is the object in the calorimeter, I will replace the word "object" with the word "metal:"

$$-q_{metal} = q_{water} + q_{calorimeter} = 1{,}400\ J + 21\ J = 1{,}421\ J$$

$$q_{metal} = -1{,}400\ J$$

Before we go on, I want you to look at the significant figures here. The "q_{water}" has its last significant figure in the hundreds place, and the "$q_{calorimeter}$" has its last significant figure in the ones place. Thus, by the rule of addition, we can only report the answer to the hundreds place. Thus, *the "q" of the calorimeter did not affect the answer*. It gained so little heat that its effect was too small to matter in terms of the significant figures that we are allowed to keep. I want to come back to this point after we finish the example.

Okay, then, what do we have now? We have "q_{metal}." How does that help? Well, we can calculate the ΔT of the metal, and we know its mass. Thus, now that we know its "q," we can use Equation (2.3) to calculate its specific heat! Since the metal was in contact with the water and the calorimeter after the experiment ended, the metal's final temperature was the same as the water's final temperature. Its initial temperature, however, was quite different from that of the water and calorimeter. Thus, the ΔT of the metal will be different from the ΔT of the water and calorimeter:

$$\Delta T_{metal} = T_{final} - T_{initial} = 24.2\ {}^{o}C - 151.7\ {}^{o}C = -127.5\ {}^{o}C$$

Notice this ΔT is negative, because the object cooled down. Now we can rearrange Equation (2.3) using algebra and solve for the specific heat of the metal:

$$c = \frac{q}{m \cdot \Delta T}$$

$$c = \frac{-1{,}400\ J}{28.5 g \cdot (-127.5\ ^{\circ}C)}$$

$$c = 0.39\ \frac{J}{g \cdot ^{\circ}C}$$

Looking at Table 2.1, we can see that the metal must have been copper, because the specific heat that we determined is the same (to two significant figures) as the specific heat of copper.

That was a long problem, so let's review how we were able to solve it. When we read the problem, we noticed it was a calorimetry problem, so we knew that Equations (2.3) - (2.5) would be involved. Based on the data, we recognized that we could calculate the heat absorbed by the water and the calorimeter. Once we calculated those heats, we plugged them into Equation (2.5) so that we could solve for the heat lost by the metal. Once we did that, we simply used Equation (2.3) again with all of the metal's data in it. Since the only thing we did not know in that equation was the metal's specific heat, we could solve for it.

Notice one more thing about this example problem: The heat absorbed by the calorimeter was so small (21 Joules), that it did not even "show up" when we added it to the heat absorbed by the water (1,400 J). This happens often in calorimetry experiments. As a result, we can often disregard the heat absorbed by the calorimeter when doing calorimeter experiments. If that is the case, then, we can treat $q_{calorimeter}$ in Equation (2.5) as being equal to zero. This simplifies the analysis of calorimetry experiments quite a bit. Thus, if I tell you to ignore the calorimeter in a calorimetry problem, just set $q_{calorimeter}$ equal to zero. However, if I give you information about the calorimeter (its mass or specific heat, for example), then you cannot ignore the calorimeter, and you must solve the problem the way we did in Example 2.5.

In order to help you fully understand the process of calorimetry, do the following experiment:

EXPERIMENT 2.2

Measuring the Specific Heat of a Metal

Note: A sample set of calculations is available in the solutions and tests guide. It is with the solutions to the practice problems.

Supplies:

- A calibrated thermometer
- A scale that reads mass, preferably in grams
- Two Styrofoam cups
- A chunk of metal that has mass of at least 30 grams (a lead sinker or a very large steel nut, for example)
- Boiling water (either in a pot or a beaker)
- Kitchen tongs
- Safety goggles

1. Heat water in a pan or beaker until it is boiling vigorously.
2. While you are waiting for the water to boil, measure the mass of the metal with your scale. Remember to report your answer to one more decimal place than what is marked off by the scale.

3. Once the water on the stove is boiling, drop the metal into the pot and let it sit there for about five minutes. This will heat the metal to the temperature of the boiling water (100.0 °C).
4. While the metal is heating, take the two Styrofoam cups and nest one inside the other. This will be your calorimeter. Since Styrofoam is a good insulator, two Styrofoam cups make an excellent calorimeter.
5. Measure the mass of this calorimeter with your scale. After you do this, fill the calorimeter about three-quarters full of room-temperature water. Measure the mass again, and determine the mass of the water by difference.
6. Place your thermometer in the water for three minutes, then read the temperature. Use the calibration determined in Experiment 2.1 to correct for any errors in your thermometer. Once the metal has been in the boiling water for five minutes, quickly pull it out of the boiling water with the kitchen tongs and transfer it to the calorimeter.
7. Stir the water carefully with your thermometer and periodically read the temperature without lifting the thermometer out of the water. Continue this process until the temperature stops increasing. Write down the final temperature.
8. Now we can use Equation (2.3) to determine how much heat the metal transferred to the water. In this case, we will ignore the heat absorbed by the calorimeter. Use the change in temperature in °C, the mass of the water in grams, and the specific heat of water as $4.184 \frac{J}{g \cdot {}^{o}C}$. This will then put your answer in Joules. Be sure to use the correct number of significant figures in determining the heat absorbed by the water.
9. Now you can use Equation (2.5). Since we are ignoring the calorimeter in the experiment, $q_{calorimeter} = 0$. You calculated q_{water} in the previous step, so you can determine q_{metal}. It will come out negative, because the metal *lost* energy.
10. Once you have q_{metal}, you can rearrange Equation 2.3 to calculate the specific heat of the metal. To do that, however, you need to know the mass and ΔT of the metal. You measured its mass. What is its ΔT? Since the metal was in boiling water, its initial temperature was 100.0 °C. Since it was in contact with the water and the calorimeter at the end of the experiment, its final temperature was the same as the final temperature of the water. With those two numbers, you can calculate ΔT. This number will also turn out negative, canceling the negative sign on the "q."
11. Now that you have q, m, and ΔT for the metal, you can use Equation (2.3) to calculate c, just as I did in Example 2.5. Be sure to keep track of significant figures!
12. What should your specific heat be? Well, metals have small specific heats, so it should probably be less than $1 \frac{J}{g \cdot {}^{o}C}$.
13. Clean up your mess.

Now that you've seen two examples of calorimetry problems and you've done an experiment on calorimetry, you should be able to do the following problem:

ON YOUR OWN

2.8 A calorimeter holds 175.0 grams of water at an initial temperature of 25.3 °C. A 54.3 g piece of metal at 100.0 °C is dropped into the calorimeter, and the final temperature of the water is 27.1 °C. What is the specific heat of the metal? Ignore the calorimeter in this problem.

2.9 A calorimeter (m = 9.0 g, c = 2.31 $\frac{J}{g \cdot {}^{o}C}$) holds 100.0 grams of water at an initial temperature of 25.1 °C. A 43.2 g piece of metal at 111.0 °C is dropped into the calorimeter, and the final temperature of the water is 29.1 °C. What is the specific heat of the metal?

2.10 A calorimetry experiment is performed with a calorimeter that has a specific heat of 3.77 $\frac{J}{g \cdot {}^{o}C}$. The calorimeter has a mass of 9.5 g and is filled with 117 g of an unknown liquid that is initially at a temperature of 25.0 °C. If a 50.0 g chunk of copper at 112 °C is dropped in the calorimeter and the liquid's temperature increases by 2.3 °C, what is the specific heat of the unknown liquid?

This concludes your first module on energy. However, don't think we've left this topic for good. What you have learned in this module is just an introduction to the fascinating topic of energy. We will revisit this subject again in later modules and delve deeper into the wonders of how energy is used in God's creation. For right now, however, be sure you understand everything you've read by answering the review questions and the practice problems at the end of this module.

ANSWERS TO THE "ON YOUR OWN" PROBLEMS

2.1 To solve this one, we simply need to use Equation (2.1):

$$^{\circ}C = \frac{5}{9}(^{\circ}F - 32)$$

$$^{\circ}C = \frac{5}{9}(98.6 - 32)$$

$$^{\circ}C = 37$$

Our measurement starts out with three significant figures, so we must end up with three significant figures, since the other numbers in this equation are exact. Therefore, the answer is 37.0 °C.

2.2 This one requires that we use algebra to rearrange Equation (2.1) so that we can solve for °F:

$$^{\circ}F = \frac{9}{5}(^{\circ}C) + 32.0$$

$$^{\circ}F = \frac{9}{5}(100.0) + 32.0$$

$$^{\circ}F = 212$$

Since 100.0 has four significant figures and everything else in the equation is exact, our answer is 212.0 °F.

2.3 The only way we can convert to Fahrenheit is if we have a temperature in Celsius. So, before we can get the answer, we must first convert 0.00 K to degrees Celsius. We do this by rearranging Equation (2.2):

$$^{\circ}C = K - 273.15$$

$$^{\circ}C = 0.00 - 273.15 = -273.15$$

Since we are subtracting, we look at precision. The original measurement goes out to the hundredths place, as does 273.15. Therefore, our answer should go out to the hundredths place. Now we can convert to Fahrenheit:

$$^{\circ}F = \frac{9}{5}(^{\circ}C) + 32$$

$$^{\circ}F = \frac{9}{5}(-273.15) + 32$$

$$^{\circ}F = -459.67$$

Since 273.15 is the only number in the equation that is not exact, the answer must have the same number of significant figures. The answer, then, is -459.67 °F.

2.4 Copper has a lower specific heat than glass. This means that copper is easier to heat up and therefore takes the least energy.

2.5 Since both kettles are on the same fire for the same amount of time, they are each getting the same amount of heat. Therefore, the kettle made out of the substance which is more difficult to heat up will be the coolest one. Since iron has a larger specific heat, it is more difficult to heat up, so it will be cooler than the copper kettle. That means the copper kettle's temperature will be higher than 100.0 °C.

2.6 As we look at this problem, we can see that we are given q, m, and (from Table 2.1) c. Thus, if we use Equation (2.3), we can calculate ΔT. Once we know ΔT, we can use it and the final temperature (22 °C) to get the unknown initial temperature. So our first step is to rearrange Equation (2.3) to solve for ΔT:

$$\Delta T = \frac{q}{m \cdot c}$$

As mentioned in the hint, since the iron rod releases energy, its "q" is negative. The mass is given, and since we know that the rod is iron, we can use Table 2.1 to get the specific heat:

$$\Delta T = \frac{-597.5 \not{J}}{502 \not{g} \cdot 0.4521 \frac{\not{J}}{\not{g} \cdot {}^{\circ}C}}$$

$$\Delta T = -2.63\ {}^{\circ}C$$

Now that we have ΔT, we can get the initial temperature by rearranging Equation (2.4):

$$T_{initial} = T_{final} - \Delta T = 22\ {}^{\circ}C - -2.63\ {}^{\circ}C = 25\ {}^{\circ}C$$

Since our final temperature had the lowest precision, our answer must also have the same precision. Thus, the initial temperature was 25 °C.

2.7 In this problem we are given q, m, and ΔT and asked to solve for c. So this is simply a rearrangement of Equation (2.3):

$$c = \frac{q}{m \cdot \Delta T}$$

$$c = \frac{836.8\ J}{15.6\ g \cdot 21\ {}^{\circ}C}$$

$$c = 2.6\ \frac{J}{g \cdot {}^{\circ}C}$$

2.8 We can ignore the calorimeter in this problem, so that makes it a bit easier. We have all of the information that we need to calculate q_{water}, so we might as well start there:

$$q_{water} = m \cdot c \cdot \Delta T$$

$$q_{water} = (175.0 \text{ g}) \cdot \left(4.184 \frac{\text{J}}{\text{g} \cdot {}^{\circ}\text{C}}\right) \cdot (27.1 \text{ }^{\circ}\text{C} - 25.3 \text{ }^{\circ}\text{C})$$

$$q_{water} = (175.0 \cancel{\text{g}}) \cdot \left(4.184 \frac{\text{J}}{\cancel{\text{g}} \cdot {}^{\circ}\cancel{\text{C}}}\right) \cdot (1.8 \text{ }^{\circ}\cancel{\text{C}})$$

$$q_{water} = 1,300 \text{ J}$$

Notice that we have to determine the significant figures of ΔT by looking at precision, because we are subtracting at that point. Since both temperatures go out to the tenths place, ΔT must go out to the tenths place as well. When we then multiply by ΔT, the two significant figures in ΔT limit the answer to two significant figures.

Since we can assume $q_{calorimeter} = 0$, we can now determine q_{metal}:

$$-q_{metal} = q_{water} + q_{calorimeter} = 1,300 \text{ J} + 0$$

$$q_{metal} = -1,300 \text{ J}$$

The value for q_{metal} is negative because the metal *lost* energy. We can calculate the ΔT of the metal. Since the metal started out at 100.0 °C and ended up at the same final temperature as the water, ΔT is:

$$\Delta T_{metal} = T_{final} - T_{initial} = 27.1 \text{ }^{\circ}\text{C} - 100.0 {}^{\circ}\text{C} = -72.9 \text{ }^{\circ}\text{C}$$

Once again, because we are subtracting, the significant figures here are determined by precision. We now have all the information we need to calculate the specific heat of the metal:

$$c = \frac{q}{m \cdot \Delta T}$$

$$c = \frac{1,300 \text{ J}}{54.3 \text{ g} \cdot 72.9 \text{ }^{\circ}\text{C}}$$

$$\underline{c = 0.33 \frac{\text{J}}{\text{g} \cdot {}^{\circ}\text{C}}}$$

2.9 We are given the details of the calorimeter in this problem and we are not told that we can ignore it. Thus, to solve this problem, we will need to determine the heat gained by the water *and* the heat gained by the calorimeter. Let's start with the calorimeter:

$$q_{calorimeter} = m \cdot c \cdot \Delta T$$

$$q_{calorimeter} = (9.0\ g) \cdot \left(2.31 \frac{J}{g \cdot {}^{o}C}\right) \cdot (29.1\ {}^{\circ}C - 25.1\ {}^{\circ}C)$$

$$q_{calorimeter} = (9.0\ \cancel{g}) \cdot \left(2.31 \frac{J}{\cancel{g} \cdot {}^{o}\cancel{C}}\right) \cdot (4.0\ {}^{\circ}\cancel{C})$$

$$q_{calorimeter} = 83\ J$$

Now we can move on to the water:

$$q_{water} = m \cdot c \cdot \Delta T$$

$$q_{water} = (100.0\ \cancel{g}) \cdot \left(4.184 \frac{J}{\cancel{g} \cdot {}^{o}\cancel{C}}\right) \cdot (4.0\ {}^{\circ}\cancel{C})$$

$$q_{water} = 1{,}700\ J$$

Now that we have calculated the "q's" of the water and calorimeter, we can use Equation (2.5) to determine the "q" of the metal.

$$-q_{metal} = q_{water} + q_{calorimeter} = 1{,}700\ J + 83\ J = 1{,}783\ J$$

$$q_{metal} = -1{,}800\ J$$

Since the "q" of the water has its last significant figure in the hundreds place, the answer must be reported to the hundreds place. That's why 1,783 was rounded to 1800. Now that we have "q_{metal}," we can determine ΔT for the metal and then use Equation (2.3) to determine the specific heat:

$$\Delta T_{metal} = T_{final} - T_{initial} = 29.1\ {}^{\circ}C - 111.0\ {}^{\circ}C = -81.9\ {}^{\circ}C$$

Notice that this ΔT is negative, because the object cooled down. Now we can rearrange Equation (2.3) using algebra and solve for the specific heat of the metal:

$$c = \frac{q}{m \cdot \Delta T}$$

$$c = \frac{-1{,}800\ J}{43.2\ g \cdot (-81.9\ {}^{\circ}C)}$$

$$\underline{c = 0.51 \frac{J}{g \cdot {}^{o}C}}$$

2.10 When we look at this problem, we see that we're trying to discover the specific heat of the liquid in the calorimeter, not the specific heat of the metal, as is usually the case. Instead, the specific heat of the metal can be found in Table 2.1 ($c = 0.3851 \frac{J}{g \cdot {}^oC}$). In order to find the liquid's specific heat, we need to find out "q_{liquid}". How can we do that? Well, we have enough information to calculate "q_{metal}" and "$q_{calorimeter}$", so we can start there.

To calculate "q_{metal}," I need to know the "ΔT" for the metal. To get that, I need to know the initial and final temperatures. The initial temperature is given, but what is the final temperature? Well, the water started out at 25.0 °C and increased a total of 2.3 °C as a result of the experiment. Thus, the final temperature of the water, which is also the final temperature of the metal, is 27.3 °C.

$$q_{metal} = m \cdot c \cdot \Delta T$$

$$q_{metal} = (50.0\ g) \cdot \left(0.3851 \frac{J}{g \cdot {}^oC}\right) \cdot (27.3\ {}^oC - 112\ {}^oC)$$

$$q_{metal} = (50.0\ \cancel{g}) \cdot \left(0.3851 \frac{J}{\cancel{g} \cdot {}^o\cancel{C}}\right) \cdot (-85\ {}^o\cancel{C})$$

$$q_{metal} = -1{,}600\ J$$

Notice that the heat is negative. You should expect that because the metal lost energy. We can now determine the heat gained by the calorimeter. Notice that "ΔT" for the water and calorimeter is given, since the problem told us how much the temperature *increased.* Since it increased by 2.3 °C, ΔT = 2.3 °C:

$$q_{calorimeter} = m \cdot c \cdot \Delta T$$

$$q_{calorimeter} = (9.5\ \cancel{g}) \cdot \left(3.77 \frac{J}{\cancel{g} \cdot {}^o\cancel{C}}\right) \cdot (2.3\ {}^o\cancel{C})$$

$$q_{calorimeter} = 82\ J$$

Now we can use Equation (2.5) to determine "q_{liquid}". Since the unknown liquid is replacing water in the calorimeter, we can replace "q_{water}" with "q_{liquid}" in the equation:

$$-q_{metal} = q_{liquid} + q_{calorimeter}$$

$$q_{liquid} = -q_{metal} - q_{calorimeter} = -(-1{,}600\ J) - 82\ J = 1{,}500\ J$$

Since the "q" of the metal has its last significant figure in the hundreds place, the answer must be reported to the hundreds place. Thus, the result of the equation above (1,518 J) was rounded down.

We were given the mass of the liquid (117 g) and the "ΔT" of the liquid (2.3 °C), so we can use the "q_{liquid}" that we just calculated to determine the specific heat of the liquid:

$$c = \frac{q}{m \cdot \Delta T}$$

$$c = \frac{1{,}500 \text{ J}}{(117 \text{ g}) \cdot (2.3\ ^{\circ}\text{C})}$$

$$\underline{c = 5.6 \frac{\text{J}}{\text{g} \cdot {}^{\circ}\text{C}}}$$

REVIEW QUESTIONS FOR MODULE #2

1. A man pushes a stalled car for 2 minutes. During the first 1.5 minutes, the car does not move, but for the last 0.5 minutes the man pushes even harder and succeeds in getting the car to move. Is the man doing work for the entire 2 minutes? If not, does he ever succeed in doing work? If so, when?

2. Classify each of the following as having either potential or kinetic energy, or both:
    a. A lump of coal
    b. A flash of lightning
    c. A candle flame
    d. A tornado

3. A chemist watches the temperature of a vat of liquid and notices that it decreases over time. What can the chemist conclude from this observation?

4. If a hot object is totally insulated from the outside world (totally unable to interact with anything else), will its temperature ever be able to change?

5. Why should you never look to science as an ultimate source of truth?

6. Heat is added to two identical objects that are initially at the same temperature. If the first absorbs 100.0 cal and the second absorbs 100.0 J, which gets hotter?

7. Some diet books tell you that an excellent way to lose weight is to drink ice-cold water. Why?

8. Examine the heating curve for the following unknown substance:

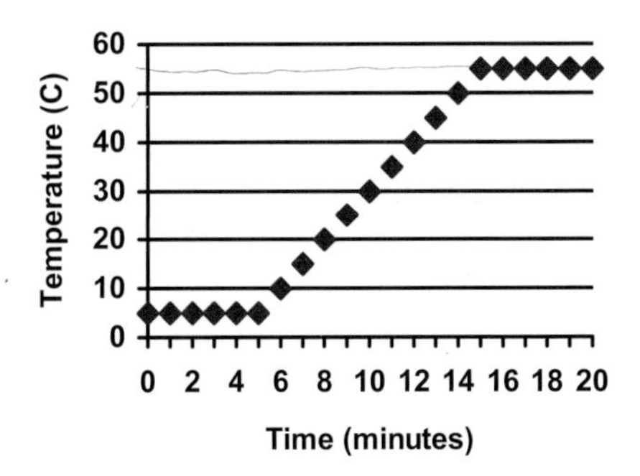

At what temperature does this substance boil?

9. The specific heat of iron is roughly three times that of gold. Equal masses of gold and iron start out at room temperature. Heat is added to each mass at the same rate for the same amount of time. When it is all over, the gold's temperature increased by 900 °C. What was the temperature increase of the iron?

10. Why should a calorimeter always be made from an insulating material?

PRACTICE PROBLEMS FOR MODULE #2

1. Convert 15.0 °C into Fahrenheit.

2. The temperature of deep space is 3.5 K. What is that in Celsius? In Fahrenheit?

3. Some parts of the western United States regularly reach temperatures of 115 °F in the summer. What is that in °C?

4. The average person uses 2,500.0 Cal of energy per day. How many Joules is that? Remember that a food calorie (Cal) is 1,000 chemistry calories (cal).

5. How many Joules does it take to heat up 15.1 kg of glass from 15 °C to 45 °C? (refer to Table 2.1)

6. 124.1 g of an unknown substance absorbs 50.0 kJ of heat and increases its temperature by 36.3 °C. What is its specific heat? (Remember that "k" is the abbreviation for "kilo," so "kJ" stands for kiloJoules.)

7. A 245 g piece of copper at room temperature (25 °C) loses 456.7 Joules of heat. What will its new temperature be?

8. Review question #7 mentioned that drinking ice-cold water is a way of burning excess Calories. Calculate how many Calories are burned when a 12-ounce (3.40×10^2 g) glass of water at 0.0 °C is warmed up to body temperature (37.0 °C).

9. A calorimeter is filled with 150.0 g of water at 24.1 °C. A 50.0 g sample of a metal at 100.0 °C is dropped in this calorimeter and causes the temperature to increase a total of 5.4 °C. What is the specific heat of the metal? Ignore the calorimeter in this problem.

10. A 345.1 g sample of copper at 100.0 °C is dropped into a 4.5 g calorimeter made of an unknown substance. If the calorimeter has 150.0 grams of water in it and the temperature changed from 24.2 °C to 25.1 °C, what is the specific heat of the calorimeter? (refer to Table 2.1)

MODULE #3: Atoms and Molecules

Introduction

In the last module, we began to explore matter by examining the energy contained in it. We learned that, under certain circumstances, this energy can be released or absorbed. We need to return to that subject in more detail, when we are ready to learn *how* matter can release or absorb energy. Before we can learn that, however, we need to learn some of the most fundamental aspects of matter. We need to know what matter is composed of, why matter exists in such a wide variety of substances, and how matter is able to change from one substance to another.

Now if I were to give you a radio and ask you to tell me what it is made of, what would you do? Most likely, you would begin to take it apart. You would find a way to remove the back cover and look inside. Of course, you would find a lot of things inside the radio, so you would probably have to take it apart piece by piece until you had broken it down into all of its individual components. You could then learn what each component was and make a list of everything that makes up the radio. Thus, by taking the radio apart, you would be able to learn about its composition.

Taking a radio apart tells you more than just what it's made of, however. If you wanted to learn *how* that radio works, you could learn how each individual component works, and that knowledge would help you immensely in understanding how the entire radio works. Once you learn how each component works, you also gain some understanding regarding the way in which other electronic devices, such as televisions, work. In the end, then, to truly understand the nature of something, we have to break it down into all of its component parts. Once we have done that, we can learn about each individual component, and that will allow us to understand whatever thing we are investigating.

This has long been the chemist's method for trying to understand the nature of matter. Chemists believe that if you can break matter down into its fundamental components and learn everything there is to learn about those components, you will then know everything there is to know about matter. In this module, we will begin to explore how chemists over the ages have pursued this method and what they have learned as a result.

Early Attempts to Understand Matter

During the earliest years of scientific investigation, ancient Greek philosophers thought that the matter inside a substance was continuous. By this they meant that you could take any substance and divide it in half over and over again. You would eventually have to get some tweezers and a very powerful magnifying glass, but no matter how small the substance became, you could still divide it in half again. In other words, they thought that substances were composed of long, unbroken blobs of matter. This concept was known as the **continuous theory of matter**.

About four centuries before Jesus Christ walked this earth, however, the Greek philosopher Democritus was walking towards the shore of a beach and was suddenly struck with an amazing thought. He noticed that from a distance, the sand along the shore looked like it was one long, continuous blob of yellow. As he walked closer to the beach, however, he could see that the sand was not at all continuous. Instead, it was made up of hundreds of thousands of individual particles called "grains."

Democritus then looked out at the ocean and wondered to himself whether or not water was made exactly the same way. From the shore, the water seemed to be composed of a huge mass of unbroken matter. But suppose you could magnify the water. Would you eventually see "grains" of water? Could it possibly be that the continuous theory of matter was based on an illusion? Perhaps matter was not really continuous. Maybe it just appeared that way because we could not magnify it enough to see the individual particles that make it up. Democritus wrote a great deal about this idea, which is referred to as the **discontinuous theory of matter**. Instead of being continuous, he argued, matter is, in fact, composed of tiny, individual particles. Our eyes simply cannot see these particles, so the apparent continuity of matter is, in fact, just an illusion. Democritus' musings represent man's first attempt to think in terms of atoms.

Since no experiments could be imagined to test the validity of either the continuous or the discontinuous theory of matter, chemists could not come to any firm conclusions regarding which theory was more scientifically sound. Each theory had its fervent supporters and its ardent critics. Unfortunately, neither side had any empirical evidence (facts gained by experiment and observation) to support its theory, so scientific progress on this point stalled for many centuries.

The Law of Mass Conservation

Progress in understanding the nature of matter picked up in the late 1700s. During that time, Antoine Lavoisier (lah vwah zyay'), considered by some to be the founder of chemistry, developed a theory that stunned the scientific world. This theory is now considered a scientific law and is called **The Law of Mass Conservation**.

The Law of Mass Conservation - Matter cannot be created or destroyed; it can only change forms.

Thus, when you burn wood in the fireplace, the matter in the wood is not destroyed. Instead, it changes from wood into other substances. Today, we know that those substances are carbon dioxide, water, and ash.

In light of what we learned about energy, this may not sound very surprising to you. However, to the chemistry community in Lavoisier's day, this was a startling revelation. You see, early chemists were guided by what they observed, and, since the human eye cannot observe everything in nature, this led them to some very wrong conclusions.

For example, early chemists would watch wood burning in the fire and would see the wood slowly disappear. They were unable to see the carbon dioxide and water vapor that was forming from the wood, because those substances are colorless gases. As a result, early chemists simply thought that the matter in the wood was disappearing. Thus, they thought that matter was being destroyed. Once again, because their experiments were not accurate, early chemists developed scientific laws that were quite wrong.

Today, we have developed methods that allow us to detect the presence of colorless gases, so we can measure the mass of the carbon dioxide and water vapor produced when wood burns. We can also determine that when wood burns, the oxygen in the air around the wood is used up. This is why a fire cannot burn unless there is a plentiful supply of oxygen. If we measure the mass of the wood and the oxygen before a fire starts, and we trap all of the carbon dioxide and water produced by the fire after it starts, we will find out that the mass of the wood and oxygen used by the fire is exactly equal to the mass of the ash, water vapor, and carbon dioxide produced by the fire. The following experiment allows you to see that even when matter goes through substantial change, the total mass is not affected.

EXPERIMENT 3.1

The Conservation of Mass

Supplies:

- A 100-mL beaker (A juice glass can be used instead.)
- A 250-mL beaker (A juice glass can be used instead.)
- A watch glass (A small saucer can be used instead. It must cover the mouth of the 100-mL beaker or juice glass listed above.)
- A teaspoon
- Lye (This is commonly sold in supermarkets with the drain cleaners. A popular brand is Red Devil® Lye. If you cannot find lye, any *powdered* drain cleaner ought to work.)
- White vinegar
- Several leaves of red (often called "purple") cabbage
- Water
- A small pot for boiling water
- A measuring cup
- A stove
- Mass scale
- Stirring rod (A spoon will work.)
- Safety goggles
- Rubber cleaning gloves

1. Add two cups of water to the pot.
2. Place the cabbage leaves into the water.
3. Place the pot on the stove and turn on the burner. You want the water to boil for five minutes.
4. While you are waiting for the water to boil, use the scale to measure the mass of your beaker.
5. Add approximately 60 mL of vinegar to the beaker. The beaker should have volume levels marked, so you can just fill the beaker to the 60 mL mark. You needn't add exactly 60 mL. Anywhere from 50 to 60 will be fine. If you are not using a beaker, add about ¼ of a cup. When chemists make measurements like these, where they don't need to get an exact amount, they call it a **qualitative** measurement. On the other hand, when chemists make measurements as exactly as possible, they call them **quantitative** measurements. Remember these terms, because we will use them again.
6. Put on the gloves.
7. Measure out one teaspoon of lye and put it on the watch glass. You can take the gloves off, but whenever you are touching *anything* that has lye on it (as you will do later), you need to put the gloves back on, because lye is caustic (it can cause chemical burns).

8. Once the water in the pot has been boiling for five minutes, take the pot off of the stove and allow it to cool for a few minutes.
9. Carefully pour about 50 mL of the solution in the pot into the 250-mL beaker. If some cabbage leaves get into the beaker, that is fine.
10. Allow the solution in the 250-mL beaker to cool.
11. Once the solution is cool enough that you can comfortably pick up the beaker in your hand, pour about 10 mL of the solution into the 100-mL beaker. The vinegar in the beaker should turn pink. This is due to chemicals (anthocyanins) that come from the cabbage leaves. As you will learn in Module #10, anthocyanins turn pink in the presence of acids such as vinegar.
12. Place the stirring rod in the beaker so that it stands in the beaker.
13. Place the watch glass on top of the beaker so that it covers the mouth of the beaker.
14. Place the beaker (with the stirring rod in it and the watch glass covering it) on the scale, and measure the mass of the total assembly. Your setup should look like the picture shown below:

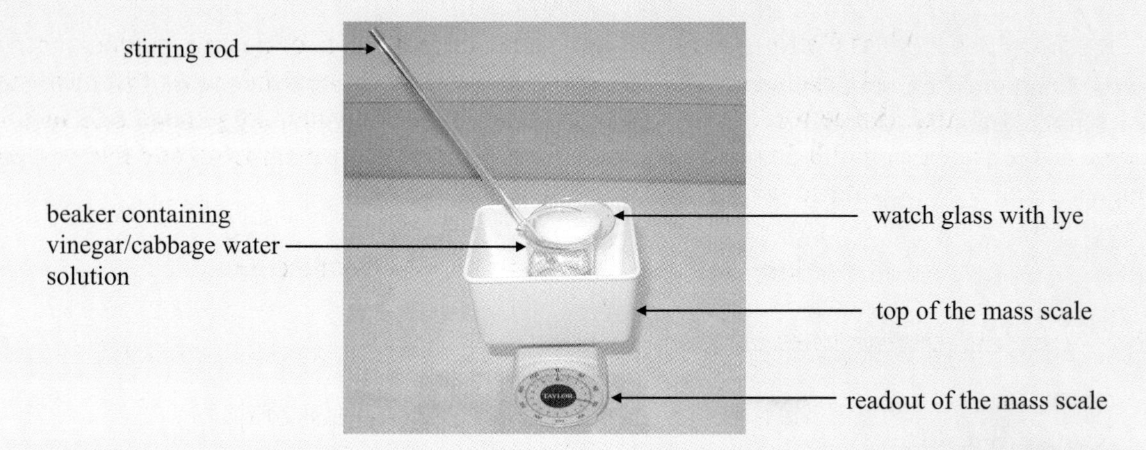

15. After you have measured the total mass of the experimental setup, dump the lye into the vinegar.
16. Stir the solution with the stirring rod. Do not lift the stirring rod out of the solution, as you run the risk of losing some of the solution.
17. You should quickly observe a color change, because the lye (a base) neutralizes the vinegar (an acid). Since anthocyanins are pink in the presence of acids, you know that the vinegar is being changed, because the pink color is replaced by a yellow color. You will learn a lot more about this process in Module #10. For now, just realize that you are observing a chemical reaction, where the vinegar and lye are forming two completely different chemicals. The fact that the solution turns from pink to yellow tells you that this is happening.
18. Notice also that most of the lye seems to have "disappeared." Of course, it didn't really disappear. It simply reacted with the vinegar and changed into a chemical that is now dissolved in the water.
19. Rest the stirring rod against the side of the beaker and place the watch glass back on the top of the beaker so that (aside from the lye and the color change), the setup looks like it did in step #14.
20. Carefully touch the sides of the beaker. The beaker should be noticeably warmer than when you put it on the scale. This is further evidence that a chemical reaction occurred. Potential energy in the vinegar and lye was released via the chemical reaction and converted to kinetic energy in the form of heat.
21. Read the mass of the entire assembly again. It should be the same as what you read in step #14. This demonstrates the Law of Mass Conservation. Despite the fact that vinegar and lye were changed into two different substances; despite the fact that energy was released; and despite the fact that the lye seemed to "disappear," the mass of the entire assembly did not change.
22. Clean up your mess.

Once the Law of Mass Conservation was accepted by chemists, they started using it to analyze many of the changes they observed in matter. Example 3.1 illustrates one way that the Law of Mass Conservation can be used in analyzing how matter changes.

EXAMPLE 3.1

A chemist notices that when given enough energy, a certain white powder changes into two different substances: tin and oxygen. A chemist watches 151 grams of powder undergo this change. The chemist easily collects the tin that was formed and measures its mass to be 119 grams. Unfortunately, the chemist could not collect the oxygen that was formed and therefore could not determine its mass. Use the Law of Mass Conservation to help the chemist determine how much oxygen was made.

The Law of Mass Conservation tells us that we cannot create or destroy matter. Thus, if the chemist started with 151 grams of matter, no matter what change occurred, he or she must end up with 151 grams of matter. Since we know that only two substances, tin and oxygen, were formed in the course of the experiment, all 151 grams must be accounted for between the tin and the oxygen. Mathematically, we would say:

Mass of powder = Mass of tin + Mass of oxygen

Rearranging the equation gives us:

Mass of oxygen = Mass of powder - Mass of tin

Plugging in our numbers:

Mass of oxygen = 151 grams - 119 grams = 32 grams

Notice the significant figures here. Despite the fact that both 151 and 119 have three significant figures, the answer can have only two. This is because we are subtracting, and in addition and subtraction, we do not count significant figures. Instead, we look at the decimal place. Since both numbers have their last significant figure in the ones place, the answer must also have its last significant figure in the ones place. Thus, 32 grams of oxygen were formed when 151 grams of the white powder underwent the observed change.

Make sure that you understand how to use the Law of Mass Conservation by performing the following problem:

ON YOUR OWN

3.1 In a careful experiment, a chemist burns 15.4 kg of wood in a fireplace. The chemist collects all of the substances produced by the fire. He measures the mass of ash produced to be 925 grams, the mass of carbon dioxide produced to be 15.14 kg, and the mass of water produced to be 3.12 kg. How much oxygen was used up by the wood while it burned?

Elements: The Basic Building Blocks of Matter

Once chemists began using the Law of Mass Conservation, they noticed something quite interesting: There are some substances that can undergo change in such a way as to make many other substances, each less massive than the original. For example, they noticed that a black powder called iron sulfide can undergo a change that forms iron and sulfur. If they started out with 88 grams of iron sulfide, they would make 56 grams of iron and 32 grams of sulfur. Similarly, 85 grams of a white powder known as sodium nitrate could undergo a change that would produce 23 grams of sodium, 48 grams of oxygen, and 14 grams of nitrogen. In both of these cases, the change that occurred seemed to divide up the matter that was in the powder into smaller bundles that each became a new substance. Chemists called this type of change **decomposition**.

On the other hand, chemists noticed that there were some substances that would not undergo decomposition in any way. For example, once a chemist had made sulfur from iron sulfide, there was nothing the chemist could do to force the sulfur to decompose into several less massive substances. No matter what these chemists tried to do, the sulfur would not change into different substances that were less massive than the original sample of sulfur. It seemed that the matter in the sulfur could not be made into smaller substances. This told chemists that there was something special about sulfur.

Remember the example of the radio I used earlier? If you began to take apart a radio, you would eventually get to the point where you had a lot of parts (screws, wires, and dials, for example) that you could not disassemble any further. When you got to that point, you would say that you had broken the radio down into all of its component parts.

When chemists saw that iron sulfide could be decomposed into iron and sulfur, but that the iron and sulfur could not be decomposed any further, they assumed that this meant iron and sulfur were the component parts of iron sulfide. These component parts could not be decomposed any further, so they represented the simplest form of the matter within the iron sulfide. Chemists called these component parts **elements.**

Element - Any substance that cannot be decomposed into less massive substances

Chemists soon discovered that there were many, many substances that could not be decomposed into smaller substances; thus, there were many, many elements. Since elements were considered to be the component parts of all forms of matter, chemists wanted to catalog them carefully. They compiled the most important information concerning each element into a table that was called the **Periodic Table of Elements**. As time went on, this table became the chemist's most important tool. It remains so even to this day. The modern form of The Periodic Table of Elements is shown on the next page.

The Periodic Table of Elements can also be found on the very first page of this book. Feel free to photocopy it and keep it handy so you can easily look at it as you read the rest of this and other modules. In addition, you will want to use the Periodic Table of Elements heavily while doing all problems and tests. You might want to put it in a clear plastic sheet protector and use it as a bookmark so you will always have it while you read. The Periodic Table of Elements contains a huge amount of information, so I will be referring to it over and over again. It will help you quite a bit to look at the chart every time I refer to it so that you will see how to use it.

THE PERIODIC TABLE OF ELEMENTS

1A	2A	3B	4B	5B	6B	7B	8B	8B	8B	1B	2B	3A	4A	5A	6A	7A	8A
1 **H** 1.01																	2 **He** 4.0
3 **Li** 6.94	4 **Be** 9.01											5 **B** 10.8	6 **C** 12.0	7 **N** 14.0	8 **O** 16.0	9 **F** 19.0	10 **Ne** 20.2
11 **Na** 23.0	12 **Mg** 24.3											13 **Al** 27.0	14 **Si** 28.1	15 **P** 31.0	16 **S** 32.1	17 **Cl** 35.5	18 **Ar** 39.9
19 **K** 39.1	20 **Ca** 40.1	21 **Sc** 45.0	22 **Ti** 47.9	23 **V** 50.9	24 **Cr** 52.0	25 **Mn** 54.9	26 **Fe** 55.8	27 **Co** 58.9	28 **Ni** 58.7	29 **Cu** 63.5	30 **Zn** 65.4	31 **Ga** 69.7	32 **Ge** 72.6	33 **As** 74.9	34 **Se** 79.0	35 **Br** 79.9	36 **Kr** 83.8
37 **Rb** 85.5	38 **Sr** 87.6	39 **Y** 88.9	40 **Zr** 91.2	41 **Nb** 92.9	42 **Mo** 95.9	43 **Tc** (98)	44 **Ru** 101.1	45 **Rh** 102.9	46 **Pd** 106.4	47 **Ag** 107.9	48 **Cd** 112.4	49 **In** 114.8	50 **Sn** 118.7	51 **Sb** 121.8	52 **Te** 127.6	53 **I** 126.9	54 **Xe** 131.3
55 **Cs** 132.9	56 **Ba** 137.3	57 **La** 138.9	72 **Hf** 178.5	73 **Ta** 180.9	74 **W** 183.9	75 **Re** 186.2	76 **Os** 190.2	77 **Ir** 192.2	78 **Pt** 195.1	79 **Au** 197.0	80 **Hg** 200.6	81 **Tl** 204.4	82 **Pb** 207.2	83 **Bi** 209.0	84 **Po** (209)	85 **At** (210)	86 **Rn** (222)
87 **Fr** (223)	88 **Ra** 226.0	89 **Ac** (227)	104 **Rf** (261)	105 **Db** (262)	106 **Sg** (266)	107 **Bh** (264)	108 **Hs** (269)	109 **Mt** (268)	110 **Uun** (271)	111 **Uuu** (272)	112 **Uub** (285)						

58 **Ce** 140.1	59 **Pr** 140.9	60 **Nd** 144.2	61 **Pm** (145)	62 **Sm** 150.4	63 **Eu** 152.0	64 **Gd** 157.3	65 **Tb** 158.9	66 **Dy** 162.5	67 **Ho** 164.9	68 **Er** 167.3	69 **Tm** 168.9	70 **Yb** 173.0	71 **Lu** 175.0
90 **Th** 232.0	91 **Pa** 231.0	92 **U** 238.0	93 **Np** (237)	94 **Pu** (244)	95 **Am** (243)	96 **Cm** (247)	97 **Bk** (247)	98 **Cf** (251)	99 **Es** (252)	100 **Fm** (257)	101 **Md** (258)	102 **No** (259)	103 **Lr** (262)

Nonmetals

Standard metals

Transition metals

Inner transition metals

(The multimedia CD has a funny song about the Periodic Table of Elements.)

The Periodic Table of Elements is often called the "periodic chart." I usually shorten that even further and simply call it "the chart." Notice that it has 112 boxes on it. Each box represents an element; thus, there are currently 112 elements. Believe it or not, only 92 of these elements (those with boxes numbered 1-92) appear naturally in creation. The 20 other elements (numbered 93-112) have been manufactured by scientists in the lab. I say that there are "currently" 112 elements because every now and again, a scientist will figure out how to manufacture a new one. Element number 112, for example, was manufactured by a research group in Darmstadt, Germany, in 1996.

Think for a minute about what this means: Since elements are the building blocks of matter, everything you see around you is composed of elements which are listed on the chart. If you were to take anything in nature and break it down into its individual components of matter, then each of those components would be one of the elements listed on the chart.

Each box in the chart contains one or two letters. These letters make up the chemical abbreviation that we use for each element. For example, box #6 has a "C" in it. This is the chemical abbreviation for the element carbon. In the same way, box #20 contains a "Ca." This is the chemical abbreviation for the element calcium. So we already see the first inconsistency on the chart. Sometimes the abbreviation for an element is simply the first letter in the element's name, and sometimes it is the first two letters in the element's name. It gets even worse with elements 110-112, because they have three letters in their abbreviations. If an abbreviation contains more than one letter, the first letter is always capitalized and the others are not capitalized.

Now one thing you will quickly find out in chemistry is that every rule has exceptions. This is true of nearly every rule in chemistry, including the rule for how to abbreviate elements. For example, what would you expect the abbreviation for sodium to be? Probably "S" or "So," right? Well, that would be consistent with the rule I stated above, but it turns out that it's not right. The chemical abbreviation for sodium is "Na." Now where did that come from?

Well, it turns out that sometimes chemists do not use the English name of the element to determine its abbreviation; they use its Latin name. The Latin name for sodium is "natrium," so its abbreviation is "Na." In addition, chemists sometimes abbreviate elements with just the first letter of the Latin name. For example, the abbreviation for potassium is "K," because the Latin name for potassium is "kalium." This unruly system has come about because chemists were working all over the world discovering elements. Since in some countries chemists used Latin as the scientific language, they used Latin to name the elements. In other countries, chemists used English. As a generalization, then, we can say:

The element abbreviation which appears on the chart is usually either the first or the first two letters of the element's English or Latin name.

As time goes on, you will begin to learn the names that go with the abbreviations on the chart. For right now, however, you will be expected to memorize a few of them so that you can understand some of the things you read in this module. By the end of this module, you will be expected to know the first 20 elements that appear on the chart: (H through Ca). You need not know the numbers that appear with them on the periodic chart. You need only know that "H" is the abbreviation for "hydrogen," "He" is the abbreviation for "helium," etc. The last page of this module has a list of all element names and their corresponding abbreviations so that you can begin to learn them.

The next thing you might notice is that each box on the chart contains two numbers. These numbers tell us some very important things about each element. We will not get to the meaning of those numbers for a couple of modules, so right now you can simply ignore them. Also, you might think that the chart is structured in a funny way. The boxes are not simply piled on top of each other or arranged in nice, even rows. Well, once again, we will learn in a later module that the chart is laid out in this way in order to arrange the elements in a very descriptive fashion. For right now, however, don't worry about it.

Finally, notice the heavy, jagged line that starts at the element boron (B) and moves diagonally down the chart to astatine (At). This very important line separates the elements into two classes: **metals** and **nonmetals**. Elements that are called metals are usually malleable (can be easily bent and shaped), have luster (are shiny), and are able to conduct electricity. On the other hand, nonmetals are typically brittle (break easily when you try to bend or shape them), lack luster, and do not conduct electricity.

Elements that appear on the chart to the left of the jagged line (Zn, Ca, and Pt, for example) are classified as metals. Elements that appear on the chart to the right of the jagged line (B, N, and Br, for example) are nonmetals. Now as usual, this rule has a few exceptions. The most important exception (and one you will be expected to remember) is hydrogen:

Even though hydrogen is left of the jagged line, it is always considered a nonmetal.

You will have to wait for a while before you have enough information to understand *why* hydrogen is an exception to this rule. For right now, you just need to memorize the fact that it is.

The other exception to this rule is a class of elements known as **metalloids**. These elements have some metal properties and some nonmetal properties. Silicon (Si), for example, is brittle but it has luster. In addition, it will conduct electricity under certain conditions but not under other conditions. For right now, however, you do not have to worry about metalloids. As far as you are concerned, there are only two classes of elements, metals and nonmetals, and these classes are separated by the heavy jagged line that appears on the chart.

The classification of elements into metals and nonmetals is important because we will use it again and again to identify the nature of different substances. It is important, therefore, that you can recognize whether an element is a metal or a nonmetal by looking at the chart. In addition, it is important for you to know the three properties that describe metals and nonmetals. Try the following problems to make sure you understand this classification scheme:

ON YOUR OWN

3.2 Which of the following elements is (are) metal(s)? O, Mg, I, In, Os, Rn, H

3.3 If you want to make a wire for an electrical system and you only have the following elements to work with, which should you choose? Ni, S, Se

3.4 Which of the following would you expect to break when it is bent? Rh, Ba, P

Compounds

Since chemists classify as elements all substances that cannot be decomposed into simpler substances, we also need to classify the substances that *can* be decomposed into simpler substances. These we call **compounds**:

Compounds - Substances that can be decomposed into elements by chemical means

Based on these two definitions, then, we can say that matter comes in two forms: elements and compounds. Elements are the basic building blocks of all substances. Compounds are formed when elements group together to form a new substance. Alternatively, you could say that elements are formed when compounds are broken down into their component parts. Either way of looking at these two forms of matter is correct.

In 1794, French chemist Joseph Proust (proost) demonstrated a fact about compounds that later became known as the **Law of Definite Proportions**:

The Law of Definite Proportions - The proportion of elements in any compound is always the same.

What Proust discovered was that when several elements come together to form a compound, they always come together in the same proportion. For example, water is made when the element hydrogen (H) and the element oxygen (O) react together. Whenever this happens, it always takes 8 grams of oxygen for every 1 gram of hydrogen. If you started with 10 grams of hydrogen, you would need 80 grams of oxygen to react with it. When the reaction was done, 90 grams of water would be formed, in agreement with the Law of Mass Conservation. No matter how much hydrogen you started with, the mass of oxygen had to be eight times greater, or there would be hydrogen left over.

You can think about the Law of Definite Proportions in the same way that you think about recipes. If a cake recipe says that the cake serves eight people, you could make a cake to serve 40 people by using five times the ingredients called for in the recipe. What the Law of Definite Proportions tells us is that there is a single "recipe" that governs how much of each element is needed to make a particular compound. If you don't add elements together in exactly the proportion the recipe calls for, then there will be some "leftover" ingredients at the end. Figure 3.1 tries to illustrate this idea:

Illustrations by Megan Whitaker

FIGURE 3.1
The Law of Definite Proportions

A chemist decides that she wants to make table salt. She has already determined that table salt is a compound composed of two elements: sodium (Na) and chlorine (Cl). What she doesn't know is what proportion of each element she must use. To determine this, she does a few experiments. First, she reacts 10.0 grams of sodium with 10.0 grams of chlorine. This is actually a very dangerous reaction, because sodium (a solid metal) is very reactive and chlorine (a gas) is poisonous. When the reaction is over, she finds that she has 16.5 grams of table salt and 3.5 grams of sodium left over.

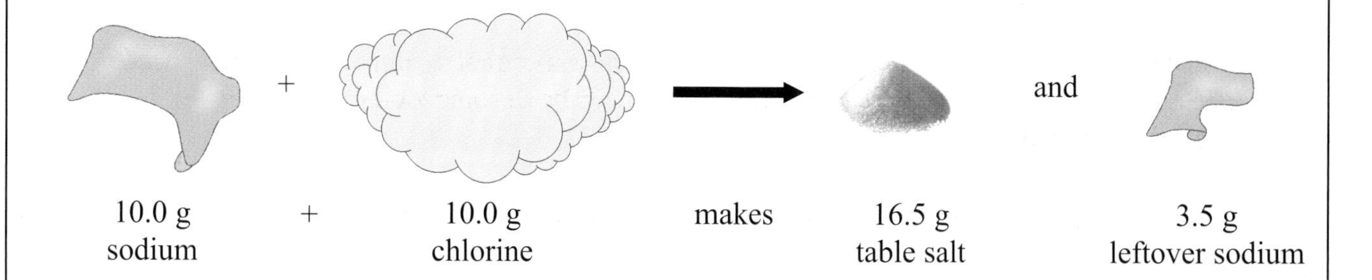

First of all, we see that this experiment confirms the Law of Mass Conservation. Since the chemist started out with 20.0 grams (10.0 g + 10.0 g) of matter, she also had to end up with 20.0 grams (16.5 g + 3.5 g) of matter. Since the sodium and chlorine were not added in the correct proportion, however, the 20.0 grams that she ended up with was not pure table salt. Instead, only 16.5 grams of table salt could be made. The rest of the 20.0 grams, then, is accounted for by sodium that could not react, because too much had been added in the first place.

Second, we see that the *chlorine* must have run out. There was sodium left over; thus, there must have been plenty of it. However, since there was no leftover chlorine, that must mean that the chlorine was completely used up in the reaction. This little fact allows us to determine the proper proportions (the proper "recipe") of sodium and chlorine a chemist must have to make table salt. Since 3.5 grams of sodium were left over, then that mass of sodium was not used in the reaction. To find out how much sodium *was* used, we simply subtract the leftover amount from the amount we started with:

$$\text{Amount of sodium used} = \text{Starting amount} - \text{Amount left over}$$

$$\text{Amount of sodium used} = 10.0 \text{ g} - 3.5 \text{ g} = 6.5 \text{ g}$$

So we see that only 6.5 grams of sodium were actually used when reacted with 10.0 grams of chlorine. Thus, the proper proportion in which to add sodium and chlorine is 6.5 grams of sodium for every 10.0 grams of chlorine.

To test this recipe, our chemist then takes 10.0 grams of chlorine and reacts it with 6.5 grams of sodium. According to the Law of Definite Proportions, these two elements should now react to form table salt with no leftovers, since we are adding the elements together in the proper proportions:

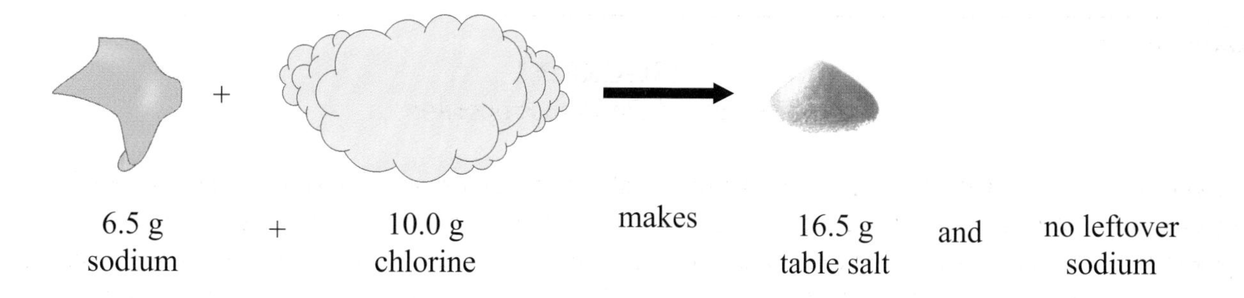

So she did, indeed, determine the proper recipe for making table salt: 10.0 grams chlorine plus 6.5 grams sodium makes 16.5 grams table salt. To convince herself that the Law of Definite Proportions really works, she tripled the masses of sodium and chlorine and tried again:

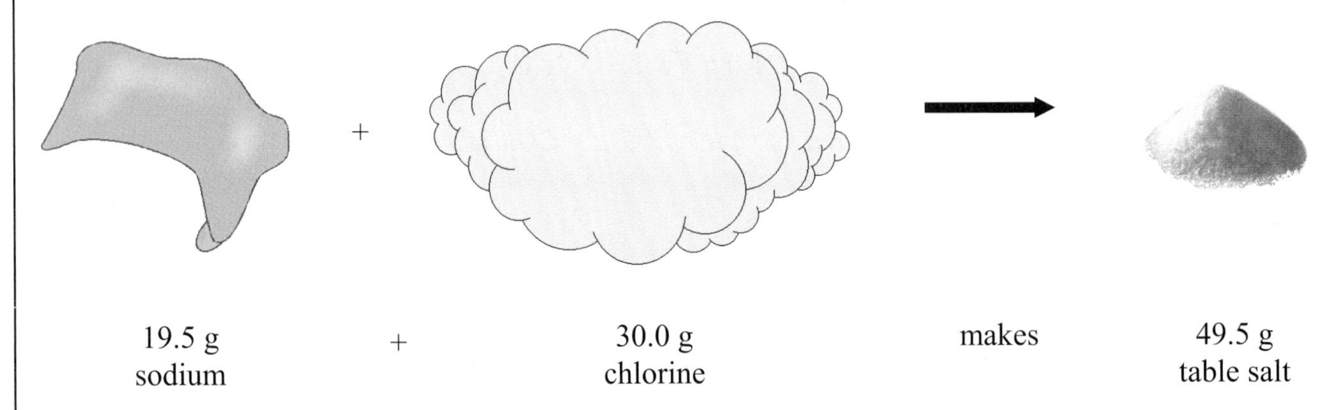

Once again, the experiment ended with no leftovers, because the elements were added together in the proper proportion.

As a final experiment, the chemist adds 6.5 grams of sodium to 15.0 grams of chlorine. In this case, the Law of Definite Proportions would tell you that the chemist used too much chlorine, because 6.5 grams of sodium needs only 10.0 grams of chlorine. There should, therefore, be 5.0 grams of chlorine left over in the end:

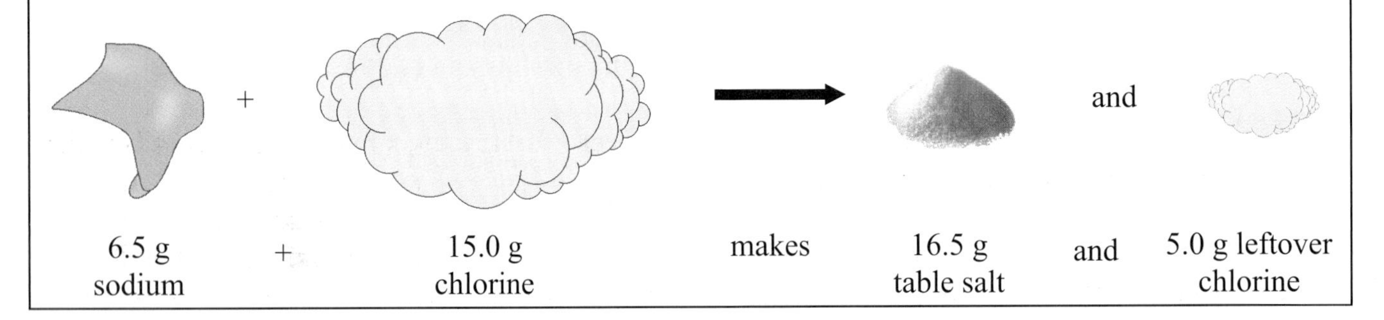

The Law of Definite Proportions, coupled with the Law of Mass Conservation, can be used to solve problems like the following example:

EXAMPLE 3.2

A chemist reacts 15.0 grams of calcium (Ca) with 15.0 grams of oxygen (O). This reaction makes 21.0 grams of a compound known as lime. Along with the lime, there is also some leftover oxygen. If the chemist wants to make 55 grams of lime and have no leftover oxygen or calcium, how much of each element should he use?

To solve this problem, we first need to determine the recipe for making lime. In order to be able to do that, we need to figure out how much oxygen was actually used in the reaction. Since there was no leftover calcium, we know that all 15.0 grams of calcium were used. However, there was some leftover oxygen. How much leftover oxygen was there? We can find out by using the Law of Mass Conservation:

$$\text{Total mass before reaction} = \text{Total mass after reaction}$$

$$15.0 \text{ g} + 15.0 \text{ g} = 21.0 \text{ g} + \text{Mass of leftover oxygen}$$

$$\text{Mass of leftover oxygen} = 30.0 \text{ g} - 21.0 \text{ g} = 9.0 \text{ g}$$

Since 9.0 grams of oxygen were left over:

$$\text{Mass of oxygen used} = \text{Starting mass} - \text{Mass left over}$$

$$\text{Mass of oxygen used} = 15.0 \text{ g} - 9.0 \text{ g} = 6.0 \text{ g}$$

So the proper recipe to use in making lime is: 15.0 grams of calcium plus 6.0 grams of oxygen makes 21.0 grams of lime. The problem, however, asks how much of each element we must use in order to make 55 grams of lime. So, we obviously need to increase the recipe, but by how much? Well, what we have to do is find what number, when multiplied by 21.0, gives us 55. That will tell us by what factor we need to increase our recipe:

$$21.0 \text{ g} \cdot x = 55 \text{ g}$$

$$x = \frac{55 \cancel{\text{g}}}{21.0 \cancel{\text{g}}} = 2.6$$

Notice that in the expression for x, the unit of grams cancels. As a result, x has no units. Numbers like this are called **dimensionless quantities.**

Now we know that in order to make 55 grams of lime, we must increase our recipe by a factor of 2.6. Thus, if the original recipe was 15.0 grams calcium + 6.0 grams oxygen, then the new recipe is:

$$\text{Mass of calcium} = 15.0 \text{ g} \cdot 2.6 = 39 \text{ g}$$

$$\text{Mass of oxygen} = 6.0 \text{ g} \cdot 2.6 = 16 \text{ g}$$

Thus, to make 55 grams of lime, you need to react <u>39 grams of calcium with 16 grams of oxygen</u>.

ON YOUR OWN

3.5 When 12.0 grams of the element carbon (C) react with 4.00 grams of the element hydrogen (H), natural gas (the same stuff you burn in a gas stove or a gas furnace) is produced. There is no leftover carbon or hydrogen when these quantities are used. How much carbon and hydrogen would you need to make 100.0 grams of natural gas?

3.6 A chemist decomposes 30.0 grams of a purple powder into 13.6 grams of the element cobalt (Co) and an unmeasured amount of chlorine gas. The chemist uses the results of this experiment to help him determine the recipe for making this purple powder from its elements. He then decides to make 1.00 kg of it. How much cobalt and how much chlorine gas will he need?

The Law of Multiple Proportions

One more law that deals with the nature of compounds is the **Law of Multiple Proportions,** which can be stated as follows:

The Law of Multiple Proportions - If two elements combine to form different compounds, the ratio of masses of the *second* element that react with a fixed mass of the *first* element will be a simple, whole-number ratio.

This law says that sometimes you can find different ways of combining elements. If this is the case, there is a simple relationship between the "recipes" for the two compounds. Although the law sounds a bit hard to understand when you read it, an example will make it very clear.

When 12.0 grams of carbon (C) and 32.0 grams of oxygen (O) combine, they make 44.0 grams of carbon dioxide, a gas that humans exhale. On the other hand, under the right conditions, 12.0 grams of carbon can combine with just 16.0 grams of oxygen to make 28.0 grams of carbon monoxide, a completely different compound. To illustrate just how different these compounds are, carbon monoxide is poisonous to humans. Thus, when 12.0 grams of carbon combine with 32.0 grams of oxygen, they form a compound that humans exhale each time they breathe. However, when carbon and oxygen combine in a different proportion (12.0 grams and 16.0 grams), they make a compound that is lethal to humans.

Now, the Law of Multiple Proportions says that there is a simple relationship between the amount of oxygen in the "recipe" for carbon dioxide and the amount of oxygen in the "recipe" for carbon monoxide. It first says that we must use a *fixed* amount of the first element (carbon). We have already done that. In each recipe, the amount of carbon is the same (12.0 grams). Thus, the amount of carbon is fixed. Then we determine the ratio of the mass of oxygen used to make the first compound and the mass of oxygen used to make the second compound. It took 32.0 grams of oxygen to make carbon dioxide, but only 16.0 grams of oxygen to make carbon monoxide. The ratio of the mass of oxygen in carbon dioxide to the mass of oxygen in carbon monoxide, then, is 32.0:16.0, or 2:1. That's a simple, whole-number ratio.

Let's look at one more example just to make sure you understand this law. When 36.0 grams of carbon react with 12.0 grams of hydrogen, you get 48.0 grams of methane, which is also called "natural gas." On the other hand, when 36.0 grams of carbon react with 8.0 grams of hydrogen, you get 44.0 grams of propane, which is the gas typically burned by outdoor gas grills. Once again, we have two elements (carbon and hydrogen) reacting in different proportions to make different gases. We have a fixed amount of carbon (36.0 grams). If we then take the ratio of the masses of hydrogen that react with this fixed mass of carbon, we get 12.0:8.0, or 3:2. Once again, a simple, whole-number ratio. That's what the Law of Multiple Proportions says should happen.

What's the big deal about the Law of Multiple Proportions? It was the main evidence for a "modern" atomic theory proposed by John Dalton. Dalton used his theory to *predict* the Law of Multiple Proportions, and scientists then did experiments to demonstrate that the law was, indeed, correct. This provided excellent evidence for the validity of Dalton's theory.

Dalton's Atomic Theory

The Law of Definite Proportions, combined with the Law of Mass Conservation, led John Dalton, an English chemist who worked in the early 1800s, to propose a theory he thought would help chemists better understand the nature of matter. With this theory, Dalton thought that he could explain *why* these laws were always obeyed. He called it his "atomic theory," and it forms the basis for our modern-day interpretation of the nature of matter. Dalton's theory contained four vital assumptions:

1. All elements are composed of small, indivisible particles called "atoms."

2. All atoms of the same element have exactly the same properties.

3. Atoms of different elements have different properties.

4. Compounds are formed when atoms are joined together. Since atoms are indivisible, they can only join together in simple, whole-number ratios.

Remember, these four ideas were assumptions. He could not offer definitive proof that they were true. However, he said that if you assume that they are true, you can explain *why* the laws of mass conservation and definite proportions are always obeyed. Also, he predicted the Law of Multiple Proportions (as you see in statement #4), which was later demonstrated to be true.

How do these assumptions explain the laws of mass conservation and definite proportions? Well, consider the Law of Mass Conservation. According to Dalton's assumptions, all compounds and elements that undergo change are simply rearranging their atoms. Since those atoms are indivisible, they obviously cannot be destroyed, so the total number of atoms in the entire system must stay the same. If the total number of atoms stays the same, the total amount of matter stays the same, and therefore, the mass stays the same.

Now let's consider The Law of Definite Proportions. Dalton's theory assumes that atoms can only join together to make compounds in simple, whole-number ratios. In other words, a compound might have 1 carbon atom and 2 oxygen atoms, while another compound might have 1 carbon atom and 1 oxygen atom. A compound cannot, however, have 1 carbon atom and 1.5 oxygen atoms, because atoms are indivisible. You can either have 1 atom or 2 atoms, never 1.5 atoms. Well, if atoms combine together in this way, their masses would always combine in the same proportion, which is exactly what the Law of Definite Proportions says.

Of course, the real evidence for Dalton's theory came with the fact that he *predicted* the Law of Multiple Proportions. Since Dalton believed that atoms combine in simple, whole-number ratios, he predicted that you should be able to see those ratios in the "recipes" that exist for different compounds. When scientists studied this, they found out that he was right. This helped Dalton's atomic theory gain acceptance in the scientific community. After all, it is nice for a theory to be able to explain facts that are already known. However, if a theory *predicts* facts that have not already been determined, it adds a lot of weight to the theory.

As time has gone on, chemists have seen that a great many of the facts that we know about nature can be explained quite nicely if we assume that Dalton's atoms do indeed exist. It turns out that atoms are far too small to be seen, so we can never really *prove* that they exist. However, if we assume they exist, we can explain a wealth of facts that we have accumulated over the years. Thus, chemists today simply accept the idea that atoms exist, regardless of the fact that their existence can never be proven.

If we could see these atoms, what would they look like? Figure 3.2 gives you some idea of what a group of atoms might look like if we were able to actually see them.

FIGURE 3.2
A Scanning Tunneling Electron Microscope Image of the Surface of a Nickel Foil

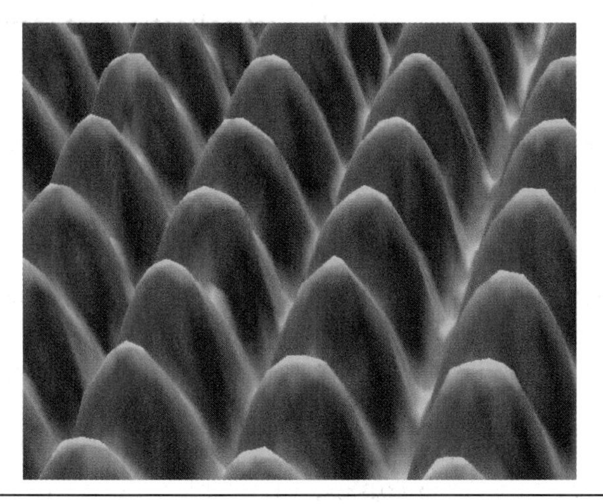

Image courtesy of the IBM research division

The first thing I must say about this figure is that you are *not* looking at a picture of atoms. In order for us to see an object (or take a picture of it), visible light must bounce off of the object and hit our eyes (or the camera's film). That cannot happen with atoms, because atoms are too small for visible light to bounce off of them. Thus, you are not seeing a picture here. You are seeing a *computer image* that is based on some data.

In this figure, a scanning tunneling electron microscope was used to examine the surface of a nickel foil. In this kind of microscope, an electrified probe is placed very close to the surface of the metal, and the flow of electricity from the probe to the metal is monitored by a computer. Based on a theory called "quantum mechanical tunneling," the computer then draws what the theory says the surface of the metal must look like in order for the electricity to flow the way that it did. Thus, if the theory is correct (and if the computer used the theory correctly), this is probably a good representation of what a group of nickel atoms might look like on the surface of the nickel foil. If the theory is incorrect, however, then who knows what this image represents? Thus, this image is certainly not proof that atoms exist. However, it does give more evidence for their existence.

It turns out that the image in Figure 3.2 is pretty much how Dalton pictured atoms. Today we are fairly sure that atoms have a more detailed structure than what is shown in the figure. The scanning tunneling electron microscope, however, does not have enough resolution to show that detailed structure. However, as you will learn in an upcoming module, other experiments have been done to probe the detailed structure of the atom, and today we have a theory that helps us picture the detailed structure of an atom. Before we can get to that, however, we need to learn a little more about matter and how it changes.

One final note regarding Dalton's atomic theory: Although chemists do accept the fact that atoms exist, we have also come to realize that two of Dalton's assumptions are not quite right. Assumption #1, for example, contains a small mistake. It turns out that under the right circumstances, atoms can, indeed, be split apart. Thus, atoms are not truly indivisible. Luckily, atoms are never split in chemistry, so in this class, we can assume that Dalton was right on this point. Assumption #2 is not quite right either. Certain atoms within an element can be heavier than other atoms in that same element. These atoms are called **isotopes** and will be discussed in detail in an upcoming module.

Molecules: The Basic Building Blocks of Compounds

I have already said that elements are simply vast collections of identical atoms. Thus, if I were able to see a sample of the element helium (He), the gas we put in balloons to make them float, at the atomic level, I would see a bunch of identical particles jumbled together much like grains of sand on a beach. Thus, Democritus (way back in the fourth century B.C.) was, indeed, right. Matter is discontinuous: It contains little "grains" that we call atoms.

What about compounds, however? If I were to take the water in the ocean and look at it on the atomic level, would I see a bunch of atoms jumbled together? No, not exactly. Water is a compound that is formed when 2 hydrogen (H) atoms join with 1 oxygen (O) atom. Thus, if you looked at water on an atomic level, you would not see just a bunch of hydrogen and oxygen atoms jumbled together; you would instead see little groups of atoms that each contained 2 hydrogens and 1 oxygen. Those *groups* would be jumbled up together like grains of sand. The groups that are formed when atoms join together to form compounds are called **molecules**. In other words, while all elements are composed of identical atoms, all compounds are composed of identical molecules.

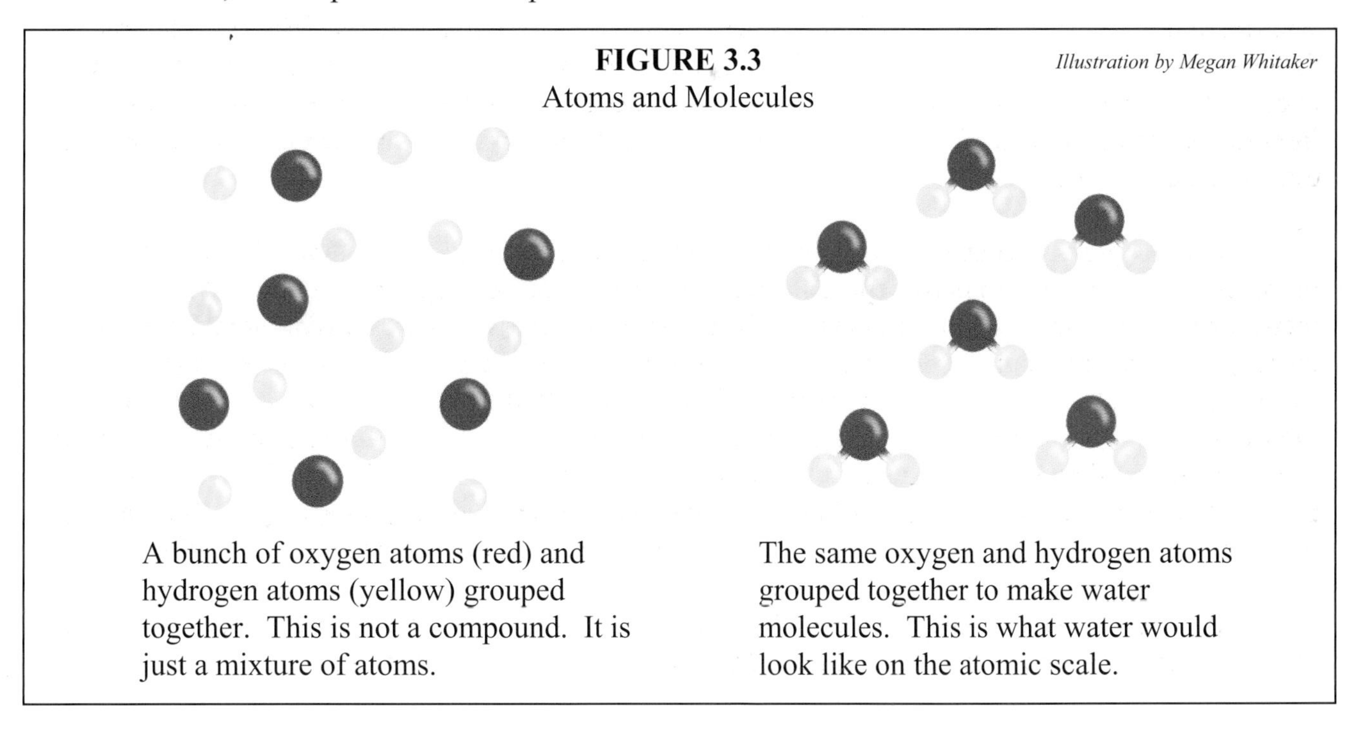

FIGURE 3.3
Atoms and Molecules

Illustration by Megan Whitaker

A bunch of oxygen atoms (red) and hydrogen atoms (yellow) grouped together. This is not a compound. It is just a mixture of atoms.

The same oxygen and hydrogen atoms grouped together to make water molecules. This is what water would look like on the atomic scale.

The basic building blocks of matter, then, are atoms. When a substance is made up of identical atoms, that substance is called an element. On the other hand, when atoms join together, they form molecules. When a substance is made up of molecules, it is called a compound.

Abbreviating and Classifying Compounds

We've already seen that each element has its own name and abbreviation. What about compounds? Each compound needs to have a name, and it would be nice to be able to abbreviate that name. How do we do this? Well, we name elements based on the atoms that make them up. For example, the element fluorine (F) is a substance made up of individual fluorine atoms. Thus, an element's name is the same as the atoms that make it up.

Compounds, then, should be named after the molecules that make them up. So how do we name a molecule? Well, in order to learn how to name molecules, we first have to learn how to abbreviate them. Since molecules are made up of atoms, we can use the abbreviations on the chart to help us abbreviate molecules as well. For example, I have already told you that a water molecule is formed when 2 hydrogen (H) atoms join together with 1 oxygen (O) atom. The abbreviation for a water molecule, then, is H_2O.

How do we get this abbreviation? The letters in the abbreviation come from the atoms which make up the molecule. Since water has hydrogen atoms and oxygen atoms in it, the abbreviations for these two atoms, H and O, must be in the abbreviation for the molecule. In addition to the letters in the molecule's abbreviation, there are also numbers. These numbers, written as subscripts, tell us how many of each atom are in the molecule. The subscript "2" after the "H" tells us that water contains 2 hydrogen atoms. Since there is 1 oxygen atom in a water molecule, why isn't there a "1" subscript after the oxygen? That's because chemists love to leave off ones. In chemistry, if a number should exist but is not written down, then it is assumed to be 1. So in order to abbreviate a molecule, we simply take the abbreviation for each atom that makes up the molecule and follow it by a subscript that indicates the number of those atoms that are in the molecule. If the number is "1," we don't write it down.

Before we do some examples and problems, there are two terms I must introduce. Chemists do not really use the term "abbreviation" when dealing with atoms and molecules. Instead, they use the terms **chemical symbol** and **chemical formula**. A chemical symbol is the abbreviation for an atom. Thus, "H" is the chemical symbol for hydrogen. A chemical formula simply tells you how many of each atom make up the molecule you are interested in. Thus, "H_2O" is the chemical formula for a water molecule. To make things even a bit more confusing, sometimes chemists drop the word "chemical" and simply use the terms "symbol" and "formula." You will have to get used to these terms, so we will begin using them now.

EXAMPLE 3.3

What is the chemical formula for a molecule that contains 1 atom of sodium, 1 atom of nitrogen, and 3 atoms of oxygen?

The chemical symbols for sodium, nitrogen, and oxygen are Na, N, and O, respectively. Since there is only 1 Na and 1 N, then we needn't put number subscripts after them. However, there are 3 O atoms, so a subscript of 3 must follow the O:

$$\underline{NaNO_3}$$

How many atoms are in the molecule whose formula is Na_2HPO_4?

The formula indicates that there are 2 Na's, 1 H, 1 P, and 4 O's. How did I know that H was one atom's symbol, P was another atom's symbol, and O was another's? Remember, each atom's symbol only contains one capital letter. Thus, every time you see a capital letter, you know you are dealing with a new atom. The total number of atoms, then, is simply the sum of all the individual atoms:

$$\text{Total number of atoms} = 2+1+1+4 = 8$$

So there are a total of 8 atoms in this molecule.

ON YOUR OWN

3.7 Give the formula for each of the following molecules:

a. A molecule that has 1 atom of sodium and 1 atom of chlorine
b. A molecule that is made up of 6 carbon atoms, 12 hydrogen atoms, and 6 oxygen atoms
c. A molecule with 2 sodium atoms, 1 carbon atom, and 3 oxygen atoms

3.8 How many of each atom exist in the following molecules?

a. $KMnO_4$ b. Na_3PO_4 c. RbCl

Classifying Matter as Ionic or Covalent

Once you have determined a compound's chemical formula, you are well on your way to figuring out its name. There is, however, one other thing you must do in order to name a compound properly. You must classify it. Just as elements are classified into metals and nonmetals, compounds are classified based on their properties as well. There are many, many different classification schemes for compounds, however, and we will learn several of them as this course continues. For right now, we will look at the broadest, simplest classification scheme.

Just as elements can be classified into two groups, compounds can be classified as either **ionic** or **covalent**. We will learn as time goes on that there are subclasses within the covalent class, but this is a good start for right now. A compound is considered ionic if, when dissolved in water, it conducts electricity. Pure water is not able to conduct electricity, but when an ionic compound is dissolved in water, it can.

"Now wait a minute," you might be saying, "aren't we always told not to use water around electricity? If water cannot conduct electricity, then why are we worried about it?" Well, as I said, *pure* water cannot conduct electricity. However, the water that comes out of your sink tap is not pure water. It has several ionic compounds dissolved in it. The most notable ionic compound in water is sodium fluoride, which helps protect your teeth against decay. Almost every city in the United States adds sodium fluoride (or something equivalent) to its drinking water for precisely that reason. It turns out that even rainwater is not pure water. It also has ionic compounds dissolved in it. About the only way to get pure water is to buy or make **distilled** water. Distilled water is about 99% pure water.

If a compound does not allow water to conduct electricity after it is dissolved, it is a covalent compound. How can you tell whether a compound is ionic or covalent? Well, one way would be to dissolve it in water and see if it conducts electricity. That's what you will do in the following experiment.

If you purchased the MicroChem kit discussed in the introduction, you can perform Experiment #3 in that kit in addition to the experiment below.

EXPERIMENT 3.2

Electrical Conductivity of Compounds Dissolved in Water

<u>Supplies</u>

- Distilled water (available at grocery stores - half a gallon is plenty)
- Baking soda
- Sugar
- 2 pieces of wire (preferably insulated), each of which is at least 15 cm long
- Scissors or wire cutters to strip insulation from wire (if it is insulated)
- Tape (preferably black electrical tape)
- Teaspoon
- 100 mL beaker or small glass
- 9-volt battery (***<u>DO NOT</u>*** use an electrical outlet in place of the battery. The energy contained in a wall socket can easily kill you!)
- Safety goggles

1. Rinse out your beaker or glass with tap water to get rid of any detergent residues left in it from washing. You must then wash it twice with distilled water to get out any impurities that were left behind from the tap water.
2. Add 80 mL of distilled water to the beaker. Treat this volume as a qualitative measurement.
3. Take your wires and, if they are insulated, strip the insulation off of each wire end so that you have about 2 cm of bare wire at all ends. When that is done, attach one wire's end to one post of the battery and the other wire's end to the other post. The best way to do this is to wad up the bare wire into a ball and shove the ball inside the battery post. Then, while pressing firmly, tape the insulated part of the wire down at the top of the post. The main concern is that you must have the bare wire touching the battery post firmly so that it makes good electrical contact.
4. Once each wire is attached to its own battery post, immerse the other ends of the wires into the water contained in your glass or beaker. Make sure the ends do not touch each other. In the end,

your experiment should look something like the drawing below:

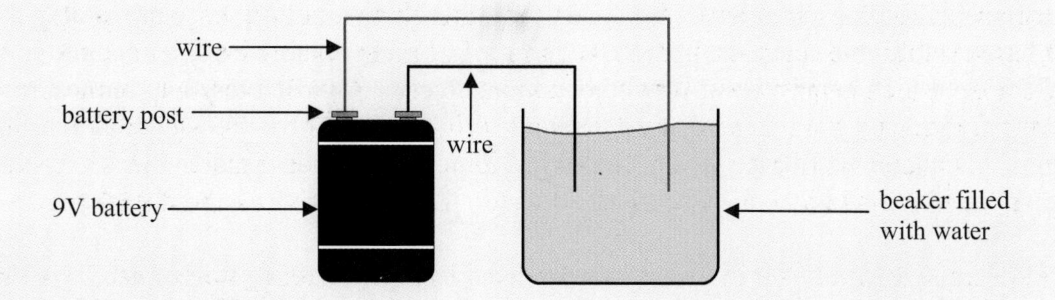

5. Watch the bare ends of the wire closely for a little while. Nothing exciting happens, right?
6. Now remove the wires from the water and add half a teaspoon of baking soda to the water. Mix it around a bit to get it to dissolve. The water should be cloudy, but the undissolved baking soda will eventually settle and the solution will be relatively clear again.
7. Immerse the ends of the wire as you had done previously. Now what's happening? If the other ends of your wires are firmly touching their individual battery posts, you should see bubbles forming on each of the wire ends that are immersed in the solution.
8. Why is the bubbling happening? You'll learn in physics that when electricity is conducted, particles called electrons (ee lek' trons) move from one end of the battery to the other. This motion contains a lot of kinetic energy. The way we have set the experiment up, the electrons must travel through one wire, through the water, and through the other wire to reach the other end of the battery. When this motion takes place, some of that energy can be given to individual water molecules. Under the conditions of this experiment, the kinetic energy generated by the electricity is enough to cause the water to decompose into hydrogen and oxygen, its constituent elements. Since hydrogen and oxygen are gases, they form bubbles at the ends of the wires. In order for this energy to exist to begin with, however, electricity must be conducted. Since pure water cannot conduct electricity, nothing happened when you immersed the wires in the pure water. However, when you dissolved baking soda in the water, it became able to conduct electricity. This means that baking soda must be an ionic compound.
9. Now rinse the glass or beaker with tap water and distilled water as you did before.
10. Repeat the experiment with table sugar. You will find that no bubbles form in this case. This must mean that no electricity can be conducted, which means that table sugar must be a covalent compound.
11. Clean up your mess.

In this experiment, then, we used the decomposition of water as an indicator for electrical conductivity. If bubbles appeared, that told us water was being decomposed, which meant that electricity was being conducted. This is another example of measuring something indirectly, as we discussed in Module #2.

Now in order to classify compounds as ionic or covalent, you could go through an experiment like this for each compound you want to classify. Fortunately, there is an easier way. To classify compounds, all you have to do is look at the chemical formula and follow the following rules:

1. **If a compound contains at least one metal atom and at least one nonmetal atom, the compound is ionic.**

2. If a compound is made up of solely nonmetal atoms, the compound is covalent.

Following these rules, you can now see why our experiment turned out the way that it did. The chemical formula for table sugar is $C_{12}H_{22}O_{11}$. The atoms C, H, and O are all nonmetals, so table sugar is a covalent compound. Thus, it will not conduct electricity when dissolved in water. On the other hand, the chemical formula for baking soda is $NaHCO_3$. The atom Na is a metal while H, C, and O are nonmetals; therefore, baking soda is an ionic compound and will conduct electricity when dissolved in water.

I've already told you that there are exceptions to nearly every rule in chemistry. Although all ionic compounds conduct electricity when dissolved in water, there are compounds that are not ionic but will nevertheless conduct electricity when mixed with water. HCl is an excellent example. It is clearly a covalent compound, because both hydrogen and chlorine are nonmetals. When mixed with water, however, HCl *will* conduct electricity, because it chemically reacts with the water to produce ions. You will learn about this in Module #10. At this point in your chemistry education, however, you can ignore such exceptions. As far as you are concerned, if a compound conducts electricity when dissolved in water, it is an ionic compound.

In the same way, although the vast majority of compounds made up of solely nonmetal atoms are covalent, there are some exceptions. Ammonium hydroxide (NH_4OH), for example, is made up solely of nonmetal atoms. However, it is an ionic compound. You will learn why this is the case in Module #9. Once again, at this point in your chemistry education, you can ignore such exceptions and just follow the rules as discussed in this module. Make sure you understand these rules by classifying the molecules in the following problem:

ON YOUR OWN

3.9 Classify the following compounds as ionic or covalent:

a. CH_4 b. KCl c. $RbC_2H_3O_2$ d. $C_2H_4Br_2$

Naming Compounds

Now that you can interpret a molecule's chemical formula as well as classify it, you are finally ready to learn how to name compounds. The first thing to remember about naming compounds is that you use a different system to name ionic compounds than you do to name covalent compounds. Since the ionic-compound naming system is a little easier, we will begin there. To name an ionic compound:

1. **Start with the name of the first atom in the molecule.**
2. **Take the next atom in the molecule and replace its ending with an "ide" suffix.**
3. **Putting those two names together gives you the compound's name.**

This is why sodium chloride is the name for NaCl. The symbol Na means "sodium" while Cl stands for "chlorine." According to the rules, then, we replace the "ine" in chorine with "ide," making "chloride." Putting these two names together, we get "sodium chloride." In the same way, we would call K_2O "potassium oxide." Pretty easy, huh? In a later module, we will see that this gets more complicated when you have more than two different kinds of atoms in your ionic compound. For right now, we will stick with ionic compounds that have just two different kinds of atoms in them, so naming ionic compounds will be pretty simple at first.

Naming covalent compounds is a little trickier because covalent compounds are more complicated than ionic compounds. When metal atoms and nonmetal atoms form ionic compounds, they can only form one type of molecule. For example, when Na and Cl get together, the only molecule that they can form is NaCl. They cannot form $NaCl_2$ or Na_2Cl or anything like that. The only molecule that will ever be formed between sodium and chlorine is NaCl. Thus, when we say the name "sodium chloride," we know it means only one possible compound: NaCl.

Compare this to the situation with covalent molecules. When carbon (a nonmetal) joins with oxygen (a nonmetal), two possible compounds can be formed: CO or CO_2. In this case, if we used the ionic compound naming system, we would call both of these molecules "carbon oxide." Clearly, you can't have the same name for two different molecules. To fix this, we add prefixes in front of the name of each atom in the compound. This way, the number of atoms in each molecule is explicitly stated in its name. The prefixes we use in this naming system are summarized in the following table. You will be expected to have them memorized!

TABLE 3.1
Prefixes for Naming Covalent Compounds

Prefix	Meaning	Prefix	Meaning
mono	one	**hexa**	six
di	two	**hepta**	seven
tri	three	**octa**	eight
tetra	four	**nona**	nine
penta	five	**deca**	ten

These prefixes are inserted before each name in the compound to indicate how many of each type of atom there are in the molecule. Once again, however, this general rule has one exception. If the prefix used on the first atom is "mono," we drop it. For example, CO_2 has the name "carbon dioxide." Since there is only one carbon atom in the molecule, we should use the "mono" prefix; however, since carbon is the first atom, we drop it. There are 2 oxygen atoms, so we must use the "di" prefix in front of oxygen. Just like naming ionic compounds, we change the last atom's name to an "ide" ending.

We can only drop the "mono" prefix if it is on the first atom in the molecule. Thus, CO is named "carbon monoxide." Since there is only one oxygen atom, we need to use the "mono" prefix, and since oxygen isn't the first atom in the molecule, we cannot drop it. Finally, we can never drop any prefix unless it is "mono." So, the molecule C_2H_6 is named "dicarbon hexahydride." See if you understand both of these naming systems by studying the following examples and then completing the "On Your Own" problems.

EXAMPLE 3.4

Name the compound whose chemical formula is Al_2O_3.

In order to name a compound, we must first determine whether it is ionic or covalent. Since aluminum (Al) lies on the left of the jagged line, it is a metal. Thus, this is an ionic compound. We therefore simply change "oxygen" to "oxide" and put the names together: aluminum oxide.

What is the name of the molecule PH_3?

Since P and H are both nonmetals, this is a covalent compound. Thus, we have to use the prefixes in the name. The prefix for phosphorus (P) is "mono," but since it is the first atom, we can drop it. The prefix for hydrogen is "tri," and we must add the "ide" ending to "hydrogen": phosphorus trihydride.

ON YOUR OWN

3.10 Name the following compounds:

a. SF_6 b. K_3N c. P_2O_3 d. $CaCl_2$

3.11 Give the chemical formulas for the following covalent compounds:

a. oxygen dichloride b. dicarbon dihydride

Before we leave this module, I think I must leave you with a word of warning: This lesson in naming compounds might lead you to believe that the process of giving names to compounds is easy. Don't be fooled into thinking this. We have merely scratched the surface of naming compounds. Notice that the only compounds we have named are those that contain only two different types of atoms. That's because compounds with more than two types of atoms are more difficult to name. In addition, some ionic molecules break the rules, and we have to develop a new naming system for them. Finally, there is a class of covalent compounds called "organic compounds" that have a whole new naming scheme to them. We will revisit some of these tougher topics later on, so don't dismiss as easy the very difficult process of naming compounds.

One other complication that crops up when we start naming compounds is the problem of common names. Think for a minute about why we don't call water by its real chemical name: dihydrogen monoxide. This is because many compounds (like water) were named long before chemists knew about atoms and molecules. Thus, these compounds were named without following any of the rules we just studied. Since old habits die hard, many compounds are still called by their common names rather than their proper chemical names. Ammonia (NH_3) is another example. The proper name for ammonia is nitrogen trihydride, but chemists called it ammonia long before our naming system was developed. Unfortunately, common names like these are here to stay, so we just have to get used to recognizing these names and associating them with the proper compound.

In the end, then, we can see that Democritus' musings on the nature of matter were correct. Chemists have shown throughout the years that matter does not come in continuous sheets; it comes in little packages known as atoms. When identical atoms are jumbled together, the substance that results is called an element. When atoms join together, however, they make molecules. When these molecules are collected together, they form substances called compounds. Even though he didn't know any of these details, Democritus' ideas were essentially right. Not bad for someone who lived more than 2,000 years ago, is it?

LIST OF ELEMENT NAMES AND SYMBOLS

The list is compiled in the order that they appear on the chart

Name	Symbol	Name	Symbol	Name	Symbol	Name	Symbol
Hydrogen	H	Copper	Cu	Lanthanum	La	Astatine	At
Helium	He	Zinc	Zn	Cerium	Ce	Radon	Rn
Lithium	Li	Gallium	Ga	Praseodymium	Pr	Francium	Fr
Beryllium	Be	Germanium	Ge	Neodymium	Nd	Radium	Ra
Boron	B	Arsenic	As	Promethium	Pm	Actinium	Ac
Carbon	C	Selenium	Se	Samarium	Sm	Thorium	Th
Nitrogen	N	Bromine	Br	Europium	Eu	Protactinium	Pa
Oxygen	O	Krypton	Kr	Gadolinium	Gd	Uranium	U
Fluorine	F	Rubidium	Rb	Terbium	Tb	Neptunium	Np
Neon	Ne	Strontium	Sr	Dysprosium	Dy	Plutonium	Pu
Sodium	Na	Yttrium	Y	Holmium	Ho	Americium	Am
Magnesium	Mg	Zirconium	Zr	Erbium	Er	Curium	Cm
Aluminum	Al	Niobium	Nb	Thulium	Tm	Berkelium	Bk
Silicon	Si	Molybdenum	Mo	Ytterbium	Yb	Californium	Cf
Phosphorus	P	Technetium	Tc	Lutetium	Lu	Einsteinium	Es
Sulfur	S	Ruthenium	Ru	Hafnium	Hf	Fermium	Fm
Chlorine	Cl	Rhodium	Rh	Tantalum	Ta	Mendelevium	Md
Argon	Ar	Palladium	Pd	Tungsten	W	Nobelium	No
Potassium	K	Silver	Ag	Rhenium	Re	Lawrencium	Lr
Calcium	Ca	Cadmium	Cd	Osmium	Os	Rutherfordium	Rf
Scandium	Sc	Indium	In	Iridium	Ir	Dubnium	Db
Titanium	Ti	Tin	Sn	Platinum	Pt	Seaborgium	Sg
Vanadium	V	Antimony	Sb	Gold	Au	Bohrium	Bh
Chromium	Cr	Tellurium	Te	Mercury	Hg	Hassium	Hs
Manganese	Mn	Iodine	I	Thallium	Tl	Meitnerium	Mt
Iron	Fe	Xenon	Xe	Lead	Pb	Ununnilium	Uun
Cobalt	Co	Cesium	Cs	Bismuth	Bi	Unununium	Uuu
Nickel	Ni	Barium	Ba	Polonium	Po	Ununbium	Uub

ANSWERS TO THE "ON YOUR OWN" PROBLEMS

3.1 As was discussed previously, when wood burns, it uses up oxygen and produces carbon dioxide, ash, and water. Since we know the total amount of matter produced during the fire, we can determine how much oxygen had to be used up. To do this, we first recognize that the chemist measured the masses of all substances produced in the fire. Thus:

$$\text{Total mass} = \text{Mass of ash} + \text{Mass of water} + \text{Mass of carbon dioxide}$$

Before we can do this, however, we must change the mass of the ash from grams to kilograms, because that's what the other two masses are measured in. Based on our rules of conversion, 925 grams = 0.925 kg. Thus,

$$\text{Total mass} = 0.925 \text{ kg} + 3.12 \text{ kg} + 15.14 \text{ kg} = 19.19 \text{ kg}$$

Notice the significant figures. Because we are adding numbers, we must look at decimal place. The least precise numbers have their last significant figure in the hundredths place, so the answer must have its last significant figure in the hundredths place. Thus, the total mass produced in the fire is 19.19 kg. By the Law of Mass Conservation, then, the total amount of mass prior to the fire must also be 19.19 kg.

$$19.19 \text{ kg} = \text{Mass of wood} + \text{Mass of oxygen}$$

Rearranging the equation gives us:

$$\text{Mass of oxygen} = 19.19 \text{ kg} - \text{Mass of wood}$$

Plugging in the mass of the wood:

$$\text{Mass of oxygen} = 19.19 \text{ kg} - 15.4 \text{ kg} = 3.8 \text{ kg}$$

In this case, the least precise number has its last significant figure in the tenths place. Since we are subtracting, the answer must also have its last significant figure in the tenths place. Thus, <u>3.8 kg</u> of oxygen had to be used up in the fire.

3.2 Since O, I, and Rn are to the right of the jagged line, they are nonmetals. In addition, H is always a nonmetal because it is an exception to the rule. Thus, <u>Mg, In, and Os are metals</u>.

3.3 Electrical wires have to be able to conduct electricity and should bend relatively easily. Thus, you should choose a metal to work with. Metals lie to the left of the jagged line, and the only one that does is <u>Ni</u>.

3.4 Something that breaks easily when bent is brittle. Thus, the question is asking you to identify nonmetals. The only nonmetal on the list is <u>P</u>.

3.5 According to the Law of Mass Conservation, 12.0 grams of carbon + 4.0 grams of hydrogen must make 16.0 grams of natural gas since there are no leftover amounts of carbon or hydrogen. In order to make 100.0 grams of natural gas, then, all we have to do is determine the factor we need to increase our recipe by:

$$16.0 \text{ g} \cdot \text{x} = 100.0 \text{ g}$$

$$\text{x} = \frac{100.0 \cancel{\text{g}}}{16.0 \cancel{\text{g}}} = 6.25$$

If we need to increase our recipe by a factor of 6.25, then all we need to do is multiply all of our ingredient amounts by 6.25:

$$\text{Mass of carbon} = 12.0 \text{ g} \cdot 6.25 = 75.0 \text{ g}$$

$$\text{Mass of hydrogen} = 4.00 \text{ g} \cdot 6.25 = 25.0 \text{ g}$$

Thus, to make 100.0 grams of natural gas, we must add 75.0 grams of carbon to 25.0 grams of hydrogen.

3.6 Before we can answer this question, we need to determine how much chlorine gas was in the original sample. That will then tell us the recipe. For this, we use the Law of Mass Conservation. Since we started with 30.0 g, we must end with 30.0 grams as well. If 13.6 grams of mass is accounted for by the cobalt, then the rest must have gone to the chlorine, thus:

$$\text{Mass of chlorine} = 30.0 \text{ g} - 13.6 \text{ g} = 16.4 \text{ g}$$

So the recipe for making 30.0 grams of the purple powder is to take 13.6 grams of cobalt and react it with 16.4 grams of chlorine. But the chemist wants to make 1.00 kg of the stuff. Well, first we have to convert that to grams so all of our units are the same. 1.00 kg is the same as $1.00 \text{ x } 10^3$ g. Now we need to determine what to multiply our recipe by so it will make $1.00 \text{ x } 10^3$ g:

$$30.0 \text{ g} \cdot \text{x} = 1.00 \times 10^3 \text{ g}$$

$$\text{x} = \frac{1.00 \times 10^3 \cancel{\text{g}}}{30.0 \cancel{\text{g}}} = 33.3$$

So now we just multiply our list of ingredients by this number, and we'll have the answer:

$$\text{Mass of cobalt} = 13.6 \text{ g} \cdot 33.3 = 453 \text{ g}$$

$$\text{Mass of chlorine} = 16.4 \text{ g} \cdot 33.3 = 546 \text{ g}$$

Thus, to make 1.00 kg of the purple powder, you react 453 grams of cobalt with 546 grams of chlorine. *NOTE: You might notice that these two numbers do not add up to $1.00 \text{ x } 10^3$ grams as they should by the Law of Mass Conservation. Instead, they add to $0.999 \text{ x } 10^3$ g. This is a result of the rounding that we must do in order to keep track of significant figures.*

3.7 a. The chemical symbols for sodium and chlorine are Na and Cl. Since there is only one atom of each, then there is no need to put any numbers in the formula. Thus, NaCl is the chemical formula.

b. The symbols for carbon, hydrogen, and oxygen are C, H, and O. Since there are 6 carbons and 6 oxygens, there must be a "6" subscript after the C and O. Since there are 12 hydrogens, then a "12" must follow the H. So $C_6H_{12}O_6$ is the proper chemical formula.

c. The proper formula is Na_2CO_3.

3.8 a. Since there is no number after the K (potassium), we assume there is 1 potassium atom. The same assumption holds true for the Mn, so there is 1 manganese atom. The "4" after the oxygen means there are 4 oxygen atoms.

b. In this molecule, there are 3 sodium atoms, 1 phosphorus atom, and 4 oxygen atoms.

c. This molecule is made up of 1 rubidium atom and 1 chlorine atom.

3.9 a. Since C is on the right of the jagged line and H is always a nonmetal, this molecule is comprised solely of nonmetals. This makes it covalent.

b. Because K (potassium) is on the left of the jagged line, it is a metal, while Cl, on the right of the jagged line, is a nonmetal. Thus, this means the compound is ionic.

c. This is also ionic because Rb is a metal.

d. All atoms in this molecule are nonmetals, so the compound is covalent.

3.10 a. This molecule is covalent, so we need to use prefixes. The prefix for sulfur (S) is "mono," but we drop it because it is first. The prefix for fluorine (F) is "hexa," so we have sulfur hexafluoride.

b. K is a metal, so this is an ionic compound. Thus, we use no prefixes: potassium nitride.

c. diphosphorus trioxide

d. calcium chloride

3.11 a. OCl_2 b. C_2H_2

REVIEW QUESTIONS FOR MODULE #3

1. What is the difference between the continuous theory of matter and the discontinuous theory of matter?

2. What two laws were instrumental in the development of Dalton's atomic theory? What law did Dalton predict using his theory?

3. Write the Law of Mass Conservation in your own words.

4. List the four assumptions of Dalton's atomic theory.

5. What is the difference between an atom and a molecule?

6. What are the physical characteristics that distinguish metals from nonmetals?

7. How can you determine whether an atom is a metal or a nonmetal from the periodic chart?

8. How can you experimentally determine whether a compound is ionic or covalent?

9. How can you determine from the periodic chart whether a compound is ionic or covalent?

10. Why do chemists use two different naming systems for compounds?

PRACTICE PROBLEMS FOR MODULE #3

1. A chemist uses an experiment similar to Experiment 3.2 to decompose 150.0 grams of water (H_2O). If the chemist finds that he makes 16.7 grams of hydrogen from the decomposition, how much oxygen must he have produced?

2. A compound is decomposed into its constituent elements of carbon, nitrogen, and hydrogen. If, in the decomposition, 12.0 grams of carbon, 14.0 grams of nitrogen, and 3.0 grams of hydrogen were formed, what was the mass of the compound before it was decomposed?

3. Identify the following as metal or nonmetal: I, Y, N, Re.

4. In an experiment to make hydrogen peroxide, a chemist mixes 20.0 grams of hydrogen with 20.0 grams of oxygen. She finds that 21.3 grams of hydrogen peroxide are produced. In addition, there is a lot of hydrogen left over at the end of the experiment. How many grams of hydrogen should have been added to the 20.0 grams of oxygen so that there would be no leftovers at the end of the experiment?

5. In making calcium nitride, 100.0 grams of calcium plus 100.0 grams of nitrogen makes 152.5 grams of product along with some leftover nitrogen. How much nitrogen and calcium should be added together in order to make 1.0 kg of calcium nitride without any leftovers?

6. What is the chemical formula of a molecule that is made up of 3 carbon atoms, 6 hydrogen atoms, 2 chlorine atoms, and one oxygen atom?

7. How many total atoms are in one molecule of $C_{12}H_{24}O_{10}S_2$?

8. Which of the following molecules are covalent?

a. SO_3 b. P_2H_6 c. $CaSO_4$ d. KF

9. Name the compounds in problems 8a, 8b, and 8d.

10. What are the chemical formulas for the following compounds?

a. tetranitrogen hexahydride b. nitrogen monoxide

MODULE #4: Classifying Matter and Its Changes

Introduction

In the last module, we learned about the basic building blocks of matter: atoms and molecules. We learned that elements result when atoms of the same type are collected together, and compounds result when molecules of the same type are collected together. It turns out, however, that there are other ways atoms and molecules can combine to form other types of matter. One of the things we will do in this module is classify matter into several groups, depending on the ways that atoms and molecules exist in it.

In addition, we will start to study the most interesting aspect of matter: how it can change. Changes in matter also need to be classified, and we will learn the criteria for this classification. In addition, we will start learning the basis of the most fundamental type of change matter can undergo: the chemical reaction.

Classifying Matter

In order to determine how matter changes, we first need to classify it. If we can properly classify the various types of matter we see in God's creation, it will be easier for us to determine how that matter is changing. First of all, every bit of matter that you see can be classified into two large groups: **mixtures** and **pure substances**:

Pure substance - A substance that contains only one element or compound

Mixture - A substance that contains different compounds and/or elements

These definitions may sound a bit confusing at first, but don't worry; you'll soon get the hang of them. In the last module, you used distilled water in one of your experiments. I said that the distilled water was 99% pure. What I meant by that was 99% of all molecules in distilled water are H_2O molecules. If 100% of all the molecules in distilled water were H_2O molecules, then distilled water could be labeled a pure substance. This is what I mean when I say that pure substances contain only one element or compound.

Examples of pure substances would include sulfur, iron, sugar, and table salt. In each of these substances, only one element or compound exists. The first two examples are elements; thus, they are made up of atoms. In a sample of sulfur, only one type of atom exists: sulfur (S) atoms. In a sample of iron, only one type of atom exists: iron (Fe) atoms. The last two examples are compounds; thus, they are made up of molecules. Sugar is made up of only one type of molecule, $C_{12}H_{22}O_{11}$. Likewise, table salt is composed entirely of NaCl molecules. As a point of review, you should recognize that sugar is a covalent compound (entirely made up of nonmetals), whereas table salt is ionic (made up of a metal and a nonmetal).

Mixtures, on the other hand, contain more than one compound and/or element. If I were to take table salt and dissolve it in pure water, the result would be a mixture. This mixture would have two compounds: H_2O and NaCl. This is what I mean when I say that mixtures have different compounds and/or elements. The H_2O molecules and the NaCl molecules in a salt water mixture do not combine to make new molecules. They just occupy the same general area. That's what a mixture is. The important thing to realize about mixtures is that, while in a mixture, individual elements or compounds retain their individual properties. For example, pure water is tasteless. Table salt, on the other hand, has a distinct, sharp taste. A mixture of salt water, then, tastes sharp like salt, but not as sharp as pure salt. Each compound retains its taste, and the tasteless water dilutes the sharp taste of the salt in the mixture.

This leads to another way we can distinguish between pure substances and mixtures: The individual components in a mixture can be physically separated based on their individual properties. Consider, for example, Figure 4.1.

Photos by Flanders Video Productions

FIGURE 4.1
Mixtures Can be Separated Based on the Properties of Their Components

Sulfur (the yellow powder) and iron (the black powder) are mixed together.

A magnet (the white bar) is brought near the mixture.

The magnet pulls the iron out of the mixture and leaves the sulfur behind.

Sulfur and iron are mixed together and then heated so that they form the compound iron sulfide (the black powder).

A magnet (the white bar) is brought near the iron sulfide.

The magnet doesn't pick up anything, because the iron no longer has its individual properties.

In the top portion of the figure, we are working with a *mixture* of sulfur and iron. Since this is a mixture, the sulfur retains its properties (its yellow color, for example), and the iron retains its properties (its black color, for example). These properties can be used to separate the two elements. If a magnet is brought near the mixture, the iron will cling to the magnet, but the sulfur will not. Thus, we could separate the mixture back into its components (sulfur and iron) by using the magnet to pull out all of the iron. In the end, we would have two pure substances, iron and sulfur. A mixture, then, can be separated using the individual properties of the components which make up the mixture.

Compare this to the lower portion of the figure, where the same two elements (sulfur and iron) are heated together so that they form the *compound* iron sulfide. Although iron sulfide contains both sulfur and iron, it is *not* a mixture: it is a pure substance. You can see that in several ways. For example, notice that there is no yellow in iron sulfide: it is black. Thus, the sulfur does not retain its characteristic yellow color. Notice also that when the magnet is brought near the iron sulfide, nothing clings to it. That's because the iron does not retain its ability to respond to a magnet. Both the sulfur and the iron have "given up" their individual properties to become a compound. This, then, is a pure substance. Its elements cannot be separated based on their individual properties, because they do not retain their individual properties when they form a compound.

The only way you could get the iron and sulfur out of iron sulfide is to *decompose* the iron sulfide back into a mixture of iron and sulfur *and then* separate the two. Thus, mixtures can be *physically separated* into their individual components while pure substances must be *decomposed* before their individual components can be physically separated. Experiment 4.1 gives you the opportunity to find another way of separating the components in a mixture.

EXPERIMENT 4.1
Separating a Mixture of Sand and Salt

Supplies:

- Two beakers - 100 mL and 250 mL (or two glasses - one large and one small)
- Sand (Kitty litter is an acceptable substitute, but don't use the kind that clumps.)
- Table salt
- Funnel
- Water
- Filter paper (You can cut a circle out of the bottom of a coffee-maker filter.)
- Stirring rod (or small spoon)
- Teaspoon
- A heat source such as a stove or alcohol burner
- Safety goggles

1. Take a teaspoon of sand and a teaspoon of salt and pour them into the 100 mL beaker.
2. Use your stirring rod to mix them up well. At this point, you have a mixture of sand and salt in your beaker. The molecules of sand (SiO_2) do not combine with the molecules of salt (NaCl) in any way. These two different types of molecules simply occupy the same general space now. In this experiment, you will learn how to separate this mixture back into sand and salt.
3. One way we could accomplish this separation would be to get a fine set of tweezers and a powerful magnifying glass. If we looked at the mixture through the magnifying glass, we could see little grains of salt and little grains of sand. We could then use the tweezers to slowly pick the salt out of the sand. With enough patience and time, we could separate the mixture in that way. Unfortunately, this method is quite tedious and would take an enormous amount of time. In order

to speed up the procedure, we will use a different method. Remember that in a mixture, the individual molecules still retain their own, unique properties. One property of salt is that it dissolves in water. Sand, however, does not. We will use this property to separate the two.

4. Add about 25 mL (1/8 cup) of water (this is a qualitative measurement) to the mixture and use your stirring rod to mix it in well. As the water mixes with the sand and salt, the salt will dissolve in water and the sand will not. Now you have an ugly-looking mixture of sand, salt, and water in your beaker. How does this help separate the sand and the salt? Well, now you will filter your mixture, and the salt water will separate from the sand.
5. To do this, take a circle of filter paper and fold it as shown below:

Illustration by Megan Whitaker

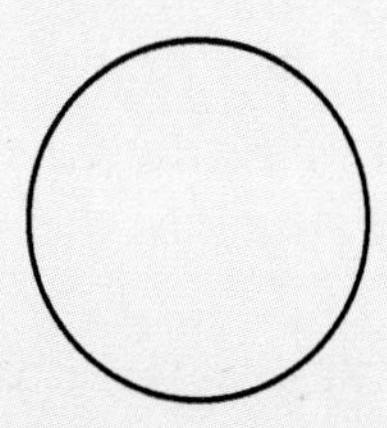

Start with a large piece of circular filter paper.

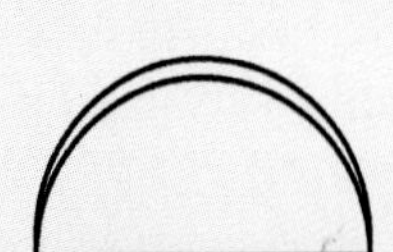

Fold it in half.

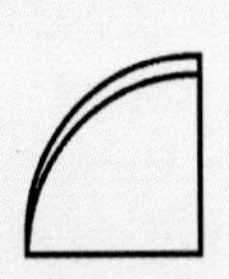

Fold it in half again so it makes a triangle.

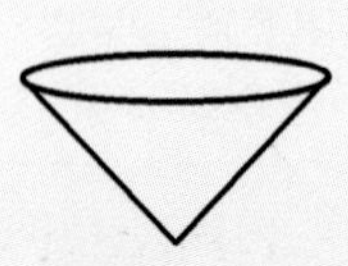

Open the triangle to make a cone.

6. Take the resulting cone of filter paper and place it in the funnel. In order to get the filter paper to retain its cone shape, wet it down with some water. This will cause the filter paper to stick to the funnel.
7. Hold the funnel above the empty 250mL beaker so that anything falling through the funnel will land in the 250 mL beaker.
8. Now pour the mixture of water, salt, and sand into the filter paper. Make sure that the water level never rises above the top of the filter-paper cone. In addition, for best results, try to pour off just the liquid first and leave the majority of the sand in your beaker until the very end. As the liquid filters through the paper, you will see it fall into the 250 mL beaker. As long as you do not allow the liquid level to rise above the top of the filter-paper cone, the liquid in the beaker will be clear. Continue this process until everything in the beaker (including the sand) has been poured into the funnel.
9. Once all of the liquid has filtered through the filter paper (this may take a while!), rinse the 100 mL beaker with 10 mL (1 tablespoon) of water and dump it all into the filter paper. Allow all of the liquid to drain through the filter paper.
10. Now what you have is pure sand on the filter paper and a mixture of salt and water in your beaker. To get rid of the water, simply start heating it up so that it will boil. **WARNING: If you are using glasses and not beakers for this experiment, do not heat the glass. Instead, pour the salt water into a pot and boil it there. Regular glasses tend to break when they are heated to 100 ^{o}C!**
11. Allow all of the water to boil away. What's left behind? Pure table salt is left as a residue in the beaker. So now you have separated your mixture of sand and salt back into its components. The sand is on the filter paper, and the salt is in the 250 mL beaker.
12. Clean up your mess.

If you purchased the MicroChem kit discussed in the introduction, you can learn another method used to separate mixtures by performing Experiment #1.

This is the idea behind the concept of mixtures. Since the individual molecules within a mixture retain their unique properties, we can use those properties to separate them. Since salt dissolves in water and sand does not, we used that particular property as a means of separation in the experiment.

Mixtures and pure substances, then, are the two broad classifications of matter. However, they can both be broken down into two subclassifications as well. Pure substances can be either elements or compounds. Thus, the classifications you learned before are really subclasses of the pure substance classification. Mixtures can be further classified as either **heterogeneous** (het' uh roh jee' nee us) or **homogeneous** (ho' moh jee' nee us).

Homogeneous mixture - A mixture with a composition that is always the same no matter what part of the sample you are observing

Heterogeneous mixture - A mixture with a composition that is different depending on what part of the sample you are observing

These definitions make sense when you think about the prefixes "homo" and "hetero." The prefix "homo" means "the same" while the prefix "hetero" means "different." Thus, homogeneous solutions are the same throughout the sample, whereas heterogeneous solutions are different in different parts of the sample.

The best way to explain the difference between heterogeneous and homogeneous mixtures is by example. In Experiment 4.1, you made up a mixture of sand and salt. When you first dumped the sand and the salt together, it was a heterogeneous mixture. If you looked at one portion of the mixture, you might see mostly sand, whereas another portion of the mixture might be mostly salt. As you stirred the mixture, however, it started to become homogeneous, because the sand and salt were mixing evenly throughout the sample. If you stirred the sand and salt enough, you would end up with a mixture in which the salt was evenly spread throughout the sand. That would be a homogeneous mixture.

Examples of heterogeneous mixtures include Italian salad dressing, a bowl of cereal and milk, and a cake with icing on it. In each case, these mixtures have a different composition depending on where you observe the sample. When you take Italian dressing out of the refrigerator, there are seeds, herbs, and spices in the bottle, but they are mostly at the bottom. Thus, the seeds, herbs, and spices are more concentrated in one part of the sample than the other. A bowl of cereal and milk typically has mostly cereal on the top and milk at the bottom, because the cereal tends to float on top of the milk. Finally, an iced cake is a mixture of cake and icing, but the icing is completely on the top while the cake is underneath.

Examples of homogeneous mixtures would be salt water, a Coke®, or steel. In each case, these mixtures have the same composition throughout. A glass of salt water has the same amount of salt and water at the top of the glass as at the bottom. Coke® is a mixture of several ingredients, but there is no difference in taste or composition from the top of the Coke® can to the bottom. Finally, steel is a mixture of iron (Fe) and carbon (C). The iron gives steel its strength while the carbon helps the steel stay rigid. The amount of carbon and iron is the same no matter where you observe a sample of steel, so it is also a homogeneous mixture. Figure 4.2 sums up these classifications of matter.

FIGURE 4.2
A Classification Scheme for Matter

ALL MATTER

Pure Substances
Composed of only one compound or element

Examples:

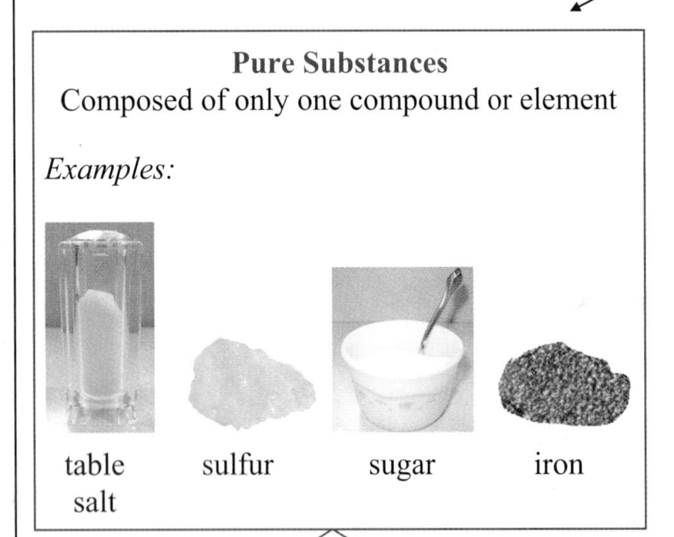

Mixtures
Composed of at least two different compounds and/or elements

Examples:

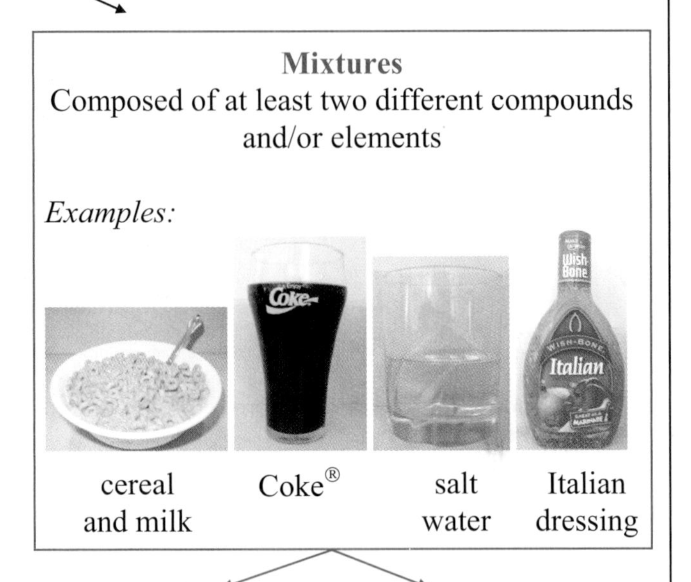

Elements
Composed of only one type of atom

Examples:

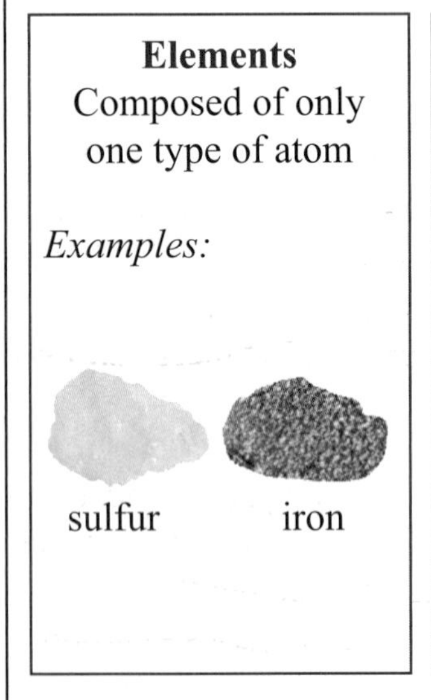

Compounds
Composed of only one type of molecule

Examples:

Homogeneous
Mixtures that have the same composition throughout

Examples:

Heterogeneous
Mixtures with a varied composition throughout

Examples:

There are a couple of interesting facts related to the concept of mixtures that are worth discussing. First, if you buy a carton of milk, it says that the milk is "homogenized." Now that you know this classification scheme, you can understand what that word means. When milk is taken from a cow, it is a heterogeneous mixture. The cream in the milk tends to rise to the top, while the water tends to fall to the bottom. Thus, the milk is thick and creamy if it comes off the top of the bottle, whereas it is thin and watery if it comes off the bottom of the bottle. Before you buy the milk, however, it is treated so that the cream stays evenly distributed throughout the milk. Thus, the formerly heterogeneous mixture is turned into a homogeneous mixture. This is what the term "homogenized" means.

Second, did you know that air is a mixture? Most people think that since animals breathe in oxygen, the air around them must be pure oxygen. Nothing could be farther from the truth. In fact, air is a mixture of several gases, including nitrogen, argon, water vapor, carbon dioxide, and oxygen. Interestingly enough, oxygen actually makes up less than ¼ of the air that we breathe. Assuming there is no humidity, the air we breathe is 78% nitrogen, 21% oxygen, 0.9 % argon, 0.03% carbon dioxide, and 0.07% other gases. Figure 4.3 shows this relationship in pie graph form.

FIGURE 4.3
The Composition of Dry Air

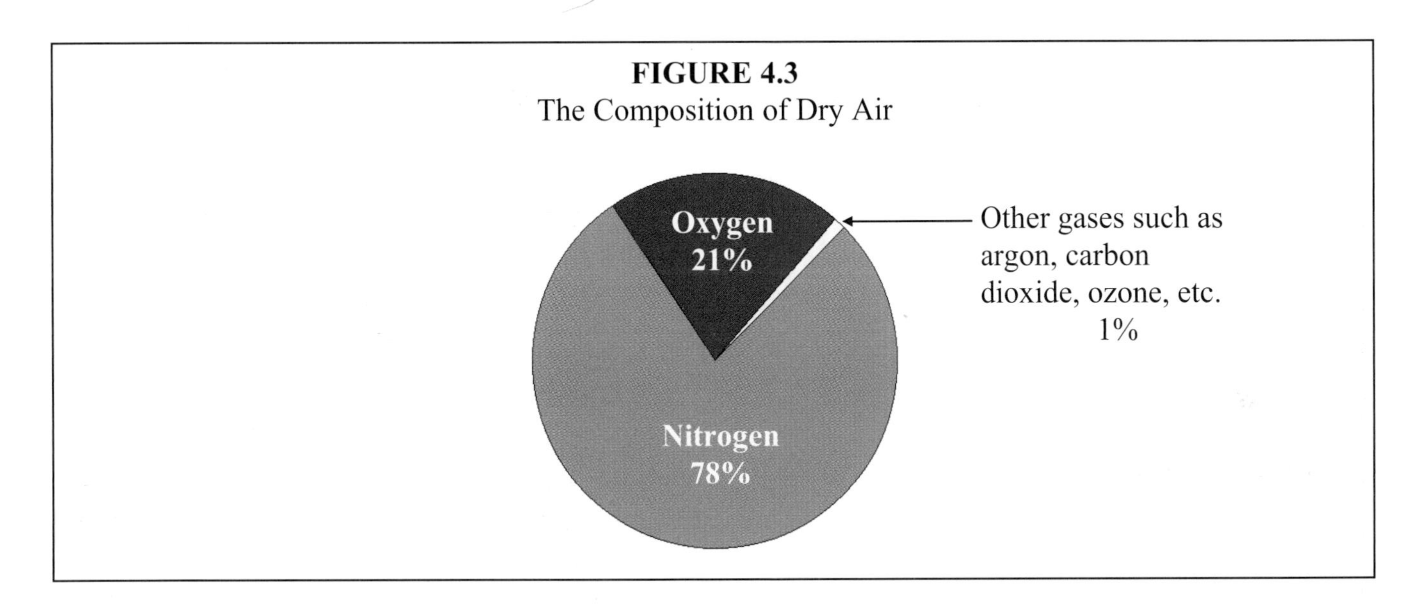

It turns out that this mixture of gases is ideal for human life. Even though we need oxygen in order to live, we cannot have too much oxygen. Breathing elevated levels of oxygen for prolonged periods of time can have serious consequences such as lung damage, chest pains, and even blindness! In addition, the frequency of natural forest fires increases significantly when the amount of oxygen in the air increases. For every 1 percent increase in the amount of oxygen in the air, the chance of a natural forest fire occurring goes up by 70 percent! Thus, we need oxygen in our air, but there can't be too much of it.

The air in our atmosphere does not have too much oxygen in it because the oxygen has been "diluted" with nitrogen. It turns out that nitrogen is the perfect gas to accomplish this dilution, because it does not react with our bodies in any way. We breathe in nitrogen with every breath, but then we breathe it right out again because it does not interact with our bodies. Almost all other gases - carbon monoxide, sulfur dioxide, etc. - are poisonous to human life, so we are very lucky that those gases do not make up any significant fraction of the air we breathe.

Doesn't this seem like an amazing coincidence? The air on this planet just happens to have oxygen, but not too much oxygen. It has the perfect amount of oxygen for human life. The other gas that makes up the vast majority of the rest of our planet's air just happens to be one of the few gases that does not affect our bodies in any way! Clearly, such a perfect mixture of gases would never have come about by chance. The air around us has been made by God to be perfect for sustaining human life. As we go on in this course, you will see that there are thousands and thousands of "coincidences" like this one that clearly show that this world has a Creator. This is why I say that real science points unerringly to God.

See if you understand our matter classification scheme by answering the following question.

ON YOUR OWN

4.1 Classify the following as elements, compounds, homogeneous mixtures, or heterogeneous mixtures:

a. copper b. carbon dioxide c. cake icing d. tomato e. sugar water
f. sodium g. fruitcake h. sodium fluoride i. tap water

Classifying Changes That Occur in Matter

Now that we know how to classify matter, we can finally begin to study how matter changes. Chemists classify the changes that matter undergoes into two large groups: **physical change** and **chemical change.**

Chemical change - A change that affects the type of molecules or atoms in a substance

Physical change - A change in which the atoms or molecules in a substance stay the same

The way to distinguish between these two types of changes is to think about the atoms or molecules involved. For example, if you rip a piece of paper in half, have you changed the molecules in the paper into different molecules? Obviously not. Both halves of the paper are still paper. All you did was take the sample of molecules and split it in half. The molecules stayed the same; some were just separated from others. This is an example of physical change. Whenever you change a substance without altering its molecules or atoms, you have made a physical change.

On the other hand, suppose you took that piece of paper and burned it. The molecules in the paper would slowly change into carbon dioxide, water vapor, and ash. This is an example of chemical change. If the type of atoms or molecules in a substance changes, the substance has undergone a chemical change.

Distinguishing between chemical and physical change is very important, so we are going to spend some time on it. Suppose I dissolve sugar in water. Is this a chemical change or a physical change? Well, when the sugar dissolves in the water, it seems to disappear, but it hasn't changed its chemical makeup in any way. The sugar molecules are still sugar molecules; they are now just evenly dispersed around the water. Thus, dissolving one substance in another is a physical change, because the act of dissolving things does not affect the type of atoms or molecules involved.

On the other hand, remember what happened when you put lye into vinegar during Experiment 3.1? The lye did not *dissolve* in the vinegar; it *reacted* with the vinegar. You saw evidence of this because a color change occurred and heat was released. The color change was caused by the fact that the lye (NaOH) reacted with the active ingredient in vinegar ($C_2H_4O_2$) to make water (H_2O) and sodium acetate ($NaC_2H_3O_2$). Thus, the molecules involved actually changed. Before you put the lye in the vinegar, you had NaOH and $C_2H_4O_2$. After you mixed the two and allowed them to react, you got H_2O and $NaC_2H_3O_2$. That, then, was a chemical change.

Another way you can think about chemical and physical change is to think about whether a change can be easily reversed or not. Suppose you dissolved salt into water. Could you reverse that process? Could you recover the salt? Of course. You did so in the previous experiment. To recover the salt, all you would have to do is boil off the water. In the same way, you could easily reverse the change that occurred when you ripped the paper in half. All you would need to do is put the two halves back together. Physical changes, then, can usually be reversed relatively easily.

On the other hand, is there any way you could "unburn" paper? No. Once you have burned paper, there is no way to get the paper back. In the same way, once you reacted the lye with the vinegar in Experiment 3.1, you could not easily undo that change. If you boiled the liquid away, you would not find lye left over. Instead you would find sodium acetate. Chemical changes, then, are not easily reversed.

Although this way of distinguishing between chemical and physical change is convenient, it is not 100% correct. For example, chemical changes *can* be reversed, but it takes a chemical reaction to do so. In fact, in Module #15, you will see that in principle, all chemical reactions are reversible. However, for right now, thinking about whether or not a change is *easily* reversible is a convenient way to distinguish between chemical and physical change. In order to become proficient at understanding the changes that occur in matter, you must be able to make this crucial distinction. To help you get more experience doing so, perform the following experiment:

EXPERIMENT 4.2

Distinguishing Between Chemical and Physical Change

Supplies:

- Water
- A **plastic** graduated cylinder (A rain gauge will work as well. A measuring cup may work, but if you have a frost-free freezer, the opening of whatever you use should be very small.)
- An egg in its shell
- Toilet bowl cleaner (The list of ingredients must include hydrochloric acid or hydrogen chloride. The Works® was the brand used when this experiment was tested. You can use vinegar if you cannot find a proper toilet bowl cleaner, but the experiment will have to sit overnight in order for it to work. However, if you *do* use vinegar, the egg will be *much* more interesting to observe!)
- A tall glass
- A sink
- A spoon
- Rubber cleaning gloves
- Safety goggles

1. The first part of this experiment will take quite a while to finish, so ideally you should start it now and finish the rest tomorrow. To start the first part of the experiment, simply fill your graduated cylinder with 30.0 mL of water (a quantitative measurement), and then put it in the freezer.

(NOTE: If you have a frost-free freezer, cover the cylinder loosely with plastic wrap.) If you are using a measuring cup or a rain gauge, it doesn't really matter how much water you use, as long as the meniscus is sitting right on one of the measuring lines, so that you know exactly what volume of water is in there.

2. You need to let the graduated cylinder remain in the freezer for at least several hours.
3. Take the tall glass and place it in the sink. You need to do this because the experiment might cause suds to overflow the glass, and you will want the suds to fall right into the sink. ***DO NOT use aluminum foil around this experiment, as aluminum will react with the toilet bowl cleaner and produce a nasty gas. Also, keep the glass in the sink throughout the entire experiment!***
4. Put on the rubber gloves and fill the glass about one-third full of toilet bowl cleaner. ***Be careful not to get any of the toilet bowl cleaner on your skin; it is caustic and will burn.*** *If you do get it on your skin, just rinse it off with plenty of water.*
5. Using the spoon, gently lower the egg into the toilet bowl cleaner and hold it completely under the liquid. Observe what is happening. There should be a lot of bubbling going on. Since most toilet bowl cleaners contain detergent, there might also be some suds forming. If you are using vinegar, you will not see the bubbling, and you will have to let it sit overnight rather than the few minutes discussed in the next step.
6. Allow this to go on for about two minutes, then use the spoon to remove the egg from the glass. Rinse the egg with water and examine it. You should see that its shell is much thinner than it was before. This is because the acid in the toilet bowl cleaner is reacting with the egg shell. When this happens, the calcium carbonate ($CaCO_3$) which makes up the egg shell is turned into calcium chloride, water, and carbon dioxide (which is making the bubbles you see). This is obviously a *chemical* change, because the $CaCO_3$ is being changed into different molecules. This change is also not reversible. There is just no way to get the eggshell back once you've done this experiment.
7. If you do this experiment just right, you can actually strip away the shell of the egg and end up with just a thin, semi-transparent membrane holding the egg together. This is much more likely if you use vinegar and wait overnight. If your egg still has a lot of shell on it, just put it back in the toilet bowl cleaner for a while and then pull it out again and rinse it off. Continue to do this until the shell is gone and a thin membrane is all that exists to hold the egg together. This can be a little tricky, however, because the egg gets pretty fragile and may well break as you try to take it out of the toilet bowl cleaner.
8. After the graduated cylinder has been in the freezer at least 12 hours, pull it out and observe what has happened. As expected, the water has turned to ice, but read the volume. What has happened? The volume has increased. The water actually expanded as it froze. As we will learn a little later in this module, water is an exception to the general rule. Most substances decrease in volume when they freeze, but the volume of water increases upon freezing. This is, indeed, a change, but what kind of change is it? It is a *physical* change. Think about the molecules involved. When water is liquid, its chemical formula is H_2O. After it freezes, it is still H_2O; it is just in solid form. Also, this change is reversible. You can allow the graduated cylinder to stand out at room temperature until the ice melts. Then, the volume will be back to 30.0 mL. This kind of change, called a **phase change**, will be discussed in great detail in a moment. For right now, the important thing to realize is that phase changes are physical changes and do not affect the identity of the molecules or atoms that make up the substance being studied.
9. Clean up your mess.

Now that you've had a bit of experience at identifying physical and chemical changes, try this exercise:

ON YOUR OWN

4.2 Classify the following changes as physical or chemical:

a. Steam condenses into water.
b. Gasoline burns in an automobile engine.
c. A barber cuts your hair.
d. Milk sours.
e. Sugar is dissolved in water.
f. Milk is poured on cereal.
g. An iron nail rusts.

Phase Changes

As I mentioned in the previous experiment, when a substance changes from solid to liquid (or vice versa), that change is referred to as a **phase change**. A change from liquid to gas (or vice versa) is also a phase change. These phase changes are quite interesting and worth discussing in some detail. The first thing to realize about phase changes is that they require only one thing: energy. If I want to change something from a solid to a liquid, all I need to do is add heat. Solid butter becomes liquid if we just heat it up a little. Ice becomes water if it is allowed to warm up to room temperature. We call the transition from solid to liquid **melting**. The reverse process, liquid turning into solid, we call **freezing**. Instead of adding energy to a substance, if I take energy away from it, I can make the substance freeze. Likewise, going from liquid to gas, called **boiling**, requires energy, whereas going from gas to liquid, called **condensing**, requires that energy be removed. We could schematically depict this relationship as follows:

The three states of matter (solid, liquid, and gas) are called **phases** of matter, and that is why changes between them are called phase changes. Now, as I said before, phase changes are physical changes. Thus, the molecules (or atoms) that make up the substance undergoing a phase change are not altered in any way. But if the molecules (or atoms) are not being changed, what is? Something obviously must be changing in order for a phase change to take place. What can it be? Performing the next experiment might help you find out:

EXPERIMENT 4.3
Condensing Steam in an Enclosed Vessel

Supplies:

- A rectangular metal can with a lid (Turpentine and paint thinner are usually sold in such cans. The can must be empty and thoroughly rinsed out.)
- Water
- A stove
- Hot pads
- Safety goggles

1. Thoroughly rinse the can several times with water.
2. After you have rinsed the can, put about 3 cups (a qualitative measurement) of water in it.
3. Leaving the lid off, heat the can on the stove. You will eventually hear the water boiling. Allow it to boil for quite some time, until steam is pouring out of the hole in the top where the lid should be.
4. Use a hot pad to remove the can from the stove.
5. Quickly put the lid on the can and close it as tightly as possible.
6. Now watch the can for a while. You will start to hear it creak and groan, and then the can will start collapsing in on itself. If this doesn't happen, it means that your can has a leak in it somewhere, so you must start over with a new can.
7. Clean up your mess.

Why did the can collapse on itself? Well, while the water was boiling, water vapor began filling up the can, pushing any air that was in the can out through the hole in the top. Thus, by the time you pulled the can off of the stove and placed the lid on it, the can was full of water in its gas phase. When you took it off of the stove, however, the water vapor began to cool. When water vapor cools, it condenses back into liquid water. When this happened in your experiment, the can collapsed in on itself because water vapor takes up considerably more volume than water liquid. Thus, in its vapor form, the water was able to fill the can. When it turned into liquid, however, the volume it required was significantly less. Since the lid was on the can, no air could rush in and fill the void left behind by the condensing vapor, so the can had to collapse in reaction to the fact that the condensing vapor was taking up significantly less volume than it had before.

I hope that this experiment convinced you of the main difference between phases of matter: the volume they occupy. With the exception of water (which will be discussed in a moment), all substances take up the least amount of volume in their solid phase, a little more volume in their liquid phase, and the largest amount of volume in their gas phase. Why? Remember, all substances are made up of individual atoms or molecules that come together and occupy the same general area. The reason these atoms or molecules come together in the first place is that they are attracted to one another. Thus, sulfur atoms tend to stay with other sulfur atoms because they attract one another. In the same way, sugar molecules tend to hang around other sugar molecules because they are attracted to one another.

When things are attracted to one another, they tend to get close together. The closer atoms or molecules get to one another, the less volume is occupied by the substance that they comprise. In other words, when sugar molecules get close together, the sugar occupies little volume. When the sugar molecules get far apart, the substance occupies quite a bit of volume. Since solids occupy the least volume, that means that when a substance is in its solid phase, the atoms or molecules that make up that substance are very close together. On the other hand, when a substance is in its gas phase, the atoms or molecules that make it up are very far away from one another.

For example, carbon dioxide (like all matter) can exist in any of the three phases we have discussed. Its most common phase is the gas phase. When we breathe out carbon dioxide, it is in its gas phase. In this phase, the carbon dioxide molecules are far away from one another. Under the right conditions, however, carbon dioxide can also become a liquid. This happens when the carbon dioxide molecules get closer together. As they start to move closer to one another, the gas starts to condense, and the carbon dioxide becomes a liquid. When this happens, the amount of space occupied by the same mass of carbon dioxide decreases significantly. Finally, carbon dioxide can also exist in the solid phase. We call solid carbon dioxide "dry ice." Dry ice is formed when the molecules of carbon dioxide get as close as possible to one another. This further reduces the volume that the same mass of carbon dioxide occupies, making it a solid.

Now the obvious question is, why do the molecules in carbon dioxide get closer together or farther apart? If carbon dioxide molecules are attracted to one another, why don't they all just get as close as possible to one another, making carbon dioxide always a solid? In order to answer this question, we have to discuss the **kinetic theory of matter**. We will do that in the next section.

ON YOUR OWN

4.3 Cubes of frozen alcohol are put into liquid alcohol. Will the cubes float or sink?

4.4 Natural gas companies store most of their natural gas in liquid form rather than in its gas phase, even though they must deliver it to their customers as a gas. Why do they do this?

The Kinetic Theory of Matter

At this point, I need to tell you something that will probably surprise you a little. Did you know that right now, as you read this sentence, the molecules that make up the paper in this module and the molecules that make up the ink in these words are all moving very rapidly? You may not believe it, but it's true! The molecules or atoms that make up all of the matter we observe are always in constant motion. That is the basis of the kinetic theory of matter.

"Wait a minute," you might ask, "how in the world can the molecules which make up this paper be moving when the paper is sitting still?" That's a reasonable question. It turns out that atoms and molecules are very, very small. As I mentioned in a previous module, they are so small that they cannot be seen, even with a very powerful microscope. As a result, any matter that you can actually see is made up of billions and billions and billions of atoms or molecules. These atoms or molecules are in random motion. In the solid phase, for example, molecules (or atoms) vibrate back and forth a few billion times every second. The reason that matter in the solid phase doesn't appear to be moving is that for every molecule (or atom) that is vibrating one way, there is another molecule (or atom) vibrating another way. Thus, on average, the solid matter cannot move anywhere, because all its molecules or atoms are moving in different directions.

In the liquid phase, molecules and atoms do not just vibrate; they move around. If you were to look at a glass of water and track a single molecule, you would find that the molecule travels throughout the glass. It may start out on the bottom of the glass, but it will quickly move in one direction until it collides with the wall of the glass or with another water molecule. It will then switch directions during the collision and speed off in another direction until it collides with something else. Thus, the motion of a water molecule is a lot like a bumper car that continues moving in one direction until it hits something. It then changes direction and speeds off again until it hits something else. In the same way, a molecule of water will travel throughout the entire volume of the glass, continually running into things, changing directions, and speeding off again.

In the gas phase, things are just like the liquid phase, only the molecules travel faster and they make fewer collisions with other molecules because they are farther apart from one another. The gas molecules simply speed around, collide into things, change directions, and speed off again. Whether we are talking about the solid phase, liquid phase, or gas phase, the molecules that make up matter are in some kind of random motion. Figure 4.4 tries to illustrate this concept:

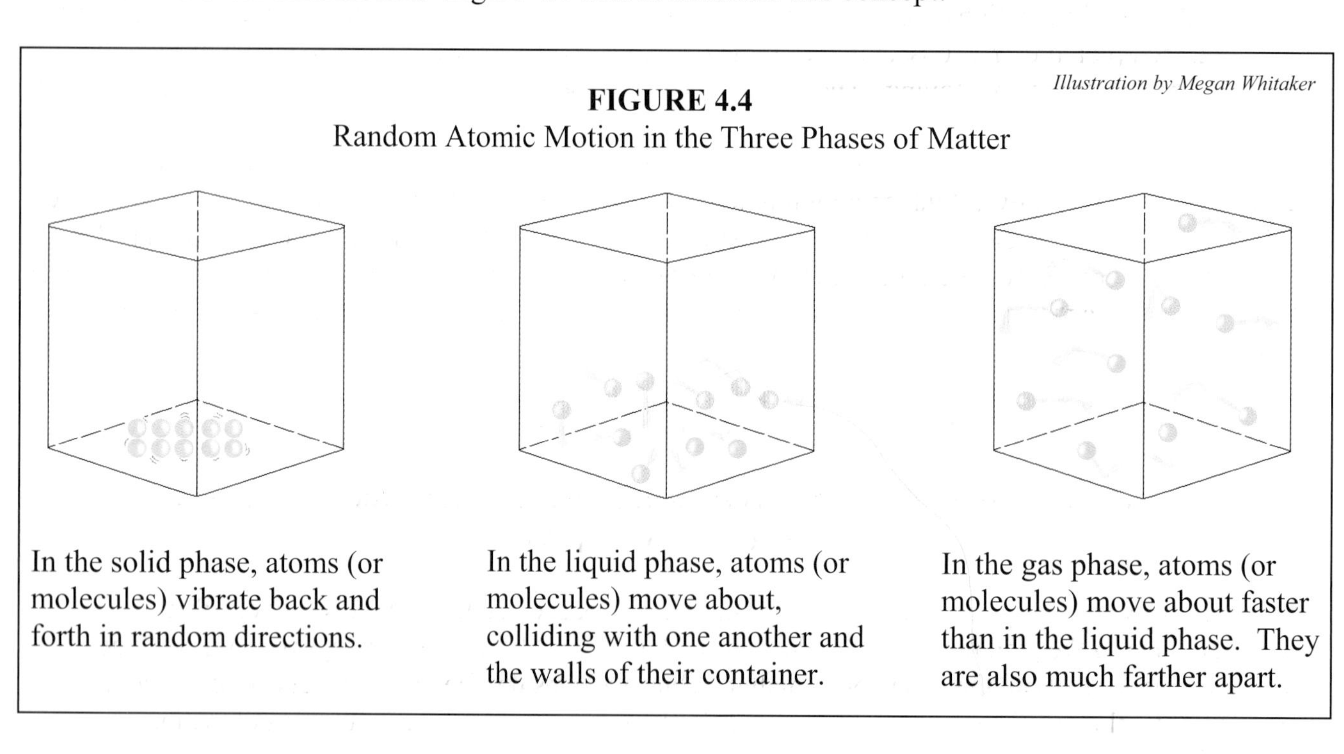

Illustration by Megan Whitaker

FIGURE 4.4
Random Atomic Motion in the Three Phases of Matter

In the solid phase, atoms (or molecules) vibrate back and forth in random directions.

In the liquid phase, atoms (or molecules) move about, colliding with one another and the walls of their container.

In the gas phase, atoms (or molecules) move about faster than in the liquid phase. They are also much farther apart.

Since motion requires work, and work requires energy, the molecules in a substance must have energy in order to move around as they do. Where do they get this energy? They get it from the surroundings. The higher the temperature of the surroundings, the faster the molecules move. The faster the molecules move, the farther they get away from the other molecules in the substance. Thus, when the temperature of a substance's surroundings increases, the molecules (or atoms) in that substance start to absorb energy. As they absorb energy, they start moving faster. As they move faster, they get farther away from their neighbors. As the molecules (or atoms) get farther away from their neighbors, the substance eventually changes phase. This is why a substance changes from solid to liquid to gas as the temperature is increased. The temperature at which a substance changes from solid to liquid is usually called the **melting point** of the substance. The temperature at which a substance goes from its liquid phase to its gas phase is called the **boiling point**.

If you purchased the MicroChem kit discussed in the introduction, you can get some experience measuring melting points by performing Experiment #2.

As a substance's surroundings decrease in temperature, energy travels out of the molecules (or atoms) in the substance and into the surroundings. This is because heat always travels from a hot body to a cold body. Thus, if the surroundings decrease in temperature, the molecules (or atoms) in a substance will lose energy to their surroundings. Since they lose energy, their motion begins to decrease. As their motion decreases, the molecules (or atoms) get closer to their neighbors. This causes a substance to change from gas to liquid to solid as temperature decreases.

Remember in Module #2 when I said you needed a little more information before you could understand the real definition of temperature? Well, now you have all the information you need. It turns out that the temperature of a substance is actually determined by the speed at which the molecules (or atoms) in the substance are moving. The faster the molecules (or atoms) move, the higher the temperature.

If you have a hard time understanding how a thermometer can measure the speed at which molecules (or atoms) move, think about it this way: When a thermometer is immersed in a substance, the molecules (or atoms) in the substance can collide with the thermometer as they are moving around. The harder they hit the thermometer, the hotter the thermometer gets. This causes the liquid in the thermometer to expand. Thus, a thermometer really measures the speed at which molecules (or atoms) move in a substance.

The kinetic theory of matter, then, can be summarized as follows:

Molecules and atoms are in constant motion, and the higher the temperature, the greater their speed.

Since we can never really see atoms or molecules, we cannot see that they actually move, either. However, making the assumption that molecules or atoms in a substance are always moving helps us explain a great deal about the behavior of matter. Thus, we consider the kinetic theory of matter a very good scientific theory. Perform the following experiment to get a visual understanding of the kinetic theory of matter:

EXPERIMENT 4.4

The Kinetic Theory of Matter

Supplies:

- Two glass canning jars or peanut butter jars, both the same size
- Food coloring (any color)
- A pan and stove to boil water, and a hot pad to hold the pan
- Safety goggles

1. Boil some water. You need to boil enough water so that the boiled water will fill one of the jars about halfway.
2. Once the water is boiling, take it off of the stove (use the hotpad!) and pour it into one of the jars. Pour in enough water so that the jar is filled about halfway.
3. Take the other jar and fill it about halfway with cold water from the tap.
4. Wait a few minutes so that the water in each jar is still.
5. Drop a single drop of food coloring into each jar. Observe what happens over the next several minutes.
6. Clean up your mess.

What did you see in the experiment? If everything went well, you should have seen the drop of food coloring mix rapidly with the hot water, coloring the entire jar of water relatively quickly. In the jar of cold water, however, the food coloring should not have mixed nearly as well. This is an excellent illustration of the kinetic theory of matter. The molecules in the cold water were moving, but not nearly as rapidly as those in the hot water. When the food coloring was added to both jars, the water molecules began colliding with the molecules in the food coloring. This began spreading out the molecules in the food coloring, mixing them with the water. Since the water molecules in the hot water were moving faster than those in the cold water, however, the food coloring spread out much more rapidly in the hot water. Thus, the hot water changed color much more quickly than did the cold water because its molecules were moving more rapidly.

ON YOUR OWN

4.5 If a substance is cooled down 10 degrees, will its molecules be moving faster or slower? Will the molecules be closer together or farther apart?

Phase Changes in Water

Before we leave the subject of phase changes, we must spend a little time discussing water. As I have said before, although almost all natural substances occupy less volume in the solid phase than in the liquid phase, solid water (ice) actually occupies more volume than liquid water. This is an exception to the phase change rule I discussed, and we are very lucky that this exception exists.

Because water occupies more volume in its solid phase than in its liquid phase, the density of ice is *less* than the density of liquid water. You learned in Module #1 that less dense substances float on top of more dense substances. This means that ice floats on top of liquid water, since its density is lower. This is why ice cubes float in a drink. This is also why ice forms on the TOP of a freezing lake. A layer of ice on top of a lake actually insulates the water below it, keeping it from freezing further. As a result, a lake never freezes completely. There is always plenty of water underneath the ice, allowing the fish and other organisms to stay alive.

If ice did not float on top of water, a lake would freeze from the bottom up. There would be nothing to insulate the unfrozen water, so most lakes would freeze solid each winter, killing most (if not all) of the life in the lake. Water-living organisms provide food for many different creatures, and water-living algae are responsible for a great deal of the photosynthesis that keeps the oxygen levels in the atmosphere constant. Thus, it is hard to imagine how life could exist on earth without the organisms that live in lakes.

In addition to the fact that water's exception to the rule makes it possible for fresh-water organisms to survive, many biochemists believe that it also makes it possible for life as we know it to exist. The same physical effect that causes water to be an exception to the phase change rule also keeps water a liquid at normal temperatures. This effect is called "hydrogen bonding." Amazingly enough, all of the substances that are chemically similar to water (H_2S, H_2Se, H_2Te) are gases at normal temperatures. Water, on the other hand, is a liquid at those temperatures because it can participate in hydrogen bonding. Most biochemists agree that if water were not a liquid at normal temperatures, life as we know it could not even exist!

Once again, isn't this an amazing "coincidence?" There are only a handful of natural substances that expand when turning into a solid. If it weren't for water's special trait, life as we know it would not exist. I do not understand how any real scientist can continue to believe that events like these are "coincidences." Water is an exception to the rule because God designed it that way in order to allow life as we know it to exist. The mere existence of water as an exception to the phase change rule is powerful evidence that the universe was designed to support life!

Chemical Reactions and Chemical Equations

Now that we have taken an in-depth look at physical change, it is time to turn our attention to chemical change. Remember, chemical change occurs when the molecules within a substance change into different molecules. For the rest of this module, we will examine how this change (called a **chemical reaction**) occurs and how to write these changes down in an abbreviated form called a **chemical equation**.

We will begin our examination of chemical change by looking at a process that is very important to today's economy: the burning of coal. When coal, which is mostly carbon, is burned, the energy released is used to make electricity. Currently, a little more than half the electricity produced in the United States is produced through coal burning, so it is a pretty important chemical change and deserves close examination. Before we can do this, however, we must discuss a concept that has been put off since we first discussed atoms and molecules.

I said in Module #3 that elements are composed of individual atoms whereas compounds are composed of individual molecules. This rule is not completely correct. As I have told you over and over, nearly every rule in chemistry has its exceptions. There are elements in nature that are not composed of individual atoms. These elements, called **homonuclear** (ho' moh new' klee er) **diatomics** (dye uh tom' iks), are composed of individual molecules, not individual atoms.

As their name implies, homonuclear diatomic molecules are made up of two atoms (diatomic) which are the same type (homonuclear).

Nitrogen, oxygen, chlorine, fluorine, bromine, iodine, astatine, and hydrogen are all elements that are made up of homonuclear diatomic molecules rather than individual atoms.

The chemical formulas for these elements are N_2, O_2, Cl_2, F_2, Br_2, I_2, At_2, and H_2. These exceptions are so important that you will have to memorize them. Notice that five of these eight homonuclear diatomics are in group 7A on the periodic chart. That should help you in memorizing them.

Now please understand that the concept of a homonuclear diatomic applies only to an *element*, not to an atom within a molecule. For example, oxygen is a homonuclear diatomic. Thus, the element oxygen exists as O_2, not as O. However, when oxygen is a part of a molecule, it can exist as one oxygen atom, two oxygen atoms, three oxygen atoms, etc., depending on the molecule. Water (H_2O) molecules, for example, contain only one oxygen atom each. Thus, the fact that oxygen is a homonuclear diatomic is irrelevant when discussing the number of oxygen atoms in a water molecule. In the same way, nitric acid (HNO_3) contains three oxygen atoms. Once again, since the oxygen atoms are a part of a molecule, the fact that oxygen is a homonuclear diatomic is irrelevant. The *only time* you worry about the fact that oxygen is a homonuclear diatomic is when you are dealing with the *element* oxygen.

Now that we have this new knowledge under our belts, we can continue our discussion of burning coal. When coal is burned, it combines with oxygen to make carbon dioxide. The energy that is released when this chemical change occurs is 30 kJ for every 1 gram of carbon that is burned. This energy output is sufficient to make electricity in an efficient manner.

This chemical change, called a **chemical reaction**, could be written as follows:

Carbon plus oxygen yields carbon dioxide.

Although this method of writing the chemical reaction does tell us what happened, it is rather bulky. Thus, we would like to abbreviate it. We will do so by writing the chemical formulas of the substances involved rather than their names. We will also abbreviate the word "yields" with an arrow, "→." Thus, our abbreviated chemical reaction is:

$$C + O_2 \rightarrow CO_2 \quad (4.1)$$

This gives us the same information as the previous representation, but in a much more condensed fashion. Notice that since oxygen is a homonuclear diatomic element, its chemical formula is "O_2," not "O." Carbon, however, is not a homonuclear diatomic, so its chemical formula is simply "C."

The abbreviations for chemical reactions, such as the one written above, are called **chemical equations**. Chemical equations form the basis for most of the work done in chemistry, so it is critically important that you understand what they are and how they are used. Of course, chemical equations get much more complicated than Equation (4.1), so we will need to spend a great deal of time learning about them.

First of all, you need to understand the framework of a chemical equation. The atoms and molecules that appear on the left side of the arrow represent the substances that exist before the chemical change takes place. We call these substances **reactants**, because they are the substances that will react together. The atoms and molecules on the right side of the equation are called **products**, because they represent the substances which are produced by the chemical reaction. Thus, a chemical equation is always written with reactants on the left side of the arrow and products on the right side.

To give you an idea of what really goes on during a chemical reaction, Figure 4.5 attempts to illustrate, on the atomic level, the steps that take place when carbon reacts with oxygen to make carbon dioxide:

Illustration by Megan Whitaker

FIGURE 4.5
Burning Carbon at the Atomic Level

1. The chemical reaction starts with just the reactants: a carbon atom and a homonuclear diatomic oxygen molecule.

carbon atom oxygen molecule

2. The oxygen molecule splits apart to make two independent oxygen atoms.

carbon atom two oxygen atoms

3. The carbon then joins with both oxygen atoms, leaving just the product: a carbon dioxide molecule.

carbon dioxide molecule

This, then, is how a chemical reaction proceeds. The molecules must disassemble into their constituent atoms, and then the atoms recombine to form the products.

Now, we can make Equation (4.1) a little more meaningful by adding some information about the phase of each substance in the equation. When we burn coal, for example, the carbon is in its solid state, the oxygen is in its gaseous state, and the carbon dioxide produced is also in its gaseous state. We can add these facts in the equation by putting symbols in parentheses next to each substance:

$$C\ (s) + O_2\ (g) \rightarrow CO_2\ (g) \qquad (4.2)$$

Whenever you see a small letter enclosed with parentheses in a chemical equation, you know that it is a symbol telling you the phase of the substance. An "(s)" is used to denote solid, an "(l)" means liquid, a "(g)" stands for gas, and an "(aq)" represents a substance that has been dissolved in water. The letters "aq" are an abbreviation for the Latin term *aqueous*, which means "in water." Chemists do not always use phase symbols in their chemical reactions. Sometimes they are there; sometimes they are not. It is always preferable to include them, however, because they provide us with important information. Don't be confused if you do not see them, though.

Now if this were as complicated as things got, chemistry would be a breeze. Unfortunately, things get a little tougher from here. Let's look at another important chemical reaction: the burning of methane (CH_4). Methane is the principal component of natural gas. If you have a gas stove, gas water heater, or gas furnace, this chemical reaction occurs quite frequently in your home. Once again, the combustion (burning) of methane releases quite a bit of energy. This energy can be used to cook, to heat water, or to heat your home. There are even some electrical plants that run on methane. More than 45% of the electricity in Texas, for example, is generated by methane-burning power plants. Thus, the combustion of methane is another important chemical reaction.

The combustion of methane is a little more complicated than the combustion of coal, because it makes two products: carbon dioxide and water. We could write this reaction as:

Methane + oxygen yields carbon dioxide + water.

Using chemical formulas and phase symbols to turn this into a chemical equation, we come up with:

$$CH_4\ (g) + O_2\ (g) \rightarrow CO_2\ (g) + H_2O\ (g) \qquad (4.3)$$

There are several things to note about Equation (4.3). First, you might wonder about how I came up with the phase symbols. How did I know that CH_4 is a gas, for example? Well, this knowledge just comes from experience. At this stage, I would not expect you to know what phase the reactants and products are in during a chemical reaction. The exceptions to this are oxygen and carbon dioxide. We have talked about these two substances so much that you should know by now that they are gases. If I expect you to put phase symbols with any other substances, however, I will tell you what phase they are in.

Second, you might notice that I listed water as a gas. You might think that I should have listed water as a liquid. After all, at normal temperatures, water is, indeed, a liquid. This particular reaction, however, releases a lot of energy, making its surroundings very hot. When the surroundings get hot, the water boils. So, because this reaction is hot, it puts water into its gas phase, and that's why I listed it as a gas. In chemical reactions, sometimes water will be a liquid, sometimes it will be a gas, and sometimes it will be a solid. Once again, only experience can help you make that distinction.

Finally, Equation (4.3), as it is written now, is *wrong*. To see why it is wrong, study Figure 4.6.

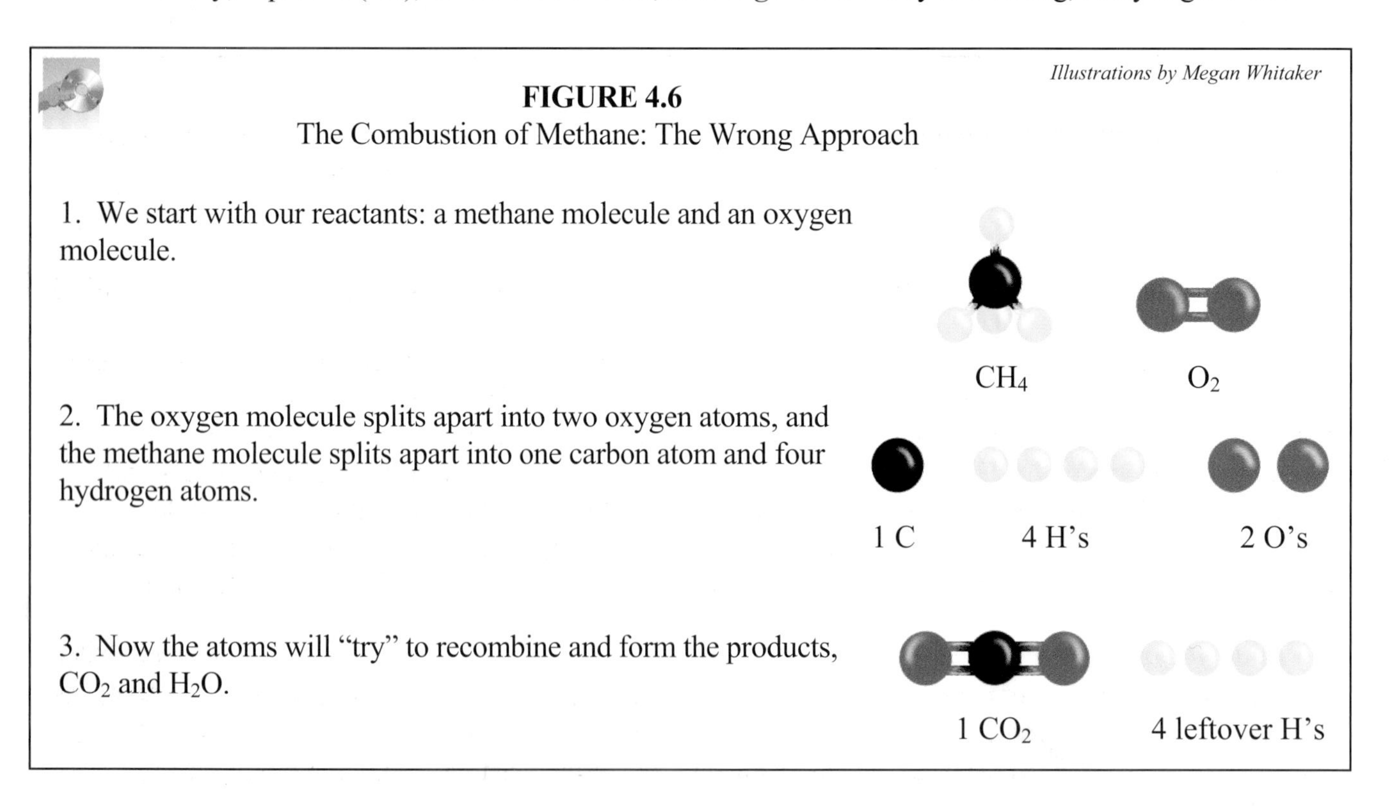

Illustrations by Megan Whitaker
1. We start with our reactants: a methane molecule and an oxygen molecule.
CH4
O2
2. The oxygen molecule splits apart into two oxygen atoms, and the methane molecule splits apart into one carbon atom and four hydrogen atoms.
1 C
4 H's
2 O's
3. Now the atoms will "try" to recombine and form the products, CO2 and H2O.
1 CO2
4 leftover H's

FIGURE 4.6
The Combustion of Methane: The Wrong Approach

Do you see the problem in the figure? There are not enough oxygen atoms to get the job done. The chemical reaction makes both carbon dioxide and water. We were able to make the carbon dioxide molecule, but then we ran into a problem. In order to make H_2O, I need both hydrogen atoms *and* oxygen atoms. I have enough hydrogen atoms, but I don't have any oxygen atoms left after the CO_2 was formed. In order to solve this problem, I have to use another oxygen molecule. Figure 4.7 goes through this procedure again, but it starts with two oxygen molecules. As you will see, this works out much better!

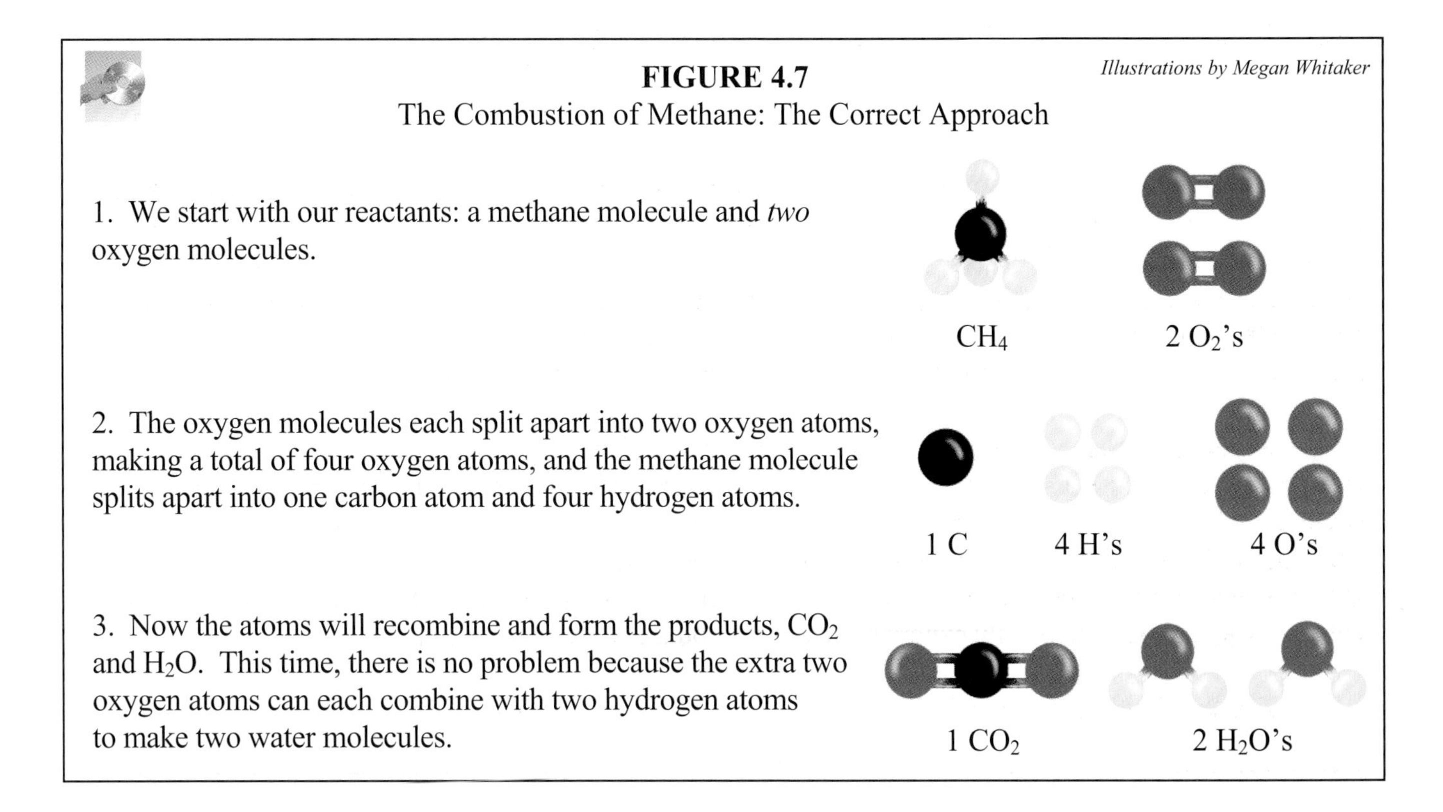

Illustrations by Megan Whitaker
FIGURE 4.7
The Combustion of Methane: The Correct Approach
1. We start with our reactants: a methane molecule and two oxygen molecules.
CH4
2 O2's
2. The oxygen molecules each split apart into two oxygen atoms, making a total of four oxygen atoms, and the methane molecule splits apart into one carbon atom and four hydrogen atoms.
1 C
4 H's
4 O's
3. Now the atoms will recombine and form the products, CO2 and H2O. This time, there is no problem because the extra two oxygen atoms can each combine with two hydrogen atoms to make two water molecules.
1 CO2
2 H2O's

In the combustion of methane, then, *two* oxygen molecules must combine with *one* methane molecule. When this happens, *one* carbon dioxide molecule and *two* water molecules will be produced. We can add this information to Equation (4.3) so that it will be correct:

$$CH_4\ (g) + O_2\ (g) + O_2\ (g) \rightarrow CO_2\ (g) + H_2O\ (g) + H_2O\ (g) \qquad (4.4)$$

Now, at least, the chemical equation is correct. It tells us that our reactants are one methane molecule and two oxygen molecules, and our products are one carbon dioxide molecule and two water molecules. It is critically important for you to understand why Equation (4.3) is not correct and Equation (4.4) is correct. The way Equation (4.3) was written, we could not form the products that were listed, and we had extra atoms left over at the end. Neither of these two things is possible in nature. A chemical reaction must start with its reactants, end up with its products, and have no extra atoms left over. The chemical equation you use to describe a chemical reaction must reflect this fact.

We can clean up this chemical equation a little more by combining like terms. Remember from algebra that $x + x = 2x$. It's the same in chemistry. When we have several identical terms on the same side of a chemical equation, we can group them together. Thus, $O_2\ (g) + O_2\ (g) = 2O_2\ (g)$ and $H_2O\ (g) + H_2O\ (g) = 2H_2O\ (g)$. Remember, just as in algebra, the terms must be identical before you can combine them. Thus, the chemical equation that describes the combustion of methane is:

$$CH_4\ (g) + 2O_2\ (g) \rightarrow CO_2\ (g) + 2H_2O\ (g) \qquad (4.5)$$

Equation (4.5) is called a **balanced chemical equation.** If a chemical equation is not balanced, then it is not correct. An unbalanced chemical equation will result in not all products being made or in leftover atoms (or both). Both of those results are impossible in nature, so unbalanced chemical equations are just not correct. Also, an unbalanced chemical equation violates the Law of Mass

Conservation. Since atoms cannot be created or destroyed, the total mass before a chemical reaction takes place is dependent on the number of each type of atom in the reactants. Thus, when the reaction is over, we must have the same number of each type of atom in the products, or the mass will not be the same. Since balancing chemical equations is so important, the rest of this module will be devoted to teaching you how to balance chemical equations. You cannot proceed to the next module until you learn this important technique, so please concentrate and practice.

Before we begin practicing how to balance chemical equations, let's make sure we understand everything that Equation (4.5) tells us. First, we know this equation tells us the reactants and products in the reaction. It also tells us what phase each substance is in during the reaction. Equally important, however, the chemical equation tells us the recipe we need to get this reaction to go. The big numbers to the left of each chemical formula tell you how many molecules of each substance are necessary in order for the reaction to work. As usual, if there is no number to the left of a chemical formula, we must assume it is "1." Thus, Equation (4.5) tells us that *one* CH_4 molecule will react with *two* O_2 molecules. These reactants will produce *one* CO_2 molecule and *two* H_2O molecules.

Remember not to confuse the big numbers on the left of the chemical formulas with the numbers that appear in the subscripts. The numbers in the subscripts are part of the molecule's chemical formula. They tell you *what* molecules you are dealing with. The big numbers to the left of the chemical formulas (the coefficients), on the other hand, tell you *how many* of those molecules you are dealing with. This is an important distinction and one that you must understand in order to be able to balance chemical equations.

Determining Whether or Not a Chemical Equation Is Balanced

In order to balance chemical equations, we must first learn to keep track of all atoms in the equation. It turns out that an equation will always be balanced if the number of each type of atom on the reactants side of the equation is the same as the number of that type of atom on the products side of the chemical equation. For example, if a chemical equation has five H atoms on its reactants side, it must have five H atoms on its products side. There is a trick to counting the atoms, though, so we will go back to Equation (4.5) to teach you this trick.

Remember that the number to the left of each molecule tells you how many molecules there are in the equation. The numbers in subscripts tell you how many of each atom are in the molecule. Thus, to count the number of atoms on each side of the equation, we must take the number to the left of the molecule and multiply it by the subscript number that follows each atom. This will tell us the total number of each atom on each side of the equation. Let's see how that works in Equation (4.5). First, I want to rewrite the equation so that you can refer to it easily:

$$CH_4\ (g) + 2O_2\ (g) \rightarrow CO_2\ (g) + 2H_2O\ (g) \tag{4.5}$$

Now let's count the number of carbon atoms on the reactants side of this equation. To do this, you must take the number to the left of the CH_4 and multiply it by the subscript that comes directly after the "C." Since there is no number to the left of the CH_4, we assume it is 1. Also, since there is no subscript after the C, we assume it is 1 as well. Thus, there is 1 x 1 = 1 carbon atom on the reactants side of the equation. To count the carbon atoms on the products side of the equation, we look at the only molecule that contains carbon on the products side: CO_2. Since there is no number to the left of CO_2 and no subscript after the C, they are both 1. Thus, there is 1 x 1 = 1 carbon atom on the products side of the equation. Since the number of carbon atoms is the same on each side of the equation, the equation is balanced with respect to carbon.

Now we must count the H's to see if the equation is balanced with respect to hydrogen. On the reactants side, the only molecule that has H in it is CH_4. There is no number to the left of CH_4, so it must be 1. There is a subscript of 4 after the H, so the number of H atoms on the reactants side of the equation is 1 x 4 = 4 hydrogen atoms. On the products side, the only molecule with H in it is H_2O. There is a 2 to the left of the H_2O and there is a subscript of 2 after the H. Therefore, there are 2 x 2 = 4 hydrogen atoms on the products side of the equation. Since there are the same number (4) of hydrogen atoms on each side of the equation, the equation is balanced with respect to hydrogen.

Finally, we must determine whether or not the equation is balanced with respect to oxygen. On the reactants side, only the O_2 molecule has O's in it. There is a 2 to the left of the O_2 and a subscript of 2 after the O, so there are 2 x 2 = 4 oxygen atoms on the reactants side of the equation. What about the products side? This will be a little tougher because there are two molecules, CO_2 and H_2O, that each have O's in them. To count the total number of O's, we figure out how many O's come from each molecule, and then we add those numbers together. This will give us the total number of O's on the products side of the equation. There is no number to the left of the CO_2, so it must be 1. There is a subscript of 2 after the O, so there are 1 x 2 = 2 oxygen atoms coming from the CO_2 molecule. There is a 2 to the left of the H_2O, and no subscript number after the O, so it must be 1. Therefore, there are 2 x 1 = 2 oxygen atoms coming from the H_2O molecule. The total number of oxygen atoms, then, is the number of oxygens that come from CO_2 plus the number of oxygens that come from H_2O, or 2 + 2 = 4. Since there are the same number of oxygen atoms (4) on both sides of the equation, it is balanced with respect to oxygen.

Since the equation is balanced with respect to all of the atoms, it is completely balanced. Although this might seem a little tough at first, it gets easier with practice. Try to follow the next example and then complete the "On Your Own" problems after it. Please do not go on to the last part of the module until you have mastered this technique.

EXAMPLE 4.1

Determine whether or not the following equation is balanced:

$$2HCl\ (aq) + CaBr_2\ (s) \rightarrow 2HBr\ (aq) + 2CaCl_2\ (s)$$

To determine whether or not this equation is balanced, we must count the number of atoms on each side of the equation:

Reactants Side	Products Side	
#H : 2x1 = 2	#H: 2x1 = 2	Since the number of H's is the same, the equation is balanced with respect to H.
#Cl: 2x1 = 2	#Cl: 2x2 = 4	There are more Cl's on the products side than the reactants side, so this equation is not balanced. We could stop here, but I will complete the process for the sake of illustration.

Reactants Side	Products Side	
#Ca: 1x1 = 1	#Ca: 2x1 = 2	The equation is also not balanced with respect to Ca.
#Br: 1x2 = 2	#Br: 2x1 = 2	The equation is balanced with respect to Br.

If an equation has even one type of atom that is not balanced, then the whole equation is not balanced. Therefore, this equation is not balanced.

Determine whether or not the following equation is balanced:

$$6CO_2 + 6H_2O \rightarrow C_6H_{12}O_6 + 6O_2$$

Once again, we must count up the atoms to see if they balance:

Reactants Side	Products Side	
#C: 6x1 = 6	#C: 1x6 = 6	The equation is balanced with respect to carbon.
#O: 6x2 + 6x1 = 18	#O: 1x6 + 6x2 = 18	Remember, since O appears in two molecules, we add the number of O's that come from each molecule. This makes the equation balanced with respect to O.
#H: 6x2=12	#H:1x12 = 12	The equation is balanced with respect to H.

Since the equation is balanced with respect to all its atoms, it is a balanced chemical equation.

ON YOUR OWN

4.6 Determine whether or not the following equation is balanced:

$$2C_2H_6\ (g) + 7O_2\ (g) \rightarrow 4CO_2\ (g) + 6H_2O\ (g)$$

4.7 Determine whether or not the following equation is balanced :

$$6NH_4Cl + Al_2S_3 \rightarrow AlCl_3 + 3N_2H_8S$$

Balancing Chemical Equations

Now that you can count the atoms in a chemical equation, you are finally ready to balance chemical equations. When presented with a chemical equation that is not balanced, you must try to balance it by altering the number of molecules involved. You can alter the number of molecules involved by changing the numbers that appear to the left of each molecule. Those numbers can be changed in any way, as long as they are not made negative. ***You cannot, however, change any of the numbers that appear in the subscripts.***

If you think about what these numbers mean, this rule should make sense. When balancing chemical equations, all we are trying to do is get the number of molecules involved in the reaction to work out. The number of (how many) molecules is determined by the numbers that appear to the left of the molecules. The numbers in the subscripts tell you *which* molecules are involved. If you change any of those numbers, you are changing the substances in the chemical reaction, and that is not allowed. So remember, you can only alter the numbers that appear to the left of the molecules, not the subscript numbers. See if you understand these rules by following the next example:

EXAMPLE 4.2

Balance the following equation:

$$N_2 (g) + H_2 (g) \rightarrow NH_3 (g)$$

To balance this equation, we must get the same number of each type of atom on both sides of the equation. To see where we are, we will first count the number of atoms on each side:

Reactants Side	Products Side
#N: 1x2 = 2	#N: 1x1 = 1
#H: 1x2 = 2	#H: 1x3 = 3

Clearly, this equation is not balanced. How can we balance it? We must change the numbers in the equation so that each individual type of atom balances. Remember, we can't change the numbers in the subscripts; we can only change the numbers to the left of the molecules. Let's concentrate on the N's first. There are 2 N's on the reactants side and only 1 on the products side. To fix this, we must double the number of nitrogens on the products side. The only way we are allowed to do this is to change the number to the left of the molecule. Since there is no number to the left of NH_3, it must be 1. To double the number of nitrogens, we must double that number, so we change the equation to look like this:

$$N_2 (g) + H_2 (g) \rightarrow 2NH_3 (g)$$

Once again, I did not change the subscripted numbers at all. I just put a "2" in front of the NH_3. I have emphasized this by making the "2" red. This simply means there are two molecules of NH_3. Now when we count up atoms again, we will see that the nitrogens balance:

Reactants Side	Products Side
#N : 1x2 = 2	#N: 2x1 = 2
#H: 1x2 = 2	#H: 2x3 = 6

Now the nitrogens are balanced. Of course, the hydrogens are not balanced, so we must balance them next. Since there are three times as many hydrogens on the products side, we must triple the number of hydrogens on the reactants side. We can do this by tripling the number that's to the left of H_2:

$$N_2 (g) + 3H_2 (g) \rightarrow 2NH_3 (g)$$

Now look what happens when we count up the atoms:

Reactants Side	Products Side
#N : 1x2 = 2	#N: 2x1 = 2
#H: 3x2 = 6	#H: 2x3 = 6

We now have a balanced chemical equation. The answer, then, is:

$$\underline{N_2 (g) + 3H_2 (g) \rightarrow 2NH_3 (g)}$$

Write a balanced chemical equation for the following reaction:

Propane (C_3H_8) reacts with oxygen to make carbon dioxide and water.

Don't let the fact that the reaction is written with words bother you. Just convert everything into equation form. You are given the chemical formula of propane, and you know that oxygen is a homonuclear diatomic, so its chemical formula is O_2. You also know how to go from a proper chemical name like carbon dioxide to the chemical formula CO_2. Finally, you know that the chemical formula for water is H_2O. Thus, the unbalanced chemical equation is:

$$C_3H_8 + O_2 \rightarrow CO_2 + H_2O$$

Since phases were not given, you cannot write the phase symbols in this equation. Now we must balance it. We start by counting the atoms:

Reactants Side	Products Side
#C: 1x3= 3	#C: 1x1 = 1
#H: 1x8 = 8	#H: 1x2 = 2
#O: 1x2 = 2	#O: 1x2 + 1x1 = 3

Since this equation is a little more complicated, we must use some strategy. If I were to start by balancing the O's, my job would be difficult because there are O's in two molecules on the products side. I would not know what numbers to put where. The general rule of thumb to follow when you are balancing equations is *start with atoms that appear in only one molecule on each side of the equation.* These atoms will be easier to balance. So let's balance the C's first. Carbon appears only once on the reactants side (in C_3H_8) and only once on the products side (in CO_2). That makes it an excellent atom with which to start. There are 3 C's on the reactants side and only 1 on the products side, so I must triple the number of C's on the products side:

$$C_3H_8 + O_2 \rightarrow 3CO_2 + H_2O$$

Now we count the atoms again:

Reactants Side	Products Side
#C: 1x3= 3	#C: 3x1 = 3
#H: 1x8 = 8	#H: 1x2 = 2
#O: 1x2 = 2	#O: 3x2 + 1x1 = 7

Now we can balance the H's, because they also appear only once on each side of the equation. There are 8 on the reactants side of the equation and 2 on the products side, so we must quadruple the number of hydrogens on the products side:

$$C_3H_8 + O_2 \rightarrow 3CO_2 + 4H_2O$$

Reactants Side	Products Side
#C: 1x3 = 3	#C: 3x1 = 3
#H: 1x8 = 8	#H: 4x2 = 8
#O: 1x2 = 2	#O: 3x2 + 4x1 = 10

The only unbalanced atom left is O. Since oxygen appears in several molecules, we need to think about where we want to start changing the equation. If we changed the numbers in front of C_3H_8, CO_2, or H_2O, it would throw off the balancing that we already did. Thus, the only molecule we really want to work with is the O_2 molecule. That's fine, however, since we can balance the entire equation by multiplying O_2 by 5:

$$C_3H_8 + 5O_2 \rightarrow 3CO_2 + 4H_2O$$

Now, as you can see, the equation is balanced:

Reactants Side	Products Side
#C: 1x3 = 3	#C: 3x1 = 3
#H: 1x8 = 8	#H: 4x2 = 8
#O: 5x2 = 10	#O: 3x2 + 4x1 = 10

So the balanced chemical equation is:

$$\underline{C_3H_8 + 5O_2 \rightarrow 3CO_2 + 4H_2O}$$

Try your hand at balancing chemical equations with the following problems:

ON YOUR OWN

4.8 Hydrogen and oxygen can react together violently to make water. Write a balanced chemical equation for this reaction.

4.9 Solid calcium carbonate ($CaCO_3$) is made when solid calcium reacts with gaseous oxygen and solid carbon. What is the balanced chemical equation for this reaction?

4.10 Balance the following equation:

$$C_5H_{12}\,(l) + O_2\,(g) \rightarrow CO_2\,(g) + H_2O\,(g)$$

This represents your first introduction to the wonderful world of balancing chemical equations. Like most of what we've discussed so far, this can get harder, as we will see in later modules. It is, however, critically important that you understand things up to this point. If you have problems balancing the equations given to you in the "On Your Own" problems and the practice problems, you can look at the extra practice problems given in Appendix B. Also, you can go to the library and check out another chemistry book. See how that book explains balancing chemical equations and try your hand at some of their problems and examples. You absolutely must be able to balance chemical equations before you can continue to Module #5!

ANSWERS TO THE "ON YOUR OWN" PROBLEMS

4.1 a. Copper is on the periodic chart, so it is an element.

b. Carbon dioxide is the name for the molecule CO_2, and molecules form a compound.

c. Icing has sugar, butter, and flavorings in it, and they are mixed together so the composition is the same throughout. Thus, icing is a homogeneous mixture.

d. If you cut a tomato open, you can see that there is a skin on the outside and pulp and seeds on the inside. Since the composition is different inside and outside, a tomato is a heterogeneous mixture.

e. Sugar water is sugar dissolved in water. The composition is the same throughout, so it is a homogeneous mixture.

f. Sodium is on the periodic table, so it is an element.

g. Fruitcake has several different things in it, and the composition varies throughout, so it is a heterogeneous mixture.

h. Sodium fluoride is the name for the molecule NaF, so it is a compound.

i. Tap water, as was mentioned in the previous module, has more than just water in it. There is also sodium fluoride and chlorine in tap water if it comes from a city water supply. Since tap water has the same composition throughout, it is a homogeneous mixture.

4.2 a. Steam is simply water vapor. Thus, when steam condenses back into liquid water, there is no change in the molecules. Also, this change can be easily reversed by simply boiling the water again. Either way you look at it, this is a physical change.

b. When gasoline burns, it combines with oxygen to make carbon dioxide and water, just like when wood burns. Also, it is impossible to get the gasoline back once it's burned, so this change is irreversible. Either way you look at it, this is a chemical change.

c. When a barber cuts your hair, he or she does not change the chemical makeup of the hair. He or she simply changes its length. This could also be reversed by gluing the hair back in place. It might look a little funny, but it still could be reversed. Either way you look at it, this is a physical change.

d. Milk has its distinctive taste because of the molecules that make it up. The change in taste that occurs in sour milk is because the molecules have changed. Also, you cannot "unsour" milk. Either way you look at it, this is a chemical change.

e. When sugar is dissolved in water, the water tastes slightly sweet. That is because the sugar is still there; it is just evenly distributed around the water. Also, this change can be reversed by just boiling off the water. Either way you look at it, this is a physical change.

f. When milk is poured on cereal, the only change is that the cereal gets soggy. No molecules have changed. Also, you can reverse this change by picking the cereal out of the milk and letting it dry. Either way you look at it, this is a physical change.

g. When iron (Fe) rusts, the red color appears because the iron has reacted with the oxygen in the air to make Fe_2O_3, what we call rust. Also, there is no way to "unrust" a nail. You can scrape the rust away, but it will not turn back into iron. It will simply fall on the ground as a fine, red powder. Either way you look at it, this is a chemical change.

4.3 Since all substances, except water, occupy less volume in the solid phase than in the liquid phase, frozen cubes of alcohol would be *more dense* than liquid alcohol. As a result, the frozen cubes of alcohol would sink in liquid alcohol.

4.4 Since liquid natural gas occupies significantly less volume than natural gas in its gas phase, the gas company can store significantly larger amounts of natural gas when it is in its liquid form. The money natural gas companies save with the increased storage capacity makes up for the cost of cooling natural gas down to its liquid phase so that it can be stored.

4.5 When substances are cooled, the atoms or molecules which make them up begin to lose energy and thus move slower. As they move slower, they get closer together.

4.6 To determine whether or not the equation is balanced, we must count the atoms on both sides of the equation:

Reactants Side	Products Side	
#C: 2x2 = 4	#C: 4x1 = 4	Balanced with respect to carbon.
#H: 2x6 = 12	#H: 6x2 = 12	Balanced with respect to hydrogen.
#O: 7x2 = 14	#O: 4x2 + 6x1 = 14	Balanced with respect to oxygen.

This is a balanced chemical equation.

4.7 Don't let the fact that there are not any phase symbols bother you. We still just count up the atoms on each side of the equation.

Reactants Side	Products Side	
#N: 6x1 = 6	#N: 3x2 = 6	Balanced with respect to nitrogen
#H: 6x4 = 24	#H: 3x8 = 24	Balanced with respect to hydrogen
#Cl: 6x1 = 6	#Cl: 1x3 = 3	NOT balanced with respect to chlorine
#Al: 1x2 = 2	#Al: 1x1 = 1	NOT balanced with respect to aluminum
#S: 1x3 = 3	#S: 3x1 = 3	Balanced with respect to sulfur

This is not a balanced chemical equation because it is not balanced with respect to Cl or Al.

4.8 This reaction is written in words, but we can easily convert it to equation form. Remember, both hydrogen and oxygen are homonuclear diatomics.

$$H_2 + O_2 \rightarrow H_2O$$

Since no phases were given, we cannot put phase symbols in the equation. Now we count up the atoms:

Reactants Side	Products Side
#H: 1x2 = 2	#H: 1x2 = 2
#O: 1x2 = 2	#O: 1x1 = 1

The H's are already balanced, but the O's are not. The only way to balance the O's is to double them on the products side of the equation:

$$H_2 + O_2 \rightarrow 2H_2O$$

Reactants Side	Products Side
#H: 1x2 = 2	#H: 2x2 = 4
#O: 1x2 = 2	#O: 2x1 = 2

This, of course, throws the H's off balance, but that can be fixed by doubling the number of H's on the reactants side:

$$2H_2 + O_2 \rightarrow 2H_2O$$

Now, as you can see, this equation is balanced:

Reactants Side	Products Side
#H: 2x2 = 4	#H: 2x2 = 4
#O: 1x2 = 2	#O: 2x1 = 2

The balanced chemical equation is:

$$\underline{2H_2 + O_2 \rightarrow 2H_2O}$$

4.9 Remembering that oxygen is a homonuclear diatomic, the words translate into the following chemical equation:

$$Ca\ (s) + O_2\ (g) + C\ (s) \rightarrow CaCO_3\ (s)$$

This looks easy at first. Counting up the atoms:

Reactants Side	Products Side
#Ca: 1x1 = 1	#Ca: 1x1 = 1
#O: 1x2 = 2	#O: 1x3 = 3
#C: 1x1 = 1	#C: 1x1 = 1

Everything is balanced except the O's. The problem is, how do you balance the O's? There are 2 on the reactants side and 3 on the products side. The way to do this is to find the least common multiple of each: 6. Thus, we need to get 6 O's on both sides of the equation. We do this by tripling the O's on the reactants side and doubling the O's on the products side:

$$Ca\ (s)\ +\ 3O_2\ (g) + C\ (s)\ \rightarrow\ 2CaCO_3\ (s)$$

Now we count up atoms again:

Reactants Side	Products Side
#Ca: 1x1 = 1	#Ca: 2x1 = 2
#O: 3x2 = 6	#O: 2x3 = 6
#C: 1x1 = 1	#C: 2x1 = 2

Now the O's are balanced, but the C's and Ca's are not. This is easy to fix, however, since we need to just double the number of each on the reactants side:

$$2Ca\ (s)\ +\ 3O_2\ (g)\ +\ 2C\ (s)\ \rightarrow\ 2CaCO_3\ (s)$$

Now everything is balanced:

Reactants Side	Products Side
#Ca: 2x1 = 2	#Ca: 2x1 = 2
#O: 3x2 = 6	#O: 2x3 = 6
#C: 2x1 = 2	#C: 2x1 = 2

The balanced chemical equation is:

$$\underline{2Ca\ (s)\ +\ 3O_2\ (g)\ +\ 2C\ (s)\ \rightarrow\ 2CaCO_3\ (s)}$$

4.10 First we count up atoms:

$$C_5H_{12}\ (l)\ +\ O_2\ (g)\ \rightarrow\ CO_2\ (g)\ + H_2O\ (g)$$

Reactants Side	Products Side
#C: 1x5 = 5	#C: 1x1 = 1
#H: 1x12 = 12	#H: 1x2 = 2
#O: 1x2 = 2	#O: 1x2 + 1x1 = 3

Once again, try to start with atoms that appear in only one molecule on each side of the equation. In this case, both C and H appear only once on each side, so we will start with them. To balance C's, we need to multiply the number of C's on the products side by 5:

$$C_5H_{12}\ (l)\ +\ O_2\ (g)\ \rightarrow\ 5CO_2\ (g) + H_2O\ (g)$$

Now the C's are balanced:

Reactants Side	Products Side
#C: 1x5 = 5	#C: 5x1 = 5
#H: 1x12 = 12	#H: 1x2 = 2
#O: 1x2 = 2	#O: 5x2 + 1x1 = 11

To balance the H's, we need to multiply the H's on the products side by 6:

$$C_5H_{12}\ (l) + O_2\ (g) \rightarrow 5CO_2\ (g) + 6H_2O\ (g)$$

Now the H's balance:

Reactants Side	Products Side
#C: 1x5 = 5	#C: 5x1 = 5
#H: 1x12 = 12	#H: 6x2 = 12
#O: 1x2 = 2	#O: 5x2 + 6x1 = 16

To balance the O's we just need to multiply the O's on the reactants side by 8:

$$C_5H_{12}\ (l) + 8O_2\ (g) \rightarrow 5CO_2\ (g) + 6H_2O\ (g)$$

Now the equation is completely balanced:

Reactants Side	Products Side
#C: 1x5 = 5	#C: 5x1 = 5
#H: 1x12 = 12	#H: 6x2 = 12
#O: 8x2 = 16	#O: 5x2 + 6x1 = 16

The balanced chemical equation is:

$$\underline{C_5H_{12}\ (l) + 8O_2\ (g) \rightarrow 5CO_2\ (g) + 6H_2O\ (g)}$$

REVIEW QUESTIONS FOR MODULE #4

1. If a substance can be physically separated into its components, is it a pure substance or a mixture?

2. If a substance can be given a single chemical name, e.g. calcium bromide, is it a pure substance or a mixture?

3. What element makes up the majority of the air we inhale?

4. What element makes up the majority of the air we exhale?

5. One type of phase change we did not discuss is sublimation. Sublimation occurs when a solid changes directly into a gas, without passing through a liquid phase. Does sublimation occur when something is heated or when something is cooled?

6. If a liquid goes through a phase change and all you know is that the molecules slowed down and moved closer together, what phase did the liquid turn into?

7. What makes water an exception to the phase change rule?

8. What is the difference between a chemical change and a physical change?

9. List the chemical formulas of the homonuclear diatomic elements.

10. What makes a chemical equation balanced?

PRACTICE PROBLEMS FOR MODULE #4

1. Classify the following as either a mixture or a pure substance:

a. An egg b. A gold nugget c. A bottle of HNO_3 d. Lemonade dissolved in water

2. Reclassify everything in question #1 as an element, compound, homogeneous mixture, or heterogeneous mixture.

3. Classify the following as physical or chemical changes:

a. A precious vase is smashed to bits.
b. Eggs, flour, sugar, milk, and vanilla are baked, making a cake.
c. Dry ice turns into gas.
d. Charcoal is burned in a grill.
e. Kool-Aid® is dissolved in water.

4. What is the chemical formula of nitrogen gas?

5. Is the following equation balanced?

$$CaF_2 \text{ (aq)} + 2NH_4Cl \text{ (aq)} \rightarrow CaCl_2 \text{ (aq)} + NH_4F \text{ (aq)}$$

6. Balance the following equation:

$$HCl \text{ (aq)} + Zn \text{ (s)} \rightarrow ZnCl_2 \text{ (aq)} + H_2 \text{ (g)}$$

7. In an automobile, nitrogen from the air reacts with oxygen in the engine to make nitrogen monoxide, a toxic pollutant. Write the balanced chemical equation for this reaction.

8. Liquid heptene, C_7H_{10}, can react with gaseous hydrogen to make liquid heptane, C_7H_{16}. Write the balanced chemical equation for this reaction.

9. Balance the following equation:

$$C_7H_{16} \text{ (l)} + O_2 \text{ (g)} \rightarrow CO_2 \text{ (g)} + H_2O \text{ (g)}$$

10. Balance the following equation:

$$CO_2 \text{ (g)} + H_2O \text{ (l)} \rightarrow C_{12}H_{24}O_{12} \text{ (s)} + O_2 \text{ (g)}$$

MODULE #5: Counting Molecules and Atoms in Chemical Equations

Introduction

In the last module, you learned the basics of chemical reactions. You learned how to recognize the difference between chemical and physical change, and how to write chemical change in terms of a chemical equation. You were then given a brief introduction into the wonderful world of balancing chemical equations.

In this module, we will continue our discussion of chemical equations by showing you how to predict certain types of chemical reactions. You will learn how to figure out the products of certain chemical reactions when you are given only the reactants, as well as how to determine the reactants when you are given only the products. In addition, you will learn how to count the atoms and molecules in a sample of matter. Counting the atoms and molecules in a substance is the first step in fully understanding the mathematics of a chemical equation.

Three Basic Types of Chemical Reactions

As we go through this course, you will learn many, many different types of chemical reactions. To get you started, however, we will first concentrate on three of the most basic types of chemical reactions: **decomposition, formation,** and **combustion** reactions. These reactions are relatively simple to understand, and the equations that describe them are reasonably easy to predict and balance.

Decomposition Reactions

Probably the easiest chemical reactions to predict are **decomposition reactions**. Consider the reaction pictured in Figure 5.1:

FIGURE 5.1
The Decomposition of Water

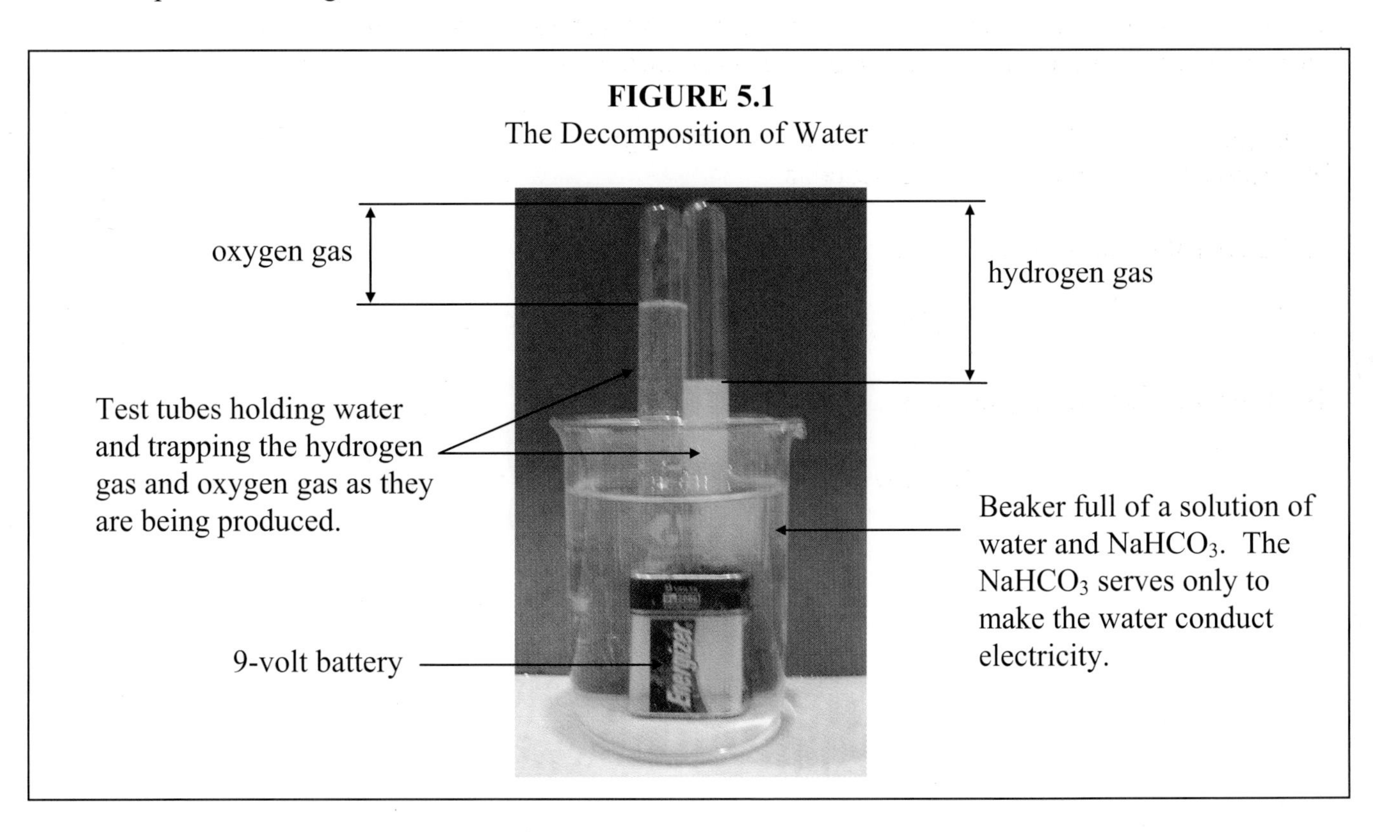
oxygen gas
hydrogen gas
Test tubes holding water and trapping the hydrogen gas and oxygen gas as they are being produced.
Beaker full of a solution of water and $NaHCO_3$. The $NaHCO_3$ serves only to make the water conduct electricity.
9-volt battery
Energizer

In the reaction shown, a battery is immersed in a solution of water and baking soda ($NaHCO_3$). Test tubes full of water are inverted over the posts of the battery. The test tube on the left is over the positive post of the battery, while the test tube on the right is over the negative post. Bubbles of oxygen gas form on the positive post of the battery, while bubbles of hydrogen gas form on the negative post of the battery. The bubbles then rise up into the test tubes, where they are trapped. The amount of space between the top of the inverted test tube and the water in the test tube is an indication of how much gas is collected. Notice that twice as much hydrogen has been produced as compared to oxygen. This is because, as the chemical formula tells us, there are two hydrogen atoms for every one oxygen atom in water.

The chemical reaction taking place in Figure 5.1 changes water into hydrogen and oxygen, according to the following equation:

$$2H_2O\ (l) \rightarrow 2H_2\ (g) + O_2\ (g) \qquad (5.1)$$

If you are wondering why H_2 and O_2 are products rather than H and O, you have to remember that hydrogen and oxygen are homonuclear diatomic elements. Thus, whenever a molecule has H or O (or any of the other homonuclear diatomics listed in the previous module), the decomposition reaction will produce homonuclear diatomic molecules. This reaction is an example of a decomposition reaction.

Decomposition reaction - A reaction that changes a compound into its constituent elements

Notice how Equation (5.1) fits the definition. A compound (H_2O) is changed into hydrogen (H_2) and oxygen (O_2), which are its constituent elements.

The nice thing about decomposition reactions is that their equations are relatively easy to predict and balance, given the compound that is undergoing the reaction. All you need to do is look at the chemical formula of the compound, split it up into its individual elements (being careful to recognize which ones are homonuclear diatomics), and then balance the equation. Example 5.1 shows you how this is done. Be sure to follow this example closely, because a new facet of balancing chemical equations is presented in the example.

EXAMPLE 5.1

Write the chemical equation that describes the decomposition of ethanol (C_2H_6O).

To solve this problem, we first look at the molecule which is undergoing decomposition. According to its chemical formula, it is made up of C, H, and O atoms. Thus, its elements are carbon, hydrogen, and oxygen. Hydrogen and oxygen are homonuclear diatomics; therefore, H_2 and O_2 will be produced. Carbon is not a homonuclear diatomic, so the decomposition will produce C. This gives us the following unbalanced equation:

$$C_2H_6O \rightarrow C + H_2 + O_2$$

This equation is not balanced, so we need to balance it. First of all, we see 2 C's on the reactants side of the equation but only 1 C on the products side. We therefore must double the C's on the products side:

$$C_2H_6O \rightarrow 2C + H_2 + O_2$$

Now the carbons are balanced, but the hydrogens are not. There are 6 on the reactants side and only 2 on the products side, so we need to triple the number of H's on the products side:

$$C_2H_6O \rightarrow 2C + 3H_2 + O_2$$

Now we need to address the problem with oxygen. This is a tricky one because there is 1 O on the reactants side and 2 O's on the products side. The way we would usually solve this problem is to double the number of O's on the reactants side. The problem is that in order to double the O's on the reactants side, we would have to multiply the C_2H_6O molecule by 2. That would end up throwing the C's and H's back out of balance. So...what do we do?

To take care of this problem, we have to employ a bit of a trick. To use this trick, we have to recognize that there is another way to balance the O's. Instead of *doubling* the O's on the *reactants* side of the equation, I could *halve* the O's on the *products* side of the equation:

$$C_2H_6O \rightarrow 2C + 3H_2 + \tfrac{1}{2}O_2$$

This balances the equation, because now there is 1 x 1 = 1 oxygen on the reactants side and $\tfrac{1}{2}$ x 2 = 1 oxygen on the products side. The problem with this solution is that a fraction in the chemical equation doesn't make sense. After all, you can't have a fraction of an atom. You can only have 0 atoms or 1 atom--not one half of an atom.

To get rid of the fraction, we must do something we learned in algebra. In algebra, we learned that you can do anything you want to one side of an equation as long as you do exactly the same thing to the other side of the equation. Thus, I could multiply the products side of the equation by 2 as long as I also multiply the reactants side of the equation by 2. What would this accomplish? Watch:

$$2 \text{ x } [C_2H_6O] \rightarrow 2 \text{ x } [2C + 3H_2 + \tfrac{1}{2}O_2]$$

Now we distribute the 2 by multiplying it with every term on both sides of the equation:

$$2C_2H_6O \rightarrow 4C + 6H_2 + 1O_2$$

When the 2 multiplies the one-half, you get 1, eliminating the fraction. If you check the equation, it is still perfectly balanced. Thus, our balanced equation for the decomposition of ethanol is:

$$\underline{2C_2H_6O \rightarrow 4C + 6H_2 + O_2}$$

This little trick actually brings up an important fact regarding chemical equations. Chemical equations are just like mathematical equations: Anything you can do to a mathematical equation you can also do to a chemical equation. You can add, subtract, multiply, or divide both sides of a chemical equation by any number. This trick is such an important aspect of balancing equations that I will present one more problem to make sure you understand it.

Write the balanced chemical equation for the decomposition of SiH_3F.

The elements in this molecule are Si, H, and F. H and F are homonuclear diatomics; thus, the decomposition reaction is:

$$SiH_3F \rightarrow Si + H_2 + F_2$$

To balance this equation, we don't need to do anything to the silicons, since there is 1 Si on each side of the equation. However, there are 3 H's on the reactants side and only 2 on the products side. We can solve this as follows:

$$SiH_3F \rightarrow Si + \frac{3}{2}H_2 + F_2$$

I chose to balance the H's this way so that I wouldn't have to multiply SiH_3F by anything, because that would destroy the balance between the Si's. Now we can balance the F's in a similar fashion:

$$SiH_3F \rightarrow Si + \frac{3}{2}H_2 + \frac{1}{2}F_2$$

Now the equation is balanced:

Reactants Side	Products Side
Si: 1x1 = 1	Si: 1x1 = 1
H: 1x3 = 3	H: $\frac{3}{2}$x2 = 3
F: 1x1 = 1	F: $\frac{1}{2}$x2 = 1

The only problem remaining is that the fractions must be eliminated. To do this, we simply have to multiply both sides of the equation by some number that will get rid of the fractions. Since both fraction denominators are 2, we can eliminate both fractions by multiplying both sides of the equation by 2:

$$2 \times [SiH_3F] \rightarrow 2 \times [Si + \frac{3}{2}H_2 + \frac{1}{2}F_2]$$

$$\underline{2SiH_3F \rightarrow 2Si + 3H_2 + F_2}$$

Now see if you can do this:

ON YOUR OWN

5.1 Write the chemical equation for the decomposition of phosphorus trihydride.

5.2 Write the chemical equation for the decomposition of $C_{12}H_{23}O_{11}$.

Before we leave the topic of decomposition reactions, I must add a word of warning. The definition that I gave for decomposition reactions is very narrow. Most chemists say that any reaction that has a single compound as the reactant and several elements or compounds as products is, in fact, a decomposition reaction. For example, consider the reaction represented by Equation (5.2):

$$CaCO_3 (s) \rightarrow CaO (s) + CO_2 (g) \tag{5.2}$$

Chemists consider this a decomposition reaction since it starts with a single compound and breaks it into two different compounds. We will also consider reactions like the one shown in Equation (5.2) to be decomposition reactions. However, when I ask you to write the equation for a decomposition reaction, always write the equation for the reaction that produces only elements, like those given in the examples and the "On Your Own" problems.

(The multimedia CD has a video demonstration of a decomposition reaction.)

If you purchased the MicroChem kit discussed in the introduction, you can perform Experiment #7 to get some experience with decomposition reactions.

Formation Reactions

Formation reactions are really just the opposite of decomposition reactions. The definition is as follows:

<u>Formation reaction</u> - A reaction that starts with two or more elements and produces one compound

Thus, a formation reaction has elements as its reactants and a compound as its product. An example of a formation reaction would be the reverse of Equation (5.1). Suppose I had hydrogen gas and oxygen gas and let them react. They would form water according to the equation:

$$2H_2 (g) + O_2 (g) \rightarrow 2H_2O (g) \tag{5.3}$$

Now please understand that this is just the opposite of what was done in Figure 5.1. In that reaction, we started with water and added energy to it so that it would absorb the energy and decompose into hydrogen and oxygen. In this reaction, we are starting with hydrogen and oxygen and reacting them. The resulting reaction, shown in Figure 5.2, produces water and releases quite a bit of energy.

Photos by Flanders Video Productions

FIGURE 5.2
The Formation of Water

The balloon is full of hydrogen gas. There is oxygen in the air surrounding the balloon. A flame is brought near the balloon.

When the flame pops the balloon, the oxygen and hydrogen are allowed to mix. The heat from the flame starts the reaction, which produces water and energy in the form of an explosion.

Just like the definition I gave for a decomposition reaction, the definition I gave for a formation reaction is a bit narrow. Chemists consider any reaction that starts with several reactants and forms a single product to be a formation reaction. Once again, however, when I ask you to write chemical equations for formation reactions, write them for a reaction that has only elements as its reactants, as illustrated in the following example:

EXAMPLE 5.2

Write the chemical equation for the formation of $MgCO_3$.

This compound is made up of Mg, C, and O (a homonuclear diatomic). Thus, the reaction is:

$$Mg + C + O_2 \rightarrow MgCO_3$$

Now we just have to balance the equation. The Mg's and C's are already balanced. To balance the O's without disturbing the Mg's and C's, we will have to use a fraction again:

$$Mg + C + \tfrac{3}{2}O_2 \rightarrow MgCO_3$$

Now we can eliminate the fraction by multiplying both sides by 2:

$$2 \times [Mg + C + \tfrac{3}{2}O_2] \rightarrow 2 \times [MgCO_3]$$

$$\underline{2Mg + 2C + 3O_2 \rightarrow 2MgCO_3}$$

Make sure you understand formation reactions by completing the following problem:

ON YOUR OWN

5.3 Write a balanced chemical equation for the formation of $KHCO_3$.

Complete Combustion Reactions

Combustion is another word for burning; thus, combustion reactions are those chemical reactions which govern the process of burning. In general, to burn something, it must be chemically reacted with oxygen. About 90% of the time, such a reaction starts with a compound that contains carbon and hydrogen, and it produces carbon dioxide and water. That kind of chemical reaction is called a **complete combustion reaction**. Why do we add the word "complete"? You will see momentarily.

Complete combustion reaction - A reaction in which O_2 is added to a compound containing carbon and hydrogen, producing CO_2 and H_2O

An example of a complete combustion reaction would be:

$$CH_2O\ (g) + O_2\ (g) \rightarrow CO_2\ (g) + H_2O\ (g) \qquad (5.4)$$

This is the standard form of a complete combustion reaction: O_2 is reacted with a compound that contains carbon and hydrogen (CH_2O in this case) to make CO_2 and H_2O. When you burn something, a reaction similar to this one usually takes place.

I need to warn you that there are other combustion reactions which are not *complete* combustion reactions. In a moment, we will discuss a type of combustion called *incomplete* combustion. In addition, there are some other forms of combustion. For example, you may not realize this, but metals can burn. If you were to heat the metal magnesium to a high enough temperature in the presence of oxygen, the following reaction would occur:

$$2Mg\ (s) + O_2\ (g) \rightarrow 2MgO\ (s) \qquad (5.5)$$

This is a combustion reaction, but it does not fit our definition of a *complete* combustion reaction, as it does not start with a compound which contains carbon and hydrogen, and it does not produce carbon dioxide and water.

(The multimedia CD has a video demonstration of the combustion of magnesium.)

EXAMPLE 5.3

A butane pocket lighter works by evaporating and then burning the liquid butane (C_4H_{10}) stored in it. Write the balanced chemical equation for this complete combustion reaction.

According to our definition of complete combustion, oxygen is added to butane so that it produces carbon dioxide and water:

$$C_4H_{10} + O_2 \rightarrow CO_2 + H_2O$$

We could go ahead and add phase symbols at this point. The problem told us that the butane was evaporated before it was burned, so the butane is a gas. We know oxygen and carbon dioxide are gases, and we also know that the temperature associated with combustion is large enough to evaporate the water produced. Therefore:

$$C_4H_{10}\ (g) + O_2\ (g) \rightarrow CO_2\ (g) + H_2O\ (g)$$

Now the only thing left to do is balance the equation. I need to quadruple the C's on the products side to balance them, and I must increase the number of H's on the products side by a factor of 5:

$$C_4H_{10}\ (g) + O_2\ (g) \rightarrow 4CO_2\ (g) + 5H_2O\ (g)$$

In counting the O's, we see that there are 1 x 2 = 2 on the reactants side and 4 x 2 + 5 x 1 = 13 on the products side. To balance the O's without disturbing the C and H balance, I must use a fraction again:

$$C_4H_{10}\,(g) + \tfrac{13}{2}O_2\,(g) \rightarrow 4CO_2\,(g) + 5H_2O\,(g)$$

This gives $\frac{13}{2}$ x 2 = 13 O's on the reactants side and 4x2 + 5x1 = 13 O's on the products side. Now that the equation is balanced, all we have to do is get rid of the fraction:

$$2 \times [C_4H_{10}\,(g) + \tfrac{13}{2}\,O_2\,(g)] \rightarrow 2 \times [4CO_2\,(g) + 5H_2O\,(g)]$$

$$\underline{2C_4H_{10}\,(g) + 13O_2\,(g) \rightarrow 8CO_2\,(g) + 10H_2O\,(g)}$$

ON YOUR OWN

5.4 An automobile uses the energy gained from burning evaporated gasoline in order to run. Write a balanced chemical equation for the combustion of C_8H_{18}, an important component of gasoline. Include phase symbols in your equation.

Incomplete Combustion Reactions

Before we leave the subject of combustion, there is one more thing we need to learn. We all know that in order to burn things, we need plenty of oxygen. If I start a candle burning and then cover the candle with a glass, the candle will eventually go out, because it used up all of the oxygen in its surroundings. Thus, if there is plenty of oxygen, combustion takes place. If there is no oxygen, no combustion takes place. What happens, however, when we are somewhere in between? Let's suppose we try to burn something in an area where there is *some* oxygen, but not *plenty* of oxygen. What happens then?

When a substance containing carbon and hydrogen burns in an environment where there is limited oxygen, one of two chemical reactions known as **incomplete combustion** occurs. In these reactions, carbon dioxide is not produced. Instead, either carbon monoxide or just plain carbon is produced. For example, let's assume we try to burn natural gas (CH_4) in an area where there is limited oxygen. One of two reactions occurs:

$$2CH_4\,(g) + 3O_2\,(g) \rightarrow 2CO\,(g) + 4H_2O\,(g) \tag{5.6}$$

or

$$CH_4\,(g) + O_2\,(g) \rightarrow C\,(s) + 2H_2O\,(g) \tag{5.7}$$

The amount of oxygen in the area determines which reaction occurs. If there is still a reasonable amount of oxygen around, the first reaction [Equation (5.6)] will occur, forming carbon monoxide and water. If there is a serious shortage of oxygen in the area, the second reaction [(Equation (5.7)] will occur, forming carbon (which we usually call "soot") and water.

As I mentioned in Module #3, while carbon dioxide is a substance that is generally not harmful to humans, carbon monoxide is poisonous to humans. Thus, whenever you try to burn something in a limited supply of oxygen, you run the risk of producing a poisonous gas! Believe it or not, this happens all of the time in everyday life. The best example is that of an automobile.

In order for an automobile to run, energy must be generated. In the vast majority of automobiles today, this is done by burning gasoline. In order to burn gasoline, the automobile must have a steady supply of oxygen. To get this oxygen, the automobile engine constantly pulls in air from the outside. The oxygen in the air is then used to burn the gasoline. The problem with this process is that an engine uses an *enormous* amount of energy; thus, it must burn an *enormous* amount of gasoline. The gasoline usually burns so quickly that it depletes the oxygen in the air faster than the engine can pull in more oxygen to replace it. Thus, when an automobile runs, it often burns gasoline in a limited supply of oxygen, causing incomplete combustion.

This incomplete combustion produces the poisonous gas carbon monoxide, as shown in the Equation (5.6) above. This is why you must always avoid being behind a car when it is running. Some of the combustion that occurs in any automobile engine is incomplete combustion. This produces carbon monoxide, which is ejected through the tailpipe of the car. Standing behind a car and breathing in the exhaust puts carbon monoxide, a poison, into your body. This is also why you should *never, ever* run a car in an enclosed area, such as a closed garage. The car exhaust contains carbon monoxide, which will build up in an enclosed area to the point that breathing the air could kill you.

The amount of carbon monoxide produced by automobiles got to be so large in the early 1970s that the U.S. government became concerned that too many people were breathing poisonous air. Thus, in 1975, the government required all new automobiles to be fitted with **catalytic** (kat' uh lih' tik) **converters**. A catalytic converter helps reduce the amount of carbon monoxide in automobile exhaust by converting the carbon monoxide into carbon dioxide. We will see exactly how this works in a later module. For right now, you just have to understand that a catalytic converter helps convert the carbon monoxide produced in an automobile into carbon dioxide, a non-poisonous gas. Unfortunately, however, even the best catalytic converter is only about 90% efficient, so automobiles still do produce carbon monoxide, making automobiles a big source of pollution.

Carbon monoxide is not just produced in automobiles, however. Often, when it is very cold, natural gas furnaces have to burn their gas faster than the oxygen supply can be replenished. Under those conditions, carbon monoxide can also be produced. Of course, a large fraction of that carbon monoxide leaves the house through the chimney, but some of it can find its way inside.

Fireplaces are also a significant source of carbon monoxide. Often there is very poor airflow around a fireplace. When the airflow is weak, the fireplace will deplete its oxygen supply to the point where incomplete combustion occurs. In fact, many fireplaces deplete the oxygen supply so much that the second reaction [Equation (5.7)] occurs. The oxygen supply gets so limited that not even carbon monoxide can be made. In the end, only carbon is formed, leaving soot in the fireplace.

The more soot you see in your fireplace, the more concerned you should be about the airflow around it. If a large amount of soot is being produced, then the airflow around the fireplace must be very poor. Most likely, this means that a large amount of the combustion that takes place in the fireplace is incomplete combustion. This means that a significant amount of carbon monoxide is being made as well. Once again, a good fraction of that carbon monoxide leaves through the chimney, but some of it can get into your house.

Every year, there are a handful of people in the United States who die because their furnaces or fireplaces produce too much carbon monoxide. Since carbon monoxide is odorless and tasteless, it is impossible to tell whether or not you are breathing it. Although this is a serious problem, there is no reason to get paranoid about it. If your furnace and fireplace are both well-ventilated, then there should be plenty of oxygen getting to them. If you are concerned about the ventilation, however, there is a simple way to find out whether carbon monoxide production is a problem in your house.

You can go to almost any hardware or department store and pick up a battery-powered carbon monoxide detector. There are several brands you can choose from, and they all emit an alarm when the carbon monoxide levels get dangerously high. One word of caution: If you put your carbon monoxide detector in a garage, it will likely go off every time you start your car. This is not a problem, however, because as soon as you pull the car out of the garage, the carbon monoxide level will decrease.

Incomplete combustion is a real problem today. It makes cars a serious source of pollution, and it has the capability of making a furnace or a fireplace dangerous. It is important for any educated person to know the basics of incomplete combustion and its effects on humans. That is why I spent time on it in this module.

Atomic Mass

Now that you are firmly grounded in the concepts behind chemical equations, it is time to learn how chemists count atoms and molecules. Once you learn this new technique, another powerful aspect of chemical equations will be revealed. Let's not get too far ahead of ourselves, however. We first must learn how to count atoms and molecules.

Since we cannot see atoms or molecules, it might sound odd to you that they can actually be counted. Nevertheless, they can. Even though we cannot point to each atom or molecule and assign it a number, we can figure out how many atoms or molecules are in a sample of matter. We do this by measuring the mass of the sample and determining the types of atoms or molecules in that sample. Those two pieces of information can tell us how many atoms or molecules exist in the sample.

In order to see how this works, we must first look at the periodic chart and learn some more features of this important table. Notice that each atom is represented by a box. Each box contains the atom's symbol and two numbers. The symbol tells us what atom the box represents, but what do the two numbers tell us?

The number that appears *above* the atomic symbol is called the **atomic number**. The atomic number tells you the place where the atom appears on the chart. If you start at the top left of the chart and work your way across, you will see that the atomic number in each box is one higher than the atomic number in the box before it. We will learn later that the atomic number tells us something else very important. For now, however, I want to concentrate on the other number in the box.

The number that appears *below* the atomic symbol is called the **atomic mass**, and it tells us how heavy the atom is. The bigger the atomic mass, the heavier the atom. If you look at the atomic masses listed on the chart, you will notice that they increase as you read the chart from left to right. Thus, as you read the periodic chart from left to right, the atoms listed on the chart increase in mass.

Of course, the first question that should come to your mind is: What are the units? After all, the chart says that a carbon atom has a mass of 12.0. What does that 12.0 mean? Is it 12.0 g? 12.0 kg? 12.0 mg? 12.0 slugs? Well, it turns out that the units used on the periodic chart are special. The masses listed there are in **atomic mass units**. We usually abbreviate this as "amu."

What is an amu? As you might expect, an atomic mass unit is very, very small. After all, atoms are so small that you cannot see them. They obviously cannot have too much mass, since they are so small. It turns out that the relationship between atomic mass units and grams is as follows:

$$1.00 \text{ amu} = 1.66 \times 10^{-24} \text{ g}$$

This is a very important relationship, but it is not one that you need to memorize, because I will give you this relationship on any assignment or test in which you need to use it. One thing you must remember about this relationship, however, is that it is not exact. When you use this relationship in conversion problems, you must consider the three significant figures it has.

Even though you needn't memorize this relationship, it is important that you understand what it means. A gram is about the mass of a housefly. An atomic mass unit is 1.66×10^{-24} of that. This is an incredibly small mass! You might have thought that atoms and molecules were small, but did you have any idea that they were 1 trillion trillionths the mass of a single housefly? Since atoms and molecules have such small masses, the calculations we will perform on them will often use numbers expressed in scientific notation, as is the case with the number of grams in the above relationship. We dealt with scientific notation in Module #1, and you need to be familiar with it before you can proceed in this module. Thus, if you don't feel comfortable with scientific notation, either review Module #1 or find another book and learn it.

Now that we know the relationship between atomic mass units and grams, let's get an idea of exactly how much mass certain atoms have.

EXAMPLE 5.4

What is the mass of a nitrogen atom in grams? (1.00 amu = 1.66 x 10^{-24} g)

In order to figure out the mass of an atom, we must first look at the periodic chart. According to the chart, a nitrogen atom has a mass of 14.0 amu. In order to determine the mass in grams, all we have to do is a conversion. We know from the relationship given that 1.00 amu = 1.66×10^{-24} g. Thus, according to what we learned in Module #1:

$$\frac{14.0 \, \cancel{\text{amu}}}{1} \times \frac{1.66 \times 10^{-24} \text{ g}}{1.00 \, \cancel{\text{amu}}} = 2.32 \times 10^{-23} \text{ g}$$

So the mass of one nitrogen atom is 14.0 amu, or 2.32×10^{-23} g. If you can't figure out how to do the arithmetic to get the right answer, refer to a math book or your calculator's instruction book. We will be using scientific notation in equations like this quite a bit, so you need to know how it's done.

ON YOUR OWN

5.5 What is the mass of a calcium atom in kg? (1.00 amu = 1.66×10^{-24} g)

Molecular Mass

What about molecules? If we can find the mass of an atom by simply looking at the periodic chart, how do we get the mass of a molecule? Well, if you think about it, the answer becomes pretty obvious. A molecule is made when atoms join together. According to the Law of Mass Conservation, the mass of the atoms before they join together should be the same as the mass of the atoms after they join together. Thus, if you just add up all the masses of all of the atoms in a molecule, you'll end up with the mass of that molecule, which is called the **molecular mass**. See if you can follow this logic in the example problems below:

EXAMPLE 5.5

What is the molecular mass of $CHCl_3$ in amu?

According to the chemical formula, a $CHCl_3$ molecule has 1 carbon atom (which has a mass of 12.0 amu according to the chart), 1 hydrogen atom (which has a mass of 1.01 amu), and 3 chlorine atoms (each of which has a mass of 35.5 amu). In order to figure out the mass of the molecule, we simply add the masses of 1 carbon, 1 hydrogen, and 3 chlorines:

$$\text{Mass of } CHCl_3 = 1 \times 12.0 \text{ amu} + 1 \times 1.01 \text{ amu} + 3 \times 35.5 \text{ amu}$$

$$\text{Mass of } CHCl_3 = 119.5 \text{ amu}$$

Notice that all we really had to do was take the mass of each atom in the molecule, multiply by the number of that atom in the chemical formula, and then add it all up.

Let me spend a moment on significant figures here. In reality, we are just adding numbers in this problem. It is true that we are multiplying 35.5 by 3, but that 3 is exact. You cannot have some fraction of an atom. Thus, there are exactly 3 chlorine atoms in a $CHCl_3$ molecule. In reality, then, we are really just adding 35.5 three times ($3 \times 35.5 = 35.5 + 35.5 + 35.5$). In this problem, then, we must follow the rule of addition and subtraction when determining the significant figures. According to that rule, we keep the lowest *precision* when we add. Although 1.01 has its last significant figure in the hundredths place, both 12.0 and 35.5 have their last significant figure in the tenths place We therefore must report the answer to the tenths place. Therefore, a $CHCl_3$ molecule has a mass of 119.5 amu. Notice that this has more significant figures than any of the atomic masses used in the equation, but that does not matter. In addition and subtraction, we worry only about precision.

What is the mass of a $Si_6F_{12}O_6$ molecule in grams? (1.00 amu = 1.66×10^{-24} g)

According to the chart, Si has a mass of 28.1 amu, F has a mass of 19.0 amu, and O has a mass of 16.0 amu. Thus, the mass of the molecule in amu is:

$$\text{Mass of } Si_6F_{12}O_6 = 6 \times 28.1 \text{ amu} + 12 \times 19.0 \text{ amu} + 6 \times 16.0 \text{ amu} = 492.6 \text{ amu}$$

Once again, we report our answer to the tenths place because the masses that we are adding together all have their last significant figure in the tenths place. Converting to grams:

$$\frac{492.6 \,\cancel{\text{amu}}}{1} \times \frac{1.66 \times 10^{-24} \text{ g}}{1.00 \,\cancel{\text{amu}}} = 8.18 \times 10^{-22} \text{ g}$$

At this point, we must count significant figures, because we are multiplying. The conversion relationship has three significant figures, so the answer can have only three significant figures. Thus, a $Si_6F_{12}O_6$ molecule has a mass of 492.6 amu, or $\underline{8.18 \times 10^{-22} \text{ g}}$.

ON YOUR OWN

5.6 What is the mass of a Na_2SO_4 molecule in g? (1.00 amu = 1.66×10^{-24} g)

The Mole Concept

It is now time to learn how chemists can count the number of atoms or molecules that exist in a sample of matter. To do this, chemists use a concept known as the **mole**. Now, the mole that we use in chemistry is not a furry little creature that lives underground. Instead, chemists use the term "mole" to refer to a specific number of atoms. If I have 6.02×10^{23} atoms, I have a mole of atoms. Alternatively, if I have 6.02×10^{23} molecules, I have a mole of molecules.

This kind of terminology might sound strange at first, but we use it all of the time in our day-to-day language. For example, if I go to the doughnut shop to get some doughnuts, I usually buy them in groups that we call a "dozen." When we say "one dozen doughnuts," we really mean 12 doughnuts. Likewise, if I ask the person at the doughnut shop for three dozen doughnuts, he or she will pick out 36 doughnuts for me. Thus, "dozen" is just a word that means "12."

In the same way, if I have a party and tell you that 12 couples attended the party, you would know that I meant there were 24 people who came to my party. The word "couple" means "two." A set of 12 couples, then, is 24 people. The word "mole" is a word that simply means "6.02×10^{23}". All of these words--"dozen," "couple," and "mole"-- refer to a group of things. The only difference is what number of things they refer to. The word "couple" refers to only two things, the word "dozen" refers to 12 things, and the word "mole" refers to 6.02×10^{23} things. Now, of course, the words "couple" and "dozen" refer to manageable numbers (two and 12). Why, then, does "mole" refer to such a big number (6.02×10^{23})?

Remember why we use the word "mole." We use it to count atoms and molecules in a sample of matter. Since atoms and molecules are so small, there obviously must be a huge number of them in any given sample of matter. In a 12-ounce glass of water, for example, there are 1.14×10^{25} molecules of water. This is, of course, a *huge* number. Thus, the word we use to count molecules or atoms must refer to a huge number as well. One mole, therefore, represents 602,000,000,000,000,000,000,000 (6.02×10^{23}) atoms or molecules.

Therefore, if I wanted to tell you how many molecules of water were in a 12-ounce glass, I could do it one of two ways. I could tell you that there are 1.14×10^{25} molecules, as I did above, or I could tell you that there are 18.9 moles of water molecules in that glass $\left(\frac{1.14 \times 10^{25}}{6.02 \times 10^{23}}\right)$. This is exactly the same as telling you how many doughnuts I bought at the store. I could tell you that I bought 36 doughnuts, or I could tell you that I bought three dozen doughnuts $\left(\frac{36}{12}\right)$. In each case, I could either tell you the number of objects directly, or I could use a word which then could be used to figure out how many objects there were. It turns out that in chemistry, using the word "mole" is a much easier way to report how many atoms or molecules exist in a sample, so that's what we will most often use.

This rather odd number (6.02×10^{23}) is often called **Avogadro's** (ah vuh gah' drohs) **number**, because the chemist whose work led to its discovery was an Italian scientist named Amedeo Avogadro. Avogadro's number represents how many atoms are in a sample of an element when that sample has the same mass in grams as the element's mass in amu. In other words, the mass of a carbon atom is 12.0 amu. A 12.0-*gram* sample of carbon has 6.02×10^{23} atoms in it. Similarly, the atomic mass of sulfur is 32.1 amu, and a 32.1-gram sample of sulfur has 6.02×10^{23} atoms of sulfur in it.

Regardless of the atom you choose, if you gather a sample with a mass in grams equal to the mass of the atom in amu, you have 6.02×10^{23} atoms in that sample. Thus (look at the periodic chart to see where I'm getting these numbers), 4.0 grams of helium, 52.0 grams of Cr, 55.8 grams of Fe, and 197.0 grams of Au will all have exactly 6.02×10^{23} atoms in each sample. Or, to put it another way, each of those samples contains one mole of atoms.

Incredibly enough, this same relationship works for molecules as well. A water molecule has a molecular formula of H_2O. This means that its molecular mass is $2 \times 1.01 + 16.0 = 18.0$ amu. If I have an 18.0-*gram* sample of water, there are 6.02×10^{23} molecules (one mole) of water in it. Similarly, the molecular mass of table salt (NaCl) is 58.5 amu. Thus, a 58.5-gram sample of table salt contains 6.02×10^{23} molecules (one mole). This leads us to a very, very important fact:

An atom's mass in amu tells you how many grams it takes to have one mole of those atoms.

In the same way:

A molecule's mass in amu tells you how many grams it takes to have one mole of those molecules.

These are the two most important concepts presented in this module. Commit them to memory and be sure that you understand what they mean.

FIGURE 5.3
An Illustration of the Mole Concept

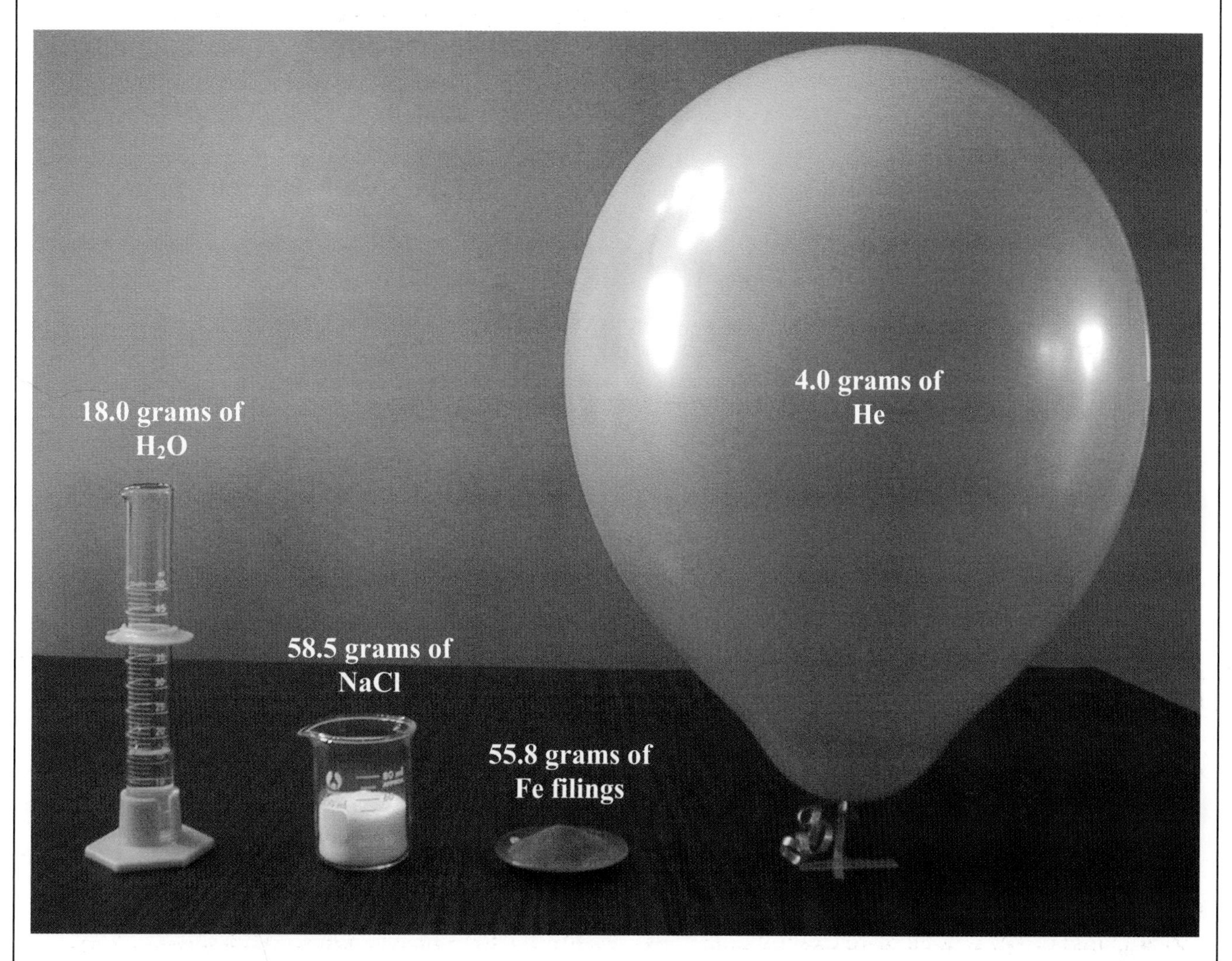

Each of these samples has exactly the same number of atoms or molecules in them. The balloon contains 4.0 g of helium. That means there is a mole of helium atoms in the balloon. There are 55.8 grams of iron filings, meaning that there is one mole of iron atoms. Since there are 58.5 grams of table salt, there is one mole of table salt. There are 18.0 grams of water, which means there is one mole of water. Thus, there are 6.02×10^{23} atoms of helium, 6.02×10^{23} atoms of iron, 6.02×10^{23} molecules of NaCl, and 6.02×10^{23} molecules of water.

If you are still having trouble grasping the mole concept, think about it this way: An atom of boron has a mass of 10.8 amu. This also means that it takes 10.8 grams of that element to get a mole of boron atoms. In other words, 10.8 grams boron = 1 mole boron. Thus, once I know the mass of an atom, I know a conversion relationship that relates grams to moles. For another atom, this conversion relationship exists, but the number of grams is different. If I need to know how many moles of copper (Cu) atoms are in a sample of copper, I must use the conversion relationship between grams of copper and moles of copper. Since the chart tells me that the mass of a copper atom is 63.5 amu, then I know that 63.5 grams Cu = 1 mole Cu. Try to follow this logic in these problems:

EXAMPLE 5.6

How many moles of fluorine atoms exist in a 50.0 g sample of fluorine?

In order to solve this problem, we first need to find out a relationship between grams and moles. We do this by looking at the periodic chart. The chart tells us that a fluorine atom has a mass of 19.0 amu. As we just learned, this also means:

19.0 grams fluorine = 1 mole fluorine

Thus, to convert between grams and moles, we simply have to use that relationship:

$$\frac{50.0\ \cancel{\text{g fluorine}}}{1} \times \frac{1\ \text{mole fluorine}}{19.0\ \cancel{\text{g fluorine}}} = 2.63\ \text{moles fluorine}$$

So a 50.0 g sample of fluorine contains <u>2.63 moles</u> of fluorine atoms. Notice that I kept three significant figures in the answer. This is because the "1" in "1 mole fluorine" is an integer and is therefore exact, giving it an infinite number of significant figures.

People are encouraged to limit their table salt (NaCl) intake to 5.000 grams per day. How many moles of NaCl is that?

To get the conversion relationship between grams and moles in this case, we must first determine the mass of an NaCl molecule. According to the periodic chart:

NaCl mass = 1 x 23.0 amu + 1 x 35.5 amu = 58.5 amu

According to what we just learned, this also means:

58.5 g NaCl = 1 mole NaCl

Now we can do our conversion :

$$\frac{5.000\ \cancel{\text{g NaCl}}}{1} \times \frac{1\ \text{mole NaCl}}{58.5\ \cancel{\text{g NaCl}}} = 0.0855\ \text{moles NaCl}$$

In the end, then, people are to limit their intake of NaCl to <u>0.0855 moles</u>. Notice that in this case, I could only keep three significant figures because of the "58.5 g NaCl." Thus, even though the "1 mole" part of the conversion relationship is considered exact, the other half of the relationship is not, and therefore we must count *its* significant figures.

Now don't get lost in the math and forget what we are doing here. These may seem like pointless conversions to you, but they are far from it! We are taking something that is very easy to measure (mass) and converting it into something that tells us how many atoms or molecules are in our sample (moles). This is a very important and useful thing to be able to do. Every time we can report the number of moles of atoms (or molecules) in a sample, we are really reporting the number of atoms (or molecules) that exist in that sample. Why is this so useful? Perform the following experiment to find out:

EXPERIMENT 5.1
Measuring the Width of a Molecule

Note: A sample set of calculations is available in the solutions and tests guide. It is with the solutions to the practice problems.

Supplies:
- ~~Safety goggles~~
- Eyedropper
- Stirring rod (or spoon)
- Water
- Large glass (at least 16 ounces)
- Graduated cylinder (Measuring cups and measuring spoons can be used, but they will be less precise.)
- Dishwashing liquid (It must be the kind used for washing dishes by hand, NOT the kind used in automatic dishwashers. Preferably, the brand should be Joy® or Sunlight®.)
- Large bowl (It should have a diameter larger than 10 inches but smaller than 12 inches. If you don't have a bowl in that size range, then use one larger than 12 inches; it will just make it a little harder to use the ruler.)
- Pepper
- Ruler

1. We've already learned that atoms and molecules are small, but exactly how small are they? We're going to use the mole concept we just learned in order to get a rough idea. To do this, you first need to fill your large bowl almost full of water. Leave only a centimeter or so from the top of the water to the rim of the bowl. Put the bowl of water where it will be easy to work with but won't be disturbed while you perform the next few steps.
2. Quantitatively measure out 5.0 mL (1 teaspoon) of the dishwashing liquid into your graduated cylinder.
3. Leaving the dishwashing liquid in the graduated cylinder, fill the cylinder up to the 50.0 mL (1/4 cup) mark with water. Do it slowly to reduce the formation of bubbles.
4. Pour the entire contents of the cylinder into the large glass.
5. Now that the cylinder is empty, fill it up to the 50.0 mL (1/4 cup) mark with water and pour it into the same glass.
6. Repeat this step six more times, so that a total of 395.0 mL (2 cups) of water and 5 mL (1 teaspoon) of dishwashing liquid are in the tall glass.
7. Stir the contents of the glass slowly but thoroughly. By performing the preceding steps, you have taken 5.0 mL of dishwashing liquid and **diluted** it to 400.0 mL.
8. Next, you need to calibrate your eyedropper. To do this, clean and dry your graduated cylinder.
9. Use your eyedropper to transfer 10.0 mL of the dilute dishwashing liquid (*drop by drop)* into the graduated cylinder. Count how many drops it takes to fill the cylinder up to the 10.0 mL mark.

10. Take the number 10.0 and divide it by the number of drops you counted. The result represents how many mL are in each drop of your eyedropper. Your answer should be somewhere between 0.060 mL and 0.020 mL. In this way, you have used your graduated cylinder to calibrate your eyedropper so that you can figure out how many mL of liquid are in each drop of the eyedropper.
11. Take the pepper over to the bowl of water. Since it has just been sitting for a while, the water in the bowl should be very calm.
12. Lightly sprinkle the pepper onto the surface of the water in the bowl. You should sprinkle the pepper on the water so that there are a few dots of pepper on the surface of the of water in each part of the bowl. It is very important that you don't sprinkle *too much* pepper on the water's surface, however. You want only a slight dusting of pepper on the surface of the water.
13. Use your eyedropper to put *one single drop* of diluted dishwashing liquid into the very center of the bowl. Do this by holding the dropper a few centimeters above the surface of the water at the center of the bowl and carefully allowing one drop to fall into the bowl. Observe what happens. The pepper that is in the center of the bowl is pushed outwards in a roughly circular form. Thus, you will see a circle of clear water form, outlined by the pepper that has been pushed away.
14. Quickly use your ruler to measure the diameter of this circle. You can do this by laying the ruler across the top of the bowl and looking over it at the circle. Try to do this quickly, because the circle will tend to collapse, and you want to measure its largest diameter.
15. Why did the circle form in the pepper? The pepper was floating on top of the water, evenly distributed around the surface of the water. When the dilute dishwashing liquid (a mixture that has several compounds in it) was dropped into the bowl, most of the compounds in the mixture started to sink into the water and mix with it. One component of the dishwashing liquid (sodium stearate), however, did not sink. Its chemical nature caused it to float on top of the water. While it floated on the water, it began to spread out in a circle, pushing the pepper out of its way. This is why you saw a clear circle form with the pepper at the edge of the circle. As long as there wasn't too much pepper on top of the water, the sodium stearate would continue to spread out until it formed a single, circular layer of molecules sitting on top of the water.
16. We can now use the mole concept and the size of the circle to determine the width of the molecules. After all, we can use the size of the circle to determine its area. Then we can use the mole concept to determine how many molecules were in the circle. That way, we can determine the area that each molecule must have occupied. We can then take that area and use it to determine the width of the molecule.
17. First, we need to know how many molecules were in the circle. We have all the information we need to calculate that. First, take the volume of one drop that you calculated when you calibrated your eyedropper. Multiply that volume by the density of your diluted dishwashing liquid, which is 1.00 grams/mL. This gives you the number of grams of solution you dropped into the water.
18. Of course, only a tiny fraction of the drop that you put in the water was actually sodium stearate, the chemical responsible for making the circle. As long as you did your dilution right, about 0.0125% of the solution was sodium stearate. Thus, take the number of grams you just calculated, and multiply it by 0.000125. This will tell you how many grams of sodium stearate you added to the bowl. Your answer should be somewhere between 2.5×10^{-6} grams and 7.5×10^{-6} grams.
19. The chemical formula for sodium stearate is $NaC_{18}H_{35}O_2$. Use this information to calculate the molecular mass of sodium stearate, just like we did in the first part of Example 5.5.
20. Use the molecular mass of sodium stearate to convert the grams of sodium stearate into moles of sodium stearate, just like we did in Example 5.6. Your answer should be somewhere between 8.0×10^{-9} moles and 2.9×10^{-8} moles.
21. To determine how many molecules were in the circle, take the number of moles you just calculated and multiply it by 6.02×10^{23}. After all, each mole contains 6.02×10^{23} molecules. If you

multilply the number of moles by 6.02×10^{23}, you get the number of molecules in the sample. Your answer should be somewhere between 4.8×10^{15} and 2.0×10^{16} molecules within your circle. If you don't exactly understand every step we took to get this number, don't worry. You just need to understand the concept: By knowing the mass and using the mole concept, we can calculate the number of molecules that were in the circle we observed.

22. Now we need to know the area of the circle. That's easy. Take the diameter of the circle (in centimeters) and divide it by 2. This gives you the radius of the circle. Now put that value in the equation for the area of a circle, Area = πr^2. You do this by squaring the radius you calculated and then multiplying by 3.14. Now you have the area of the circle in cm^2.
23. Take that area and divide it by the number of molecules you calculated above. The result gives you the area occupied by each individual molecule.
24. Now that we have the area occupied by each individual molecule, we can estimate the molecules' width. To do this, we will assume that the molecules are roughly squares. This is not really true, but it at least gives us an easy way to determine the width. The length of the sides of a square is equal to the square root of the area. Thus, take the square root of the area that you calculated in the previous step, and you have an estimation of the width of a molecule.
25. Look at your answer. It should be on the order of a few times 10^{-8} cm or a few times 10^{-7} cm. This is a very small number. The average human hair has a with of about 10^{-2} cm. Thus, the width of a molecule is 100,000 to 1,000,000 times smaller than the width of a single human hair!
26. Clean up your mess.

Once again, if you don't understand why we did each calculation, don't worry about it. What you should gain from this experiment is that the mole concept is a very powerful tool, allowing us to count atoms and determine things about them that cannot be determined by direct observation.

Now that you have seen some examples that deal with the mole concept and have also performed an experiment that uses the mole concept, make sure you have a firm grasp of it by performing the following problems:

ON YOUR OWN

5.7 How many moles of sulfur atoms are in a 50.0-gram sample of sulfur?

5.8 How many moles of glucose ($C_6H_{12}O_6$) are in a 1.25 kg sample?

It is important to note that chemists don't always convert from grams to moles. Often it is necessary to go the other way. Sometimes chemists have already figured out how many moles are in a sample, and they need to calculate the mass of the sample. This type of conversion uses the same concepts you have just learned, but it is useful for you to see an example of it, as well as try to solve a problem like it on your own.

EXAMPLE 5.7

A chemist determines that a sample of pure ammonia (NH_3) contains 12.331 moles of NH_3. What is the mass of the sample?

Notice that in this case, we already have the number of moles, but we are looking for the number of grams. In order to solve this, we once again have to get the conversion relationship between grams and moles. The molecular mass of NH_3 = 1 x 14.0 amu + 3 x 1.01 amu = 17.0 amu. This means:

$$17.0 \text{ g } NH_3 = 1 \text{ mole } NH_3$$

We can then use this relationship in our standard factor-label conversion:

$$\frac{12.331 \text{ \sout{moles NH}}_3}{1} \times \frac{17.0 \text{ g } NH_3}{1 \text{ \sout{mole NH}}_3} = 2.10 \times 10^2 \text{ g } NH_3$$

Thus, a sample that contains 12.331 moles of NH_3 molecules has a mass of <u>2.10×10^2 g</u>. Notice that this problem is really no different from the problems we did previously. We just used our conversion relationship in a different way. Notice also that since we only know the molecular mass to three significant figures, we can only report the answer to three significant figures. Scientific notation is the only way we can report this answer properly, as 210 has only two significant figures.

So you see that whether a chemist is converting from grams to moles or from moles to grams, the first step is to determine the mass of the molecule or atom involved. Then the chemist must use that mass as a conversion relationship. Make sure you understand this process by doing one more problem:

ON YOUR OWN

5.9 What is the mass of a NO_3 sample if it contains 0.23 moles?

Before you leave this section, make sure you understand the importance of what you have just learned. When you convert from grams to moles, you are taking a measurement that is very easy to obtain (mass) and converting it into something that actually tells you how many molecules or atoms are present in a sample of matter (moles). This is an enormously useful thing to be able to do. Thus, these conversion problems are not useless exercises that are designed to make sure you understand the factor-label method! Instead, these problems are designed to make sure you understand one of the most powerful tools in all of chemistry. We will use this tool over and over and over again, so you must become intimately familiar with it.

The mole is such an important concept in chemistry that some have actually tried to make an official holiday celebrating this event. It is called "National Mole Day," and starts at 6:02 a.m. on October 23 (10/23). If you think about Avogadro's number, you should see why National Mole Day is defined that way. You can learn more about this "holiday" at www.moleday.org.

Using the Mole Concept in Chemical Equations

Now that you have a reasonable grasp of the mole concept, you are ready to see why it is considered such a powerful tool. The real importance of the mole concept is revealed when you look at how it relates to chemical equations. Remember that a chemical equation is really a recipe. By looking at the coefficients which appear to the left of the substances in the equation, we can learn how much of each substance we need to make a reaction occur. For example, the following reaction:

$$CH_4\ (g)\ + 2O_2\ (g) \rightarrow CO_2\ (g)\ + 2H_2O\ (g) \qquad (5.8)$$

tells us that when *1 molecule of* CH_4 is reacted with *2 molecules of* O_2, the products will be *1 molecule of* CO_2 and *2 molecules of* H_2O. Thus, the coefficients in the chemical equation tell us the quantity of substance that we either use or make.

Now suppose I ran the reaction represented by equation (5.8) twice. I could say that *2 molecules of* CH_4 reacted with *4 molecules of* O_2, producing 2 *molecules of* CO_2 and *4 molecules of* H_2O. Since the reaction ran twice, I can simply double my list of ingredient and products. There is another way to say this, though. I could also say that *1 couple of* CH_4 *molecules* reacted with *2 couples of* O_2 *molecules*, producing *1 couple of* CO_2 *molecules* and *2 couples of* H_2O *molecules*. When I state what happened in this way, I am simply using the fact that the word "couple" means "2."

Let's suppose that instead of running this reaction twice, I run it a total of 12 times. Then I could say that *12 molecules of* CH_4 reacted with *24 molecules of* O_2, producing *12 molecules of* CO_2 and *24 molecules of* H_2O. Alternatively, I could say that *1 dozen* CH_4 *molecules* reacted with *2 dozen* O_2 *molecules*, producing *1 dozen* CO_2 *molecules* and *2 dozen* H_2O *molecules*. This way, I am using the word "dozen" to refer to groups of 12 molecules.

Finally, suppose I ran this reaction 6.02×10^{23} times. If I did this, I could say that *1 mole of* CH_4 *molecules* reacted with *2 moles of* O_2 *molecules*, producing *1 mole of* CO_2 *molecules* and *2 moles of* H_2O *molecules*. This is why the mole concept is so important:

In chemical equations, the coefficients which appear to the left of each substance tell you how many moles of that substance must be used in the chemical equation.

Think about this fact for a minute: As long as you can balance chemical equations, you can determine the recipe necessary for making that chemical reaction occur! That's a very powerful technique to understand.

Now please understand that the coefficients in a chemical equation *do not* give you the recipe in grams. Thus, I cannot use Equation (5.8) to say that 1 gram of CH_4 reacts with 2 grams of O_2, producing 1 gram of CO_2 and 2 grams of H_2O. That is just not the case. The coefficients in the chemical equation tell you how many *molecules* react and are produced. Since the word "mole" just represents a group of molecules, they also tell you how many *moles* react and are produced. Thus, a chemical equation is a wonderfully useful thing, but to use it properly, you can only use *moles*, not grams. If you stick with the concept of moles, however, you can do quite a lot. In fact, once you have a balanced chemical equation, you can convert from one substance in the equation to any other substance in that equation at will! Let's see how to do that:

EXAMPLE 5.8

The decomposition of hydrogen peroxide into water and oxygen is described by the following chemical equation:

$$2H_2O_2\ (l) \rightarrow 2H_2O\ (l) + O_2\ (g)$$

If a chemist begins with 2.4 moles of H_2O_2, how many moles of O_2 will be produced?

The chemical equation tells us that *2 moles of H_2O_2* will decompose into *1 mole of O_2*. We can think of this fact as a conversion relationship. For this particular chemical reaction,

$$2 \text{ moles } H_2O_2 = 1 \text{ mole } O_2$$

I have to stress that this relationship is only good for *this particular reaction*. Another reaction that has hydrogen peroxide and oxygen in it will, most likely, have a completely different relationship between these two substances. However, for this particular equation, the above relationship does indeed hold.

Thus, to find out how many moles of oxygen are formed, all I have to do is *convert* moles of hydrogen peroxide into moles of oxygen. The equation above represents a conversion relationship like any other conversion relationship. I can therefore use the factor-label method to convert between moles of H_2O_2 and moles of O_2:

$$\frac{2.4 \cancel{\text{moles } H_2O_2}}{1} \times \frac{1 \text{ mole } O_2}{2 \cancel{\text{moles } H_2O_2}} = 1.2 \text{ moles } O_2$$

Thus, when 2.4 moles of hydrogen peroxide decompose, they produce 1.2 moles of oxygen gas. The relationship that we develop from the chemical equation is considered to be exact. After all, it comes from the coefficients in the chemical equation, and those are integers. Therefore, both numbers in the conversion relationship have an infinite number of significant figures. As a result, the only number with error in it is the original amount of H_2O_2. Thus, its number of significant figures determines the number of significant figures we can report in the answer.

It may seem odd to you that we can convert from one substance to another in a chemical equation. Nevertheless, we can. After all, what does a chemical reaction do? It *converts* one or more substances into different substances. Thus, you can use the factor-label method in this type of conversion, just like you can in unit conversions.

You might also find it odd that I did not have to use any information about the water produced. Water is, indeed, produced in the reaction, but the amount of water produced does not really affect the amount of oxygen produced, and vice versa. Both of them are dependent on the amount of hydrogen peroxide that decomposes, but neither is dependent on the other. This is an important aspect of chemical equations. In general, you needn't know everything about the substances in a chemical equation in order to arrive at an answer to a problem. Usually, you can relate any one substance in a chemical equation to any other substance in that equation. There are some exceptions to this general rule (of course), and we will discuss those exceptions in detail in the next module.

Let's do one more problem just to make sure you really understand how to use the coefficients in a chemical equation:

If a chemist has 0.234 moles of C_2H_6O, how many moles of carbon will the chemist get by decomposing the sample?

We are not given the chemical equation in this problem, but we can figure it out. The problem says that the chemist is decomposing C_2H_6O. The chemical reaction, then, is:

$$C_2H_6O \rightarrow C + H_2 + O_2$$

Remember, both hydrogen and oxygen are homonuclear diatomics, but carbon is not. Thus, the chemical formula for carbon is C, but the chemical formula for hydrogen is H_2, and the chemical formula for oxygen is O_2. Balancing the equation gives us:

$$2C_2H_6O \rightarrow 4C + 6H_2 + O_2$$

This chemical equation tells us that 2 moles of C_2H_6O will decompose into 4 moles of C. Thus,

$$2 \text{ moles } C_2H_6O = 4 \text{ moles C}$$

Now that we have our conversion relationship for this particular reaction, we can simply convert:

$$\frac{0.234 \cancel{\text{moles } C_2H_6O}}{1} \times \frac{4 \text{ moles C}}{2 \cancel{\text{moles } C_2H_6O}} = 0.468 \text{ moles C}$$

Our chemical equation, then, tells us that when 0.234 moles of C_2H_6O decompose, they produce 0.468 moles of C. Once again, the numbers in the conversion relationship are exact, since they come from the chemical equation. Therefore, the number of significant figures in the original measurement determines the number of significant figures in the answer.

Hopefully you are beginning to appreciate the power behind the mole concept. It not only allows us to count the number of atoms or molecules that exist in a sample of matter, but it also allows us to predict the quantities of substances that are produced in a chemical reaction. See if you can perform the latter of these two techniques:

ON YOUR OWN

5.10 A chemist obtains a 122.3-mole sample of ammonia (NH_3). If the chemist allows this sample to decompose, how many moles of hydrogen gas will be produced?

The section you have just completed is an introduction into the most important aspect of chemistry: relating the quantities of chemical substances in a chemical reaction. This subject, commonly called **stoichiometry** (stoy' kee ahm' uh tree), will be the main subject of the next module. There isn't a chemist in the world who doesn't use stoichiometry all of the time, so, needless to say, you must learn this subject inside and out.

If you have a good handle on how to convert from grams to moles, how to convert from moles to grams, and how to convert from moles of one substance to moles of another substance by using a chemical equation, then you are well on your way to understanding stoichiometry. If, on the other hand, you are unsure of one of these three techniques, you need to review what you have read, look at the extra problems in Appendix B, or seek out alternative resources. You cannot really be successful in the next module (not to mention the rest of the course) unless you are comfortable with these three techniques.

Cartoon by Speartoons, Inc.

ANSWERS TO THE "ON YOUR OWN" PROBLEMS

5.1 Phosphorus trihydride is PH_3, which contains two elements: P and H. Hydrogen is a homonuclear diatomic, and phosphorus is not. The chemical formula for phosphorus, then, is just P, while the chemical formula for hydrogen is H_2.

$$PH_3 \rightarrow P + H_2$$

Now we just have to balance the equation. The P's are already balanced, but there are 3 H's on one side and only 2 on the other. To balance the H's without disturbing the P's:

$$PH_3 \rightarrow P + \tfrac{3}{2}H_2$$

Now the H's balance, but we must get rid of the fraction:

$$2 \text{ x } [PH_3] \rightarrow 2 \text{ x } [P + \tfrac{3}{2}H_2]$$

$$\underline{2PH_3 \rightarrow 2P + 3H_2}$$

5.2 This molecule has C's, H's, and O's in it. Hydrogen and oxygen are both homonuclear diatomics, but carbon is not.

$$C_{12}H_{23}O_{11} \rightarrow C + H_2 + O_2$$

To balance, we start by making 12 C's on the products side:

$$C_{12}H_{23}O_{11} \rightarrow 12C + H_2 + O_2$$

To balance the H's without disturbing the C's, we need to get 23 H's on the products side:

$$C_{12}H_{23}O_{11} \rightarrow 12C + \tfrac{23}{2}H_2 + O_2$$

To balance the O's without disturbing the C's or the H's, we need to get 11 O's on the products side:

$$C_{12}H_{23}O_{11} \rightarrow 12C + \tfrac{23}{2}H_2 + \tfrac{11}{2}O_2$$

Now we just need to get rid of the fractions:

$$2 \text{ x } [C_{12}H_{23}O_{11}] \rightarrow 2 \text{ x } [12C + \tfrac{23}{2}H_2 + \tfrac{11}{2}O_2]$$

$$\underline{2C_{12}H_{23}O_{11} \rightarrow 24C + 23H_2 + 11O_2}$$

5.3 The compound is composed of K, H, C, and O. H and O are homonuclear diatomics, but K and C are not.

$$K + H_2 + C + O_2 \rightarrow KHCO_3$$

The K's and C's are already balanced. To balance the H's and O's without disturbing the balance of the K's and C's, use fractions:

$$K + \tfrac{1}{2}H_2 + C + \tfrac{3}{2}O_2 \rightarrow KHCO_3$$

Eliminating the fractions:

$$2 \text{ x } [K + \tfrac{1}{2}H_2 + C + \tfrac{3}{2}O_2] \rightarrow 2 \text{ x } [KHCO_3]$$

$$\underline{2K + H_2 + 2C + 3O_2 \rightarrow 2KHCO_3}$$

5.4 A complete combustion reaction involves adding oxygen while producing carbon dioxide and water:

$$C_8H_{18} + O_2 \rightarrow CO_2 + H_2O$$

We can also add phase symbols here because we were told that the gasoline was evaporated; thus, we know it is a gas. We also know that oxygen and carbon dioxide are both gases. Finally, combustion is usually hot enough to turn water into a gas, therefore:

$$C_8H_{18}\ (g) + O_2\ (g) \rightarrow CO_2\ (g) + H_2O\ (g)$$

Now all we have to do is balance the equation. Balancing the C's and H's is easy:

$$C_8H_{18}\ (g) + O_2\ (g) \rightarrow 8CO_2\ (g) + 9H_2O\ (g)$$

To balance the O's without disturbing the balance of the C's and H's, we use a fraction:

$$C_8H_{18}\ (g) + \tfrac{25}{2}O_2\ (g) \rightarrow 8CO_2\ (g) + 9H_2O\ (g)$$

To continue, we need to eliminate the fraction:

$$2 \text{ x } [C_8H_{18}\ (g) + \tfrac{25}{2}O_2\ (g)] \rightarrow 2 \text{ x } [8CO_2\ (g) + 9H_2O\ (g)]$$

$$\underline{2C_8H_{18}\ (g) + 25O_2\ (g) \rightarrow 16CO_2\ (g) + 18H_2O\ (g)}$$

5.5 First, we look at the chart and see that calcium atoms have a mass of 40.1 amu. All we need to do now is convert. This conversion is a little more difficult than the one in Example 5.4 because we are asked to make a conversion to kg. The problem is, we have a relationship between amu and g, not amu and kg. Thus, we need to make two conversions. The first will convert amu to g, and the second will convert g to kg. We will do it all on one line as explained in Module #1:

$$\frac{40.1 \cancel{\text{amu}}}{1} \times \frac{1.66 \times 10^{-24} \cancel{\text{g}}}{1.00 \cancel{\text{amu}}} \times \frac{1 \text{ kg}}{1{,}000 \cancel{\text{g}}} = 6.66 \times 10^{-26} \text{ kg}$$

So the mass of a calcium atom is 40.1 amu, or $\underline{6.66 \times 10^{-26} \text{ kg}}$.

5.6 This molecule has 2 Na's (23.0 amu), 1 S (32.1 amu), and 4 O's (16.0 amu). The molecular mass, then, is:

$$\text{Mass of } Na_2SO_4 = 2 \text{ x } 23.0 \text{ amu} + 32.1 \text{ amu} + 4 \text{ x } 16.0 \text{ amu} = 142.1 \text{ amu}$$

Converting to grams:

$$\frac{142.1 \cancel{\text{amu}}}{1} \times \frac{1.66 \times 10^{-24} \text{ g}}{1.00 \cancel{\text{amu}}} = 2.36 \times 10^{-22} \text{ g}$$

The mass is <u>$2.36 \text{ x } 10^{-22}$ g</u>. Note that the conversion relationship between grams and amu limited the number of significant figures we can have. As discussed in the module, the conversion relationship has only three significant figures, so we must count those when using it.

5.7 The periodic chart tells us that one sulfur atom has a mass of 32.1 amu. Therefore:

$$32.1 \text{ grams S} = 1 \text{ mole S}$$

This is the conversion relationship we need for converting grams of sulfur into moles of sulfur:

$$\frac{50.0 \cancel{\text{g S}}}{1} \times \frac{1 \text{ mole S}}{32.1 \cancel{\text{g S}}} = 1.56 \text{ moles S}$$

There are <u>1.56 moles</u> in the sample.

5.8 To do these conversions, we need to know the mass of the molecule:

$$\text{Mass of } C_6H_{12}O_6 = 6 \text{ x } 12.0 \text{ amu} + 12 \text{ x } 1.01 \text{ amu} + 6 \text{ x } 16.0 \text{ amu} = 180.1 \text{ amu}$$

Remember, we are adding here. When adding (or subtracting) you do not count significant figures. Instead, you report your answer to the same decimal place as the least precise number in the problem. Most of the masses have their last significant figure in the tenths place. The mass of hydrogen has its last significant figure in the hundredths place. Thus, the answer must be reported to the tenths place. This means:

$$180.1 \text{ g } C_6H_{12}O_6 = 1 \text{ mole } C_6H_{12}O_6$$

This is the conversion relationship we need for converting grams of glucose into moles of glucose. Before we can use it, though, we need to convert kg of glucose into g of glucose. By now, you should be able to do that on your own. The conversion yields $1.25 \text{ x } 10^3$ g glucose. Now we can do the grams-to-moles conversion:

$$\frac{1.25 \times 10^3 \cancel{\text{g } C_6H_{12}O_6}}{1} \times \frac{1 \text{ mole } C_6H_{12}O_6}{180.1 \cancel{\text{g } C_6H_{12}O_6}} = 6.94 \text{ moles } C_6H_{12}O_6$$

There are <u>6.94 moles</u> in the sample.

5.9 This is just another conversion problem, but this time we are converting moles into grams. We still need to determine the molecular mass of NO_3 first, however:

$$\text{Mass of } NO_3 = 1 \times 14.0 \text{ amu} + 3 \times 16.0 \text{ amu} = 62.0 \text{ amu}$$

This means:

$$62.0 \text{ grams } NO_3 = 1 \text{ mole } NO_3$$

Now we can do our conversion:

$$\frac{0.23 \text{ moles } NO_3}{1} \times \frac{62.0 \text{ g } NO_3}{1 \text{ mole } NO_3} = 14 \text{ g } NO_3$$

The mass is 14 g.

5.10 To solve this problem, we must first find the balanced chemical equation for this reaction. According to the problem, the reaction is the decomposition of NH_3:

$$NH_3 \rightarrow N_2 + H_2$$

Now we have to balance it:

$$2NH_3 \rightarrow N_2 + 3H_2$$

Remember what this tells you. This tells you the *recipe* for the reaction *in moles*. Thus, 2 moles of NH_3 decompose into 1 mole of N_2 and 3 moles of H_2. The relationship between NH_3 and H_2, then, is given by:

$$2 \text{ moles } NH_3 = 3 \text{ moles } H_2$$

This is a conversion relationship that allows us to convert from moles of one substance into moles of another:

$$\frac{122.3 \text{ moles } NH_3}{1} \times \frac{3 \text{ moles } H_2}{2 \text{ moles } NH_3} = 183.5 \text{ moles } H_2$$

The answer is 183.5 moles.

REVIEW QUESTIONS FOR MODULE #5

1. Can an element undergo a decomposition reaction?

2. Classify the following reactions as decomposition, formation, complete combustion, or none of these:

a. $C_6H_{12}O_6$ (s) + $6O_2$ (g) → $6CO_2$ (g) + $6H_2O$ (g)

b. $2H_2$ (g) + O_2 (g) → $2H_2O$ (l)

c. NaCl (s) + $AgNO_3$ (aq) → AgCl (s) + $NaNO_3$ (aq)

d. 2HBr (l) + $Ca(OH)_2$ (s) → $2H_2O$ (l) + $CaBr_2$ (aq)

e. $2NaHPO_4$ (s) → 2Na (s) + H_2 (g) + 2P (s) + $4O_2$ (g)

3. Which of the following is *not* a formation reaction?

a. H_2CO_3 + 2KOH → $2H_2O$ + K_2CO_3

b. CO_2 (g) + H_2O (g) → H_2CO_3 (l)

c. 2S (s) + Cl_2 (g) + $5F_2$ (g) → $2SF_5Cl$

4. Which of the following is *not* a complete combustion equation?

a. $2C_3H_8O$ + $9O_2$ → $6CO_2$ + $8H_2O$

b. H_2CO_3 (s) → CO_2 (g) + H_2O (g)

c. CH_4 (g) + $2O_2$ (g) → CO_2 (g) + $2H_2O$ (g)

5. What is the difference between complete combustion and incomplete combustion?

6. What does a catalytic converter do for an automobile?

7. Which has more mass: 100 hydrogen atoms, 4 sulfur atoms, or 1 lanthanum (La) atom?

8. What do the following things have in common?

32.1 g of sulfur, 40.1 g of calcium, and 60.1 g of SiO_2

9. What is Avogadro's Number and what does it represent?

10. Why must a chemical reaction be balanced before it can be used in stoichiometry?

PRACTICE PROBLEMS FOR MODULE #5

(1.00 amu = 1.66×10^{-24} g)

1. Write a balanced chemical equation for the decomposition of $RbNO_3$.

2. Write a balanced chemical equation for the formation of $NaHSO_4$.

3. Write a balanced chemical equation for the complete combustion of propionaldehyde (C_3H_6O).

4. What is the mass of a tin atom (Sn) in grams?

5. What is the mass of a K_2CrO_4 molecule in kg?

6. How many moles of Cr atoms are in a 200.0-gram sample of Cr?

7. How many moles of $NaHCO_3$ are in a 125-gram sample of the compound?

8. What is the mass of a $CuCl_2$ sample if it contains 0.172 moles of the compound?

9. How many grams are in 15.0 moles of Ce?

10. In the decomposition of 1.2 moles of dinitrogen pentaoxide, how many moles of oxygen would be formed?

MODULE #6: Stoichiometry

Introduction

In the previous module, you learned how to count atoms and molecules using the mole concept. You were then given a brief introduction to the relationship between the mole concept and the mathematics of chemical equations. You found out that, using the mole concept, you can actually calculate the amount of product formed in a chemical reaction if you are given the number of moles of reactant material.

In this module, we will expand on the mathematical relationships that exist within chemical equations. We will see that, using the mole concept, you can relate the quantity of any substance in a chemical equation to the quantity of any other substance in that same equation. We will also see that, given the proper conversion steps, you can develop relationships between the masses of those substances as well. As you learned in the previous module, relating the quantities of different substances in a chemical equation is called **stoichiometry**, and it is probably the single most important concept in the entire course. Please read this module carefully and thoroughly, because we will constantly go back to the concepts presented here. You must be firmly grounded in this subject matter to be able to proceed in the course.

Mole Relationships in Chemical Equations

In the previous module, you did a few problems in which you related the number of moles of a reactant to the number of moles of products formed in a chemical reaction. That was your first step in the process of understanding the mole relationships between substances in a chemical equation. Now you are ready for the next step. As I said in the previous module, chemical equations are really the same as mathematical equations. Anything you can do to one, you can do to the other. We will use this fact to help us take the next step.

In the stoichiometry problems you performed previously, you were always given a decomposition reaction and the amount of reactant, and you were always asked to determine the amount of product that was formed. This may have left you with the impression that stoichiometry must always relate the amount of reactant to the amount of product. This simply isn't true. Like mathematical equations, chemical equations can be used to relate the left side of the equation to the right side of the equation, *and vice versa.*

For example, consider the mathematical relationship below:

$$y + 3 = x \tag{6.1}$$

If I told you that y = 4, then you could determine the value for "x," right? You would simply replace "y" with "4" in Equation (6.1), and you would end up determining that x = 7. In this case, you related the terms on the left side of the equation to the term on the right side of the equation.

Suppose, however, that I gave you the value for "x" instead of "y." Suppose I told you that x = 7. Could you solve for "y"? Of course you could. You would simply replace "x" with "7" in Equation (6.1), and you would use your algebra skills to determine that y = 4. In this case, you related the term on the right side of the equation to those on the left side of the equation.

Chemical equations behave in exactly the same way. I do not always have to give you the amount of reactants and have you calculate the amount of products formed. Instead, I could tell you how much product was formed, and you could use that information to determine the quantities of the reactants involved. See if you can follow this reasoning through the following example:

EXAMPLE 6.1

A chemist forms ammonium sulfate, $(NH_4)_2SO_4$, an excellent fertilizer, by combining ammonia (NH_3) and sulfuric acid (H_2SO_4). If the chemist ends up making 12.2 moles of ammonium sulfate, how many moles of ammonia were used?

To be able to do any stoichiometry problem, we must first figure out the chemical equation. According to the problem, ammonium sulfate is the product, while sulfuric acid and ammonia are the reactants:

$$NH_3 + H_2SO_4 \rightarrow (NH_4)_2SO_4$$

Before we can balance the equation, I need to explain the chemical formula for ammonium sulfate (the product). Notice that it has parentheses in it. What does that mean? You will learn the *physical* meaning in Module #9. For right now, however, you need to know the *practical* meaning. The "2" that you see outside the parentheses applies to *every atom within the parentheses*. Thus, we must take that "2" and multiply it with the numbers inside the parentheses. That's the only way we can find out how many N's and H's are in the molecule. When I multiply the "2" with the numbers inside the parentheses, I get $N_2H_8SO_4$. Thus, in this molecule, there are really two N's and eight H's. You will need to get used to this, because you will see it again. Whenever you see parentheses in a molecule, you must distribute the subscript after the parentheses to every atom within the parentheses. That's the only way you will know how many of each atom is in the molecule.

Okay, now that we have that out of the way, we can go back to work. We first need to balance the equation. Remember, there are two N's and eight H's in the product. Thus, the balanced chemical equation is:

$$2NH_3 + H_2SO_4 \rightarrow (NH_4)_2SO_4$$

The problem gives us the number of moles of ammonium sulfate and asks us how many moles of ammonia were used in the reaction. Well, according to the chemical equation, the relationship between ammonia and ammonium sulfate is:

$$2 \text{ moles } NH_3 = 1 \text{ mole } (NH_4)_2SO_4$$

Using this relationship, we can convert from moles of ammonium sulfate to moles of ammonia:

$$\frac{12.2 \cancel{\text{moles } (NH_4)_2SO_4}}{1} \times \frac{2 \text{ moles } NH_3}{1 \cancel{\text{mole } (NH_4)_2SO_4}} = 24.4 \text{ moles } NH_3$$

Thus, the chemist must have started with <u>24.4 moles of NH_3</u>.

The first thing you should notice about this example is that there is nothing new here, except for the parentheses in the chemical formula of ammonium sulfate. We have used exactly the same kind of reasoning that we used in the previous module--we just went from products to reactants instead of reactants to products. Thus, the relationships between substances in a chemical equation are *independent* of which side of the equation you begin with.

The problem asked how much NH_3 was used to make the ammonium sulfate, but we could also figure out how much sulfuric acid was used. After all, the chemical equation relates moles of sulfuric acid to moles of ammonium sulfate:

$$1 \text{ mole } H_2SO_4 = 1 \text{ mole } (NH_4)_2SO_4$$

We can use that relationship to convert from moles of ammonium sulfate to moles of sulfuric acid:

$$\frac{12.2 \cancel{\text{moles } (NH_4)_2SO_4}}{1} \times \frac{1 \text{ mole } H_2SO_4}{1 \cancel{\text{mole } (NH_4)_2SO_4}} = 12.2 \text{ moles } H_2SO_4$$

Thus, we learn from the chemical equation that the chemist had to start with 24.4 moles of ammonia and 12.2 moles of sulfuric acid in order to make 12.2 moles of ammonium sulfate.

So you see that the quantity of just one substance in the equation has the potential to give you the quantities of all other substances in the chemical equation. This gives you some idea of how powerful the tool of stoichiometry can be. Check your stoichiometry skills with the following problem:

ON YOUR OWN

6.1 A chemist performs the following reaction to make sodium nitrate:

$$2NaNO_2\,(s) + O_2\,(g) \rightarrow 2NaNO_3\,(s)$$

How many moles of oxygen gas must the chemist use in order to make 15.0 moles of $NaNO_3$? How many moles of $NaNO_2$ are necessary?

Limiting Reactants and Excess Components

Now if all of this seems rather easy to you, don't worry: This subject gets a lot harder. For example, so far we have worked only with decomposition or formation reactions. It turns out that those kinds of reactions are a little easier to deal with, because for both of these reaction types, one side of the equation has only one substance. In decomposition reactions, there is only one reactant, whereas in formation reactions, there is only one product. Stoichiometry is a bit more difficult when multiple substances appear on both sides of the equation. Perform the following experiment to see why:

EXPERIMENT 6.1
Limiting Reactants

Supplies:

- Baking soda
- Vinegar
- String or tape measure
- Graduated cylinder (or measuring cups and spoons)
- Ruler
- Round balloon
- Plastic 2-liter bottle (or other large bottle)
- Mass scale
- Funnel or butter knife (see step #2)
- ~~Safety goggles~~

1. Quantitatively measure 250.0 mL (1¼ cups) of vinegar into the clean 2-liter plastic bottle.
2. Fill the balloon up with 10 grams of baking soda. You can do this with a funnel, or if you have a hard time doing it that way, use a butter knife. Use the butter knife to pick up some baking soda, and then carefully push the knife into the balloon. Once the knife is in the balloon, tip them both so that the baking soda on the knife falls into the balloon. It is best to do this over the sink because you will spill a significant amount of baking soda. As you fill the balloon, keep measuring its mass until you have 10.0 grams of baking soda.
3. Once you have 10.0 grams of baking soda in the balloon, attach the balloon to the bottle as shown below, making sure that the lip of the balloon covers the bottle opening completely so that gas cannot escape.

4. Now hold the balloon upright so that the baking soda falls into the vinegar. You will see bubbles form, and the balloon will inflate. If you hear gas escaping from the balloon, then you must redo the experiment, being more careful to fasten the lip of the balloon tightly to the opening of the bottle. In order for this experiment to work, no gas can escape from the experimental setup!
5. Carefully pick up the bottle and swirl the vinegar around so that it reacts with all of the baking soda. Also, gently shake the balloon from side to side to make sure that all of the baking soda has fallen into the vinegar. After a few minutes, the bubbles should die down.
6. Once the bubbles have died down, take the string and wrap it once around the biggest part of the balloon. Use your fingers to mark exactly how much of the string was needed to go around the balloon once, then measure the length of that part of the string. This will give you a rough value

for the circumference of the balloon. If you have a tape measure, use it to measure the circumference directly.

7. Now repeat this experiment twice. The first time you repeat it, use 30.0 grams of baking soda rather than 10.0 grams, and the second time you repeat it, use 60.0 grams of baking soda. Each time, however, you should use 250.0 mL of vinegar. *Do not re-use the old vinegar!* Clean the bottle out each time you have finished and use 250.0 mL of fresh vinegar each time.
8. Once you have all three circumference measurements, clean up your mess.

What does this experiment illustrate? In each trial you ran, the chemical reaction that occurred was:

$$HC_2H_3O_2\ (l) + NaHCO_3\ (s) \rightarrow H_2O\ (l) + CO_2\ (g) + NaC_2H_3O_2\ (aq) \qquad (6.2)$$

acetic acid in vinegar — baking soda — sodium acetate

The carbon dioxide that was formed in the reaction is the gas that filled up the balloon. According to the chemical equation, each mole of baking soda that you add to the acetic acid should produce another mole of gas. This should result in the balloon inflating more. Thus, you might expect that in each trial you ran, the balloon would get bigger. If you look at your measurements for the circumference of the balloon, however, that's not what you should see. The circumference of the balloon in the second trial should have been significantly greater than the circumference of the balloon in the first trial. However, in the third trial, despite the fact that you added twice as much baking soda as you did in the second trial, the circumference of the balloon should not be significantly larger than it was in the second trial. Why?

In the first trial, there were more moles of acetic acid in the bottle than the moles of baking soda that you added. Thus, when the reaction was over, there was leftover acetic acid waiting to react with more baking soda. However, since there was no more baking soda, the reaction stopped, and the balloon stopped filling up. In the second trial, you added slightly more moles of baking soda compared to the moles of acetic acid. Thus, the acetic acid ran out first that time. When the reaction was over, then, there was no leftover acetic acid, but there was a little bit of leftover baking soda. When you went to the third trial, adding all of that extra baking soda did not increase the amount of gas produced at all. After all, the acetic acid ran out after reacting with the first 30.0 grams of baking soda. Thus, all of that extra baking soda had nothing to react with, so it produced nothing extra.

This demonstrates the concept of **limiting reactants**. When there is more than one reactant in a chemical equation, one of those reactants will usually run out first, stopping the entire reaction. We call that reactant the *limiting* reactant, because once it runs out, the reaction cannot continue. Since the reaction cannot continue, no more products will form. Thus, the limiting reactant *limits* the amount of product that can be formed in the chemical reaction.

Looking back over this experiment, you can see that in the first trial, baking soda was the limiting reactant. In that trial, the baking soda ran out first, stopping the formation of any more product. Thus, when more baking soda was added in the second trial, the balloon got larger because there was still acetic acid to react with, and therefore more product was made. In the final trial, however, the acetic acid was the limiting reactant, stopping the production of carbon dioxide and leaving plenty of leftover baking soda. So even though there was a lot of baking soda in the reaction, the balloon could not fill with more product because the reaction ran out of acetic acid.

As we begin to do more complex stoichiometry problems, it will be necessary for me to tell you which reactant is the limiting reactant. Once you know that, you can use its quantity to predict the quantity of products formed. Thus, in more complex chemical equations, the relationships between quantities of substances depends only on the quantity of the *limiting reactant*. As a result, you can essentially ignore the quantities of the other reactants, because the limiting reactant determines how much product can be formed in the reaction.

Fully Analyzing Chemical Equations

Now that you understand the limiting-reactant concept, we can discuss how to fully analyze a chemical equation. Consider the following equation:

$$Mg\ (s) + 2HCl\ (aq) \rightarrow MgCl_2\ (aq) + H_2\ (g) \qquad (6.3)$$

In this equation, there are two reactants and two products. These reactants and products all relate to one another. For example, if I told you the number of moles *of any one substance* in the equation, you could use it to calculate the moles *of any other substance* in the equation. This may sound a little strange to you because you are used to algebra, where a single equation can only provide you with the value of a single unknown. In chemistry, however, a single equation can be used to determine many unknowns.

For example, suppose I told you that the above reaction was started with 1 mole of Mg. According to Equation (6.3), 1 mole of Mg reacts with 2 moles of HCl. Thus, one thing you can determine from the reaction is that in order for all of the Mg to be used up, 2 moles of HCl must be added. After all, if less than two moles of HCl were added, HCl would be the limiting reactant and there would be leftover Mg. Alternatively, if more than 2 moles of HCl were added, all of the Mg would be used up and there would be leftover HCl. That would make Mg the limiting reactant.

In the end, then, one thing that you can determine from a chemical equation is the amount of other reactant(s) necessary to completely use up a given quantity of the given reactant. When reactants are added in this way, so that all reactants are used up, we say that the reactants "reacted completely." This means that all reactants were added in just the right quantities necessary to be completely used up. Thus, we would say that 2 moles of HCl are needed in this reaction in order for one mole of Mg to react completely.

Of course, we can also use the quantity of Mg to determine how much product is formed. As long as we know that the Mg reacts completely, we know that it is the limiting reactant. We can then use its number of moles to determine the number of moles of products that form. According to Equation (6.3), one mole of Mg will make one mole of $MgCl_2$. Thus, we know that the reaction we have been discussing will make one mole of $MgCl_2$. In addition, however, we can also say that since Equation (6.3) tells us that one mole of Mg also produces one mole of H_2, we know that one mole of H_2 is produced in the reaction as well. Let's see how this pans out in an example problem:

EXAMPLE 6.2

The antacid tablet Tums®, with an active ingredient of $CaCO_3$, neutralizes excess HCl in your stomach according to the following reaction:

$$CaCO_3\,(s) + 2HCl\,(aq) \rightarrow CaCl_2\,(aq) + H_2O\,(l) + CO_2\,(g)$$

If 0.0075 moles of $CaCO_3$ are in a Tums® tablet, how many moles of HCl can it neutralize? When it neutralizes that amount of stomach acid, how much $CaCl_2$, H_2O, and CO_2 are produced?

In this problem, we are given the quantity of one of the reactants and are asked how much of the other reactant it can react with. This is just like asking how many moles of HCl are necessary to use up all of the $CaCO_3$. Thus, we first look at the chemical equation and develop a relationship between HCl and $CaCO_3$:

$$1\text{ mole } CaCO_3 = 2\text{ moles HCl}$$

Now we can simply use this relationship to convert from moles of $CaCO_3$ to moles of HCl:

$$\frac{0.0075\ \cancel{\text{moles } CaCO_3}}{1} \times \frac{2\text{ moles HCl}}{1\ \cancel{\text{mole } CaCO_3}} = 0.015\text{ moles HCl}$$

This tells us, then, that a Tums® tablet neutralizes <u>0.015 moles of HCl</u>. We can learn a lot more from this equation, though. Since we have determined that $CaCO_3$ will be completely used up, we know that it is the limiting reactant. The quantity of limiting reactant is also related to the quantity of all products. According to the equation:

$$1\text{ mole } CaCO_3 = 1\text{ mole } CaCl_2$$
$$1\text{ mole } CaCO_3 = 1\text{ mole } CO_2$$
$$1\text{ mole } CaCO_3 = 1\text{ mole } H_2O$$

We can use all of these relationships to convert from the given number of moles of $CaCO_3$ into moles of each product:

$$\frac{0.0075\ \cancel{\text{moles } CaCO_3}}{1} \times \frac{1\text{ mole } CaCl_2}{1\ \cancel{\text{mole } CaCO_3}} = \underline{0.0075\text{ moles } CaCl_2}$$

$$\frac{0.0075\ \cancel{\text{moles } CaCO_3}}{1} \times \frac{1\text{ mole } CO_2}{1\ \cancel{\text{mole } CaCO_3}} = \underline{0.0075\text{ moles } CO_2}$$

$$\frac{0.0075\ \cancel{\text{moles } CaCO_3}}{1} \times \frac{1\text{ mole } H_2O}{1\ \cancel{\text{mole } CaCO_3}} = \underline{0.0075\text{ moles } H_2O}$$

With just the quantity of one reactant given, then, we can learn something about all of the other substances in the chemical equation.

ON YOUR OWN

6.2 The first step in a long process that converts iron ore into pure iron involves converting iron ore into iron oxide. This step is accomplished through the following chemical reaction:

$$4FeS_2\ (s) + 11O_2\ (g) \rightarrow 2Fe_2O_3\ (s) + 8SO_2\ (g)$$

iron ore iron oxide

If an iron manufacturer starts with 1,256.0 moles of iron ore, how many moles of oxygen will be needed to react completely with the iron ore? How many moles of iron oxide will be made?

Relating Products to Reactants in Chemical Equations

There are, of course, many other ways to analyze a chemical equation. One way chemists often use stoichiometry is to determine the quantity of a reactant necessary to make a certain amount of product. For example, suppose you were told to go into a laboratory and make 2 moles of $FeCl_3$ (a chemical used in electronics manufacturing) using the following reaction:

$$2FeTiO_3\ (s) + 4HCl\ (l) + Cl_2\ (g) \rightarrow 2FeCl_3\ (aq) + 2TiO_2\ (s) + 2H_2O\ (l) \quad (6.4)$$

How many moles of each reactant would you need to start with? If you knew you had to end up with 2 moles of $FeCl_3$, then you should be able to use the chemical equation to determine how much starting material you would need.

After all, the equation, as it is written, tells us that 2 moles of $FeTiO_3$ react with 4 moles of HCl and 1 mole of Cl_2 to make 2 moles of $FeCl_3$, 2 moles of TiO_2, and 2 moles of water. Thus, you know the recipe for making 2 moles of $FeCl_3$. Therefore, you can conclude that the minimum amount of each reactant needed is 2 moles of $FeTiO_3$, 4 moles of HCl, and 1 mole of Cl_2. If only 1 mole of HCl were used, then the HCl would run out long before the other reactants, limiting the amount of product you would make. If, on the other hand, 5 moles of HCl were used, it would not affect the amount of product made because the other two reactants would run out, leaving excess HCl.

Thus, another way we can use chemical equations is to relate the amount of product we desire to the minimum quantities of reactants we would need. Let's see how this applies to a more complicated problem:

EXAMPLE 6.3

A chemist wants to make ether ($C_4H_{10}O$) for a medical anesthetic. The chemist uses the following reaction:

$$2C_2H_6O\ (l) + H_2SO_4\ (l) \rightarrow C_4H_{10}O\ (g) + H_4SO_5\ (aq)$$

If the chemist needs 134.9 moles of ether, what is the minimum number of moles of C_2H_6O and H_2SO_4 that he needs?

In this problem, we are simply trying to relate the quantity of a product to the minimum quantities of each reactant needed. Thus, as usual, all we have to do is develop relationships between the quantity we know and the quantities we want to know. The chemical equation tells us that

$$2 \text{ moles } C_2H_6O = 1 \text{ mole } C_4H_{10}O$$

We can use this relationship, then, to convert between moles of ether desired and moles of C_2H_6O required:

$$\frac{134.9 \cancel{\text{moles } C_4H_{10}O}}{1} \times \frac{2 \text{ moles } C_2H_6O}{1 \cancel{\text{mole } C_4H_{10}O}} = 269.8 \text{ moles } C_2H_6O$$

In order to determine how much H_2SO_4 is needed, we use the relationship between ether and H_2SO_4 as given in the equation:

$$1 \text{ mole } H_2SO_4 = 1 \text{ mole } C_4H_{10}O$$

Therefore:

$$\frac{134.9 \cancel{\text{moles } C_4H_{10}O}}{1} \times \frac{1 \text{ mole } H_2SO_4}{1 \cancel{\text{mole } C_4H_{10}O}} = 134.9 \text{ moles } H_2SO_4$$

Thus, in order to make 134.9 moles of ether, the chemist must have a minimum of <u>269.8 moles of C_2H_6O and 134.9 moles of H_2SO_4</u>. Note that we could also use the chemical equation to determine how many moles of H_4SO_5 would be produced. The problem did not ask us to determine this, however, so we won't.

See if you can do a similar problem:

ON YOUR OWN

6.3 Citric acid ($C_6H_8O_7$), a component of fruit drinks, jams, and jellies, is produced using the following reaction:

$$C_{12}H_{22}O_{11} + H_2O + 3O_2 \rightarrow 2C_6H_8O_7 + 4H_2O$$

If a fruit drink manufacturer decides she needs 1.4×10^5 moles of citric acid, what is the minimum amount of reactants she will need?

<u>Using Chemical Equations When the Limiting Reactant Is Identified</u>

One other way that chemists use stoichiometry is to determine the amount of products made when the limiting reactant is identified. For example, if a chemist wants to make silver chloride (a substance used in photography) he might use the following reaction:

$$Ag_2CO_3\ (s)\ +\ 2HCl\ (aq)\ \rightarrow\ 2AgCl\ (s)\ +\ H_2O\ (l)\ +\ CO_2\ (g) \qquad (6.5)$$

If the chemist were to buy the two reactants in the equation (HCl and Ag_2CO_3), he would notice a huge difference in price. While the HCl is relatively cheap, the Ag_2CO_3 is incredibly expensive. Thus, to make sure that he doesn't waste any money, the chemist might want to use a lot of extra HCl to be confident that all of the Ag_2CO_3 reacted fully. After all, the chemist wouldn't be too worried if there were some leftover HCl to throw away, since it is relatively cheap. However, if the chemist didn't use enough HCl and as a result had leftover Ag_2CO_3 to throw away, that would be a big waste of money.

When a chemist adds lots of extra reactant in order to make sure that the other reactant is the limiting reactant, we say that the chemist is adding **excess** reactant. Thus, in our example above, the chemist would make sure that Ag_2CO_3 is the limiting reactant by adding excess HCl. Since excess HCl is used, all of the Ag_2CO_3 is certain to react, making it the limiting reactant. Once you know that Ag_2CO_3 is the limiting reactant, you can use its number of moles to calculate the number of moles of each product.

Thus, in the above reaction, if the chemist bought one mole of Ag_2CO_3, he would want to buy more than two moles of HCl. Since the equation says that two moles of HCl are required to react with one mole of Ag_2CO_3, using more than two moles (2.5 moles, for example) would ensure that there was excess HCl, making Ag_2CO_3 the limiting reactant. Since the equation says that one mole of Ag_2CO_3 makes two moles of AgCl, then the chemist will make two moles of AgCl when one mole of Ag_2CO_3 is reacted with excess HCl. If you are a little confused here, follow the next example closely. It should help clear things up.

EXAMPLE 6.4

Some ores contain a small amount of precious metals that are in compounds with many other elements. For example, one common ore found on earth is $KAgC_2N_2$. Since silver (Ag) is a precious metal, it is useful to be able to extract silver from this ore. The following reaction accomplishes this task:

$$\mathbf{2KAgC_2N_2\ (aq)\ +\ Zn\ (s)\ \rightarrow\ 2Ag\ (s)\ +\ ZnC_2N_2\ (aq)\ +\ 2KCN}$$

If a chemist adds 5.61 moles of $KAgC_2N_2$ to excess Zn, how much pure silver (Ag) can be made?

Since excess Zn is added, $KAgC_2N_2$ is the limiting reactant. Since we already learned that the limiting reactant determines the amount of all products formed, all we need to do is to develop a relationship between $KAgC_2N_2$ and Ag and then use that relationship to convert from one to the other.

According to the chemical equation:

$$2\text{ moles } KAgC_2N_2 = 2\text{ moles Ag}$$

Using this relationship to convert:

$$\frac{5.61\ \cancel{\text{moles } KAgC_2N_2}}{1} \times \frac{2\text{ moles Ag}}{2\ \cancel{\text{moles } KAgC_2N_2}} = 5.61\text{ moles Ag}$$

Thus, 5.61 moles of Ag would be made. Realize that we could do the same type of conversion to determine how much KCN and ZnC_2N_2 were produced, because the number of moles of the limiting reactant can be converted into the number of moles of all products.

Now try this problem:

ON YOUR OWN

6.4 Calcium chloride 6-hydrate ($CaCl_2H_{12}O_6$) is a substance used to melt snow on streets and sidewalks. It is made from calcium carbonate ($CaCO_3$) using the following reaction:

$$CaCO_3\ (s)\ +\ 2HCl\ (aq)\ +\ 5H_2O\ (l)\ \rightarrow\ CaCl_2H_{12}O_6\ (s)\ +\ CO_2\ (g)$$

If 3.49 moles of HCl are added to excess $CaCO_3$ and excess water, how many moles of calcium chloride 6-hydrate will be made?

I hope you are beginning to see that there is really very little difference between all of the problems we have discussed so far. In each case, we are using the relationship given by the chemical equation to convert the number of moles of one substance into the number of moles of another substance. The only difference between these problems is what we are trying to learn by doing this. Sometimes we are trying to determine how much reactant is needed to react fully with another reactant, sometimes we are trying to see how much product would be formed, and sometimes we are trying to determine how much starting material we would need in order to end up with a certain amount of product. In each case, however, we used the same mathematical tools to determine the answer.

This is the way stoichiometry works. The numbers that appear to the left of the substances in a chemical equation are used to develop a conversion relationship between the substance with a quantity we know and the substance with a quantity we do not know. If done properly, the number of moles of any one substance in the equation can be converted into the number of moles of any other substance in the equation.

Since the numbers that appear to the left of each substance in the equation are so important in stoichiometry, we give them a name. We call these numbers **stoichiometric** (stoy' kee uh meh' trik) **coefficients**, and they provide a way of converting from the number of moles of one substance in the equation to the number of moles of any other substance in the equation.

If you purchased the MicroChem kit discussed in the introduction, you can perform Experiment #4 to get some more experience with stoichiometry and limiting reactants.

Volume Relationships for Gases in Chemical Equations

Now it turns out that the stoichiometric coefficients in a chemical equation do not relate just the number of moles of each substance in a chemical equation. Quite a while ago, a French chemist named Joseph Louis Gay-Lussac determined that these coefficients also relate the *volumes* of *gases* in a chemical equation. For example, consider the formation of ammonia (NH_3) according to the following equation:

$$3H_2\,(g) + N_2\,(g) \rightarrow 2NH_3\,(g) \qquad (6.6)$$

It turns out that 3 liters of hydrogen gas react with 1 liter of nitrogen gas to form 2 liters of ammonia gas. Since these numbers correspond exactly to the stoichiometric coefficients of each substance in the equation, Gay-Lussac hypothesized that the stoichiometric coefficients relate the volumes (remember, "liters" is a volume unit) of the gases in a chemical equation as well as the number of moles.

Gay-Lussac performed many experiments on many different reactions in order to confirm his hypothesis. In the end, it became generally accepted as a scientific principle and is often called **Gay-Lussac's Law**:

Gay-Lussac's Law - The stoichiometric coefficients in a chemical equation relate the volumes of gases in the equation as well as the number of moles of substances in the equation.

We will use Gay-Lussac's Law in solving problems as well, but remember this important fact:

Gay-Lussac's Law can be used in stoichiometry only if *everything you are using in the problem is a gas.*

For example, consider the following reaction:

$$C_2H_2\,(g) + 2H_2\,(g) \rightarrow C_2H_6\,(g) \qquad (6.7)$$

In this reaction, we can say that 1 mole of C_2H_2 reacts with 2 moles of H_2 to make 1 mole of C_2H_6. In addition, we could use Gay-Lussac's Law to say that 1 liter of C_2H_2 gas reacts with 2 liters of hydrogen gas to make 1 liter of C_2H_6 gas. We can use Gay-Lussac's Law in this case because every substance that we mentioned is a gas.

On the other hand, consider the following equation:

$$CaO\,(s) + 2HCl\,(g) \rightarrow CaCl_2\,(s) + H_2O\,(g) \qquad (6.8)$$

In this reaction, we would say that 1 mole of CaO reacts with 2 moles of HCl to make 1 mole of $CaCl_2$ and 1 mole of H_2O. The most we can say using Gay-Lussac's Law, however, is that 2 liters of HCl gas will make 1 liter of H_2O gas. This is because Gay-Lussac's Law can only be used on gases. Thus, if we know that excess CaO was used, we could convert the number of liters of HCl to the number of liters of H_2O. We could say nothing, however, about the number of liters of $CaCl_2$ made, because $CaCl_2$ is not a gas.

Thus, when using Gay-Lussac's Law, we must make certain that each substance we use in our conversion is a gas. If not, then we cannot use Gay-Lussac's Law; thus, we cannot relate the volumes of substances in the chemical equation. See if you can follow this reasoning in the next example.

EXAMPLE 6.5

An automobile engine burns gaseous octane (C_8H_{18}). If an automobile completely combusts 12 liters of gaseous octane in excess oxygen, how many liters of carbon dioxide will be produced?

First of all, we must write a balanced chemical equation for the complete combustion of octane. According to our definition of complete combustion, the reaction would be:

$$C_8H_{18}\,(g) + O_2\,(g) \rightarrow CO_2\,(g) + H_2O\,(g)$$

Balancing the equation gives us:

$$2C_8H_{18}\,(g) + 25O_2\,(g) \rightarrow 16CO_2\,(g) + 18H_2O\,(g)$$

Since the substances we are interested in (octane and carbon dioxide) are both gases, we can use Gay-Lussac's Law to relate their volumes:

$$2 \text{ liters of } C_8H_{18} = 16 \text{ liters } CO_2$$

Now we can use this relationship to convert:

$$\frac{12 \cancel{\text{liters } C_8H_{18}}}{1} \times \frac{16 \text{ liters } CO_2}{2 \cancel{\text{liters } C_8H_{18}}} = 96 \text{ liters } CO_2$$

So we find out that burning 12 liters of octane in excess oxygen will make <u>96 liters of carbon dioxide</u>.

Now realize that we could just have easily used moles in this problem. Had the problem started by giving us the number of moles of octane burned, we could have calculated the number of moles of carbon dioxide produced. What Gay-Lussac's Law allows us to do is *also* answer the problem if liters (or any other volume unit) is used instead of moles, so long as the substances we are interested in are all gases.

There is one more thing you must realize. Even though Gay-Lussac's Law is a useful tool, there are many things it cannot do. One thing it cannot do is convert from moles to liters. For example, if you see a problem that gives you the quantity of a substance in moles, Gay-Lussac's Law is not capable of converting it to liters. Although there is a way to accomplish this task, we will not discuss it for a while. Gay-Lussac's Law can only be used to convert liters of one gas in a chemical equation to liters of another gas in that chemical equation. Make sure you understand this by solving the following problem:

ON YOUR OWN

6.5 A chemist decides to produce carbon dioxide gas using the following reaction:

$$Na_2CO_3\ (s)\ +\ 2HCl\ (g)\ \rightarrow\ 2NaCl\ (s)\ +\ CO_2\ (g)\ +\ H_2O\ (g)$$

If the chemist wants to make 3.67 liters of CO_2 gas, how many liters of HCl gas must be used?

Mass Relationships in Chemical Equations

What we have learned so far is that the stoichiometric coefficients in chemical equations relate the number of moles of each substance in that chemical equation. In addition, if the substances are gases, the stoichiometric coefficients can also relate their volumes. This makes stoichiometry a powerful tool in chemistry.

Unfortunately, we rarely measure the quantity of substances in moles. Usually, when asked to determine "how much" of a certain substance you have, the easiest thing to do is to measure the substance's mass. In other words, we usually report the quantity of a substance in grams, not moles. Thus, to make stoichiometry really useful, we have to determine a way that we can calculate relationships between the masses of substances in a chemical equation.

For example, suppose you were asked to prepare 50.0 grams of CuI with the following chemical reaction:

$$2CuCl_2\ +\ 4KI\ \rightarrow\ 2CuI\ +\ 4KCl\ +\ I_2 \tag{6.9}$$

How would you go about determining the amount of $CuCl_2$ and KI that you need in order to accomplish your goal? You couldn't do it exactly the way we have done previously, because we can only use stoichiometry to relate the number of *moles* of product to the number of *moles* of reactant. In this case, however, we were not given the number of moles of product. We were given the mass of the product. So what should we do?

Well, in the previous module, we learned how to convert from grams to moles. If we did that, then we would have the number of moles of product we need, and that could then be used to determine the number of moles of reactants we need. When we had accomplished that, however, we would probably still not be done. It turns out that you cannot purchase chemicals by the mole. Instead, when you purchase chemicals, you do it by the gram. Thus, if you wanted to order the necessary quantities of reactants, you would have to determine how many grams of each you required. Well, once again, in the previous module we learned how to convert from moles to grams. Thus, once you determined how many moles of each reactant you needed, you could then convert to grams. See how this works in the next example:

EXAMPLE 6.6

A chemist needs to produce 100.0 grams of $GaCl_3$ using the following reaction:

$$2Ga + 6HCl \rightarrow 2GaCl_3 + 3H_2$$

How many grams of Ga and HCl are necessary to accomplish this?

The problem asks us to relate the quantity of a product to the required quantities of the reactants. Now if the quantity of $GaCl_3$ were given in moles, we could use the techniques we have already learned to start solving the problem, much like we solved Example 6.3. Unfortunately, the amount of $GaCl_3$ was not given in moles; it was given in grams. To fix this problem, the first thing we need to do is convert from grams of $GaCl_3$ to moles of $GaCl_3$, as we learned in the previous module:

$$\text{Mass of } GaCl_3 = 69.7 \text{ amu} + 3 \text{ x } 35.5 \text{ amu} = 176.2 \text{ amu}$$

Now remember, when adding and subtracting, you do not count significant figures; you look at decimal place. Both of these masses have their last significant figure in the tenths place, so the answer should have its last significant figure in the tenths place. As you should recall from the previous module, the mass also tells us:

$$1 \text{ mole } GaCl_3 = 176.2 \text{ grams } GaCl_3$$

We can use this relationship to convert from grams of $GaCl_3$ to moles of $GaCl_3$:

$$\frac{100.0 \cancel{\text{grams GaCl}_3}}{1} \times \frac{1 \text{ mole GaCl}_3}{176.2 \cancel{\text{grams GaCl}_3}} = 0.5675 \text{ moles GaCl}_3$$

Now that we have the number of moles of $GaCl_3$, we can use the chemical equation to determine how many moles of each reactant we need. According to the chemical equation:

$$2 \text{ moles Ga} = 2 \text{ moles } GaCl_3$$
$$6 \text{ moles HCl} = 2 \text{ moles } GaCl_3$$

Using these relationships to convert:

$$\frac{0.5675 \cancel{\text{moles GaCl}_3}}{1} \times \frac{2 \text{ moles Ga}}{2 \cancel{\text{moles GaCl}_3}} = 0.5675 \text{ moles Ga}$$

$$\frac{0.5675 \cancel{\text{moles GaCl}_3}}{1} \times \frac{6 \text{ moles HCl}}{2 \cancel{\text{moles GaCl}_3}} = 1.703 \text{ moles HCl}$$

As we learned before, this tells us the minimum quantity of reactants needed to make 100.0 grams of $GaCl_3$. We aren't done yet, however, because the problem asked us to calculate the number of *grams* of each reactant. Well, since we have the number of moles already, we can use the skills we learned in Module #5 to convert this answer to grams.

Let's start with Ga:

$$\text{Mass of Ga} = 69.7 \text{ amu}$$

Thus:

$$1 \text{ mole Ga} = 69.7 \text{ grams Ga}$$

Therefore:

$$\frac{0.5675 \cancel{\text{moles Ga}}}{1} \times \frac{69.7 \text{ grams Ga}}{1 \cancel{\text{mole Ga}}} = 39.6 \text{ grams Ga}$$

Now we can move on to HCl:

$$\text{Mass of HCl} = 1.01 \text{ amu} + 35.5 \text{ amu} = 36.5 \text{ amu}$$

Thus:

$$1 \text{ mole HCl} = 36.5 \text{ grams HCl}$$

Therefore:

$$\frac{1.703 \cancel{\text{moles HCl}}}{1} \times \frac{36.5 \text{ grams HCl}}{1 \cancel{\text{mole HCl}}} = 62.2 \text{ grams HCl}$$

So we learn from this complicated process that in order to make 100.0 grams of $GaCl_3$, we must start with a minimum of 39.6 grams of Ga and 62.2 grams of HCl.

Now the previous example may have seemed a little long and complicated at first, but as you examine the steps we took, it should begin to make sense. After all, we know that stoichiometry allows us to relate the number of moles of one substance in a chemical equation to the number of moles of another substance in the chemical equation. Thus, in order to use stoichiometry, *we must have our quantities in moles*. Therefore, if a problem gives us the amount of a substance in grams, we must first convert to moles before we can use stoichiometry.

Once we have the substance's quantity in moles, we can then use the chemical equation and its stoichiometric coefficients to convert the number of moles of that substance into the number of moles of any other substance in the chemical equation. If the problem asks us to put our answer into grams, then we must convert from the moles we just calculated back into grams.

In case you are having a little trouble picturing all of this in your mind, I want to illustrate the procedure. Consider Figure 6.1:

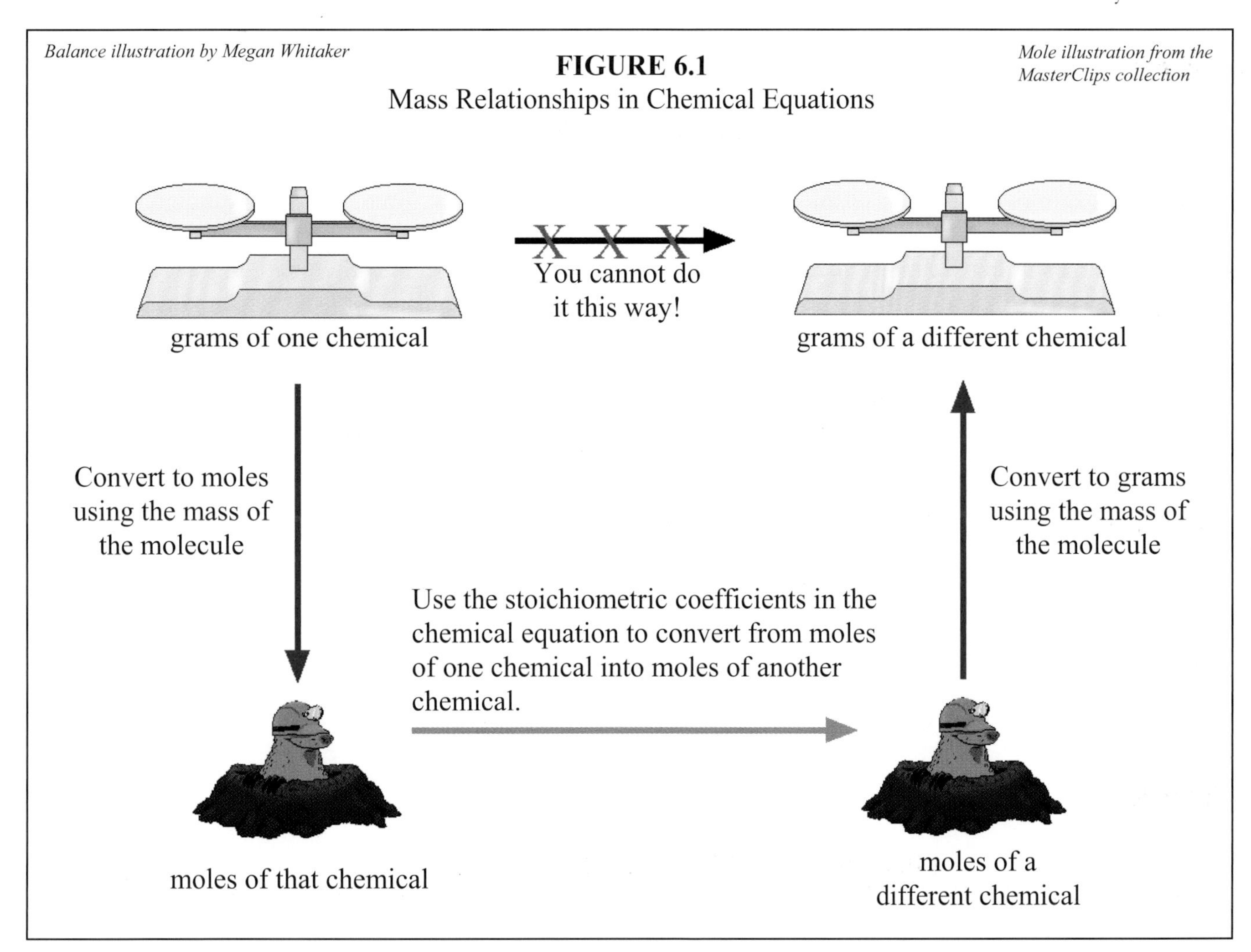
Balance illustration by Megan Whitaker
FIGURE 6.1
Mass Relationships in Chemical Equations
Mole illustration from the MasterClips collection
You cannot do it this way!
grams of one chemical
grams of a different chemical
Convert to moles using the mass of the molecule
Convert to grams using the mass of the molecule
Use the stoichiometric coefficients in the chemical equation to convert from moles of one chemical into moles of another chemical.
moles of that chemical
moles of a different chemical

In this figure, the balance represents the mass of a chemical, while the cartoon mole represents (what else?) the number of moles of a chemical. Since we can measure the mass of a sample pretty easily, we would like to relate the mass of one chemical in a chemical equation to the mass of another chemical in that same chemical equation. However, we cannot just convert from the mass of one chemical into the mass of another. There is nothing that will give us a conversion relationship between the masses of the two different chemicals. However, we *do* have something that will give us the conversion relationship in *moles*. The chemical equation does that. So, in order to determine the mass relationship between one chemical and another in a chemical equation, we must *convert to moles*.

We do this by starting with the chemical whose mass is given. Using the chemical formula, we can calculate the mass of a single molecule, and that tells us how many grams it takes to make a mole. We can use that information to convert from grams into moles. Once we have the amount of the chemical in *moles*, we can use the chemical equation. The chemical equation gives us a conversion relationship between moles of one chemical in the equation and moles of any other chemical in the equation. That conversion relationship is given by the stoichiometric coefficients in the equation. We can use that information to convert from moles of the given chemical to moles of any other chemical in the chemical equation.

Of course, knowing the number of moles of the other chemical is nice, but in the laboratory, it is hard to measure the number of moles in a sample. Thus, we usually need to convert moles back into grams. We can do that, of course, by using the chemical formula to calculate the mass of a single molecule. That tells us the number of grams it takes to make a mole, which will allow us to convert from moles back into grams.

Thus, to relate the masses of chemicals in a chemical equation, I must go via an indirect route. Since the chemical equation gives me the conversion relationship *in moles*, I must first convert from grams to moles. Then I can use the chemical equation to convert from one chemical into another. At that point, I have the moles of the new chemical, not the mass. Therefore, I must then convert from moles back into grams. I can relate the masses of chemicals in a chemical equation only in this "roundabout" way. I want to really make sure that you have this idea down pat, so here are two more example problems:

EXAMPLE 6.7

The human body produces the energy it requires to live by burning simple sugars such as $C_6H_{12}O_6$ according to the following chemical equation:

$$C_6H_{12}O_6 + 6O_2 \rightarrow 6CO_2 + 6H_2O$$

This reaction produces a lot of heat, which is used to power all of the body's motion. When the sugar is burned, the CO_2 produced is exhaled, while the H_2O produced is mostly eliminated through urine. If the average adult human burns about 1.2 kg of simple sugars each day in excess oxygen, how many grams of water does the average human eliminate every day as a result of this process?

This problem asks us to relate the quantity of $C_6H_{12}O_6$ burned in the body to the quantity of H_2O produced. It also tells us that the O_2 is in excess, indicating that $C_6H_{12}O_6$ is the limiting reactant. Thus, all we need to do is convert from the quantity of the limiting reactant to the quantity of product. Unfortunately, we can only do this if our quantities are in moles. So the first thing we must do is convert the quantity of $C_6H_{12}O_6$ from kilograms to moles. To do this, we must first convert 1.2 kilograms into 1.2×10^3 grams. Then, we can convert to moles:

$$\text{Mass of } C_6H_{12}O_6 = 6 \times 12.0 \text{ amu} + 12 \times 1.01 \text{ amu} + 6 \times 16.0 \text{ amu} = 180.1 \text{ amu}$$

This means:

$$1 \text{ mole } C_6H_{12}O_6 = 180.1 \text{ grams } C_6H_{12}O_6$$

Which allows us to convert to moles :

$$\frac{1.2 \times 10^3 \text{ \sout{grams } C_6H_{12}O_6}}{1} \times \frac{1 \text{ mole } C_6H_{12}O_6}{180.1 \text{ \sout{grams } C_6H_{12}O_6}} = 6.7 \text{ moles } C_6H_{12}O_6$$

Now that we have moles of $C_6H_{12}O_6$, we can use stoichiometry to determine how much water will be produced:

$$1 \text{ mole } C_6H_{12}O_6 = 6 \text{ moles } H_2O$$

$$\frac{6.7 \text{ moles } C_6H_{12}O_6}{1} \times \frac{6 \text{ moles } H_2O}{1 \text{ mole } C_6H_{12}O_6} = 4.0 \times 10^1 \text{ moles } H_2O$$

So we now know that burning 1.2 kg of $C_6H_{12}O_6$ makes 4.0×10^1 moles of H_2O. Note that since the number of moles must have two significant figures, the answer must be expressed in scientific notation. This is not the final answer; however, because the problem asks how many *grams* of water are produced, not how many *moles*. So now we have to convert from moles to grams:

Mass of H_2O = 2 x 1.01 amu + 16.0 amu = 18.0 amu

1 mole H_2O = 18.0 grams H_2O

$$\frac{4.0 \times 10^1 \text{ moles } H_2O}{1} \times \frac{18.0 \text{ grams } H_2O}{1 \text{ mole } H_2O} = 7.2 \times 10^2 \text{ grams } H_2O$$

In the end, then, we find out that the average human eliminates $7.2\text{x}10^2$ grams of water each day as a result of burning 1.2 kg of simple sugars.

Toxic acid spills, which occur regularly in industrial settings, are usually cleaned up with a substance known as barium hydroxide, $Ba(OH)_2$. If, for example, HCl (a toxic acid) is spilled, barium hydroxide neutralizes it with the following reaction:

$$2HCl\ (l) + Ba(OH)_2\ (s) \rightarrow BaCl_2\ (aq) + 2H_2O\ (l)$$

The products of this reaction are not nearly as toxic as the reactants, so as long as enough barium hydroxide is used so that all of the HCl reacts, a spill that is potentially very toxic gets neutralized. If 540.0 grams of HCl are spilled, what is the minimum number of grams of barium hydroxide necessary to completely neutralize the spill?

This problem simply gives us the mass of one reactant and asks what mass of the other reactant is necessary for a complete reaction. Once again, we can use stoichiometry to figure this out, but only if the quantity of reactant is given in moles. Our first job, then, is to convert from grams of HCl to moles of HCl:

Mass of HCl = 1.01 amu + 35.5 amu = 36.5 amu

1 mole HCl = 36.5 grams HCl

$$\frac{540.0 \text{ grams HCl}}{1} \times \frac{1 \text{ mole HCl}}{36.5 \text{ grams HCl}} = 14.8 \text{ moles HCl}$$

Now that we have the quantity of HCl in moles, we can do stoichiometry:

2 moles HCl = 1 mole $Ba(OH)_2$

$$\frac{14.8\ \cancel{\text{moles HCl}}}{1} \times \frac{1\ \text{mole Ba(OH)}_2}{2\ \cancel{\text{moles HCl}}} = 7.40\ \text{moles Ba(OH)}_2$$

This tells us how much barium hydroxide is needed. Unfortunately, the mean guy who wrote this problem specifically asked how many grams of barium hydroxide are needed, so now we have to convert from moles to grams. Remember, since there are parentheses in this chemical formula, we must multiply the "2" after the parentheses with every atom inside the parentheses. Thus, this molecule contains one barium atom, but *two* oxygen atoms and *two* hydrogen atoms.

Mass of $Ba(OH)_2$ = 137.3 amu + 2 x 16.0 amu + 2 x 1.01 amu = 171.3 amu

1 mole $Ba(OH)_2$ = 171.3 grams $Ba(OH)_2$

$$\frac{7.40\ \cancel{\text{moles Ba(OH)}_2}}{1} \times \frac{171.3\ \text{grams Ba(OH)}_2}{1\ \cancel{\text{mole Ba(OH)}_2}} = 1.27 \times 10^3\ \text{grams Ba(OH)}_2$$

In the end, then, this problem shows us that to clean up 540.0 grams of HCl, at least <u>1.27×10^3 grams of barium hydroxide</u> are necessary.

Problems such as these are the toughest stoichiometry problems that you will be asked to perform. If you understand the examples, all you need to do is cement your stoichiometry skills by doing the "On Your Own" problems found below. If you are not comfortable with these types of problems, you need to explore stoichiometry further with another resource. Check the course website to see if one of the links related to this module explains stoichiometry in a way that is easier for you to understand. If you need more practice solving problems like this, you can find more problems in Appendix B.

I cannot stress enough how important it is for you to understand stoichiometry so that you can solve problems like these. Since all of chemistry is built upon developing relationships between substances in chemical equations, you could say that all of chemistry is based on stoichiometry! Every college chemistry course stresses this subject for that very reason, so it is important for you to learn it now.

ON YOUR OWN

6.6 The insecticide DDT ($C_{14}H_9Cl_5$), which is now banned because of its deleterious environmental effects, was manufactured with this chemical reaction:

$$2C_6H_5Cl + C_2HCl_3O \rightarrow C_{14}H_9Cl_5 + H_2O$$

If 486.1 grams of C_6H_5Cl are reacted with excess C_2HCl_3O, how many grams of DDT will be produced?

6.7 One type of rocket engine uses the following reaction to develop its thrust:

$$7H_2O_2 \text{ (l)} + N_2H_4 \text{ (s)} \rightarrow 2HNO_3 \text{ (aq)} + 8H_2O \text{ (l)}$$

If a rocket engine has 11.2 kg of N_2H_4 in it, what is the minimum mass of H_2O_2 necessary to allow the N_2H_4 to react completely?

6.8 Welding torches burn a chemical called acetylene (C_2H_2), which is made with the following reaction:

$$CaC_2 + 2H_2O \rightarrow Ca(OH)_2 + C_2H_2$$

How many grams of water are necessary to make at least 100.0 grams of acetylene?

Using Stoichiometry to Determine Chemical Formulas

In Module #3 you were introduced to the use of chemical formulas, and since that time you have become intimately familiar with them. You can use them to name compounds, figure out the mass of a compound, count the molecules in a compound, and perform stoichiometric calculations. Have you ever wondered, however, how we got these chemical formulas?

After all, when I say that the chemical formula for water is H_2O, I am saying that there are two hydrogen atoms and one oxygen atom linked together to make a water molecule. How do I know that? As I said in a previous module, it's really not possible to see atoms, so there is no way that I could look at a water molecule and point out the two hydrogen atoms and the one oxygen atom. So how do I know that they are there?

Well, remember in Module #5 when we measured the width of a molecule without ever seeing it? We used the mole concept to count the number of molecules in a circle, and we used the size of the circle to determine the width of the molecules. In other words, we measured the width of a molecule *indirectly*. That same basic principle has guided the way in which chemists have determined chemical formulas over the years.

When a compound (molecule) is decomposed, all of the elements (atoms) in that compound are separated. Once they are separated, their mass can be measured. Using the mole concept, the mass allows you to calculate how many atoms of each element existed in the compound. Once you have counted all of the atoms that were in the compound, you have essentially determined its chemical formula. In a nutshell, that's how we can determine a molecule's chemical formula without ever seeing the atoms that make it up. If this all sounds a little confusing right now, don't worry about it. The whole concept will get a lot easier when you see a few examples:

EXAMPLE 6.8

A chemist decomposes an unknown compound. The products of this decomposition are 12.00 g of carbon and 4.04 grams of hydrogen. What is the chemical formula of the unknown compound?

In looking at the information given, the first thing you should notice is that the unknown compound contains only carbon and hydrogen. If it contained other types of atoms, there would have been more products in the decomposition. Thus, this particular compound has only C's and H's, so its chemical formula must be C_xH_y, where x and y are numbers that we must determine. The decomposition reaction, then, could be written:

$$C_xH_y \rightarrow C + H_2$$

Remember, we don't know x and y, but we do know that elemental carbon has a chemical formula of "C," and that hydrogen (a homonuclear diatomic) has a formula of H_2. Thus, regardless of what values exist for x and y, the products must be C and H_2.

Now, in order for this to be a proper equation for a decomposition reaction, it must be balanced. The problem is, since we do not have x or y, balancing will be a little difficult. That's where the experimental data comes in. The problem says that 12.00 grams of carbon were produced. Well, the stoichiometric coefficients in a chemical formula simply represent the number of moles, so let's convert the grams of carbon into moles of carbon:

$$\text{Mass of C} = 12.0 \text{ amu}$$

$$1 \text{ mole C} = 12.0 \text{ g C}$$

$$\frac{12.00 \text{ \sout{g C}}}{1} \times \frac{1 \text{ mole C}}{12.0 \text{ \sout{g C}}} = 1.00 \text{ moles C}$$

Since the stoichiometric coefficients simply represent the number of moles of the substance they multiply, and we also know that in this decomposition, 1 mole of carbon was produced, we could say that the stoichiometric coefficient for "C" is 1.

In the same way, we could calculate the number of moles of hydrogen produced:

$$\text{Mass of } H_2 = 2 \text{ x } 1.01 \text{ amu} = 2.02 \text{ amu}$$

$$1 \text{ mole } H_2 = 2.02 \text{ g } H_2$$

$$\frac{4.04 \text{ \sout{g H}}_2}{1} \times \frac{1 \text{ mole } H_2}{2.02 \text{ \sout{g H}}_2} = 2.00 \text{ moles } H_2$$

Once again, since the stoichiometric coefficients really just represent the number of moles, and since we know that there were 2.00 moles of H_2 produced, the stoichiometric coefficient for "H_2" is 2. Therefore, our chemical equation becomes:

$$C_xH_y \rightarrow C + 2H_2$$

Now the problem of figuring out x and y is a snap! In order for the equation to balance, there must be 1 C on both sides of the equation. Therefore, x = 1. In order for the equation to balance, there must be 4 H's on both sides of the equation, making y = 4. Therefore, the equation is:

$$CH_4 \rightarrow C + 2H_2$$

Thus, the chemical formula of the unknown compound is $\underline{CH_4}$!

Let's review what just happened so that we understand why this method works. First of all, a decomposition reaction is carried out which separates a compound into its constituent elements. This tells us what atoms go in the chemical formula and allows us to write an unbalanced chemical equation for the decomposition. The masses of each individual element then allow us to calculate the moles of each element, and that tells us the stoichiometric coefficients which go next to each element on the products side of the equation. Once we have that, then we can figure out the chemical formula necessary to make the equation balance.

Empirical and Molecular Formulas

Now before you start thinking that this is all a little too easy, there are two wrinkles that get thrown into this mess. First, there is an inherent assumption we make in this type of experiment which you probably didn't notice. The assumption was that the stoichiometric coefficient next to the C_xH_y on the reactants side of the equation is a "1." After all, I said the only way we could balance the equation was if x=1 and y=4, but that is simply not true. There are other possibilities. Consider the following equations:

$$\tfrac{1}{2}C_2H_8 \rightarrow C + 2H_2$$

$$\tfrac{1}{3}C_3H_{12} \rightarrow C + 2H_2$$

$$\tfrac{1}{4}C_4H_{16} \rightarrow C + 2H_2$$

Each one of these equations are balanced; thus, any of the chemical formulas listed on the reactants side of the equation represents a *possible* chemical formula for our unknown compound. Now of course we would have to get rid of the fractions to make these *proper* chemical equations, but nevertheless, they do balance. So it turns out that the example we just did doesn't really measure the *exact* chemical formula of the compound. It just determines that the compound must have a formula like CH_4, C_2H_8, C_3H_{12}, C_4H_{16}...

Now if you notice, these chemical formulas are all multiples of each other. Each one of them has 1 "C" for every 4 "H's". Therefore, our experiment did help us learn something about the chemical formula. It told us that the chemical formula of this compound must have 1 "C" for every 4 "H's". This rules out a host of other possibilities and leads us to only a few possible formulas, which is more than we knew in the beginning.

This leads us to developing a distinction between chemical formulas. When we know exactly how many atoms of each type exist in a molecule, we will say that we know the **molecular formula** of the compound. On the other hand, when we only know the ratio of atoms in a compound (1 "C" for every 4 "H"'s), we call that an **empirical** (im peer' uh kul) **formula**. Just to make sure you understand the distinction, I will write out formal definitions:

Molecular formula - A chemical formula that provides the number of each type of atom in a molecule

Empirical formula - A chemical formula that tells you a simple, whole-number ratio for the atoms in a molecule

For example, if I tell you that the molecular formula for benzene is C_6H_6, you know that there are 6 C's and 6 H's in each molecule of benzene. However, instead of telling you its molecular formula, I could just tell you that benzene's empirical formula is CH (1 C for every 1 H). This would tell you that a benzene molecule has 1 C and 1 H *or* 2 C's and 2 H's *or* 3 C's and 3 H's *or* 4 C's and 4 H's *or* 5 C's and 5 H's *or* 6 C's and 6 H's, and so on. Study the next example problem and do the "On Your Own" problem that follows to make sure you understand the difference between empirical and molecular formulas.

EXAMPLE 6.9

The molecular formulas of three compounds are listed below. Determine the empirical formula for each compound:

a. C_4H_{12} **b.** $Na_2S_2O_8$ **c.** $K_2Cr_2O_7$

The empirical formula of a compound represents the simplest, whole-number ratio of the atoms in the molecule. Thus, to get the simplest, whole-number ratio, we must divide the molecular formulas by any common factor that exists between the subscripts:

a. 4 and 12 have a common factor of 4; thus, we divide both subscripts by 4 and get CH_3. This represents the simplest, whole-number ratio of C's and H's in the molecule.

b. 2, 2, and 8 have a common factor of 2, so we must divide all subscripts by 2 and get $NaSO_4$.

c. 2, 2, and 7 have no common factors, so $K_2Cr_2O_7$ is already an empirical formula. This particular example shows that empirical formulas *can* be molecular formulas.

ON YOUR OWN

6.9 Which of the following molecular formulas are also empirical formulas? If the formula is not an empirical formula, write the empirical formula for that molecule.

a. Al_6O_9 b. PbO_2 c. $MgCO_2$ d. $C_8H_{18}O_2$ e. $Si_3H_9O_2$

In the end, experiments such as the one in Example 6.8 only give you empirical formulas, not molecular formulas. Even though empirical formulas do not contain as much detailed information about a molecule as do molecular formulas, they are better than nothing. It is possible, however, to convert from a molecule's empirical formula to its molecular formula, if you happen to know the molecule's mass. Think about it: To go from a molecular formula to an empirical formula, what do you have to do? You have to find the common factor and divide by it. In Example 6.9a, for example, we noted that the common factor in the molecular formula was 4. When we divided by 4, we turned the molecular formula into an empirical formula.

Well, if we want to turn an empirical formula into a molecular formula, what will we have to do? We will have to *multiply* by the common factor. How do we know what that common factor is? We can determine it from the molecule's mass. Example 6.10 shows you how to do that.

EXAMPLE 6.10

A compound's empirical formula is determined by experiment to be CH_2O. In a separate experiment, the molecule's mass was determined to be 180.1 amu. What is the compound's molecular formula?

Now remember, to go from an empirical formula to a molecular formula, we must multiply by the common factor. So this problem really boils down to figuring out what the common factor is. We figure that out by looking at the mass. If the molecular formula really were CH_2O, the mass would be:

$$\text{mass of } CH_2O = 1 \times 12.0 \text{ amu} + 2 \times 1.01 \text{ amu} + 1 \times 16.0 \text{ amu} = 30.0 \text{ amu}$$

This isn't anywhere close to the correct mass of 180.1 amu. Thus, I need to multiply the numbers in the empirical formula by some factor. How do I find that factor? Think about it. I know that I need a mass of 180.1 amu. The empirical formula has a mass of 30.0 amu. To determine the common factor, I simply divide the mass that I need by the mass of the empirical formula:

$$\text{common factor} = \frac{\text{mass of molecule}}{\text{mass of empirical formula}} = \frac{180.1 \cancel{\text{amu}}}{30.0 \cancel{\text{amu}}} = 6.00$$

This tells me that if I multiply each number in the empirical formula by 6, I will get a molecule whose empirical formula is CH_2O and whose mass is 180.1 amu:

$$\text{molecular formula} = C_{1x6}H_{2x6}O_{1x6} = C_6H_{12}O_6$$

The molecular formula, then, is $\underline{C_6H_{12}O_6}$.

If you are a little shaky on this, don't worry. We will go over another problem just like it in a moment. However, before we do that, I want to define a term that you will often see in problems like this one. Sometimes, chemists talk about a compound's **molar mass** instead of its molecular mass. It is important to know the definition of that term.

<u>Molar mass</u> - The mass of one mole of a given compound

Now think about what this definition means. Suppose I have a mole of $C_6H_{12}O_6$. What would its mass be? Well, the molecule has a mass of 180.1 amu. That also means it takes 180.1 grams to make a mole. The molar mass, then, is 180.1 grams.

In other words, the molar mass is just the conversion relationship that goes from grams to moles. If we know the molecular mass of a compound, we know its molar mass. For example, the molecular mass of CH_4 is 16.0 amu. That means it takes 16.0 grams to make one mole. The molar mass, therefore, is 16.0 grams. In the same way, if we know the molar mass of a compound, we know its molecular mass. The molar mass of KCl, for example, is 74.6 grams. That means its molecular mass is 74.6 amu.

Why do chemists use the term "molar mass"? We use it because molar mass is typically what we measure in the lab. It is hard to measure the mass of a single molecule, because a single molecule has very little mass (on the order of 10^{-24} grams). Thus, the molecular mass is hard to measure. It *can* be measured; it just requires some pretty sophisticated equipment. However, it is pretty easy to measure the mass of a mole of molecules, because a mole of molecules has a mass that is somewhere between several grams and a few hundred grams. That's something you can measure on a typical lab scale. Thus, we typically *measure* the molar mass of a compound and then *infer* from that its molecular mass.

See if you can put this concept together with the kind of problem we just did.

EXAMPLE 6.11

A compound's empirical formula is PO_3. If its molar mass is 158 g, what is its molecular formula?

The problem tells us that the mass of one mole of this compound is 158 grams. This tells us that the molecular mass of the compound must be 158 amu. To get the molecular formula from the empirical formula, we must determine the common factor. To do that, we divide the molecular mass by the mass of the empirical formula. Thus, we must first determine the mass of the empirical formula:

$$\text{Mass of } PO_3 = 1 \times 31.0 + 3 \times 16.0 = 79.0 \text{ amu}$$

To get the common factor, we divide this into the molecular mass:

$$\text{common factor} = \frac{\text{mass of molecule}}{\text{mass of empirical formula}} = \frac{158 \cancel{\text{amu}}}{79 \cancel{\text{amu}}} = 2.0$$

Therefore, to get the molecular formula, we must multiply each number in the formula by 2:

$$\text{molecular formula} = P_{1x2}O_{3x2} = \underline{P_2O_6}$$

Do the following problem to make sure you understand how to convert empirical formulas into molecular formulas:

ON YOUR OWN

6.10 In one experiment, the empirical formula of a compound is determined to be $NaPO_4$. In a separate experiment, its molar mass is determined to be 354 grams. What is its molecular formula?

More Complicated Experiments for Determining Chemical Formulas

Remember I said that there were two wrinkles in the problem of determining chemical formulas. The first was the distinction between molecular and empirical formulas. The second is the fact that most experiments that determine empirical formulas are much harder to analyze than the experiment presented in Example 6.8. That particular example had very easy numbers to work with. In the next two examples, you will see that the numbers don't always work out so nicely:

EXAMPLE 6.12

A sample of brown-colored gas is decomposed into 2.34 g of nitrogen and 5.34 g of oxygen. What is the empirical formula of the compound?

Since only nitrogen and oxygen are products in the decomposition, then the formula of the compound must be N_xO_y. Nitrogen and oxygen are both homonuclear diatomics, so the decomposition reaction looks like:

$$N_xO_y \rightarrow N_2 + O_2$$

Now, to determine the stoichiometric coefficients on the products side, we calculate the number of moles of the elements produced. Let's start with nitrogen:

$$\text{Mass of } N_2 = 2 \text{ x } 14.0 \text{ amu} = 28.0 \text{ amu}$$

$$1 \text{ mole } N_2 = 28.0 \text{ g } N_2$$

$$\frac{2.34 \cancel{\text{g } N_2}}{1} \times \frac{1 \text{ mole } N_2}{28.0 \cancel{\text{g } N_2}} = 0.0836 \text{ moles } N_2$$

Now let's figure out the number of moles of oxygen produced:

$$\text{Mass of } O_2 = 2 \text{ x } 16.0 \text{ amu} = 32.0 \text{ amu}$$

$$1 \text{ mole } O_2 = 32.0 \text{ g } O_2$$

$$\frac{5.34 \cancel{\text{g } O_2}}{1} \times \frac{1 \text{ mole } O_2}{32.0 \cancel{\text{g } O_2}} = 0.167 \text{ moles } O_2$$

Since the number of moles of each product is its stoichiometric coefficient, our equation looks like:

$$N_xO_y \rightarrow 0.0836\ N_2 + 0.167\ O_2$$

Now, of course, this is where the wrinkle occurs. Unlike in Example 6.8, these numbers did not work out "nicely." This time, the stoichiometric coefficients are not integers. In order to be able to determine an empirical formula, we need integer numbers of atoms. How do we get that? We get that by dividing both of the coefficients by the smallest number present. If we do that, we will get integers:

$$N_xO_y \rightarrow \frac{0.0836}{0.0836}\ N_2 + \frac{0.167}{0.0836}\ O_2$$

Do you see how this works? When I divide 0.0836 by 0.0836, I get 1! When I divide 0.167 by 0.0836, I get 2. Thus, dividing by the smallest number has turned these decimals into integers:

$$N_xO_y \rightarrow N_2 + 2O_2$$

Thus, the only way to balance the equation is if x=2 and y=4. The formula, then, is N_2O_4. This, however, is not the answer. Remember, these kinds of problems can only determine the *empirical* formula. They cannot determine the molecular formula. Thus, even though the answer is a *possible* molecular formula, we cannot be sure that it is the *correct* molecular formula. Since these experiments can only determine empirical formulas, we must convert this into an empirical formula by dividing by the common factor, which is 2. Thus, the empirical formula is $\underline{NO_2}$.

This example shows that when you solve problems like this one, if the stoichiometric coefficients turn out not to be integers, you can simply divide by the smallest of them, and that will give you a reasonable chemical equation. Balancing the equation then gives you the chemical formula. If the chemical formula is not an empirical formula, you can then convert it to one. See if you can do this on your own:

ON YOUR OWN

6.11 A compound is decomposed into 1.40 g sulfur and 1.39 g oxygen. What is its empirical formula?

I hope you now have an appreciation for how useful stoichiometry is. This powerful tool allows us to analyze chemical equations and determine empirical formulas. We will see in future modules that there are many, many other uses of stoichiometry, and that's why I say you must understand the concepts presented in this module to be successful in chemistry.

ANSWERS TO THE "ON YOUR OWN" PROBLEMS

6.1 The chemical equation tells us that

$$1 \text{ mole } O_2 = 2 \text{ moles } NaNO_3$$

Thus, we can use it as a conversion relationship to convert moles of $NaNO_3$ into moles of O_2:

$$\frac{15.0 \cancel{\text{ moles } NaNO_3}}{1} \times \frac{1 \text{ mole } O_2}{2 \cancel{\text{ moles } NaNO_3}} = 7.50 \text{ moles } O_2$$

Note that the numbers which come from the chemical equation are integers and thus are exact. Therefore, 15.0 is the only number with a finite amount of significant figures. Since there are three significant figures in 15.0, we must have three in the answer. To make 15.0 moles of sodium nitrate, then, the chemist must start with <u>7.50 moles of oxygen</u>.

We can also use the chemical equation to develop a relationship between moles of $NaNO_2$ and moles of $NaNO_3$:

$$2 \text{ moles } NaNO_2 = 2 \text{ moles } NaNO_3$$

Using this in a conversion:

$$\frac{15.0 \cancel{\text{ moles } NaNO_3}}{1} \times \frac{2 \text{ moles } NaNO_2}{2 \cancel{\text{ moles } NaNO_3}} = 15.0 \text{ moles } NaNO_2$$

Once again, the 15.0 determines the significant figures. Thus, in order to make 15.0 moles of sodium nitrate, a chemist must have 7.50 moles of oxygen and <u>15.0 moles of $NaNO_2$</u>.

6.2 The first thing we are asked to do is calculate how many moles of oxygen are necessary to use up all of the iron ore. According to the equation:

$$4 \text{ moles } FeS_2 = 11 \text{ moles } O_2$$

Thus, converting from iron ore to oxygen :

$$\frac{1{,}256.0 \cancel{\text{ moles } FeS_2}}{1} \times \frac{11 \text{ moles } O_2}{4 \cancel{\text{ moles } FeS_2}} = 3{,}454.0 \text{ moles } O_2$$

In addition, the problem asks us to calculate how much iron oxide will be produced. Once again, since all substances in a chemical equation relate to all other substances in that equation, we can simply convert from moles of iron ore to moles of iron oxide. The equation tells us

$$4 \text{ moles } FeS_2 = 2 \text{ moles } Fe_2O_3$$

Thus:

$$\frac{1256.0 \text{ ~~moles FeS$_2$~~}}{1} \times \frac{2 \text{ moles Fe}_2\text{O}_3}{4 \text{ ~~moles FeS$_2$~~}} = 628.00 \text{ moles Fe}_2\text{O}_3$$

In the end, then, the reaction will require 3,454.0 moles of oxygen and will produce 628.00 moles of iron oxide. We could also have calculated how much sulfur dioxide was produced, but the problem did not ask us to, so we won't.

6.3 The problem asks us to calculate the minimum amount of each reactant needed to produce 1.4×10^5 moles of citric acid. Thus, all we have to do is determine the relationship between citric acid and each of the reactants and then use those relationships to convert:

To determine the amount of $C_{12}H_{22}O_{11}$ needed:

$$1 \text{ mole } C_{12}H_{22}O_{11} = 2 \text{ moles } C_6H_8O_7$$

$$\frac{1.4 \times 10^5 \text{ ~~moles C$_6$H$_8$O$_7$~~}}{1} \times \frac{1 \text{ mole } C_{12}H_{22}O_{11}}{2 \text{ ~~moles C$_6$H$_8$O$_7$~~}} = 7.0 \times 10^4 \text{ moles } C_{12}H_{22}O_{11}$$

To determine the amount of water needed:

$$1 \text{ mole } H_2O = 2 \text{ moles } C_6H_8O_7$$

$$\frac{1.4 \times 10^5 \text{ ~~moles C$_6$H$_8$O$_7$~~}}{1} \times \frac{1 \text{ mole } H_2O}{2 \text{ ~~moles C$_6$H$_8$O$_7$~~}} = 7.0 \times 10^4 \text{ moles } H_2O$$

To determine the amount of oxygen needed:

$$3 \text{ moles } O_2 = 2 \text{ moles } C_6H_8O_7$$

$$\frac{1.4 \times 10^5 \text{ ~~moles C$_6$H$_8$O$_7$~~}}{1} \times \frac{3 \text{ moles } O_2}{2 \text{ ~~moles C$_6$H$_8$O$_7$~~}} = 2.1 \times 10^5 \text{ moles } O_2$$

To make the citric acid, then, the manufacturer must use 7.0×10^4 moles of $C_{12}H_{22}O_{11}$, 7.0×10^4 moles of water, and 2.1×10^5 moles of oxygen.

6.4 Since the problem tells us that both $CaCO_3$ and water are added in excess, HCl is the limiting reactant. Thus, the amount of HCl determines the amount of product. In order to solve the problem, then, we need to convert from the amount of HCl to the amount of product using the relationship given in the chemical equation.

The chemical equation tells us:

$$2 \text{ moles HCl} = 1 \text{ mole } CaCl_2H_{12}O_6$$

Using this relationship to convert:

$$\frac{3.49\ \cancel{\text{moles HCl}}}{1} \times \frac{1\ \text{mole}\ CaCl_2H_{12}O_6}{2\ \cancel{\text{moles HCl}}} = 1.75\ \text{moles}\ CaCl_2H_{12}O_6$$

Thus, 1.75 moles will be produced. Once again, we *could* also use the information above to determine how much CO_2 is produced, because the number of moles of limiting reactant can be used to determine the number of moles of all products.

6.5 The problem asks us to relate the volumes of two gases in a chemical equation. Since both are gases, we can use Gay-Lussac's Law. Thus, the stoichiometric coefficients in the chemical equation relate the volumes of the gases:

$$2\ \text{liters of HCl} = 1\ \text{liter of}\ CO_2$$

Using this relationship to convert :

$$\frac{3.67\ \cancel{\text{liters}\ CO_2}}{1} \times \frac{2\ \text{liters HCl}}{1\ \cancel{\text{liters}\ CO_2}} = 7.34\ \text{liters HCl}$$

Thus, so long as enough Na_2CO_3 is added, you need 7.34 liters of HCl to make 3.67 liters of CO_2.

6.6 In reading this problem, the first thing you should notice is that you were told one reactant is in excess. This means that the other reactant (C_6H_5Cl) is the limiting reactant. Once you recognize that, you can relate its number of moles to the number of moles of any other substance in the chemical equation. Of course, you were not given the number of moles; you were given the number of grams. There is no conversion relationship between grams of C_6H_5Cl and grams of $C_{14}H_9Cl_5$. So the first thing we must do is convert from grams to moles. Only then can we use the chemical equation.

Mass of C_6H_5Cl = 6 x 12.0 amu + 5 x 1.01 amu + 1 x 35.5 amu = 112.6 amu

1 mole of C_6H_5Cl = 112.6 g C_6H_5Cl

$$\frac{486.1\ \cancel{\text{g}\ C_6H_5Cl}}{1} \times \frac{1\ \text{mole}\ C_6H_5Cl}{112.6\ \cancel{\text{g}\ C_6H_5Cl}} = 4.317\ \text{moles}\ C_6H_5Cl$$

Now that you have the number of moles of limiting reactant, you can relate that to any substance in the equation using stoichiometry :

2 moles C_6H_5Cl = 1 mole $C_{14}H_9Cl_5$

$$\frac{4.317\ \cancel{\text{moles}\ C_6H_5Cl}}{1} \times \frac{1\ \text{mole}\ C_{14}H_9Cl_5}{2\ \cancel{\text{moles}\ C_6H_5Cl}} = 2.159\ \text{moles}\ C_{14}H_9Cl_5$$

This gives us the amount of product, but the mean person who wrote this problem asked us to calculate the grams of product, not the moles. So now we have to take the number of moles and convert to grams:

Mass of $C_{14}H_9Cl_5$ = 14 x 12.0 amu + 9 x 1.01 amu + 5 x 35.5 amu = 354.6 amu

1 mole $C_{14}H_9Cl_5$ = 354.6 g $C_{14}H_9Cl_5$

$$\frac{2.159 \cancel{\text{moles } C_{14}H_9Cl_5}}{1} \times \frac{354.6 \text{ g } C_{14}H_9Cl_5}{1 \cancel{\text{mole } C_{14}H_9Cl_5}} = 765.6 \text{ g } C_{14}H_9Cl_5$$

There will be 765.6 g of product.

6.7 This problem asks you to relate the quantity of one reactant to the quantity of the other reactant that will allow for everything to be used up. This is easy to do with stoichiometry, but first you must convert the number of grams of reactant to moles of reactant, because stoichiometry can only be done with moles:

$$\frac{11.2 \cancel{\text{kg}}}{1} \times \frac{1{,}000 \text{ g}}{1 \cancel{\text{kg}}} = 1.12 \times 10^4 \text{ g}$$

Mass of N_2H_4 = 2 x 14.0 amu + 4 x 1.01 amu = 32.0 amu

1 mole of N_2H_4 = 32.0 g N_2H_4

$$\frac{1.12 \times 10^4 \cancel{\text{g } N_2H_4}}{1} \times \frac{1 \text{ mole } N_2H_4}{32.0 \cancel{\text{g } N_2H_4}} = 3.50 \times 10^2 \text{ moles } N_2H_4$$

Note that the number of moles must be expressed in scientific notation, as there must be three significant figures. That can only be done in scientific notation, as the zero needs a decimal point in order to make it significant. Now that you have the number of moles of reactant, you can relate that to the minimum amount of the other reactant in the equation using stoichiometry:

1 mole N_2H_4 = 7 moles H_2O_2

$$\frac{3.50 \times 10^2 \cancel{\text{moles } N_2H_4}}{1} \times \frac{7 \text{ moles } H_2O_2}{1 \cancel{\text{mole } N_2H_4}} = 2.45 \times 10^3 \text{ moles } H_2O_2$$

This gives us the amount of reactant, but now we have to take the number of moles and convert to grams:

Mass of H_2O_2 = 2 x 1.01 amu + 2 x 16.0 amu = 34.0 amu

1 mole H_2O_2 = 34.0 g H_2O_2

$$\frac{2.45 \times 10^3 \cancel{\text{moles } H_2O_2}}{1} \times \frac{34.0 \text{ g } H_2O_2}{1 \cancel{\text{mole } H_2O_2}} = 8.33 \times 10^4 \text{ g } H_2O_2$$

The answer is 8.33×10^4 g.

6.8 In this problem, you are asked to calculate the minimum amount of reactant necessary to make a certain amount of product. Thus, you must assume that the reactant you have to calculate will be the limiting reactant. You can do this with stoichiometry, once you convert from grams of product to moles of product:

Mass of C_2H_2 = 2 x 12.0 amu + 2 x 1.01 amu = 26.0 amu

1 mole of C_2H_2 = 26.0 g C_2H_2

$$\frac{100.0 \cancel{\text{g } C_2H_2}}{1} \times \frac{1 \text{ mole } C_2H_2}{26.0 \cancel{\text{g } C_2H_2}} = 3.85 \text{ moles } C_2H_2$$

Now that you have the number of moles of product, you can relate that to any substance in the equation using stoichiometry:

1 mole C_2H_2 = 2 mole H_2O

$$\frac{3.85 \cancel{\text{moles } C_2H_2}}{1} \times \frac{2 \text{ moles } H_2O}{1 \cancel{\text{mole } C_2H_2}} = 7.70 \text{ moles } H_2O$$

This gives us the amount of water, but now we have to take the number of moles and convert to grams:

Mass of H_2O = 2 x 1.01 amu + 1 x 16.0 amu = 18.0 amu

1 mole H_2O = 18.0 g H_2O

$$\frac{7.70 \cancel{\text{moles } H_2O}}{1} \times \frac{18.0 \text{ g } H_2O}{1 \cancel{\text{mole } H_2O}} = 139 \text{ g } H_2O$$

The minimum mass of water necessary is 139 g.

6.9 In order for a chemical formula to be empirical, it must represent a simple, whole-number ratio of atoms. Therefore, if a chemical formula is an empirical formula, the subscripts cannot have a common factor.

a. This is NOT an empirical formula, because 6 and 9 have a common factor of 3. In order to get the empirical formula, you would divide by that common factor and get Al_2O_3.

b. This IS an empirical formula, because 1 and 2 have no common factors.

c. This IS an empirical formula, because 1, 1, and 2 have no common factors.

d. This is NOT an empirical formula, because 8, 18 and 2 have a common factor of 2. In order to get the empirical formula, you would divide by that common factor and get C_4H_9O.

e. This IS an empirical formula, because 3, 9, and 2 have no common factors.

6.10 The molar mass is 354 grams. That means it takes 354 grams to make a mole, so the molecular mass must be 354 amu. Now, to get from the empirical formula to the molecular formula, we must determine the common factor by which to multiply. To do that, we need to divide the molecular mass by the mass of the empirical formula. Thus, we need to start by determining the mass of the empirical formula:

$$\text{Mass of } NaPO_4 = 1 \times 23.0 \text{ amu} + 1 \times 31.0 \text{ amu} + 4 \times 16.0 \text{ amu} = 118.0 \text{ amu}$$

Now we can determine the common factor:

$$\text{common factor} = \frac{\text{mass of molecule}}{\text{mass of empirical formula}} = \frac{354 \text{ amu}}{118.0 \text{ amu}} = 3.00$$

To get from the empirical formula to the molecular formula, then, we must multiply by 3:

$$Na_{1x3}P_{1x3}O_{4x3} = \underline{Na_3P_3O_{12}}$$

6.11 Since only sulfur and oxygen were products in the decomposition, the chemical formula is S_xO_y. Oxygen is a homonuclear diatomic, so the decomposition reaction looks like this:

$$S_xO_y \rightarrow S + O_2$$

In order to get the stoichiometric coefficients, we determine the number of moles produced in the decomposition reaction. Let's start with sulfur:

$$\text{mass of S} = 32.1 \text{ amu}$$

$$1 \text{ mole S} = 32.1 \text{ g S}$$

$$\frac{1.40 \text{ g S}}{1} \times \frac{1 \text{ mole S}}{32.1 \text{ g S}} = 0.0436 \text{ moles S}$$

Now we can determine the number of moles of oxygen produced:

$$\text{mass of } O_2 = 2 \times 16.0 \text{ amu} = 32.0 \text{ amu}$$

$$1 \text{ mole } O_2 = 32.0 \text{ g } O_2$$

$$\frac{1.39 \text{ g } O_2}{1} \times \frac{1 \text{ mole } O_2}{32.0 \text{ g } O_2} = 0.0434 \text{ moles } O_2$$

So the equation looks like this:

$$S_xO_y \rightarrow 0.0436 \text{ S} + 0.0434 \text{ } O_2$$

The problem with this equation is that the coefficients are not integers. We need to turn them into integers in order to be able to determine the empirical formula of the reactant. How do we do that? We divide by the smallest number:

$$S_xO_y \rightarrow \frac{0.0436}{0.0434}\,S + \frac{0.0434}{0.0434}\,O_2$$

When you divide 0.0436 by 0.0434, you get 1.004608.... Since we have only three significant figures, however, the answer is 1.00. Since 0.0434 divided by 0.0434 is also 1.00, the chemical equation now looks like this:

$$S_xO_y \rightarrow S + O_2$$

The formula that balances the equation, then, is $\underline{SO_2}$. This is already an empirical formula, since 1 and 2 have no common factors.

REVIEW QUESTIONS FOR MODULE #6

1. Define stoichiometry and explain why it is such a useful tool for chemists.

2. Explain what a limiting reactant is and why it is important in stoichiometry.

3. A chemist experiments with the following reaction:

$$14HCl + K_2Cr_2O_7 \rightarrow 2KCl + 2CrCl_3 + 3Cl_2 + 7H_2O$$

If the chemist adds 15 moles of HCl to 1 mole of $K_2Cr_2O_7$, what is the limiting reactant?

4. State Gay-Lussac's Law and explain how it is used.

5. A chemist wants to perform the following reaction:

$$CaCO_3\ (s) + 2HCl\ (g) \rightarrow CaCl_2\ (aq) + CO_2\ (g) + H_2O\ (l)$$

Which substances can she use Gay-Lussac's Law to relate to one another?

6. If a stoichiometry problem gives you the quantity of a substance in grams, what is the first thing you must do to solve the problem?

7. What is the difference between molecular formulas and empirical formulas?

8. Which of the following are empirical formulas?

a. $C_{12}H_{24}O_{10}$ b. $K_2S_2O_3$ c. CaN_2O_6

9. A compound has a molecular formula of $C_{14}H_{21}O_7$. What is its empirical formula?

10. A compound has a molecular formula of H_3PO_4. What is its empirical formula?

PRACTICE PROBLEMS FOR MODULE #6

1. When 2.13 moles of C_3H_8 burn in excess oxygen, how many moles of CO_2 will be formed? Assume this is complete combustion.

2. Silver tarnishes when exposed to oxygen and dihydrogen monosulfide. The chemical reaction is as follows:

$$\underset{\text{silver}}{4Ag\ (s)} + 2H_2S\ (g) + O_2 \rightarrow \underset{\text{tarnish}}{2Ag_2S\ (s)} + 2H_2O\ (l)$$

Every household has some H_2S in the air (it smells like rotting eggs), but there is usually only a small quantity. Thus, H_2S is almost always the limiting reactant in this reaction. If a silver spoon has 0.0012 moles of tarnish on it, how many moles of H_2S was it exposed to?

3. Freon®, a very useful refrigerant, is produced in the following reaction:

$$3CCl_4\ (g) + 2SbF_3\ (s) \rightarrow \underset{\text{Freon®}}{3CCl_2F_2\ (g)} + 2SbCl_3\ (s)$$

If a chemist wants to make 1.0×10^4 moles of Freon® using excess carbon tetrachloride, how many moles of antimony triflouride will the chemist need?

4. In the reaction from problem #3, suppose the chemist wanted to make 100.0 liters of Freon® using excess antimony triflouride. How many liters of carbon tetrachloride would the chemist need?

5. To make a phosphorus fertilizer, agricultural companies use the following reaction:

$$Ca_3P_2O_8 + 2H_2SO_4 + 4H_2O \rightarrow \underset{\text{fertilizer}}{CaH_4P_2O_8} + 2CaH_4SO_6$$

If 1.50×10^4 grams of H_2SO_4 are reacted with excess $Ca_3P_2O_8$ and H_2O, how many grams of fertilizer can be made?

6. Phosgene, $COCl_2$, was once used as a chemical weapon in war. When humans breathe in the gas, it reacts with the water in their lungs in the following way:

$$COCl_2 + H_2O \rightarrow CO_2 + 2HCl$$

The product HCl burns the lining in the lungs, making them unable to function. Thus, by breathing in phosgene, a person will eventually suffocate. Assuming that 10.0 grams of HCl in the lungs is deadly, how many grams of phosgene would a person need to breathe in order to die?

7. $Mg(OH)_2$ is the active ingredient in many laxatives. It can be made with the following reaction:

$$Mg\ (s) + 2H_2O\ (l) \rightarrow Mg(OH)_2\ (s) + H_2\ (g)$$

If a laxative manufacturer has 3.6×10^5 grams of Mg, how many grams of water must be used to have all of the Mg react to form $Mg(OH)_2$?

8. A compound has an empirical formula of $CHBr_2$. If its molar mass is 345.6 grams, what is its molecular formula?

9. An unknown compound was decomposed into 63.2 g carbon, 5.26 g hydrogen, and 41.6 g oxygen. What is its empirical formula?

10. A dry-cleaning compound was decomposed into 14.5 g carbon and 85.5 g chlorine. In a separate experiment, its molar mass was determined to be 166.0 grams. What is the molecular formula of the compound?

MODULE #7: Atomic Structure

Introduction

In the past few modules, we have been working with atoms quite a bit. We have used them in chemical formulas, counted them in order to balance equations, determined how to find out their masses, and used the mole concept to count the number of atoms in a sample. After using atoms so much, you might have wondered what they look like. I have already told you that it's not possible for us to see atoms, but for a moment, let's suppose we could. If we could see an atom, what would it look like? Well, that's what we're going to learn about in this module.

Now you might wonder how in the world we can determine what atoms look like if we cannot see them. Remember, as I have told you several times already, chemists often use things that we can see to learn about things we cannot see. Just as we used water, pepper, and soap in Module #5 to determine the width of a molecule, and just as decomposition reactions can help chemists determine chemical formulas, we can use experiments to help us learn what atoms look like. The way that matter responds to certain situations can help us learn more about the structure of the atom.

Historical Overview

Ever since Democritus first theorized about atoms some 2,400 years ago, scientists have wondered what they might look like. Democritus thought that each type of atom had its own unique shape: some were triangles; some were pentagons; and some were squares. These shapes, he thought, fitted together like jigsaw puzzle pieces to form molecules. As time went on, however, chemists began to do experiments that started to unlock the mysteries of the atom.

One of the first experiments that shed some light on the question of the atom's structure was performed by an English chemist named William Crookes (krooks). In the second half of the 1800s, Crookes studied how certain gases behaved when exposed to electricity. He studied this by filling a glass tube (which we often call a **Crookes tube**) with a tiny amount of gas. He would then hook the tube up to a battery, allowing electricity to pass through it. Crookes noticed that when the battery was hooked up, the end of the tube would glow a faint greenish-yellow color:

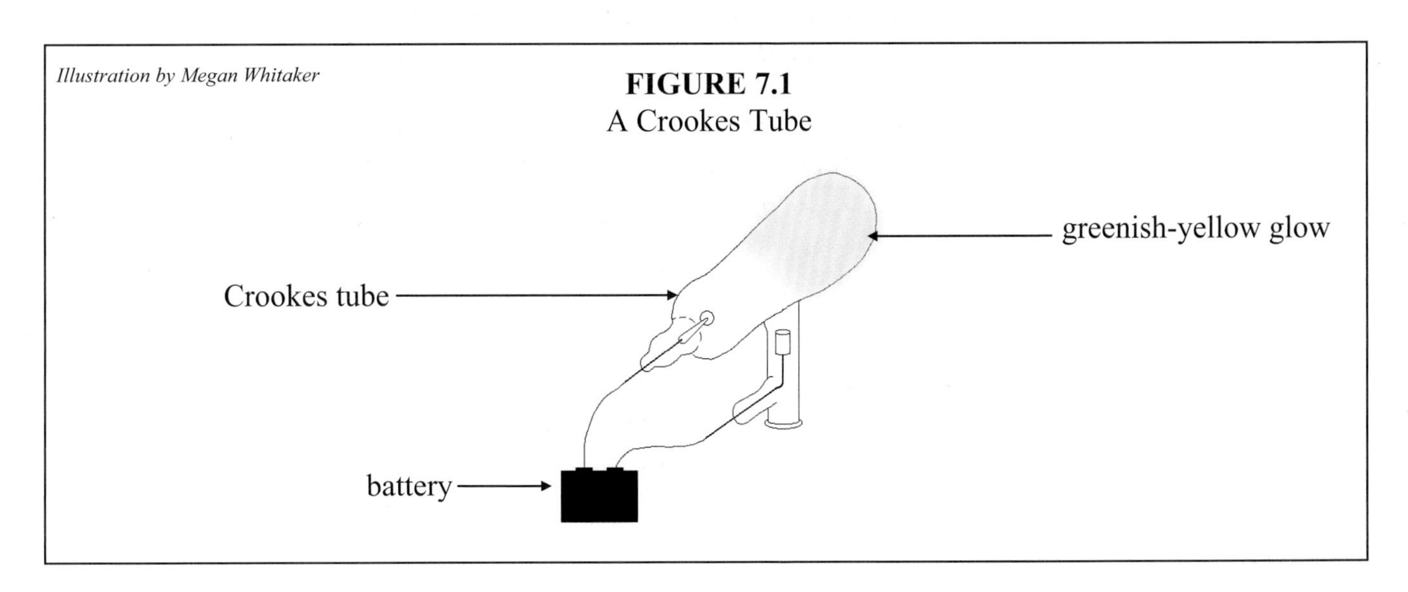

FIGURE 7.1
A Crookes Tube

When the battery was not hooked up, there would be no glow. He concluded from this experiment that particles were being produced in the gas by the battery. These particles, he theorized, traveled through the gas and hit the end of the tube, causing the glow that he saw. Since these particles always seemed to travel from the post on the battery called the "cathode" (kath' ohd), Crookes called them "cathode rays." As a result, a Crookes tube is also sometimes referred to as a **cathode ray tube**.

In order to try to confirm his theory that these cathode rays were actually particles, he put an obstacle between the cathode and the end of the tube. He reasoned that if these cathode rays were, in fact, particles, they would be blocked by the obstacle and would not hit the end of the tube. Since they could not hit the end of the tube, they would not cause a glow. This is, indeed, just what he saw, as illustrated in the figure below.

Illustration by Megan Whitaker

FIGURE 7.2
A Crookes Tube with an Obstacle

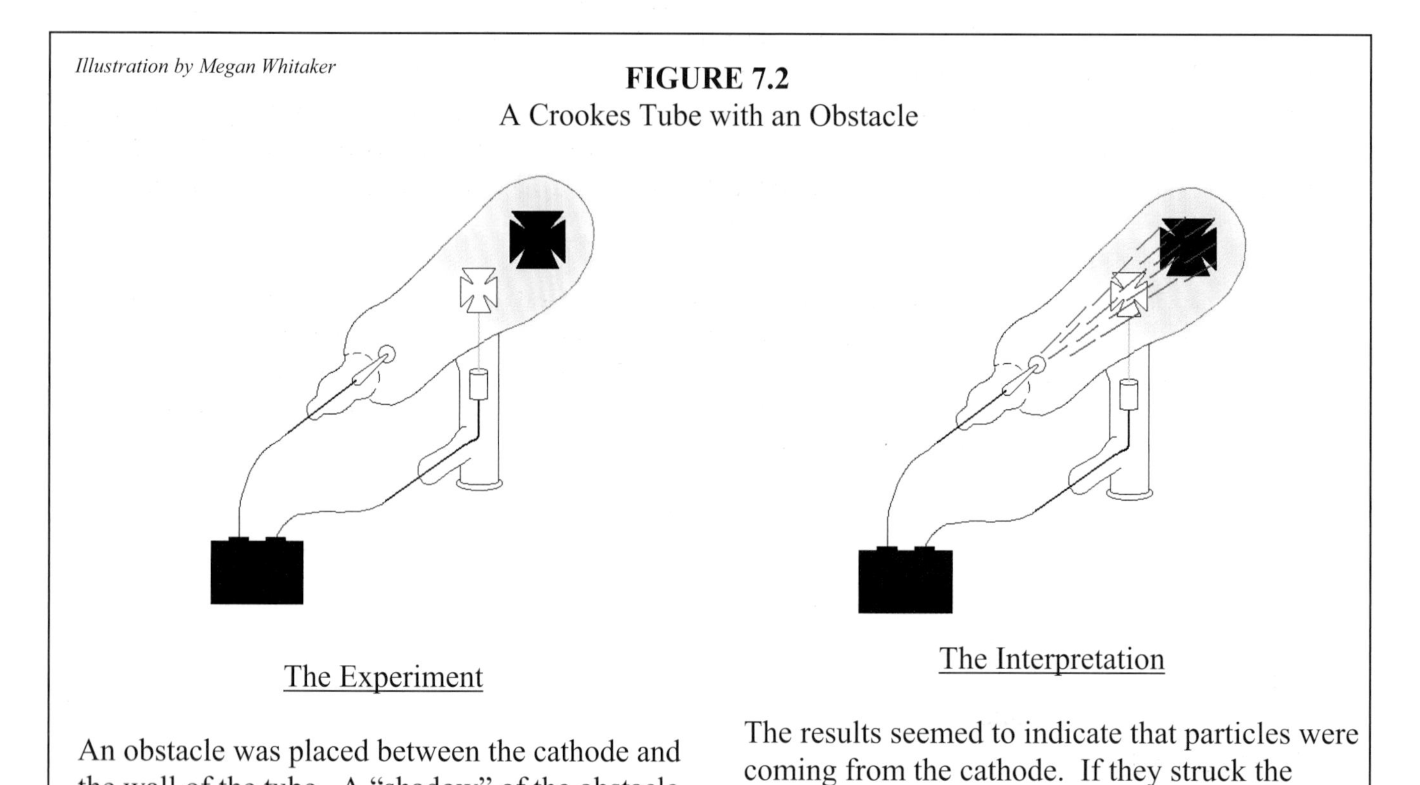

The Experiment

An obstacle was placed between the cathode and the wall of the tube. A "shadow" of the obstacle appeared at the end of the tube.

The Interpretation

The results seemed to indicate that particles were coming from the cathode. If they struck the obstacle, they could not make it to the wall, and that explains the "shadow."

About 20 years after Crookes did his experiments, another English chemist, J. J. Thomson, revisited them to try to determine more about the nature of these cathode rays. Thomson's work ended up demonstrating that one of John Dalton's main assumptions concerning atoms was wrong. Remember from Module #3 that Dalton developed a theory that assumed that atoms were indivisible. He reasoned that since atoms were the smallest unit of matter, they could not be broken down into smaller pieces. However, Thomson showed that regardless of what type of gas he put in the Crookes tube, cathode rays were always formed. He further showed that these particles were precisely the same, whether he filled the Crookes tube with helium, argon, nitrogen, or any other gas. This led Thomson to believe that these cathode rays were a part of every atom and that his experiments were breaking them away from the atom.

Thomson's idea that cathode rays were actually small parts of the atom was further boosted by his discovery that these rays were electrically charged. He determined this by seeing how they reacted in the presence of magnetic and electric fields. The way they behaved in the presence of these fields indicated that cathode rays were, indeed, electrically charged. In fact, he determined that they had negative electrical charge. To see why this fact was critical in tearing down Dalton's assumption that atoms are indivisible, you first have to understand a few things about electrical charge.

Electrical Charge

Everyone knows that there is something called electricity. We know that it can be used to operate lights, televisions, motors, etc. Most people, however, do not understand what electricity is. If you are one of these people, don't worry about it. In fact, most scientists don't have a really good understanding of electricity. That's because understanding electricity begins with understanding the concept of electrical charge. Unfortunately, scientists do not have a clear understanding of what electrical charge is or why it exists. We know how electrical charges behave and how they generate electricity, but we really don't know much about what they are or where they come from. Thus, don't worry if you don't understand everything in the discussion that follows; I will point out the concepts that are important for you to understand at the end.

There are two types of electrical charge: **positive** and **negative**. We don't know what makes something negative or positive in charge, but we know that these charges do indeed exist. We also know that the vast majority of matter on earth does not have any overall electrical charge. We call these things electrically **neutral.** To learn a little more about what it means to be electrically positive or negative, perform the following experiment:

EXPERIMENT 7.1
Electrical Charge

Supplies
- Comb
- Aluminum foil
- Cellophane (Scotch®) tape
- Safety goggles

1. Take two pieces of cellophane tape, each approximately 5 inches long, and attach them to the top of a table. Attach them so that about one-half of an inch of tape hangs off the table.
2. Once the tape is firmly attached to the table, grab the ends of the tape that hang over the table edge and quickly pull both pieces so that they detach from the table.
3. Hold the tape so that each piece hangs down.
4. Turn the pieces of tape so that their sticky sides face each other, and slowly bring them together. As you move the sticky sides towards each other, you will see them begin repel each other. The pieces of tape will actually float away from each other as you try to move them together.
5. Now, take some aluminum foil and cut it into several very small pieces and lay those pieces in a small pile on the table.
6. Take the comb and pass it over the foil pieces. Bring the comb as close as you can to the foil without actually touching it. Nothing should happen.
7. Now take the comb and comb your hair with it. In order for this experiment to work, your hair must be reasonably clean. If it is too greasy, the next step of the experiment will not work.

8. After combing your hair, pass the comb over the aluminum pieces as you did before. This time, the aluminum pieces should move in response to the comb. The aluminum pieces should actually move on their own towards the comb. If this didn't happen, your hair was probably a bit too greasy. Wash your hair, let it dry completely, then try again.
9. Clean up your mess.

How do we explain what happened in the experiment? Well, let's start with what happened in the first part. To explain the results, you must understand that when you pulled the tape off the table, the sticky substance pulled some electrical charges off the table. Since each piece was pulling off the same type of charge, each piece of tape ended up having the same type of electrical charge. As those like charges were brought together, they repelled each other. Thus, we come to our first rule of electrical charge:

Like charges repel one another.

What about the second part of the experiment? In that part, you started with a comb and some aluminum foil, neither of which was electrically charged. In fact, both the comb and the aluminum have electrical charges in them, but there are as many positive electrical charges as there are negative ones. Just as $-1 + 1 = 0$, the negative and positive charges cancel each other out to give no charge at all.

However, when you combed your hair, the comb picked up some stray negative electrical charges that were in your hair. This caused the comb to have more negative charges than positive charges, so it became negatively charged. When the comb was then placed near the aluminum foil, the negative charge of the comb repelled all of the negative charges in the aluminum foil. Thus, the negative charges in the aluminum foil got as far away from the comb as possible. But when that happened, the parts of the foil near the comb had more positive charges than negative ones, and those positive charges were attracted to the comb. Thus, the foil pieces began to move towards the comb. Therefore, we come to the second rule of electrical charge:

Opposite charges attract each other.

The other important thing to remember about electrical charge was mentioned in the previous discussion:

Every substance on earth has electrical charges. However, if there are as many positive charges as there are negative charges, the substance has no overall charge ($-1 + 1 = 0$). If there is more of one charge than the other, the substance takes on that electrical charge ($-2 + 1 = -1$).

Thus, the comb you used started with no overall electrical charge because it had an equal number of positive and negative charges in it. However, when it picked up a few extra negative charges from your hair, it became negatively charged because there were more negative charges than positive charges in it.

If you don't understand everything you've just read, don't worry. The concepts of electrical charge are difficult to comprehend. Even the most brilliant chemist in the world still doesn't fully understand them. What you have to be able to understand are the three boldface statements above. Those are the foundational concepts of electrical charge that we will use throughout this module.

Electrical Charge and Atomic Structure

We now have enough information under our belts to understand why Crookes' and Thomson's experiments destroyed Dalton's postulate that atoms are indivisible. When the Crookes tube was filled with gas, the gas was always electrically neutral. Thus, the gas had as many negative charges as positive charges. However, when the battery was hooked up to the Crookes tube, negatively charged particles were produced. Thomson reasoned that those negatively charged particles must have been taken from the atoms of gas, much like the comb you used in the previous experiment pulled negative charges from your hair.

Thus, if negative charges could be pulled from an electrically neutral atom, then the atom must be composed of smaller components. Some of those components must be negatively charged, and the other components must be positively charged. Since the gas atoms started out neutral, there must be an equal number of positively and negatively charged components in atoms. Crookes and Thomson showed that these charged components could be separated; thus, atoms could be broken down into smaller components. This destroyed Dalton's assumption that atoms are indivisible.

Thomson later did a famous experiment which actually measured the ratio of the charge and mass of these negatively charged particles. He showed that no matter what atoms they were pulled from, these particles always had the same ratio of charge to mass. He thus considered these particles to be something common to all atoms, and he called them **electrons** (ee lek' trahnz). As I have already said, since electrons are pulled from electrically neutral atoms, there must also be positive components in the atom. Thomson's student, Ernest Rutherford, discovered those components and called them **protons** (pro' tahns). In 1932, another English scientist, James Chadwick, discovered that there was an additional component to the atom. This component was electrically neutral (it had no charge), so Chadwick called it the **neutron** (new' trahn).

Thus, by 1932, scientists had determined that atoms were made up of three components: protons, electrons, and neutrons. What makes one atom different from another atom is the number of protons, neutrons, and electrons contained in it. For example, look at your periodic chart again. The first box on the periodic table represents the atom "hydrogen." A hydrogen atom is composed of 1 proton, 1 electron, and no neutrons. The second box on the chart represents the atom "helium," which contains 2 protons, 2 electrons, and 2 neutrons. Hydrogen and helium have totally different properties. For example, hydrogen is explosive when placed in contact with a flame or even a spark. On the other hand, helium will never burn, no matter how hot a flame it comes into contact with. This difference is the result of the different number of protons, neutrons, and electrons that make up the atoms.

Determining the Number of Protons and Electrons in an Atom

You might wonder how I know that a hydrogen atom has 1 proton and a helium atom has 2. It turns out that this is what the top number in each box of the periodic table tells you. We call that number the **atomic number**, and it tells you how many protons are in an atom:

An atom's atomic number tells you how many protons it contains.

In addition, since all atoms are electrically neutral, we know they must contain an equal number of protons (positive charges) and electrons (negative charges):

All atoms have an equal number of electrons and protons.

Thus, if you look at your periodic chart, you will see that sulfur (S) has an atomic number of 16. This tells you that a sulfur atom has 16 protons and 16 electrons in it. Make sure you understand this idea by doing the following problem:

ON YOUR OWN

7.1 How many protons and electrons are in the following atoms?

a. Chromium (Cr) b. Beryllium (Be) c. Lanthanum (La)

Determining the Number of Neutrons in an Atom

As you can see, when you are armed with the periodic chart, it is very simple to determine the number of protons and electrons in an atom. Determining the number of neutrons in an atom is a little more difficult, however. To understand why this is the case, I must first tell you that so far, we have been a bit sloppy in our discussion of the periodic table. All along, I have told you that each box in the periodic table represents a single type of atom. That's not really true. In fact, each box on the periodic chart represents several different atoms, but all of those atoms are part of the same element. For example, the sixth box on the chart represents carbon. It turns out that there are three different types of carbon atoms. Each has the same number of protons and electrons (6), but they have different numbers of neutrons. When atoms have the same number of protons and electrons but different numbers of neutrons, we call them **isotopes** (eye' suh tohps).

Isotopes - Atoms with the same number of protons but different numbers of neutrons

The most common type of carbon atom is composed of 6 protons, 6 electrons, and 6 neutrons. There are, however, two other types of atoms which belong to the element we call carbon. The second type of carbon atom still has 6 protons and 6 electrons, but it has 7 neutrons. Finally, the last type of carbon atoms has 6 protons, 6 electrons, and 8 neutrons. This is what we mean by the term "isotope." All three of these types of atoms are made up of the same number of protons and electrons. They are comprised of different numbers of neutrons, however. Thus, they are all isotopes.

What is the difference between these carbon isotopes? Not much. The one with 8 neutrons is a little heavier than the one with 7 neutrons. Similarly, the one with 7 neutrons is a bit heavier than the one with 6 neutrons. Other than that, their chemical natures are precisely the same. Any reaction that one of the isotopes can participate in, they all can participate in. In fact, if you were to have a lump of carbon (a black solid) in your hand, all three isotopes would be present. The first isotope would make up about 98.9% of all atoms in the sample. The second isotope would make up about 1.1% of the sample, and the last isotope would make up less than 1.0×10^{-12}% of the sample. In almost every way, however, these isotopes would appear identical. They would all participate exactly the same in any chemical reaction. You could not find a way to separate them from one another chemically. This is a very important point:

Isotopes behave identically in their chemistry; the main difference between them is their mass.

It turns out that every element is composed of several isotopes. This means that every box on the periodic chart actually represents several different isotopes. Since that is the case, it is impossible to look at the periodic chart and determine how many neutrons are in a given atom. Thus, in order to determine the number of neutrons in an atom, you must be given a little more information. This information is provided by something we call the **mass number:**

<u>Mass number</u> - The total number of neutrons and protons in an atom

This mass number is not always provided when discussing elements, because often it is not necessary to worry about the number of neutrons in an atom. However, if it is given, then it is easy to determine how many neutrons are in that particular atom.

Remember I said that the most common isotope of carbon has 6 protons and 6 neutrons? Thus, its mass number is 6 + 6 = 12. You can include this mass number in one of two ways. When writing out the full name of the element (carbon), you can include the mass number by adding a hyphen followed by the mass number. Thus, we could call this isotope "carbon-12." On the other hand, if you just want to write the symbol for carbon (C), then the mass number must appear as a superscript which *precedes* the symbol. Thus, the symbol for this isotope would be ^{12}C. What about the other isotopes of carbon? The one with 6 protons and 7 neutrons would have a mass number of 6 + 7 = 13. Thus, we could call it carbon-13 or ^{13}C. Similarly, the isotope with 6 protons and 8 neutrons would be called carbon-14 or ^{14}C.

Before you go any further, I want to make a point in order to clear up some confusion you might have. Some students interpret the term "isotope" incorrectly, thinking that there is one "normal" version of an atom and that all the others are isotopes. For example, they think that carbon-12 is the "normal" version of carbon and that carbon-13 and carbon-14 are isotopes. That's *not* the proper use of the term. *All three types of carbon are isotopes.* Look back at the definition. Carbon-12, carbon-13, and carbon-14 all have 6 protons, but they have different numbers of neutrons. Thus, they are *all* isotopes of carbon. Think about it this way: The term "isotope" is a *relational* term. It refers to how atoms relate to one another. Consider three men who each have the same father. We would say that these men are brothers. The term "brother" is a relational term. It tells you how the men relate to one another. There is not one "normal" man, while the other two men are the brothers! They are *all* brothers. In the same way, if several atoms all have the same number of protons (the same atomic number) but different numbers of neutrons, they are *all* isotopes.

Now that you have learned about isotopes, you have what you need to determine the makeup of any atom on the periodic chart, provided I give you enough information. Study the following examples and then tackle the "On Your Own" problems that follow in order to be certain you understand how this is done.

EXAMPLE 7.1

What is the name and symbol of an atom which is made up of 10 protons, 10 electrons, and 11 neutrons?

Since the atom has 10 protons, its atomic number is 10. Looking at the periodic chart, neon (Ne) is the element with that atomic number. To give the full name, we must also include its mass number. Since it has 10 protons and 11 neutrons, its mass number is 10 + 11 = 21. Therefore, the name of the atom is neon-21, and its symbol is ^{21}Ne.

How many protons, electrons, and neutrons make up a ^{129}I atom?

To determine the number of protons and electrons, all we have to do is find I (iodine) on the periodic chart. We see that its atomic number is 53. This means it has 53 protons and 53 electrons. In addition, we are given its mass number, which represents the total number of protons and neutrons. If the atom has 53 protons, and its total number of protons + neutrons is 129, then the number of neutrons it has is 129-53 = 76 neutrons.

ON YOUR OWN

7.2 Give the number of protons, electrons, and neutrons that make up the following atoms:

a. ^{40}Ar b. Chlorine-37 c. ^{139}La

7.3 An atom has 34 protons, 34 electrons, and 41 neutrons. What is its full name and symbol?

Isotopes and Nuclear Bombs

Before we leave this discussion of protons, electrons, neutrons, and isotopes, there is one thing I would like to point out. If you pay attention to the news at all, you have probably heard about how the United States and other nuclear powers in the world are trying to keep nuclear technology away from certain countries and are concerned about those countries that are developing nuclear technology. Have you ever wondered, however, exactly what "nuclear technology" is? Most people, when asked to define "nuclear technology," would say that it refers to one's ability to make a nuclear bomb. In fact, that's not true at all! Any person who has studied nuclear physics at the college level has all of the knowledge needed to build a nuclear bomb! It turns out that the trick is not in making the nuclear bomb; almost anyone (with a little bit of training) can do that! The real trick to making a nuclear bomb is in making the bomb's fuel!

Uranium is the usual fuel used to make nuclear bombs. However, the element uranium has three naturally-occurring isotopes: ^{238}U, ^{235}U, and ^{234}U. A sample of natural uranium is about 99.3% ^{238}U, about 0.7% ^{235}U, and less than 0.01% ^{234}U. Only the ^{235}U isotope, however, can be used to make a bomb. Thus, for a nuclear bomb to work, the uranium used in the bomb must have a significantly larger percentage of ^{235}U than natural uranium has. In other words, in order to be able to produce a nuclear bomb, you must be able to artificially increase the amount of ^{235}U in the uranium that you are using. The way to do this, of course, is to get rid of some of the ^{238}U that is in your uranium. This process, called **isotopic** (eye' suh top' ik) **enrichment**, is very tricky. After all, the only real difference between ^{235}U and ^{238}U is the mass. Chemically, they behave in exactly the same way. This makes it nearly impossible to separate the two for isotopic enrichment.

Of course, "nearly impossible" and "impossible" are two different things. It turns out that during World War II, scientists in a group called the "Manhattan Project" were able to figure out a way to partially separate ^{238}U from ^{235}U. The process is incredibly inefficient, but it *is* able to take natural uranium and enrich the ^{235}U content from 0.7% to a little more than 7%, which is enough to fuel a nuclear bomb. *This technique for isotopic enrichment is the secret that must be kept from those countries trying to become nuclear powers.* In other words, making a nuclear bomb is simple; however, getting the fuel to power it is nearly impossible, and the technique for making this fuel is a carefully guarded secret.

Atomic Structure in More Detail

Now that you know how to determine the number of protons, neutrons, and electrons in an atom, you might ask how these particles arrange themselves in an atom. Are they jumbled up together like a bunch of marbles in a sack? Do they arrange themselves in groups? Do they form any neat arrangements or designs? These are the same questions that chemists were asking themselves in the early 1900s.

After electrons were first discovered (and before neutrons were discovered), chemists thought that an atom looked a lot like a dish of plum pudding. This concept, called the "plum pudding model," is illustrated in Figure 7.3:

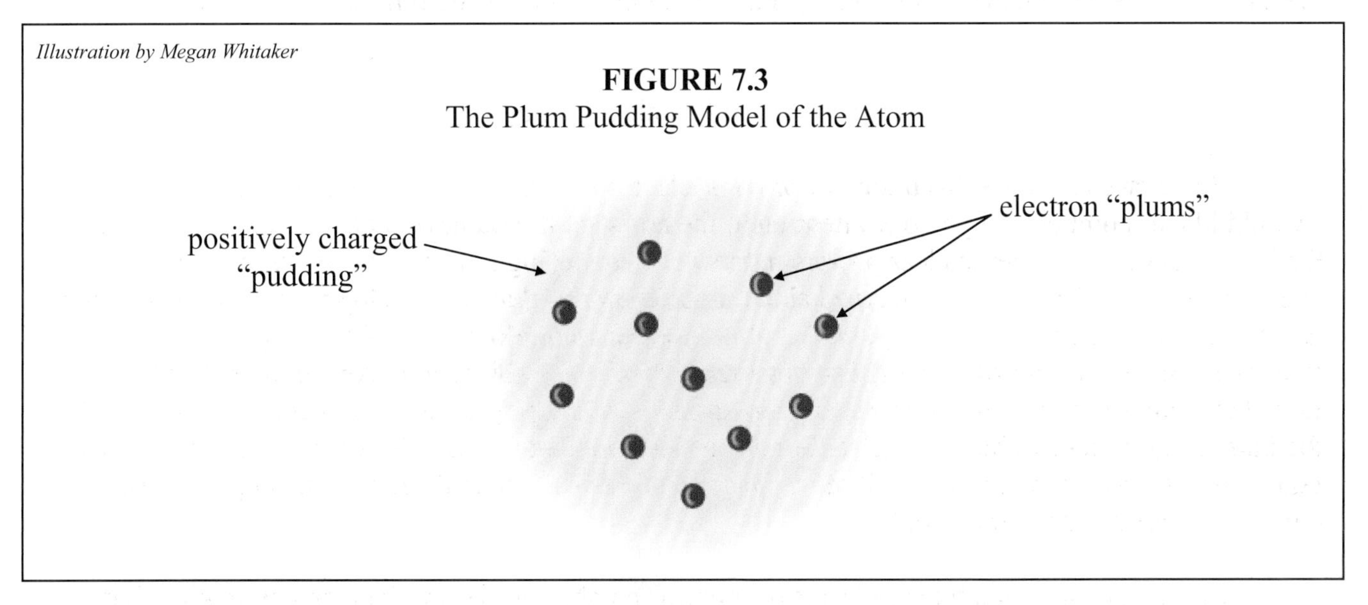

FIGURE 7.3
The Plum Pudding Model of the Atom

In essence, chemists thought that the atom was made up of a pudding-like substance which had positive electrical charge. Electrons, they thought, were suspended in this "pudding" like plums, making the entire atom electrically neutral. Remember, it's impossible to see atoms; thus, chemists just had to imagine what an atom might look like. That's why we call this concept a **model** of the atom. A model is simply a constructed image of something that we cannot see with our eyes.

Since we can't see atoms, how do we know whether or not the plum pudding model is correct? Well, the best way to determine that is to compare it to some experimental data. If the model is inconsistent with the experimental data, then the model is obviously incorrect. If, however, the model is consistent with the data, then we don't necessarily know that the model is right, but at least we know that it might not be wrong. The trick, however, is in designing an experiment that will produce data that really tests all aspects of the model.

Ernest Rutherford (Thomson's student) designed the perfect experiment to test the validity of the plum pudding model. In earlier experiments, Rutherford had identified a type of radiation called "alpha particles." He determined that these alpha particles have positive electrical charges and are emitted by certain isotopes that were called "radioactive." You will learn a great deal more about radioactive isotopes if you take advanced chemistry. For right now, just remember that alpha particles are positive particles emitted by certain radioactive isotopes.

Rutherford decided that what he needed to do was to shoot some of the alpha particles at a thin metal foil. He reasoned that if the plum pudding model of the atom were correct, the alpha particles should pass through the foil without changing their direction much. After all, the positive alpha particles would be attracted by the negative "plums" (the electrons) but repelled by the positive "pudding." Since the plums were randomly distributed around the pudding, the alpha particles would be attracted and repelled in many different directions, and the result would be that they would travel straight through the atom.

If you have trouble understanding why the alpha particles should travel straight through a plum pudding atom, think about it this way. Suppose you were trying to move a large boulder by tying many ropes to it and having several people pull on the ropes while others push on the boulder. If everyone pulled and pushed in different, random directions, the boulder wouldn't go anywhere. In order to move the boulder, everyone would need to pull and push in roughly the same direction. The same principle applies here. When traveling through a plum pudding atom, the alpha particles would be pulled and pushed in different, random directions. This would result in the alpha particles not being moved from their original path.

I hope that you can see the ingenuity behind Rutherford's experiment. The data provided would be an excellent test of the model in question. If the alpha particles did not behave as Rutherford expected, then clearly the plum pudding model was wrong. If, however, they did behave as expected, then that would be one more bit of evidence in favor of the plum pudding model. In the end, the experiment looked something like Figure 7.4:

Illustration by Megan Whitaker

FIGURE 7.4
Rutherford's Experiment

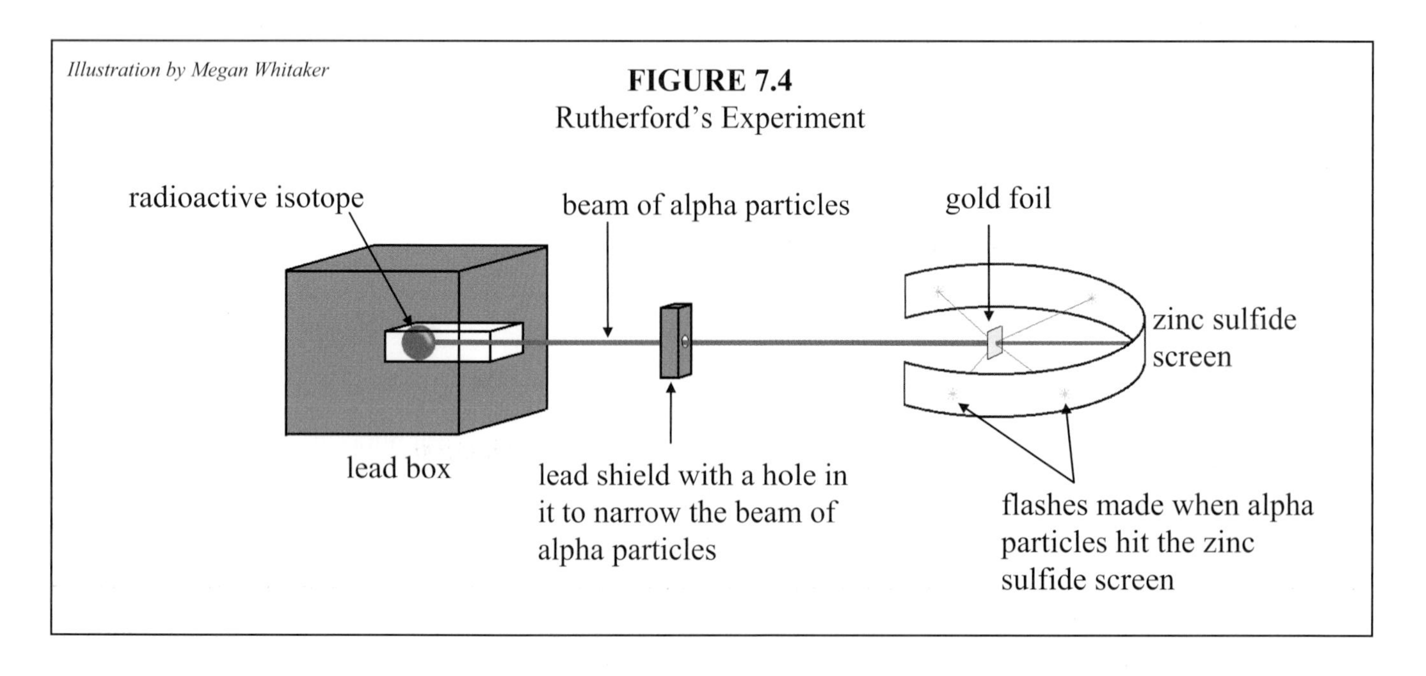

In this experiment, Rutherford put a radioactive isotope in a thick lead box that had a single hole in it. The radioactive isotope inside the box was emitting alpha particles (a type of nuclear radiation). Most of the alpha particles were stopped inside the box, but any that were emitted in the direction of the hole escaped. This made a "beam" of alpha particles that could be pointed towards a target. The beam was narrowed by another piece of lead with a hole in it. This thin beam of alpha particles then hit a thin gold foil.

Rutherford surrounded the target with a zinc sulfide screen. When an alpha particle hits zinc sulfide, a bright green glow is emitted. Thus, Rutherford could see where the alpha particles went after they collided with the gold foil by simply looking at where the zinc sulfide screen was glowing. Wherever it was glowing very brightly, many alpha particles were hitting it. Wherever it was glowing faintly, only a few alpha particles were hitting it.

Rutherford ran his experiment and, to his astonishment, the zinc sulfide screen glowed everywhere! To be sure, the glow was the brightest directly in front of the target, but alpha particles were hitting the zinc sulfide screen at all other positions as well. Rutherford noticed that the farther away from the front of the target, the dimmer the glow was. Nevertheless, alpha particles were colliding with the target and bouncing in every conceivable direction.

This, of course, spelled the end of the plum pudding model of the atom. After all, there was no way to imagine how the alpha particles could possibly bounce off of the target and change direction if the plum pudding model of the atom were correct. In the words of Rutherford himself, it would be like "...firing shells at a piece of paper handkerchief and having them bounce back at you." Rutherford tried his experiment with several different types of targets, to make sure that his conclusion applied to all atoms. In the end, however, the results were the same, regardless of what target was used.

Of course, it wasn't enough for Rutherford to simply determine that the plum pudding model of the atom was wrong. Now that he had all this experimental data, he needed to try to develop a new model of the atom that was consistent with all of the information he had up to that point. He did come up with such a model. Rutherford called it the **planetary model** of the atom, but it is usually referred to today as the **Rutherford model** of the atom. This model is illustrated in Figure 7.5:

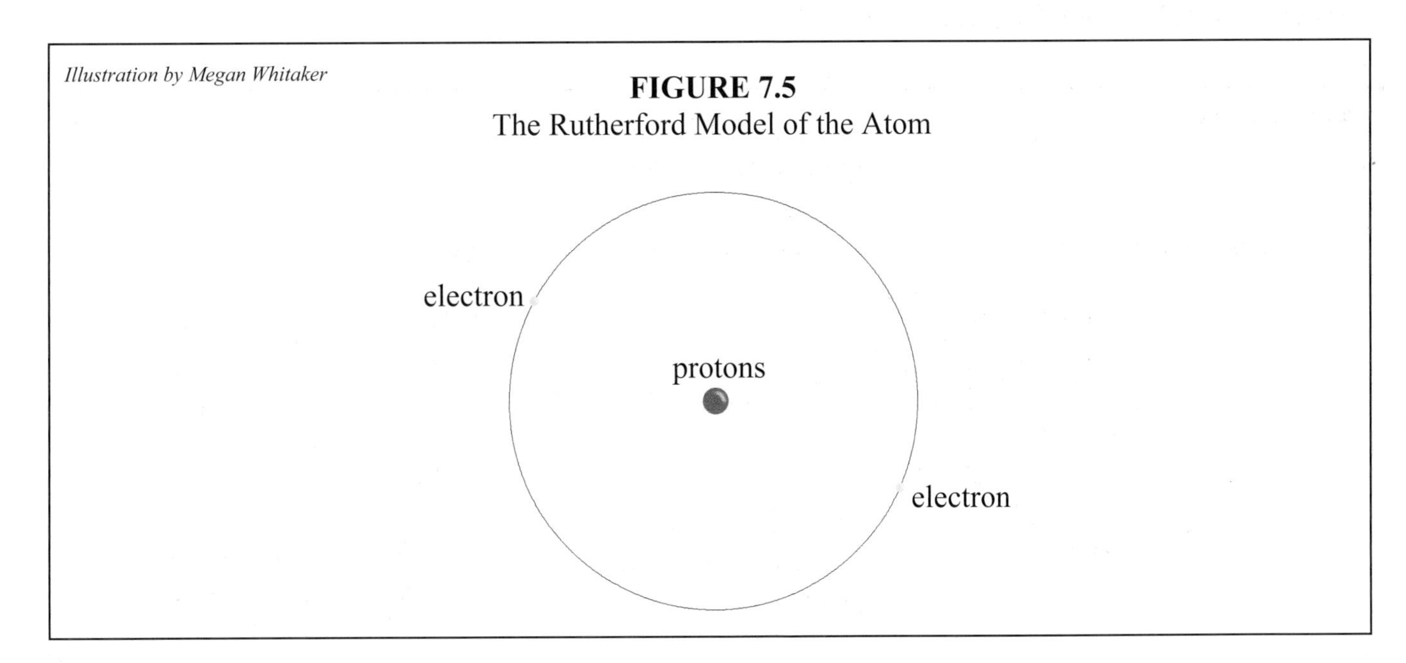

Illustration by Megan Whitaker

FIGURE 7.5
The Rutherford Model of the Atom

In Rutherford's model, the protons were clustered together at the center of the atom, called the **nucleus** (new' klee us), while the electrons orbited around the nucleus in the same way that the planets in our solar system orbit around the sun. This similarity between his view of the atom and our solar system is why Rutherford called it the "planetary model" of the atom. Notice that Rutherford's model does not contain neutrons. This is because neutrons had not been discovered yet.

Realize that this idea of what the atom looks like was not something Rutherford just dreamed up. It was the only feasible model that explained the data he got from his experiment. Remember, Rutherford's experiment showed that when positive particles were shot at atoms, most of those particles passed through the atoms undisturbed. However, some were deflected away from their original path. Sometimes the deflection was enormous, and the alpha particle would bounce backwards after hitting an atom.

Rutherford's model is consistent with all of this data. To help you understand why, take a look at Figure 7.6:

Illustration by Megan Whitaker

FIGURE 7.6
Why Rutherford's Model Is Consistent With His Data

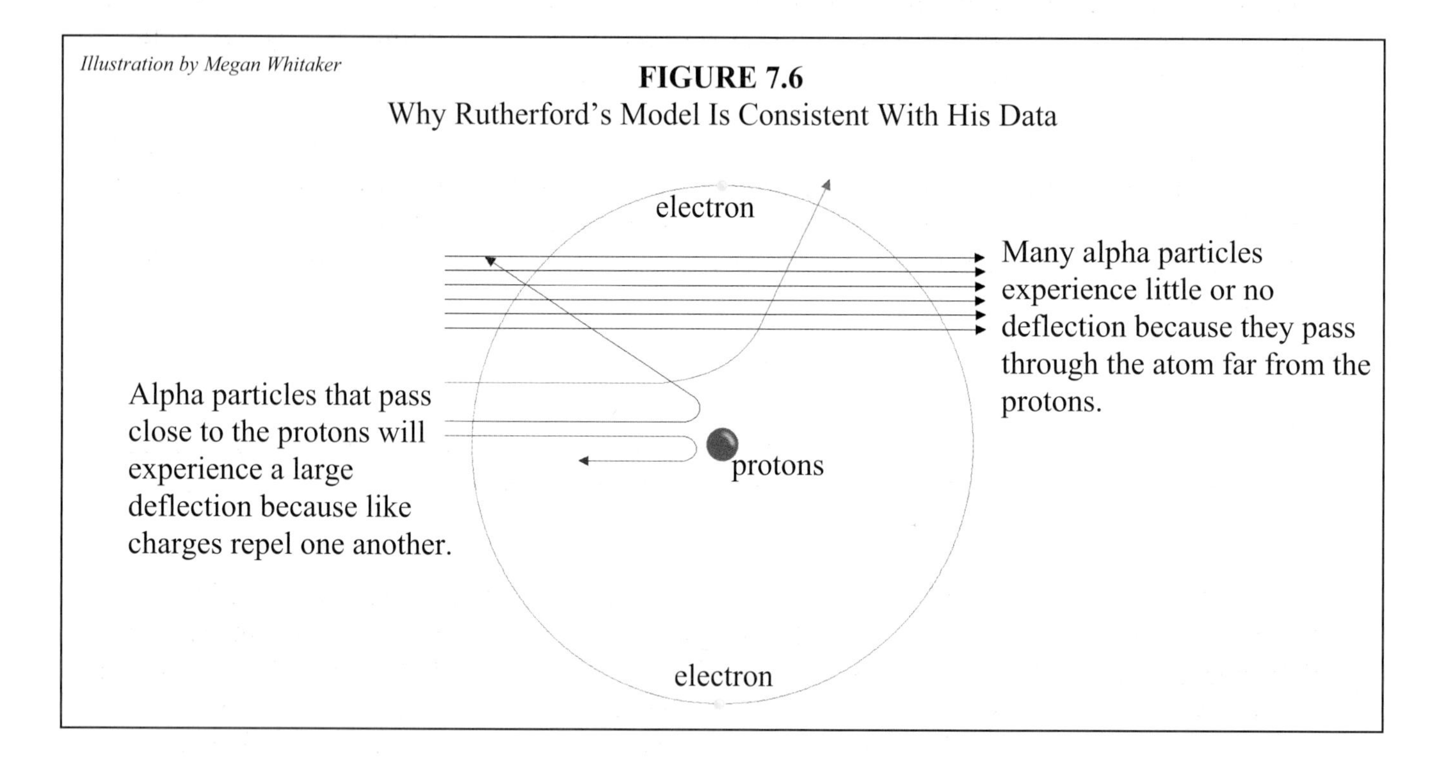

electron
Many alpha particles experience little or no deflection because they pass through the atom far from the protons.
Alpha particles that pass close to the protons will experience a large deflection because like charges repel one another.
protons
electron

In the figure, the arrows represent alpha particles that are passing through the atom. Unlike the plum pudding model, this model has the positive electrical charge concentrated at the center of the atom. The negative charges (the electrons) are orbiting around the protons. Since the electrons are not concentrated in one spot, they do not affect the path of the alpha particles greatly. Thus, the protons cause most of the deflection of the alpha particles. The vast majority of alpha particles that pass through the atom, then, will not be deflected, because they will pass through the atom far from the protons. These alpha particles, represented by the straight arrows, will continue to travel straight, hitting the zinc sulfide screen in front of the target.

However, if an alpha particle passes through the atom close to the protons, it will be repelled by the protons' positive charge, and it will be deflected from its original path. The closer it passes to the protons, the more bent its path will become. If it is headed directly towards the proton, it will be repelled so much that it will actually turn around and head in the opposite direction. These deflected alpha particles are represented by the curved arrows.

Since the vast majority of alpha particles are not deflected because they do not pass very close to the protons, they strike the zinc sulfide screen directly in front of the target, as Rutherford saw in his experiment. In addition, his observation that the number of alpha particles that were deflected decreases as you begin looking at large deflections also makes sense in terms of this model, as the number of particles that pass close to the protons is quite small. In the end, then, Rutherford's model was consistent with all of the data produced in his experiment.

The one thing Rutherford's model was not consistent with, however, was a current theory in physics. Physicists during Rutherford's time had noticed that whenever electrically charged particles moved in circles, they emitted light. When they emitted this light, they would lose energy and thus slow down. According to this theory, then, the Rutherford model of the atom would never be stable. This theory would predict that the electrons moving in a circle would continue to emit light until they lost all of their energy. This theory would additionally predict that the electrons, as they lost energy, would spiral towards the protons in the atom until they collided with them.

Even though Rutherford's model was not consistent with a current theory in physics, it was the only model of the atom that could be devised that was consistent with his experimental data. Thus, it became the accepted model of the atom until it was later modified by Niels Bohr (boor). To understand Bohr's model of the atom and what experiments led up to it, however, you must first learn a little bit about light.

The Nature of Light

In Module #1, you learned that since light neither has mass nor takes up space, it is not matter. Unfortunately, you haven't yet learned what light *is* and what its properties are. That's the purpose of this discussion.

The first thing you must realize about the nature of light is that scientists don't completely understand it. There are currently two models for light: the particle model and the wave model. The particle model states that light is actually made up of small particles called **photons**. Thus, any beam of light that you see is actually made up of billions of individual particles (photons) that are all traveling in the same direction. The wave model, on the other hand, says that a beam of light is actually a wave that travels much like a wave on the ocean travels.

To make things even more confusing, sometimes light behaves like a particle, and sometimes it behaves like a wave. Scientists really aren't sure they understand this phenomenon, but they accept it. Currently, theory states that light is both a particle *and* a wave. The situation that light is in determines whether it will behave as a particle or a wave. Scientists call this theory the **particle/wave duality theory**, and it is used frequently in describing light.

Particle/Wave Duality Theory - The theory that light sometimes behaves as a particle and sometimes behaves as a wave

If you don't understand the particle/wave duality theory, don't worry. Scientists don't understand how light can be both a particle and a wave, but for now they are forced to believe it because that's what the experimental data indicate.

The wave model of light is by far the most useful way to picture light, so we will start there. Consider the following wave:

FIGURE 7.7
A Wave

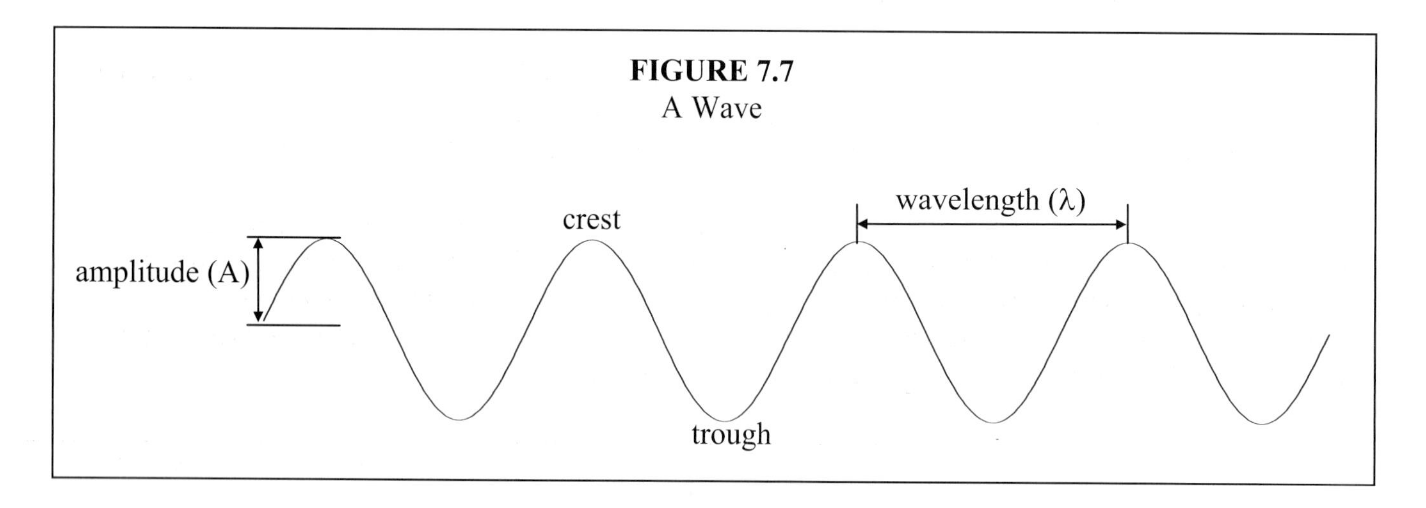

This is one way to picture light. It moves through space just like a wave moves through water. The high points of the wave are called **crests**, and the low points are called **troughs**. If I were to describe the above wave, I would need to specify two things. First, I have to say how far apart the waves are. We use the term **wavelength** to accomplish this.

Wavelength - The distance between the crests (or troughs) of a wave

Since wavelength is a measurement of distance, its units will be meters, centimeters, etc. The wavelength of a wave is usually given the symbol "λ" (the lower-case Greek letter "lambda"), as illustrated above.

In order to describe the wave completely, however, I not only have to tell you how far apart the waves are; I also need to tell you how big the waves are. This is referred to as the **amplitude** of the wave.

Amplitude - A measure of the height of the crests or the depths of the troughs on a wave

Amplitude is usually given the symbol "A," as illustrated above. Amplitude is not a distance measurement, however. The units for amplitude vary from wave to wave, and we will not get into those distinctions here. If you take physics, you should learn about the units of amplitude. For right now, just remember that amplitude means how big the wave is. The larger the amplitude, the bigger the wave.

How does all of this apply to light? Well, the wavelength of light corresponds to its *color*. Perhaps you have seen what happens when sunlight hits a prism. The apparently white light, once it passes through the prism, turns into a rainbow of colors. This is because a prism, through laws you will learn in physics, breaks light into its different wavelengths. Each color you see represents a range of wavelengths.

The same thing happens to make a rainbow. In Genesis 9:11-17, God promised Noah that the rainbow would symbolize His promise to never again destroy the earth with a flood. God accomplishes this by allowing little water droplets to be suspended in the air after a rainfall. These water droplets act like little prisms, separating light into different wavelengths.

Think about the last time you saw a rainbow. What colors did you see? Red was at the top of the rainbow; orange was next; yellow followed orange; and then there was green, blue, indigo, and violet. These are the basic colors of light, and they always appear in a rainbow in that order. Why?

Red light, for example, is made up of waves whose wavelengths are between 650 and 700 nanometers long. The prefix "nano" means 10^{-9}, thus red light's wavelengths are very small. However, of all the colors, red has the longest wavelength. Figure 7.8 summarizes each basic color and the approximate range of wavelengths that make up that color. This range of light wavelengths is typically called the "**visible spectrum**." You will learn why it is called this a bit later.

Rainbow photo ©GoodShoot/SuperStock, Inc.

Visible Spectrum Illustration by Megan Whitaker

FIGURE 7.8
The Visible Spectrum

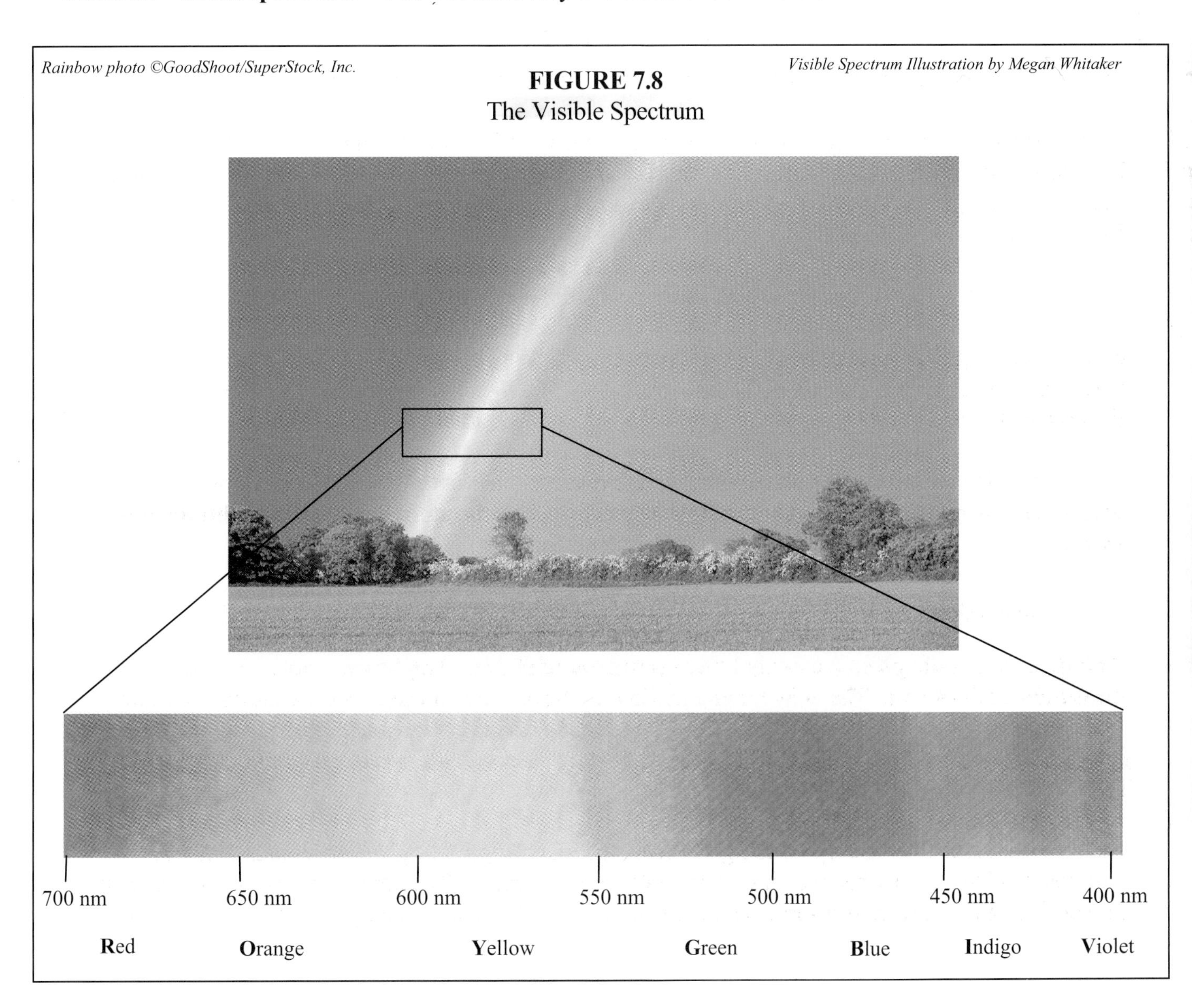

As you will learn in physics, the wavelength of light determines how the light passes through a prism. Since wavelength means color, the colors of light in a rainbow always appear in the same place, because they always pass through the water droplets in the air in exactly the same way.

Although you needn't memorize the wavelengths that each color corresponds to, you *do* need to memorize the *relative* size of the wavelengths in question. What I mean by this is that you need to know that red light has the largest wavelengths, orange light has smaller wavelengths, and so on. This is easy to do if you think about the colors as a single name. Notice how I have put the first letter of each color listed in the figure in boldfaced type. This is to emphasize that when you put them together in the proper order, you come up with a man's name: ROY G. BIV. Thus, if you think of ROY G. BIV every time you think of the colors in the visible spectrum, you will always know that red light has the longest wavelength, violet light has the smallest, and you will also know the order of all colors in between.

Now that we know what the wavelength of light means, what does the amplitude of light mean? Well, simply put, the amplitude of a light wave tells us how bright the light wave is. A dim light bulb produces light waves that have a small amplitude, while a bright light bulb produces light with a large amplitude. Thus, if I had two light bulbs of the same color, but one was dimmer than the other, I could say that the light bulbs were both producing light of the same wavelength but of different amplitudes. Similarly, if I had two light bulbs of different color but the same brightness, I could say that the light bulbs produced light waves of equal amplitude but of different wavelengths.

There's a bit more to describing a light wave than just its amplitude and wavelength; the speed of the light wave is also important. After all, we know that light travels from place to place. If I turn on a light bulb, the light rushes from the light bulb and fills up the room. Thus, light does travel. But how fast does it travel? Have you ever watched a thunderstorm and seen a lightning flash but heard the thundercrash a few moments later? The reason this happens is that the light generated by the lightning flash travels more quickly than the sound it makes. Thus, the light reaches your eyes before the sound reaches your ears. It turns out that for every 5 seconds between the time you see the lightning flash and the time you hear the thunderclap, the lightning is striking 1 mile away from you.

Light, then, travels fast: much faster than sound. In fact, light travels in air with a speed of 3.0×10^8 meters per second (m/s). In units you are more familiar with, that translates to 9.8×10^8 feet per second (ft/s)! That's really moving! To put it another way, light travels 190,000 miles every second! One thing that's very interesting about this number is that it does not change with the wavelength or amplitude of the light wave. Regardless of the color or the brightness of light, its speed in air is always 3.0×10^8 m/s. Since this value doesn't change, we call it a **physical constant**.

<u>Physical constant</u> - A measurable quantity in nature that does not change

There are several physical constants you will be introduced to in this course, and the speed of light is the first one. Each physical constant has its own symbol. The symbol for the speed of light is "c." Thus, when I say that $c = 3.0 \times 10^8$ m/s, I am giving you the speed of light. To put your mind at ease, do not worry about trying to memorize this number. I will always provide you with any physical constants you need to solve problems that I assign.

Now that we know about the speed of light, we can finally speak of another very important means of describing a wave: **frequency.**

<u>Frequency</u> - The number of wave crests (or troughs) that pass a given point each second

To best understand frequency, suppose you are standing in the ocean. If you look out at the waves, the distance between each crest is the wavelength. On the other hand, if you stand still and count the number of wave crests that hit you every second, that is frequency.

Think for a moment about the unit associated with frequency. Suppose you stand still in the water at the beach and count that two ocean wave crests hit you every second. You would report your measurement as "2 wave crests per second." However, since everyone at the beach can see the fact that you are counting waves, you might just say "2 per second." What does "per second" mean? Well, the unit "meters per second" is mathematically expressed as "meters/second." Thus, the "per second" part just means "1/second." This means that the standard unit for frequency is simply 1/second. This unit is usually called the "**Hertz**" (abbreviated "**Hz**"), in honor of Heinrich Rudolf Hertz, a German physicist who greatly expanded our knowledge of light and its relationship to electricity.

Now it turns out that the frequency, wavelength, and speed of light are all related to each other by a simple equation:

$$f = \frac{c}{\lambda} \tag{7.1}$$

In this equation, "f" stands for frequency, "c" is the speed of light, and "λ" stands for wavelength. Notice how the units work out. Speed is in m/s, while wavelength is in m. When I divide the speed by the wavelength, meters cancel, and I am left with 1/s, which is also called "Hertz." This is a very important equation. I expect you to memorize it. See how we use it in the two examples that follow:

EXAMPLE 7.2

A violet light wave has a wavelength of 405 nm. What is its frequency? ($c = 3.0 \times 10^8$ m/s, and the prefix "nano" means 10^{-9}).

To solve this problem, we first must recognize that we use Equation (7.1):

$$f = \frac{c}{\lambda}$$

Thus, to find the frequency, we just need to divide the speed, which is given, by the wavelength, which is also given. Unfortunately, before we can use the equation, we must make sure our units work out. The wavelength is in nm, while the speed is in m/s. In order for the equation to work, however, our units must agree. Thus, either m needs to change to nm or vice versa. I choose to change nm into m. According to what's given:

$$1 \text{ nm} = 10^{-9} \text{ m}$$

$$\frac{405 \cancel{\text{nm}}}{1} \times \frac{10^{-9} \text{ m}}{1 \cancel{\text{nm}}} = 4.05 \times 10^{-7} \text{ m}$$

You should try to put that equation into your calculator to see if you can get the right answer. If your calculator is a scientific calculator, you can put the numbers into the calculator in scientific notation. However, most calculators have a very specific way in which they want you to input scientific notation. If you don't do it correctly, your answers will be off by a factor of 10 or more.

Now that our units agree, we can put the numbers into the equation:

$$f = \frac{3.0 \times 10^{8} \ \frac{\cancel{m}}{s}}{4.05 \times 10^{-7} \ \cancel{m}} = 7.4 \times 10^{14} \ \frac{1}{s}$$

Remember, since the speed of light as given has only 2 significant figures, I can only report my answer to 2 significant figures as well; thus, the frequency is $\underline{7.4 \times 10^{14}\ \text{Hz}}$. What does the answer mean? It means that when you look at violet light, 7.4×10^{14} crests of the light wave hit you every second! Once again, make sure you can put this equation into your calculator and get the right result. If you cannot, consult your calculator manual or contact me using any of the means listed on the "NEED HELP?" page near the front of this book.

A light wave has a frequency of 4.70×10^{14} Hz. What is its wavelength and color according to the chart in Figure 7.8? ($c = 3.0 \times 10^{8}$ m/s, "nano" = 10^{-9})

Once again, we use Equation (7.1):

$$f = \frac{c}{\lambda}$$

But now we have to rearrange it so that we are solving for wavelength:

$$\lambda = \frac{c}{f}$$

Now we can plug in the numbers:

$$\lambda = \frac{3.0 \times 10^{8} \ \frac{m}{\cancel{s}}}{4.70 \times 10^{14} \ \frac{1}{\cancel{s}}} = 6.4 \times 10^{-7} \ m$$

The chart in Figure 7.8, however, lists wavelength in nm. Thus, if we are to refer to the chart, we must convert to nm:

$$\frac{6.4 \times 10^{-7} \ \cancel{m}}{1} \times \frac{1 \ nm}{10^{-9} \ \cancel{m}} = 640 \ nm$$

The wavelength is <u>640 nm</u>. According to Figure 7.8, then, this light is <u>orange light</u>.

Make sure you understand how to use Equation (7.1) by completing the following problems:

ON YOUR OWN
($c = 3.0 \times 10^8$ m/s, "nano" = 10^{-9})

7.4 What color is light that has a frequency of 6.4×10^{14} Hz?

7.5 If red light has wavelengths from 7.00×10^2 nm to 6.50×10^2 nm, what is the frequency range for red light?

The Electromagnetic Spectrum

Now that you've had some experience *using* Equation (7.1), go back and just look at it for a moment. Notice how the equation tells us that frequency and wavelength are related. As wavelength gets large, Equation (7.1) tells us that we will be dividing by a large number. What does that tell us about frequency? It tells us that for large wavelengths, frequency is small. Conversely, if wavelength is small, frequency is large. When two quantities behave like this, we say that they are **inversely related** to one another. When studying light, this inverse relationship between frequency and wavelength is a very important thing to remember:

When wavelength is large, frequency is small. When wavelength is small, frequency is large.

This inverse relationship between frequency and wavelength should actually make sense if you think about it. Imagine yourself once again standing in the ocean. As you look at the waves coming towards you, if their crests are far apart (they have a large wavelength), they will not hit you very often (they will have a small frequency). However, if the wave crests are close together (small wavelength), they will hit you very often (large frequency).

Now what you may not realize about light is that the light we see with our eyes (the visible spectrum) is actually only a small part of the light that comes to us from the sun. It turns out that the sun bathes our planet with light of many, many different wavelengths and frequencies. Our eyes perceive only a small fraction of the total amount of this light. Figure 7.9 is a more complete representation of all light that comes from the sun, which we call the **electromagnetic spectrum**.

Illustrations from the MasterClips collection

FIGURE 7.9
The Electromagnetic Spectrum

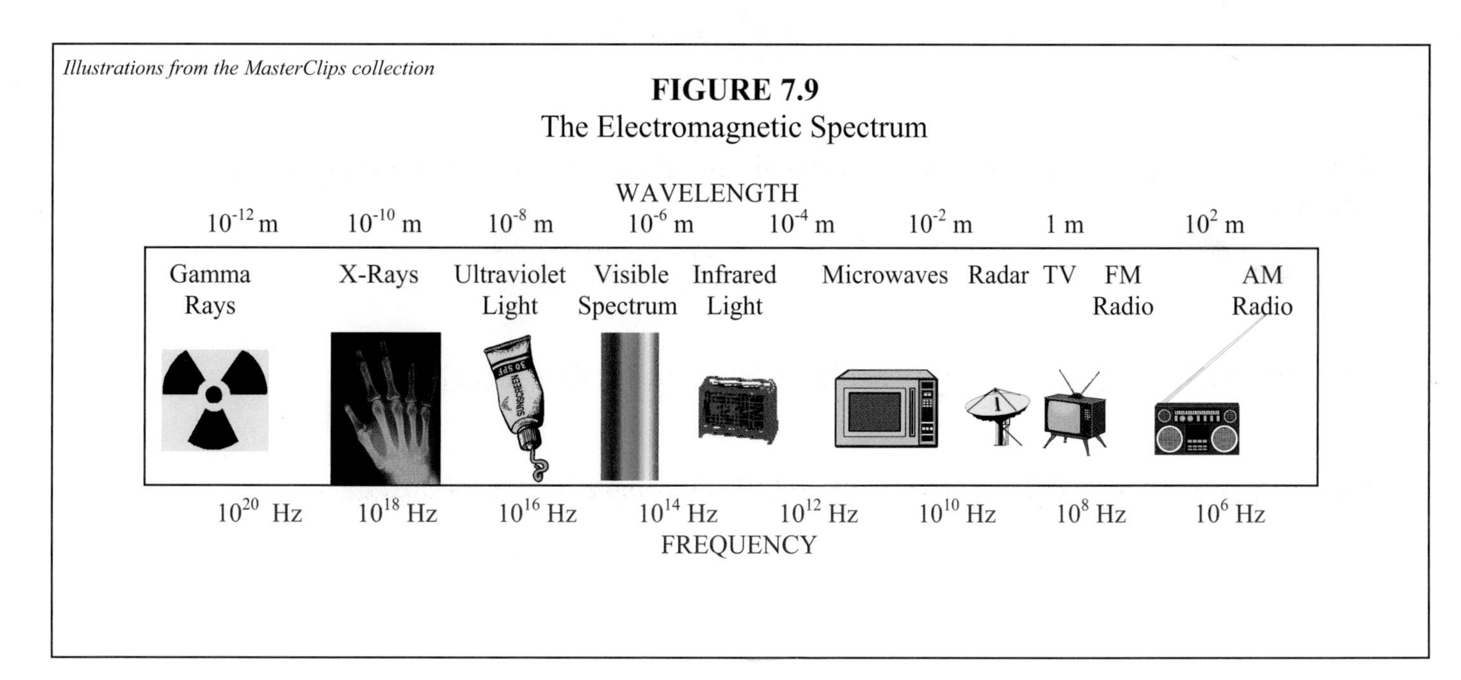

If you examine Figure 7.9, you will first realize just how small a part of the electromagnetic spectrum we actually see with our eyes. The rainbow under the words "visible spectrum" represents that part of the electromagnetic spectrum. Everything else shown in Figure 7.9, from wavelengths of 10^{-14} m all the way to wavelengths of 10^{4} m, is still light. Our eyes, however, are only sensitive to the light contained in the visible spectrum; thus, that is the only light we perceive. However, as you can tell by the figure, we find uses for much of the light we cannot see.

For example, the signals that are sent out from a radio station to your radio antennae are, in fact, waves of light. We cannot see these light waves, but they are there nevertheless. Since they are there, we can use them to carry radio signals (FM and AM radio signals, for example). The only difference between these light waves and the ones we see with our eyes is the wavelength. Wavelengths of light used to carry radio signals are longer than those of visible light. In addition, television antennae also use light with wavelengths longer than those of visible light. What may be even more surprising to you is that microwave ovens use light in order to cook food. Microwaves also have longer wavelengths than those of visible light.

Notice that while we often use several of the light waves with wavelengths *longer* than visible light, we do not frequently use any of the light waves with wavelengths *shorter* than visible light. There is a very good reason for this. Light, of course, has energy. Its energy is kinetic, since the light is moving. It turns out that the energy of a light wave is directly proportional to its frequency. This is an important enough fact that you need to memorize it:

As a light wave's frequency increases, its energy increases.
As its frequency decreases, its energy decreases.

Since frequency and wavelength are inversely related to each other, we could also say:

As a light wave's wavelength increases, its energy decreases.
As its wavelength decreases, its energy increases.

Based on these two facts, then, light waves with wavelengths shorter than those of visible light have higher frequencies and therefore higher energy. This is a very important consideration because when light strikes something (your eye, your skin, etc.) it deposits its energy into what it strikes. Thus, when visible light strikes your eye, it deposits its energy there. As you'll see in the next experiment, your eye uses that energy to transmit signals to the brain, and that's what causes you to see. It turns out that light with wavelengths shorter than visible light has enough energy to kill living tissue. Thus, when ultraviolet light (the first thing in Figure 7.9 with wavelengths shorter than those of visible light) strikes your skin, it can kill some of your cells. If your skin is exposed to too much ultraviolet light, a large number of cells will die, and you will get a burn on your skin, which we call a **sunburn**. You can avoid a sunburn by covering your skin with sunscreen, a lotion containing chemicals that absorb the ultraviolet light before it hits your skin.

Because of the lethal nature of ultraviolet light, God has built into this wonderful planet a very efficient system that filters out the majority of these rays. You may have heard some talk about the ozone layer. We will go into this in detail in the next module, but for right now I'll tell you that the ozone layer is part of the filtering system that God has designed to shield us from this destructive type of light. Although it is incredibly efficient, the ozone layer does allow some ultraviolet light through, and that is why you can get sunburned if you stay out in the sun too long without protection.

Gamma rays and X-rays are more energetic than ultraviolet light, so they are even more dangerous to living tissue. Nevertheless, we still use both of these forms of light occasionally in medical procedures. X-rays, of course, are used to examine bones and other aspects of our internal anatomy for diagnostic purposes. Although these X-rays will kill some living tissue, that risk is worth the benefit of being able to diagnose internal problems without surgery. Thus, as long as you do not get X-rays frequently, your risk associated with the nature of the light is low, and the benefit you get from the diagnosis of internal problems is high. Gamma rays are also used in some medical applications. For example, certain types of cancer can be treated by exposing the cancerous site to gamma rays. Even though this leads to the death of healthy tissue, it can also lead to the death of cancerous tissue, so once again, the risk of gamma ray exposure can be worth the medical benefit.

Before we leave this section, I want to mention one point of terminology. Notice that Figure 7.9 is titled "The Electromagnetic Spectrum." That's because we know that light is actually an electromagnetic phenomenon. It is produced by the interaction of electrically charged particles. Thus, light is often called **electromagnetic** (ee lek' troh mag net' ik) **radiation**. Although that sounds like a scary term, it just means "light," but it includes all wavelengths of light, not just the wavelengths that are visible.

The Relationship Between Frequency and Energy

You need to know one more equation that deals with light. This equation relates the frequency of a light wave to its energy. As I already mentioned, as a light wave's frequency increases, its energy does as well. The mathematical equation that governs this relationship is:

$$E = h \cdot f \qquad (7.2)$$

In this equation, "E" is the energy of the light wave and "f" is the frequency. The symbol "h" refers to another physical constant, called **Planck's** (Plahnks) **constant**. Named after its discoverer, German physicist Max Planck, this constant allows us to relate energy and frequency. Its value is 6.63×10^{-34} J/Hz. Once again, you needn't memorize the number. It will be given in any problem for which it is needed. Instead of worrying over the value, look at the units. Notice that if I stick those units into Equation (7.2), then I will get:

$$E = \frac{J}{\cancel{Hz}} \times \cancel{Hz}$$

When this happens, the Hertz unit cancels, leaving Joules, an energy unit.

Now remember what "Hertz" means. It means "1/second." With that in mind, let's look at the unit for Planck's constant once again. If I take the unit "Joule" and divide it by "1/second," I get:

$$\frac{J}{Hz} = \frac{J}{1/s} = J \cdot s$$

Thus, "Joule·second" is another way of expressing the unit for Planck's constant.

Now that you understand the units for Planck's constant, see how we use it in the two examples that follow:

EXAMPLE 7.3
(h = 6.63 x 10^{-34} J/Hz, c = 3.0 x 10^{8} m/s)

A light wave has a frequency of 2.3 x 10^{16} Hz. What is its energy?

This is easy. We simply plug our numbers into Equation (7.2):

$$E = h \cdot f$$

$$E = 6.63 \times 10^{-34} \frac{J}{\cancel{Hz}} \cdot 2.3 \times 10^{16} \cancel{Hz}$$

$$\underline{E = 1.5 \times 10^{-17} J}$$

We can only keep two significant figures because our frequency had only two. Now this might seem like a small amount of energy, but it corresponds to the energy of ultraviolet light, so it is enough to kill living tissue!

A light wave has a wavelength of 1.21 x 10^{-12} m. What is its energy?

This problem is a little harder because the only equation we can use to calculate the energy of light is Equation (7.2), and it uses *frequency*, not *wavelength*. In order to solve this, then, we must first turn the wavelength we've been given into frequency. We can do this with Equation (7.1):

$$f = \frac{c}{\lambda}$$

$$f = \frac{3.0 \times 10^{8} \frac{\cancel{m}}{s}}{1.21 \times 10^{-12} \cancel{m}} = 2.5 \times 10^{20} \frac{1}{s} = 2.5 \times 10^{20} \text{ Hz}$$

Now that we have the frequency, we can use Equation (7.2):

$$E = h \cdot f$$

$$E = 6.63 \times 10^{-34} \frac{J}{\cancel{Hz}} \cdot 2.5 \times 10^{20} \cancel{Hz}$$

$$E = 1.7 \times 10^{-13} J$$

The energy is $\underline{1.7 \times 10^{-13} J}$.

Make sure you understand how to use Equation (7.2) by completing the following exercises:

ON YOUR OWN
($h = 6.63 \times 10^{-34}$ J/Hz, $c = 3.0 \times 10^{8}$ m/s, "nano" means 10^{-9})

7.6 If a light wave has energy of 3.4×10^{-14} J, what is its frequency?

7.7 If a visible light wave has energy of 3.3×10^{-19} J, what color is it? (You can use Figure 7.8 if you need to.)

How the Eye Detects Color

Before we go on to discussing the Bohr model of the atom, I would like you to perform the following experiment. Hopefully, it will give you some insight into how humans perceive color.

EXPERIMENT 7.2
How the Eye Detects Color

Supplies

- Two plain white sheets of paper (There cannot be lines on them.)
- A bright red marker (A crayon will also work, but a marker is better.)

1. Take one of the sheets of paper and make a thick cross on it with the red marker. The cross should be about 6 inches long, and the two legs which make it up should be about 3/4 of an inch thick.
2. Color the entire cross so that you have a large, solid bright red cross in the middle of a white sheet of paper.
3. Take the clean sheet of white paper and put it underneath the sheet with the cross on it. Make sure the cross faces you so that you can see it.
4. Stare at the cross for a full 60 seconds. You can blink if you need to, but do not take your eyes off of the cross.
5. After a full 60 seconds of staring at the cross, quickly pull the top sheet of paper out of the way, so that you can only see the clean sheet of paper on the bottom. What happens? For about 90% of students, a blue-green cross will appear for a few seconds on the blank sheet of paper. After a few moments, it should vanish, however. This optical illusion will not work for some people, especially if they have a tendency towards color blindness.

What happened in the experiment? In order to see light, your eyes are equipped with cells called **rods** and **cones**. These cells transmit electrical signals to the brain whenever they are hit by certain energies of light. The brain receives the electrical transmissions and uses them to form an image in your mind. It turns out that the rods are sensitive mostly to low levels of light and are not very sensitive to color. However, the cones in your eye respond to certain specific energies of light. Some cones are sensitive only to low-energy visible light (red), while others are sensitive to medium-energy visible light (green), while still others are sensitive to the highest energy light (blue). When colored light hits these cells, they will only send signals to the brain if the light that they are sensitive to is hitting them. Thus, if blue light hits your eyes, the cones that respond to blue light will send signals to your brain, but the cones that respond to green and red light will not. As a result, your brain interprets the signals coming from your eyes and puts a blue image in your head. Similarly, if red light hits your eyes, the cones that respond to red light will send signals to your brain, but the cones that respond to green and blue will not. As a result, your brain constructs a red image.

If a mixture of colors hits your eyes, the cones send signals to your brain in proportion to the amount of light to which they are sensitive. Suppose, for example, you are looking at a purple ball. When that purple hits your eyes, the cones that respond to red light send some signals to your brain, and the cones that respond to blue light send some signals to your brain. Your brain then constructs a purple image of the ball in response. If the purple color of the ball is on the blue side of purple, then the cones that respond to blue light send more signals to the brain than the cones that respond to red light. On the other hand, if the purple is more red than blue, the cones that respond to red light send more signals to the brain than those that respond to blue light. As a result, all of the colors that you see are really the result of the brain receiving signals from three types of cone cells and adding those three primary colors (red, green, and blue) with the weight indicated by the amount of signal coming from each type of cell.

Now, what happened in the experiment? While you were looking at the red cross, your cones that respond to low-energy (red) light were sending signals to the brain, but your other cones weren't doing anything. It turns out that cones get tired pretty quickly, and when they have sent the same signal to the brain for a period of several seconds, they eventually just shut off. The brain, sensing that no more signals are coming from the cones, assumes that they have shut off simply because they are tired, and it holds the same image in your mind until new signals come along. Thus, as you were staring at the cross, your low-energy cones eventually shut off. Since no more signals were coming from the low-energy cones, and since no signals had ever come from the medium and high-energy cones, the brain was receiving no more signals. It therefore assumed you were still looking at the cross and continued to hold the image in your mind.

When you yanked the top sheet away, white light began to hit your eyes where only red light had hit them before. Since white light contains all energies, your medium and high-energy cones began to receive light and transmit signals to your brain. Your low-energy cones, however, were still shut off, so they didn't send any signals, even though they should have. Thus, the brain started receiving new signals, but only from the high- and medium-energy cones. Thus, it constructed an image of green (medium-energy) and blue (high-energy) light. Eventually, however, your low-energy cones realized that they had to start transmitting again, and, once they did, the brain realized that the eyes were seeing all energies of light and thus formed a white image in your mind.

The way we perceive color, then, is based on the energy of the light that hits our eyes. Isn't it marvelous how well designed the eye is to handle such a complex operation? That should tell you something about how marvelous its Designer is!

The Bohr Model of the Atom

During my discussion of Rutherford's model of the atom, I mentioned that it was inconsistent with electromagnetic theory. Nevertheless, since the model seemed to be the only way to explain Rutherford's experimental data, it became the accepted model of the day. As chemists sought to understand more about atomic structure, however, they did more experiments.

One of the experiments that shed some light on atomic structure involved adding energy to atoms by heating them or passing electricity through them. When this was done, the atoms emitted light, some of which was in the visible spectrum. The interesting thing, however, was that each type of atom seemed to emit its own unique color or colors of visible light. For example, when mercury atoms were heated, a pale blue glow would occur. On the other hand, when electricity was passed through neon atoms, a bright pink light was emitted.

As chemists investigated these experiments further, the mystery deepened. They learned that the colors which were being emitted during these experiments were actually the result of a handful of *individual* wavelengths of visible light being emitted by the atom. For example, when electricity is passed through hydrogen gas, the atoms emit a purple light. It turns out that this light is composed of exactly four wavelengths: 410 nm (violet), 434 nm (indigo), 486 nm (blue), and 656 nm (red-orange). These wavelengths mix together to form a purple color.

What makes these results interesting is the fact that atoms seem to emit *individual* wavelengths of light. This is very unusual. Most colors that we see are the result of a *range* of wavelengths, not *individual* ones. For example, if I were to make a red light, I would probably take a normal light bulb and paint it red. The paint would keep all of the wavelengths except the red ones from getting through, and thus I would have a red light. The light coming from this painted bulb, however, would not be made up of just one individual wavelength. Instead, all wavelengths from 700 to 650 nm would be there. The fact that atoms emit only a handful of individual wavelengths indicates that something unique is at work.

An additionally interesting fact is that each element has its own *unique* set of wavelengths that it produces when heated or electrified. Thus, while hydrogen emits visible light with wavelengths of 410 nm, 434 nm, 486 nm, and 656 nm, tantalum (Ta) emits visible light of only a single wavelength: 535 nm. This little fact is quite useful. After all, if you need to determine the atoms in a substance, one way to do it is to heat the substance up and determine the wavelengths of light that are emitted. Since each element in the substance will emit its own unique wavelengths of light, you should be able to match the elements with their respective wavelengths and thus determine the elemental composition of the substance.

In fact, this is the main way that astronomers determine the elemental makeup of distant stars. The reason stars are bright is because the substances that make them up are very hot and they therefore emit light. They emit so many different wavelengths of light that, to our eyes, the light appears white. However, a scientific instrument known as a **spectrometer** (spek trahmm' uht er) can analyze the light and determine all of the individual wavelengths that make it up. Then, by matching the elements to their expected individual wavelengths, the elemental composition of the star can be determined. This method of analysis, known as **spectroscopy** (spek trahs' kuh pee), can be applied to chemicals as easily as it is applied to stars. Thus, it is one of the most useful tools a chemist has in analyzing unknown substances.

(The multimedia CD has a video demonstration of how atoms emit individual wavelengths of light.)

The ability of atoms to emit individual wavelengths of light also helped Niels Bohr develop his own model of the atom. Bohr took the planetary model developed by Rutherford and added his own little twist. Rutherford assumed that all of the electrons orbited around the protons in a single orbit. Bohr suggested, instead, that there were several possible orbits that the electrons could be in. Furthermore, he suggested, the electrons could jump from one orbit to another. In addition, by the time Bohr had come up with his model of the atom, the neutron had been discovered by Chadwick. Bohr put the neutrons and protons together in the center of the atom and called it the **nucleus**. This model is shown in Figure 7.10.

FIGURE 7.10
The Bohr Model of the Atom

Illustration by Megan Whitaker

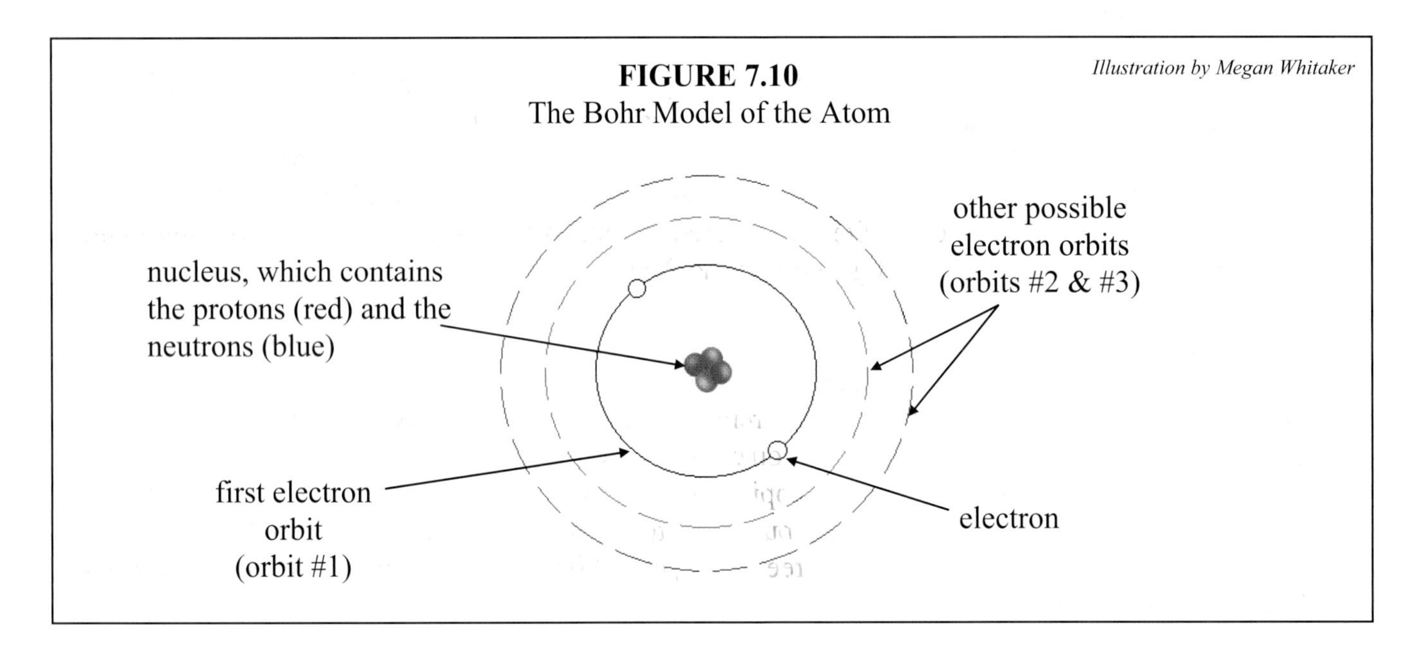

One very important advance that Bohr made in developing his model was his assumption that the electrons could only jump into and out of specific orbits. So, if an electron decided to jump out of orbit #1, it could not go just anywhere. It had to go into another allowed orbit, such as orbit #2. And even though it could be in orbit #1 and jump to orbit #2, *it could not be anywhere in between*. This type of assumption is called a **quantum** (kwahnt' uhm) assumption, and at the time was considered rather odd. After all, if the electron could go to either orbit, why couldn't it go somewhere in between? Bohr had no answer to this question (indeed, even modern chemists have only a rough answer), but it was a necessary assumption in his model.

Now, one of the first questions you might ask is: How do the electrons jump from one orbit to another? Well, the orbits that are farther away from the nucleus require more energy. Thus, if an electron wants to jump from an orbit that is close to the nucleus (orbit #1) to an orbit that is far away from the nucleus (orbit #3), it can do so by absorbing some energy. This is what happens when you heat a substance or pass electricity through it. The atoms in the substance absorb energy, allowing the electrons to move from an orbit close to the nucleus to an orbit far away from the nucleus. When an electron moves from an orbit close to the nucleus to an orbit far from the nucleus, we say that the electron has been **excited**.

If, on the other hand, an electron would like to move from an orbit far away from the nucleus to an orbit close to the nucleus, the electron must somehow get rid of its extra energy. In other words, while in the orbit that is far from the nucleus, the electron has more energy than it would in an orbit that is close to the nucleus. In order to get to that closer orbit, the electron must find a way to get rid of that excess energy. The way it does this is to emit light. When an electron moves from an orbit far from the nucleus to an orbit close to the nucleus, we say it has **de-excited**.

Thus, when you heat up a substance and its atoms emit light, you are witnessing a two-step process. Electrons in the atoms first gain energy (because of the heat) and move into orbits that are far away from the nucleus. Then, after they sit in the far orbits for a short time, they decide they would rather be back in their original orbit, so they emit light in order to get rid of the excess energy and go back to their original orbit.

This is what makes the flames of a fire. When you burn wood, the chemicals in the wood are first changed into carbon dioxide gas and water vapor via a combustion reaction, as we have discussed before. However, heat is also produced in the reaction. This heat excites the electrons in the gases surrounding the fire, and they move to orbits farther away from their nucleus. However, after being there for a while, they decide to de-excite back to their original orbit. To do this, they must emit light. This is where the pleasant yellow-orange glow of the flames comes from.

If you light a natural gas oven, you will notice that the flames are blue, not orange. This is because the natural gas company purposely contaminates its natural gas with a substance whose atoms emit a blue glow when excited. Thus, the difference in the flame color is due to the fact that there are different atoms being excited in each case. You can examine another similar phenomenon if you burn some wrapping paper. The dyes in most wrapping papers have elements in them that are different from other papers. Thus, while most papers burn with the same yellow-orange glow that wood burns with, wrapping paper typically burns with green, blue, or red flames because of the elements that make up the dyes in the wrapping paper.

(Electrons will emit light when they are excited by *any* means, because the electrons must eventually de-excite. The multimedia CD has a video demonstration of light being emitted by atoms excited by chemical means rather than with heat or electricity.)

Remember I said earlier that Bohr assumed that an atom's electrons could only be in certain orbits? Well, since each orbit has a specific energy associated with it, this means that an electron can only have certain energies. If it is in orbit #1, it has a certain energy (E_1). If it is in orbit #2 it has another, higher energy (E_2). But, since the electron can never be in between orbit #1 and orbit #2, it can also never have any of the energies between E_1 and E_2. This assumption makes no sense at all, because common sense tells us that any object should be able to attain any energy it wants, provided there is some way of transferring the necessary energy. Bohr, however, said there were only certain energies that the electrons in the atom could have. This, another quantum assumption in Bohr's model, is actually the one we now see as critical to understanding atomic structure.

Now once again, Bohr did not just dream up his model for the atom. In fact, for the hydrogen atom, he had detailed mathematical descriptions of the electron moving in its various orbits. He could actually use his mathematics to determine the energy of each electron orbit. Thus, he could determine how much energy the electron needed to jump from one orbit to another. Once he knew the energy necessary for the electron to jump from one orbit to the next, he could then determine the wavelength of light that the electrons must emit in order to de-excite. It turns out that the wavelengths he calculated were *exactly* equal to the wavelengths that chemists observed for de-exciting hydrogen atoms! This indicated that Bohr was on the right track, and his model for the atom quickly gained acceptance.

In fact, even though today's accepted model of the atom is slightly different from Bohr's model, the detailed mathematics that Bohr developed are still taught in college chemistry today. These mathematics, which are a bit too detailed to tackle in this course, provide good insight into the physical mechanisms that hold the atom together. If you end up taking advanced chemistry, you will learn the mathematics of the Bohr model of the atom.

Additionally, the stylized pictures of atoms you often see are really based on the Bohr model of the atom. Thus, although science has moved on past Bohr's model, the scientific community recognizes his achievements to this day.

The Quantum Mechanical Model of the Atom

Today's model of the atom, called the **quantum mechanical model**, is a bit more complicated than Bohr's model, but it still retains many of Bohr's concepts. For example, chemists still believe that electrons orbit around the nucleus and that there are many, many different orbits that electrons can continually jump between. In addition, we still say the electrons need energy to go into orbits that are far away from the nucleus, and they need to release energy in the form of light in order to get back to orbits close to the nucleus. Finally, Bohr's quantum assumption regarding energy is still followed. We assume that there are only certain energies that an electron in an atom can have. Thus, in one orbit, an electron might have an energy of E_1, and in the next higher orbit have an energy of E_2, but the electron can have no energy in between E_1 and E_2.

Where the quantum mechanical model mainly differs from the Bohr model is the type of orbits that the electrons can occupy. In the Bohr model, electrons orbit the nucleus in circular orbits. In the quantum mechanical model, this is not the case. The way we theorize that electrons orbit the nucleus today is a little more complex. We assume that electrons do not orbit in fixed circles, but instead orbit in "clouds" we call **orbitals** (or' buh tul). When orbiting in a fixed circle, the electron is always the same distance from the nucleus. In today's model of the atom, electrons can, at different times, be at different distances from the nucleus but still be in the same orbital. However, if you were to watch the electron for quite a while, you would see that it stayed within a certain boundary.

If you are having trouble understanding what I am saying, think about it this way. Suppose I had a dog in my backyard, and suppose I drove a stake in the center of the yard and tied a rope to it. When the dog is tied to the rope, he can be anywhere in the backyard within a circle that is determined by the length of the rope. Although the dog can be anywhere inside the circle, he cannot go outside of the circle. This is like an electron in an orbital. The orbital has a general shape, and the electron can be anywhere within that shape. On the other hand, suppose the rope got wet during the winter and froze while it was stretched out. Then, because the rope was rigid and would always stay stretched out, the dog could not move anywhere except on the edge of the circle. This is like an electron orbiting in the Bohr model. It cannot be anywhere except on the edge of the circular orbit. That's the difference between electron orbits (the Bohr model) and electron orbitals (the quantum mechanical model).

The other difference between the Bohr model and the quantum mechanical model is the shape of the orbitals. In the Bohr model, all electrons had to orbit in circles. In the quantum mechanical model, there are several differently shaped orbitals that electrons can use to orbit around the nucleus. The shapes of these orbitals can become very complex, so we will limit ourselves to a discussion of the first three types of electron orbitals. The majority of atoms in nature use only these orbitals, so that will be enough for an introduction into this rather esoteric concept.

The simplest type of orbital an electron can occupy is called the **s orbital**. This orbital, as illustrated in Figure 7.11, is spherically shaped, with the nucleus at the center. Once again, an electron that occupies an s orbital can be anywhere inside the sphere, but cannot venture outside of the sphere.

FIGURE 7.11
s orbitals

Illustration by Megan Whitaker

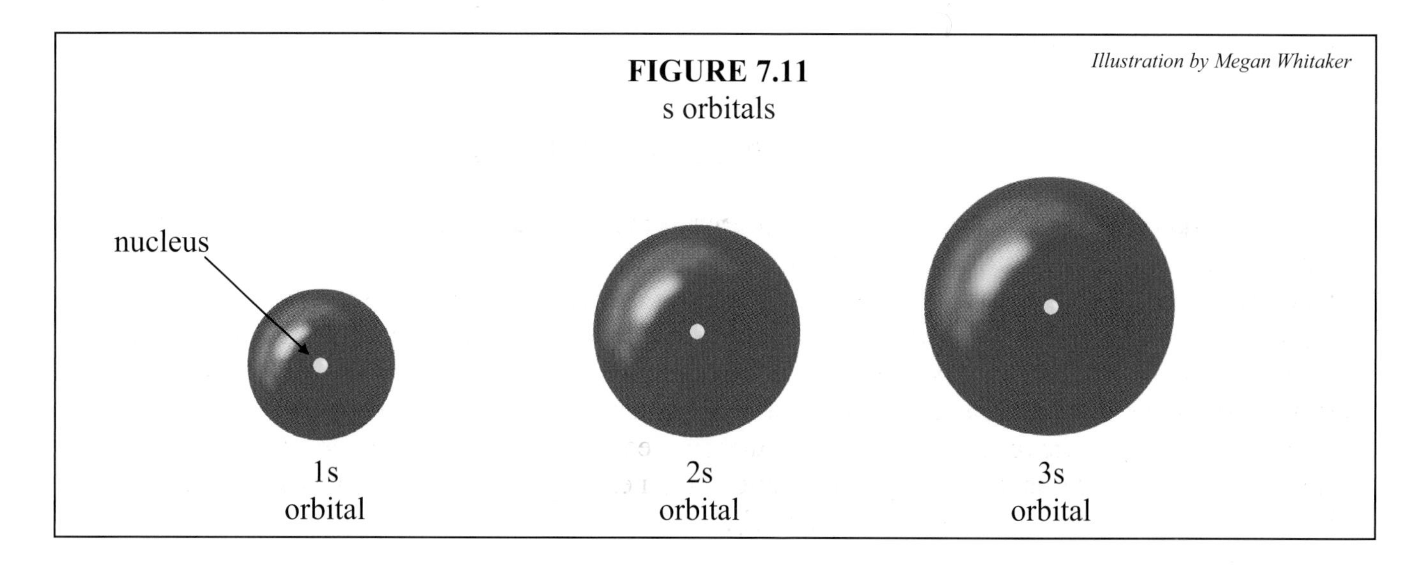

Notice that in Figure 7.11 that there are three s orbitals, each of differing size. This is because in the quantum mechanical atom, just like in the Bohr atom, electrons can orbit far away from the nucleus or close to the nucleus depending on their energy. In the Bohr atom, the orbit number (1, 2, 3, etc.) determined how far they were from the nucleus. In the same way, the quantum mechanical model of the atom allows electrons to orbit far from the nucleus or close to it. If an electron orbits far away from the nucleus in a spherical pattern, it will occupy a large s orbital. If it orbits close to the nucleus in a spherical pattern, it will occupy a small s orbital. The way we delineate small s orbitals from large ones is the number that appears next to the orbital letter. A 1s orbital is always smaller than a 2s orbital, which is always smaller than a 3s orbital.

Just like in the Bohr model, the farther away from the nucleus the electron is, the more energy it must have. Thus, the electrons in a 1s orbital have less energy than the electrons in a 2s orbital. Since the size of the orbital determines the energy of the electrons, the number that appears next to the orbital letter is often called the "energy level" of the atom.

The next orbital shape we will consider is the **p orbital**. This orbital is dumbbell-shaped, as illustrated in Figure 7.12. The nucleus is in the center of the dumbbell. Just as there are different sizes of s orbitals that correspond to different energies of electrons, there are also different sizes of p orbitals that correspond to different electron energies. One interesting thing about p orbitals is that there are none of them on the first energy level of the atom. Thus, there is no such thing as a 1p orbital. The first p orbital is the 2p orbital.

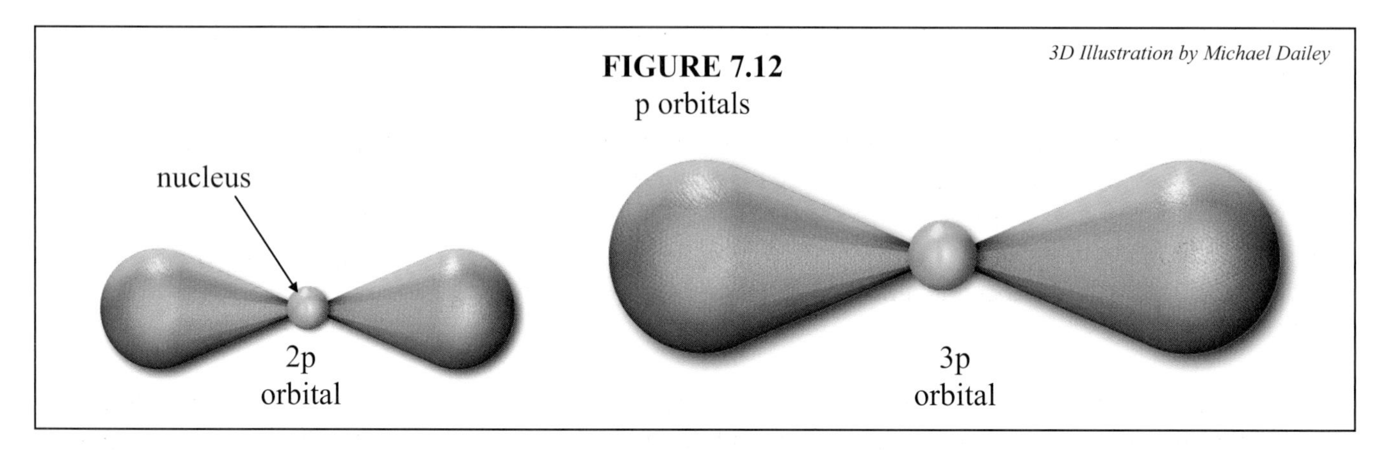

FIGURE 7.12
p orbitals

Another interesting thing about p orbitals is that for every energy level (or size of the orbital), there are actually three different p orbitals. They are all shaped the same, but they are oriented differently in space. One p orbital is oriented horizontally (as drawn in Figure 7.12), while the second one is oriented vertically. The third is oriented 90 degrees from both of the others. In other words, each of the three p orbitals is oriented along a different axis in three-dimensional space. Figure 7.13 illustrates how the three p orbitals all exist together in an atom:

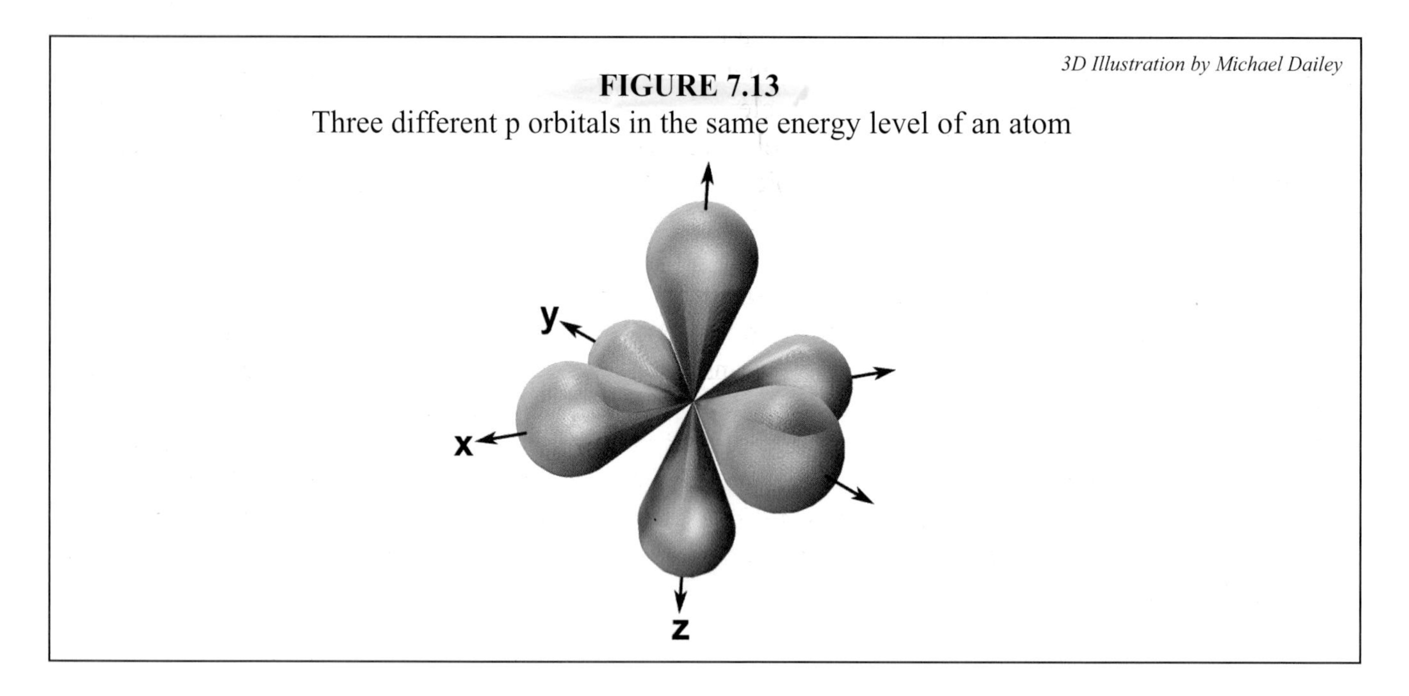

FIGURE 7.13
Three different p orbitals in the same energy level of an atom

So, based on the figure, there can be three different p orbitals for each energy level. Thus, there are three 2p orbitals, three 3p orbitals that are all bigger than the 2p orbitals, three 4p orbitals that are all bigger than the 3p orbitals, and so on. Like I said before, all of this is a little esoteric and may be hard to picture in your mind. If that is the case, don't worry too much about it. At the end of this discussion, I will summarize what you need to know about the quantum mechanical model.

Finally, the last type of orbital we will use in this course is the d orbital. The shapes of the d orbitals are very complex, so I will not attempt to illustrate them for you. What you need to know about d orbitals is that there are five different d orbitals for each energy level. Also, the lowest-energy (smallest) d orbitals are located on the third energy level; thus, the first set of d orbitals is the 3d orbitals.

One important thing to note about the different shaped orbitals is that within the same atomic energy level, the different shaped orbitals require different amounts of energy. It is rather easy for an electron to whirl around the nucleus in a sphere, so s orbitals are the lowest-energy of all orbitals. On the other hand, a dumbbell is a little more complex, so p orbitals require more energetic electrons than s orbitals. Finally, since d orbitals have even more complex shapes, they require more energy than either s or p orbitals.

One more important thing to know about orbitals is that each individual orbital can only hold two electrons. After two electrons, the orbital is full, and any new electrons must find another orbital to fill. Now, since there are three different p orbitals on an energy level, six electrons could fit into all three p orbitals on that energy level (two per orbital). In the same way, since there are five different d orbitals on an energy level, then up to 10 electrons can go into the d orbitals of a given energy level.

Now, go back in your mind and think a bit about all of these orbitals and how electrons use them. If an electron is orbiting as close as it can to the nucleus, we say that it is in the first energy level. On the first energy level, only one type of orbital exists: the s orbital. Thus, if an electron orbits as close as it possibly can to the nucleus, then it must do so in a spherically shaped orbit (1s orbital). However, if the electron goes up to the next energy level (#2), it is a little farther away from the nucleus and has the option of orbiting the nucleus in a spherical orbit (2s orbital) or in one of any three dumbbell-shaped orbits (2p orbitals). If the electron goes away from the nucleus even further into the third atomic energy level, then it has the choice of a spherically shaped orbit (3s orbital), one of three dumbbell-shaped orbits (3p orbitals), or one of five more complex orbits (3d orbitals).

This, then, is the modern-day view of the atom. The atom has a nucleus of tightly packed protons and neutrons. Its electrons whirl around that nucleus in several different energy levels (1, 2, 3...). Depending on the energy level, the electrons have several different shapes of orbitals (s, p, d, and other orbitals) that they can use to travel around the nucleus. Each specific energy level (1, 2, 3...) has its own energy requirements for the electrons. In addition, each orbital shape has additional energy requirements for the electrons. Based on the energy that an electron has, it will go into the energy level and orbital whose energy requirements it best meets. Each orbital has a maximum capacity of two electrons. This is the summary of things that you must understand about the quantum mechanical model of the atom.

Building Atoms in the Quantum Mechanical Model (Electron Configurations)

Now if all of the previous discussion confused you a bit, don't worry too much. In this section, we will try to apply what we have learned about the quantum mechanical model of the atom to describe what an atom might look like. After you have seen this application, you might gain a better understanding of what you read in the last section.

Before we start describing what atoms look like, however, there is one more fact of nature that you must learn:

All forms of matter try to stay in their lowest possible energy state.

This is a fact of nature that we will use over and over again. Think about it this way: Matter is basically lazy. It doesn't want to work any harder than it needs to. Thus, if at all possible, matter will always try to stay in its lowest-energy configuration.

For example, remember when we were discussing the light that was emitted from excited atoms? We said that when an atom is heated, its electrons absorb some of that kinetic energy and use it to travel to an orbital farther away from the nucleus. The electron then emits light to go back down to its original orbital. Why did it do that? Once it jumped to the higher-energy orbital, why didn't it just stay there? It didn't stay there because once it was in a higher-energy orbital, it was not in its lowest possible energy state. Thus, the electron released energy in the form of light in order to get back to its lowest energy state.

Since the lowest possible energy state is so important to matter, we give it a special name. We call it the **ground state** of the substance:

Ground state - The lowest possible energy state for a given substance

Be sure you understand that the ground state is different for every substance on earth. The ground state of a hydrogen atom, for example, is rather different from the ground state of a helium atom, as we will soon see.

Now that we know this very important fact about nature, we are finally ready to start building up atoms. First, let's describe the simplest atom in nature: the hydrogen-1 (^{1}H) atom. By looking at the chart, we can tell how many protons, neutrons, and electrons are in this atom. On the chart, hydrogen's atomic number is 1. This means there is 1 proton and 1 electron in a hydrogen atom. Also, since the mass number of this hydrogen atom is 1, then there are 1-1 = 0 neutrons in the atom. Thus, we already know that a hydrogen atom has 1 proton and no neutrons in its nucleus. What about the electron? Where is it?

Well, since all matter in nature tries to get to its lowest energy state, the electron will go into the lowest energy orbital available. Energy level #1 is the lowest energy level, and energy level #1 has only an s orbital in it, so the electron goes into the 1s orbital. This means that the electron is orbiting the nucleus relatively close and the orbit has the shape of a sphere. Thus, the overall picture of a hydrogen atom in its ground state is one electron orbiting close to one proton, and the electron's orbital is spherically shaped.

Now let's move to another simple atom, the helium-4 (^{4}He) atom. After looking at the chart, we see that it has 2 protons and 2 electrons. In addition, since its mass number is 4, it has 4 - 2 = 2 neutrons. The two electrons will try to go into the lowest energy orbits available. Well, since energy level #1 is the lowest energy level and since it has only an s orbital in it, then both electrons will go into the 1s orbital. Thus, our picture for the helium-4 atom in its ground state is two electrons orbiting closely to a nucleus tightly packed with two protons and two neutrons. The electrons whirl about this nucleus in a spherical orbit.

Continuing on, we can consider a lithium-6 (^{6}Li) atom. After looking at the chart and dealing with the mass number, we see that it has three protons, three electrons, and three neutrons. The three electrons will try to go into the lowest energy orbits available. Well, since energy level #1 is the lowest energy level, and since it has only an s orbital in it, two electrons will go into the 1s orbital. However, any single orbital can hold only two electrons. Thus, the third electron in the atom cannot fit into the 1s orbital. It must, therefore, find the next lowest energy orbital to go into. Since there is no more room on the first energy level, it goes into energy level #2. At this point, it can either go into a 2s orbital or a 2p orbital. Since s orbitals are lower in energy than p orbitals, it will go into the 2s orbital. Thus, our picture for the lithium-6 atom is two electrons orbiting closely to and one electron orbiting farther away from a nucleus tightly packed with three protons and three neutrons. All three electrons whirl about this nucleus in spherical orbitals.

Do you see how the thinking goes here? Let's try a much more complicated atom and we will then introduce an abbreviated notation that will make things a little easier to describe. Consider a ^{22}Ne atom. According to our rules, it has 10 protons, 10 electrons, and 12 neutrons. Where do the electrons go?

The electrons will each look for the lowest-energy orbitals available. The first two will occupy the 1s orbital since energy level #1 is the lowest energy level available. Since there is only one s orbital available on energy level #1, the rest of the electrons will have to go into higher energy levels. On energy level #2, there are both s and p orbitals available. Since the s orbitals are lower in energy, the next two electrons will fill that orbital up. So far, then, we have placed four electrons in their orbitals and we have six to go. On energy level #2, there are three p orbitals that can also be filled. Each p orbital can hold two electrons, so the six remaining electrons can go into the 2p orbitals.

So, our picture of the ^{22}Ne atom is a tightly-packed nucleus of 10 protons and 12 neutrons with 10 electrons whirling around it. Two of the electrons are orbiting the nucleus closely in a spherical pattern. There are two more electrons whirling around the nucleus in a bit farther away, also in a spherical pattern. However there are also six electrons whirling about in dumbbell-shaped clouds. This is the way a modern-day chemist would describe a ^{22}Ne atom. Figure 7.14 illustrates what this would look like:

FIGURE 7.14
The Quantum Mechanical View of a ^{22}Ne Atom

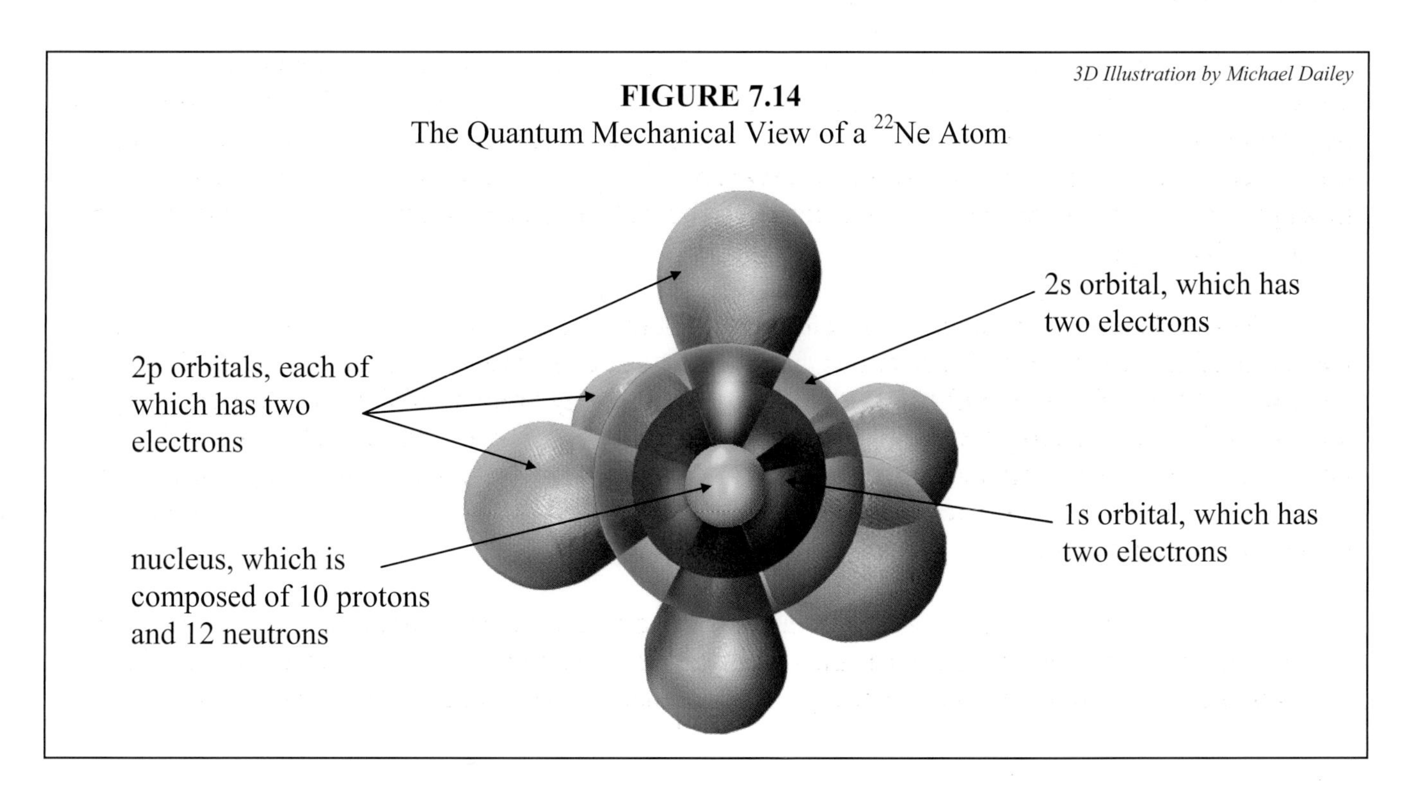

Hopefully you now have a reasonably good grasp of the quantum mechanical model of the atom. If you understand how I came up with my description of the ^{22}Ne atom, then you are all set. If you don't quite understand it, then you need to reread the material or seek out another source because, as I have said about other topics, you can't go farther in this course without understanding this important concept.

Since the quantum mechanical model of the atom is complicated, we will simplify it a bit with some notation. First of all, we will learn in the next module that all of the chemical behavior of an atom is governed solely by the number of electrons it has. As a result, the nucleus of an atom is relatively unimportant to a chemist. Thus, when giving the quantum mechanical description of an atom, chemists do not bother to mention the nucleus. Also, rather than saying in words where all of the electrons go, we develop an abbreviation to help us. The abbreviation for the neon atom we just discussed is:

$$1s^2 2s^2 2p^6$$

To a chemist, this says all one needs to know about the structure of a neon atom. The numbers to the left of the letters represent the energy level, whereas the letters represent the shape of the orbital. The superscripted numbers tell you how many electrons are in the orbital. Thus, the abbreviation above tells us that there are two electrons in the 1s orbital, two electrons in the 2s orbital, and six electrons in the 2p orbitals. This type of abbreviation is called an **electron configuration**, and you are going to become *intimately* familiar with this kind of notation in this and the next module.

Now before we do a few sample electron configurations to show you how easy this is, you need to be introduced to one more interesting fact in chemistry. Remember how we decided which electrons in an atom went into which orbitals? We did it by using energy arguments. We said that the first energy level should fill up before electrons go into the second energy level. In the same way, the second energy level should fill all of its orbitals before electrons go into the third energy level. This is all true. However, at the third energy level, things get a little messy. It turns out that the d orbitals in the third energy level take a lot of energy. In fact, the 3d orbitals require more energetic electrons than the 4s orbitals. Thus, if I had enough electrons to fill up the 3s and 3p orbitals, I would then start filling the 4s orbitals before I go back and fill up the 3d orbitals. Messy, isn't it?

As atoms get more and more electrons, the order of orbitals gets even more messy. In order to be able to determine the proper electron configuration of any atom, then, we need to have some way of determining the order in which orbitals fill up. Believe it or not, we *already have that,* because we have the periodic chart. You see, the periodic chart gets its weird shape *because* of electron configurations. By looking at the chart, you can easily determine an atom's electron configuration; thus, you can avoid all of the messy energy arguments we have been making. These energy arguments, of course, still form the *reason* that electron orbitals fill up in the way that they do. However, we needn't go through those arguments every time we want to determine an atom's electron configuration. All we have to do is look at the chart!

Of course, to do this you have to see why the chart is arranged the way it is. Examine Figure 7.15 to understand how the chart helps you with electron configurations:

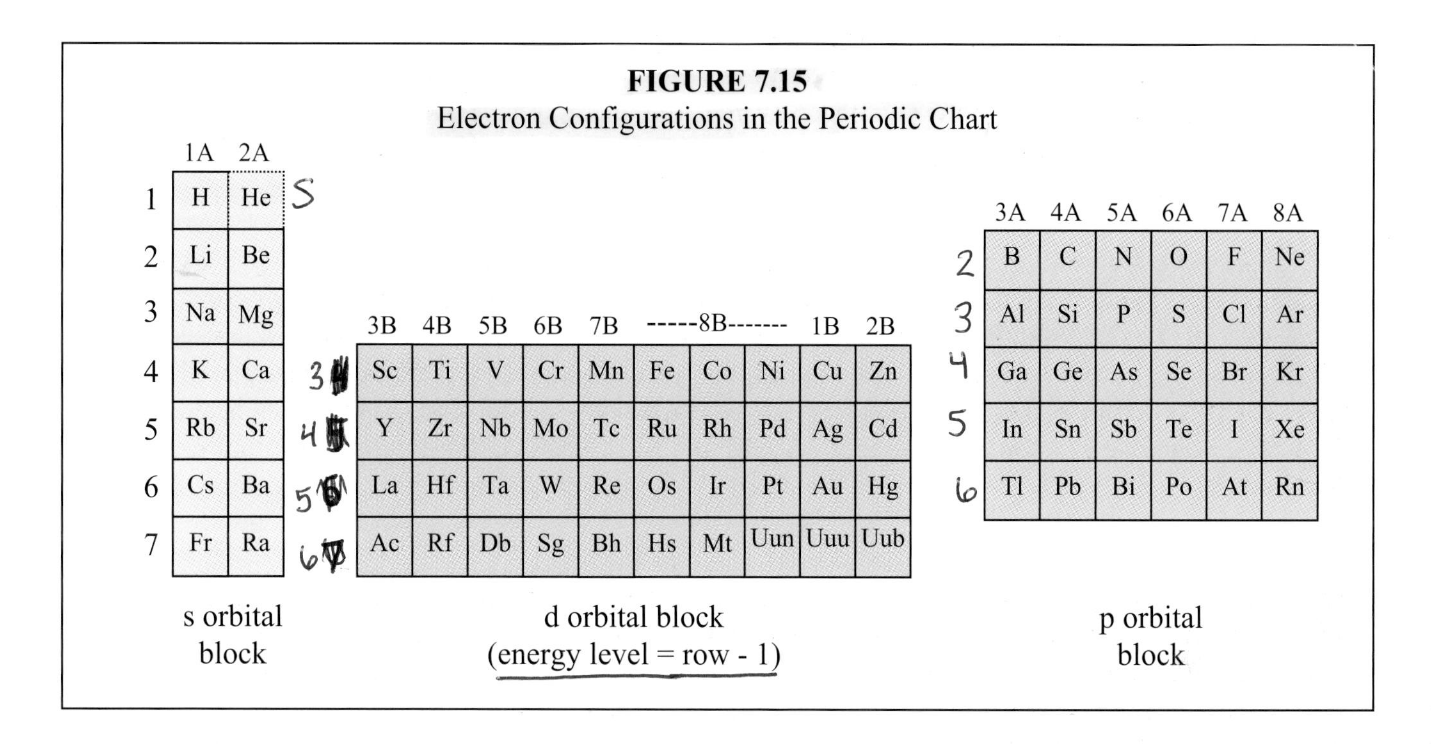

FIGURE 7.15
Electron Configurations in the Periodic Chart

Looking at Figure 7.15, you should first see that for the purposes of this discussion, we pretend that helium sits right next to hydrogen on the chart. The next thing you should see in the figure is that the columns have been separated. The first two columns on the chart we call the “s orbital” block, because the last electrons in the atoms of that block end up in an s orbital. Since each s orbital can only have two electrons in it, the s orbital block can only be two columns wide. The next group represents those atoms whose last electrons end up in d orbitals. Since there are five d orbitals, and each can have two electrons, there are 10 elements in each row that have their last electrons in d orbitals. Finally, the last block is the p orbital block. These elements have their last electrons in a p orbital. Since the three p orbitals can each hold two electrons, the p orbital block is six columns wide. Note that the two lowest rows on the periodic chart (the lanthanides and actinides) are not in the figure. That’s because we will not consider them here. However, if you are interested, their last electrons go into f orbitals, which are even more complex than d orbitals.

Since the columns represent the electron orbitals in the atoms, the rows of the chart represent the energy level that those electrons are in. Notice, for example, that Ne is in row #2. This means that the last electron in Ne goes in the second energy level. The fact that Ne is in a column that is a part of the p orbital block tells us that the last electron is in a p orbital. Thus, Ne’s last electron goes in the 2p orbital. We had figured this out earlier with energy arguments, but figuring it out this way is easier.

To use the arrangement of the periodic chart to determine electron configurations, you just have to look at the periodic chart and find the atom whose electron configuration you wish to determine. Then, starting with hydrogen, walk through the chart, assigning electrons in each orbital one box at a time. When you get to the box that represents the atom you are interested in, you’re done. The only trick you have to remember is that in the d orbital block, the row that the elements are on is actually one number higher than the energy level that the electrons are in. So, when filling up d orbitals, subtract one from the row number in order to get the energy level. If this all sounds a bit confusing, the next three examples should clear things up a bit.

EXAMPLE 7.4

Write the electron configuration for a carbon atom.

First, we look at the periodic chart and find out that carbon (C) is element #6. To get to this atom, we start walking through the chart from the beginning. The first two elements are H and He. Their boxes are in row 1 and the s orbital group. Thus, the first two electrons go into the 1s orbital. Going on through the chart, we get to Li and Be. They are in row 2 and the s orbital group, so we put two electrons in the 2s orbital. Continuing on, we hit B and then C, where we want to stop. This means that we went through two boxes in the row 2, p orbital block. Thus, there are two electrons in the 2p set of orbitals. Since we've hit carbon, we are done. So the electron configuration is:

$$\underline{1s^2 2s^2 2p^2}$$

Remember, the superscripts in the electron configuration represent the number of electrons in each orbital. Thus, the sum of the superscripts must equal the number of electrons in the atom. Carbon has six electrons, and our superscripts add up to 6. Our answer, therefore, checks out.

Write the electron configuration for a Cl atom.

In order to get to Cl (#17), we have to go through several boxes. We must walk through all of row 1, which has two boxes in the s orbital block. That gives us $1s^2$. We must also go through all of row 2, which has two boxes in the s orbital block and six boxes in the p orbital block. This gives us $2s^2 2p^6$. We also must go through the s orbital block in row 3. This gives us $3s^2$. Finally, to get to Cl, we go through five boxes in the row 3, p orbital block. This gives us $3p^5$. The electron configuration, then, is:

$$\underline{1s^2 2s^2 2p^6 3s^2 3p^5}$$

The superscripts add up to the number of electrons in a Cl atom (17), so our answer checks out.

Give the electron configuration of a Ag atom.

To get to Ag, we have to walk through several boxes on the chart. We must walk through all of row 1, which has two boxes in the s orbital block. That gives us $1s^2$. We must also go through all of row 2, which has two boxes in the s orbital block and six boxes in the p orbital block. This gives us $2s^2 2p^6$. Row 3 has two boxes in the s orbital group and six in the p orbital box, so that gives us $3s^2 3p^6$. Row 4 has two boxes in the s orbital block, 10 in the d orbital block, and six in the p orbital block. Remembering that for d orbitals, we must subtract 1 from the row number, this gives us $4s^2 3d^{10} 4p^6$. In row 5, we must go through the two boxes in the s orbital block, so that gives us $5s^2$. In row 5, we also have to go through nine boxes to get to Ag. Once again, we subtract one from the row number with d orbitals, so this gives us $4d^9$. In the end, then, our electron configuration is:

$$\underline{1s^2 2s^2 2p^6 3s^2 3p^6 4s^2 3d^{10} 4p^6 5s^2 4d^9}$$

Our superscripts add up to 47, so our answer checks out.

This is all you do to get the electron configuration of an atom. With the periodic table as a guide, it is not a hard task at all. However, I do need to give you a couple of words of advice:

(1) Some chemistry books teach a different method to get electron configurations. Their method, however, requires memorization and is not the way real chemists do electron configurations. That's why I don't teach it their way. Don't get confused if you look at other books, however. Their method might be different, but the answers should be the same.

(2) Remember that these electron configurations are not just a worthless exercise. They have very, very deep meaning to a chemist. They not only tell you the structure of the atom you are interested in, but, as we will see in the next module, they also tell you a great deal about the chemistry of the atom.

See if you can do electron configurations on your own. Feel free to use the periodic chart I gave you in Module #3, but do not use Figure 7.15. You should be able to do your electron configurations from a normal chart, not one specially enhanced for electron configurations.

ON YOUR OWN

7.8 Write the electron configuration for phosphorus.

7.9 Write the electron configuration for arsenic (As).

7.10 What is the electron configuration of ruthenium (Ru)?

Abbreviated Electron Configurations

Before we leave electron configurations, I would like to give you a handy abbreviation that will keep you from writing a lot when you give your electron configurations. It turns out that the last column in the chart, column 8A, is very important. We will see why it is so important in the next module. For right now, just realize that most chemists gauge an atom by how far it is away from this column in the chart. Thus, one way you can abbreviate electron configurations is to look for the nearest atom in 8A that has a *lower* atomic number than the atom you are interested in. Then, you simply add any extra electrons on top of that atom's configuration. See how I do this in the next two example problems:

EXAMPLE 7.5

Write the abbreviated electron configuration for Sr.

To abbreviate an electron configuration, we just find the nearest 8A element that has a lower atomic number than the atom we are interested in. In this case, Kr fits the bill. We can then say that the only difference between Sr and Kr is that there are two more boxes in the row 5, s orbital block. Thus, we can write the electron configuration as:

$\underline{[Kr]5s^2}$

Putting Kr in square brackets tells a chemist that you are abbreviating the electron configuration in terms of the 8A element Kr.

Write the abbreviated electron configuration for Se.

The nearest 8A element that has a lower atomic number than Se is Ar. Don't be fooled here. Kr is the closest 8A element, but it does not have a *lower* atomic number. The only differences between Se and Ar are that there are two boxes in the row 4, s orbital group, 10 boxes in the row 4, d orbital group, and four boxes in the row 4, p orbital group. Thus, the abbreviated electron configuration for Se is:

$$\underline{[Ar]4s^2 3d^{10} 4p^4}$$

Do the following problems to cement the idea of abbreviated electron configurations in your mind:

ON YOUR OWN

7.11 What is the abbreviated electron configuration for I?

7.12 Write the electron configuration for Ga in abbreviated form.

As I said before, you will soon learn that the meaning of electron configurations is very deep, so your ability to predict them is a necessity.

The Amazing Design of Atoms

Before we leave this module, there are a couple of interesting things that I would like to tell you about atoms. First of all, the atom is mostly made up of empty space. Since matter is made up entirely of atoms, this also means that matter is mostly empty space. This may sound like a crazy statement to you. After all, when you hit your head against the wall, the wall certainly doesn't *feel* like empty space to you. Nevertheless, the statement is true.

Nuclear chemistry experiments have demonstrated that the nucleus of a hydrogen atom has a radius of about 1×10^{-15} meters. On the other hand, the mathematics of the quantum mechanical model of the atom tell us that the radius of the 1s orbital in that same atom (the closest orbital to the nucleus) is about 5×10^{-11} meters. Thus, on the atomic scale, the electrons are *very far away* from the nucleus. If you don't understand the implications of these numbers, think about it this way: If the nucleus of the hydrogen atom were one inch in diameter, the electron would be orbiting it about 3/4 of a mile away! Between the nucleus and the electron there would be absolutely nothing. No air, no matter. Nothing. Just empty space. Thus, the vast majority of an atom is simply empty space.

Now the reason the wall you hit your head against doesn't *feel* like empty space is that the electrons in the atoms of the wall are constantly moving around, as are the electrons in the atoms of your head. These electrons repel each other, since they are of like charge. Thus, when they come into contact, this repulsion causes a force that can be quite painful!

The most interesting aspect of an atom, however, is not that it is made up of mostly empty space. The most interesting aspect of the atom is how well designed it is. You see, in order for all atoms except ^{1}H to exist, we must have neutrons. Well, it turns out that the neutron is just slightly more massive than the proton. It's a good thing, too. Nuclear chemistry experiments indicate that the neutron must be more massive than the proton, or it would not exist! Calculations indicate that if the mass difference between the neutron and the proton were changed by as little as 0.2%, all neutrons would spontaneously decay into protons, and the only atom in the universe would be ^{1}H. If that were the case, no life could exist in the universe!

The mass of the electron is also quite crucial for the existence of life. Remember, the electrons whirl around the nucleus in the orbitals that we just discussed. Well, the *size* of those orbitals depends on the strength of the electrical attraction between the protons and electrons, as well as the mass of the electron. If the electron's mass were larger, the size of the orbitals would be smaller. If the electron's mass were smaller, the size of the orbitals would be larger. Why does that matter?

Remember that there are two kinds of compounds: ionic compounds and covalent compounds. As you will learn in the next module, whether a compound is ionic or covalent depends on how the atoms in the molecules rearrange their electrons. In ionic compounds, one atom takes electrons from another atom. The atom that takes the electrons ends up being negatively charged, and the atom from which the electrons are taken becomes positively charged. In covalent compounds, however, the two atoms actually *share* each others' electrons.

If the electron's mass were larger, the electron orbitals would be smaller. As a result, atoms would hold on to their electrons more tightly, and it would be difficult to form ionic compounds. Many of the chemical processes that occur in living organisms depend on the existence of ions from ionic compounds. Without these ionic compounds, the chemical processes that depend on them would cease, and living organisms would die.

On the other hand, most of the compounds that actually make up living organisms are covalent compounds. As I told you a moment ago, atoms in covalent compounds share their electrons. Well, if the electron mass were smaller, the electron orbitals would be larger and, as a result, atoms would hold onto their electrons rather loosely. This would make it very easy to form ionic compounds, but very hard to form covalent compounds. If that were the case, living organisms could not even form, much less exist!

In the end, then, the electron mass is just right to produce electron orbitals that are neither too big nor too small. The balance between ionic and covalent molecules is perfect for the formation and existence of life, due to the mass of the electron. Estimates indicate that a variation of as little as 2% in the mass of the electron would make it impossible for life as we know it to exist!

Believe it or not, we have not even reached the most amazing design feature of the atom! As we have already noted, protons are positively charged while electrons are negatively charged. These opposite charges are what hold the electrons to the atom, keeping them from flying away from the nucleus. The problem is, in order for the atom to be stable, the charges must be precisely balanced. According to our best measurements, the proton has *exactly* as much positive charge as the electron has negative charge. That's a good thing, too, because calculations indicate that if the charges were out of balance by as little as 0.00000001%, all of the atoms in our bodies would instantaneously explode!

What is really amazing about the perfect charge balance between the proton and the electron is that these two particles are incredibly different. The proton is about 2,000 times heavier than the electron. In addition, the proton seems to be made up of three smaller particles we call "quarks," while as far as we can tell, the electron is not made up of any smaller particles. Nevertheless, despite the fact that these two particles are quite different, they have *perfectly* balanced charges!

The most important thing to realize is that these atoms are the *simplest* building blocks of matter. Even these very simple building blocks, however, have delicately balanced parts which all work together to give the atom its properties. If one of these parts were even slightly altered, the delicate balance would be destroyed, and the chemistry necessary for life would be forever lost. How is it, then, that these two incredibly different particles are able to carry *exactly* the right amount of charge to make a stable atom? How is it that the neutron has *precisely* the right amount of extra mass to make the atom stable? How is it that the electron has *precisely* the right mass to make the existence of living organisms possible? Without these three incredible "coincidences," life could never exist in the universe. Nevertheless, we are here.

As I have pointed out before, there are many such incredible "coincidences" in nature. There are so many that I find it impossible for anyone who really understands physics or chemistry to believe that all this we are studying just came about by chance! The universe is too perfectly designed for that. From the smallest atom to the largest galaxy, incredible design features abound in our universe. These design features are the fingerprints of God, making all who objectively study the universe aware of His existence.

ANSWERS TO THE "ON YOUR OWN" PROBLEMS

7.1 a. On the periodic chart, Cr has an atomic number of 24. Thus, a Cr atom has 24 protons and 24 electrons.

b. On the periodic chart, Be has an atomic number of 4. Thus, a Be atom has 4 protons and 4 electrons.

c. On the periodic chart, La has an atomic number of 57. Thus, a La atom has 57 protons and 57 electrons.

7.2 a. Looking at the chart, Ar has an atomic number of 18. This means it has 18 protons and 18 electrons. Its mass number, according to the problem, is 40. If it has 40 total protons + neutrons, and it has 18 protons, then it has 40 - 18 = 22 neutrons.

b. Looking at the chart, chlorine (Cl) has an atomic number of 17. This means it has 17 protons and 17 electrons. Its mass number, according to the problem, is 37. If it has 37 total protons + neutrons, and it has 17 protons, then it has 37 - 17 = 20 neutrons.

c. Looking at the chart, La has an atomic number of 57. This means it has 57 protons and 57 electrons. Its mass number, according to the problem, is 139. If it has 139 total protons + neutrons, and it has 57 protons, then it has 139 - 57 = 82 neutrons.

7.3 First, since it has 34 protons, we know its atomic number is 34. The element on the chart which has an atomic number of 34 is selenium (Se). The problem states that this atom also has 41 neutrons; thus, its mass number is 34 + 41 = 75. Therefore, the atom is selenium-75, which has a symbol of ^{75}Se.

7.4 To solve this, we use Equation (7.1):

$$f = \frac{c}{\lambda}$$

But now we have to rearrange it so that we are solving for wavelength:

$$\lambda = \frac{c}{f}$$

Now we can plug in the numbers:

$$\lambda = \frac{3.0 \times 10^{8}\ \frac{m}{\cancel{s}}}{6.4 \times 10^{14}\ \frac{1}{\cancel{s}}} = 4.7 \times 10^{-7}\ m$$

Figure 7.8 lists wavelength in nm. Thus, if we are to refer to the chart, we must convert to nm:

$$\frac{4.7 \times 10^{-7}\ \cancel{m}}{1} \times \frac{1\ nm}{10^{-9}\ \cancel{m}} = 4.7 \times 10^{2}\ nm$$

According to Figure 7.8, this light is blue light.

7.5 To solve this problem, we need to simply convert the wavelength range into a frequency range. To do this, we just convert the two wavelengths in the range into frequencies:

$$\frac{7.00 \times 10^{2}\ \cancel{\text{nm}}}{1} \times \frac{10^{-9}\ \text{m}}{1\ \cancel{\text{nm}}} = 7.00 \times 10^{-7}\ \text{m}$$

Now that our units agree, we can put the numbers into the equation:

$$f = \frac{3.0 \times 10^{8}\ \frac{\cancel{\text{m}}}{\text{s}}}{7.00 \times 10^{-7}\ \cancel{\text{m}}} = 4.3 \times 10^{14}\ \frac{1}{\text{s}}$$

Now we convert the other part of the range:

$$\frac{6.50 \times 10^{2}\ \cancel{\text{nm}}}{1} \times \frac{10^{-9}\ \text{m}}{1\ \cancel{\text{nm}}} = 6.50 \times 10^{-7}\ \text{m}$$

Now that our units agree, we can put the numbers into the equation:

$$f = \frac{3.0 \times 10^{8}\ \frac{\cancel{\text{m}}}{\text{s}}}{6.50 \times 10^{-7}\ \cancel{\text{m}}} = 4.6 \times 10^{14}\ \frac{1}{\text{s}}$$

So the frequency range is between 4.3×10^{14} Hz and 4.6×10^{14} Hz.

7.6 This problem is a direct application of Equation (7.2). Since we know both E and h, we simply have to use algebra and rearrange the equation to solve for f:

$$f = \frac{E}{h}$$

$$f = \frac{3.4 \times 10^{-14}\ \cancel{\text{J}}}{6.63 \times 10^{-34}\ \frac{\cancel{\text{J}}}{\text{Hz}}} = \underline{5.1 \times 10^{19}\ \text{Hz}}$$

Looking at Figure 7.9, this turns out to be the frequency of X-rays. Thus, if you have ever had a medical X-ray, light of approximately this energy was projected onto you in order to produce the picture that the doctor or dentist examined.

7.7 In this problem, we are given the energy and asked to determine the color. The only thing that can tell us color, however, is wavelength. Equation (7.2) only relates energy and frequency. So, we must first determine the frequency, and then we can use Equation (7.1) to figure out the wavelength.

To determine the frequency:

$$f = \frac{E}{h}$$

$$f = \frac{3.3 \times 10^{-19} \not{J}}{6.63 \times 10^{-34} \frac{\not{J}}{Hz}} = 5.0 \times 10^{14} \text{ Hz}$$

Now that we have the frequency, we can use algebra to rearrange Equation (7.1) so that we can determine the wavelength:

$$\lambda = \frac{c}{f}$$

$$\lambda = \frac{3.0 \times 10^{8} \frac{m}{s}}{5.0 \times 10^{14} \text{ Hz}} = \frac{3.0 \times 10^{8} \frac{m}{\not{s}}}{5.0 \times 10^{14} \frac{1}{\not{s}}} = 6.0 \times 10^{-7} \text{ m}$$

In order to use Figure 7.8, we must convert this answer to nm, because that's the unit used in the figure. Remember, 1 nm = 10^{-9} m:

$$\frac{6.0 \times 10^{-7} \not{m}}{1} \times \frac{1 \text{ nm}}{10^{-9} \not{m}} = 6.0 \times 10^{2} \text{ nm}$$

According to Figure 7.8, light with this wavelength is yellow light.

7.8 In order to get to P (#15), we have to go through several boxes. We must walk through all of row 1, which has two boxes in the s orbital block. That gives us $1s^2$. We must also go through all of row 2, which has two boxes in the s orbital block and six boxes in the p orbital block. This gives us $2s^2 2p^6$. We also must go through the s orbital block in row 3. This gives us $3s^2$. Finally, to get to P, we go through three boxes in the row 3, p orbital block. This gives us $3p^3$. The electron configuration, then, is:

$$1s^2 2s^2 2p^6 3s^2 3p^3$$

The superscripts add up to the number of electrons in a P atom (15), so our answer checks out.

7.9 To get to As, we have to walk through several boxes on the chart. We must walk through all of row 1, which has two boxes in the s orbital block. That gives us $1s^2$. We must also go through all of row 2, which has two boxes in the s orbital block and six boxes in the p orbital block. This gives us $2s^2 2p^6$. Row 3 has two boxes in the s orbital group and six in the p orbital box, so that gives us $3s^2 3p^6$. Row 4 has two boxes in the s orbital block and 10 in the d orbital block that we must go through. Remembering that for d orbitals we must subtract one from the row number, this gives us $4s^2 3d^{10}$. Finally, to get to As, we must walk through three boxes in the row 4, p orbital block. That gives us $4p^3$. Therefore, the final electron configuration is:

$$\underline{1s^2 2s^2 2p^6 3s^2 3p^6 4s^2 3d^{10} 4p^3}$$

Our superscripts add up to 33, so our answer checks out.

7.10 To get to Ru, we have to walk through several boxes on the chart. We must walk through all of row 1, which has two boxes in the s orbital block. That gives us $1s^2$. We must also go through all of row 2, which has two boxes in the s orbital block and six boxes in the p orbital block. This gives us $2s^2 2p^6$. Row 3 has two boxes in the s orbital group and six in the p orbital box, so that gives us $3s^2 3p^6$. Row 4 has two boxes in the s orbital block, 10 in the d orbital block, and six in the p orbital block. Remembering that for d orbitals we must subtract one from the row number, this gives us $4s^2 3d^{10} 4p^6$. In row 5, we must go through the two boxes in the s orbital block, so that gives us $5s^2$. In row 5, we also have to go through six d orbital boxes to get to Ru. Once again, we subtract one from the row number with d orbitals, so this gives us $4d^6$. In the end, then, our electron configuration is:

$$\underline{1s^2 2s^2 2p^6 3s^2 3p^6 4s^2 3d^{10} 4p^6 5s^2 4d^6}$$

Our superscripts add up to 44, so our answer checks out.

7.11 To abbreviate an electron configuration, we just find the nearest 8A element that has a lower atomic number than the atom we are interested in. In this case, Kr fits the bill. We can then say the only differences between I and Kr are that there are two more boxes in the row 5, s orbital block, 10 boxes in the row 5, d orbital block, and five boxes in the row 5, p orbital block. Thus, we can write the electron configuration as:

$$\underline{[Kr]5s^2 4d^{10} 5p^5}$$

7.12 The nearest 8A element that has a lower atomic number than Ga is Ar. The only differences between Ga and Ar are that there are two boxes in the row 4, s orbital group, 10 boxes in the row 4, d orbital group, and one box in the row 4, p orbital group. Thus, the abbreviated electron configuration for Ga is:

$$\underline{[Ar]4s^2 3d^{10} 4p^1}$$

REVIEW QUESTIONS FOR MODULE #7

1. What was Rutherford's main contribution to the study of atomic structure?

2. What do we call the experimental apparatus that William Crookes used in his experiments, and what did he discover with it?

3. If two electrically charged particles repel one another, what can we conclude about their charges?

4. If a substance has 15 positive charges and 11 negative charges, is it positively charged, negatively charged, or neutral?

5. Where do the protons and neutrons exist in an atom?

6. Why is it so hard to separate one isotope from another?

7. What were the differences between the plum pudding model of the atom and the planetary model of the atom?

8. If you have an orange light bulb and a violet one, which emits waves with the largest wavelength? Which emits light of higher frequency? Which emits the higher energy light? (Do NOT look at the figures in the book. Just remember the name I told you to remember!)

9. Two light bulbs emit light of the same color. One, however, is much brighter than the other. What can you say about the wavelengths and amplitudes of the waves being emitted by each bulb?

10. If an atom absorbs energy, what happens to its electrons? If the atom is emitting light, what are its electrons doing?

11. What is the concept of ground state and why is it so important in chemistry?

12. The three fundamental particles that make up the atom are the proton, neutron, and electron. Order them in terms of decreasing mass.

PRACTICE PROBLEMS FOR MODULE #7
($c = 3.0 \times 10^8$ m/s, "nano" = 10^{-9}, $h = 6.63 \times 10^{-34}$ J/Hz)

1. Give the number of protons, electrons, and neutrons in the following atoms:

   a. ^{90}Zr b. ^{202}Hg c. ^{58}Ni d. ^{222}Rn

2. Which of the following atoms are isotopes?

   ^{22}Na, ^{22}Ne, ^{23}Na, ^{22}Mg, ^{24}Na

3. What is the symbol of the atom made up of 39 protons, 45 neutrons, and 39 electrons?

4. If light has a frequency of 1.2×10^{14} Hz, what is its wavelength?

5. What is the energy of a light wave with a frequency of 5.3×10^{20} Hz?

6. If a light wave has a wavelength of 351 nm, what is its energy?

7. One electron is in a 2p orbital while another is in a 3s orbital. Which has the higher energy? What shape is each electron's orbit?

8. Give the full electron configurations of the following atoms:

   a. Ti b. S c. Rb

9. Give the abbreviated electron configurations for the following atoms:

   a. V b. Sn c. In

10. What is wrong with the following electron configurations?

    a. $1s^2 2s^2 2p^6 3s^2 3p^7 4s^2 3d^{10}$

    b. $1s^2 2s^2 2p^6 3s^2 3p^6 3d^{10} 4s^2 4p^5$

MODULE #8: Molecular Structure

Introduction

In the last module, we learned a little bit about what atoms would look like if we were able to actually see them. We learned that atomic structure is very complex and depends mostly on the way that the atom's electrons are arranged in their orbitals. Now that we have this information under our belts, we can undertake the task of determining what molecules would look like if we could actually see them.

Since you learned that atomic structure is very complex, you might think that molecular structure is complex as well. Although this is certainly true to some extent, you will be happy to know that in terms of what you will learn in this course, molecular structure is simple in comparison to atomic structure. Indeed, determining the structure of a molecule is much like putting together a jigsaw puzzle. Like most things you learn in this course, however, there is a fair amount of background material we must understand before we tackle the actual subject of molecular structure.

Electron Configurations and the Periodic Chart

In the last module, you (hopefully) became quite efficient at determining an atom's electron configuration by looking at the periodic chart. Now we need to learn a little bit about what a chemist can learn by looking at an electron configuration. To do this, let's take a look at the electron configuration of bromine:

$$1s^2 2s^2 2p^6 3s^2 3p^6 4s^2 3d^{10} 4p^5$$

We already learned what all of this means in terms of atomic structure, but what does it mean in terms of the chemical behavior of the compound?

Think for a minute about the electrons in a bromine atom. There are 35 of them, and they are all whirling around the nucleus. The electron configuration tells us that they are whirling about in three different orbital shapes (because s, p, and d orbitals are all present) and at four different distances from the nucleus (since energy levels 1-4 are present). Now when a bromine atom wants to join with another atom in order to make a molecule, what part of the bromine atom will play the most important role in this joining? Clearly the electrons will. After all, the bromine atom (and whatever atom it decides to join with) both have electrons whirling outside the nucleus. When these atoms move close to one another, their electrons will be the first things to come into contact with each other.

It turns out that when joining to form a molecule, atoms never get close enough to each other for their nuclei to interact. As a result, when forming molecules, the only important part of the atom is its electrons. The nucleus of the atom plays no role whatsoever in making the molecule. In fact, not even *all* of the electrons are important. The only electrons that significantly affect the chemistry of an atom are those that are farthest away from the nucleus. Those are the electrons that will begin to interact with another atom's electrons.

If you look back at the electron configuration for bromine, which electrons are the farthest away from the nucleus? Well, in bromine's electron configuration, energy levels 1-4 are present. According to what we learned in the last module, the larger the energy level's number, the farther it is from the nucleus. Therefore, the electrons in energy level 4 are the ones that are farthest from the nucleus. These are the ones that will affect the chemistry of bromine. Although the electrons in energy levels 1-3 are present, they do little to affect the way in which bromine behaves chemically.

Since the electrons farthest away from the nucleus are so important, we give them a name: **valence** (vay' lents) **electrons**.

Valence electrons - The electrons that exist farthest from an atom's nucleus. They are generally the electrons with the highest energy level number.

Since bromine has two electrons in the 4s orbital and five electrons in the 4p orbitals, it has seven valence electrons. Those seven electrons are responsible for almost all of bromine's chemical behavior. This is a very important point; you must remember it:

An atom's chemistry is mostly determined by the number of valence electrons it has.

Thus, one thing we learn from the electron configuration of an atom is how many valence electrons it has. This fact will help us determine the atom's chemical behavior.

Now let's look at a few other atoms and their electron configurations:

F: $1s^2 2s^2 2p^5$
Cl: $1s^2 2s^2 2p^6 3s^2 3p^5$
I: $1s^2 2s^2 2p^6 3s^2 3p^6 4s^2 3d^{10} 4p^6 5s^2 4d^{10} 5p^5$

How many valence electrons do each of the above atoms have? If you count the number of electrons in each atom's highest energy level, you will see that they all have seven valence electrons. Notice also that these atoms are all in the same column of the periodic table. This is no accident. The periodic table is arranged so as to guarantee that all atoms in a given column have the same number of valence electrons. Since they have the same number of valence electrons, they have essentially the same chemistry. This is another important point worth remembering:

Atoms in the same column of the periodic chart have the same number of valence electrons and thus have very similar chemistry.

Since the columns of the periodic chart contain atoms of similar chemistry, they are often called "groups" or "families."

There are certain groups of elements that are rather important in chemistry. The first of these is called the **noble gases**. They are the atoms that appear in column 8A of the chart. The reason that this group of atoms is important is that they have the ideal electron configurations in nature:

He: $1s^2$
Ne: $1s^2 2s^2 2p^6$
Ar: $1s^2 2s^2 2p^6 3s^2 3p^6$
Kr: $1s^2 2s^2 2p^6 3s^2 3p^6 4s^2 3d^{10} 4p^6$
Xe: $1s^2 2s^2 2p^6 3s^2 3p^6 4s^2 3d^{10} 4p^6 5s^2 4d^{10} 5p^6$

In each case, the s and p orbitals in their highest energy level are full. It turns out that this is a very low-energy situation, and, since all matter strives to get to the lowest energy possible, we consider such an electron configuration to be ideal.

Since these atoms have ideal electron configurations, they have no reason to change. As a result, they rarely undergo chemical reactions. As I have already said, the electron configuration of an atom determines its chemistry. Since these atoms already have ideal electron configurations, there is no need for them to change, so they will almost never undergo a chemical change. In other words, the noble gases rarely participate in chemical reactions. Substances like these are called **inert** substances.

Now since the electron configurations of the noble gases are ideal, all other atoms strive to attain them. This is the guiding principle behind chemistry. Chemical reactions are really just the result of atoms trying to rearrange their electrons so that they can get an ideal electron configuration. Since (with the exception of helium) all noble gases have eight valence electrons, we could say that atoms strive to get eight valence electrons. This very important principle is called the **octet rule**, and we will use it over and over again in this course:

Octet rule - Most atoms strive to attain eight valence electrons.

As usual in chemistry, there are exceptions to this rule. Hydrogen strives to get two valence electrons (so that it will be like helium). Also, there are a few others that (at times) break this rule. However, in this module, the only exception you need to worry about is hydrogen. Thus, you need to remember that all atoms except for hydrogen strive for eight valence electrons. Hydrogen strives for two.

There is one more thing that you need to know about the periodic chart and electron configurations. Since the number of valence electrons is so important to the chemistry of an atom, the periodic chart has been designed so that a chemist can get this information very quickly. Rather than determining the electron configuration of an atom each time you need to figure out how many valence electrons it has, just look at the number of the column that the atom resides in. For example, we said that bromine has seven valence electrons. What column is it in? 7A. The noble gases (with the exception of helium) have eight valence electrons. What group are they in? 8A.

It turns out that as long as the atom is in column 1A-8A, the number of valence electrons it has is the same as the column number. This shortcut does not work for the elements in columns 1B-8B, but that's okay. The chemistry of these elements is more difficult to understand, so we will not be dealing with them in this module. Thus, all you need to do is remember this important fact:

For atoms in groups 1A-8A, the number of the column in which the atom is located on the periodic chart equals the number of valence electrons the atom has.

Lewis Structures

Since the valence electrons play such a hefty role in the chemistry of an atom, chemists like to use a type of notation that emphasizes them. This notation, called **Lewis structures**, was developed by Gilbert N. Lewis, a professor of chemistry at the University of California at Berkeley, in the early

1900s. It has become a central concept in chemistry, because it forms the basis of our understanding of how molecules form.

A Lewis structure consists of an atom's symbol surrounded by dots. Each dot represents a valence electron in the atom. For example, the atom lithium is in group 1A of the periodic chart. This means that it has one valence electron. Thus, the Lewis structure for lithium looks like this:

Li·

The single dot to the right of the symbol represents the one valence electron that lithium has. The rest of the atoms in the second row of the periodic chart have Lewis structures that look like this:

Be· ·B· ·C· ·N: ·O: :F: :Ne:

Beryllium (Be) has two dots because it is in group 2A and thus has two valence electrons. Boron has three valence electrons, carbon four, nitrogen five, oxygen six, fluorine seven, and neon has eight. Realize that the dots do not represent the *total* number of electrons that the atom has. That number is given by the atomic number on the chart. The dots represent the number of *valence* electrons that the atom has, which is given by the column number on the chart.

Notice the way in which the dots are put in a Lewis structure. The first dot is placed on the right side of the elemental symbol. If the atom has two valence electrons, the second dot is placed on the bottom of the symbol. The third valence electron's dot is placed to the left of the symbol, and the fourth is placed on top. If the element has more than four valence electrons, the remainder must be paired up with the dots that are already there. Notice also that the ideal Lewis structure (shown in Ne because it has the ideal electron configuration) has a pair of dots on each side and at the top and bottom of the elemental symbol.

Since Lewis structures use dots to represent the valence electrons in an atom, they are sometimes called **electron-dot diagrams**. This name, however, fails to credit Lewis for his seminal work in molecular structure, so most chemists do not like this terminology. Do not be confused, however, if you read about electron-dot diagrams in other books. They are the same as Lewis structures. Make sure you understand how to determine an atom's Lewis structure by studying the example below and solving this problem:

EXAMPLE 8.1

Draw the Lewis structure for the following atoms:

a. Na **b. Ga** **c. Te**

a. Since Na is in group 1A, it has one valence electron. In a Lewis structure, a single electron is written on the right side of the symbol:

<u>Na·</u>

b. Since Ga is in group 3A, it has three valence electrons. We start on the right side of the symbol and put a single dot on the right, then the bottom, then the left of the symbol:

·Ga·

c. Since Te is in group 6A, it has six valence electrons. The first four are put on the sides, bottom, and top of the symbol by themselves. After that, the last two must be paired up. We pair the electrons up in the same order that we put them down singly; thus, the last two dots will pair up with the dots on the right and bottom of the symbol:

·Ṫe:

ON YOUR OWN

8.1 Draw the Lewis structures for the following atoms:

a. Ca b. Si c. At

Lewis Structures for Ionic Compounds

The whole reason we are investigating Lewis structures is so that we can learn about the structure of molecules. But remember from Module #3 that there are two different types of compounds in nature: ionic and covalent. Do you remember the distinction between these two types of compounds? Ionic compounds are made up of *both* metals and nonmetals, whereas covalent compounds are made up entirely of nonmetals. In order to make this distinction, you obviously have to remember how to tell metals from nonmetals. On the periodic chart, there is a heavy jagged line that runs down the right side of the table. All elements to the left of the line are metals (except hydrogen), and all elements to the right of the line (plus hydrogen) are nonmetals.

Now that we've reviewed all of that, we can begin to see how Lewis structures help us in determining molecular structure. We will start with ionic compounds, because they are the easiest to understand. Consider table salt, NaCl. Why do the elements Na and Cl come together and make a molecule? Also, why is there one Na for every one Cl, as the formula indicates? Why aren't there two Na's for every three Cl's so that the formula is Na_2Cl_3? The answers to these questions are readily available by looking at the Lewis structures of these two atoms:

Na· :Ċl:

First, look at chlorine's Lewis structure. It is very close to the ideal valence electron configuration. All it needs to do is get one more electron, and then it will have eight valence electrons. Well, sodium just happens to have one valence electron (one dot in its Lewis structure). Chlorine could reach the ideal electron configuration if it could just take the one valence electron from sodium and make that electron its own. Look at the Lewis structures that result when that happens:

$$\text{Na}\cdot \; :\ddot{\underset{..}{\text{Cl}}}: \longrightarrow \text{Na} \quad :\ddot{\underset{..}{\text{Cl}}}:$$

Once chlorine takes sodium's electron (the red dot) away, it now has the ideal electron configuration, because it has eight valence electrons. This is, in fact, the way nonmetal atoms try to get an ideal electron configuration when they are a part of an ionic compound. They take electrons from other atoms until they have eight valence electrons. This is a fact we must remember:

In ionic compounds, nonmetals try to gain electrons so that their Lewis structure has eight dots around it.

Once a nonmetal's Lewis structure contains eight dots, it has the ideal electron configuration and will be stable. Hydrogen, of course, is an exception to this rule. When hydrogen is a part of an ionic compound, it gains an electron to get two dots in its Lewis structure, not eight.

What about sodium, however? Well, once it loses its electron, it also has the ideal electron configuration. This may be hard to see, but if we go back to the full electron configuration, you'll understand what I mean. Sodium's complete electron configuration is:

$$1s^2 2s^2 2p^6 3s^1$$

If the chlorine takes an electron away, sodium has only 10 electrons, not 11. Thus, its new configuration is:

$$1s^2 2s^2 2p^6$$

This is the same as the electron configuration for neon, which is an ideal electron configuration.

So we see that metals behave differently than nonmetals. Whereas nonmetals tend to gain electrons until their Lewis structure has eight dots around it, metals tend to give away their electrons until their Lewis structure has no dots around it. This is, once again, a fact that we must remember:

In ionic compounds, metals try to lose electrons until their Lewis structure has no dots around it.

When a metal's Lewis structure has no dots, it has the ideal electron configuration and will be stable. This is, in fact, the fundamental difference between metals and nonmetals: Nonmetals try to gain electrons, while metals try to lose electrons.

So when a single Na atom gives up its electron and donates it to a single Cl atom, both atoms end up with ideal electron configurations. This is what makes the chemical formula of table salt NaCl. Each Na has exactly enough electrons to satisfy each Cl's desires. Thus, one Na will always react with one Cl and they will always form NaCl. In fact, this kind of consideration governs the formation of all molecules. Atoms will join together in such a way as to make sure that they all end up with ideal electron configurations.

One thing has to be made clear, however. The last two Lewis structures I drew were *wrong*. You see, when a sodium atom loses an electron, it is no longer a sodium atom. Remember, all atoms have the same number of protons and electrons. A sodium atom, according to the periodic chart, has 11 protons and thus 11 electrons. But when sodium gives up an electron, it ends up with 11 protons and only 10 electrons. Thus, it has more positive charges (protons) than negative charges (electrons). This gives it an overall positive charge. Since it has one more proton than it has electrons, we say it has a 1+ charge. Remember, however, that chemists love to drop 1's. They do not report 1's in chemical formulas or chemical equations. In the same way, they often do not report 1's with electrical charges. Thus, rather than listing the charge as "1+", it is often simply listed as "+".

In the same way, a chlorine atom has 17 protons and 17 electrons. However, when the chlorine atom gets the electron from the sodium atom, it ends up with 17 protons and 18 electrons. This gives it an overall negative charge, since it has more negative charges (electrons) than positive charges (protons). Since it has one more electron than it has protons, we say it has a "1-" charge, which is usually reported as simply "-". The correct Lewis structures in this situation, then, are:

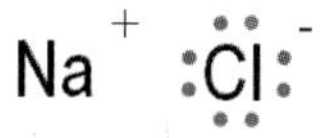

$$Na^{+} \quad :\ddot{\underset{..}{Cl}}:^{-}$$

These Lewis structures emphasize the fact that each of the elements in the molecule are electrically charged. When an atom gives up or gains one or more electrons to become electrically charged, we no longer refer to it as an atom. We call it an **ion**:

<u>Ion</u> - An atom that has gained or lost electrons and thus has become electrically charged

Thus, we would say that a molecule of table salt is composed of one sodium ion and one chloride ion.

Notice the names I just used. Positively charged ions keep the same name as the atom from which they came. Negative ions, on the other hand, have an "ide" ending after the atom from which that they came. This is an important fact to remember:

Positive ions have the same name as the atom from which they came.
Negative ions have an "ide" suffix added to the name of the atom from which they came.

This is where ionic compounds get the names we learned about in Module #3. To name an ionic compound, we simply name the ions that make it up. Thus, since table salt is made up of a sodium ion and a chloride ion, we call it sodium chloride.

Now let's consider a more difficult example: the ionic compound that is formed between magnesium and fluorine. The Lewis structures for these two atoms are:

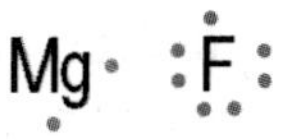

$$\underset{\cdot}{Mg}\cdot \quad :\dot{\underset{..}{F}}:$$

Since the fluorine atom has seven valence electrons (seven dots), it must gain one more electron to attain the ideal electron configuration. Thus, the fluorine takes one electron away from the magnesium:

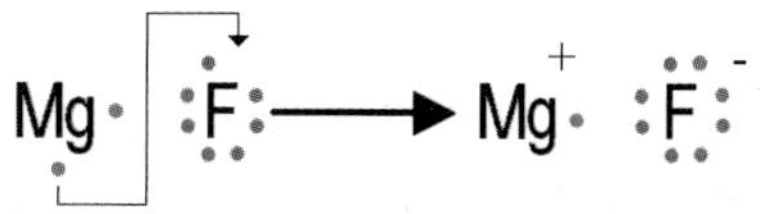

$$\underset{\cdot}{Mg}\cdot \quad :\dot{\underset{..}{F}}: \longrightarrow Mg^{+}\cdot \quad :\ddot{\underset{..}{F}}:^{-}$$

Now the fluorine has eight valence electrons (eight dots). Since this is one more electron than it normally has, we assign it a - charge. Since the magnesium now has one less electron than it normally has, we assign it a + charge. Remember, electrons are negative, so the atoms that gain electrons (nonmetals) will develop negative charges and the atoms that lose electrons (metals) will develop positive charges.

Are we finished with this molecule? The answer is no. Even though the fluorine has an ideal electron configuration, the magnesium still does not. Magnesium is a metal (on the left side of the jagged line) and thus will have its ideal electron configuration only when it has no dots in its Lewis structure. Since it still has one dot in the Lewis structure above, it does not have an ideal electron configuration. This means we still have work to do!

In order to get no dots in magnesium's Lewis structure, we need to give that last electron away. The problem, however, is that the fluorine has all the electrons it wants. Thus, it will not take magnesium's last electron. However, there are, most likely, other fluorine atoms around. Thus, *another* fluorine atom (one that hasn't gained any electrons yet) could come up and take magnesium's last electron:

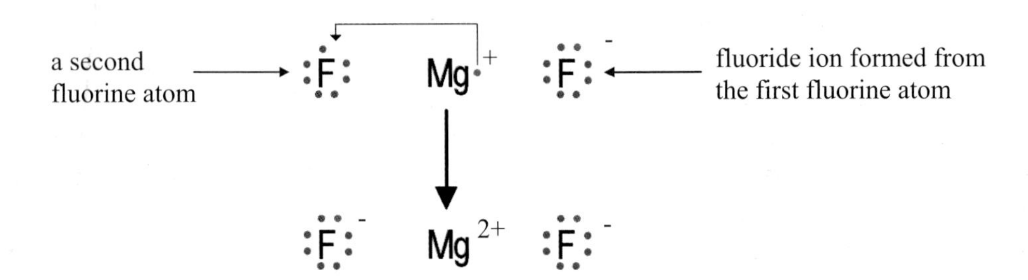

Now we see that all atoms have their ideal electron configurations. Notice that each individual fluorine atom gained one electron, so each individual fluoride ion has a 1- charge. On the other hand, a single magnesium lost two electrons. Thus, it has a 2+ charge. We cannot drop the 2 like we drop 1's, so we must list the charge as "2+". Since it took two fluorines to gain the electrons and one magnesium to lose them so that every atom had an ideal configuration, the chemical formula for this compound must be MgF_2.

This is how ionic compounds get their chemical formulas. The atoms start exchanging electrons until all of them attain an ideal electron configuration. Once they all have achieved this, they are attracted to one another because of the electrical charges that they have gained. Thus, in magnesium fluoride, the magnesium ion will always be close to the two fluoride ions because the positive charge of the magnesium is attracted to the negative charges of the fluorides. This electrical attraction is what holds ionic compounds together.

Now that you have seen how ionic compounds get their chemical formulas, I will show you a simple shortcut that will allow you to determine the chemical formula of an ionic compound. The first step in this shortcut is to determine what charge an atom needs to have in order to attain an ideal electron configuration. For example, in the discussion above, we found that fluorine needed to gain one electron to attain an ideal configuration; thus, its ion's charge was 1-. Magnesium, on the other hand, needed to lose two electrons to attain an ideal configuration, so its ion developed a 2+ charge.

Since the number of valence electrons an atom has determines how many electrons it must gain or lose to attain an ideal configuration, all atoms in a given column on the periodic chart tend to develop the same charge. In other words, we already determined that since fluorine has seven valence electrons, it needs one more electron in order to attain its ideal configuration. Thus, the fluoride ion develops a charge of 1-. Well, all of the other atoms in group 7A also have seven valence electrons, so their ions will all develop a charge of 1-. In the same way, since Mg develops a 2+ charge in order to gain an ideal electron configuration, all of the atoms in its column (group 2A) develop a 2+ charge.

Looking at groups 1A-8A:

- Atoms in group 1A have one valence electron. Since they are metals, they want to lose all of their valence electrons, so they will all lose that one valence electron. This gives their ions a charge of 1+ in ionic compounds.
- Atoms in group 2A, as we have found out, will develop a 2+ charge in ionic compounds.
- Group 3A is a little more difficult. Boron is a nonmetal and is therefore different from the other elements in this column. We will look at boron in a moment. All others in this column will develop a 3+ charge, because they have 3 valence electrons to lose.
- Group 4A is also difficult, because C and Si are nonmetals while the rest of the atoms in the column are metals. As a result, C and Si want to *gain* electrons to get to their ideal configuration, while Ge, Sn, and Pb would like to lose electrons. Thus, C and Si have four valence electrons and need to get eight, so they need to gain four electrons. This gives them a 4- charge in ionic compounds. Ge, Sn, and Pb want to be positive, but they don't follow the rules very well. We will discuss them in a moment.
- Group 5A has mostly nonmetals, which want to gain three electrons in order to get to their ideal configuration. They develop a 3- charge in ionic compounds. Don't worry about the metals in Group 5A.
- Group 6A also has mostly nonmetals, which will develop a 2- charge in ionic compounds. Once again, don't worry about the metals.
- Group 7A, as we have seen, has nonmetals which develop a 1- charge in ionic compounds.

Using these rules, you need to be able to determine the charge that any atom in groups 1A-8A (minus the exceptions) develops in an ionic compound. Now at first this may seem like a huge memorization task, but it isn't. Remember, you will always have a periodic chart with you. The column number on the chart tells you how many valence electrons the atom has. All you have to do is remember that nonmetals (those that lie to the right of the jagged line) will gain electrons in order to get eight, while metals (left of the jagged line) will lose electrons to get none. This determines the charge. So really, you needn't memorize any of the rules I just listed. Just remember the principles discussed in this paragraph and, with the aid of the periodic chart, you can determine the charges on your own.

Once you have determined the charges, determining the formula is a snap. There are just two rules to remember:

- **If the charges have the same numerical value, then the subscript for each ion is "1" and can therefore be ignored.**
- **If the charges have different numerical values, drop the + and - signs, switch the numbers, and use them as subscripts.**

This is all you do in order to determine an ionic molecule's chemical formula. If it sounds a little confusing now, don't worry. The following three example problems will clear things up.

EXAMPLE 8.2

What is the chemical formula for the ionic compound formed by aluminum and sulfur? What is its name?

To solve this problem, we first need to determine the charges that Al and S develop in an ionic compound. Al is in group 3A, so it wants to be 3+. S is in group 6A, so it wants to be 2-. The charges do not have the same numerical values, so we drop the + and - signs, switch the numbers, and make them subscripts. Thus, the 3 gets switched to the S and the 2 goes to the Al:

$$\underline{Al_2S_3}$$

Al is the metal, so its ion name is aluminum. S is the nonmetal, so its ion name is sulfide. Therefore, the compound's name is <u>aluminum sulfide</u>.

What is the chemical formula of barium oxide?

According to the name given, this ionic compound is composed of a barium ion, which comes from the metal barium, and an oxide ion, which comes from the nonmetal oxygen. Barium is in group 2A, so it wants a 2+ charge. Oxygen is in group 6A, so it wants a 2- charge. The numerical values for the charges are the same, so we ignore them:

$$\underline{BaO}$$

Notice that we did not have to be told that this is an ionic compound. Because it has a metal and a nonmetal in it, we know it is ionic and thus know that we are using ionic rules to determine the formula.

What is the chemical formula for the compound that forms when Na and N are reacted together? What is its name?

The metal Na is in Group 1A, so it develops a charge of 1+. The nonmetal N is in group 5A, however, so it develops a charge of 3-. Ignoring the charges and switching the numbers gives us:

$$\underline{Na_3N}$$

The ion that comes from sodium is named sodium, since it is positive. Since the ion that comes from nitrogen is negative, it is called nitride. Thus, the name is:

<u>sodium nitride</u>

Once again, we did not have to be told that this is an ionic compound. The fact that it is ionic is noticeable from the fact that it contains a metal and a nonmetal.

This shortcut allows you to get the chemical formula of an ionic compound very quickly, but don't let that distract you from what is really going on here. Remember that this method is just a shortcut to skip writing out the Lewis structures and moving the dots around. The chemical formulas of these compounds are, in nature, determined the way I showed you with Lewis structures. This method is simply a shortcut that saves time. Make sure you understand it by doing the following problem:

ON YOUR OWN

8.2 Give the chemical formulas for the following compounds:

a. calcium chloride b. potassium sulfide c. aluminum nitride d. magnesium phosphide

Handling the Exceptions in Ionic Compounds

Now you might wonder about all of the exceptions that are on the periodic table. In the list of rules on determining charge, I mentioned that B, Ge, Sn, and Pb are all exceptions. Why are they exceptions? Well, the answer to that question is too complex for this course, but, as you might expect, it has to do with electron orbitals. For now, you will just have to take my word that they are exceptions. Even though they are exceptions, we can still use them in ionic compounds as long as we know the way chemists name ionic compounds that have exceptions in them.

First of all, you must realize that the vast majority of the exceptions on the periodic chart are metals. After all, the entire middle of the chart (groups 1B-8B) is called the "**transition metal**" region simply because it doesn't follow our standard rules. Since these exceptions are mostly metals, we can still figure out the chemical formula of any ionic molecule that contains them if we just modify the way we name an ionic compound with such a metal in it.

Remember, the tough part about using these exceptions is that we don't know their charges. Well, the way we get around this is to state the metal's charge explicitly in the name of the compound. We do this by inserting a Roman numeral in parentheses right after the metal ion's name. This Roman numeral represents the positive charge of the metal. Thus, a name like:

tin(II) chloride

tells us two things. First, it tells us that the metal we are dealing with is an exception to the rules. We know this because of the Roman numeral in the name. Second, it tells us that in this particular compound, tin has a charge of 2+. Chlorine is not an exception, so chloride has a charge of 1-. Thus, the formula is:

$$SnCl_2$$

The reason we must name ionic compounds that have tin in them this way is because 2+ is not the only charge that tin can have. Because of reasons too complex to explain here, tin also can have a charge of 4+. For example:

tin(IV) chloride

is also an ionic compound found in nature. Because of tin's 4+ charge, this compound's formula is:

$SnCl_4$

Thus, when dealing with ions of atoms which we call exceptions to the rules, you will always be given a Roman numeral which signifies the positive charge on the metal. With that information, you can use the same skill you just learned to determine the chemical formulas of those ionic compounds as well. Make sure you understand this by doing the following problem:

ON YOUR OWN

8.3 Give the chemical formulas of the following compounds:

a. iron(III) fluoride b. copper(I) iodide c. manganese(III) oxide

Ionization Potential and Periodic Properties

It is important to remember that although most atoms want to become charged in order to attain the ideal electron configuration, they never start out that way. When an atom is formed (we only have theories as to *how* this happens), it is always formed as an atom. Thus, it always has just as many electrons as it has protons.

However, depending on how much an atom wishes to attain an ideal configuration, it may gain or lose electrons rather quickly. When this happens, we called the process **ionization** (eye' uh nuh zay' shun).

Ionization - The process by which an atom turns into an ion by gaining or losing electrons

Other atoms, however, may remain as atoms for a long time, simply because they do not have a strong enough desire for an ideal electron configuration. We measure an atom's desire to become a positive ion by measuring its **ionization potential**.

Ionization potential - The amount of energy needed in order to take an electron away from an atom

It turns out that each atom holds onto its valence electrons with a different strength. Some hold them tightly, while others hold them rather loosely. For those atoms that hold their valence electrons tightly, it takes a lot of energy to take an electron away. On the other hand, if an atom holds on to its valence electrons loosely, not much energy is required in order to remove one.

But how do we know whether or not an atom holds on to its electrons tightly or loosely? Well, as you might expect, it depends on the atom's electron configuration and can be determined from the periodic chart. Consider, for example, the elements in group 1A. These atoms all have one valence electron. In order to get to an ideal electron configuration, all they have to do is lose that one electron. Thus, as you might expect, the ionization potential for these atoms is rather low. In other words, these atoms will be so much better off by the loss of one electron that they give that electron up willingly.

On the other hand, consider the atoms in group 7A. These atoms have seven valence electrons and need only one more to attain an ideal electron configuration. How willingly do you think that these atoms will *give up* one of their electrons? Not at all. Since these atoms need to *gain* an electron to attain an ideal configuration, they are not very likely to give up an electron, taking them farther away from an ideal configuration.

Thus, one thing we can say is that as you look at the periodic chart, the elements on the left side of the chart have a rather low ionization potential (they give up their electrons easily), while the atoms on the right side of the periodic chart have high ionization potentials (they give up their electrons grudgingly). This is a fact that is worth remembering:

In general, the ionization potential of atoms increases from the left of the periodic chart to the right of the periodic chart.

In other words, we could say that since Mg lies left of P on the periodic chart, Mg has a lower ionization potential than P.

Any atomic property that varies regularly across the periodic chart is called a **periodic property**.

Periodic property - A characteristic of atoms that varies regularly across the periodic chart

Periodic properties are nice because they allow you to make predictions about the relative behavior of atoms by simply looking at the chart. For example, suppose a chemist was working on an experiment and realized that a source of electrons was needed to make the experiment a success. Looking at his inventory, the chemist sees that he has carbon and lithium. Looking at the periodic chart, he sees that lithium lies to the left of carbon on the chart. This tells the chemist that lithium gives up its electron much more easily than carbon, so the chemist knows to choose lithium for the experiment.

Not only does ionization potential change from left to right on the periodic chart, it also changes from the top to the bottom. It turns out that atoms near the top of the periodic chart hold on to their electrons more tightly than those on the bottom on the chart. In other words:

In general, the ionization potential of atoms decreases from the top of the periodic chart to the bottom of the periodic chart.

Thus, if two atoms lie in the same column of the periodic chart, their relative ionization potential can be determined by their vertical position on the chart. For example, O and Se are both in the same column of the chart, so neither lies to the left or right of the other. However, since O is above Se on the periodic chart, O has a higher ionization potential than Se.

Why does ionization potential decrease from the top to the bottom of the periodic chart? Well, look at the difference between O and Se. The valence electrons in oxygen are on energy level 2 (determined by the electron configuration), while selenium's valence electrons are in energy level 4. This means that selenium's valence electrons are farther away from the nucleus than oxygen's. As a result, the protons in the nucleus cannot attract selenium's valence electrons as strongly. This means that selenium holds on to its valence electrons more loosely, making its ionization potential lower.

The way in which ionization potential changes relative to an atom's position in the chart is summarized in Figure 8.1. Such periodic trends are important and worth remembering.

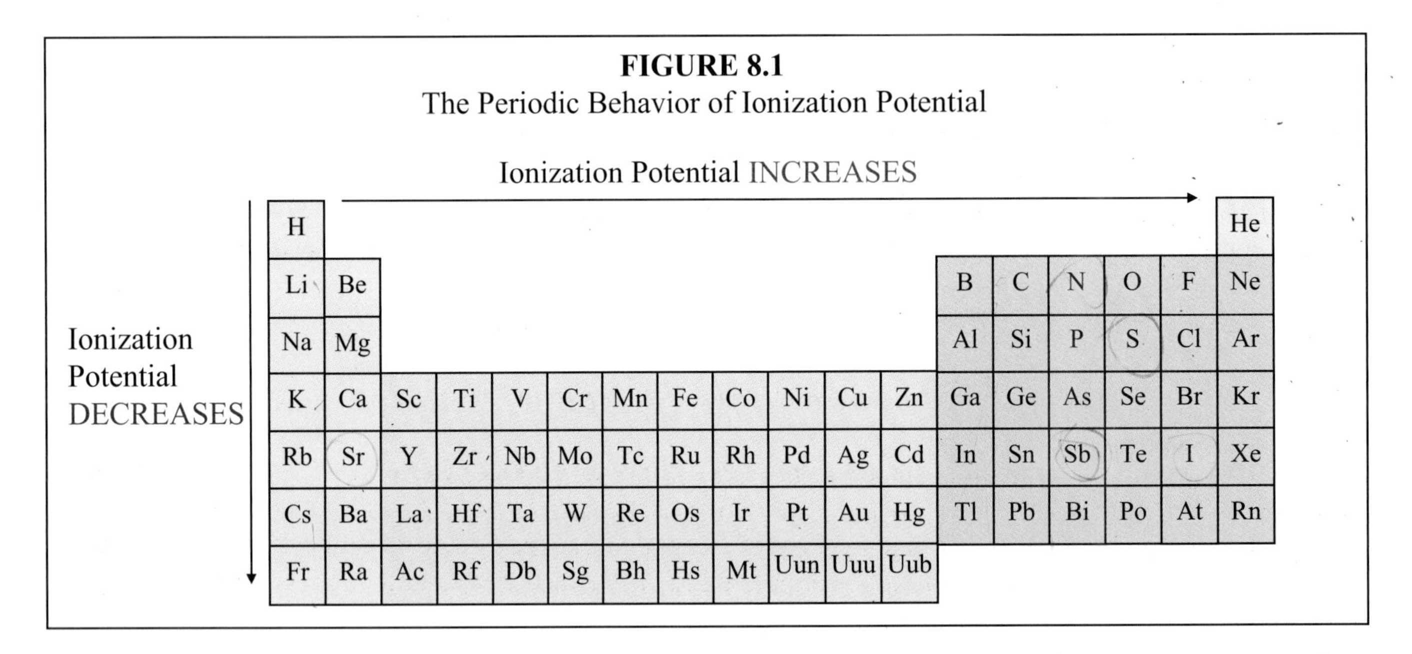

FIGURE 8.1
The Periodic Behavior of Ionization Potential

Make sure you understand how to use this periodic property by studying the next example and then performing the problems that follow:

EXAMPLE 8.3

A chemist needs to find an element that readily gives up an electron. Which of the following would suit the chemist's needs best: Ca, As, or Br?

Since all of these atoms lie in the same row of the periodic chart, they do not lie above or below each other. Thus, the fact that ionization potential decreases from the top to the bottom of the chart is useless here. However, Ca lies to the left of As, which lies to the left of Br. Since ionization potential increases from left to right on the periodic chart, Ca has the lowest ionization potential. Thus, Ca gives up its electrons most readily and is the best choice for the experiment.

Which of the following holds onto its electrons the tightest: Al, In, or B?

All of these elements lie in the same column, so the fact that ionization potential increases from left to right on the chart is of no use. However, B lies above Al, which lies above In. Since ionization potential decreases from the top to the bottom of the chart, B has the highest ionization potential. This means that B holds on to its electrons the tightest.

ON YOUR OWN

8.4 Order the following atoms in terms of increasing ionization potential: I, Ru, Sn, Rb.

8.5 Which atom most readily gives up its electrons: K, Li, Cs, or Fr?

 (The multimedia CD has a video demonstration on the periodicity of ionization potential.)

Electronegativity: Another Periodic Property

The way in which we have defined ionization potential deals only with positive ions. After all, we defined ionization potential as the energy needed to take an electron away from an atom. Well, when electrons are taken away from atoms, the atoms become positive ions. What about negative ions? Is there a way of determining how much an atom desires to gain electrons? Of course there is. The property that determines an atom's desire to gain electrons is called **electronegativity**.

Electronegativity - A measure of how strongly an atom attracts extra electrons to itself

As the electronegativity of an atom increases, its desire to become a negative ion increases.

Let's determine how electronegativity varies across the periodic chart. First, consider atoms in group 1A. Do they desire extra electrons? No. They need to lose an electron to attain the ideal electron configuration, so they certainly have little desire to gain electrons. On the other hand, atoms in group 7A have a strong desire to gain an extra electron, because that will give them the ideal electron configuration. Thus, we can conclude:

In general, the electronegativity of atoms increases from left to right on the periodic chart.

What about an atom's vertical position on the chart? How does that affect its electronegativity? Well, as we said before, atoms that are higher on the periodic chart tend to have their valence electrons closer to the nucleus as compared to atoms that appear near the bottom of the chart. Thus, the attraction of an atom for electrons should decrease from top to bottom on the chart, because the farther the valence electrons are from the nucleus, the less attraction the protons can exert on them.

In general, the electronegativity of atoms decreases from top to bottom on the periodic chart.

Figure 8.2 summarizes the way in which both ionization potential and electronegativity vary across the periodic chart.

FIGURE 8.2
The Periodic Behavior of Ionization Potential and Electronegativity

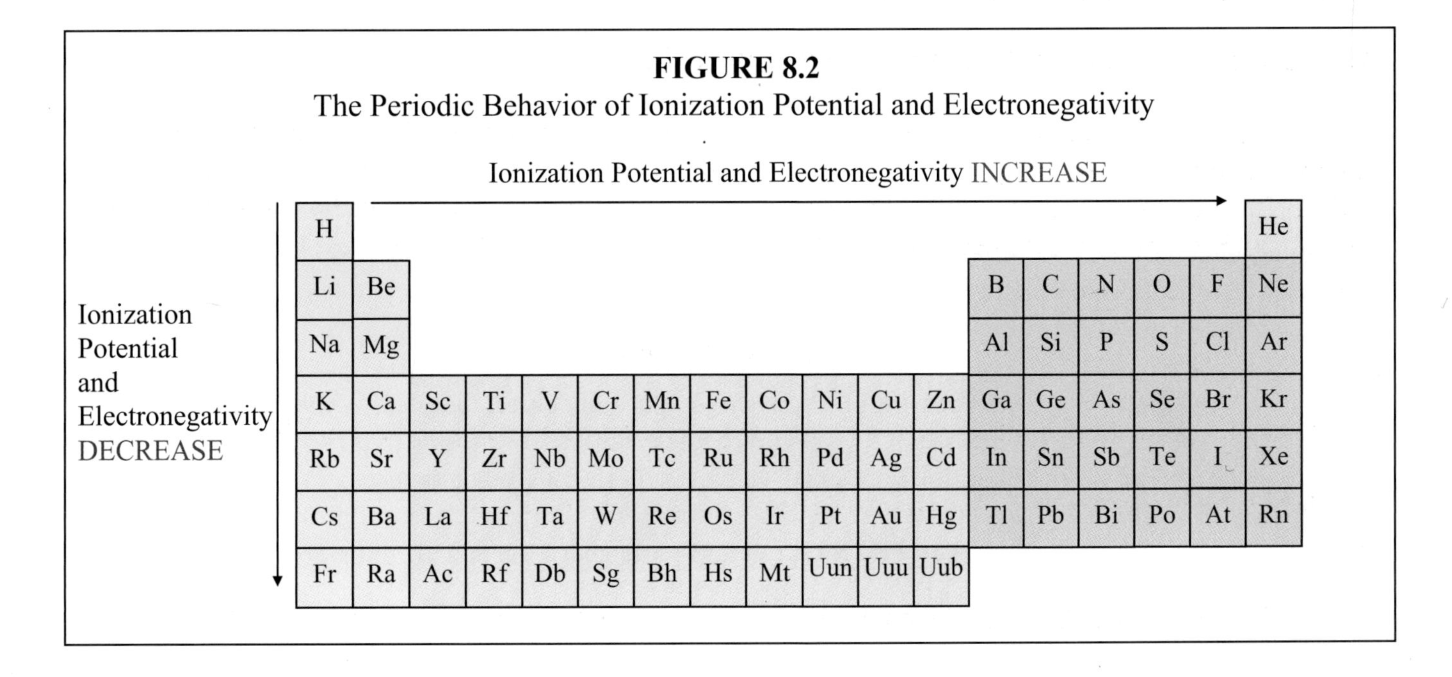

Use the same reasoning you learned in the previous section to solve the following problems:

ON YOUR OWN

8.6 Which element desires electrons most: Mg, P, Na, or S?

8.7 Order the following in terms of increasing desire for electrons: F, At, Cl, I, and Br.

Atomic Radius: Another Periodic Property

Before we move on to the Lewis structures of covalent compounds, I want to discuss one more periodic property: the **atomic radius**. Even though atoms have rather complex shapes due to the shapes of the electron orbitals, to some extent, we can approximate atoms as tiny spheres. If we do that, then, the radius of an atom is an indicator of its size. The larger the atom's radius, the larger the atom.

If you think about it, you should quickly realize that (once again) the electron configuration of an atom is what determines its size. After all, the protons and neutrons in an atom are tightly packed at the center, but the electrons whirl around the nucleus far from the center. Thus, the electrons are the "outer part" of an atom. Since electron configuration determines the atom's size, it is not surprising that atomic radius varies according to the placement of the atom on the periodic chart.

Think for a moment about how the electron configuration changes as you move *down* the chart. Well, as I said before, atoms that are higher on the periodic chart tend to have their valence electrons closer to the nucleus as compared to atoms that appear near the bottom of the chart. Thus, it should be pretty obvious that the lower an atom is on the periodic chart, the bigger the atom is, because its valence electrons are farther from the nucleus. In other words:

In general, the atomic radius increases from top to bottom on the periodic chart.

Okay, that's not too bad. What about an atom's horizontal position on the periodic chart? How does that affect its atomic radius? Well, as you move across the periodic chart, the electrons are not filling up higher energy levels. Instead, they are filling orbitals in the same energy level or below. For example, look at row 4 on the periodic chart. As you travel across row 4, the electrons first fill up the 4s orbital, then the 3d orbitals, then the 4p orbitals. Thus, the energy level of the orbitals does not increase. As a result, you might think that the size of the atom doesn't vary as you travel across the periodic chart. However, if you are thinking that, you are forgetting something.

Remember, as you travel from left to right on the periodic chart, the *atomic number* of the atom is increasing. What does that tell you? It tells you that the number of protons in an atom increases as you travel across the chart. Thus, the total positive charge in the nucleus increases. What does that mean? It means that the nucleus can exert more electrical force on the electrons, *pulling them closer*. As a result, when you travel from left to right on the periodic chart, the size of the atoms actually *decreases*!

In general, the atomic radius decreases from left to right on the periodic chart.

Figure 8.3 summarizes the way in which atomic radius varies across the periodic chart.

FIGURE 8.3
The Periodic Behavior of the Atomic Radius

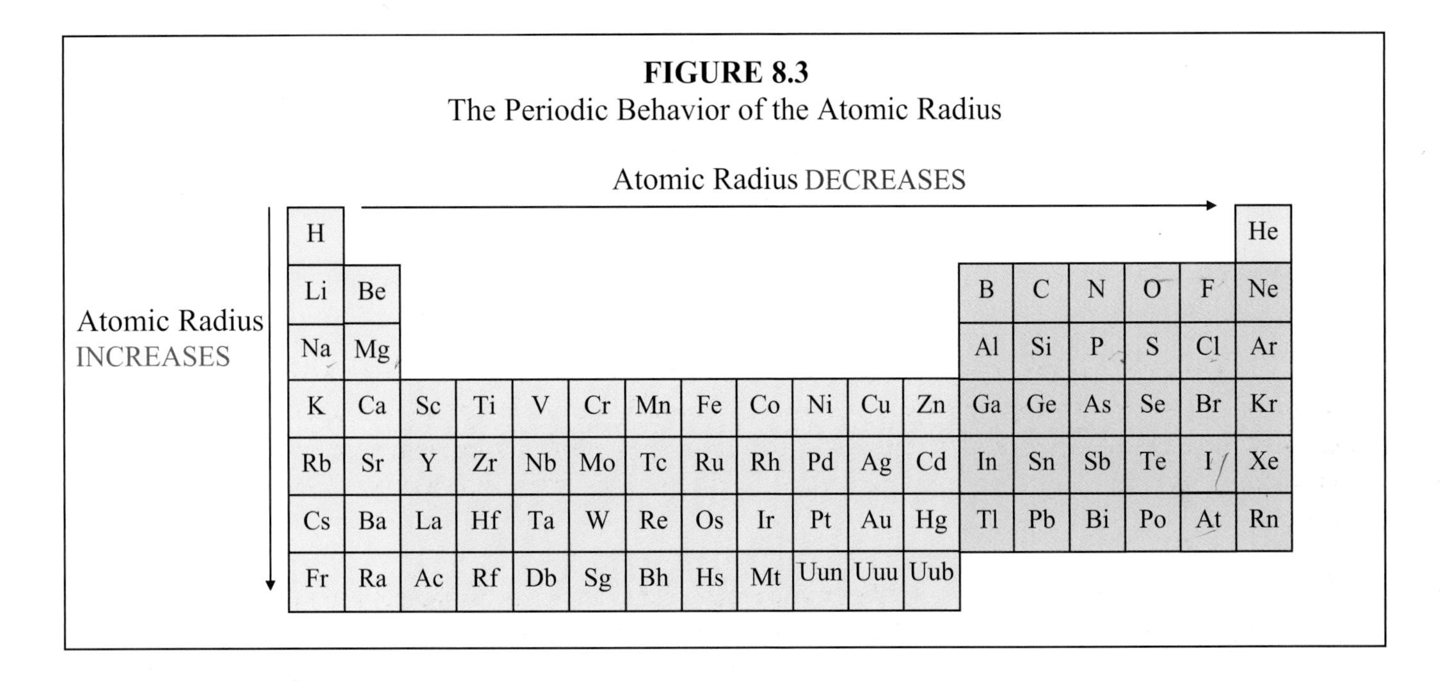

ON YOUR OWN

8.8 Which element is the largest: Mg, P, Na, or S?

8.9 Order the following in terms of increasing atomic radius: F, At, Cl, I, and Br.

<u>Lewis Structures of Covalent Compounds</u>

Now that we have learned a great deal about the nature of ionic compounds, we turn our attention to covalent compounds. As you learned in Module #3, the difference between ionic and covalent compounds revolves around the type of elements that make them up. Ionic compounds contain both metals and nonmetals, while covalent compounds contain only nonmetals. Now that you understand Lewis structures and how atoms desire to attain an ideal electron configuration, you can finally learn the *fundamental* difference between these two types of compounds.

Now, if we are to believe that the main driving force for the formation of molecules is the desire for each atom to attain a perfect electron configuration, there seems to be a major problem in explaining the formation of covalent compounds. After all, we already learned that covalent compounds contain only nonmetals. We also learned that nonmetals always gain electrons. They never give up electrons. Thus, how do nonmetals combine together to get ideal electron configurations if none of them are willing to give up electrons so that others can gain them?

Think about what happens when a friend of yours has something that you want but is unwilling to give it to you. What do you usually do? Hopefully you end up *sharing* the item. Well, the same goes for atoms. If two nonmetals come together and neither will give up its electrons, the two atoms will *share* electrons in order to help them both attain ideal electron configurations. The best way to understand this is by looking at some Lewis structures.

In Module #4 we learned that the elements H, N, O, F, Cl, Br, I, and At all exist as homonuclear diatomics. In other words, no F atoms exist as a stable form of matter. The element fluorine exists as the molecule F_2. Now we can see why. Consider the Lewis structure of an individual fluorine atom:

:F:

In order to attain an ideal electron configuration, fluorine needs to gain one electron. If a metal atom were around, it could take that electron away from the metal and become an F^- (fluoride) ion. However, suppose no metal were around. What could the fluorine atom do?

Well, wherever there is one fluorine atom, there are probably a few more. If another fluorine atom were to enter into the picture, a possibility would begin to emerge:

:F: :F:

I made the second fluorine's electrons red so you would be able to tell one fluorine from another. Now notice that each atom has a space for one electron, and each atom also has one unpaired electron. Suppose each atom's unpaired electron were to fit into the space that the other atom has:

:F: :F: ⟶ :F:
:F:

Make sure you understand what happened here. I did *not* transfer electrons from one atom to the other. I brought the atoms close enough so that each atom's unpaired electron was able to sit next to the other atom's unpaired electron. To make it more readable, I also turned the top "F" (the one with the blue electron dots) over, so that both atomic symbols were right side up.

Now look at what has happened. Although each atom still only has seven valence electrons, each electron has another electron to pair up with. Because the two fluorines share one of each other's electrons, they actually think they each have eight valence electrons, because all of the electrons seem to be paired up. When all of an atom's electrons are paired up, the atom thinks that it has eight electrons.

If you don't quite understand why sharing electrons like this makes an atom think it has an ideal electron configuration, think of it this way. Each fluorine is sharing an electron with the other. This means that, for certain moments in time, the electron pair pictured between the two atomic symbols above is actually whirling around the first fluorine. For those moments, the first fluorine has eight valence electrons. During other moments, the shared electron pair is whirling around the second fluorine, making it have eight valence electrons. Thus, even though neither of the fluorines *always* has eight valence electrons, each of them has eight valence electrons *sometimes*.

Since both fluorines gain a little bit in this arrangement, they stay together, sharing the pair of electrons that are pictured in between the two atomic symbols in the Lewis structure above. This shared electron pair is what we call a **covalent bond**.

Covalent bond - A shared pair of valence electrons that holds atoms together in covalent compounds

This covalent bond is the fundamental thing that separates an ionic compound from a covalent one:

The atoms in covalent compounds share electrons in order to form molecules, while the atoms in ionic compounds give and take electrons to form ions.

Notice the difference here. The atoms in ionic molecules stay together because of the fact that the positively charged ions are attracted to the negatively charged ions. In covalent compounds, however, there is a physical bond (a pair of shared electrons) which holds the atoms together.

Since the covalent bond created by shared electron pairs is the thing that holds the atoms in a molecule together, we often emphasize this in the Lewis structure by replacing the pair of dots with a dash. This illustrates that there is something physically attaching one atom to the other. It also helps to distinguish the electron pairs that are not being shared from the ones that are being shared. Thus, the best way to represent the Lewis structure for a fluorine molecule is:

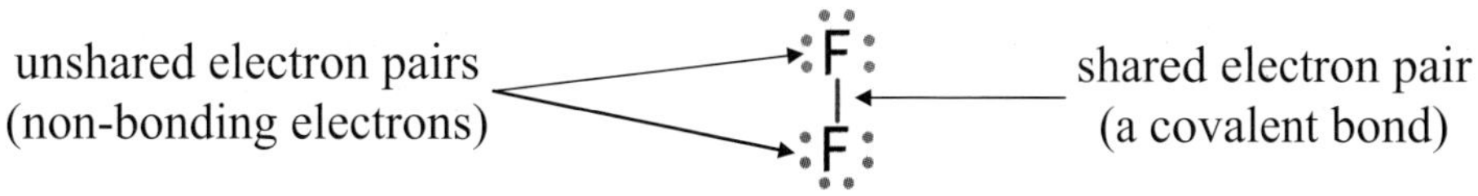

unshared electron pairs (non-bonding electrons)

F
|
F

shared electron pair (a covalent bond)

Now do you see why I said that determining molecular structure is a lot like putting together a jigsaw puzzle? In order to determine what a molecule of any covalent compound looks like, all you have to do is find out how you can "fit" the atoms together in such a way as to give every atom its ideal electron configuration.

Let's look at another covalent compound so that we can better see how to determine the Lewis structure of molecules. Consider the compound carbon tetrachloride. According to the nomenclature rules you learned in Module #3, this compound is composed of one carbon atom and four chlorine atoms. In order to determine its Lewis structure, we need to look at the Lewis structures for these atoms:

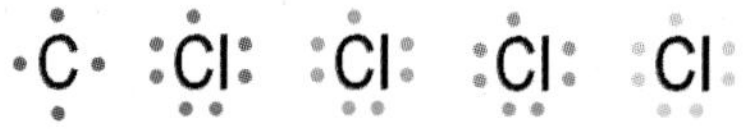

C Cl Cl Cl Cl

This is roughly equivalent to laying out all of the puzzle pieces before actually starting the puzzle. It helps you see what you have to work with.

Now the first thing you have to realize about working with these Lewis structures is that the atoms given in the chemical formula are all you have to work with. You cannot use any more or any less than the atoms listed in the chemical formula. Thus, no matter what, we must use exactly one C and four Cl's to make this molecule.

Secondly, the dots in the Lewis structures for each of these atoms are all of the dots that you have to work with. You cannot use any more or any less than the dots drawn. Thus, since carbon has four valence electrons (four dots) and the four chlorines each have seven valence electrons (for a total of 28 dots), we have exactly 32 dots to work with. No more, no less. We must construct a molecule in which each atom has eight dots around it, using only these 32 dots. How do we go about doing this?

When putting together a jigsaw puzzle, experts will tell you to construct the frame of the puzzle first. Since the frame pieces are generally the easiest to find, it tends to make things go more quickly if you start there. In putting together Lewis structures, however, it is best to start at the *center* and work your way out. Generally, you take the atom with the *most unpaired electrons* and put it in the center. You then try to attach the remaining atoms to this one central atom. If you happen to have more than one of the atom with the most unpaired electrons, you generally hook them all together and put the whole chain in the center.

Since carbon has four unpaired electrons in its Lewis structure and the four chlorines have only one each, our rule tells us that we should put the carbon in the center and try to attach all of the chlorines to it:

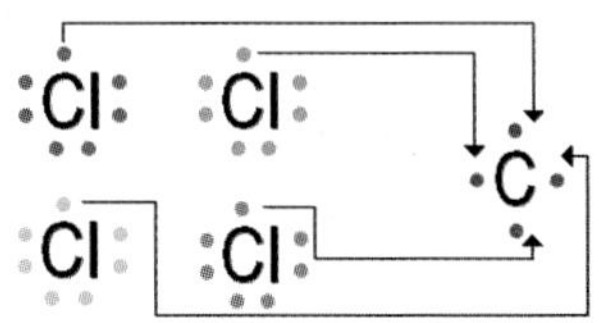

Well, since each chlorine has one spot open for an electron, and since there are four of them, we can have each chlorine share one of carbon's unpaired electrons:

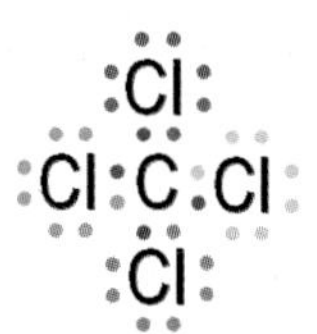

Now we can see that each atom has eight electrons around it, giving each atom an ideal electron configuration. This means we have a good Lewis structure for this molecule.

If you are having problems seeing why each atom has eight electrons around it, look at the Lewis structure this way. Any dots that are next to an atom's symbol belong to that atom. In addition, all dots between two atomic symbols belong to *both* atoms. Thus, the blue box below encloses all dots that belong to the carbon:

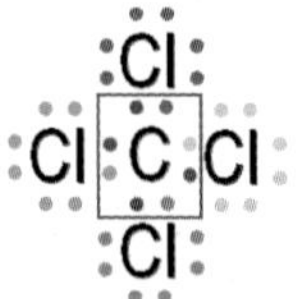

You can see that the box contains all dots between the carbon and the chlorines. Since the box has eight electrons in it, carbon has its ideal electron configuration.

Now look at the chlorines. If we draw boxes around all dots that are next to a chlorine or between a chlorine and the carbon, we get the following:

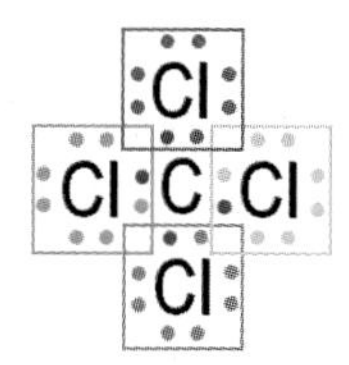

Since each box contains eight dots, we can say that all chlorines have their ideal electron configurations. Since every atom has its ideal electron configuration, we know we have a good Lewis structure, and we can finish by replacing the shared electron pairs with dashes:

:Cl:
:Cl—C—Cl:
:Cl:

This, then, is the final electron configuration for CCl_4.

In summary, determining the Lewis structure for a molecule is rather like putting together a puzzle, as long as we follow these general guidelines:

1. **Write out the Lewis structures for all atoms in the chemical formula. This gives you an idea of what you have to work with. You must work with the number of dots in those Lewis structures: no more, no less.**

2. **Put the atom with most unpaired electrons in the center. If there is more than one of those atoms, put them all in the center, linked together.**

3. **Try to arrange the other atoms around the atom(s) you put in the center, making sure that they all end up with eight dots around them, except for H atoms, which should only have two dots.**

Study the next example so that you see how this works, and then try the problem that follows:

EXAMPLE 8.4

What is the Lewis structure for a water molecule?

Water has the chemical formula H_2O. This means that a water molecule has two hydrogen atoms and one oxygen atom. In order to determine the molecule's Lewis structure, we first have to look at the Lewis structures for the individual atoms:

H· H· ·O:

Now remember that unlike most atoms, H needs only two valence electrons. Therefore, we must use the dots pictured above and arrange them in such a way as to give the hydrogen atoms two valence electrons and the oxygen atom eight. We start by putting the atom with the most unpaired electrons in the middle. That would be oxygen:

```
H·———→ ··
       :O:
H·———→ ··
```

Now we have to attach the H's to the O. This isn't hard, since each H has one space open for an electron, and the oxygen has two spaces open:

```
  H
  ··
H:O:
  ··
```

This gives us eight electrons around the oxygen and two around each hydrogen. Thus, we have a good Lewis structure. Now all we need to do is replace the shared electron pairs with dashes to indicate that they are covalent bonds.

```
  H
  |
H-O:
  ··
```

This is the final Lewis structure for water.

Now please realize that the actual position of the H's around the O is irrelevant. Because molecules are free to rotate in three-dimensional space, any of the following Lewis structures would be acceptable for water:

```
                ··                        ··          H
  ··           H-O:        H             :O-H         |
H-O-H            |         |              |          :O:
  ··             H        :O-H            H           |
                           ··                         H
```

Thus, your Lewis structure does not need to be *identical* to the one that I draw. It just needs to have the same central atom as mine, the same atoms bonded to that central atom as mine, and the same number of non-bonding electron pairs for each atom as mine. In the case of water, for example, as long as your Lewis structure has oxygen at the center with two non-bonding electron pairs, and as long as it has two hydrogens bonded to the oxygen with one shared electron pair each, your Lewis structure is just as valid as mine.

ON YOUR OWN

8.10 Draw the Lewis structures for the following molecules:

a. NH_3 b. $SiBr_4$ c. H_2S d. ICl

More Complicated Lewis Structures

Now if Lewis structures were always this simple, anyone could do them. It turns out that there are much more complicated molecules in nature, calling for more complicated Lewis structures. For example, consider the very important molecule, oxygen. Oxygen is a homonuclear diatomic; thus, its chemical formula is O_2, which means that we have two oxygen atoms to work with:

Each O has two unpaired electrons, so we could start by pairing up one of the unpaired electrons on the first O with one of the unpaired electrons on the second O:

Now we have a problem. Each O has seven electrons around it, but they each need eight. There are no more electrons, so what can we do? Well, we just need to remember that a dot counts for an atom if it is next to the atom's symbol *or in between the atom's symbol and another atom's symbol.* So, we just need to take the unpaired electron from the first O and move it in between the two:

How does this help? Well, count the number of atoms that belong to the second oxygen now:

The dots inside the box all belong to the second O, because they are all right next to the atomic symbol or in between the two atomic symbols. If you count, there are eight of them. This gives the second oxygen its ideal electron configuration! What about the first O, however?

The first O still has only seven dots. This is easy to fix, however, because there is an unpaired electron on the second O. If we move it down in between the two O's, it will count for both atoms:

If you count the dots now, each O has eight electrons either next to it or in between it and the other O. Thus, each atom has its ideal electron configuration. All we have to do is to replace each shared electron pair with a dash to indicate that it is a covalent bond. Since there are two shared electron pairs, however, we must use two dashes.

unshared electron pair
(non-bonding electrons)

double bond

Since two shared electron pairs make up this bond, we call it a **double bond**. As you might expect, a double bond is much stronger than a single bond. Thus, comparing F_2 (which we saw earlier is linked by a single bond) to O_2, we could say that it would be more difficult to break down an O_2 molecule than an F_2 molecule, because of the stronger bond in O_2.

Now realize what went on when we constructed the double bond in our Lewis structure above. We moved electrons from one place in the Lewis structure to another place. This is perfectly legal, as long as we do not actually get rid of any electrons. Also, notice that when we move an electron from one side of an atom to a spot in between the two atoms, the electron suddenly counts for both atoms. This will be a very useful technique in determining the Lewis structure of complex molecules.

Let's look at one more complex Lewis structure before we go on to examples and "On Your Own" problems. We learned in Module #4 that the two principle components of the air we breathe are N_2 and O_2. We just learned about the Lewis structure for O_2, so now let's look at N_2. As the chemical formula indicates, N_2 is composed of two nitrogen atoms:

·N: ·N:

Both of these atoms have three unpaired electrons. The first thing we can do with them, then, is pair up one of the unpaired electrons on the first N with one of the unpaired electrons on the second N:

:N:N:

This is a good start, but we are far from done. Each N has only six electrons right now. They both need eight. Well, we can give the second N eight electrons by taking the two unpaired electrons on the first N and putting them in between the two N's. This allows the electron to count for both atoms.

:N:N: ⟶ :N::N:

Now the N on the right has eight electrons, but the N on the left still has only six. We can fix that, however, by moving the two unpaired electrons on the second N in between the two N's.

:N::N: ⟶ :N:::N:

Once again, since the electrons between the two N's count for both of them, we now have eight electrons with each atom. Thus, the only thing left to do is replace the shared electron pairs with dashes. Since there are three shared electron pairs, however, we must have three dashes:

:N≡N: ⟵ unshared electron pair (non-bonding electrons)

triple bond

This type of bond is, naturally, called a **triple bond**. As you might expect, it is stronger than a double bond. Thus, N_2 would be more difficult to break apart than O_2, because the triple bond that holds N_2 together is stronger than the double bond that holds O_2 together.

The real trick to figuring out Lewis structures is just to keep trying. Start working on an idea and see if it pans out. If it doesn't, don't worry. Start over and try it again with a new idea. In other words, don't be afraid to make some mistakes before you come up with the correct answer. Make sure you understand the way to make double and triple bonds in Lewis structures by studying the following examples and doing the problems that follow. The next module uses Lewis structures extensively, so it is important that you know how to work with them.

EXAMPLE 8.5

Determine the Lewis structure for carbon dioxide.

The name tells us that the chemical formula is CO_2, which means that we have one carbon and two oxygens to work with:

·C· ·O: ·O:

According to the procedure we are supposed to follow, we must put the carbon in the middle, since it has the most unpaired electrons, and then we must try to attach the oxygens to it:

·O: ·C: ·O: ⟶ :O:C:O:

This arrangement gives the carbon six electrons and each of the oxygens seven. We can give the carbon eight electrons by taking the unpaired electrons on each oxygen and putting them in between the atoms:

:O·:C:·O: ⟶ :O·:C:·O:

Now the carbon atom has eight electrons. The oxygens only have seven, but we can fix that. There are two unpaired electrons on the carbon. If each of them is placed between each oxygen and the carbon, each oxygen will have eight dots:

:O::C::O: ⟶ :O::C::O:

Now we see that all atoms in the molecule have eight electrons, so our Lewis structure is complete. All we have to do now is replace the shared electron pairs with dashes, to indicate that they are covalent bonds:

:O=C=O:

A carbon dioxide molecule, therefore, consists of two oxygen atoms each double-bonded to a carbon atom.

What is the Lewis structure for HCN?

The formula tells us that we have one hydrogen, one carbon, and one nitrogen to work with:

H· ·C· ·N:

According to the procedure we are supposed to follow, we must put the carbon in the middle, since it has the most unpaired electrons, and try to attach the others to it:

H· ·C· ·N: → H:C:N:

This arrangement gives the carbon six electrons, the hydrogen two, and the nitrogen six. The hydrogen, then, has its ideal electron configuration. We can give the nitrogen eight electrons by taking the unpaired electrons on the carbon and putting them in between the two atoms:

H:C:N: → H:C::N:

The carbon only has six electrons, but we can fix that. There are two unpaired electrons on the nitrogen. If each of them is placed between the carbon and nitrogen, the carbon will get eight dots:

H:C::N: → H:C:::N:

Now we see that all atoms in the molecule have their ideal electron configurations, so our Lewis structure is complete. All we have to do now is replace the shared electron pairs with dashes to indicate that they are covalent bonds:

H–C≡N:

ON YOUR OWN

8.11 Determine the Lewis structure for CH_2O.

8.12 Draw the Lewis structure for C_2H_2.

An Application of Lewis Structures

Lewis structures are not only helpful in determining what a molecule looks like; they are also helpful in understanding a molecule's chemical behavior. For example, I mentioned in the previous module that the surface of our planet is protected from the ultraviolet rays of the sun by something called the "ozone layer." Your knowledge of Lewis structures will now allow you to understand how this system works.

Recall from the previous module that the sun is bathing the earth with all wavelengths of light. Our eyes use the visible light in order to see. We use light with longer wavelengths (infrared, microwave, etc.) for many diverse purposes, including microwave ovens, television, and radio. Light waves with wavelengths smaller than visible light (gamma rays, X-rays, ultraviolet rays), however, are quite dangerous to living organisms. These light waves have enough energy to destroy any living tissue that they contact. If the surface of our planet were not protected from these light waves, nothing could survive. Any living organism would immediately be fried by these energetic light waves.

How is the earth protected from these light waves? Well, there is actually a three-tiered system that blocks high-energy light that comes from the sun. First of all, the highest-energy light that comes from the sun consists of gamma rays. These light waves have very small wavelengths, indicating very large energy. As these light waves strike our atmosphere, they encounter an enormous amount of nitrogen. As you recall from our previous discussions, nitrogen's Lewis structure looks like this:

:N≡N:

I told you before that this triple bond makes the nitrogen molecule very hard to break apart. These high-energy light rays, however, have plenty of energy, so when they strike the nitrogen molecule, they are able to break it apart. This, however, uses up all of the energy in the light wave. As I said in Module #1, light is pure energy. Thus, when the energy gets used up, the light wave is gone!

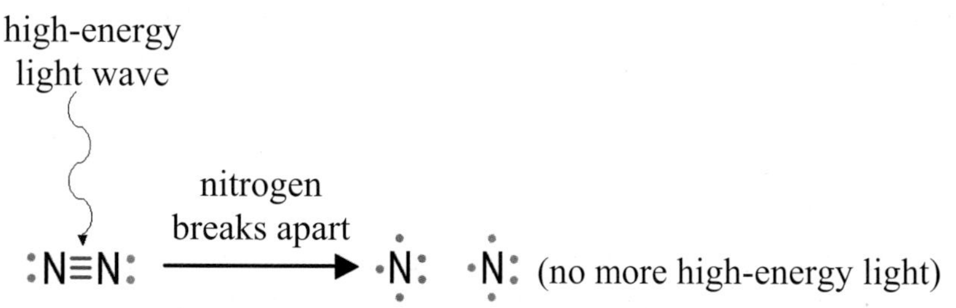

In this way, nitrogen protects the earth from harmful gamma rays.

But gamma rays are not the only high-energy light waves that come from the sun. X-rays also strike the earth's atmosphere daily. Although these light waves still have enough energy to destroy living tissue, they do not have enough energy to break apart nitrogen molecules. Thus, the nitrogen cannot block them. Fortunately, however, there is also a lot of oxygen in the atmosphere. Oxygen, as we have learned, has the following Lewis structure:

:O=O:

The double bond indicates that the oxygen molecule is held together tightly, but not as tightly as the nitrogen molecule. Although X-rays cannot break nitrogen molecules apart, they can break oxygen molecules apart. Thus, the oxygen in our atmosphere blocks the X-rays that come from the sun.

There is still one more type of high-energy light that comes from the sun, however. Ultraviolet rays have enough energy to destroy living tissue, but not enough energy to break apart oxygen molecules. Thus, oxygen cannot block ultraviolet rays. Instead, a substance known as ozone takes care of them. In order to see why ozone can block ultraviolet rays, we need to determine its Lewis structure.

Ozone is a strange substance. Its chemical formula is O_3. Thus, to determine its Lewis structure, we start by writing out the three O's that we have to work with:

We have no choice but to link them all together, so that's where we'll start:

The center oxygen now has eight electrons, so it's all set. The others only have seven. We cannot split up paired electrons, so we must do something with unpaired electrons. I will do something you might not have thought about up to now. I will remove the unpaired electron from one O and move it to the other:

Now two of the O's have eight, and the last one has only six. We can take care of the last O by taking one of the unshared electron pairs on the center O and moving it between the two O's:

Now all oxygens have eight dots, so we have a valid Lewis structure. All we have to do now is replace the shared electron pairs with dashes:

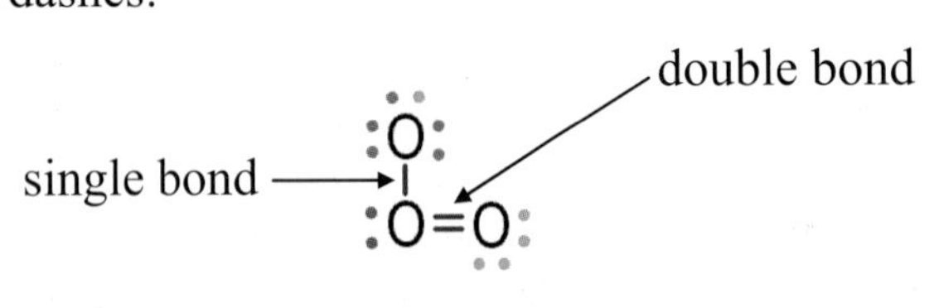

Before we continue, let me assure you that you will not be asked to construct a Lewis structure this complicated. I merely went through its construction to give you more practice at Lewis structures.

Now that we have the Lewis structure, we can see why ozone is uniquely designed to block ultraviolet light that comes from the sun. Ultraviolet light has less energy than gamma rays, so it cannot break apart a molecule with a double bond. However, ozone has a single bond, which is weaker than a double bond. Ultraviolet light has just enough energy to break this bond. Once again, when the bond is broken, the energy of the light is used up, so the light disappears.

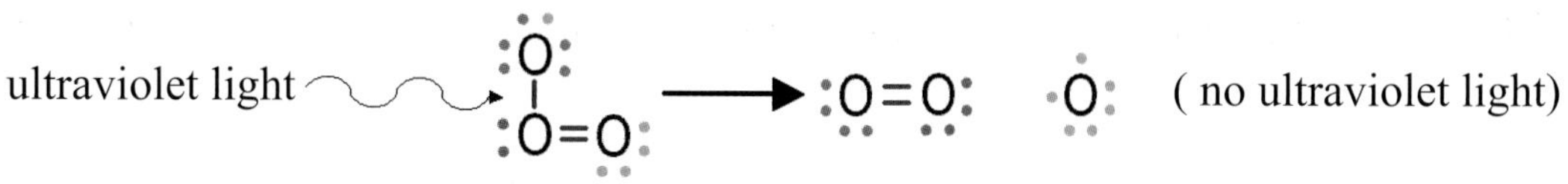

ultraviolet light → O_3 → O_2 + O (no ultraviolet light)

This is how ozone blocks ultraviolet light.

Do you see the marvelous amount of engineering that went into this system? Since many different types of high-energy light from the sun must be blocked, the Designer of this planet installed a three-tiered protection system. The nitrogen in the atmosphere protects us from the most highly energetic light; the oxygen in the atmosphere protects us from the intermediately high-energy light; and the ozone in the atmosphere protects us from the moderately high-energy light. This complex system for protecting living organisms from harmful light rays is just one more piece of evidence that the earth was created, not formed by chance.

If you are impressed with this system so far, hang on! The most incredible aspect of this light-blocking system hasn't even been discussed yet! Although its ability to block ultraviolet light makes ozone necessary for life as we know it, ozone is also incredibly poisonous to living organisms. If living organisms breathe in too much ozone, they die. So, we must have ozone to protect us from the sun's ultraviolet rays, but we cannot breathe too much of it, or it would kill us. Seems like a contradiction, doesn't it? Well, it would be, except the *Designer* of our planet is a little smarter than you and me.

Earth's atmosphere contains plenty of ozone, but the vast majority of it exists in a layer of the atmosphere ("the ozone layer") which is 20 to 30 kilometers (12.4 to 18.6 miles) above sea level, where no living organism breathes! Think about all of this for a moment. Earth just *happens* to have all of the gases necessary to be protected from the sun's harmful rays, and they just *happen* to be in the right place. If nitrogen weren't in the atmosphere, life could never exist because of the gamma rays hitting the planet. If oxygen weren't present, living organisms would suffocate as well as die from exposure to X-rays. Finally, if ozone were not in the atmosphere, living organisms would die from exposure to ultraviolet rays. Having ozone in the atmosphere, however, is not all that is necessary. If ozone were evenly distributed around the atmosphere, as are nitrogen and oxygen, living organisms would die from the poisonous effects of ozone! So we must have ozone, but it must also be somewhere in the atmosphere where no living organisms breathe!

Once again, design features such as this one provide ample evidence that this earth did not appear by chance. The probability of a planet forming with just the right gases in just the right places in the atmosphere is simply too small to consider. Add to this the fact that the design features inherent in our atmosphere constitute only a minuscule fraction of the design features of the entire planet, and an unbiased scientist can only come to one conclusion: This planet is a system that has been designed and created by God.

Before we leave this module, I feel compelled to comment on the media coverage concerning earth's ozone layer. You have probably heard that the ozone layer is being eaten away by certain man-made chemicals called "CFCs." This is, in fact, true. CFCs (chlorofluorocarbons) are chemicals that we use for refrigeration, sterilization, and fire extinguishing. They are incredibly useful chemicals that cost little money to produce. Unfortunately, evidence has been mounting that these chemicals are, indeed, eating away at the ozone layer.

What you are probably not aware of is the *insignificance* of this problem. The media, egged on by environmental extremists, have blown the problem way out of proportion. The following facts may put the debate in a little bit of perspective:

1. The *only* places where CFCs are eating away at the ozone layer are at the North and South Poles, principally the South Pole. The weather conditions at these sites are ideal for CFC-induced destruction of the ozone layer. These weather conditions do not exist elsewhere on the planet, so there is little chance of the problem spreading.

2. The total destruction of the ozone layer by CFCs is, by all best estimates, about 5% at the South Pole. To put this number in perspective, the amount of ozone in the ozone layer fluctuates *naturally* by about 50%! Thus, the natural ups and downs in the ozone layer are 10 times the amount of destruction that has been caused by CFCs.

3. Not a single illness (or death) can be traced to ozone destruction. The effect is simply too small to affect living organisms.

Despite these facts, world governments (egged on by environmental extremists) have called for the elimination of CFCs. The United States banned production of CFCs at the end of 1995. While these acts will restore what little of the ozone layer has been damaged by CFCs, the cost in human lives will not be negligible. You see, while no illnesses or deaths can be traced to ozone depletion, thousands of lives are saved yearly by CFCs. Professional firefighters call them the most miraculous of all fire-extinguishing chemicals. They are the best refrigerants in the world, making food distribution more efficient than ever thought possible. Surgical sterilizers that use CFCs are more efficient than all others, providing the best protection against deadly infection. Unfortunately, other chemicals used in these capacities are not nearly as good as CFCs.

When CFCs are finally banned by all governments, food distribution will be less efficient, causing starvation in Third World countries. Estimates indicate that a worldwide CFC ban could result in as many as 40 million deaths due to hunger, starvation, and food-borne diseases *(Environmental Overkill: Whatever Happened to Common Sense?* by Dixy Lee Ray with Lou R. Guzzo). Despite these measurable effects on human life, CFCs will be banned to solve a problem that cannot yet be traced to a single human calamity. This is just one example of the world's need for leaders who truly understand science and are not swayed by propaganda. If you would like to learn more about this controversy, check out the following books in your library:

Two excellent books on how environmentalist propaganda succeeds over real science:

Eco-Sanity: A Common-Sense Guide to Environmentalism by Joseph L. Bast and others

Ecoscam: The False Prophets of Ecological Apocalypse by Ronald Bailey

Two books that give you the environmental extremist point of view:

Betrayal of Science and Reason: How Anti-Environmental Rhetoric Threatens Our Future by Paul R. Ehrlich and Anne H. Ehrlich

Earth in the Balance by Albert Gore

ANSWERS TO THE "ON YOUR OWN" PROBLEMS

8.1 a. Since Ca is in group 2A, it has two valence electrons:

Ca

b. Since Si is in group 4A, it has four valence electrons:

c. Since At is in group 7A, it has seven valence electrons:

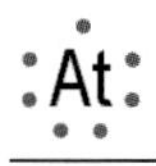

8.2 a. Since calcium is in group 2A, it has a 2+ charge. Chlorine is in group 7A, so chloride has a 1- charge. Ignoring the + and - signs and switching the numbers gives us:

$CaCl_2$

b. Since potassium is in group 1A, it has a 1+ charge. Sulfur is in group 6A, so sulfide has a 2- charge. Ignoring the + and - signs and switching the numbers gives us:

K_2S

c. Since aluminum is in group 3A, it has a 3+ charge. Nitrogen is in group 5A, so nitride has a 3- charge. The numerical values of the charge are identical, so we ignore them:

AlN

d. Since magnesium is in group 2A, it has a 2+ charge. Phosphorus is in group 5A, so phosphide has a 3- charge. Ignoring the + and - signs and switching the numbers gives us:

Mg_3P_2

8.3 a. In this problem, the Roman numeral after iron tells you that it has a 3+ charge. Fluorine is in group 7A, and thus fluoride has a 1- charge. Ignoring the signs and switching the numbers gives us FeF_3.

b. In this problem, the Roman numeral after copper tells you that it has a 1+ charge. Iodine is in group 7A, and thus iodide has a 1- charge. The numerical values of the charges are the same, so we ignore them, giving us CuI.

c. In this problem, the Roman numeral after manganese tells you that it has a 3+ charge. Oxygen is in group 6A, and therefore oxide has a 2- charge. Ignoring the signs and switching the numbers gives us Mn_2O_3.

8.4 These elements all lie in the same row. This means that the only difference in their position on the chart is left to right. Based on the fact that ionization potential increases from left to right on the chart, the correct order is:

Rb < Ru < Sn < I

8.5 The only difference in position for these elements is vertical. Based on the fact that ionization potential decreases from top to bottom on the chart, Fr has the lowest ionization potential. This means that Fr gives up its electrons most readily.

8.6 All of these atoms are in the same row on the periodic chart, so the only difference between their positions is left to right. Since S lies farthest to the right, it has the highest electronegativity. This means that S has the highest desire for electrons.

8.7 These elements all lie in the same column on the chart, so their only difference in position is top to bottom. Based on the fact that electronegativity decreases from top to bottom of the chart, the order is:

At < I < Br < Cl < F

8.8 All of these elements are in the same row. Since the atomic radius decreases from left to right on the periodic chart, the leftmost element will be largest. That means Na is the largest.

8.9 All of these elements are in the same column. Since the atomic radius increases as you move down the chart, the order is:

F < Cl < Br < I < At

8.10 a. The chemical formula tells us that we have one N and three H's to work with:

```
 •
•N:  H•  H•  H•
 •
```

Because N has the most unpaired electrons, it goes in the center and we try to attach the H's to it:

```
H•————→ •
H•———→•N:
H•————→ •
```

This is easy, since each H has a space for an unpaired electron, and the N has three unpaired electrons. The Lewis structure, then, looks like this:

```
  H
  ••
H:N:
  ••
  H
```

Remember, the ideal electron configuration for hydrogen is two valence electrons, while for nitrogen it is eight. This is what each atom has in this Lewis structure, so we are all set. Now we just have to replace the shared electron pairs with dashes:

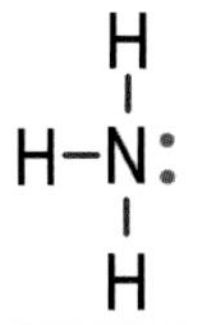

b. The chemical formula tells us that we have one Si and four Br's to work with:

Because Si has the most unpaired electrons, it goes in the center and we try to attach the Br's to it:

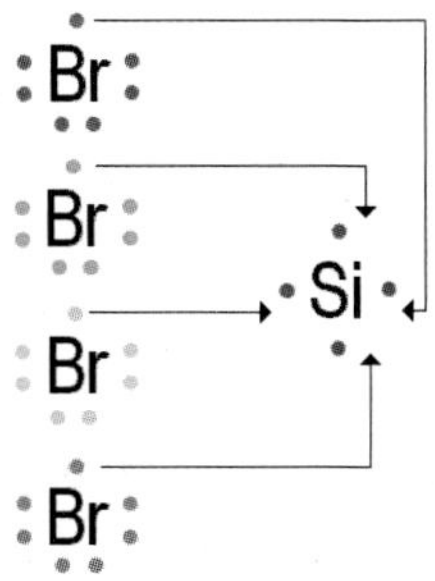

This is easy, since each Br has a space for an unpaired electron, and the Si has four unpaired electrons. The Lewis structure, then, looks like this:

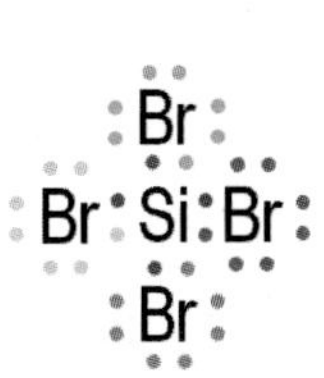

All atoms have eight valence electrons now, so we are all set. Now we just have to replace the shared electron pairs with dashes:

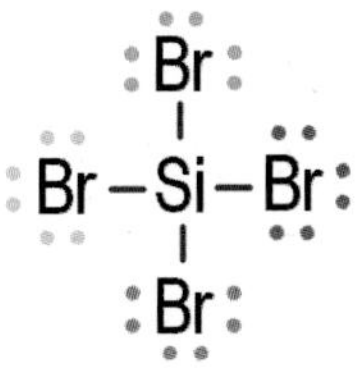

c. The chemical formula tells us that we have one S and two H's to work with:

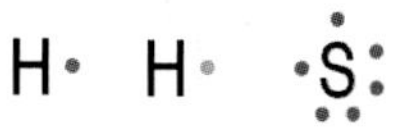

Because S has the most unpaired electrons, it goes in the center and we try to attach the H's to it:

H
H S

This is easy since each H has a space for an unpaired electron, and the S has two unpaired electrons. The Lewis structure, then, looks like this:

H
H:S:

Remember, the ideal electron configuration for hydrogen is two valence electrons, while for sulfur it is eight. This is what each atom has in this Lewis structure, so we are all set. Now we just have to replace the shared electron pairs with dashes:

H
|
H–S:

d. The chemical formula tells us that we have one I and one Cl to work with:

:I: :Cl:

This isn't bad. Its just like F_2, because each atom has one unpaired electron. We just pair these together:

:I:
:Cl:

Both atoms have eight valence electrons now, so we are all set. Now we just have to replace the shared electron pair with a dash:

:I:
|
:Cl:

8.11 The chemical formula tells us that we have one C, two H's, and one O to work with:

H· H· ·C· ·O:

According to our procedure, we put the C in the middle (since it has the most unpaired electrons) and try to attach everything to it:

O
H C
H

This gives us:

H:C:O:
H

Now we see that the H's already have their ideal electron configuration, since they each have two dots. The carbon and oxygen, however, both have only seven. We can take care of the oxygen by taking the unpaired electron on the carbon and putting it between the carbon and the oxygen. That way it will count for both.

H:C:O: ⟶ H:C:.O:
H H

Now the oxygen has eight, but the carbon still has only seven. We can fix that, however, by putting the unpaired electron on the oxygen between the carbon and oxygen. This will, once again, allow the electron to count for both atoms.

H:C:.O: ⟶ H:C::O:
H H

Now all atoms have their ideal electron configuration, so we just have to replace the shared electron pairs with dashes:

H–C=O:
H

8.12 The chemical formula tells us that we have two C's and two H's to work with:

·C· ·C· H· H·

The procedure that we have been following tells us that if we have two (or more) of the atom with the most unpaired electrons, we link them together and put them in the center. We then try to hang everything else off of them:

H· :C:C:
H·

This gives us:

H:C:C:H

You might wonder how I decided to put one H on one C and that other H on the other C. There are several unpaired electrons, so why didn't I put both H's on one C? Well, there really is no way to know this when you start out. It just turns out that this is the only way you can get a good Lewis structure. You might have made several attempts before you came across this solution. One thing you might consider, however: Creation tends to like symmetry (balance). Putting an H on each carbon "balances" the molecule.

Now we can see that the H's have their ideal electron configuration, but both C's have only six electrons. We can fix the second C by taking the unpaired electrons on the first C and putting them in between the two:

H:C::C:H ⟶ H:C::C:H

This gives the second C eight electrons, but the first C still has only six. We can fix that by taking the two unpaired electrons on the second C and putting them in between the two as well:

H:C::C:H ⟶ H:C:::C:H

This gives all atoms their ideal electron configuration. The only thing left to do, then, is replace the shared electron pairs with dashes:

H–C≡C–H

NOTE: All Lewis structure illustrations in this module by Megan Whitaker.

REVIEW QUESTIONS FOR MODULE #8

1. What are valence electrons and why are they so important in chemistry?

2. How many valence electrons are in the following electron configuration?

$$1s^22s^22p^63s^23p^64s^23d^{10}4p^65s^24d^{10}5p^3$$

3. The chemistry of life is based on carbon. This means that all of the molecules which govern life-giving chemistry have carbon atoms as their principal component. Chemists have often speculated that there could be a life form with a chemistry based on silicon. Why?

4. What are the noble gases and why are they important in chemistry?

5. What is the fundamental difference between metals and nonmetals?

6. What is the difference between an atom and an ion?

7. What is ionization potential?

8. Name three periodic properties of atoms.

9. What is the fundamental difference between ionic compounds and covalent compounds?

10. What gases are involved in earth's three-tiered protection system against high-energy light that comes from the sun?

PRACTICE PROBLEMS FOR MODULE #8

1. Draw the Lewis structures for the following atoms:

    a. Ge   b. Te   c. Ba

2. Give the chemical formulas for the following molecules:

    a. aluminum sulfide b. cesium nitride   c. magnesium oxide   d. chromium(III) oxide

3. Which atom gives up its electrons most readily: Al, B, In, or Ga?

4. Order the following atoms in terms of increasing ionization potential: Sr, Sb, I.

5. Which atom has the greatest desire for extra electrons: N, Sb, or As?

6. Order the following in terms of increasing atomic radius: As, K, Br, Se.

7. Draw a Lewis structure for CH_4.

8. What is the Lewis structure for PCl_3?

9. What is the Lewis structure for FNO?

10. Based on their Lewis structures, which molecule is easiest to break apart:

    C_2H_4, PN, or H_2?

MODULE #9: Polyatomic Ions and Molecular Geometry

Introduction

In the last module, we learned all about Lewis structures and how they relate to both ionic and covalent compounds. As promised, this module still deals with Lewis structures, but it applies them to brand new concepts. First, you will learn about more complicated ionic compounds. By design, the ionic compounds we have dealt with so far have been rather simple. In creation, however, the most important ionic compounds tend to be a little more complicated. Thus, to get a good understanding of chemistry, we must improve our knowledge of ionic compounds.

In addition to learning about more complicated ionic compounds, we will also explore the three-dimensional aspect of molecules. After all, the Lewis structures we learned in the last module are very informative, but they have a big drawback. They do not allow us to see what the molecules look like in three-dimensional space. In other words, all Lewis structures are flat (two-dimensional) because they are drawn on a flat piece of paper. In reality, molecules are not flat; they are three-dimensional. We will learn some concepts in this module that will help us visualize the three-dimensional character of molecules.

Finally, the three-dimensional nature of molecules will help us understand why certain liquids mix together easily (like milk and water) while other liquids do not (like gasoline and water). To learn all of these new concepts, we will have to be comfortable with the skills discussed in the last module. Thus, if you feel tentative about any of the material presented in the last module, it might be a good idea to go back and look at that material again or look at another resource that deals with such material before plunging into this module.

Polyatomic Ions

So far, you have learned that when a single atom gains or loses electrons, an ion forms. That is not, however, the only way that an ion can form in nature. It turns out that several atoms can come together and gain or lose electrons as a group. Since these ions require more than one atom, we call them **polyatomic** (pahl' ee uh tom' ik) **ions**.

Polyatomic ions - Ions which are formed when a group of atoms gains or loses electrons

Although you might not understand how this can happen, a quick look at the Lewis structures of a couple of polyatomic ions might help.

Consider, for example, what might happen if a single hydrogen atom tried to form a molecule with a single oxygen atom. If we tried to construct the Lewis structure for such a molecule, we would start by drawing the Lewis structures of the individual atoms:

H·   ·Ö:

We would then link the two atoms together to see what we got:

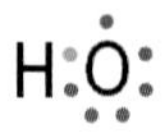

H:O:

In this Lewis structure, the H has its ideal electron configuration, but the O has only seven electrons. You can play around with this Lewis structure as long as you like, but there is no way to move the electrons around so that both H and O get their ideal electron configurations. Since we cannot draw a good Lewis structure for the OH molecule, we have to assume that it doesn't exist.

Despite the fact that the OH molecule doesn't exist, the OH^- *ion* does exist. After all, if you look at the Lewis structure above, all that the O needs to attain its ideal electron configuration is one more electron. Suppose a sodium atom was around and was willing to donate its electron to the oxygen in the OH:

H:O: Na· ⟶ $H{:}O{:}^-$ Na^+

This situation is now ideal for all atoms concerned. The hydrogen has two electrons, the oxygen has eight electrons, and the sodium has no electrons in its Lewis structure. This gives each atom its ideal electron configuration. But look what has happened. The oxygen and hydrogen are linked together by shared electrons:

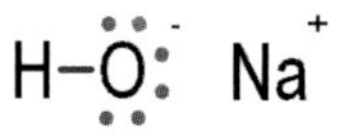

$H{-}O{:}^-$ Na^+

So despite the fact that OH^- is an ion, it also contains a shared electron pair. These two characteristics make OH^- a polyatomic ion. Both the O and the H share the negative charge brought about by the extra electron taken from the sodium atom. Since both the O and the H share the negative charge, this is neither an oxide ion nor a hydride ion. It is something else. We call this ion a "hydroxide" ion. The resulting ionic compound, sodium hydroxide, has a chemical formula of NaOH. The O and the H are kept next to each other in the chemical formula to emphasize that they are hooked together by a pair of shared electrons.

Thus, any ion that has more than one atom is a polyatomic ion. Another example of such a substance would be the carbonate ion. This ion is composed of one carbon atom and three oxygen atoms:

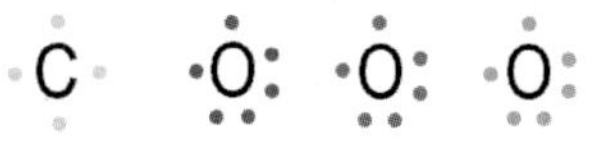

·C· ·O: ·O: ·O:

If we link these atoms together according to the rules we followed in the previous module, we get:

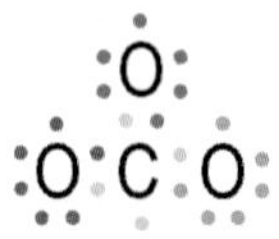

:O:
:O:C:O:

If we play with the Lewis structure for a while, about the best we can do is as follows:

This gives the carbon and one of the oxygens their ideal electron configurations, but two of the oxygens are each one electron short. The only way that those oxygens will come up with eight electrons each is for them to each get an extra electron. Well, suppose two potassium atoms were around and willing to donate their electrons:

K· K·

This would lead to:

K^{+} $2-$ K^{+}

Since each potassium lost one electron, they each have a 1+ charge. The polyatomic ion, on the other hand, gained two electrons, so it has a 2- charge.

K^{+} $2-$ K^{+}

The resulting CO_3^{2-} ion is called the "carbonate" ion. The ionic compound that results from the above Lewis structure, K_2CO_3, is called potassium carbonate. Once again, the CO_3 is written together as a group to emphasize that these atoms are all linked together as a polyatomic ion.

Now that you know about polyatomic ions and why they exist, you need to become familiar with several of the most common ones. You need to know their chemical formulas, charges, and how to determine the chemical formulas of the ionic compounds of which they are a part. The good news is that you do not have to learn their Lewis structures. You will never, in this course, be asked to construct a Lewis structure for a polyatomic ion, because it involves adding (or subtracting) electrons like I did in the previous two Lewis structures for the hydroxide ion and the carbonate ion. I don't want you to do this, because it breaks one of the rules we established in the last module. If you go on to advanced chemistry, however, you will be expected to draw polyatomic ion Lewis structures.

Although you do not need to know their Lewis structures, you do need to know the names, formulas, and charges of the polyatomic ions listed in Table 9.1. Unfortunately, there is not a whole lot of rhyme or reason to these names and charges, so you will simply have to memorize them.

TABLE 9.1
Important Polyatomic Ions

Ion Name	Formula	Ion Name	Formula
ammonium (uh moh' nee uhm)	NH_4^+	cyanide (sigh' uh nide)	CN^-
hydroxide (hye drox' ide)	OH^-	carbonate (kar' bun ate)	CO_3^{2-}
chlorate (klor' ate)	ClO_3^-	chromate (krohm' ate)	CrO_4^{2-}
chlorite (klor' ite)	ClO_2^-	dichromate (dye krohm' ate)	$Cr_2O_7^{2-}$
nitrate (nye' trate)	NO_3^-	sulfate (suhl' fate)	SO_4^{2-}
nitrite (nye' trite)	NO_2^-	sulfite (suhl' fite)	SO_3^{2-}
acetate (as' uh tate)	$C_2H_3O_2^-$	phosphate (fahs' fate)	PO_4^{3-}

This is a significant amount of memorization work, but it will get easier as time goes on, because we will use these polyatomic ions quite a bit. Notice that even though there is not a lot of rhyme or reason behind these names, there is a little. When an ion ends with an oxygen, its name usually starts with the first atom in the ion and ends in "-ate" or "-ite." In addition, for a certain starting atom, the ion with more oxygens has the "-ate" ending, while the one with fewer oxygen atoms has the "-ite" ending. These facts may help you in memorizing these ions.

Not only do you need to know the names, charges, and formulas of polyatomic ions, you also have to be able to determine the chemical formulas of ionic compounds that contain polyatomic ions. This doesn't require learning anything new, however, because the technique you used in the previous module for determining the chemical formula of ionic compounds still applies here. You just have to be careful in how you write the final result. Study the examples below and then perform the problems that follow to make sure you understand how to do this:

EXAMPLE 9.1

What is the chemical formula for calcium nitrate?

According to the name, calcium nitrate is made up of the calcium ion and the nitrate ion. Calcium is in group 2A, so the calcium ion must have a charge of 2+. The nitrate ion is a polyatomic ion. According to the table, it has a charge of 1-. To determine the chemical formula using the shortcut, we ignore the charges and switch the numbers. Thus, the "1" goes with the calcium ion and the "2" goes with the nitrate ion. We write the resulting formula like this:

$$\underline{Ca(NO_3)_2}$$

Let's make sure we understand why the chemical formula is written this way. I said that the "2" went with the nitrate ion; thus, there are two nitrate ions in the chemical formula. The only way we can show that the "2" goes with all of the atoms in the nitrate ion is to use the parentheses. This tells us that the entire nitrate ion appears twice in the formula for calcium nitrate. Now you see why some compounds have parentheses in their chemical formulas. Back in Module #6, you used compounds like $(NH_4)_2SO_4$ in stoichiometry problems, and you learned how to deal with the parentheses in the chemical formula at that time. Now you can see *why* there are parentheses in the chemical formula: They emphasize that this ionic compound has two ammonium (NH_4^+) ions in it.

What is the chemical formula of magnesium phosphate?

According to the name, this molecule is composed of the magnesium ion and the phosphate ion. According to the periodic chart, the magnesium ion will have a charge of 2+. According to the table, the phosphate ion has a charge of 3-. To determine the chemical formula, we ignore the signs and switch the numbers. Thus, we have three magnesium ions and two phosphate ions. Using our parentheses notation, we get:

$$\underline{Mg_3(PO_4)_2}$$

Once again, realize that the "2" outside of the parentheses goes with both atoms inside the parentheses. This means that the molecule is composed of three magnesiums, two phosphoruses, and eight oxygens.

What is the chemical formula for copper(I) acetate?

According to the name, this substance is composed of the copper ion and the acetate ion. Since there is a Roman numeral after the copper, we know that copper is an exception to our rules. The "(I)" tells you that in this molecule, it has a charge of 1+. According to the table, the acetate ion has a charge of 1-. Since the numerical values of the charges are the same, we ignore them. This gives us:

$$\underline{CuC_2H_3O_2}$$

Notice that in this chemical formula, parentheses are not needed for the polyatomic ion, because we have only one acetate ion.

ON YOUR OWN

9.1 Balance the following chemical equation:

$$Ba(NO_3)_2\ (aq) + NaOH\ (aq) \rightarrow Ba(OH)_2\ (s) + NaNO_3\ (aq)$$

9.2 Give the chemical formula for each of the following compounds:

a. potassium sulfate b. ammonium carbonate c. aluminum acetate

9.3 When aluminum nitrite reacts with sodium carbonate, aluminum carbonate and sodium nitrite are produced. Write a balanced chemical equation for this reaction.

Molecular Geometry: The VSEPR Theory

Now that we have spent a considerable amount of time on ionic compounds, we turn back to the subject of covalent compounds. In the last module, we learned how to use Lewis structures to help us picture what a molecule might look like. The problem is, this picture is still a bit wrong. After all, we were using Lewis structures that were written on a flat (two-dimensional) piece of paper.

Molecules, however, are rarely flat. They generally have a three-dimensional structure to them. In other words, molecules not only have length and width, but they also have *depth*. Lewis structures do not allow us to picture the depth of a molecule. We must learn something else before we can fully picture what a molecule really looks like.

In order to do this, we must first recall something about Lewis structures. Consider, for example, the Lewis structure for CH_4:

```
    H
    |
H — C — H
    |
    H
```

Remember that the dashes in this Lewis structure represent covalent bonds. Remember also that these covalent bonds are just electron pairs that are shared between two atoms. Well, since each of these covalent bonds is made up of electron pairs, we can easily predict something about their behavior: They will tend to repel one another, because they each have the same type of electrical charge.

If these electron pairs repel each other, it stands to reason that they will try to move as far apart from each other as possible. How will they do that? If the molecule were to stay flat, then the electron pairs could not get any farther apart from each other than what is pictured in the Lewis structure above. In geometry terms, we could say that the largest angle between the chemical bonds would be 90 degrees. On the other hand, if the molecule took advantage of three-dimensional space, the bonds could get farther apart from each other. Figure 9.1 illustrates this phenomenon.

FIGURE 9.1
CH_4 Puts the Maximum Distance Between Its Bonds by Forming a Tetrahedron

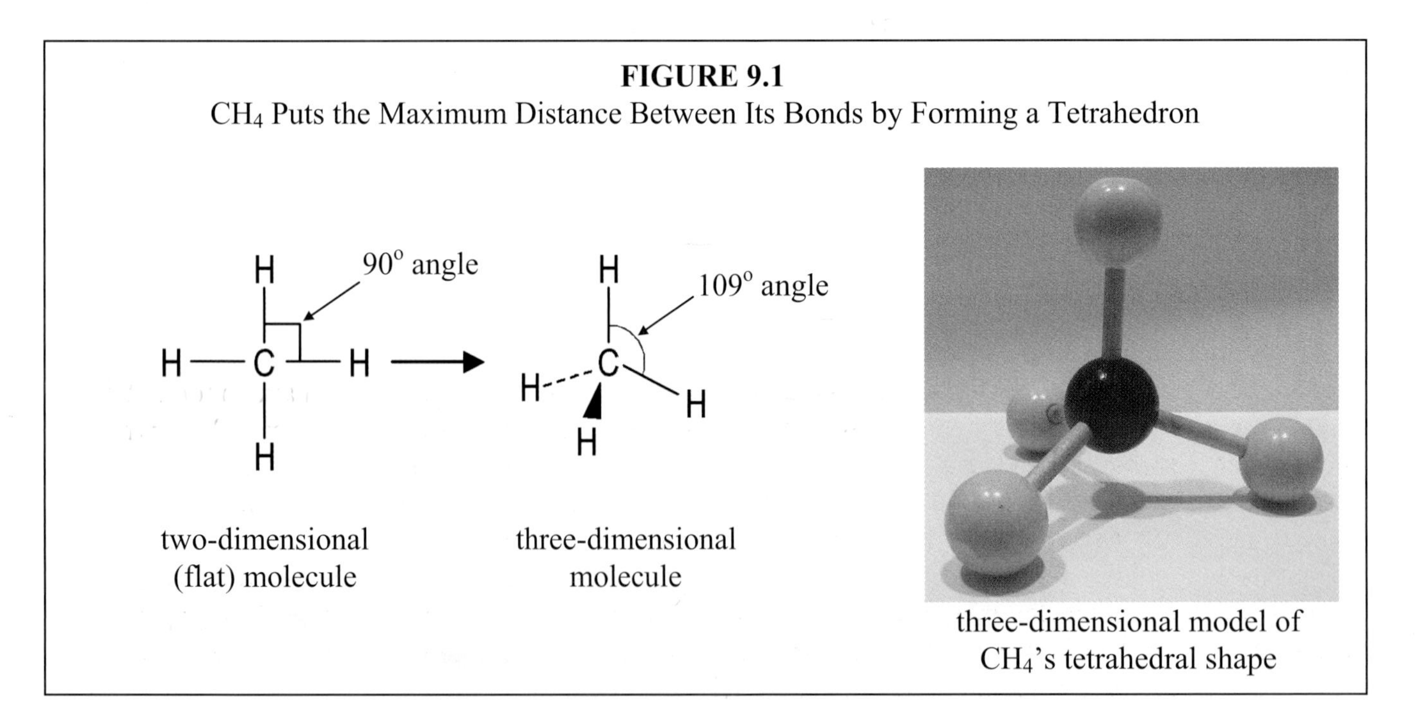

three-dimensional model of CH_4's tetrahedral shape

As a flat, two-dimensional structure, the bonds between the C and H cannot get any farther apart from one another than 90 degrees. However, if the molecule takes advantage of three-dimensional space, the bonds can spread out a bit more. It turns out that the bonds can get as far as

109 degrees from each other in a three-dimensional structure. We try to picture this three-dimensional structure by using solid, dashed, and triangular lines. The solid lines represent chemical bonds that are in the plane of the paper. The dashed line, however, is supposed to indicate a bond that goes behind the plane of the paper. Finally, the heavy, triangular line represents a chemical bond that is out in front of the plane of the paper. Thus, the hydrogen atom attached to the dashed line sits behind the paper, while the hydrogen atom attached to the heavy, triangular line is sitting in front of the paper. The other two hydrogens are sitting on the paper. This shape, called a **tetrahedron** (teh truh he' drun), is one of the fundamental shapes that a molecule can attain. This shape is easiest to understand by looking at the photograph, where the black sphere represents the carbon atom, and the yellow spheres represent the hydrogen atoms. The wooden sticks represent the bonds between the atoms.

The theory that allows us to predict that CH_4 has a tetrahedral shape is called **VSEPR theory**. VSEPR stands for **V**alence **S**hell **E**lectron **P**air **R**epulsion. The VSEPR theory states that molecules will attain whatever shape keeps the valence electrons of the central atom as far apart from one another as possible. Thus, we know that CH_4 has a tetrahedral shape because that shape allows carbon's valence electrons to be as far apart from each other as possible.

If you have a hard time picturing all of this in your mind, don't worry too much. Trying to picture three-dimensional shapes can be very difficult for some, while it is perfectly natural for others. As we go through more molecular shapes, I will give them all names. In addition, I will mention the angles that exist between the chemical bonds. All you need to be able to do is learn those two things. It would be nice if you could also picture it in your mind, but I realize that this is a little too hard for some people. Hopefully, you will get the hang of this as we go along.

The tetrahedral shape is not the only shape a molecule might posses. Consider, for example, an ammonia molecule: NH_3. The Lewis structure for this molecule, which you have already drawn, is pictured below:

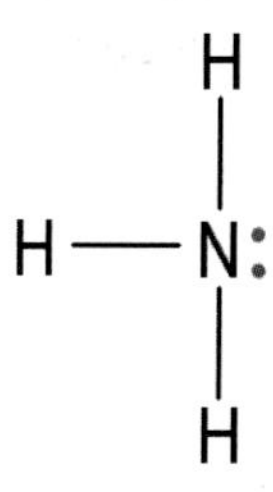

In trying to determine what kind of three-dimensional shape a molecule attains, we have to look at the electron *groups* that surround the central atom. In this case, the central atom is nitrogen. How many groups of electrons surround the N? There are a total of four. Three of them are bonds, represented by the dashes in the Lewis diagram. The fourth, however, is a pair of non-bonding electrons. Even though these electrons do not form a bond, we still have to worry about them because they are still electrons and will repel any other electron pair that comes close to them. Thus, the four electron groups will try to get as far away from each other as possible. As we saw above, when four groups of electrons try to get as far away from each other as possible, a tetrahedron is formed. This tetrahedron, however, is a bit misshapen, as shown in Figure 9.2.

FIGURE 9.2
How NH_3 Acquires a Pyramidal Shape

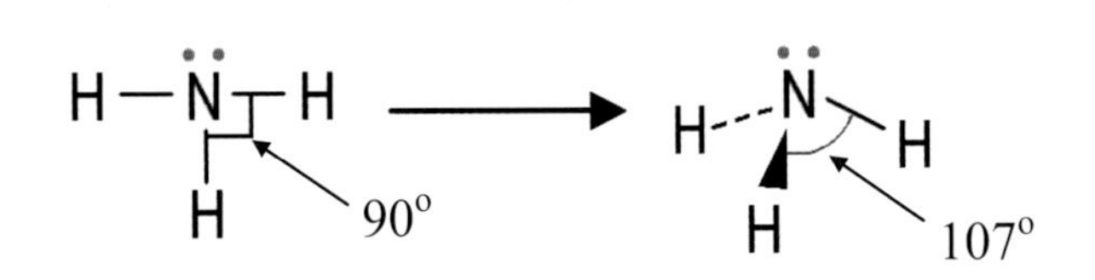

Lewis structure rotated to make 3-D drawing easier to see

three-dimensional structure

three-dimensional model of NH_3's pyramidal shape

In the figure, I took the original Lewis structure and rotated it so that the non-bonding electron pair is on top of the nitrogen atom. This is done simply to make the next drawing a little easier to understand. In that drawing, I tried to illustrate the three-dimensional nature of the molecule. Once again, the dashed line indicates that the bond is behind the plane of the paper, while the heavy, triangular line indicates that the bond lies in front of the plane of the paper. The non-bonding electron pair and the bond indicated by the solid line lie in the plane of the paper.

You may be tempted to call this a tetrahedron as well, because it looks very similar to the situation with CH_4, which I said formed a tetrahedron. However, there is one big difference. The top of the tetrahedron is not a bond in this case. It is a pair of non-bonding electrons. Thus, the top "leg" of the tetrahedron is missing, deforming the tetrahedron. The resulting shape looks a lot like a pyramid, with the H's as the pyramid's base and the N as its apex. Thus, we say that NH_3 has a **pyramidal** (pih ram' uh dul) shape. This pyramidal shape is easiest to understand by looking at the photograph, where the blue sphere represents the nitrogen atom and the yellow spheres represent hydrogen atoms. Once again, the wooden sticks represent covalent bonds.

In a tetrahedron, the legs are 109° apart. In this shape, the legs are slightly closer together because the non-bonding electron pair tends to repel the bonds just a little more than the bonds repel one another. As a result, the bonds stay a little farther away from the non-bonding electron pair and a little closer to one another. Thus, the bond angle in this case is about 107°.

As you might imagine, we aren't done looking at the different shapes that a molecule can have. Next, we turn our attention to water:

H
|
H-O:

According to VSEPR theory, we must look at the electrons surrounding the central atom and determine what shape will allow the electron groups to stay as far apart from each other as possible. Once again, there are four groups of electrons around the central oxygen atom. Two of the groups are chemical bonds, while the other two are non-bonding electron pairs. Since a tetrahedron allows the four electron groups to stay as far apart from each other as possible, the basic shape of this molecule is also a tetrahedron. Once again, however, since two of the groups are non-bonding electron pairs, the tetrahedron is deformed.

FIGURE 9.3

How H_2O Gets Its Bent Shape

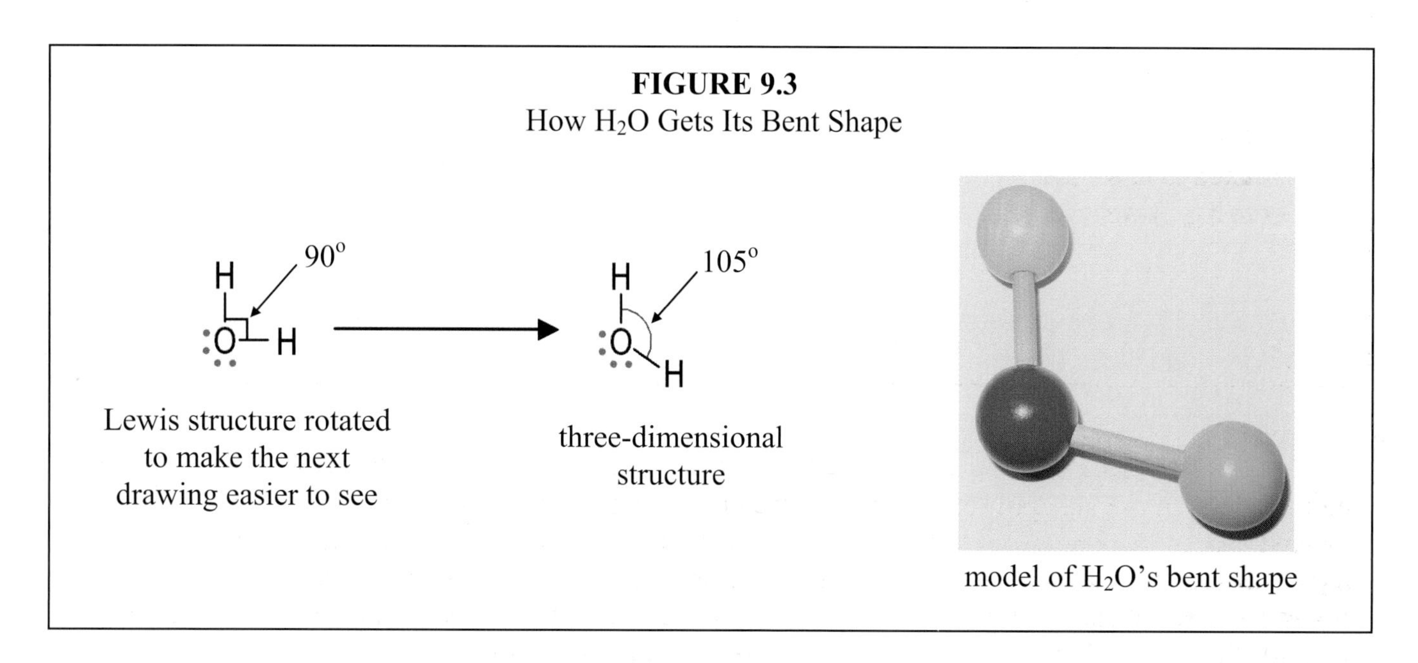

model of H_2O's bent shape

It is easier to understand water's molecular shape if you first rotate the Lewis structure. When that is done, we once again put each group of electrons on a leg of the tetrahedron. In the second drawing above, the H's are each on a leg that is in the plane of the paper. One of the non-bonding electron pairs is on the tetrahedron leg that juts out in front of the paper, while the other is on the leg that sticks out behind the paper. Since these two non-bonding electron pairs are in the molecule, the tetrahedron looks like it is missing those two legs. What is left, then, is a shape that looks **bent**. As I mentioned with NH_3, non-bonding electron pairs tend to repel more than bonding electron pairs. As a result, the two bonds in the water molecule are closer together than they would be in a tetrahedron or a pyramidal shape, so the bond angle is about 105°.

The three shapes we have considered so far are all built on the framework of a tetrahedron. The pyramidal shape is just a tetrahedron with one leg removed, while the bent shape is a tetrahedron with two legs removed. This tetrahedron framework exists because in the three molecules we have considered so far, the central atom has four groups of electrons around it. However, four groups of electrons around the central atom is not the only possibility in a Lewis structure. As a result, we have more shapes to consider.

Let's look at a more complicated Lewis structure. In the previous module, you constructed the Lewis structure for CH_2O:

H-C=O:
H

Notice that in this molecule, there are only *three* groups of electrons around the central carbon atom. There are two single bonds linking the H's to the C. That's two groups of electrons. There is also one double bond linking the O to the C. Although a double bond represents four electrons, it is still only one *group* of electrons. After all, a double bond is really only *one* bond; it is simply twice as strong as a single bond. Thus, in this molecule, there are only three groups of electrons surrounding the central carbon: two single bonds and one double bond. What happens in a situation like this?

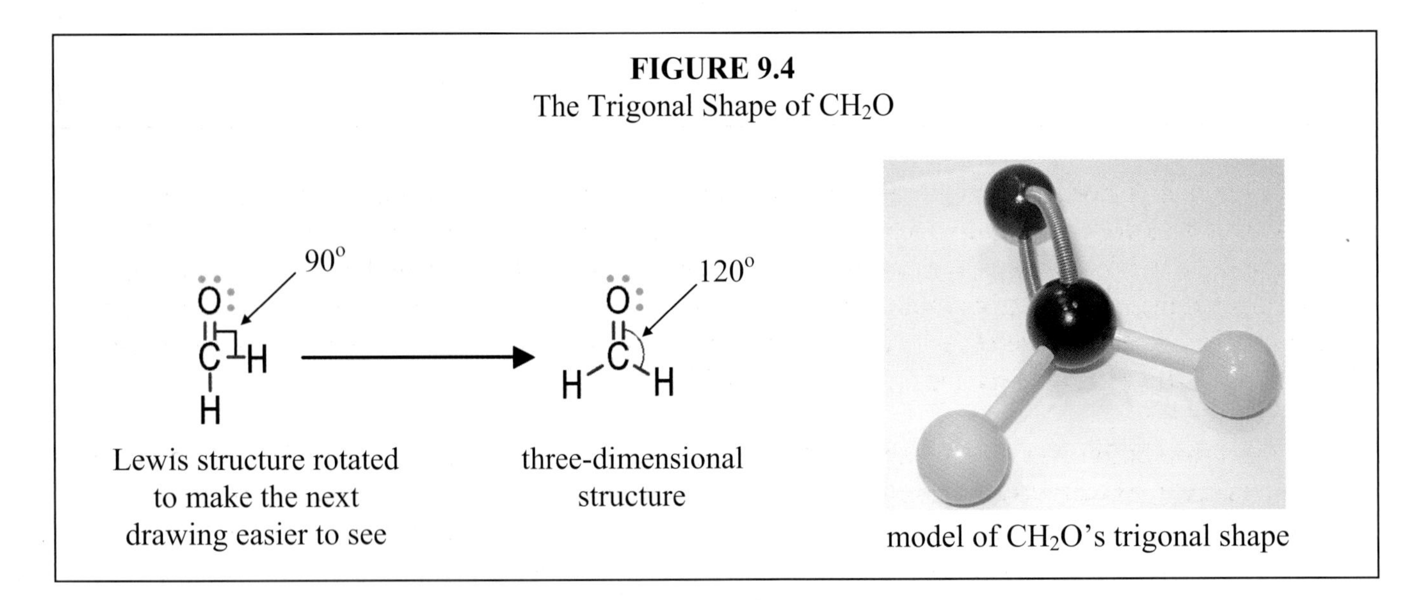

FIGURE 9.4
The Trigonal Shape of CH_2O

Lewis structure rotated to make the next drawing easier to see

three-dimensional structure

model of CH_2O's trigonal shape

Now remember, the guiding principle behind the VSEPR theory is that the groups of electrons try to get as far apart from one another as possible. Well, the farthest that three groups of electrons can get from each other is 120°. Thus, the shape of this molecule is triangular, with the hydrogen and oxygen atoms each occupying a vertex of the triangle. We'll call this a **trigonal** (trg' uh nul) shape. Since all of the atoms lie in the same plane, it is sometimes called a "trigonal planar" shape. The picture shows a model of this geometry, with the black sphere representing carbon, the red sphere representing oxygen, and the yellow spheres representing hydrogen atoms. Notice that the model uses two springs to represent the double bond that exists between oxygen and carbon. You will see that again in the next figure.

Believe it or not, there is really only one more molecular shape that we are going to consider. There are, in fact, several more shapes that molecules can take, but the ones we are covering here are the most popular. If you take another year of chemistry, you will learn additional molecular shapes then. For right now, we have only one more shape to consider: linear. It turns out that there are two ways for a molecule to become linear. The first should be pretty obvious. If a molecule consists of only two atoms, the molecule must be linear. After all, with only two points, there is no other shape possible than a line. Thus, a molecule like F_2:

must be linear. However, there is another way that a molecule might attain a linear shape. Figure 9.5 illustrates how this can happen.

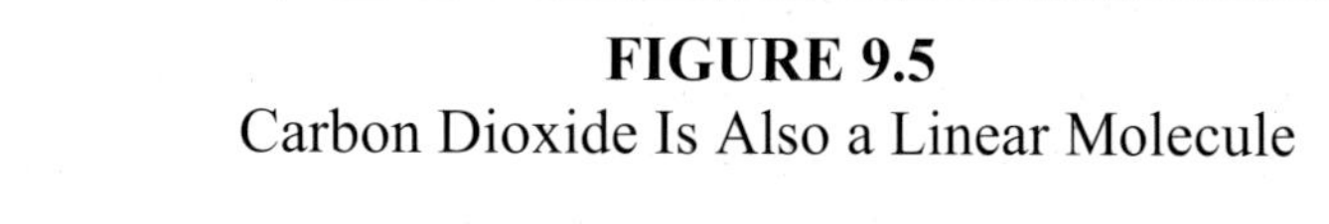

FIGURE 9.5
Carbon Dioxide Is Also a Linear Molecule

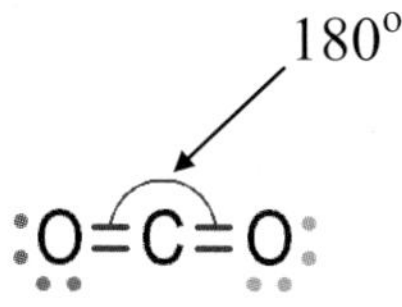

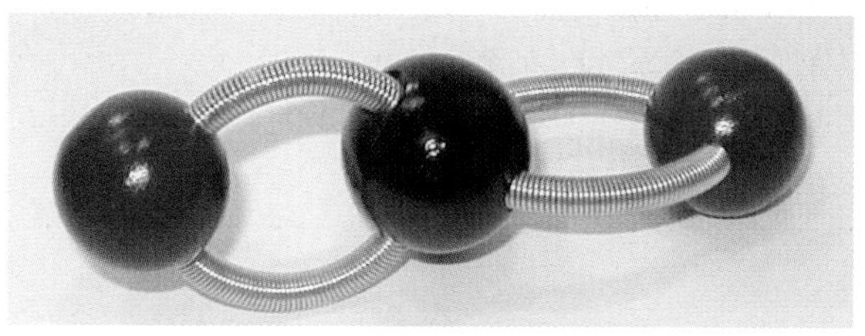

In this molecule, there are only two groups of electrons around the central atom. Even though it looks like there are four bonds around the carbon, there are not. As I said before, a double bond is really only one bond; it is simply twice as strong as a single bond. Thus, even though there are four *dashes* around the central carbon, there are really only two *bonds*. That means there are two groups of electrons. Two groups will get as far away from each other as possible when they form a straight line. Thus, this is a **linear** shape, with a bond angle of 180°.

As I said before, if you are having trouble visualizing all of this, don't worry. I will give you a set of rules that will help you determine these molecular shapes. Thus, you need not be able to visualize this in order to solve the problems. You will be better off, however if you are able to visualize the molecules. It will help you as you go to the last section of this module. In addition, much of the chemistry of life is based on molecular shape. Thus, if you wish to pursue chemistry as it applies to living creatures (i.e., organic chemistry, biochemistry, or pharmaceutical chemistry), you will need to be able to "see" the shapes of molecules in your head.

To determine the shape of a molecule:

- **Determine the Lewis structure of the molecule.**
- **If there are only two atoms in the molecule or if there are only two groups of electrons around the central atom, then the molecule must be *linear* and have a bond angle of *180°*.**
- **If there are three groups of electrons around the central atom, then the shape is *trigonal* with a bond angle of *120°*.**
- **If none of the above-listed conditions exists, then there must be four groups of electrons around the central atom, so the basic shape of the molecule is a tetrahedron.**
- **For every non-bonding pair of electrons, you must take a "leg" of the tetrahedron away. Thus, if there are no non-bonding electrons around the central atom, the shape is *tetrahedral* with a bond angle of *109°*.**
- **If the central atom has one pair of non-bonding electrons, then you have a tetrahedron with one leg removed. This results in a *pyramidal* shape, with a bond angle of about *107^{0}*.**
- **If two non-bonding electron pairs surround the central atom, then two legs of the tetrahedron are missing. This ends up giving the molecule a *bent* shape with a bond angle of approximately *105°*.**

Let's go through a few examples of determining the shape of a molecule so that you will see how we apply the rules you just read.

EXAMPLE 9.2

What are the shape and bond angle of a $PHCl_2$ molecule?

To answer this question, we must first determine the molecule's Lewis structure. Hopefully you are familiar enough with Lewis structures to be able to get that far on your own:

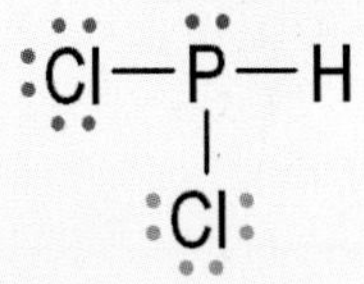

:Cl—P—H
|
:Cl:

Now that we see the Lewis structure, we can tell that there are four groups of electrons around the central phosphorus atom. This means that the basic shape of the molecule is a tetrahedron. However, since there is one pair of non-bonding electrons on the phosphorus, this takes one leg away from the tetrahedron. Thus, the shape of the molecule is pyramidal with a bond angle of 107°. The molecule, then, would look something like this:

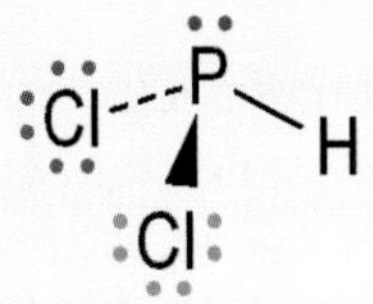

:Cl---P—H
:Cl:

Notice, before we leave this example problem, that the non-bonding electron pairs that surround the chlorines played no role in determining the shape of the molecule. This is because the shape of a molecule is based on its central atom. Thus, only the electrons around the central atom come into play here. If you go on in chemistry, you will eventually start dealing with molecules that have so many atoms in them that it is difficult to determine the central atom. At that point, determining molecular shape can be very difficult. For this course, however, we will continue to concentrate on relatively simple molecules when talking about molecular shapes.

What are the shape and bond angle of a SiO_2 molecule?

First we must determine the Lewis structure:

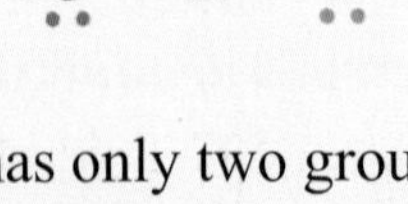

:O=Si=O:

This is a simple molecule to deal with. It has only two groups of electrons (two double bonds) around the central atom. This makes it a linear molecule with a bond angle of 180°.

What are the shape and bond angle for an OCl_2 molecule?

The Lewis structure is:

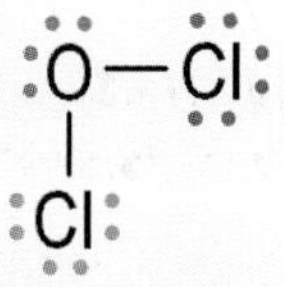

```
 ..   ..
:O — Cl:
 |    ..
:Cl:
 ..
```

The central atom has four groups of electrons around it, making the basic shape a tetrahedron. Two of the groups of electrons, however, are non-bonding electron pairs. This means that two "legs" of the tetrahedron are gone, resulting in a bent shape with a bond angle of 105°. The molecule would look something like this:

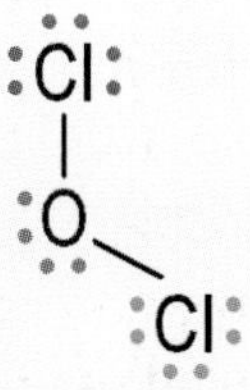

```
 ..
:Cl:
  |
:O
 ..\
   :Cl:
    ..
```

What are the shape and bond angle for a BF_3 molecule? (Note: Boron is an exception to the octet rule. It wants only six valence electrons, not eight.)

The note says that boron wants only six valence electrons. Thus, the Lewis structure is pretty easy. Boron has three valence electrons, and each fluorine needs an extra valance electron. Thus, the Lewis structure is:

```
 ..      ..
:F — B — F:
 ..  |   ..
    :F:
     ..
```

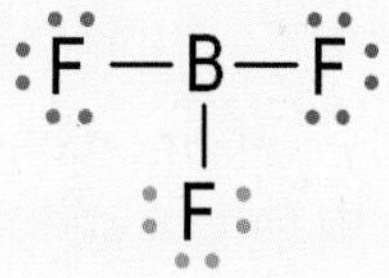

The central atom has three groups of electrons around it, making a trigonal shape with a bond angle of 120°. The molecule would look something like this:

```
     ..
    :F:
     |
 ..  B  ..
:F /   \ F:
 ..      ..
```

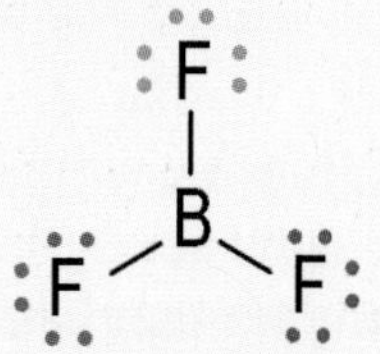

Make sure you understand the concept of molecular shape by solving the following problems:

ON YOUR OWN

9.4 Determine the shape and bond angle of an oxygen molecule.

9.5 What are the shape and bond angle of $CHCl_3$?

9.6 What shape and bond angle are possessed by an arsenic trichloride molecule?

9.7 Determine the shape and bond angle of a sulfur dibromide molecule.

9.8 What are the shape and bond angle of a $SiCl_2O$ molecule?

Purely Covalent and Polar Covalent Bonds

In Module #3, I told you that covalent compounds could be further divided into two important subclasses. Now that you understand Lewis structures and molecular shapes, it is time to learn these two subclasses and their distinctions. In order to learn this, however, we must first go back to Lewis structures and think a bit about what they mean. For example, consider the Lewis structure for F_2:

The dash between the fluorine atoms indicates a covalent bond, which is, in reality, a shared pair of electrons. But how do atoms share electrons?

As we learned in the last module, the fluorine atoms share electrons because they would each like to have eight valence electrons. If there were metal atoms around, they would try to take the metal's electrons away so that they could each have their own eight valence electrons. However, for whatever reason, no metal atoms could be found. As a result, the atoms decided to share electrons in order to have eight valence electrons for at least some fraction of time.

The reason they each have eight valence electrons for a fraction of time is because the shared electron pair is whirling around both atoms. For those moments in which the electrons are whirling around the first fluorine atom, that atom has eight valence electrons. Of course, when that happens, the second atom has only six valence electrons. This is a problem for the second atom, so it tugs on the electrons, forcing them to whirl around it. Of course, at that point, the first fluorine atom wants the electron pair back, so it tugs on them, trying to get them to come back.

Thus, a tug-of-war is established over this shared electron pair. Now, since both of the atoms in this molecule are the same (they are both fluorine atoms), neither of them can win the tug-of-war. They each pull on the electron pair with the same strength. As a result, one fluorine has them exactly half of the time, while the other has them for the other half of the time. We could say then, that the molecular bond in this molecule is made up of a pair of electrons which are shared *equally* between the two atoms. When electrons in a chemical bond are shared equally, we call it a **purely covalent bond**.

Purely covalent bond - A covalent bond in which the electrons are shared equally between the atoms involved

Since the only bond in the fluorine atom is covalent, we call fluorine a **purely covalent molecule**.

Contrast the situation in a fluorine molecule to that in a hydrochloric acid molecule:

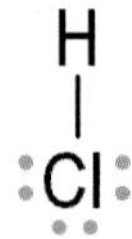

In this case, the electron pair that makes up the chemical bond is being shared between two different atoms: a hydrogen atom and a chlorine atom. One of these atoms is, most likely, a little stronger at tugging on electrons than the other. Can you predict which atom that would be? Well, in the last module, we learned about a concept called electronegativity, which determines the desire an atom has for extra electrons. Doesn't it make sense, then, that the atom with the higher electronegativity will also be able to tug on electrons with more force?

Since chlorine is to the right of hydrogen on the periodic chart, we can assume that it has a higher electronegativity. You might think that since chlorine is also lower on the chart than hydrogen, its electronegativity might actually be less than hydrogen's. Well, if you look at the chart, chlorine is much farther to the right of hydrogen than it is lower. Thus, the fact that electronegativity increases as you travel to the right on the chart outweighs the fact that it decreases as you move down the chart. Thus, chlorine has a higher electronegativity than hydrogen and can tug on the electron pair more strongly than hydrogen can.

What is the result of this fact? Well, if the chlorine is stronger, it can hold onto the electrons for a longer period of time than the hydrogen atom can. Thus, the electrons are no longer shared equally; they are shared *unequally*. When the electrons that make up a chemical bond are shared unequally, we call it a **polar bond**.

Polar bond - A covalent bond in which the electrons are shared unequally between the atoms involved

Since the only bond in a hydrochloric acid molecule is polar, we call it a **polar covalent molecule**.

So what's the big deal? Why worry about whether or not electrons are shared equally between atoms in a molecule? Well, if you think about it, there is a rather important consequence if electrons are not shared equally in a molecule. For example, in the case of the HCl molecule, the electrons spend more than half of their time around the chlorine, and only a small fraction of time around the hydrogen. As a result, the chlorine gets more than its fair share of electrons. Thus, in a way, it has an extra electron, at least for part of the time. What happens when an atom gets extra electrons? It becomes negatively charged. Thus, for some fraction of time, the chlorine is actually negatively charged. In the same way, the hydrogen atom doesn't get its fair share of electrons. As a result, it has a deficit of electrons, at least part of the time. When atoms have a deficit of electrons, they become positively charged. Thus, for some fraction of time, the hydrogen atom is positively charged.

Now remember, these charges only exist for a fraction of time. As a result, the chlorine never develops a *full* negative charge, and neither does the hydrogen develop a *full* positive charge. We could picture this situation in the following way:

$$\overset{\delta+}{H}-\overset{\delta-}{Cl}$$

The symbol "δ" in the picture above is the lower case Greek letter "delta." Scientists and mathematicians often use this symbol to indicate a small fraction. Thus, the Lewis structure pictured above indicates that the hydrogen in the HCl molecule has a small fraction of a positive charge (a *partial* positive charge) whereas the chlorine has a small fraction of negative charge (a *partial* negative charge).

This is, in fact, the situation in an HCl molecule. Even though the molecule has no true ions in it, there are still small electrical charges within the molecule itself. In order to see exactly what the consequences of this fact are, perform the following experiment:

EXPERIMENT 9.1

Polar Covalent Versus Purely Covalent Compounds

Supplies

- Glass of water
- Vegetable oil
- Styrofoam or paper cup
- ~~Comb~~ balloon
- Pen
- Safety goggles

1. Take the pen and punch a small hole into the bottom of the cup. The smaller the hole, the better.
2. While holding it over the sink, pour some water into the cup from the glass. Water should start running out of the hole in the bottom of the cup. Make sure that the water is pouring out of the hole in a steady stream, not dripping. If it is dripping, make your hole just a little bigger.
3. Once the water is pouring out of the cup in a steady stream, take the comb in your other hand and vigorously comb your hair. This is meant to accomplish the same thing it did in Experiment 7.1: to make the comb develop an electrical charge. As I mentioned in that experiment, if your hair is greasy, this may not work too well.
4. Once you have combed your hair for a few seconds, bring the comb (bristles first) near the stream of water. You should let the comb get very close to, but not actually touch, the water stream. What happens? The water stream should bend towards the comb. The more you combed your hair, the stronger the bend in the water should be.
5. Repeat this same experiment using vegetable oil instead of water. You may have to make the hole a little bigger this time, because vegetable oil doesn't flow as easily as water does. Once you do the experiment, however, you should observe that the stream of vegetable oil doesn't bend like the water did. You may see the stream sputter a bit in reaction to the nearness of the comb, but the stream of oil does not bend.
6. Clean up your mess.

Why did the water stream bend while the oil stream didn't? The answer lies in the fact that water is a polar covalent compound, while vegetable oil is purely covalent. A water molecule looks like this:

In this molecule, electrons are being shared between hydrogen atoms and an oxygen atom. The oxygen, because it is far to the right of hydrogen on the periodic chart, is more electronegative. As a result, it wins the tug-of-war for electrons and has more than its fair share of electrons. As a result, it develops a partial negative charge. The hydrogen atoms, then, have a slight electron deficit, giving them a partial positive charge. The comb, once you combed your hair with it, became negatively charged, attracting the partial positive charges on the hydrogen atoms. This caused the water molecules to move towards the comb, with their hydrogen atoms facing the comb and their oxygen atoms facing away from it:

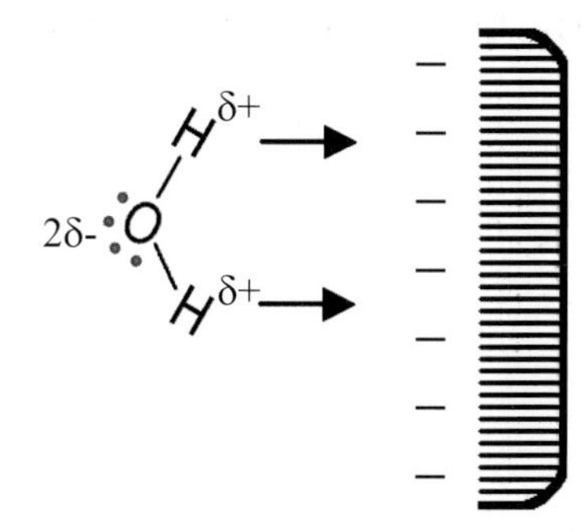

That's why the stream of water bent towards the comb. It was electrically attracted to it.

The vegetable oil, on the other hand, is made up of purely covalent molecules. There are no electrical charges in those molecules. Thus, they were not attracted to the comb, and therefore the stream of oil didn't bend towards it. If you happened to see the stream of oil sputter a little bit when the comb got near it, that's due to the fact that there might have been a small amount of water in your vegetable oil. When the water was attracted to the comb, it started moving towards it. The rest of the stream (which was oil) was not attracted to the comb, so it still moved straight down. Since one part of the stream wanted to go one way and the other part wanted to go another, the stream sputtered. This might not have happened in your experiment, indicating that there was no water contaminating your vegetable oil.

In order to fully understand the difference between purely covalent molecules and polar covalent molecules, you must be able to identify polar bonds as well as determine which direction the unequally shared electrons are pulled toward. In order to do this, chemists have developed a symbolic way of labeling polar bonds. When you are determining which bonds in a molecule are polar, you represent them with the following symbol:

The arrowhead in the symbol should point towards the atom with the largest electronegativity, to indicate in what direction the unequally shared electrons are being pulled. The "plus" sign on the other end of the arrow, then, represents the electron deficit that results on the other atom.

Now that you know how to *label* a polar bond, how do you *determine* what bonds in a molecule are polar? Fortunately, that task is really easy. If a bond exists between two identical atoms (two C's, for example), then it is not polar. If it exists between two different types of atoms (a C and an F, for example), then it is polar. Study the following examples and then solve the "On Your Own" problems in order to make sure you understand how to identify polar bonds and the direction in which the unequally shared electrons are pulled.

EXAMPLE 9.3

Identify and label the polar bonds in NH_3 (ammonia).

In order to identify the polar bonds and determine the charges that result, we must first look at the Lewis structure:

```
    H
    |
  H-N:
    |
    H
```

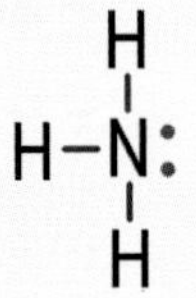

In this molecule, there are three bonds that will be polar, since there are three bonds that exist between two different types of atoms. To determine how to label the bonds, we must look at the electronegativity of the atoms involved. According to our rules, the nitrogen must have a larger electronegativity because it lies far to the right of hydrogen on the periodic table and only a little lower than hydrogen. Since it gains a lot of electronegativity from being far to the right and only loses a little by being just a bit lower, the net result is that nitrogen has more electronegativity than hydrogen. Thus, nitrogen gets more than its fair share of electrons from each of the three hydrogens. This means that the arrows on the polar bond labels need to point towards the nitrogen:

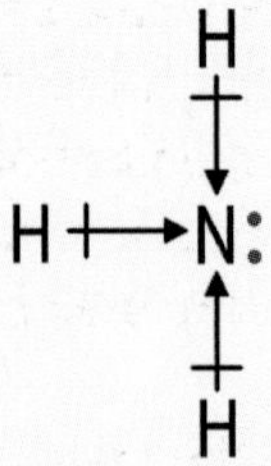

```
      H
      +
      ↓
  H+→N:
      ↑
      +
      H
```

Identify and label the polar bonds in the C_2F_4 molecule.

First, we must look at the Lewis structure:

```
 ..        ..
:F-C=C-F:
 ..  |   |  ..
    :F: :F:
     ..  ..
```

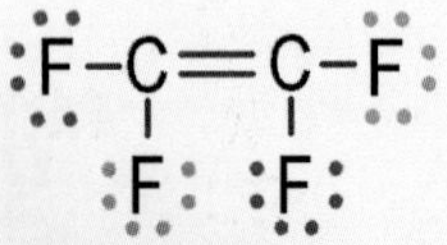

According to the way we defined polar bonds, the carbon-to-carbon double bond is obviously not polar. After all, the bond is between two identical atoms. Thus, neither of them can win the tug-of-war for electrons. The four carbon-to-fluorine bonds, however, are polar because they exist between different types of atoms. Since fluorine is more electronegative than carbon (due to its position on the periodic table), the arrows on our polar bond label must point towards the fluorines:

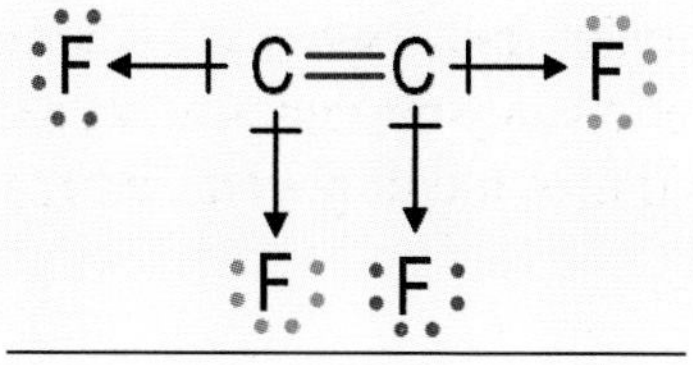

ON YOUR OWN

9.9 Identify and label the polar bonds in the following molecules:

a. CO_2 b. $CClF_3$

Purely Covalent and Polar Covalent Molecules

Now remember, the important thing about polar covalent bonds and purely covalent bonds is their resulting effect on the molecule that they are in. If a molecule is polar covalent, it contains electrical charges. On the other hand, a purely covalent molecule has no electrical charges. Since this is a relatively important thing to know about a molecule, it is important to be able to determine whether a given molecule is polar covalent or purely covalent.

So what makes a molecule polar covalent? Well, it stands to reason that a molecule must contain polar bonds in order to be a polar covalent molecule. After all, if the bonds do not result in fractional charges, then there is no way that a molecule can become polar covalent. It might surprise you, however, that polar bonds are not enough to ensure that a molecule is, in fact, polar covalent. This is because, under certain circumstances, the polar bonds in a molecule can all work against each other and end up canceling each other out. When this happens, the molecule has no charges in it, despite the fact that it has polar bonds, and thus the molecule itself is purely covalent.

The best way to explain why this is the case is to give you an example of a molecule that has polar bonds but is, in fact, purely covalent. The perfect example of this is the CCl_4 molecule. The shape of this molecule is as follows:

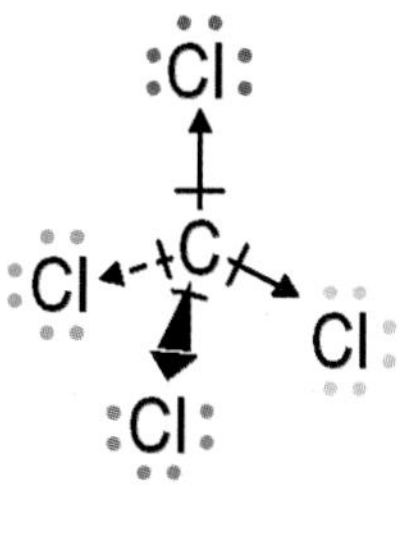

As I have indicated, the chlorines are more electronegative than the carbon. As a result, electrons are being pulled away from the carbon and towards the chlorines. The problem, however, is that the chlorines are all pulling from exactly opposite directions. As a result, the electrons are being pulled evenly in opposite directions.

What happens when something is pulled evenly in opposite directions? It ends up sitting still, of course. That's what happens to the electrons around the carbon in this molecule. Since they are all being pulled evenly in opposite directions, they end up staying put. Thus, despite the fact that this molecule has four polar bonds, they end up canceling each other out, making the molecule purely covalent. In other words, since the electrons around the carbon cannot move, there are no resulting charges in the molecule.

Notice that this is not what happens in the case of $AsCl_3$. This molecule has the following shape:

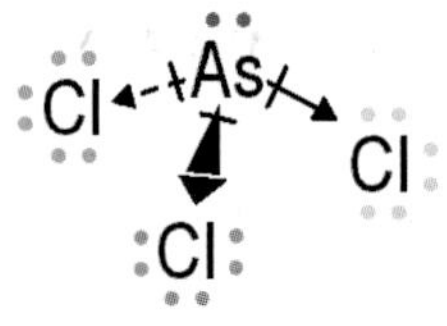

In this case, it should be obvious that the polar bonds do not cancel each other out. Since there is no bond on top of the As, there is nothing pulling electrons up. As a result, the three bonds exert a net pull downwards. Thus, since there is a net pull on the As atom's electrons, there will be charges in this molecule. We can therefore determine that the molecule is polar covalent.

Thus, in order for a molecule to be polar, two requirements must be met. First, the molecule must have polar bonds in it. Second, those polar bonds cannot cancel each other out. How can we determine whether or not polar bonds cancel each other out? Well, we have to determine the *shape* of the molecule. If there is an equal pull on the central atom's electrons from opposite directions, then the polar bonds will end up canceling each other out, and the molecule will be purely covalent. On the other hand, if there is not an equal pull on the central atom's electrons from opposite directions, then the molecule will end up having a net fractional charge, making it polar covalent.

You must be careful in making this determination, because, as I have stated, there must be an *equal* pull on the central atom's electrons in order for the polar bonds to end up canceling each other out. This can be a little confusing unless you are paying close attention. Consider, for example, the molecule $CFCl_3$. The polar bond situation in this case looks like this:

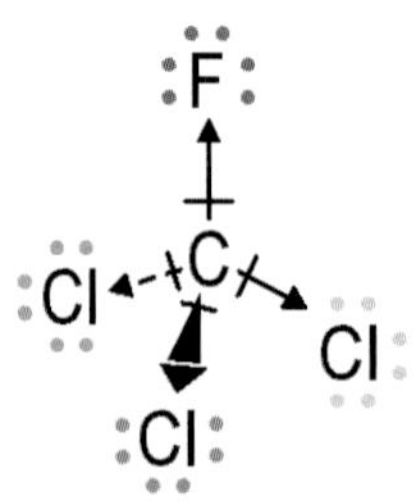

Looking at this molecule, you see that the electrons on the central carbon are, indeed, being pulled in opposite directions. But are the pulls equal? No, they are not. According to our rules of electronegativity, fluorine is more electronegative than chlorine, because it is higher on the periodic chart. Thus, the pull from fluorine is harder than the pull from the chlorines. Thus, while the carbon's electrons are being pulled in opposite directions, they are not being pulled equally, so the molecule is polar covalent.

To reiterate the conditions under which a molecule is polar covalent, these following conditions must be met:

- **The molecule must have polar bonds in it.**
- **Those polar bonds cannot be of equal strength and pull in opposite directions.**

Try to see how we apply these rules in the example problems that follow:

EXAMPLE 9.4

Classify SiO_2 as polar covalent or purely covalent.

To classify this molecule, we must first draw its Lewis structure:

:O=Si=O:

Since there are two groups of atoms around the silicon, this is a linear molecule. Thus, the shape I have drawn above is correct. Now that we know the shape, we can look at the bonds. According to their relative positions on the periodic table, oxygen is more electronegative than silicon. As a result, the silicon atom's electrons are being pulled toward the oxygen atoms:

:O ⟵ Si ⟶ O:

Notice, however, that the electrons are being pulled in exactly opposite directions. In addition, they are all being pulled by oxygen atoms, so the pulls are equal. Thus, these polar bonds cancel each other out, and the molecule is therefore purely covalent.

Is OF_2 polar covalent or purely covalent?

Once again, to solve this problem, we must look at the Lewis structure:

:O–F:
|
:F:

In this Lewis structure, there are four groups of electrons around the central atom. Thus, the base shape is a tetrahedron. However, this tetrahedron has two legs removed, making the shape of this molecule bent. Now that we know the shape, we can consider the bonds. Both bonds are polar, and since fluorine lies to the right of oxygen on the periodic table, the direction of the electron pull is as follows:

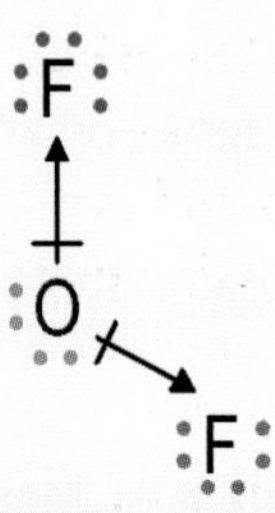

The electrons are not being pulled in exactly opposite directions, so there is, indeed, a net pull on the oxygen atom's electrons. Thus, the molecule is polar covalent.

Is the HBr molecule polar covalent or purely covalent?

The Lewis structure of HBr is:

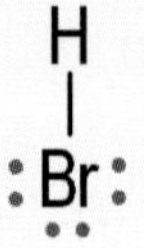

Since Br is far to the right of and only a little lower than H on the periodic chart, then the electron pull looks like this:

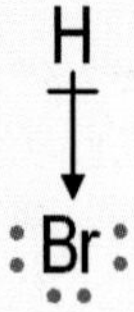

Since there is only one polar bond in this molecule, there is nothing to cancel it out. As a result, the molecule is, indeed, polar covalent.

Before you try your hand at determining whether a molecule is polar covalent or purely covalent, I want to make a point about terminology. Some chemistry books use the terms "polar" and "non-polar" to describe covalent molecules. Thus, a molecule that is polar covalent may be called a "polar" molecule, while a molecule that is purely covalent may be called a "non-polar" molecule. Although I like the complete terminology (polar covalent or purely covalent) better, it is important to note that the abbreviated terminology (polar or non-polar) exists as well.

ON YOUR OWN

9.10 Classify the following molecules as polar covalent or purely covalent:

a. Br_2 b. CO_2 c. H_2S d. CF_2Cl_2 e. SiF_4 f. PH_3

The Practical Consequence of Whether or Not a Molecule Is Polar Covalent

Now you might be saying to yourself, "So what? Now I can tell whether or not a molecule is purely covalent (non-polar) or polar covalent (polar). What's the big deal?" Well, it turns out that the difference between polar covalent compounds and purely covalent compounds is of great practical significance. To understand why, do the following experiment:

EXPERIMENT 9.2
Solubility of Ionic Compounds in Polar Covalent and Purely Covalent Compounds

Supplies
- Two test tubes (Thin glasses will work, but they must be transparent)
- Table salt
- Water
- Vegetable oil
- Safety goggles

1. Take both test tubes and put enough salt into each of them to cover the bottom of each tube. Put in enough so that it is easy to see that the salt is there, but do not put in too much.
2. Fill one of the tubes ¾ of the way with water and the other ¾ of the way with vegetable oil.
3. Cover the tubes with your thumbs and shake them vigorously for several minutes.
4. Allow them to stand for about a minute.
5. Hold the test tubes up to a light and look through them. In the one that you filled with water, the salt should be gone. It should have dissolved into the water. What about the one with vegetable oil, however? If you look closely, you should still see all of the salt either lying at the bottom of the tube or suspended in the oil. None of it dissolved.
6. Clean up your mess.

Why did the table salt dissolve in water but not in vegetable oil? Well, table salt, as we learned several modules ago, is an ionic compound. Therefore, we know that it is composed of ions, which possess electrical charge. Water, since it is polar, also has charges in it, even if they are only partial charges. Vegetable oil, as we learned in the first experiment, is purely covalent. It has no electrical charges. Thus, since it has no electrical charges, it cannot interact with (and therefore cannot dissolve) the table salt. This leads us to an important point regarding these two subclasses of covalent compounds:

Polar covalent compounds can dissolve other polar covalent compounds and ionic compounds. This is because both types of compounds contain electrical charges.

Purely covalent compounds, on the other hand, can only dissolve other purely covalent compounds, because no electrical charges are present.

These two chemical principles explain many household cleaning woes. For example, do you know why it is so hard to get grass and blood stains out of clothes? Well, the reason is that grass and blood are both made of purely covalent compounds. When you wash the clothes, you are washing them in water, which is polar covalent. As a result, the water cannot dissolve the grass or blood stains, so they tend to stay on the clothing. Adding soap to the water helps, because soap contains long molecules that are ionic on one end and purely covalent on the other. The reason that such a molecule can exist is a bit beyond the scope of this course, but at least you can see the advantage gained by adding such a molecule to your wash water. The ionic part of the soap molecule is attracted to the polar covalent water molecules, while the purely covalent part of the soap molecule dissolves the grass or blood stain.

When you pretreat a stain with something like Shout®, all you are doing is adding a lot of soap directly to the stain. The hope is that the purely covalent part of the soap molecules will dissolve the entire stain by virtue of the fact that you are using so much of it right at the stain. Then, when the clothing is in the wash, the water will attract the ionic part of the soap molecule away from the fabric, pulling the stain away with it.

So you see that the classification of compounds as either ionic, polar covalent (polar), or purely covalent (non-polar) can help us determine a lot about the behavior of certain compounds. Make sure you grasp this by completing this problem:

ON YOUR OWN

9.11 Which of the following substances dissolve in water?

a. Br_2 b. CO_2 c. H_2S d. CF_2Cl_2 e. SiF_4 f. PH_3

Cartoon by Speartoons, Inc.

ANSWERS TO THE "ON YOUR OWN" PROBLEMS

9.1 This problem is designed to remind you how to count atoms when the chemical formulas have parentheses in them. As written, each side has the following number of atoms:

Reactants side	Products side
Ba: 1x1=1	Ba: 1x1 = 1
N: 1x1x2 = 2	N: 1x1 = 1
O: 1x3x2 + 1x1 = 7	O: 1x1x2 + 1x3 = 5
Na: 1x1 = 1	Na: 1x1 = 1
H: 1x1 = 1	H: 1x1x2 = 2

To balance the N's, we need to multiply the $NaNO_3$ on the products side by 2:

$$Ba(NO_3)_2\ (aq)\ +\ NaOH\ (aq)\ \rightarrow\ Ba(OH)_2\ (s)\ +\ 2NaNO_3\ (aq)$$

Now the atom count looks like this:

Reactants side	Products side
Ba: 1x1=1	Ba: 1x1 = 1
N: 1x1x2 = 2	N: 2x1 = 2
O: 1x3x2 + 1x1 = 7	O: 1x1x2 + 2x3 = 8
Na: 1x1 = 1	Na: 2x1 = 2
H: 1x1 = 1	H: 1x1x2 = 2

Now Ba's and N's are balanced. We should skip oxygen because when we balance equations, we should first deal with atoms that appear only once on each side of the equation. Thus, we will balance the Na's next. To do this, we need to multiply the NaOH on the reactants side by 2:

$$\underline{Ba(NO_3)_2\ (aq)\ +\ 2NaOH\ (aq)\ \rightarrow\ Ba(OH)_2\ (s)\ +\ 2NaNO_3\ (aq)}$$

Now the atom count looks like this:

Reactants side	Products side
Ba: 1x1=1	Ba: 1x1 = 1
N: 1x1x2 = 2	N: 2x1 = 2
O: 1x3x2 + 2x1 = 8	O: 1x1x2 + 2x3 = 8
Na : 2x1 = 2	Na: 2x1 = 2
H: 2x1 = 2	H: 1x1x2 = 2

Now we see that the equation is balanced.

9.2 a. The name tells us that this substance is composed of the potassium ion and the sulfate ion. The periodic chart tells us that the potassium ion has a charge of 1+, while we should have the sulfate ion memorized as having a 2- charge. Ignoring the signs and switching the numbers, we get:

$$\underline{K_2SO_4}$$

No parentheses are needed here because there is only one sulfate ion.

b. The name tells us that this substance is composed of the ammonium ion and the carbonate ion. Our memory tells us that the ammonium ion has a 1+ charge, and the carbonate ion has a 2- charge. Ignoring the signs and switching the numbers, we get:

$$\underline{(NH_4)_2CO_3}$$

No parentheses are needed for the carbonate ion because there is only one. However, since there are two ammonium ions, we must have parentheses around the NH_4.

c. The name tells us that this substance is composed of the aluminum ion and the acetate ion. The periodic chart tells us that the aluminum ion has a charge of 3+, while we should have the acetate ion memorized as having a 1- charge. Ignoring the signs and switching the numbers, we get:

$$\underline{Al(C_2H_3O_2)_3}$$

9.3 This problem combines everything you need to know about ionic compounds that have polyatomic ions in them. To get the chemical equation, we first have to determine the chemical formulas:

aluminum nitrite: Al is 3+, NO_2 is 1-; therefore, the chemical formula is $Al(NO_2)_3$.
sodium carbonate: Na is 1+, CO_3 is 2-; therefore, the chemical formula is Na_2CO_3.
aluminum carbonate: Al is 3+, CO_3 is 2-; therefore, the chemical formula is $Al_2(CO_3)_3$.
sodium nitrite: Na is 1+, NO_2 is 1-; therefore, the chemical formula is $NaNO_2$.

According to the problem, sodium carbonate and aluminum nitrite are the reactants while the other two are products, thus:

$$Al(NO_2)_3 + Na_2CO_3 \rightarrow Al_2(CO_3)_3 + NaNO_2$$

We can start by balancing the Al's. There is only 1 on the reactants side and 2 on the products side; thus, we must multiply the $Al(NO_2)_3$ by 2:

$$2Al(NO_2)_3 + Na_2CO_3 \rightarrow Al_2(CO_3)_3 + NaNO_2$$

This balances the Al's, but the N's still aren't balanced. There are 6 (2x1x3) on the reactants side and only 1 on the products side. We therefore have to multiply $NaNO_2$ by 6:

$$2Al(NO_2)_3 + Na_2CO_3 \rightarrow Al_2(CO_3)_3 + 6NaNO_2$$

Now the Al's and N's are balanced. We will skip the O's because they appear in all substances. The Na's are not balanced because there are 2 on the reactants side and 6 on the products side. Thus, we must multiply Na_2CO_3 by 3:

$$\underline{2Al(NO_2)_3 + 3Na_2CO_3 \rightarrow Al_2(CO_3)_3 + 6NaNO_2}$$

Now we can see that all atoms are balanced.

9.4 The Lewis structure for an oxygen molecule is:

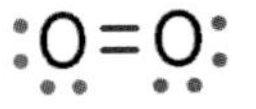

:O=O:

This, then, is a simple molecule. Because it is made up of only two atoms, it is <u>linear with a bond angle of 180°</u>.

9.5 The Lewis structure of $CHCl_3$ is:

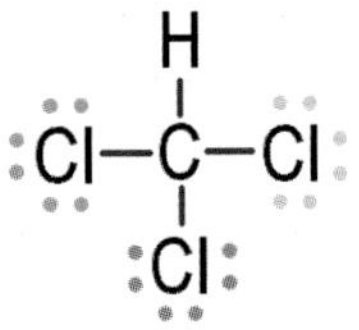

H
:Cl—C—Cl:
:Cl:

Looking at this structure, we have four groups of electrons around the central carbon atom. This means that the basic shape is that of a tetrahedron. Since there are no non-bonding electrons around the central atom, the final shape is <u>tetrahedral with a bond angle of 109°</u>. The molecule, then, would look something like this:

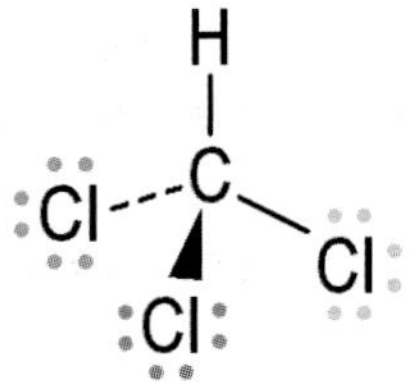

H
:Cl---C
Cl:
:Cl:

9.6 The name of this molecule indicates that its molecular formula is $AsCl_3$. The Lewis structure, then, is:

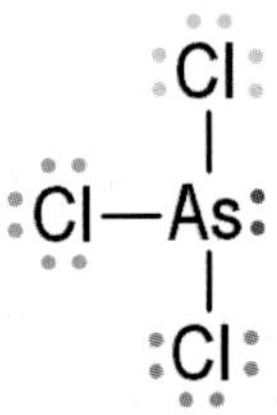

:Cl:
:Cl—As:
:Cl:

Based on this Lewis structure, we can see that there are four groups of electrons around the central atom, making the basic shape a tetrahedron. However, there is one pair of non-bonding electrons, taking one "leg" from the tetrahedron. This results in a pyramidal shape with a bond angle of 107°. The molecule would end up looking something like this:

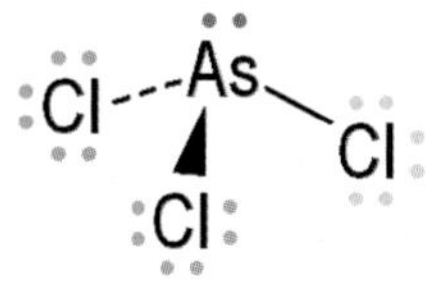

9.7 Sulfur dibromide has a formula of SBr_2, making its Lewis structure:

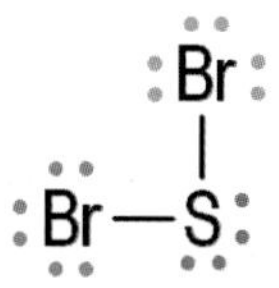

The Lewis structure indicates that the central sulfur atom is surrounded by four groups of electrons, making the basic shape a tetrahedron. Two of the groups of electrons, however, are non-bonding electrons, taking two "legs" from the tetrahedron. The resulting shape is bent with a bond angle of 105°. The molecule, then, looks like this:

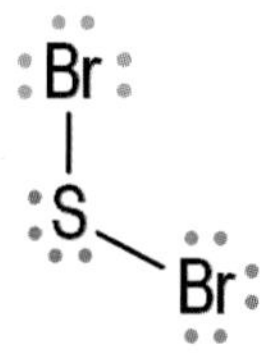

9.8 An $SiCl_2O$ molecule has the following Lewis structure:

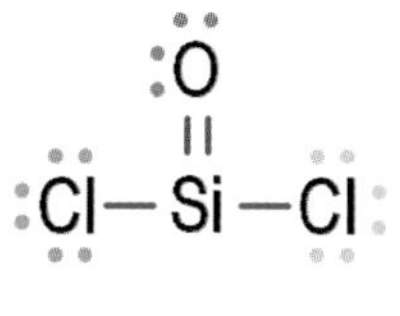

In this Lewis structure, there are only three groups of electrons around the central atom. Sure, one of those groups is a double bond, but a double bond is only *one* bond; it is just a lot stronger than a single bond. Thus, there are three bonds around the silicon, which means three groups of electrons. That makes this molecule trigonal with a bond angle of 120°. The molecule, then, looks like this:

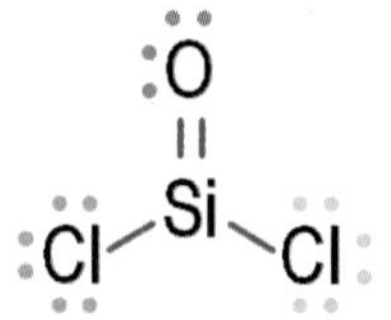

9.9 a. To solve this problem, we must first look at the Lewis structure of the molecule:

:O=C=O:

This molecule is linear, since there are only two groups of electrons around the central atom. Those two double bonds both exist between atoms of different types; thus, they are both polar. Since oxygen lies to the right of carbon, it is more electronegative, and therefore the arrows point toward the oxygens:

:O ⟵ C ⟶ O:

This drawing, then, shows that the carbon atom's electrons are being pulled toward the oxygen atoms.

b. The Lewis structure for $CClF_3$ is:

:Cl:
|
:F—C—F:
|
:F:

In this molecule, there are four groups of electrons around the central atom. Thus, the base geometry is a tetrahedron. Since all four groups are bonds, all legs of the tetrahedron exist, and therefore the molecule is tetrahedral. All four bonds are polar because each of them links different types of atoms. The fluorines and chlorine are more electronegative than the carbon, so the arrows point toward the fluorines and chlorine:

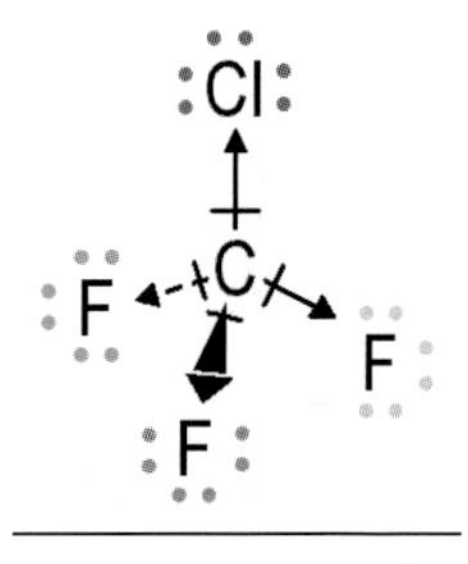

9.10 a. We start each of these problems by looking at the Lewis structure:

:Br–Br:

Since the only bond in this molecule is between identical atoms, there are no polar bonds in the molecule. As a result, the molecule cannot be polar covalent. Therefore, it is <u>purely covalent</u>.

b. We already determined the electron pull of this molecule in an earlier problem:

:O⇇C⇉O:

There are two groups of electrons around the central atom (two double bonds). Thus, this molecule is linear. Although there are polar bonds in this molecule, they pull the electrons in precisely opposite directions. As a result, the molecule cannot be polar covalent. It is therefore <u>purely covalent</u>.

c. The Lewis structure for this molecule is:

H
|
H–S:

There are four groups of electrons surrounding the central atom (two bonds, two non-bonding electron pairs). This means that the base geometry of the molecule is a tetrahedron. However, this tetrahedon has two legs removed, so the resulting shape is bent. Since the bonds exist between different types of atoms, they are polar. Sulfur is more electronegative than hydrogen, due to its position on the periodic table. The electron pull, then, looks like this:

H
↓
:S ← H

In this molecule, the polar bonds do not pull in exactly opposite directions, so they do not cancel each other out. As a result, this molecule is <u>polar covalent</u>.

d. The Lewis structure for this molecule is:

:F:
|
:Cl–C–Cl:
|
:F:

In this molecule, there are four groups of electrons around the central atom, making the base geometry tetrahedral. Since all four groups are bonds, all legs of the tetrahedron are present, so this is, indeed, a tetrahedral molecule. All four bonds in this molecule are polar. In addition, both the Cl's and the F's are more electronegative than the C, so the electron pull looks like:

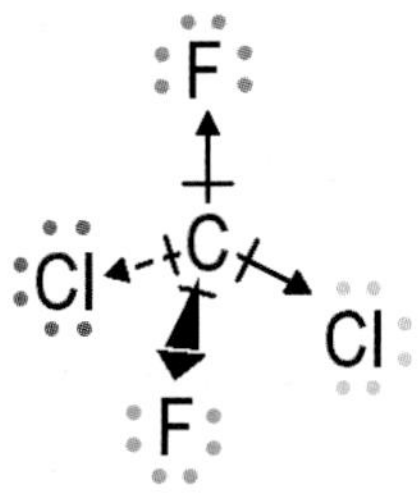

In this case, the carbon atom's electrons are being pulled in exactly opposite directions, but the pull is not equal. Since fluorine is more electronegative than chlorine, the pull from the fluorines is a little bigger. As a result, there is a net pull on the carbon atom's electrons. This makes the molecule polar covalent.

I want to emphasize here how important it is to consider the *shape* of the molecule in determining the electron pull, not the Lewis structure. If you look at the Lewis structure I drew at the beginning of the problem, it looks like the electron pulls cancel out. After all, the F's are opposite each other, so they look like they may pull in opposite directions (one up and one down). In addition, the Cl's are opposite each other, so it looks like they may pull in opposite directions as well (one left and one right). However, the Lewis structure is *not* an accurate representation of the molecule's geometry. You have to figure out the geometry with VSEPR theory. Sure, you use the Lewis structure in VSEPR theory, so the Lewis structure is important. However, it is not the final word on the molecule's shape.

When we use VSEPR theory to determine that the molecule is tetrahedral, we see that the F's are not exactly opposite one another. One is in the plane of the paper, while the other one is pointed out of the plane of the paper towards you. The same is true of the Cl's. They are not exactly opposite, because one is in the plane of the paper and the other is pointed out of the plane behind the paper. If you cannot visualize this well, go back to Figure 9.1 and look at the model of CH_4. Imagine two of the yellow balls as fluorines and the other two as chlorines. There is no way you can get the fluorines to cancel each other out, and there is no way you can get the chlorines to cancel each other out. The only way that bonds in a tetrahedral molecule can cancel is if they are all identical.

e. The Lewis structure for this molecule is:

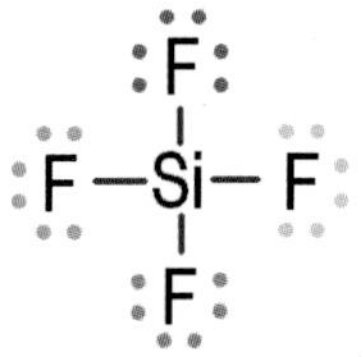

There are four groups of electrons around the central atom, and all of them are bonds. Thus, this is a tetrahedral molecule. All of the bonds in this molecule are polar, and, since the fluorine is both to the right of and above silicon on the periodic chart, fluorine is more electronegative than silicon. As a result, the electron pull looks like this:

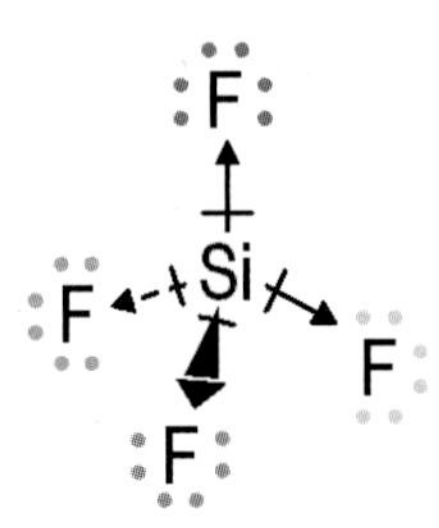

In this molecule, the silicon atom's electrons are all being pulled in opposite directions with equal strength. As a result, the polar bonds cancel each other out. This makes the molecule purely covalent.

f. The Lewis structure for PH_3 is:

The bonds are all polar, and P is more electronegative than H. Thus, the electron pull looks like:

The bonds do not pull from opposite directions; thus, this molecule is polar covalent.

9.11 We already classified each of these substances as either polar covalent or purely covalent. The polar covalent substances will dissolve in water; the purely covalent substances will not. Thus,

Br_2, CO_2, and SiF_4 will not dissolve readily in water because they are purely covalent compounds. The rest will dissolve in water.

NOTE: All Lewis structure illustrations in this module by Megan Whitaker.

REVIEW QUESTIONS FOR MODULE #9

1. What is the difference between the ions you learned about in previous modules and the polyatomic ions you learned about in this module?

2. List the chemical formula (including charge) for each of the following ions:

a. sulfate b. nitrate c. acetate d. sulfide e. hydroxide f. phosphate

3. Give the names of the following ions:

a. NO_2^- b. O^{2-} c. ClO_3^- d. CO_3^{2-} e. SO_3^{2-} f. CrO_4^{2-}

4. What does VSEPR stand for?

5. Explain VSEPR theory in your own words.

6. Why are the bond angles in a pyramidal shape smaller than the bond angles in a tetrahedral shape?

7. Explain the difference between polar covalent bonds and purely covalent bonds.

8. If two different atoms are bonded together with a purely covalent bond, what can we say about the electronegativities of the two atoms?

9. "Oil and water do not mix" is a popular phrase used to explain why very different people tend to not get along. Why do oil and water not mix?

10. Why does soap help to wash away stains that water cannot wash away by itself?

PRACTICE PROBLEMS FOR MODULE #9

1. Give the chemical formulas for the following compounds:

   a. potassium sulfate   b. calcium nitrate   c. magnesium carbonate   d. aluminum chromate

2. Name the following compounds:

   a. $(NH_4)_2O$ b. KNO_2 c. $Ca_3(PO_4)_2$ d. $AlPO_4$

3. Give the balanced chemical equation for the reaction in which aqueous calcium nitrate reacts with aqueous sodium carbonate to produce solid calcium carbonate (chalk) and aqueous sodium nitrate.

4. Determine the shape of a PCl_3 molecule. Give its bond angle and draw a picture of it.

5. Determine the shape of an H_2 molecule. Give its bond angle and draw a picture of it.

6. Determine the shape of a $SiCl_4$ molecule. Give its bond angle and draw a picture of it.

7. Determine the shape of an H_2S molecule. Give its bond angle and draw a picture of it.

8. Determine the shape of a CS_2 molecule. Give its bond angle and draw a picture of it.

9. Classify each of the following molecules as ionic, polar covalent, or purely covalent:

   a. $MgCl_2$ b. CF_3Cl c. CS_2 d. H_2 e. $SiCl_4$ f. PCl_3

10. Which of the substances in problem #9 would you expect to dissolve in water?

MODULE #10: Acid/Base Chemistry

Introduction

In the last three modules, we took a break from the subject of chemical reactions so that we could spend some time learning what atoms and molecules might look like if we were able to see them. The vast majority of chemistry, however, revolves around chemical reactions, so it is time we got back to studying them. In order to study reactions, we must remember the skills we learned in Modules 1-6. We will be using the mole concept, stoichiometry, unit conversions, ionic compound naming, and our chemical equation balancing skills in this module. Thus, if you are having trouble understanding some of the things in this module, you might need to go back and review some of those concepts.

You have already learned about three different classes of chemical reactions: formation, decomposition, and combustion. There are, as you might expect, several more categories of chemical reactions. As we go on through this course, we will touch on almost all of them. This module, however, concentrates on one of the most basic, fundamental classes of chemical reaction: the acid/base reaction.

Acids and Bases

Of course, in order to be able to understand acid/base reactions, the first thing we must do is understand what an acid is and what a base is. Now, when you read the word "acid," an image probably pops up in your mind. You probably think of an acid as some liquid chemical that can eat up anything it touches. Although it's true that some acids can eat up certain materials, it is not true of the vast majority of acids. In fact, many of the things we eat and drink contain acids. So acid is not necessarily a caustic, dangerous chemical. It is often a beneficial and even tasty substance.

There are typically two methods that chemists use to define acids. Acids have the same general properties, so one way to define an acid is by its characteristics. There is also a strict chemical definition for acids. In this module, you need to learn both types of definitions. We will start with the characteristics of an acid, and later on we will discuss the chemical definition.

In general, acids have the following properties:

1. Acids taste sour.
2. Acids are covalent compounds that conduct electricity when added to water.
3. Acids turn blue litmus paper red.

First of all, even though we include the taste of an acid in this list, let me make it clear that you should *never, ever* taste a substance to determine what it is. Many man-made substances, as well as many naturally occurring substances, are lethal poisons for human beings. In general, you should never taste something that you cannot identify. Additionally, even though all acids taste sour, some acids will cause severe damage to your body if you try to taste them. As a result, you should *never, ever* taste something in a chemistry experiment unless you are expressly instructed to do so by your teacher!

The second characteristic of acids might seem to contradict something we learned in Module #3. In that module, we said that *ionic* compounds conduct electricity when dissolved in water, while *covalent* compounds *do not* conduct electricity when dissolved in water. In general, that's true.

However, acids are an exception to that general rule. Even though acids are covalent compounds (meaning they are composed of atoms that share electrons rather than form ions), they can conduct electricity when added to water. Later on in this module we will see *how* this happens.

You won't really be able to understand the third characteristic of acids until you know what litmus (lit' mus) is. Litmus is part of a broad class of compounds known as **indicators**.

Indicator - A substance that turns one color in the presence of acids and another color in the presence of bases

Litmus has the property of turning red whenever it is in the presence of an acid. Thus, we could say that litmus is an excellent indicator of the presence of an acid. This is where the political term "litmus test" comes from. Often, there is a single issue that can define an entire political outlook. For many conservatives, that issue is abortion. If you are pro-life, you are most likely a conservative, whereas if you are pro-abortion, you are most likely a liberal. Thus, the abortion issue is often called a "litmus test," because it easily indicates a person's political beliefs. You will be doing your own litmus tests on household chemicals shortly.

It turns out that acids are rarely discussed in chemistry without also discussing bases. When you see the chemical definitions of these two classes of compounds, you will see why. In general, bases have the following characteristics:

1. **Bases taste bitter.**
2. **Bases are slippery to the touch when dissolved in water.**
3. **Bases turn red litmus paper blue.**

Of course, the same warnings about tasting acids also apply to bases. It turns out that some bases are just as caustic and dangerous as the most dangerous acids, so you should never taste any chemical in an experiment unless you are expressly instructed to do so.

The characteristics of acids and bases are important enough that you should remember them. In order to help you do this, as well as to help you learn about the nature of certain household chemicals, perform the following experiment:

EXPERIMENT 10.1

Common Household Examples of Acids and Bases

Supplies
- Red litmus paper
- Blue litmus paper
- Apple
- Orange juice or soda pop
- Toilet bowl cleaner (Both The Works® and Lime-Away® have been tested, but any toilet bowl cleaner designed to combat lime should work.)
- Bar soap (make sure it doesn't say "pH balanced")
- All-purpose cleaner (Windex® and 409® have been tested, but any spray cleaner not specifically designed for toilets should work.)

- Powdered drain unclogger (like Dran-O® or Red Devil® Lye) or scouring powder (like Comet®).
- 4 test tubes (small cups will work)
- Watch glass (a small saucer will work)
- Stirring rod
- Rubber gloves (Gloves are recommended whenever you use powdered drain uncloggers and toilet bowl cleaners, because these chemicals are caustic.)
- Safety goggles

1. Take a nice bite out of the apple and eat it. Would you say that the taste was more bitter or sour? Based on that, would you say that the apple has acid or base in it?
2. Confirm your guess by pressing some blue litmus paper against the part of the apple that was exposed by your bite. Do the same with some red litmus paper. You should see that the red litmus paper does not change colors, but the blue litmus paper changes to red. Thus, apples contain acid.
3. Rinse your hands to get off any apple juice that might be on them.
4. Lay out your watch glass and spray some of the all-purpose cleaner on it.
5. Take two fingers and rub them against the wet watch glass. Feel the consistency of the all-purpose cleaner on your fingers. Slippery, isn't it? What kind of substance does that mean it is?
6. Confirm your guess with red and blue litmus paper by placing the papers directly on the watch glass. You should see that the blue litmus paper does not change colors, but the red litmus paper changes to blue. This means that the all-purpose cleaner contains base.
7. Now rinse your hands to get off the all-purpose cleaner.
8. Shave some soap off the bar of soap and put the shavings into a test tube.
9. Fill the test tube about ¾ of the way with the shavings, then add just enough water to fill the tube almost to the top.
10. Cover the tube with your thumb and shake it vigorously for about 2 minutes. Based on your experience with soap, would you think it is an acid or base?
11. Confirm your guess with red and blue litmus paper by dipping the paper into the test tube. The red litmus should turn blue and the blue litmus should not change colors. Thus, soap contains base. You should have guessed this based on the fact that soap is slippery when exposed to water.
12. Once again, thoroughly rinse your hands and fill another test tube with the orange juice or soda pop. Based on your previous experience with this drink (think about the taste), do you expect it to be a base or an acid?
13. Confirm your guess by dipping red and blue litmus paper into the test tube. The blue litmus should turn red and the red litmus should not change colors, indicating that the drink contains acid.
14. Rinse off your hands and put the gloves on.
15. Fill a test tube with the toilet bowl cleaner. **Be careful when you do this**, because toilet bowl cleaner can be very caustic. **Do not get any on your skin!**
16. Test this substance by dipping red and blue litmus paper into the test tube. The blue litmus should turn red, and the red litmus should not change colors. Thus, the toilet bowl cleaner contains acid.
17. With your gloves still on, add a few crystals of powdered drain unclogger to a test tube. If you are using scouring powder, fill a test tube ¾ of the way with it.
18. Add enough water to fill the test tube almost full.
19. Stir the test tube contents with a stirring rod.
20. Test the solution you just made with red and blue litmus paper. The red paper should turn blue and the blue paper should not change colors. This means that powdered drain uncloggers contain base.
21. Clean up your mess.

Although you have identified the fact that some of these household chemicals contain acid and some contain base, you still don't know *why* they contain acid or base. Well, don't get too far ahead of yourself. As you begin to understand the chemistry of acids and bases, you will begin to understand why they are so useful for household items.

(The multimedia CD has a video demonstration about acid/base indicators.)

The Chemical Definitions of Acids and Bases

Now that we know something about the *properties* of acids and bases, it is time to learn how chemists *define* these substances. According to chemists, an acid is defined as follows:

Acid - A molecule that donates H^+ ions

A base, on the other hand, has this definition:

Base - A molecule that accepts H^+ ions

Now do you see why acids and bases are always discussed together? If an acid is a molecule that donates H^+, it won't be able to do anything unless there is a base around to accept the H^+. Thus, acids cannot do their job without the help of bases, and bases cannot do their jobs unless they have an acid to work with.

Before I actually show you how to determine whether or not a substance is an acid or a base, I want you to think for a moment about what is being donated and accepted in an acid/base reaction. Acids donate H^+ ions, and bases accept them. What is an H^+ ion? Well, the hydrogen *atom* consists of a single proton and a single electron. The hydrogen *ion* (H^+) is a hydrogen atom that has *lost an electron*. What is left when a hydrogen atom (which has one proton and one electron) loses an electron? Only a proton is left. Thus, an H^+ ion is actually just a single proton. Well, since acids donate H^+ ions, they really donate protons. In the same way, since bases accept H^+ ions, they really accept protons. As a result, some chemists call acids **proton donors** and bases **proton acceptors**. We will use the terms H^+ donor and H^+ acceptor, but you need to know that those other, equivalent terms do exist.

If you're not quite sure what it means to be a molecule that "donates H^+ ions" or "accepts H^+ ions," a few examples should clear things up a bit. Consider hydrochloric acid (HCl). When this acid is mixed with water, the following reaction occurs:

$$HCl\ (aq) + H_2O\ (l) \rightarrow H_3O^+\ (aq) + Cl^-\ (aq)$$

What happened in this reaction? Well, it might help to look at it from a Lewis structure point of view. When we mix HCl and water, it looks like this:

H—Cl: + H—O: (with an H bonded above the O)

Remember, the bonds pictured here are really just shared electrons:

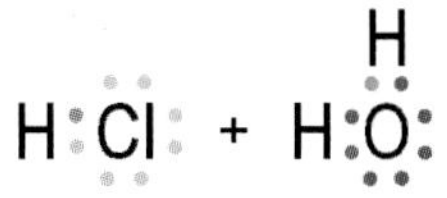

H:Cl: + H:O:H

Now think for a moment about the electron pair that is being shared by the hydrogen atom and the chlorine atom in the HCl molecule. The electronegativity of the Cl is significantly larger than that of the hydrogen. As a result, these shared electrons spend most of their time around the Cl. Suppose, then, that the hydrogen wanted a fairer share of electrons. Is there any way it could attain this goal?

It turns out that there is. You see, the oxygen atom has two extra pairs of electrons. The hydrogen in the HCl molecule can leave the electron pair it is currently sharing and give it entirely to the chlorine atom:

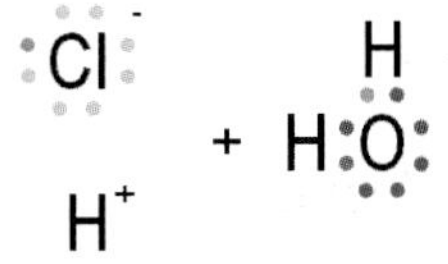

:Cl:$^-$ + H:O:H

H^+

Since the hydrogen is leaving the electron pair behind, it now has no valence electrons. It usually has one valence electron, so it is one electron shy of its normal state. This makes it a 1+ ion. In the same way, the chlorine now has sole possession of the electron pair, so it now has eight valence electrons. Compared to its normal number of valence electrons (seven), this is one too many, making it a 1- ion.

Under normal circumstances, of course, the hydrogen would not do this. It will only do it because it can get a better share of electrons if it attaches to the oxygen on the water molecule:

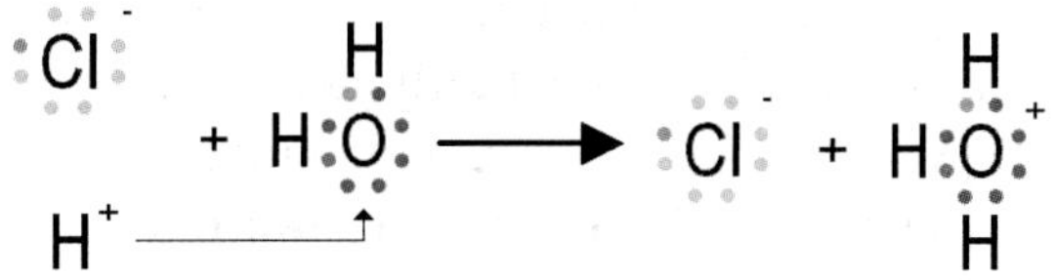

:Cl:$^-$ + H:O:H ⟶ :Cl:$^-$ + $[H_3O]^+$

H^+

In the end, then, the hydrogen that was formerly in the HCl molecule left its shared electron pair with the chlorine, making an H^+ ion. This ion then attached itself to a non-bonding electron pair on the oxygen of the water molecule. Of course, since the hydrogen in question possessed a 1+ charge, that charge is now imposed on the molecule of which it became a part. This new polyatomic ion, H_3O^+, is called the **hydronium** (hy droh' nee uhm) **ion**, and it plays a crucial role in most of acid/base chemistry. Thus, this is another polyatomic ion you must memorize!

Looking back on our original chemical equation, we should now be able to determine which compound was the acid in the reaction and which was the base:

$$\underset{\text{(acid)}}{HCl\ (aq)} + \underset{\text{(base)}}{H_2O\ (l)} \rightarrow H_3O^+\ (aq) + Cl^-\ (aq) \qquad (10.1)$$

Since the H^+ *left* the HCl molecule, we could say that the H^+ was *donated* by the HCl. Well, things that donate H^+ are called acids. Thus, HCl acted as an acid in this reaction. In the same way, the H^+ ended up *joining* the H_2O molecule, making the H_2O molecule an H^+ *acceptor*. This means that the H_2O was acting as a base.

In a chemical equation, then, it is relatively simple to determine which reactant is the acid and which is the base. All you have to do is look at where the H^+ went. The molecule that lost one or more H^+ ions is the acid, and the molecule that gained one or more H^+ ions is the base. Before we see some examples of how this is done, I want to introduce two other concepts. First, consider the following chemical equation:

$$H_2O\ (l)\ +\ NH_3\ (aq)\ \rightarrow\ NH_4^+\ (aq)\ +\ OH^-\ (aq) \qquad (10.2)$$

Which reactant is the acid in this case and which is the base? Well, the NH_3 in the reaction above was turned into NH_4^+. The only way that could happen would be for the NH_3 to gain (or accept) an H^+. Thus, NH_3 is the base. In the same way, H_2O became OH^- by losing (or donating) an H^+, making it the acid. But wait a minute, wasn't H_2O a base in the reaction represented by Equation (10.1)? How is it possible for H_2O to be an acid in one reaction and a base in another?

Well, there are certain compounds that chemists call **amphiprotic** (am' fuh proh' tik). These compounds have the interesting ability to act as a base in some reactions and an acid in others.

Amphiprotic compounds - Compounds that can act as either an acid or a base, depending on the situation

Water is the most common amphiprotic substance, but there are others that you will see from time to time in this course. So, don't let it confuse you if we discuss a compound in one context where it is an acid and then later talk about it behaving like a base. If that happens, you know you have run into an amphiprotic compound. Please note that some chemists use the term **amphoteric** (am' foh tehr' ik) instead of amphiprotic. The two terms mean essentially the same thing, so you might see one in some textbooks and the other in other textbooks.

The Behavior of Ionic Compounds in Aqueous Solutions

The second concept that must be introduced before you can begin analyzing acid/base reactions is something that we touched on way back in Module #3. Remember we said that a compound that conducts electricity when dissolved in water is an ionic compound? Well, in order to understand acid/base reactions, we must learn *why* ionic compounds conduct electricity when dissolved in water.

In order for a substance to conduct electricity, it must contain electrical charges that are free to move. Pure water, since it is a polar covalent compound, does have electrical charges, but they are not free to move. Remember, because it is polar, a water molecule has fractional positive charges on its hydrogen atoms and a fractional negative charge on its oxygen atom. These charges, however, are not free to move, because the hydrogen atoms and oxygen atom are all bonded together. Thus, although water does have electrical charges, they are not free to move. They are stuck together.

When an ionic compound is dissolved in water, however, the situation changes. The way that ionic compounds dissolve in water is to split up into their individual ions. This turns out to be an important fact worth remembering:

When ionic compounds dissolve, they split into their constituent ions.

Thus, when dissolved in water, an NaCl molecule will split up into a Na^+ ion and a Cl^- ion. These ions, because they separate, are free to move and thus can conduct electricity. That's why ionic compounds conduct electricity when dissolved in water.

We can now see why acids are covalent compounds that conduct electricity when dissolved in water. Look back at the reaction described by Equation (10.1). Notice how the acid reacted with water to make two ions: H_3O^+ and Cl^-. Once those two ions are formed, they are able to move freely, because they are separate ions. Thus, the result when acids are mixed with water is a solution that has ions that are free to move, and the solution therefore conducts electricity.

Identifying Acids and Bases in Chemical Reactions

Now that we have fully discussed the ins and outs of ions and their relationship to acids and bases, let's see how all of this pans out in a few examples problems:

EXAMPLE 10.1

Determine which reactant is the acid and which is the base in the following chemical equation:

$$H_2SO_4\,(aq) + CaCO_3\,(aq) \rightarrow Ca^{2+}\,(aq) + HCO_3^-\,(aq) + HSO_4^-\,(aq)$$

Now this looks like a very complicated chemical equation, but if we notice that $CaCO_3$ is an ionic compound, and if we remember that the phase symbol (aq) means "dissolved in water," things become a lot easier.

First, since $CaCO_3$ is an ionic compound dissolved in water, we know that it actually splits up into its constituent ions. You should be able to name this compound. It is called calcium carbonate. This means it is composed of the calcium ion and the carbonate ion. You should have the carbonate ion memorized as CO_3^{2-}. The calcium ion is a single-atom ion. Since Ca is in group 2A of the periodic chart, the ion is Ca^{2+}. In the end, then, the equation actually looks like this:

$$H_2SO_4\,(aq) + Ca^{2+}\,(aq) + CO_3^{2-}\,(aq) \rightarrow Ca^{2+}\,(aq) + HCO_3^-\,(aq) + HSO_4^-\,(aq)$$

Even though there are more components in this equation, it is actually easier to analyze now. We see that the H_2SO_4 lost an H^+ ion to become HSO_4^-. This means that H_2SO_4 is the acid, since it donated an H^+. We also see that nothing happened to the Ca^{2+}, because it is the same on both sides of the equation. We call ions like this **spectator ions**, because they don't actually do anything in the reaction. They just seem to sit and "watch" the other compounds react. Finally, we see that CO_3^{2-} gained an H^+ to become HCO_3^-. Since CO_3^{2-} gained an H^+, we could call it the base. However, our original equation did not have CO_3^{2-} written in it. Our original equation had $CaCO_3$ in it. Since $CaCO_3$ is the source of CO_3^{2-}, we can say that $CaCO_3$ is the base.

This, then, is how we figure out which reactant is the acid and which is the base in an acid/base reaction. First, we examine the chemical equation and look for any ionic compounds. If they exist, we split them up into their ions. We then see what molecule lost an H^+ ion and what molecule gained one. The molecule that gained the ion is the base, and the one that lost it is the acid. If an ion gained or lost an H^+ ion, the ionic compound from which it came is identified as the acid or base. Follow this method in two more examples.

Identify the acid and base in the following reaction:

$$\mathbf{HCl\ (aq) + NaOH\ (aq) \rightarrow H_2O\ (l) + NaCl\ (aq)}$$

To solve this problem, we first take all ionic compounds and split them up into their ions. HCl is covalent, but NaOH is ionic. You should recognize the "OH" part of the compound. That's the hydroxide ion, OH^-. Thus, this is sodium hydroxide. That means it is comprised of the sodium ion (Na^+ because Na is in group 1A) and the hydroxide ion (OH^-, which you should have memorized from the previous module). Thus, NaOH splits up into an Na^+ ion and an OH^- ion when it dissolves in water. In the same way, H_2O is covalent, but NaCl (sodium chloride) is ionic. There are no polyatomic ions in NaCl, so it just splits into a sodium ion (Na^+) and a chloride ion (Cl^-):

$$HCl\ (aq) + Na^+\ (aq) + OH^-\ (aq) \rightarrow H_2O\ (l) + Na^+\ (aq) + Cl^-\ (aq)$$

Now that we have all ionic compounds split into their ions, we can analyze the equation to see where the H^+ ion went. We can see that the Na^+ ion didn't change from the reactants side to the products side. Thus, Na^+ is a spectator ion and has no real role in the reaction. On the other hand, the HCl molecule turned into Cl^-. This means it lost an H^+ ion, and therefore HCl is the acid. In addition, the OH^- changed into H_2O. The only way that can happen is for it to gain an H^+. Thus, we could call OH^- the base. Unfortunately, OH^- is not present in the original equation. However, we know that the OH^- came from NaOH, so NaOH is the base.

Determine the acid and base in the following reaction:

$$\mathbf{2HNO_3\ (aq) + Mg(OH)_2\ (aq) \rightarrow Mg(NO_3)_2\ (aq) + 2H_2O\ (l)}$$

To analyze this equation, we must split up any dissolved ionic compounds into their components. The ionic compounds in this equation are $Mg(OH)_2$ and $Mg(NO_3)_2$. $Mg(OH)_2$ is magnesium hydroxide; thus, it is composed of the magnesium ion (Mg^{2+} because Mg is in group 2A) and the hydroxide ion (OH^-, which you should have memorized). We must be a little careful here, however. How many magnesium ions are in this molecule? Only one. But how many hydroxide ions are in the molecule? There are two, because there is a subscript of "2" after the parentheses that contain the polyatomic ion. Well, since there are two hydroxide ions in the molecule, we must be sure to put that information into our new chemical equation:

$$2HNO_3\ (aq) + Mg^{2+}\ (aq) + 2OH^-\ (aq) \rightarrow Mg(NO_3)_2\ (aq) + 2H_2O\ (l)$$

Since the magnesium hydroxide molecule has one magnesium ion and two hydroxide ions, we used a coefficient of 2 next to the hydroxide ion to show that. Of course, we are not done dealing with the ionic compounds, because we have not split $Mg(NO_3)_2$ into its constituent ions yet.

You should recognize the "NO_3" in $Mg(NO_3)_2$. It represents the nitrate ion, NO_3^-, which you should have memorized from the previous module. Thus, this is the magnesium nitrate ion, which is composed of one Mg^{2+} ion and two NO_3^- ions. When we split the ions up, then, we must put a coefficient of 2 next to the nitrate ion:

$$2HNO_3\ (aq)\ +\ Mg^{2+}\ (aq)\ +\ 2OH^-\ (aq)\ \rightarrow\ Mg^{2+}\ (aq)\ +\ 2NO_3^-\ (aq)\ +\ 2H_2O\ (l)$$

Now we can finally analyze the reaction. The Mg^{2+} ion did not change, so it is a spectator. The HNO_3, however, turned into NO_3^-. This means it lost an H^+, and therefore HNO_3 is the acid. The OH^- ended up becoming H_2O. The only way this can happen is for it to gain an H^+. Since the OH^- came from the $Mg(OH)_2$ in the original equation, $Mg(OH)_2$ is the base.

Even though this may seem like a long, drawn-out process, it is important to be able to analyze acid/base equations in this way because it builds on so many of the concepts we have learned. First, you have to be able to name ionic compounds, including those that contain polyatomic ions. That means you still must have all of the polyatomic ions from Table 9.1 memorized. In addition, you have one more polyatomic ion (the hydronium ion) to memorize. Second, you have to be able to interpret chemical formulas of ionic compounds in terms of charge. Third, you have to be able to recognize what subscripts and parentheses in a chemical formula mean. This is all review. If you are having trouble with these skills, review the modules that explained them. The only new skill you need here is the ability to determine where the H^+ ion went. Try to see if you can do these kinds of problems with the following exercises:

ON YOUR OWN

10.1 Determine the acid and base in the following chemical reaction:

$$HBr\ (aq)\ +\ NH_3\ (aq)\ \rightarrow\ NH_4^+\ (aq)\ +\ Br^-\ (aq)$$

10.2 Determine the acid and base in the following chemical reaction:

$$3C_2H_4O_2\ (aq)\ +\ Al(OH)_3\ (aq)\ \rightarrow\ Al(C_2H_3O_2)_3\ (aq) +\ 3H_2O\ (l)$$

10.3 Determine the acid and base in the following chemical reaction:

$$H_3PO_4\ (aq)\ +3KOH\ (aq)\ \rightarrow\ 3H_2O\ (l)\ +\ K_3PO_4\ (aq)$$

Recognizing Acids and Bases From Their Chemical Formulas

Hopefully, you are beginning to see a pattern. With the exception of one ($C_2H_4O_2$), all of the acids we have used so far have an "H" as the very first atom in their chemical formula. In addition, with the exception of one (NH_3), all of the bases we have used so far contain the hydroxide ion. This is a general rule. With a few notable exceptions, most acids have an "H" as the first atom in their chemical formula. Thus, while we might expect HF to be an acid (because its hydrogen atom is the first in the chemical formula), we would not expect CH_4 to be an acid (because its hydrogens are not first in the chemical formula). In the same way, most bases have the hydroxide ion in them. Thus, we would expect RbOH to be a base (because it contains the hydroxide ion), whereas we would predict that CH_3OH is not a base (because, although there is an OH in it, there is not a hydroxide ion, since the compound is not ionic and can therefore not contain ions).

If there is such a simple way to identify acids and bases, why did I show you such a complicated way? Well, first you must be able to handle the exceptions to these general rules. Second, the next section asks you to *predict* the products of a reaction between any given acid and any given base. The steps you have gone through will help you immensely in this undertaking. Finally, the only way you truly can appreciate the chemical definitions of acids and bases is to see how they pan out in chemical reactions, as you did in the section above. Now that you know the easier way to predict whether a compound is an acid or a base, you are free to use it whenever it applies.

Before we leave this section, there is one more thing you need to know about acids. They have their own names that do not follow the rules for naming compounds that you learned in Module #3. For example, according to the standard means of naming compounds, the molecule HCl would be named hydrogen monochloride. We would name it that way because it is a covalent compound, and covalent compounds use the prefix system to get their names. However, since it is an acid, chemists usually call it hydrochloric acid.

There are several acids like this. Although you needn't memorize these, the following table summarizes several of the more popular acids and their names. If I ever refer to an acid by its name in a problem, I will be sure to put its chemical formula in parentheses, so please do not spend any time trying to memorize them. However, as time goes on, you will probably get to know some of these acids by their names, because you will see them again and again.

TABLE 10.1
Some Common Acids

Formula	Name	Formula	Name
HCl	hydrochloric (hy' droh klor' ik) acid	**H_2SO_4**	sulfuric (sul fyur' ik) acid
HBr	hydrobromic (hy' droh brohm' ik) acid	**H_3PO_4**	phosphoric (fos for' ik) acid
HF	hydrofluoric (hy' droh flor' ik) acid	**H_2CO_3**	carbonic (kar bon' ik) acid
HNO_3	nitric (nye' trik) acid	**$C_2H_4O_2$**	acetic (uh see' tik) acid

Just to give you an idea of where you might find these acids, a diluted form of hydrochloric acid is sold at hardware stores for heavy-duty cleaning. It is often used to get tough lime stains (you'll see why later on) out of bathtubs, toilets, and swimming pools. In the hardware store, it is called "muriatic acid," and it is about 30% HCl and 70% water. Also, very dilute hydrochloric acid is often used in toilet bowl cleaners. Carbonic acid is a compound in soda pop that results from the carbonation process. It is mostly responsible for soda's acidic taste. In addition, acetic acid is the main component of vinegar, giving vinegar its nasty smell. Finally, sulfuric acid is the biggest chemical export for the United States, as it is synthesized in the U.S. and sold to both domestic and foreign industries.

You will be happy to know that while chemists do not follow the standard naming rules when it comes to acids, they usually do when it comes to bases. Thus, the base $Sr(OH)_2$ is simply called "strontium hydroxide," which is the proper name for this ionic compound. The only time this isn't the case is when a base is not an ionic compound. For example, we already learned that NH_3 is a base, despite the fact that it does not have a hydroxide ion. This covalent compound would have the proper name "nitrogen trihydride," but chemists usually call it by its nickname, "ammonia."

Predicting the Reactions That Occur Between Acids and Bases

Now that you've had some experience working with acid/base reactions, it is time for you to start predicting the results of mixing acids with bases. This shouldn't be too hard, if you've understood the previous section. After all, you now know how to recognize most acids and bases. In addition, you know what acids and bases do. There is really only one more thing that you need to learn before you can begin predicting the results of such reactions on your own.

The last "On Your Own" problem actually touched on this subject. In that problem, you worked with phosphoric acid (H_3PO_4). You saw in that problem that this acid actually donated three H^+ ions. Many acids have the ability to donate more than one H^+ ion. Acids that have this ability are generally called **polyprotic** (pah' lee proh' tik) **acids**:

Polyprotic acid - An acid that can donate more than one H^+ ion

If you remember my earlier discussion about what an H^+ ion is, you will understand the term "polyprotic." After all, an H^+ ion is simply a proton. Thus, "polyprotic" refers to the ability to donate more than one (poly) proton (protic).

Often, chemists get even more specific in referring to these acids. Phosphoric acid (H_3PO_4) is often called a **triprotic** (try' proh tik) acid because it can donate three protons (H^+ ions). On the other hand, sulfuric acid (H_2SO_4) is often referred to as a **diprotic** (dye' proh tik) acid, because it can donate two protons (H^+ ions). For polyprotic acids, the number of hydrogen atoms that begin the chemical formula tells you how many H^+ ions the acid will donate. Thus, H_2CO_3 (carbonic acid) is a diprotic acid.

Now that we have this new information under our belts, let's see how to predict the products of chemical reactions involving acids and bases. For example, suppose we had hydrofluoric acid (HF) and potassium hydroxide. Clearly, HF is an acid since it has a hydrogen atom as the first atom in its chemical formula, and potassium hydroxide is a base because it contains an hydroxide ion. Thus, we know we have an acid/base reaction on our hands.

The first thing we do, then, is take these two reactants and split up the ionic compound. This will allow us to see exactly what we're dealing with:

$$HF + K^+ + OH^- \rightarrow ?$$

The acid (HF) is supposed to donate an H^+ ion. Now where do you suppose that ion would like to go? Since it is positive, it will be attracted to the negatively charged hydroxide ion. Thus, the H^+ will be donated to the OH^-. When that happens, the 1+ charge on the hydrogen ion will cancel the 1- charge on the hydroxide ion, and the result will be a neutral molecule with two hydrogens and one oxygen: H_2O. The H^+ ion came from a neutral HF molecule, however. If a positive charge is taken from a neutral substance, the substance becomes negatively charged. Thus, when HF donates its H^+, it becomes F^-. In the end, then, something like this happened:

$$HF + K^+ + OH^- \rightarrow F^- + K^+ + H_2O$$

H^+

That's the actual reaction. However, chemists often don't care to write out all of those ions, even though they actually exist in solution. Instead, we often regroup the ions into the ionic compounds that formed them. Thus, the more commonly used version of this equation would be:

$$HF + KOH \rightarrow KF + H_2O$$

This is how we can predict the reactions that occur between acids and bases.

This method for determining the reactions that occur between acids and bases can actually become very cumbersome when you are dealing with polyprotic acids or bases that contain more than one hydroxide ion [$Mg(OH)_2$, for example]. Thus, we will use the results of the above discussion to develop a shortcut approach to determining the products of an acid/base reaction.

Notice the two products formed in the above discussion. In the end, the acid and base reacted to form an ionic compound and water. It turns out that when acids react with bases that are composed of hydroxide ions (which represent the vast majority of bases in nature), they will always produce water. This should make sense to you. If an acid always donates H^+ ions, and if those ions have the opportunity to react with OH^- ions, clearly the product will be water. There will most likely be another product as well.

If a base has a hydroxide ion in it, it must also have a positive ion. This positive ion will be "left behind" once the hydroxide ion reacts with the hydrogen ion. In addition, a negative ion will be "left behind" once the H^+ ion leaves the acid. In the end, then, the positive ion left over from the base and the negative ion left over from the acid can form a new ionic compound. Thus, the products of acid/base reactions are (almost always) water and some ionic compound. Chemists use the general term "salt" to refer to ionic compounds. Although you use the term "salt" to refer to table salt (the ionic compound NaCl), chemists use the term much more broadly. For chemists, "salt" is an all-inclusive term that refers to most ionic compounds. Thus, chemists often say that you can use the following general guideline to determine the products of an acid/base reaction:

ACID + BASE → SALT + WATER

Looking at the acid/base reaction we went through above, you can see that it fits this general pattern:

$$HF + KOH \rightarrow KF + H_2O$$
$$\text{acid} + \text{base} \rightarrow \text{salt} + \text{water}$$

In general, most acid/base reactions will fall into this general pattern. Thus, to predict the products of an acid/base reaction, all we have to realize is that one of them will be water. The other, then, will be an ionic compound made up of the positive ion from the base and the negative ion left over from the acid. This type of acid/base reaction is often called a **neutralization reaction**, because the acid and base neutralize each other, producing water and a salt. Let's see how this pans out with a few examples problems:

EXAMPLE 10.2

Give the balanced chemical equation that represents the reaction between HCl and RbOH.

We know that when dealing with bases that contain the hydroxide ion (as RbOH does), the general reaction is:

$$\text{ACID} + \text{BASE} \rightarrow \text{SALT} + \text{WATER}$$

The only difficulty is determining what the salt is. According to our rules, we take the positive ion from the base (Rb^+) and the negative ion left over when the H^+ leaves the acid (Cl^-) and form an ionic compound. Since the charges are the same on these two ions, we can simply ignore them in the chemical formula. Thus, the ionic compound that forms between these two ions is RbCl. The reaction, then, is:

$$\underline{HCl + RbOH \rightarrow RbCl + H_2O}$$

Notice the skills we have to use to solve these problems. First we have to recognize the acid and the base. Then we have to be able to break the base into its ions. Then we need to determine what ion is left behind when the acid donates its H^+. Finally, we have to be able to determine the chemical formula of the resulting ionic compound. Now remember, you learned how to do that last step in Module #8, and you practiced it again in Module #9 with polyatomic ions. If you have forgotten how to do that, you might want to go back and review those modules.

Write the chemical equation that describes the reaction between sulfuric acid (H_2SO_4) and NaOH.

Once again, we know that one of the products in this reaction is water. What is the other? The positive ion from the base is Na^+. The acid in question is diprotic, so it wants to donate *both* of its H^+ ions. Once H_2SO_4 donates both of its H^+ ions, the SO_4^{2-} ion (which you should recognize as the sulfate ion) remains. Thus, the other product is the ionic compound that forms between Na^+ and SO_4^{2-}. To get the formula of that ionic compound, we switch the charges and drop the signs, giving us Na_2SO_4. The equation, then, is:

$$H_2SO_4 + NaOH \rightarrow Na_2SO_4 + H_2O$$

If you look at this equation carefully, you will see that it is not balanced. Therefore, to answer this question properly, we must balance it.

$$\underline{H_2SO_4 + 2NaOH \rightarrow Na_2SO_4 + 2H_2O}$$

This, then, is the reaction that occurs between sulfuric acid and sodium hydroxide.

What is the reaction that occurs between HNO_3 and $Al(OH)_3$?

Once again, we know that one of the products in this reaction will be water. In addition, the positive ion from the base is Al^{3+}, while the negative ion left over from the acid will be NO_3^- (which you should recognize as the nitrate ion). Switching the charges and dropping the signs gives us $Al(NO_3)_3$ as the chemical formula of the salt:

$$HNO_3 + Al(OH)_3 \rightarrow Al(NO_3)_3 + H_2O$$

Now all we have to do is balance the equation:

$$\underline{3HNO_3 + Al(OH)_3 \rightarrow Al(NO_3)_3 + 3H_2O}$$

Now try the following problems to be sure you understand this concept:

ON YOUR OWN

10.4 Give the chemical equation that represents the reaction between each of the following:

a. $HClO_3$ and KOH b. HBr and $Ca(OH)_2$ c. H_3PO_4 and $Mg(OH)_2$

Now that you know how to predict the reactions between acids and bases, you can begin to see why acids and bases are so useful as household chemicals. For example, you found out in Experiment 10.1 that the all-purpose cleaner you tested contained a base. The vast majority of messes that need to be cleaned up are made of food and other substances that contain acids. When you spray a mess with all-purpose cleaner, the acids that make up the mess are turned into salt and water, which can then be picked up with a rag or paper towel.

In Experiment 10.1, you also found out that the toilet bowl cleaner was made of acid. If all-purpose cleaners are bases, why are toilet bowl cleaners acid? Well, the major problem in toilet bowls is lime stains. Lime (CaO) becomes a base when mixed with water. So in order to clean up a stain involving base, what do you need? An acid, of course. That's why toilet bowl cleaners are generally made up of acids.

<u>The Reactions Between Acids and Covalent Bases</u>

Now the method you just learned is adequate for predicting the reactions that occur between acids and bases, as long as the bases contain a hydroxide ion. But what if an acid wants to react with a base that doesn't have a hydroxide ion? What can we do to predict the subsequent reaction? Well, predicting these kinds of reactions is actually a bit easier than what you just learned.

For example, suppose we mix sulfuric acid with water. We know that water is an amphiprotic substance. Therefore, it can act as an acid or a base. Well, in the presence of an acid, water will act as a base. The problem is, water does not contain the hydroxide ion. In fact, it has no ions, because water is a covalent compound. How, then, do we determine the reaction that occurs? Well, all we need to do is fall back on the definitions of acids and bases.

Since H_2SO_4 is an acid, it wants to donate H^+ ions, and since H_2O will act as a base in the presence of acids, it wants to accept an H^+ ion. Thus, H_2SO_4 will donate an H^+ ion to H_2O and form H_3O^+. The problem is that H_2SO_4 wants to donate *two* H^+ ions. Water, however, can only accept one H^+ ion. Thus, the sulfuric acid will react with *two* H_2O molecules to form *two* H_3O^+ ions. After donating *two* H^+ ions, the H_2SO_4 becomes SO_4^{2-}. Thus, the overall reaction is:

$$H_2SO_4 + 2H_2O \rightarrow 2H_3O^+ + SO_4^{2-}$$

Notice the difference between this kind of acid/base reaction and the ones we did in the previous section. In the previous section, a covalent acid reacted with an ionic base to form an ionic compound (a salt) and a covalent compound (water). In this kind of reaction, however, a covalent acid reacted with a covalent base to form two ions. Make sure you understand this concept by studying the next example and performing the problems that follow.

EXAMPLE 10.3

What reaction occurs between H_3PO_4 and ammonia (NH_3), which is a base?

In this case, we are reacting a triprotic acid with a covalent base. Thus, we just transfer an H^+ ion to the base. However, since the acid wants to donate *three* H^+ ions but the base can accept only one, we have to do this *three* times. Once H_3PO_4 donates its *three* H^+ ions, it becomes PO_4^{3-}. The reaction, then, is:

$$\underline{H_3PO_4 + 3NH_3 \rightarrow 3NH_4^+ + PO_4^{3-}}$$

What reaction occurs between $H_2Cr_2O_7$ and water?

Water is an amphiprotic compound; therefore, in the presence of an acid, it will act as a base. This particular acid is diprotic, so it will donate *two* H^+ ions to *two* water molecules and make *two* H_3O^+ ions. Once it does this, it will become $Cr_2O_7^{2-}$, so the reaction is:

$$\underline{H_2Cr_2O_7 + 2H_2O \rightarrow 2H_3O^+ + Cr_2O_7^{2-}}$$

ON YOUR OWN

10.5 What is the chemical equation that describes the reaction between H_2SO_4 and ammonia (NH_3)?

10.6 What reaction occurs between HI and water?

Before we leave this section, step back and look at what you have accomplished. Given only the reactants in an acid/base reaction, you can write down the chemical equation and therefore determine the products that will be formed. You already learned how to do this for three other types of reactions: combustion, formation, and decomposition. Thus, at this point in your chemistry education, you can already predict the products of four different types of chemical reactions, given just the reactants! That's quite a feat! It took human beings about 2,500 years to get to the point where they could do such a thing. After less than a year of study, you now have that ability!

Molarity

Now that we know what reactions occur between acids and bases, we need to understand another very important concept that relates not only acids and bases but all chemistry that occurs in combination with water. The Works® is a specific toilet bowl cleaner that is used to get rid of lime stains in a toilet bowl. However, if you go to the hardware store, you can find a product called muriatic acid which claims (accurately) to get rid of lime stains that normal toilet bowl cleaners cannot get rid of. If you look at the active ingredient on each of these products, you will find that they each have only one active ingredient: hydrochloric acid (HCl). If they both have the same active ingredient, why is one much more powerful than the other?

The answer is quite simple. Muriatic acid is about 30% HCl and about 70% water. The Works®, on the other hand, is only about 3% HCl and about 97% water. Thus, even though they both have HCl, muriatic acid is a stronger cleaner because the HCl that makes it up is *more concentrated.* Since the acid is more concentrated, it can be more effective at its chemical job. Thus, the concentration of a chemical in a product is just as important as the type of chemical that is there.

In chemistry, we can express concentration in several different ways, depending on what we need to know. In the example above, I used percentage as an expression of concentration. The higher the percentage of HCl, the more concentrated the HCl. Usually, however, concentration is discussed in terms of amount divided by volume. For example, the unit of grams/mL (grams per milliliter) is a concentration unit, because it tells you how much (grams) of a substance is in a certain volume (mL). The most popular way of expressing concentration for chemists is with the unit moles/liter (moles per liter), which is called **molarity** (moh lehr' uh tee).

Molarity (M) - A concentration unit that tells how many moles of a substance are in a liter of solution. It is determined by taking the number of moles of a substance and dividing by the number of liters of solution.

$$M = \frac{\#\text{moles}}{\#\text{ liters}} \qquad (10.3)$$

It should make sense to you why molarity is the most popular concentration unit for chemists. After all, the most convenient way for a chemist to express the quantity of a substance is in moles, since that way it can be directly used in stoichiometry problems.

How do we use molarity? Well, let's go back to our discussion of muriatic acid. If you go to the hardware store to buy muriatic acid, you will usually find it in 4-liter bottles. A 4-liter bottle of muriatic acid contains 1,400 grams of HCl dissolved in enough water to fill up a 4-liter bottle. Since we know that the molecular mass of HCl is 36.5 amu, we know that it takes 36.5 grams of HCl to make one mole. Thus, we can figure out how many moles of HCl are in that 4-liter bottle:

$$\frac{1{,}400\ \cancel{\text{g HCl}}}{1} \times \frac{1\ \text{mole HCl}}{36.5\ \cancel{\text{g HCl}}} = 38\ \text{moles HCl}$$

Now that we know how many moles of HCl are in the bottle of muriatic acid, we can calculate the concentration of HCl in terms of molarity by dividing that number of moles by the volume of the solution (4.00 liters):

$$\text{Concentration of HCl} = \frac{\#\ \text{moles}}{\#\ \text{Liters}} = \frac{38\ \text{moles}}{4.00\ \text{liters}} = 9.5\ \text{M}$$

Now, of course, you might ask what good this does. What does a concentration of 9.5 M actually mean?

Well, knowing the concentration of an acid (or base) tells you how strong of an acid or base you might be dealing with. For example, the concentration of HCl in The Works® toilet bowl cleaner is about 0.8 M. What does that tell you? It tells you that muriatic acid is over ten times more acidic than The Works®. This means that if you are trying to get out lime stains, muriatic acid is more than 10 times better at getting the job done than The Works®. Of course, the flip side to this is that since muriatic acid is over 10 times more acidic than The Works®, it is also much more caustic and can therefore be more dangerous to work with. Spilling some of The Works® on your hands is not so bad, but spilling muriatic acid on your hands will burn your skin quite painfully.

We will see another use of the concept of concentration in a moment, but first you need to be sure that you truly understand how to calculate molarity with a couple of examples followed by some "On Your Own" problems.

EXAMPLE 10.4

What is the concentration (in units of molarity) of NaOH if there are 2.6 moles of the substance in a solution with a volume of 534 mL?

To get molarity, I must take the number of moles of NaOH and divide it by the number of liters of solution. I already have the number of moles of NaOH, but I do not have the number of liters of solution; instead, I have the volume in mL. Thus, first I must convert from mL to liters:

$$\frac{534\ \cancel{\text{mL}}}{1} \times \frac{0.001\ \text{L}}{1\ \cancel{\text{mL}}} = 0.534\ \text{L}$$

Now that I have both moles of NaOH and liters of solution, I can just use Equation (10.3) to get concentration in units of molarity:

$$\text{M} = \frac{\#\text{moles}}{\#\text{liters}} = \frac{2.6\ \text{moles NaOH}}{0.534\ \text{L}} = 4.9\ \frac{\text{moles NaOH}}{\text{L}} = 4.9\ \text{M}$$

Thus, our NaOH solution has a concentration of <u>4.9 M</u>.

If a student dissolves 124.3 grams of HNO_3 in enough water to make 250.0 mL of solution, what is the resulting molarity of HNO_3?

To get to molarity, I must have moles of acid and liters of solution. Right now I have neither, so I need to convert grams into moles and mL into liters. First, I'll convert grams of HNO_3 to moles:

$$\text{Molecular mass of } HNO_3 = 1 \times 1.01 \text{ amu} + 1 \times 14.0 \text{ amu} + 3 \times 16.0 \text{ amu} = 63.0 \text{ amu}$$

$$1 \text{ mole } HNO_3 = 63.0 \text{ grams } HNO_3$$

$$\frac{124.3 \cancel{\text{g } HNO_3}}{1} \times \frac{1 \text{ mole } HNO_3}{63.0 \cancel{\text{g } HNO_3}} = 1.97 \text{ moles } HNO_3$$

Now, we'll convert mL to liters:

$$\frac{250.0 \cancel{\text{mL}}}{1} \times \frac{0.001 \text{ liters}}{1 \cancel{\text{mL}}} = 0.2500 \text{ liters}$$

Now that we have moles and liters, we can finally calculate molarity using Equation (10.3):

$$M = \frac{\#\text{moles}}{\#\text{liters}} = \frac{1.97 \text{ moles } HNO_3}{0.2500 \text{ liters}} = 7.88 \frac{\text{moles } HNO_3}{\text{liter}} = 7.88 \text{ M}$$

In the end, then, our HNO_3 has a concentration of 7.88 M.

ON YOUR OWN

10.7 Determine the concentration of acid or base for each of the following solutions:

a. 2.1 moles HCl in 346 mL of solution
b. 6.78 grams of KOH are dissolved in enough water to make 150.0 mL of solution
c. 20.1 grams of H_2SO_4 are dissolved in enough water to make 3.4 liters of solution

The Dilution Equation

So far, you have learned to calculate the concentration of an acid or base when you are given the number of grams (or moles) of the acid or base and the total volume of the solution that contains it. This is a very useful way to calculate concentration, but it is not adequate by itself. It turns out that most of the time, chemists do not make up acid or base solutions by weighing out a certain number of grams and then dissolving the acid or base into a solution. A much more common way for chemists to make acid and base solutions is by **dilution.**

Dilution - Adding water to a solution in order to decrease the concentration

For example, when chemists buy HCl, they usually buy "concentrated HCl," which is a liquid that comes in bottles. The HCl in these bottles has a concentration of 12.0 M. Thus, if a chemist wants to make a 2.0 M solution of HCl, it is easier to just dilute the 12.0 M until its concentration decreases to 2.0 M, rather than measure out the correct number of grams of HCl.

In order to be able do this, the chemist takes a small volume of the highly concentrated solution and dilutes it with water. For this to be successful, however, the chemist must know *how much* of the original solution to take and *how* much water to add. Chemists can calculate this information with a very simple equation, called the **dilution equation:**

$$M_1V_1 = M_2V_2 \tag{10.4}$$

In this equation, "M_1" refers to the original concentration of the solution, and "V_1" stands for the volume of the original solution that is going to be diluted. On the other hand, "M_2" refers to the desired concentration, while "V_2" refers to the final volume of the new, diluted solution. Let's see how this equation is used in a couple of example problems:

EXAMPLE 10.5

A chemist wants to dilute 12.0 M HCl to make 500.0 mL of a 3.5 M HCl solution. How will he do this?

This is a dilution, so we obviously must use the dilution equation. The original concentration of the HCl is 12.0 M, so that is M_1. The chemist wants the diluted solution to have a concentration of 3.5 M, so that is M_2. The chemist also knows that he needs to have 500.0 mL of this 3.5 M solution. Since the 500.0 mL is the desired volume of the final solution, that is V_2. In the end, then, the only thing we don't know in the dilution equation is V_1, so we can solve for it:

$$M_1V_1 = M_2V_2$$

$$(12.0\text{ M})\cdot V_1 = (3.5\text{ M})\cdot(500.0\text{ mL})$$

$$V_1 = \frac{(3.5\text{ }\cancel{\text{M}})\cdot(500.0\text{ mL})}{12.0\text{ }\cancel{\text{M}}} = 1.5\times10^2\text{ mL}$$

Notice that since we are dividing and multiplying by molarity, it cancels, leaving us with the unit mL. This is good, since we are solving for volume. So what does this answer mean? It means that the chemist needs to take 1.5 x 10^2 mL of the original solution and dilute it with water until it reaches 500.0 mL of volume. This will give the chemist 500.0 mL of a 3.5 M solution.

A chemist takes 2.5 liters of 5.6 M NaOH and dilutes it with enough water to make 10.0 liters. What is the new concentration of the NaOH?

In this case, we are not told what the chemist wants, we are told what the chemist does. The chemist dilutes a solution of NaOH. We are told the original concentration ($M_1 = 5.6$ M) and volume ($V_1 = 2.5$ liters) as well as the final volume ($V_2 = 10.0$ liters). To determine the final concentration, we just use the dilution equation:

$$M_1V_1 = M_2V_2$$

$$(5.6\text{ M}) \cdot (2.5\text{ L}) = M_2 \cdot (10.0\text{ L})$$

$$M_2 = \frac{(5.6\text{ M}) \cdot (2.5\,\cancel{\text{L}})}{10.0\,\cancel{\text{L}}} = \underline{1.4\text{ M}}$$

Once again, the units work out. Since we are both multiplying and dividing by liters, they cancel out, leaving molarity, which is the unit of concentration.

ON YOUR OWN

10.8 A chemist needs to make 750.0 mL of a 3.5 M solution of sulfuric acid. Looking on the shelves, she finds a bottle of 10.0 M sulfuric acid. How can the desired solution be made?

<u>The Importance of Concentration in Chemistry</u>

Before we move on to applying the concept of concentration to stoichiometry, let me make one point very clear. I have used the subject of acids and bases to introduce the concept of concentration, but don't think for a moment that this concept is limited to acids and bases. Concentration is a very important topic that will come up time and time again throughout this course. It is important because the concentration of a chemical often determines its practical effect.

The topic in which concentration probably plays its most crucial role is that of chemical toxicity. For example, do you remember learning about ozone in Module #8? I told you then that ozone is a toxic poison for human beings and most other forms of life. Would it surprise you to know, then, that you are breathing ozone *right now, as you read this page*? If ozone is an extremely toxic poison and you are breathing ozone right now, why isn't it affecting you? The reason it doesn't affect you is that the concentration of it is too low for its poisonous effects to occur. Thus, despite the fact that ozone is a poison, you can still breathe it, as long as the concentration of the ozone is very low. This keeps it from being able to perform effectively enough, making it safe for you to inhale.

In fact, you are breathing in several different poisons right now. No matter where you are, the air that you are breathing contains nitrogen monoxide, nitrogen dioxide, ozone, sulfur dioxide, sulfur trioxide, and carbon monoxide. These are all toxic poisons, but, luckily for you, their concentration is low in most areas of the country. Thus, even though the poisons exist, they are not dangerous, because their concentration is not large enough to allow them to perform effectively. If, however, you live in places like Los Angeles, some air pollutants (especially nitrogen monoxide) can reach high enough concentrations that they can begin to harm your body. That's when you should begin getting concerned.

You might be surprised to learn that the pendulum can swing the other way. I'm sure that at one time or another you learned about vitamins. You were probably told that your body needs certain vitamins in order to continue to function properly. This is quite true. Did you also know, however, that some vitamins (especially vitamins A, D, E, and K) can actually become *poisonous* to your body if their concentrations become too high? Indeed, people have become very sick and have even died because they have taken *too many* of these vitamins! This is how important the subject of concentration is. At low concentrations, vitamins are beneficial (indeed, necessary) for your body. However, if their concentrations build up, they begin working against the body and can kill you, or at least make you very sick.

Now, of course, there is no reason to get paranoid about this. Even if you take vitamin supplements, you are probably not getting anywhere near toxic levels. It is important, however, to make sure that you are not taking *more* than the recommended dosage of your vitamin supplement. This is where people start to get into trouble. They think that if they double or triple the recommended dosage, they are doubling or tripling their bodies' benefits. That may be true of some vitamins (vitamin C, for example), but not most vitamins (especially vitamins A, D, E and K). There has been a lot of study in the area of vitamins, and that's where the recommended dosages on your vitamin supplements come from. As long as you follow those directions, you should be fine.

Using Concentration in Stoichiometry

Another place where concentration plays a critical role is in stoichiometry. This is because the concentration of an acid or a base gives you a way of determining the number of moles of acid or base that you are adding to a chemical reaction. Suppose, for example, you wanted to perform the following chemical reaction:

$$2HCl + Ba(OH)_2 \rightarrow BaCl_2 + 2H_2O$$

in order to make some barium chloride. If you have 100.0 mL of muriatic acid and an excess of barium hydroxide, you should be able to use stoichiometry to determine how much barium chloride would be produced in the reaction. The fact that you added 100.0 mL of acid tells you "how much" acid you added, but it doesn't help you in terms of stoichiometry. For stoichiometry to work, you need to know how many *moles* of HCl you added.

Well, since you know that the concentration of HCl in muriatic acid is 9.5 M (we calculated it earlier), you can use that information to convert the 100.0 mL into moles. After all, the molarity tells you the number of moles of HCl per liter of muriatic acid. You already have the number of mL of muriatic acid you used. If you were to convert that into liters, you could then multiply the concentration by the volume to get the number of moles of acid you added:

$$\frac{100.0\ \text{mL}}{1} \times \frac{0.001\ \text{L}}{1\ \text{mL}} = 0.1000\ \text{L}$$

$$\frac{9.5\ \text{moles HCl}}{1\ \text{L}} \times \frac{0.100\ \text{L}}{1} = 0.95\ \text{moles HCl}$$

Now that you have the number of moles of HCl added, you can figure out how many grams of $BaCl_2$ you produced by using stoichiometry:

$$\frac{0.95\ \text{moles HCl}}{1} \times \frac{1\ \text{mole BaCl}_2}{2\ \text{moles HCl}} = 0.48\ \text{moles BaCl}_2$$

$$\frac{0.48\ \text{moles BaCl}_2}{1} \times \frac{208.3\ \text{g BaCl}_2}{1\ \text{mole BaCl}_2} = 1.0 \times 10^2\ \text{g BaCl}_2$$

In the end, then, when 100.0 mL of muriatic acid reacts with an excess of barium hydroxide, 1.0×10^2 g of barium chloride is produced.

It is important that you see the relationship between molarity and moles:

When *number of moles* is *divided* by volume (in liters), you get *molarity*.

When *molarity* is *multiplied* by volume (in liters), you get *number of moles*.

Getting this relationship straight in your mind will help you to use concentration, especially in the stoichiometry problems that follow:

EXAMPLE 10.6

One way to make sodium sulfate is through the reaction of sulfuric acid (H_2SO_4) with sodium hydroxide. How many grams of sodium sulfate can be produced if 500.0 mL of 1.6 M sodium hydroxide is reacted with an excess of sulfuric acid?

In order to solve any stoichiometry problem, we must first figure out the balanced chemical equation. According to the problem, our reactants are H_2SO_4 and NaOH. When they react, we learned that a salt and water will be produced. The salt will be composed of the positive ion from the base (Na^+) and the negative ion left over when the acid gets rid of all of its H^+ ions. In this case, that will be the SO_4^{2-} ion. Thus, the reaction is:

$$H_2SO_4 + NaOH \rightarrow Na_2SO_4 + H_2O$$

Of course, this equation is not balanced, so that is the first thing to do:

$$H_2SO_4 + 2NaOH \rightarrow Na_2SO_4 + 2H_2O$$

Now that we have a balanced equation, we can start the stoichiometry. We are told that the sulfuric acid is in excess, so we know that NaOH is the limiting reactant. Therefore, we need to know the number of moles of NaOH in order to be able to predict how much sodium sulfate is formed. To do this, we will take concentration times volume:

$$\frac{1.6 \text{ moles NaOH}}{1 \text{ \sout{L NaOH}}} \times \frac{0.5000 \text{ \sout{L NaOH}}}{1} = 0.80 \text{ moles NaOH}$$

We can now use this information to calculate the number of moles of sodium sulfate produced:

$$\frac{0.80 \text{ \sout{moles NaOH}}}{1} \times \frac{1 \text{ mole } Na_2SO_4}{2 \text{ \sout{mole NaOH}}} = 0.40 \text{ moles } Na_2SO_4$$

Now that we have the number of moles of sodium sulfate produced, we can convert back to grams:

$$\frac{0.40 \text{ \sout{moles } \sout{Na_2SO_4}}}{1} \times \frac{142.1 \text{ g } Na_2SO_4}{1 \text{ \sout{mole } \sout{Na_2SO_4}}} = \underline{57 \text{ g } Na_2SO_4}$$

Now remember, the vast majority of this should be review. You should already know how to do the stoichiometry part of this problem. The only new things here are determining the equation (which you learned in an earlier section) and using concentration to figure out the number of moles of limiting reactant. Make sure you understand all of this by performing the following problem:

ON YOUR OWN

10.9 Barium hydroxide is often used to clean up toxic acid spills in industrial settings. If a 10.0 liter container of 12.0 M nitric acid (HNO_3) is broken, and the acid spills all over the floor, how many grams of barium hydroxide will be needed to clean up the spill?

Acid/Base Titrations

One specific application of stoichiometry to acid/base reactions is the process known as **titration**.

Titration - The process of slowly reacting a base of unknown concentration with an acid of known concentration (or vice versa) until just enough acid has been added to react with all of the base. This process determines the concentration of the unknown base (or acid).

In other words, if you have an acid of unknown concentration, you can figure out the concentration by titrating it against a base of known concentration. This is done by slowly adding the base to the acid until you have added just enough base to react completely with all of the acid present. Once you know that this has happened, you can use stoichiometry to determine the concentration of the acid. In order to be able to do this, of course, there must be a way to determine when you have added just enough base to react completely with all of the acid present. This is accomplished through the use of indicators.

At the beginning of this module, I introduced the concept of indicators, and you used litmus as an acid/base indicator. Well, there are several different types of indicators (you will make your own in the experiment below) that turn one color in the presence of a base and a different color in the presence of an acid. Let's suppose, for the sake of argument, that you were using litmus as an indicator. Litmus is red in the presence of an acid, so if you added litmus to your acid of unknown concentration, it would turn red.

Now let's suppose you slowly started adding base to your acid. At first, the litmus would stay red, because as you added base, it would immediately react with the acid. The litmus, then, would still see acid in the solution and would stay red. However, what would happen after all of the acid had reacted with the base? There would be no acid left. Therefore, the *very next drop* of base that you added would turn the litmus from red to blue. This would indicate that all of the acid was gone, and you would finish your titration. When the indicator changed color, you would call this the **endpoint of the titration**.

If this sounds a bit confusing to you at this point, don't worry. Just perform the experiment that follows, and you will understand much better after you are finished.

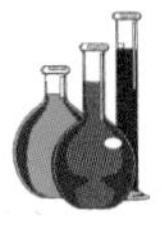

If you purchased the MicroChem kit discussed in the introduction, you can perform Experiment #12 in place of or in addition to this experiment.

EXPERIMENT 10.2
Acid/Base Titration

Note: A sample set of calculations is available in the solutions and tests guide. It is with the solutions to the practice problems.

Supplies

- Eyedropper
- Mass scale
- Distilled water (available at any grocery store)
- White sheet of paper (no lines)
- Stirring rod (or a small spoon)
- ~~A few leaves of red cabbage~~ (it must be red cabbage, not regular cabbage)
- 2 beakers (If you don't have beakers, one should be a short, fat glass that is transparent, and the other can be a small pot to boil water in.)
- ~~Flame or stove for heating~~
- Graduated cylinder (Measuring cups and spoons will work, but the experiment will be much harder.)
- Clear ammonia solution (This is sold with the cleaning supplies in most supermarkets. It must be clear. A colored solution will mess up the endpoint.)
- Clear vinegar (Once again, colored vinegar will mess up the endpoint.)
- Safety goggles

1. Rinse your beakers, graduated cylinders, and medicine dropper thoroughly with distilled water. Any contamination in them will mess up the endpoint of the experiment.
2. Take your small beaker (or boiling pot) and rinse it out thoroughly with distilled water.
3. Put the red cabbage leaves in the small beaker and fill it with about 70 mL (a qualitative measurement) of distilled water. If you are using a boiling pot, you will probably need to add three times as much water, because the water will evaporate more quickly in a pot than in the small beaker.
4. Boil the cabbage leaves and water in the beaker for about three minutes.
5. Allow the beaker to cool, and then carefully remove the leaves from the beaker. The water should now be a pinkish-red color. It might be blue-green, but that indicates that you contaminated your water a bit. This is your indicator. In the presence of acid (or neutral water) it is a pinkish-red color. In the presence of base, it is a blue-green color.
6. Take the large beaker and use the graduated cylinder to add 10.0 mL (a quantitative measurement) of the clear ammonia to the beaker. (If you are using measuring spoons, this is 2 teaspoons.)
7. Add 90.0 mL of distilled water (about ½ cup) to the beaker as well. Stir the solution.
8. Add a little less than half of the indicator to the beaker. The indicator should immediately turn greenish-blue. This indicates the presence of a base, ammonia (NH_3). You will now do a titration to determine the concentration of the ammonia in the solution that you bought.
9. Rinse your graduated cylinder thoroughly with distilled water.
10. Measure the mass of the graduated cylinder.
11. Fill the graduated cylinder with 50.0 mL (a quantitative measurement) of vinegar and measure the new mass of the cylinder plus the vinegar.
12. Determine the mass of the vinegar by difference.
13. Now use the mass and the volume (50.0 mL) to calculate the density of vinegar. (If you are using measuring cups you can do the same thing with ¼ of a cup. You must also calibrate your medicine dropper the way we did in Experiment 5.1, using the ¼ cup as a good approximation of 50.0 mL.)
14. Place the beaker full of ammonia solution on the white piece of paper so that it is easy to see the color of the solution.
15. Take your medicine dropper and fill it up from the graduated cylinder full of vinegar.
16. Squirt the vinegar into the ammonia solution and stir. Watch for any color change. (If you are using measuring cups, you cannot do it this way. Read the following instructions, but perform only the instructions in step #27). Continue to do this until the color of the solution turns pinkish. It will not turn as vividly pinkish-red as the original color of the indicator, but it should turn noticeably pink. This tells you that you passed the endpoint of the titration.
17. Read the number of mL of vinegar left in the graduated cylinder and, by difference, determine how much vinegar you added in the titration.
18. What you just did is referred to as a "rough titration." You ended up adding more acid than necessary, but you did it fairly quickly. Now you need to start over and do a careful titration.
19. Pour the titration solution down the drain, and thoroughly rinse the beaker with distilled water.
20. Fill it back up with 10.0 mL of ammonia and 90.0 mL of distilled water.
21. Stir the solution, and then add most of what is left of the indicator solution.
22. Put the beaker back on the white piece of paper and fill the graduated cylinder up again. Once again, make sure the graduated cylinder has precisely 50.0 mL of vinegar in it.
23. Now you can begin the careful titration. Use your medicine dropper to quickly add vinegar to the beaker until you get to within 5 mL of the amount you added during the rough titration.
24. Now you need to slow down. Add the vinegar drop by drop and stir thoroughly between drops. Keep careful watch on the color. The color should change from green to blue, to a very light blue, and then to a pinkish violet. The pinkish violet color tells you that you have reached the endpoint.

The best thing to look for is the very first hint of pink color that continues to exist after you have stirred the solution. As soon as you see that, you know you have reached the endpoint.

25. Squirt the remainder of the medicine dropper *back into the graduated cylinder.*
26. Read the graduated cylinder and, by difference, calculate the number of mL you added to reach the true endpoint of the titration.
27. If you are not using measuring cups, skip this step. If you are using measuring cups, you must do a careful titration the first time. Add the vinegar drop by drop, making sure to keep track of the number of drops that you added. Watch for the endpoint as detailed in the previous steps. When you reach it, the number of drops times your dropper calibration will tell you the number of mL you needed to add to get to the endpoint.
28. Clean up your mess.

Before I tell you how to analyze your experimental results, I want to review exactly what you did so that you did not get lost in the details of the experiment. You took an ammonia solution (a base) with an unknown concentration and titrated it with a vinegar solution (an acid), whose concentration is known. By using an indicator and adding the vinegar drop-by-drop, you were able to determine exactly when you had added enough acid (vinegar) to neutralize all of the base (ammonia). That's what you do in a titration. Now that you have done that, you can do the calculation that will tell you the concentration of the ammonia.

First, realize that the reaction going on in this experiment is:

$$C_2H_4O_2 + NH_3 \rightarrow NH_4^+ + C_2H_3O_2^-$$

This reaction tells us that for every one mole of acid you added, one mole of base reacted. When you reached the endpoint, you also knew that you had added just enough acid to react with all of the base present. You can use this information to calculate the concentration of the ammonia in the reaction.

To do this, you need to start with the number of moles of acid you added. This is easy. Multiply the density of the vinegar (which you measured in the first part of the experiment) by the number of mL you added to reach the endpoint. This will tell you how many grams of vinegar you added. The acid, however, makes up only 5.00% of the vinegar, so multiply your answer by 0.0500. This will tell you how many grams of acid you added. Use the chemical formula of the acid in the equation above to convert the number of grams of acid added into the number of moles of acid added.

Now that you know how many moles of acid you added, you can figure out how many moles of base were originally present. In the equation, we see that every one mole of acid requires one mole of base. The fact that we reached the endpoint of the titration tells us that all of the acid added reacted with all of the base present. Therefore, the number of moles of base present are equal to the number of moles of acid added. We are almost done.

The concentration of a base is defined as the number of moles of base (which you just figured out) divided by the volume of the solution (10.0 mL or 0.0100 liters). You divide by 10.0 mL because that's the volume of the ammonia solution. The 90.0 mL of water you added was just to make the color easier to see. You want to know the concentration of the ammonia solution that you bought. The volume of *that* solution was 10.0 mL. Divide the number of moles by the number of liters, and you have the concentration of the ammonia solution. Check your calculations against the sample calculations in the solutions and tests guide to make sure you did everything correctly.

Titrations like the one you just performed are a very common tool in laboratory chemistry. Now, of course, most laboratories use slightly more sophisticated equipment than you did. Instead of a graduated cylinder and an eyedropper, for example, most titrations are carried out with a **buret** (byoor eht'). This piece of laboratory glassware is a long, glass tube that has a volume scale on it and a dropper at the bottom. It is more precise than a graduated cylinder because it is able to measure volume to 0.01 mL. A buret is quite expensive, however.

Notice that there is *absolutely nothing new* to learn in order to understand titrations. Titrations are simply stoichiometry experiments that involve acids and bases. The special thing about these stoichiometry problems, however, is that there are not really any limiting reactants. By being able to identify the endpoint of the titration, you have identified the point at which exactly enough of one reactant is added to react completely with the other reactant. Thus, you can use either of the reactants to perform any stoichiometry you wish to perform. Using the titration you just did as a guide, finish this module by performing this last problem:

ON YOUR OWN

10.10 125 mL of nitric acid (HNO_3) with unknown concentration is titrated against magnesium hydroxide with a concentration of 2.3 M. If 35.4 mL of base are required in order to reach the endpoint, what was the concentration of the acid?

ANSWERS TO THE "ON YOUR OWN" PROBLEMS

10.1 We usually split up the ionic compounds into their component ions. However, since there are no metals in the entire reaction, there are also no ionic compounds. Looking at the equation, then, it is rather easy to see what has happened. The HBr has lost its H^+ to become Br^-, while the NH_3 has gained an H^+ to become NH_4^+. This means, then, that HBr is the acid and NH_3 is the base.

10.2 To solve this problem, we must first split up the ionic compounds into their constituent ions. The aluminum hydroxide ($Al(OH)_3$) on the reactants side is composed of one aluminum (Al^{3+}) ion and three hydroxide (OH^-) ions. The aluminum acetate on the products side is composed of one aluminum (Al^{3+}) ion and three acetate ($C_2H_3O_2^-$) ions. Notice that to be able to split both of these ionic compounds into their constituent ions, you needed to be able to recognize the polyatomic hydroxide and acetate ions. This will be true throughout this module. Thus, if you have forgotten the polyatomic ions in Table 9.1, go back and rememorize them. After splitting up the ionic compounds, the equation looks like this:

$$3C_2H_4O_2\ (aq) + Al^{3+}\ (aq) + 3OH^-\ (aq) \rightarrow Al^{3+}\ (aq) + 3C_2H_3O_2^-\ (aq) + 3H_2O\ (l)$$

Now we can see what happened. The $C_2H_4O_2$ lost an H^+ to become $C_2H_3O_2^-$, while the OH^- gained an H^+ to become H_2O. The OH^- came from the $Al(OH)_3$; thus, $C_2H_4O_2$ is the acid, and $Al(OH)_3$ is the base.

10.3 To solve this problem, we must first split up the ionic compounds into their constituent ions. The potassium hydroxide (KOH) on the reactants side is composed of one potassium (K^+) ion and one hydroxide (OH^-) ion. The potassium phosphate on the products side is composed of three potassium (K^+) ions and one phosphate (PO_4^{3-}) ion. Thus, the equation looks like this:

$$H_3PO_4\ (aq) + 3K^+\ (aq) + 3OH^-\ (aq) \rightarrow 3H_2O\ (l) + 3K^+\ (aq) + PO_4^{3-}\ (aq)$$

Notice how we got $3K^+$ and $3OH^-$ on the reactants side of the equation. The original equation tells us that we had three KOH molecules. When we split up the KOH into its two ions, we must carry that 3 to each of the ions.

Now we can see what happened. The H_3PO_4 molecule lost three H^+ ions to become PO_4^{3-}, while the three OH^- ions each gained an H^+ to become three H_2O molecules. The OH^- came from the KOH; thus, H_3PO_4 is the acid, and KOH is the base. Don't let it bother you that more than one H^+ was traded between the acid and base. This can (and often does) happen.

10.4 a. One of the products of this reaction will be water. The other will be the ionic compound that forms between the positive ion of the base (K^+) and the negative ion left behind from the acid. When an H^+ leaves $HClO_3$, a ClO_3^- ion remains. Since the charges of these ions are of equal magnitude, we ignore the numbers in the chemical formula, and therefore the salt formed will be $KClO_3$.

$$HClO_3 + KOH \rightarrow KClO_3 + H_2O$$

This chemical equation is already balanced, so we are finished.

b. This reaction will produce water and the ionic compound that forms between Ca^{2+} and Br^-. To get that chemical formula, we switch the charges and drop the signs to get $CaBr_2$.

$$HBr + Ca(OH)_2 \rightarrow CaBr_2 + H_2O$$

This equation is not balanced, of course, so now we have to balance it.

$$\underline{2HBr + Ca(OH)_2 \rightarrow CaBr_2 + 2H_2O}$$

c. Once again, this reaction will produce water as well as the ionic compound that forms between the positive ion from the base and the negative ion formed when the three H^+ ions leave this triprotic acid. The positive ion from the base is Mg^{2+}. When three H^+ ions leave the H_3PO_4 acid, they leave behind a PO_4^{3-} ion. The ionic compound that forms between Mg^{2+} and PO_4^{3-} is $Mg_3(PO_4)_2$. The reaction, therefore, is:

$$H_3PO_4 + Mg(OH)_2 \rightarrow Mg_3(PO_4)_2 + H_2O$$

This equation is not balanced, so we must balance it now:

$$\underline{2H_3PO_4 + 3Mg(OH)_2 \rightarrow Mg_3(PO_4)_2 + 6H_2O}$$

10.5 Sulfuric acid is a diprotic acid that wants to donate *two* H^+ ions. Those ions will be donated to *two* NH_3 molecules, leaving behind a SO_4^{2-} ion:

$$\underline{H_2SO_4 + 2NH_3 \rightarrow 2NH_4^+ + SO_4^{2-}}$$

10.6 Hydroiodic acid will force the amphiprotic water to act as a base. The HI will donate an H^+ ion to the water, forming an H_3O^+ ion. When it does that, HI will become I^-:

$$\underline{HI + H_2O \rightarrow H_3O^+ + I^-}$$

10.7 a. To get concentration, we must have moles divided by liters. We already have moles, but the volume of the solution is in mL, not liters. Thus, we must first convert to liters:

$$\frac{346 \cancel{mL}}{1} \times \frac{0.001 \text{ L}}{1 \cancel{mL}} = 0.346 \text{ L}$$

Now we have moles and liters, so we can calculate concentration:

$$M = \frac{\#moles}{\#liters} = \frac{2.1 \text{ moles HCl}}{0.346 \text{ L}} = 6.1 \frac{\text{moles HCl}}{\text{L}} = 6.1 \text{M}$$

The concentration is <u>6.1 M</u>.

b. In order to get concentration, we must have moles and liters. The problem gives us grams and mL, so we must make two conversions:

$$\text{Molecular mass of KOH} = 1 \times 39.1 \text{ amu} + 1 \times 16.0 \text{ amu} + 1 \times 1.01 \text{ amu} = 56.1 \text{ amu}$$

1 mole KOH = 56.1 grams KOH

$$\frac{6.78 \text{ } \cancel{\text{g KOH}}}{1} \times \frac{1 \text{ mole KOH}}{56.1 \text{ } \cancel{\text{g KOH}}} = 0.121 \text{ moles KOH}$$

$$\frac{150.0 \text{ } \cancel{\text{mL}}}{1} \times \frac{0.001 \text{ L}}{1 \text{ } \cancel{\text{mL}}} = 0.1500 \text{ L}$$

Now we can calculate molarity:

$$M = \frac{\#\text{moles}}{\#\text{liters}} = \frac{0.121 \text{ moles KOH}}{0.1500 \text{ L}} = 0.807 \frac{\text{moles KOH}}{\text{L}} = 0.807 \text{ M}$$

The concentration is 0.807 M.

c. To get to molarity, we need moles and liters. We already have liters, so all we have to do is get grams into moles:

$$\text{Molecular mass of } H_2SO_4 = 2 \times 1.01 \text{ amu} + 1 \times 32.1 \text{ amu} + 4 \times 16.0 \text{ amu} = 98.1 \text{ amu}$$

1 mole H_2SO_4 = 98.1 grams H_2SO_4

$$\frac{20.1 \text{ } \cancel{\text{g } H_2SO_4}}{1} \times \frac{1 \text{ mole } H_2SO_4}{98.1 \text{ } \cancel{\text{g } H_2SO_4}} = 0.205 \text{ moles } H_2SO_4$$

Now we can calculate concentration:

$$M = \frac{\#\text{moles}}{\#\text{liters}} = \frac{0.205 \text{ moles } H_2SO_4}{3.4 \text{ L}} = 0.060 \frac{\text{moles } H_2SO_4}{\text{L}} = 0.060 \text{ M}$$

The concentration is 0.060 M.

10.8 We can tell that this is a dilution problem because the chemist has a solution of high concentration that she wants to make into a solution of low concentration. That's what dilution does. The original concentration of the sulfuric acid is 10.0 M, so that is M_1. The chemist wants the diluted solution to have a concentration of 3.5 M, so that is M_2. The chemist also knows that she needs to have 750.0 mL of this 3.5 M solution. Since the 750.0 mL is the desired volume of the final solution, that is V_2. In the end, then, the only thing we don't know in the dilution equation is V_1, so we can solve for it:

$$M_1V_1 = M_2V_2$$

$$(10.0\text{ M})\cdot V_1 = (3.5\text{ M})\cdot(750.0\text{ mL})$$

$$V_1 = \frac{(3.5\text{ M})\cdot(750.0\text{ mL})}{10.0\text{ M}} = 260\text{ mL}$$

The chemist, then, needs to take 260 mL of the original solution and dilute it with water until it reaches 750.0 mL of volume. This will produce 750.0 mL of a 3.5 M solution.

10.9 In order to solve any stoichiometry problem, we must first figure out the balanced chemical equation. According to the problem, our reactants are HNO_3 and barium hydroxide. What is the chemical formula of barium hydroxide? Well, remember from Module #9 that the polyatomic hydroxide ion is OH^-. Barium is in group 2A of the periodic chart, so it develops a 2+ charge in ionic compounds. To determine the chemical formula of barium hydroxide, then, we must switch the charges and drop the signs. Thus, the chemical formula is $Ba(OH)_2$.

When $Ba(OH)_2$ and HNO_3 react, a salt and water will be produced. The salt will be composed of the positive ion from the base (Ba^{2+}) and the negative ion left over when the acid gets rid of all of its H^+ ions. In this case, that will be the NO_3^- ion. Once again, to get the chemical formula of the compound that forms between these two ions, we switch the charges and ignore the signs to get $Ba(NO_3)_2$. Thus, the reaction is:

$$HNO_3 + Ba(OH)_2 \rightarrow Ba(NO_3)_2 + H_2O$$

Of course, this equation is not balanced, so that is the first thing we must do:

$$2HNO_3 + Ba(OH)_2 \rightarrow Ba(NO_3)_2 + 2H_2O$$

Now that we have a balanced equation, we can start the stoichiometry. We need to clean up the nitric acid, so we know that it is the limiting reactant. Therefore, we need to know the number of moles of nitric acid in order to be able to predict how much barium hydroxide is needed. To do this, we will take concentration times volume:

$$\frac{12.0\text{ moles } HNO_3}{1\ \cancel{\text{L } HNO_3}} \times \frac{10.0\ \cancel{\text{L } HNO_3}}{1} = 1.20 \times 10^2\text{ moles } HNO_3$$

Notice that we must use scientific notation when reporting this answer, because we must report it to three significant figures. Scientific notation is the only way we can do that. We can now use this information to calculate the number of moles of barium hydroxide needed:

$$\frac{1.20 \times 10^2\ \cancel{\text{moles } HNO_3}}{1} \times \frac{1\text{ mole } Ba(OH)_2}{2\ \cancel{\text{moles } HNO_3}} = 60.0\text{ moles } Ba(OH)_2$$

Now that we have the number of moles of barium hydroxide needed, we can convert back to grams:

$$\frac{60.0\ \cancel{\text{moles Ba(OH)}_2}}{1} \times \frac{171.3\ \text{g Ba(OH)}_2}{1\ \cancel{\text{mole Ba(OH)}_2}} = 1.03 \times 10^4\ \text{g Ba(OH)}_2$$

Thus, 1.03 x 10^4 g of barium hydroxide are needed. This answer could be reported in decimal notation as well, because 10,300 g also has three significant figures.

10.10 Remember, titrations are just stoichiometry problems, so first we have to come up with a balanced chemical equation. To do that, we must first figure out the chemical formula of the magnesium hydroxide. Magnesium is in group 2A of the periodic chart, so it takes a charge of 2+ in ionic compounds. You should have memorized the hydroxide ion as OH^-. To get the chemical formula, we switch the charges and ignore the signs to get $Mg(OH)_2$. The equation, therefore, looks like this:

$$HNO_3 + Mg(OH)_2 \rightarrow Mg(NO_3)_2 + H_2O$$

We must now balance the equation:

$$2HNO_3 + Mg(OH)_2 \rightarrow Mg(NO_3)_2 + 2H_2O$$

Now we can do the stoichiometry. Since the endpoint was reached, we know that there was exactly enough base added to react with all of the acid. First, then, we calculate how many moles of base were added:

$$\frac{2.3\ \text{moles Mg(OH)}_2}{1\ \cancel{\text{L}}} \times \frac{0.0354\ \cancel{\text{L}}}{1} = 0.081\ \text{moles Mg(OH)}_2$$

We can now use the chemical equation to determine how many moles of acid were present:

$$\frac{0.081\ \cancel{\text{moles Mg(OH)}_2}}{1} \times \frac{2\ \text{moles HNO}_3}{1\ \cancel{\text{mole Mg(OH)}_2}} = 0.16\ \text{moles HNO}_3$$

Now that we have the number of moles of acid present, we simply divide by the volume of acid to get concentration:

$$M = \frac{\#\ \text{moles}}{\#\ \text{liters}} = \frac{0.16\ \text{moles HNO}_3}{0.125\ \text{L}} = 1.3\ \text{M}$$

The concentration is 1.3 M.

NOTE: All Lewis structure illustrations in this module by Megan Whitaker

REVIEW QUESTIONS FOR MODULE #10

1. State whether the substance described is most likely an acid or a base:

   a. A substance that tastes bitter
   b. A substance that feels slippery
   c. A substance that turns blue litmus paper red

2. Identify the acid in the following reaction:

$$PH_3 + CH_4O \rightarrow CH_3O^- + PH_4^+$$

3. Some books define a base as a proton acceptor. Why is this just as correct a definition as the one presented in this book?

4. Which of the following is (are) polyprotic acid(s)?

   a. HNO_3 b. H_2CO_3 c. $Ca(OH)_2$ d. H_3PO_4

5. Which of the following units could be used as concentration units?

   a. $\frac{mL}{meter}$ b. $\frac{grams}{m^3}$ c. $\frac{moles}{sec}$ d. $\frac{moles}{mL}$

6. A chemist wants to make a solution of 3.4 M HCl. There are two solutions of HCl that he can find on the shelf. One has a concentration of 6.0 M, while the other has a concentration of 2.0 M. Which solution can the chemist use to make the desired acid?

7. When the compound $Al_2(CO_3)_3$ dissolves in water, what ions does it form and how many of each ion are present?

8. Consider the following reactions:

$$HCO_3^- + NaOH \rightarrow Na^+ + CO_3^{2-} + H_2O$$

and

$$H_2O + HCO_3^- \rightarrow H_2CO_3 + OH^-$$

Which substance in the two equations is acting as an amphiprotic substance?

9. In acid/base chemistry, what two uses does an indicator serve?

10. In a titration, what is the significance of the endpoint?

PRACTICE PROBLEMS FOR MODULE #10

1. A chemist mixes sulfuric acid (H_2SO_4) with aluminum hydroxide. Give the balanced chemical equation that represents the resulting reaction.

2. Give the balanced chemical equation for the reaction that occurs between nitric acid (HNO_3) and calcium hydroxide.

3. Given that ammonia (NH_3) is a base, determine the reaction that occurs between H_2CO_3 and ammonia and give the balanced chemical equation.

4. When HBr is mixed with water, what chemical reaction occurs? Give the balanced chemical equation.

5. Give the concentration (in M) of each of the following solutions:

   a. 3.51 moles of HNO_3 are dissolved in enough water to make 1.2 liters of solution.
   b. 234.1 grams of KOH are dissolved in enough water to make 345 mL of solution.
   c. 4.1 grams of H_3PO_4 are dissolved in enough water to make 45 mL of solution.

6. A chemist has 12.0 M HBr and would like to use it to make 500.0 mL of 3.5 M HBr. Describe how she should do this.

7. A chemist makes a "stock" solution of NaOH by dissolving 950.0 grams in enough water to make 2.00 liters of solution. If, later on, he wants to use this stock solution to make 100.0 mL of 0.10 M NaOH, what would he need to do?

8. A chemist reacts 50.0 mL of 4.5 M carbonic acid (H_2CO_3) with excess potassium hydroxide. How many grams of potassium carbonate are produced?

9. A chemist needs to know the concentration of some NaOH that is in the laboratory. To find this out, she titrates a 50.0 mL sample of the solution with 3.5 M HCl. If it takes 34.3 mL of the HCl to reach the titration endpoint, what is the concentration of the NaOH solution?

10. A solution of H_2SO_4 is titrated against 1.2 M KOH. A total of 345.1 mL of KOH must be added to a 500.0 mL sample of H_2SO_4 in order to reach the endpoint. What is the concentration of the H_2SO_4?

MODULE #11: The Chemistry of Solutions

Introduction

In the previous module, we spent some time learning about acids and bases. We learned that acids and bases are often dissolved in water in order to make an acidic or basic solution. Although the word "solution" was used several times, the concept was not fully explained. It is now time to do that. It turns out that there is quite a lot of interesting chemistry involved in the simple task of making a solution. That's what we will study in this module.

One of the first things you have to know about solutions is some terminology. A chemist usually makes a solution by mixing things together. If the mixture is homogeneous, we generally say that one of the substances dissolved in the other substance. In a moment, we will learn what the word "dissolve" really means. For right now, however, we need to get some terms straight. First of all, we generally call the substance that we are dissolving the **solute** (sahl' yoot). The substance that we are dissolving the solute in is called the **solvent** (sahl' vent). When solute and solvent are mixed together, we have a **solution**. Thus, when you dissolve table salt in water, the table salt is the solute, the water is the solvent, and the salt water that you make is the solution.

It is important to note that even though we are most familiar with dissolving solids (like table salt) into liquids (like water), solutes do not have to be solids and solvents do not necessarily have to be liquids. For example, one of the steps in the manufacture of soda pop involves dissolving gaseous carbon dioxide into water. This is what gives soda pop its fizz. In this case, the solute (carbon dioxide) is a gas, not a solid.

As a sidelight, this same procedure also gives soda pop its acidic nature, which you discovered in the previous module. When carbon dioxide dissolves in water, a fraction of the molecules will undergo the following reaction:

$$CO_2\ (g) + H_2O\ (l) \rightarrow H_2CO_3\ (aq)$$

You should recognize the product of this reaction as carbonic acid. This is what turned the blue litmus paper red in Experiment 10.1.

How Solutes Dissolve in Solvents

Now that you know the terminology, you can begin to learn the science. The first thing you might ask yourself is *why* do solutes dissolve in solvents? After all, when you buy table salt (NaCl) at the grocery store, it is a solid. When you mix it with water, why does it go into the aqueous state? Why doesn't it just stay solid? The answers to these important questions involve the polar covalent nature of the water and the ionic nature of the NaCl.

First, you must remember that NaCl is an ionic compound; thus, it is composed of Na^+ ions and Cl^- ions. Unlike covalent molecules, the ionic NaCl molecule has no chemical bond, because the atoms that make it up do not share electrons. Instead, the Na has donated its electrons to the Cl. As a result, the Na is positive and the Cl is negative. *The electronic attraction between these opposite* *charges is the only thing that holds the Na^+ and Cl^- ions together.* Thus, a sample of NaCl would look something like this:

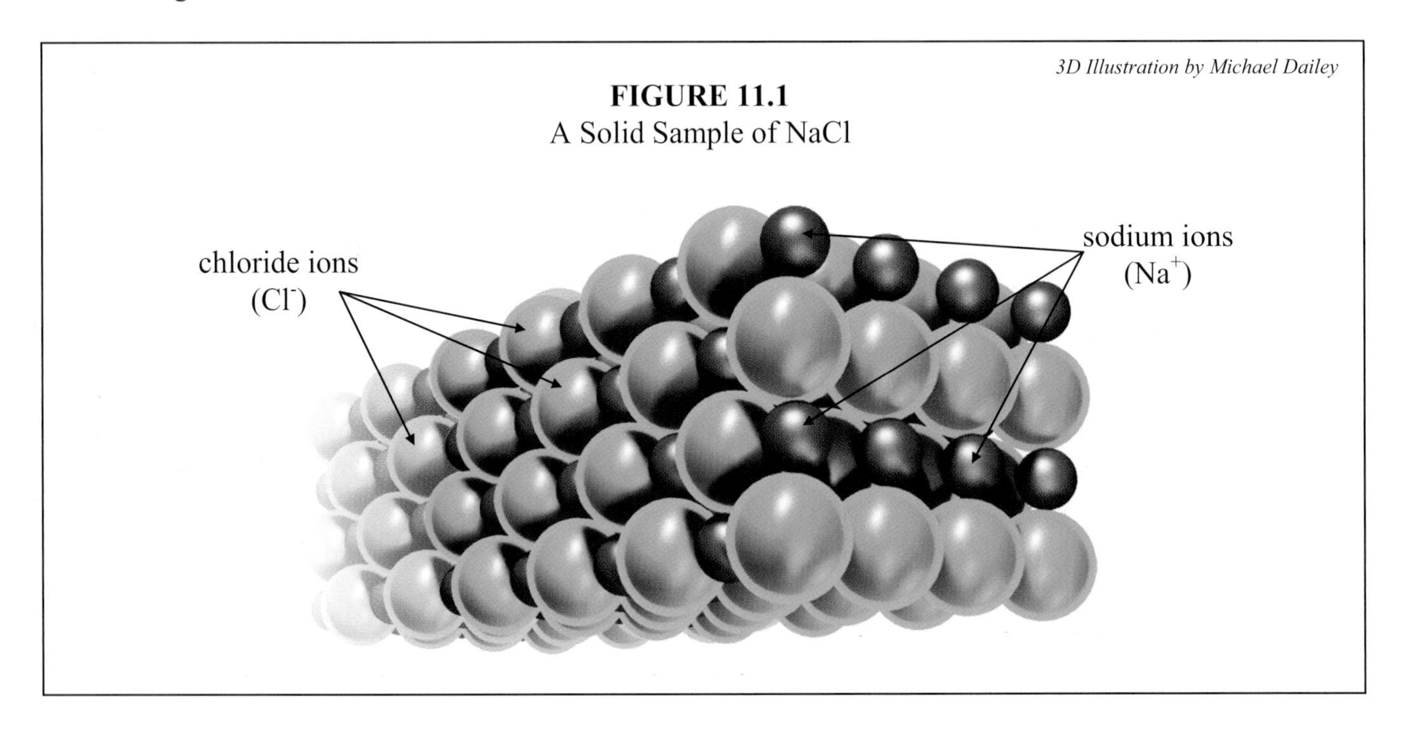

FIGURE 11.1
A Solid Sample of NaCl

Now NaCl is a solid because the Na^+ and Cl^- ions are very close to one another, due to their mutual electronic attraction. As we learned in Module #4, the only way to get NaCl out of its solid phase is to pull these ions away from each other. This is where the water comes in.

Remember, because water is a polar covalent compound, it has fractional positive charges on its hydrogen atoms and a fractional negative charge on its oxygen atom:

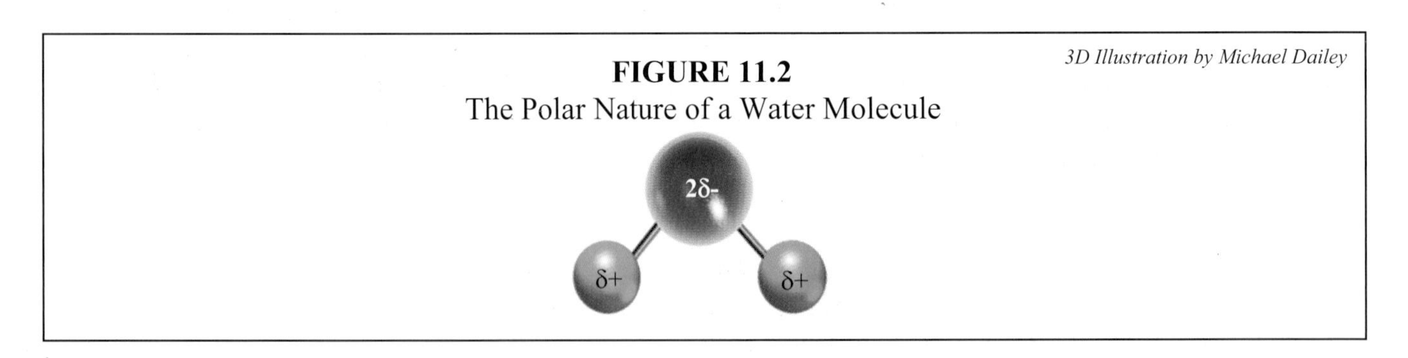

FIGURE 11.2
The Polar Nature of a Water Molecule

When water and NaCl are mixed, the fractional positive charges on the water's hydrogen atoms are attracted to the negative charges on the chloride ions. In the same way, the fractional negative charge on the water's oxygen atom is attracted to the positive charges on the sodium ions.

In order to get these charges closer to one another, the water molecules muscle in between the ions in NaCl. Because the water molecules get between the ions, the ions are not as strongly attracted to each other as they were before. This allows the water molecules to surround the ions and pull them away from each other. Figure 11.3 illustrates this.

3D Illustration by Michael Dailey

FIGURE 11.3
Water Dissolving Sodium Chloride

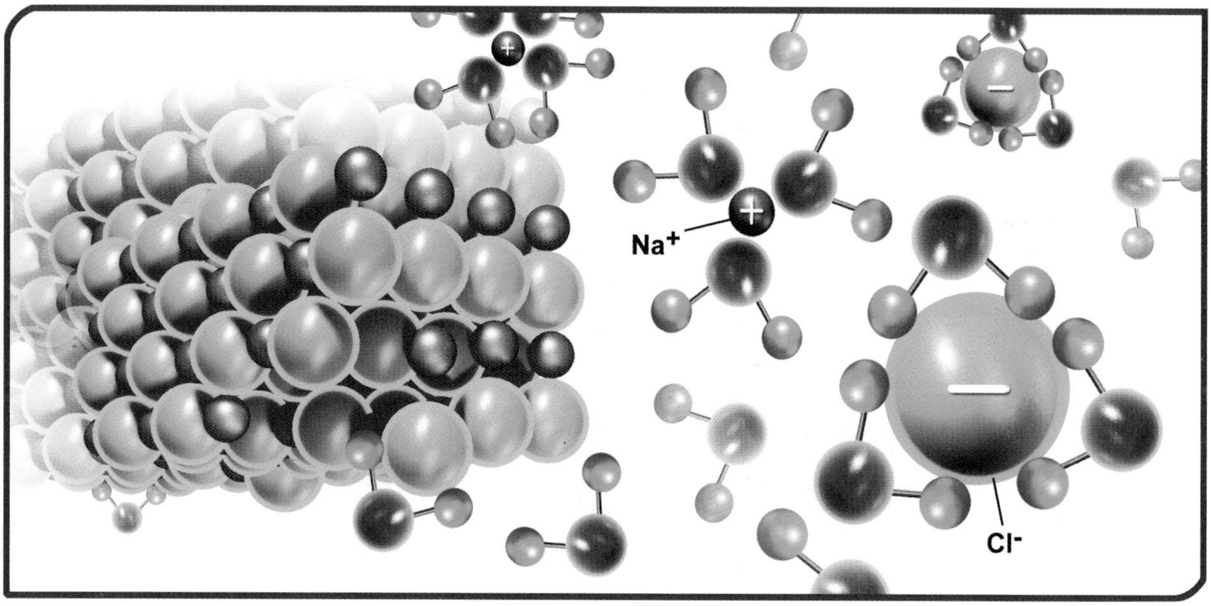

Na^+

Cl^-

Once the ions get far apart from each other, they no longer form a solid; instead, they are dissolved, and we say that they are in the aqueous state.

Notice *how* the water molecules surround the ions on the right side of the figure. The sodium ions (red spheres) are positive. The water molecules surround the sodium ions so that their oxygen atoms point towards the sodium ions. This, of course, should make sense. After all, the oxygen atom in a water molecule carries a partial negative charge. Thus, the partial negative charge on the oxygen atom is attracted to positive charge on the sodium ion, and the water molecules orient themselves so that their oxygen atoms are as close as possible to the sodium ions. In the same way, the water molecules surround the chloride ions (blue-green spheres) so that their hydrogen atoms are as close as possible to the chloride ions. This is because the partial positive charges on the hydrogen atoms are attracted to the negative charge on the chloride ion.

This, then, is how ionic solids dissolve in water. Because of the attraction that exists between the water molecules and the ions in the ionic solid, the water molecules get between the ions, pushing them away from each other. Soon the ions are so far away from each other that they can move freely. This makes the ionic solid change its phase to aqueous.

Does this mean that all ionic solids dissolve in water? No, it does not. Remember, in order for the solid to dissolve in water, the water molecules must first get between the ions. In some ionic solids, the attraction between the ions is so great that the water molecules simply cannot force their way in between the ions. As a result, we say that the compound is **insoluble** in water.

If you purchased the MicroChem kit discussed in the introduction, you can perform Experiment #5 to get some experience making water-insoluble ionic compounds.

Now that you see how ionic solids dissolve in water, you need to see how covalent solids do it. The first thing to remember is that purely covalent solids *do not* dissolve well in water. Remember, since purely covalent solids have no electrical charges in them, they cannot interact easily with water. Thus, they do not dissolve well in water. Polar covalent solids, on the other hand, can dissolve in water, because they possess fractional charges. However, these charges cannot separate, because they reside on atoms that are bonded to each other. As a result, even though polar covalent molecules can dissolve in water, they cannot split up into smaller parts. Thus, the molecules dissolve as individuals.

The way that this happens is much the same as what happens with ionic solids. Consider, for example, how solid PH_3 dissolves in water. As we learned in Module #9, PH_3 is a polar covalent solid. Thus, the atoms within the molecule possess fractional charges:

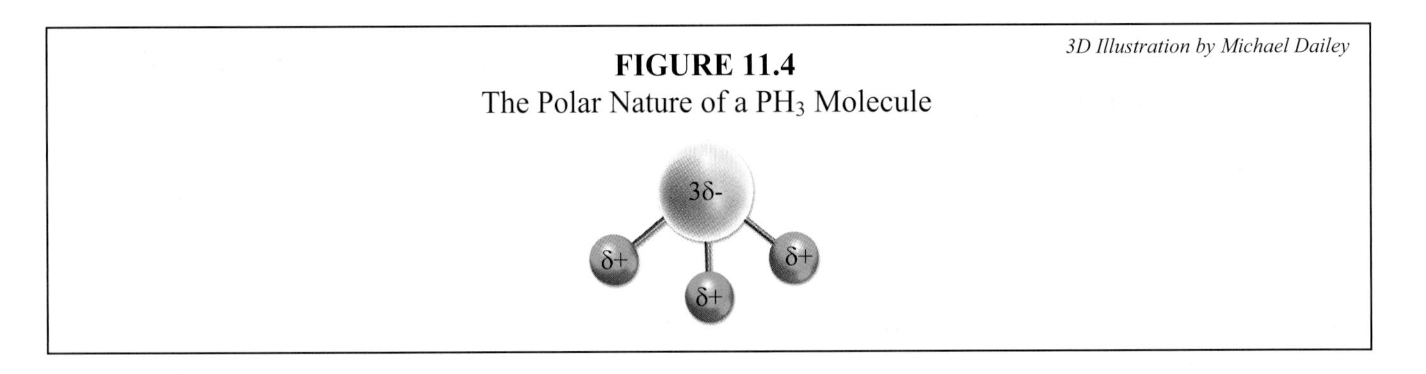

3D Illustration by Michael Dailey

FIGURE 11.4
The Polar Nature of a PH_3 Molecule

These fractional charges attract the opposite fractional charges in the water molecules. In the same way as before, then, the water molecules muscle their way in between the PH_3 molecules. They surround the individual PH_3 molecules, orienting their negatively charged oxygens towards the PH_3's positively charged hydrogens. In addition, the water's positively charged hydrogens are oriented toward the PH_3's negatively charged phosphorous atom. Once they are surrounded by the water molecules, the PH_3 molecules are pulled away from each other. When that happens, the compound changes from the solid phase into the aqueous phase:

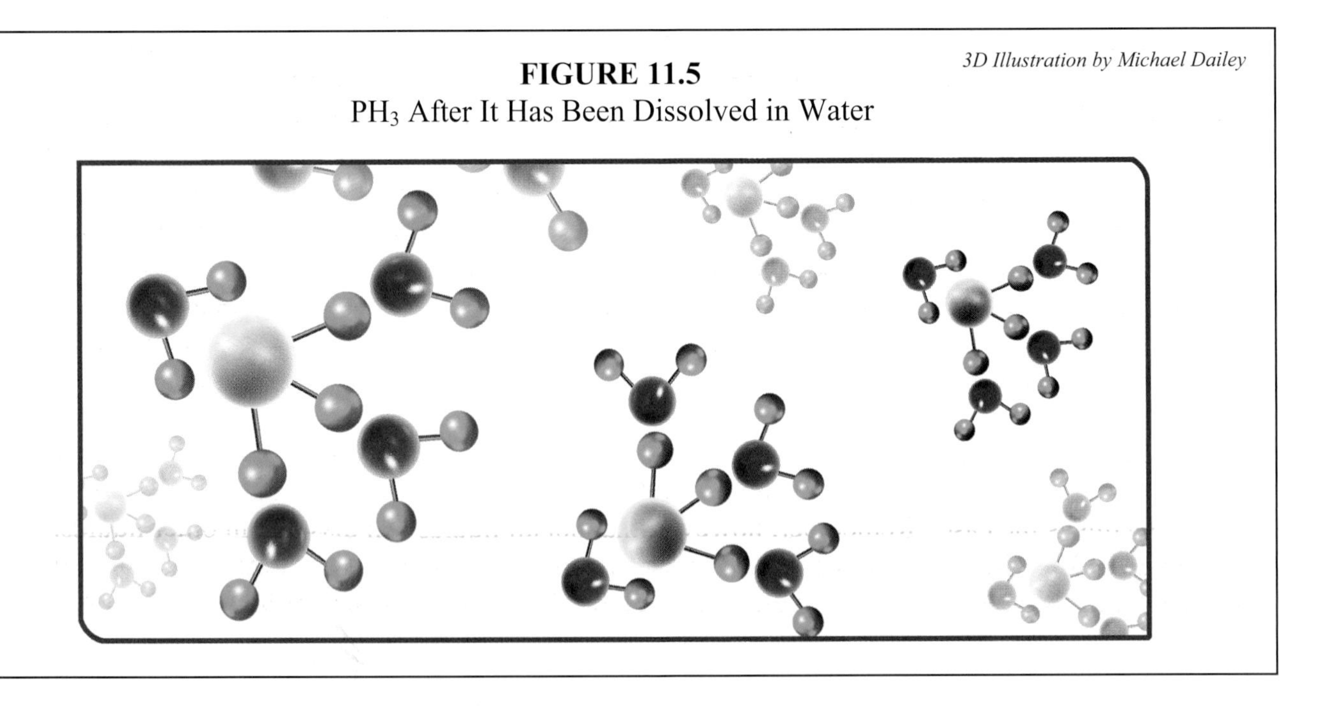

3D Illustration by Michael Dailey

FIGURE 11.5
PH_3 After It Has Been Dissolved in Water

Once again, notice *how* the water molecules surround the phosphorus trihydride molecules. The water molecules orient themselves so that their oxygen atoms point toward the hydrogen atoms in the PH_3 molecules. This is because the partial negative charge on the oxygen atom is attracted to the partial positive charges on PH_3's hydrogen atoms. In the same way, the hydrogen atoms on the water molecules are pointed toward the phosphorus atoms, because the partial positive charges on the hydrogen atoms are attracted to the partial negative charge on the phosphorus atom.

Notice the similarities and the differences between the way in which ionic solids dissolve in water and the way in which polar covalent solids dissolve in water. In both cases, water molecules are attracted to the charges that exist in the solute. Also, both types of solutes change from the solid phase to the aqueous phase because the water molecules muscle in between the solute molecules, surround them, and then pull them apart. The main difference between these two types of compounds, however, is that while ionic compounds break apart into smaller pieces (their individual ions) when dissolved, polar covalent compounds do not. Polar covalent compounds dissolve by separating each individual molecule from its neighbors.

Now this same reasoning can be used to explain how any ionic or polar covalent solid dissolves in any polar solvent. What about purely covalent compounds, however? How do purely covalent solids dissolve in other purely covalent compounds? It turns out that the explanation for this process is a bit too advanced for this course. Suffice it to say that purely covalent compounds have their own ways of being attracted to each other. If the solvent molecules are attracted to the solute molecules, then the same thing that we have just discussed can also happen. The solvent molecules will muscle in between the solute molecules, pulling them far apart from each other. The precise mechanisms for how this happens, however, are not important for right now.

Of course, everything we have discussed so far has been focused on solid solutes dissolving into liquid solvents. I have mentioned already, however, that solutes can also be gases or liquids. When dishwashing liquid, for example, is mixed with water, the dishwashing liquid is a solute that is dissolved in water (the solvent). The way in which liquid solutes dissolve in liquid solvents is similar to the way in which solid solutes dissolve in liquid solvents.

As is the case with solid solutes, the solvent molecules must be attracted to liquid solute molecules in order for the solute to dissolve. This allows the solvent molecules to muscle in between the solute molecules. However, in liquid solutes, the solute molecules are not very close to each other to begin with. Thus, it isn't very hard for the solvent molecules to get in between them. Remember, in solids, the molecules (or ions) are close together. As a result, solvent molecules must work harder to get between the solid solute molecules (or ions). Also, unlike what happens when solids dissolve, the molecules of a liquid solute are not drawn very far away from each other when the solvent gets between them. After all, the solute is a liquid already; thus, the molecules do not really have to be separated very much because they are already pretty far apart from each other. The molecules only have to be drawn apart far enough to allow the solvent molecules to get in between them.

Based on this information, you would expect that liquid solutes dissolve in liquid solvents more readily than solid solutes do. Of course, you probably knew this fact already, but at least now you know *why* this is the case. Remember, however, that not all liquids can dissolve in other liquids. As mentioned before, the solvent molecules must be attracted to the solute molecules before the solute can be dissolved. As a result, polar covalent liquid solutes will only dissolve in polar covalent liquid solvents, and purely covalent liquid solutes will only dissolve in purely covalent liquid solvents. This is why oil and water don't mix. Oil is purely covalent, while water is polar covalent.

What about gaseous solutes? How do they dissolve in liquid solvents? If you think about it, there is a fundamental difference between solid or liquid solutes and gaseous solutes. In the case of solid solutes, the solute molecules (or ions) must be pulled away from each other to convert them to the liquid state. In the case of liquid solutes, the molecules must also be pulled away from each other, but not by very much. In the case of gaseous solutes, however, the solute molecules are not pulled apart. In fact, in order to dissolve into a liquid solvent, the gaseous solute molecules must actually *be brought closer together.*

Once again, this fact should make sense if you think about it. The main difference between the phases of matter is the distance between the molecules. Molecules in the solid phase tend to be closer together; molecules in the liquid phase tend to be farther apart; and molecules in the gaseous phase tend to be very far apart. Thus, to dissolve a gas into a liquid, the gaseous molecules must be brought closer together. You can think of this as precisely the opposite of dissolving a solid solute. When dissolving a gaseous solute, the solvent molecules do not muscle in between the solute molecules. Instead, they attract the solute molecules to them, like a magnet attracts small bits of iron. As a result, the gaseous molecules come closer together.

Table 11.1 summarizes the similarities and differences in the processes that dissolve solid, liquid, and gaseous solutes. It is important to understand and remember these facts.

TABLE 11.1

A Summary of How Solutes Dissolve in Liquid Solvents

SOLUTE PHASE	REQUIREMENTS FOR DISSOLVING
SOLID	The solvent molecules must be attracted to the solute molecules so strongly that the solvent molecules can get between the solute molecules (or ions) and pull them far apart from each other.
LIQUID	The solvent molecules need only be attracted to the solute molecules a little, because the solvent does not need to separate the solute molecules very much. The solvent merely needs to get between the solute molecules.
GAS	The solvent molecules must be attracted to the solute molecules enough to pull the solute molecules closer to one another.

Solubility

Now that we see *how* solutes dissolve in solvents, the next question you might ask is *how much* solute can I dissolve in a solvent? The answer to that is a bit involved. First, let's get the vocabulary down. The term **solubility** refers to how much solute dissolves in a solvent:

Solubility - The maximum amount of solute that can dissolve in a given amount of solvent

The solubility of a solute is generally reported in terms of grams of solute that will dissolve in 100.0 g of solvent (grams per hundred). In other words, if a solute has a solubility of 10.0 grams per hundred, it means that you can dissolve as much as 10.0 grams of solute for every 100.0 grams of solvent that you use.

If you think about it, the solubility of each solute must be unique to that solute. For example, it is quite easy to dissolve table salt (NaCl) in water. At room temperature, NaCl has a solubility of 37 grams per hundred. Thus, you can dissolve 37 grams of NaCl in 100 grams of water. Under the same conditions, however, baking soda ($NaHCO_3$) has a solubility of just 12 grams per hundred. Thus, baking soda does not dissolve in water nearly as well as table salt does. In addition, the solubility of each of these solutes changes if we change the solvent. The solubility of both table salt and baking soda decreases if you try to dissolve them in alcohol rather than water, for example.

These facts should make sense to you. After all, the solubility of each solute depends fundamentally on the fact that the solvent molecules are attracted to the solute molecules. Thus, each solute molecule will have its own, unique solubility in each solvent, because the amount of attraction between solute and solvent changes depending on the identity of each. We come, then, to our first rule concerning solubility:

The solubility of any solute depends both on the identity of the solute *and* the identity of the solvent.

We will develop two more rules concerning solubility, and it will be very important for you to keep them in mind as you go through the rest of this module.

Before we learn more rules concerning solubility, however, let's learn some more terminology. As I said before, at room temperature, you can dissolve 37 grams of table salt into 100 grams of water. However, if you added any more than 37 grams of table salt, the excess would not dissolve. Thus, if I add 50.0 grams of table salt to 100.0 grams of water, 37 grams of it would dissolve, and the other 13 grams would just fall to the bottom of the container, unable to dissolve. If you were then to filter away all of that excess salt, you would have a solution that contains as much salt as can possibly be dissolved in that amount of water. We would call this a **saturated solution**:

Saturated solution - A solution in which the maximum amount of solute has been dissolved

You will make and use a saturated solution in the next experiment.

Now let's go back to developing our two additional rules about solubility. The solubility of solutes not only depends on the identity of the solute and solvent, but it also depends on certain conditions under which the solution is made. Perform the following experiment in order to see how the solubility of a solid solute depends on the temperature at which the solution is made.

EXPERIMENT 11.1

The Effect of Temperature on the Solubility of Solid Solutes

Supplies:

- 250 mL beaker (A short, fat glass or canning jar might work, but **be careful**. Glasses tend to crack when subjected to the temperature extremes of this experiment. If you don't have a beaker, you may want to just read this experiment, unless you aren't afraid of losing a glass or two.)
- 100 mL beaker (Another short, fat glass or canning jar might work.)

- Graduated cylinder (A ¼ measuring cup will work.)
- Flame heater (A stove will work.)
- Stirring rod (A spoon will work.)
- Thermometer
- Mass scale
- Salt
- Filter paper (You can cut circles out of the bottom of a coffee filter to make your own filter paper.)
- Funnel
- Oven mitt
- Safety goggles

1. Take your mass scale and measure out 50.0 grams of table salt.
2. Put the salt into the 250 mL beaker.
3. Add 50.0 mL (¼ cup) of water to the salt and stir it vigorously with your stirring rod.
4. Allow the solution to stand while you fold the filter paper and put it in the funnel the way you learned in Experiment 4.1.
5. Heat the solution with your flame heater. As it is heating, stir the solution carefully with your thermometer. You will want to wear the oven mitt on the hand that is holding the thermometer!
6. When the temperature of the solution reaches 95 ^{o}C, take the thermometer out of the solution and use your hand that has the oven mitt on it to pick up the beaker.
7. With your other hand, hold the funnel with the filter paper over the 100 mL beaker.
8. Carefully pour the solution onto the filter paper in the funnel. Make sure the level of the solution in the funnel never rises above the top of the filter paper. Continue to filter the solution in this way until you have about 20 mL of clear solution in the beaker below the funnel.
9. Look at your solution in the 100 mL beaker. It should be clear because all undissolved salt should have been filtered out by the filter paper. This is a saturated solution of table salt.
10. Place that solution in your freezer for about 20 minutes. After that, remove the solution from the freezer and examine it again. Now there should be crystals of salt lying on the bottom of the beaker.
11. Clean up your mess.

What happened in the experiment? Well, after filtering the solution, you had a saturated solution of table salt, but the solution was at a temperature near 95 ^{o}C. When you placed the solution in the freezer, it cooled to a temperature of less than 0 ^{o}C. Obviously, the table salt is less soluble in cold water than in hot water, because some of the table salt had to leave the solution and fall to the bottom of the beaker. The process by which a solid solute leaves a solution and turns back into its solid phase is called **precipitation**.

From the experiment, then, we see that the solubility of a solute also depends on temperature. For solid solutes, the solubility usually increases with increasing temperature, as we saw in the experiment. This makes sense, because in order for a solid solute to dissolve, the solvent molecules must be able to muscle in between the solute molecules (or ions) and drive them away from each other. In Module #2, we learned that heating up a compound involves adding energy to it. We can conclude, then, that the molecules in hot solvents have more energy than the molecules in cold solvents. With more energy, the solvent molecules can be more successful at muscling in between the solid molecules (or ions), making the solute dissolve more readily.

What about liquids and gases? How does the temperature of the solvent affect their solubility? In the case of liquids, the temperature of the solvent has little effect on the solubility of the solute. After all, the solvent really doesn't have to work very hard in order to dissolve a liquid solute. Thus, the amount of energy that the solvent molecules have is of little consequence to the process of dissolving a liquid solute into a liquid solvent. In the case of gaseous solutes, however, the solubility is dramatically affected by the solvent temperature, as is illustrated in the following experiment.

EXPERIMENT 11.2

The Effect of Temperature on the Solubility of a Gas

Supplies:

- Two beakers (Two saucepans will work.)
- A test tube (A tall, thin glass will work, but it must fit easily in the saucepans.)
- Flame heater (A stove will work.)
- Ice
- Cold soda pop (Pepsi®, Coke®, Sprite®, etc. It must be carbonated.)

1. Fill one beaker ¾ of the way full with hot water.
2. Put it on the flame until the water begins to boil.
3. While you are waiting for the water to boil, fill the other beaker ½ full with ice, then add enough water so that the beaker is ¾ full with the ice water mixture.
4. Once the water starts boiling, extinguish the flame so that the water is hot, but not boiling.
5. Fill the test tube ¾ full with the soda pop. Set the test tube in the beaker that contains the hot water. Observe what happens. In a few moments, the soda pop should begin bubbling.
6. Watch the soda pop bubble for a moment, then pull it out of the hot water.
7. Place it in the beaker that contains ice water. Observe what happens. In a few moments, the soda pop should stop bubbling again.
8. After a moment, return the test tube to the hot water. The test tube should start bubbling again.
9. Repeat this process a couple of times, noticing that each time the soda pop bubbles in the hot water but stops bubbling when it reaches the ice water.
10. Clean up your mess.

Why did the soda pop behave in this way? Was the soda pop boiling? No. The boiling point of soda pop is higher than that of water. If the water in the beaker was not boiling, there is no way that the soda pop was boiling. Also, your experience with boiling water should tell you that it takes quite a while to get cold water to boil. In this experiment, however, the soda pop began bubbling just moments after it hit the hot water.

If the soda pop wasn't boiling, what happened in this experiment? Well, as mentioned before, soda pop has carbon dioxide dissolved in it. Carbon dioxide is a gas; thus, soda pop is a solution made by dissolving carbon dioxide (a gaseous solute) into the pop (a liquid solvent). When you put the test tube into the beaker that contained the hot water, you were, in effect, increasing the temperature of the solvent. The bubbling that you saw was the gaseous carbon dioxide leaving the solution. Thus, when the pop got warmer, the carbon dioxide became less soluble and had to leave the solution. On the other hand, when you put the test tube into the beaker that contained the cold water, you lowered the temperature of the solvent. This caused the solubility of the carbon dioxide to increase, so no more gas had to escape the solution. As a result, the bubbling ceased. Unlike solid solutes, then, gaseous solutes are *less soluble* in hot solvents than in cold solvents!

Experiments 11.1 and 11.2 lead us to our next rule of solubility:

For solid solutes, solubility usually increases with increasing temperature.
The solubility of liquid solutes is not affected by temperature.
The solubility of gases decreases with increasing temperature.

Make sure you understand both the rule and the reasons we discussed that explain the rule. Notice also that the rule says the solubility of solid solutes "usually" increases with increasing temperature. It turns out that some solids do, indeed, dissolve better in cooler solvents. However, we will not dwell on those exceptions in this course.

The last solubility rule deals with the effect that pressure has on solute solubility. Think about what happens when you open a fresh can or bottle of soda pop. What do you hear? You usually hear a loud pop or fizz. If you look at the soda after it is opened, it bubbles. What does this tell you? The fact that the bottle or can fizzes when the lid is opened tells you that the soda pop is stored at a high pressure. When the lid is opened, the air is pushed out by that pressure, causing the fizzing sound.

The bubbles that occur after the lid is opened tell you that carbon dioxide is leaving the solution. Thus, the solubility of the carbon dioxide decreases as soon as the lid is opened. This tells us that the solubility of gaseous solutes decreases when the pressure at which the solution is stored is lowered. This is another rule of solubility:

Increasing pressure increases the solubility of gases.
Pressure does not affect the solubility of either liquids or solids.

The best way to understand this last rule is to remember how gases dissolve. In order for gases to dissolve into liquids, the gas molecules must get closer together. If the pressure surrounding the solution is large, then the molecules are forced a bit closer together, helping them dissolve. See if you can apply the three rules you just learned by solving the following problems:

ON YOUR OWN

11.1 A student is asked to dissolve 50.0 grams of potassium chromate (an orange-yellow solid) in 250 mL of water. He is having a hard time getting all of the solid to dissolve, but is not allowed to add any more water. What should he do to try to make sure all 50.0 grams dissolve?

11.2 A chemistry lab is storing several bottles of a solution made by dissolving methane gas in octane, a non-polar liquid. The laboratory chemist notices that every once in a while, the lids from the bottles pop off. She reasons that the lids are popping off because the gas that is dissolved in the liquid is escaping from solution. What can she do to stop the lids from popping off?

11.3 A student is doing an experiment that involves dissolving a gas in water. He notices that even though the temperature in the lab is exactly the same every day of the year, there are some days in which more gas can dissolve in water as compared to other days. What, most likely, is changing from day to day in order to produce this effect?

Energy Changes That Occur When Making a Solution

Now that you have learned about the physical changes that occur when solutes dissolve in solvents, it is time to investigate how the *energy* of the solute and solvent changes as well. Begin this investigation by performing the following experiment:

EXPERIMENT 11.3

Investigation of a Solute That Releases Heat When Dissolved

Supplies:

- Beaker (A short, fat glass will do.)
- Lye (This is commonly sold in supermarkets with the drain cleaners. A popular brand is Red Devil® Lye. If you cannot find lye, any *powdered* drain cleaner ought to work.)
- Rubber gloves
- Water
- Sink
- Tablespoon
- Safety goggles

1. Put the gloves on.
2. Measure out 3 tablespoons of lye into your beaker.
3. Put the beaker in the sink directly under the faucet.
4. Turn the water on slowly so that the water from the faucet begins to fill the beaker. Add enough water so that the water level in the beaker is about 1 centimeter above the lye.
5. Stir the solution with the tablespoon.
6. Take one glove off and **carefully** touch the outside of the beaker, near the bottom. Most likely, the beaker will feel warm. It might be quite hot, so be careful.
7. Continue to stir the solution with the other hand, and periodically touch the outside of the beaker near the bottom to see how hot it is getting. Once again, **be careful** when you touch the beaker with your bare hand, as it can get very hot!
8. Eventually, the solution might get so warm that you can no longer comfortably touch the beaker. At that point, put the glove back on.
9. Turn on the cold water, allowing it to fill and overflow the beaker. Keep the water running for several minutes, allowing a large amount of the lye to be flushed down the drain.
10. Tip the beaker over and rinse it out.
11. Clean up your mess.

What happened? The reason that you felt the beaker getting warm is that lye, when it dissolves, releases heat. This heat is transferred to the water, causing the water's temperature to increase. Eventually, the water might have become hot enough to boil.

The process you observed in the previous experiment is called an **exothermic** (ek' soh ther' mik) process. The prefix "exo" means out, and "therm" refers to heat; thus, exothermic refers to a process that lets out heat. The opposite of this kind of process is one that absorbs heat and is called **endothermic** (en' doh ther' mik):

<u>Exothermic process</u> - A process that releases heat

<u>Endothermic process</u> - A process that absorbs heat

These two terms will appear several times throughout this course, so you need to learn them.

Now that we know the terminology, we can discuss what happened in the experiment. Strong bases such as NaOH (the principal component of lye) dissolve exothermically. This means that they release heat. This released heat raises the temperature of the surroundings. That's why the beaker got warm. Thus, exothermic processes tend to heat up their surroundings. Endothermic processes, on the other hand, absorb heat. The only place that this heat can come from is the surroundings. Thus, an endothermic process takes energy away from its surroundings, which reduces the temperature. Endothermic processes, then, cool down their surroundings.

This relates to our discussion of solubility because most solids dissolve in an endothermic fashion. In other words, most solids must absorb energy in order to dissolve properly. Thus, when you dissolve the solid in water, the temperature of the solution actually decreases. Often, the decrease is so small that you simply do not notice it. For example, sodium chloride (table salt) dissolves in water endothermically. Thus, the temperature of the solution actually decreases once the table salt dissolves. However, the change is so small that you do not really notice the effect. However, if you were to dissolve ammonium nitrate in water, you would definitely notice the temperature change, as ammonium nitrate dissolves *very* endothermically in water.

The fact that some solids (such as ammonium nitrate) dissolve endothermically has a very practical application, illustrated in Figure 11.6.

FIGURE 11.6
Instant Cold and Hot Packs

If you have ever watched a basketball game, you might have noticed that when a player twists or sprains an ankle, the team doctor might pull out a plastic bag such as the one pictured on the left side of the figure, squeeze it, shake it, and put it on the player's ankle. This very handy bag, available at drug stores and sporting goods stores, is a temporary "ice pack."

The bag contains water and crystals of ammonium nitrate (NH_4NO_3). The water is contained in a fragile inner bag. When the outer bag is squeezed hard, the inner bag breaks, mixing the water and ammonium nitrate. As the bag is shaken, the ammonium nitrate begins to dissolve in the water. Since ammonium nitrate dissolves endothermically in water, the bag gets cold. It gets so cold that it is, in fact, a good substitute for ice, without all of the mess.

A similar product (shown on the right side of Figure 11.6) exists to provide a temporary source of heat. You can find it in most stores that sell hunting and fishing equipment. This product contains iron, water, salt, and a few other chemicals. When you open the outer plastic bag, the contents are exposed to air, and the iron begins to react with the oxygen in the air. This chemical reaction is exothermic, providing a great deal of heat. The packet pictured in the figure can stay warm up to 12 hours and is often used by hunters and fishermen to keep their hands warm on cold days.

Applying Stoichiometry to Solutions

In the last module, we learned about molarity and its applications to acid/base chemistry. In that module, I mentioned that molarity was an important concept in *any* chemistry that involved solutions. I want to emphasize this by showing you a typical example of how we would apply stoichiometry to problems involving solutions. I will then ask you to solve a couple of "On Your Own" problems as a means of making sure you understand the application.

EXAMPLE 11.1

A chemist wishes to make calcium carbonate (the principal component in Tums® antacid as well as chalk) by using the following reaction:

$$Na_2CO_3\ (aq) + Ca(OH)_2\ (aq) \rightarrow CaCO_3\ (s) + 2NaOH\ (aq)$$

If the chemist mixes 50.0 mL of 1.23 M Na_2CO_3 with an excess of calcium hydroxide, how many grams of calcium carbonate will be produced?

This is just a stoichiometry problem. We can tell this by the fact that we are being asked to determine the amount of one substance when we are given the amount of another substance. The only way to do that is by stoichiometry. Now, in order to do stoichiometry, we must first get our amount in moles. Right now, the amount is given in concentration and volume. We must turn that into moles. In order to do that, though, we must convert mL into liters, so that our volume units are consistent with our concentration unit (remember, "M" means moles/liter):

$$\frac{50.0\ \cancel{\text{mL}}}{1} \times \frac{0.001\ \text{L}}{1\ \cancel{\text{mL}}} = 0.0500\ \text{L}$$

Now we can convert from volume and concentration into moles:

$$\frac{1.23 \text{ moles } Na_2CO_3}{1 \cancel{L}} \times 0.0500 \cancel{L} = 0.0615 \text{ moles } Na_2CO_3$$

Now that we have moles, we can do stoichiometry:

$$\frac{0.0615 \cancel{\text{moles } Na_2CO_3}}{1} \times \frac{1 \text{ mole } CaCO_3}{1 \cancel{\text{mole } Na_2CO_3}} = 0.0615 \text{ moles } CaCO_3$$

This is not quite the answer we need. We were asked to figure out how many *grams* of calcium carbonate were produced, so we have to convert from moles back to grams:

$$\frac{0.0615 \cancel{\text{moles } CaCO_3}}{1} \times \frac{100.1 \text{ grams } CaCO_3}{1 \cancel{\text{mole } CaCO_3}} = 6.16 \text{ grams } CaCO_3$$

This means that 6.16 g of calcium carbonate will be produced.

How many liters of a 0.245 M solution of NaCl will be necessary to make 100.0 grams of $AlCl_3$ according to the following reaction?

$$\mathbf{3NaCl\ (aq) + Al(NO_3)_3 \rightarrow AlCl_3\ (s) + 3NaNO_3\ (aq)}$$

Once again, we can tell that this is a stoichiometry problem because we are asked to convert from the amount of one substance to the amount of another. In order to do stoichiometry, however, we must convert to moles. Since we have the number of grams of $AlCl_3$, we will use that as our starting point:

$$\frac{100.0 \cancel{\text{g } AlCl_3}}{1} \times \frac{1 \text{ mole } AlCl_3}{133.5 \cancel{\text{g } AlCl_3}} = 0.7491 \text{ moles } AlCl_3$$

Now that we have moles, we can use stoichiometry to determine how much NaCl is needed:

$$\frac{0.7491 \cancel{\text{moles } AlCl_3}}{1} \times \frac{3 \text{ moles NaCl}}{1 \cancel{\text{mole } AlCl_3}} = 2.247 \text{ moles NaCl}$$

This tells us how much NaCl we need, but it doesn't answer the question. The question asks how many *liters* of a 0.245 M solution are needed. To determine this, we must remember that "M" means moles per liter. Thus, the concentration of a solution is a conversion relationship that allows us to relate the number of moles to the number of liters. We can therefore do the following conversion:

$$\frac{2.247 \cancel{\text{moles NaCl}}}{1} \times \frac{1 \text{ L NaCl}}{0.245 \cancel{\text{moles NaCl}}} = 9.17 \text{ L}$$

Thus, 9.17 liters of the NaCl solution will be needed.

Now make sure you can do stoichiometry with solutions by performing the following problems:

ON YOUR OWN

11.4 The following reaction is carried out in the laboratory:

$$5H_2O_2\ (aq)\ +\ 2KMnO_4\ (aq)\ +\ 3H_2SO_4\ (aq)\ \rightarrow\ 5O_2\ (g)\ +\ 2MnSO_4\ (aq)\ +\ K_2SO_4\ (aq)\ +\ 8H_2O\ (l)$$

If 125 mL of 0.5 M H_2O_2 is reacted with an excess of the other reactants, how many grams of oxygen will be produced?

11.5 One way to purify a mixture of two metals is to add to the mixture something that reacts with only one of the metals and turns it into an aqueous compound. The mixture can then be filtered, isolating the other metal. A mixture of magnesium and aluminum was treated in such a way in order to isolate the magnesium. Sodium hydroxide, which reacts only with aluminum, was added:

$$2Al\ (s)\ +\ 2NaOH\ (aq)\ +\ 6H_2O\ (l)\ \rightarrow\ 2NaAl(OH)_4\ (aq)\ +\ 3H_2\ (g)$$

The aluminum was converted to an aqueous salt, and the magnesium stayed as a solid. The solution was then filtered, isolating the pure magnesium. If a mixture of these two metals contained 1.2 kg of Al, how many liters of a 6.5 M NaOH solution would be needed to make sure that all aluminum was taken out of the mixture?

Molality

When I first introduced the concept of concentration, I told you that there are many units we can use to express it. The units we choose depend on the situation we are in at the time. In a moment, you will learn two final concepts concerning solutions, and both of those concepts require a new unit of concentration. That unit is called **molality** (moh lal' ih tee).

Molality (m) - The number of moles of solute per kilogram of solvent

$$m = \frac{\#\ \text{moles solute}}{\#\ \text{kg solvent}} \tag{11.1}$$

Now do not let yourself get confused between molarity and molality. The words look very similar, but their meanings are quite different.

Remember first what molarity means. It means the number of moles of solute per *liter* of *solution*. If I take 145 mL of water and add 25 g of NaCl to it, the NaCl that I add to the water will increase the total volume of the solution. Let's suppose that adding 25 g of NaCl to 145 mL of water increases the volume to 150 mL. To get molarity, I would take the number of moles of NaCl that I added and divide it by the *total* number of liters of the solution: 0.15 liters. Thus, molarity asks you to divide the number of moles of solute by the *total volume of solution.*

Compare that to molality. In determining the molality of a solution, we instead take the number of moles of solute and divide by the number of *kilograms* of *solvent*. First, then, we are not concerned about volume when calculating molality. Instead, we are worried about mass. Second, we do not divide by the total mass of the solution. Instead, we divide only by the mass of the *solvent*. Thus, molality gives us a ratio between moles of solute and mass of solvent, whereas molarity compares the moles of solute to the volume of the entire solution. This may sound like a subtle distinction, but it is very important.

In the next section, you will learn why molality is a useful concentration unit. For right now, however, it is important to make sure that you understand how to calculate the molality of a solution. Study Example 11.2 and perform the "On Your Own" problem that follows to make sure you have a firm grasp on the concept of molality.

EXAMPLE 11.2

A chemist mixes 3.4 moles of NaCl with 4.5 kg of water. What is the molality of the resulting solution?

According to Equation (11.1), we can calculate the molality of a solution by dividing the number of moles of solute by the number of kilograms of solvent. We have both of those numbers, so we just apply Equation (11.1):

$$m = \frac{\text{\# moles solute}}{\text{\# kg solvent}} = \frac{3.4 \text{ moles NaCl}}{4.5 \text{ kg water}} = 0.76 \frac{\text{moles NaCl}}{\text{kg water}} = 0.76 \text{ m}$$

We would say, then, that the solution has a molality of 0.76 m.

If a solution is made by adding 45.0 g of $ZnCl_2$ to 150.0 g of water, what is the molality?

To use Equation (11.1), we need to have the number of moles of solute and the number of kg of solvent. In this problem, we are given neither. Thus, we have to convert to the right units. First, we will convert grams of solute to moles of solute:

$$\frac{45.0 \text{ } \cancel{\text{g ZnCl}_2}}{1} \times \frac{1 \text{ mole ZnCl}_2}{136.4 \text{ } \cancel{\text{g ZnCl}_2}} = 0.330 \text{ moles ZnCl}_2$$

Next we need to convert grams of solvent into kg of solvent:

$$\frac{150.0 \text{ } \cancel{\text{g}}}{1} \times \frac{1 \text{ kg}}{1{,}000 \text{ } \cancel{\text{g}}} = 0.1500 \text{ kg}$$

Now that we have moles of solute and kg of solvent, we simply need to apply Equation (11.1):

$$m = \frac{\text{\# moles solute}}{\text{\# kg solvent}} = \frac{0.330 \text{ moles ZnCl}_2}{0.1500 \text{ kg water}} = 2.20 \frac{\text{moles ZnCl}_2}{\text{kg water}} = 2.20 \text{ m}$$

ON YOUR OWN

11.6 Calculate the molality of the following solutions:

a. 2.3 moles of NaOH dissolved in 1.2 kg of water
b. 250.0 g KNO_3 dissolved in 2.43 kg of water
c. 34.5 g NaBr dissolved in 578 g of water

Now remember, at this point your ability to calculate molality doesn't mean very much because you have no clear indication of how molality can be used. That's okay. In the next two sections, you will see that molality is a very useful concentration unit. At that point, you will see that the examples and "On Your Own" problems which you just finished were not a series of useless exercises.

Freezing-Point Depression

Have you ever noticed that during snowy and icy conditions, street crews tend to spread salt on the roads? Perhaps you or you parents have done the same thing after you have shoveled snow off your sidewalk or driveway. Why do people do this? The answer, you think, is quite obvious. We spread salt on the roads, driveways, and sidewalks to melt the ice and snow that make them slippery. This is very true, but have you ever stopped to think about *why* salt melts the ice and snow? That's what you're going to find out by performing the following experiment:

EXPERIMENT 11.4
Freezing-Point Depression

Supplies:
- Two beakers (Two short, fat glasses will do.)
- Graduated cylinder (A ¼ measuring cup will do.)
- Table salt
- A ¼ measuring teaspoon
- Thermometer
- Safety goggles

1. Fill each of your beakers with 50 mL (¼ cup) of tap water (a qualitative measurement).
2. In one of the beakers, dissolve ¼ teaspoon of salt. Mark that beaker in some way so that you know it is the one that contains the salt-water solution.
3. Place both beakers in the freezer. Check on the beakers every few minutes to see when they begin to freeze.
4. When the salt water is half frozen, remove both beakers. If you didn't keep a careful eye on the beakers and they are both frozen solid, don't worry. Just set them out until they are about half melted. At this point, you should have a mixture of ice and liquid in both beakers.
5. Take your thermometer and carefully stir the mixture of ice and salt water. Stir for about three minutes and then read the temperature of the thermometer. It should be less than 0 °C.
6. Rinse the thermometer and stir the mixture of ice and regular tap water for the same amount of time. Read the temperature. It should be closer to 0° C.

Now remember, water freezes at 0.00 °C. If thoroughly mixed, an ice-water solution will always be at 0.00 °C, the freezing point of water. Since the temperature of the ice and salt-water mixture was *lower* than 0.00 °C, you can conclude that the *freezing point* of salt water is lower than 0.00 °C. Thus, the freezing point of a salt-water solution is lower than the freezing point of water. This turns out to be a general rule. When a solute is dissolved in a solvent, the solution always has a *lower* freezing point than the pure solvent. In this experiment, your solvent was water. Pure water has a freezing point of 0.00 °C. However, when you added salt to the water in the second beaker, you were adding a solute to the solvent. The resulting solution had a lower freezing point. This phenomenon is known as **freezing-point depression** and is a very useful concept in the chemistry of solutions.

Now, hopefully, you see why salt is spread on icy and snowy roads, driveways, and sidewalks. The salt becomes a solute, mixing with the water. This lowers the freezing point of the water that is on the roads. Thus, the weather has to get *much* colder in order to freeze the water. This makes icy conditions much less likely.

Although I told you that solutes tend to reduce the freezing points of solvents, I still haven't told you *why* this happens. That explanation requires some arguments based on energy. First, we must remember why a liquid freezes. In its liquid state, the molecules of a substance have a significant amount of energy. They are free to move around, so they are constantly traveling around their container. To get into the solid state, the molecules have to slow down and stop moving around so that they only vibrate back and forth. The only way that this can happen is to take energy away from the molecules. That's what you are doing when you cool a substance down.

As a substance cools, its molecules lose energy. As they lose energy, they start to slow down. Eventually, they lose enough energy so that they almost stop moving around altogether. At that point, the liquid begins to turn into a solid. Well, when a solute is dissolved in a solvent, the solute molecules (or ions) also begin moving around in the solution. Since the solvent molecules are attracted to the solute molecules, a chase ensues. The mutual attraction between the solvent and solute adds more energy to the chase. Thus, in order to get all of the molecules to slow down and stop moving around, *more energy must be removed.* The only way that this can happen is by lowering the temperature further.

In other words, the mutual attraction between solute and solvent tends to add energy to the molecules in the solution. Since freezing involves pulling energy out of the molecules in a liquid, the freezing temperature of a solution will have to be less than that of the solvent. This is because more energy must be pulled out of the molecules in a solution in order to freeze the solution as compared to the energy that must be pulled out of a pure solvent in order to freeze the solvent.

Hopefully, you now not only have learned the concept of freezing-point depression, but you also have a reasonable understanding of *why* it happens. In fact, this is still not quite enough to master the topic. It turns out that we can use mathematics to actually predict the amount by which the freezing point of a solvent is lowered when a solute is added. This is where the concept of molality plays a role.

When a solute is dissolved in a solvent, the resulting change in the freezing temperature is easily determined using the following equation:

$$\Delta T = -i \cdot K_f \cdot m \qquad (11.2)$$

In this equation, the "Δ" symbol is the capital Greek letter "delta." In science, the delta symbol means "change in." Thus, "ΔT" means "change in temperature." The symbol "i" refers to the number of molecules (or ions) that the solute splits into when it dissolves, while the "m" refers to the molality of the solution. The symbol "K_f" is called the **freezing-point depression constant**; it is a physical constant that depends solely on the chemical nature of the *solvent*. Like many physical constants, each solvent has its own, unique value of K_f. Water, for example, has a K_f of 1.86 °C/m while acetic acid (the active ingredient in vinegar) has a K_f of 3.90 °C/m.

Why the strange units on K_f? Look at Equation (11.2). What units should ΔT end up in? Since ΔT is a temperature measurement, its units should be °C. The symbol "i" has no units, because it is just a number, and the symbol "m" refers to molality. Thus, the K_f must cancel the molality unit and replace it with a temperature unit. That's why K_f has such a strange unit attached to it.

If you're not sure exactly what everything in Equation (11.2) means, don't worry. A couple of examples should clear everything up:

EXAMPLE 11.3

When 15.0 grams of NaCl are added to 100.0 grams of water, what is the change in the water's freezing point? (K_f of water = 1.86 °C/m)

In a freezing-point depression problem, you must use Equation (11.2). However, in order to use that equation, we must know K_f, i, and m. Right now, we only know K_f. However, we have been given enough information to determine both "i" and "m." You just got through calculating molalities in the previous section, so that shouldn't be too hard:

$$\frac{15.0 \ \cancel{\text{g NaCl}}}{1} \times \frac{1 \text{ mole NaCl}}{58.5 \ \cancel{\text{g NaCl}}} = 0.256 \text{ moles NaCl}$$

$$\frac{100.0 \ \cancel{\text{g}}}{1} \times \frac{1 \text{ kg}}{1{,}000 \ \cancel{\text{g}}} = 0.1000 \text{ kg}$$

$$m = \frac{0.256 \text{ moles NaCl}}{0.1000 \text{ kg water}} = 2.56 \text{ m}$$

To figure out "i", we just have to think about how NaCl dissolves. The symbol "i" is defined as the number of molecules or ions that the solute splits into when it dissolves. NaCl, being an ionic compound, splits up into its ions when dissolved. Each molecule of NaCl has one sodium ion and one chloride ion, so i = 2.

Now that we have all of the components of Equation (11.2), we can use it:

$$\Delta T = -i \cdot K_f \cdot m = -2 \cdot 1.86 \frac{^{\circ}C}{\cancel{m}} \cdot 2.56 \cancel{m} = -9.52\ ^{\circ}C$$

First, let's make sure that we understand what our answer means. The negative sign means that the freezing point lowered. Thus, this answer means that the freezing point of the NaCl solution is <u>9.52 °C lower than the normal freezing point of water</u>.

Notice that I kept three significant figures in my answer, despite the fact that the symbol "i" has only one significant figure. This is because the value for "i" is exact. There are exactly 2 (not 1.999 or 2.001) ions when NaCl dissolves. As a result, this number has no effect on significant figures. Thus, when you are solving these kinds of problems, ignore the value of "i" when calculating significant figures.

What is the freezing point of a solution that is made by mixing 2.5 g $Al(NO_3)_3$ in 150.0 g of water? (K_f of water = 1.86 °C/m)

In a freezing-point depression problem, you must use Equation (11.2). In order to use that equation, we must know K_f, i, and m. Right now, we only know K_f. However, we have been given enough information to determine both "i" and "m." Let's start with m:

$$\frac{2.5 \cancel{g\ Al(NO_3)_3}}{1} \times \frac{1 \text{ mole } Al(NO_3)_3}{213.0 \cancel{g\ Al(NO_3)_3}} = 0.012 \text{ moles } Al(NO_3)_3$$

$$\frac{150.0 \cancel{g}}{1} \times \frac{1 \text{ kg}}{1{,}000 \cancel{g}} = 0.1500 \text{ kg}$$

$$m = \frac{0.012 \text{ moles } Al(NO_3)_3}{0.1500 \text{ kg water}} = 0.080 \text{ m}$$

To figure out "i", we just have to think about how $Al(NO_3)_3$ dissolves. The symbol "i" is defined as the number of molecules or ions that the solute splits into when it dissolves. $Al(NO_3)_3$, being an ionic compound, splits up into its ions when dissolved. According to its chemical formula, each molecule of aluminum nitrate has one aluminum ion and three nitrate ions. Thus, the total number of ions is four, so i = 4.

Now that we have all of the components of Equation (11.2), we can use it:

$$\Delta T = -i \cdot K_f \cdot m = -4 \cdot 1.86 \frac{^{\circ}C}{\cancel{m}} \cdot 0.080 \cancel{m} = -0.60\ ^{\circ}C$$

Now realize, *this is not the answer to the problem.* The problem asks what the freezing point of the solution is. What we found was the *change in the freezing point.* You are supposed to have memorized that water freezes at 0.00 °C. Our answer indicates that the freezing point of this solution is 0.60 °C lower than that. Thus, the answer to our problem is that the freezing point of the solution is <u>-0.60 °C</u>.

What is the freezing point of a solution that is made by dissolving 100.0 grams of table sugar ($C_{12}H_{22}O_{11}$) into 950 grams of water? (K_f of water = 1.86 $\frac{^\circ C}{m}$)

We start solving this problem by calculating m:

$$\frac{100.0 \cancel{\text{g } C_{12}H_{22}O_{11}}}{1} \times \frac{1 \text{ mole } C_{12}H_{22}O_{11}}{342.2 \cancel{\text{g } C_{12}H_{22}O_{11}}} = 0.2922 \text{ moles } C_{12}H_{22}O_{11}$$

$$\frac{950 \cancel{\text{g}}}{1} \times \frac{1 \text{ kg}}{1{,}000 \cancel{\text{g}}} = 0.95 \text{ kg}$$

$$m = \frac{0.2922 \text{ moles } C_{12}H_{22}O_{11}}{0.95 \text{ kg water}} = 0.31 \text{ m}$$

To figure out "i", we just have to think about how table sugar dissolves. According to its formula, there are no metals present. This means that table sugar is not ionic. Thus, it does not split up into ions. It dissolves one molecule at a time; therefore, i = 1.

Now that we have all of the components of Equation (11.2), we can use it:

$$\Delta T = -i \cdot K_f \cdot m = -1 \cdot 1.86 \frac{^\circ C}{\cancel{m}} \cdot 0.31 \cancel{m} = -0.58\ ^\circ C$$

Thus, the solution has a freezing point that is 0.58 °C lower than the freezing point of the solvent. Since the solvent is water, which normally has a freezing point of 0.00 °C, this solution has a freezing point of -0.58 °C.

Make sure you understand how to do these problems by solving the following problems:

ON YOUR OWN

11.7 In order to prevent the water in a car's cooling system from freezing, people add "antifreeze" to the water in the radiator. This antifreeze is typically ethylene glycol ($C_2H_6O_2$). What is the freezing point of a radiator solution made from 4.00 kg water and 1.00 kg ethylene glycol? (K_f = 1.86 °C/m)

11.8 If a solution is made by dissolving 2.30 grams $Mg(OH)_2$ into 25.0 grams of acetic acid ($C_2H_4O_2$), what is the freezing point? (K_f of acetic acid = 3.90 °C/m, freezing point of pure acetic acid = 16.6 °C)

Before we leave this section, I must add a note. In the discussion so far, I have used the term "freezing point." It is important for you to realize, however, that other books may substitute the term "melting point." Even though you might think that these two terms are different, they are, in fact, the same. After all, any substance will freeze at the same temperature at which it melts! Whether you call it a freezing point or a melting point really just depends on direction. If a substance is turning from liquid to solid, we usually call it a freezing point. If, on the other hand, the substance is turning from solid to liquid, we call it a melting point. The two are equivalent, however, so some books use them interchangeably. Do not get confused if you see that terminology.

Boiling-Point Elevation

Boiling-point elevation is another property of solutions that is closely tied to freezing-point depression. When a solute is dissolved in a solvent, the boiling point of the resulting solution is *higher* than that of the pure solvent. Once again, this can be understood by looking at the way in which solutes dissolve.

The reason that solutes dissolve in solvents is that there is an attraction between the two. Well, when you are trying to boil a liquid, what are you trying to do? You are trying to pull the molecules far away from each other so that the substance will change from a liquid to a gas. If, however, a solute is present, the molecules of the solvent are attracted to the molecules of the solute. This makes the solvent molecules harder to pull apart. Since they are harder to pull apart, more energy must be added to the solution to make it boil. Thus, the boiling temperature increases.

Amazingly enough, boiling-point elevation is governed by an equation that is almost identical to that of the one for freezing-point depression:

$$\Delta T = i \cdot K_b \cdot m \qquad (11.3)$$

In this equation, "ΔT", "i", and "m" are the same as they were in Equation (11.2). The only differences between Equation (11.2) and Equation (11.3) are that there is no negative sign in Equation (11.3) and that we use the boiling-point elevation constant (K_b) instead of the freezing-point depression constant (K_f).

Both of these differences should make sense. First, the boiling point is elevated, not depressed. Thus, ΔT should be positive, not negative. Second, boiling is a completely different process from freezing. We should therefore expect substances to have a K_b that is quite different from K_f. For water, $K_b = 0.512\ ^{o}C/m$. Even though the equations for freezing-point depression and boiling-point elevation are a bit different, their application is the same. Study the example and then do the "On Your Own" problems that follow to be sure you understand this.

EXAMPLE 11.4

In cooking pasta, a chef adds 20 grams of table salt (NaCl) to 1.45 kg of water and brings the water to a boil. What is the temperature of this boiling solution? (K_b of water = 0.512 $^{o}C/m$)

The addition of a solute (NaCl) to the water will elevate the boiling point. To see how the boiling point is elevated, we must use Equation (11.3). To do this, we must calculate "i" and "m":

$$\frac{20 \text{ g NaCl}}{1} \times \frac{1 \text{ mole NaCl}}{58.5 \text{ g NaCl}} = 0.3 \text{ moles NaCl}$$

$$\frac{0.3 \text{ moles NaCl}}{1.45 \text{ kg water}} = 0.2 \text{ m}$$

Once again, "i" just refers to how the substance splits up when it dissolves. NaCl, being an ionic compound, splits up into its one sodium ion and one chloride ion. Thus, i = 2:

$$\Delta T = i \cdot K_b \cdot m = 2 \cdot 0.512 \frac{^{\circ}\text{C}}{\text{m}} \cdot 0.2 \text{ m} = 0.2 \text{ }^{\circ}\text{C}$$

We must think about what this tells us. The ΔT that we just solved for tells us how many °C *above the normal boiling point* that the solution boils. You are supposed to have memorized that water boils at 100.0 °C, so the boiling point of this solution is 100.2 °C.

ON YOUR OWN

11.9 Acetic acid has a K_b of 2.93 °C/m and a normal boiling point of 118.1 °C. What would be the boiling point of a solution made by dissolving 50.0 g of $CaCO_3$ in 500.0 g of acetic acid?

11.10 If you wanted to boil a solution made by dissolving 250.0 grams of table sugar ($C_{12}H_{22}O_{11}$) into 1.3 kg of water, to what temperature would you have to raise it? (K_b of water = 0.512 °C/m)

Now that you have a basic idea of how things dissolve and the effects of solutes on solvents, you are ready to proceed with a few different subjects in the next couple of modules. Like most of the subjects we have discussed, however, we have only scratched the surface of solubility in this module. If you pursue more study in the field of chemistry, you will find that there is much more depth to this intriguing subject than what was discussed here.

ANSWERS TO THE "ON YOUR OWN" PROBLEMS

11.1 Since the solubility of most solids increases as you increase the temperature of the solution, the student should try heating up the solvent a bit. This should increase the solid's solubility. It may, of course, decrease it because we do not know whether or not this particular solute is an exception to the general rule. However, you should assume it will follow the rule unless otherwise indicated.

11.2 If the gas is escaping the solution, the solubility of the gas is not quite large enough under the conditions of the storeroom. The chemist needs to increase the solubility of the gas. The way to do this is to reduce the temperature of the area that stores the bottles. This could be done by putting the bottles in a refrigerator or just turning off the heat to the storeroom. Note that she should NOT just tape the lids down to keep them from popping off. If the gas were really escaping the solution, the pressure would continue to build, even after the lids are taped down. Eventually, the pressure buildup could be great enough to explode the bottles!

11.3 Since temperature is not the issue, we must consider pressure. Increased pressure increases the solubility of gases in liquids. Since the atmospheric pressure varies every day, the pressure in the room probably varies with it. Thus, the pressure in the room is probably varying. On days in which the pressure is high, more gas can dissolve in the water. On days in which the pressure is low, less gas will dissolve. If the student really needs a constant amount of gas dissolved in liquid, he may have to control the pressure in the lab as well as the temperature.

11.4 This is just a stoichiometry problem. We can tell this by the fact that we are being asked to determine the amount of one substance when we are given the amount of another substance. The only way to do that is by stoichiometry. Now, in order to do stoichiometry, we must first get our amount in moles. Right now, the amount is given in concentration and volume. We must turn that into moles. In order to do that, though, we must convert mL into liters, so that our volume unit is consistent with our concentration unit (remember, "M" means moles/L).

$$\frac{125\ \cancel{\text{mL}}}{1} \times \frac{0.001\ \text{L}}{1\ \cancel{\text{mL}}} = 0.125\ \text{liters}$$

Now we can convert from volume and concentration into moles:

$$\frac{0.5\ \text{moles}\ H_2O_2}{1\ \cancel{\text{L}}} \times 0.125\ \cancel{\text{L}} = 0.06\ \text{moles}\ H_2O_2$$

Now that we have moles, we can do stoichiometry:

$$\frac{0.06\ \cancel{\text{moles}\ H_2O_2}}{1} \times \frac{5\ \text{moles}\ O_2}{5\ \cancel{\text{moles}\ H_2O_2}} = 0.06\ \text{moles}\ O_2$$

Now, of course, this is not quite the answer we need. We were asked to figure out how many grams of oxygen were produced, so we have to convert from moles back to grams:

$$\frac{0.06\ \cancel{\text{moles}\ O_2}}{1} \times \frac{32.0\ \text{grams}\ O_2}{1\ \cancel{\text{moles}\ O_2}} = 2\ \text{grams}\ O_2$$

Therefore, 2 g of oxygen will be produced.

11.5 Once again, we can tell that this is a stoichiometry problem because we are asked to convert from the amount of one substance to the amount of another. In order to do stoichiometry, however, we must convert to moles. Since we have the number of kg of Al, we will use that as our starting point. First, however, we must convert kg to g:

$$\frac{1.2\ \cancel{\text{kg}}}{1} \times \frac{1{,}000\ \text{g}}{1\ \cancel{\text{kg}}} = 1.2 \times 10^3\ \text{g Al}$$

$$\frac{1.2 \times 10^3\ \cancel{\text{g Al}}}{1} \times \frac{1\ \text{mole Al}}{27.0\ \cancel{\text{g Al}}} = 44\ \text{moles Al}$$

Now that we have moles, we can use stoichiometry to determine how much NaOH is needed:

$$\frac{44\ \cancel{\text{moles Al}}}{1} \times \frac{2\ \text{moles NaOH}}{2\ \cancel{\text{moles Al}}} = 44\ \text{moles NaOH}$$

This tells us how much NaOH we need, but it doesn't answer the question. The question asks how many liters of a 6.5 M solution is needed. To determine this, we must remember that "M" means moles per liter. Thus, the concentration of a solution is a conversion relationship that allows us to relate the number of moles to the number of liters. We can therefore do the following conversion:

$$\frac{44\ \cancel{\text{moles NaOH}}}{1} \times \frac{1\ \text{L NaOH}}{6.5\ \cancel{\text{moles NaOH}}} = 6.8\ \text{L of solution}$$

This means that 6.8 liters of the NaOH solution would be needed.

11.6 a. Since the problem already gives us the number of moles of solute and the number of kg of solvent, this problem is just a straight application of Equation (11.1):

$$\text{molality} = \frac{\#\ \text{moles solute}}{\#\ \text{kg solvent}} = \frac{2.3\ \text{moles NaOH}}{1.2\ \text{kg water}} = 1.9\ \frac{\text{moles NaOH}}{\text{kg water}} = \underline{1.9\ \text{m}}$$

b. This problem is a bit more difficult because, although we are given the number of kg of solvent, we are given the number of grams of solute. So we first must convert grams to moles:

$$\frac{250.0\ \cancel{\text{g KNO}_3}}{1} \times \frac{1\ \text{mole KNO}_3}{101.1\ \cancel{\text{g KNO}_3}} = 2.473\ \text{moles KNO}_3$$

Now that we have moles of solute and kg of solvent, we can use Equation (11.1):

$$\text{molality} = \frac{\#\text{ moles solute}}{\#\text{ kg solvent}} = \frac{2.473 \text{ moles KNO}_3}{2.43 \text{ kg water}} = 1.02 \frac{\text{moles KNO}_3}{\text{kg water}} = 1.02 \text{ m}$$

The molality is 1.02 m.

c. In this problem, we are given neither moles of solute nor kg of solvent, so we must convert both:

$$\frac{34.5 \cancel{\text{g NaBr}}}{1} \times \frac{1 \text{ mole NaBr}}{102.9 \cancel{\text{g NaBr}}} = 0.335 \text{ moles NaBr}$$

$$\frac{578 \cancel{\text{g}}}{1} \times \frac{1 \text{ kg}}{1{,}000 \cancel{\text{g}}} = 0.578 \text{ kg}$$

Now that we have moles of solute and kg of solvent, we can use Equation (11.1):

$$\text{molality} = \frac{\#\text{ moles solute}}{\#\text{ kg solvent}} = \frac{0.335 \text{ moles NaBr}}{0.578 \text{ kg water}} = 0.580 \frac{\text{moles NaBr}}{\text{kg water}} = 0.580 \text{ m}$$

The molality is 0.580 m.

11.7 In a freezing-point depression problem, you must use Equation (11.2). However, in order to use that equation, we must know K_f, i, and m. Right now, we only know K_f. However, we have been given enough information to calculate both "i" and "m." First let's calculate m:

$$\frac{1.00 \times 10^3 \cancel{\text{g C}_2\text{H}_6\text{O}_2}}{1} \times \frac{1 \text{ mole C}_2\text{H}_6\text{O}_2}{62.1 \cancel{\text{g C}_2\text{H}_6\text{O}_2}} = 16.1 \text{ moles C}_2\text{H}_6\text{O}_2$$

$$\text{m} = \frac{16.1 \text{ moles C}_2\text{H}_6\text{O}_2}{4.00 \text{ kg water}} = 4.03 \text{ m}$$

To figure out "i", we just have to realize that according to its formula, ethylene glycol is not ionic. Thus, i = 1.

Now that we have all of the components of Equation (11.2), we can use it:

$$\Delta T = -\text{i} \cdot K_f \cdot \text{m} = -1 \cdot 1.86 \frac{^\circ\text{C}}{\cancel{\text{m}}} \cdot 4.03 \cancel{\text{m}} = -7.50 \ ^\circ\text{C}$$

So the freezing point is 7.50 °C lower than that of normal water, or -7.50 °C.

11.8 In a freezing-point depression problem, you must use Equation (11.2). However, in order to use that equation, we must know K_f, i, and m. Right now, we only know K_f. However, we have been given enough information to calculate both "i" and "m." To calculate m:

$$\frac{2.30 \text{ g Mg(OH)}_2}{1} \times \frac{1 \text{ mole Mg(OH)}_2}{58.3 \text{ g Mg(OH)}_2} = 0.0395 \text{ moles Mg(OH)}_2$$

$$\frac{25.0 \text{ g}}{1} \times \frac{1 \text{ kg}}{1{,}000 \text{ g}} = 0.0250 \text{ kg}$$

$$\text{m} = \frac{0.0395 \text{ moles Mg(OH)}_2}{0.0250 \text{ kg acetic acid}} = 1.58 \text{ m}$$

To figure out "i", we just have to think about how $Mg(OH)_2$ dissolves. Being an ionic compound, it splits up into its ions when dissolved. Each molecule of magnesium hydroxide has one magnesium ion and two hydroxide ions, so i = 3.

Now that we have all of the components of Equation (11.2), we can use it:

$$\Delta T = -i \cdot K_f \cdot m = -3 \cdot 3.90 \frac{^\circ C}{m} \cdot 1.58 \text{ m} = -18.5 \ ^\circ C$$

Now realize that *this is not the answer to the problem.* The problem asks what the freezing point of the solution is. What we found was the *change in the freezing point.* You are told that pure acetic acid has a freezing point of 16.6 °C. Our answer indicates that the freezing point of this solution is 18.5 °C lower than that. Thus, the answer to our problem is that the freezing point of the solution is -1.9 °C.

11.9 To calculate boiling points, we must use Equation (11.3). To do that, however, we must know "i" and "m". To calculate "m":

$$\frac{50.0 \text{ g CaCO}_3}{1} \times \frac{1 \text{ mole CaCO}_3}{100.1 \text{ g CaCO}_3} = 0.500 \text{ moles CaCO}_3$$

$$\frac{500.0 \text{ g}}{1} \times \frac{1 \text{ kg}}{1{,}000 \text{ g}} = 0.5000 \text{ kg}$$

$$\text{m} = \frac{0.500 \text{ moles CaCO}_3}{0.5000 \text{ kg acetic acid}} = 1.00 \text{ m}$$

Since calcium carbonate is an ionic compound, it dissolves by splitting up into its calcium ion and its carbonate ion. Thus, i = 2.

$$\Delta T = i \cdot K_b \cdot m = 2 \cdot 2.93 \frac{^\circ C}{m} \cdot 1.00 \text{ m} = 5.86 \ ^\circ C$$

This means that the boiling point of the solution is 5.86 °C *higher* than that of pure acetic acid. The boiling point of pure acetic acid was given as 118.1 °C, so the boiling point of this solution is 124.0 °C.

11.10 The addition of a solute (sugar) to the water will elevate the boiling point. To see how the boiling point is elevated, we must use Equation (11.3). To do this, however, we must calculate "i" and "m":

$$\frac{250.0 \cancel{\text{g } C_{12}H_{22}O_{11}}}{1} \times \frac{1 \text{ mole } C_{12}H_{22}O_{11}}{342.2 \cancel{\text{g } C_{12}H_{22}O_{11}}} = 0.7306 \text{ moles } C_{12}H_{22}O_{11}$$

$$\frac{0.7306 \text{ moles } C_{12}H_{22}O_{11}}{1.3 \text{ kg water}} = 0.56 \text{ m}$$

Once again, "i" just refers to how the substance splits up when it dissolves. Since table sugar is not ionic, it splits up into one molecule at a time when it dissolves. Thus, i = 1.

$$\Delta T = i \cdot K_b \cdot m = 1 \cdot 0.512 \frac{^{\circ}C}{\cancel{m}} \cdot 0.56 \cancel{m} = 0.29 \ ^{\circ}C$$

We must think about what this tells us. The ΔT that we just solved for tells us how many °C *above the normal boiling point* that the solution boils. You are supposed to have memorized that water boils at 100.0 °C, so the boiling point of this solution is 100.3 °C.

REVIEW QUESTIONS FOR MODULE #11

1. When you dissolve salt in water, which compound is the solute and which is the solvent?

2. Contrast the way ionic compounds dissolve in water with the way that polar covalent compounds dissolve in water.

3. What is a saturated solution?

4. What kind of solute (solid, liquid, or gas) dissolves best under high temperature conditions?

5. What kind of solute (solid, liquid, or gas) dissolves best under high pressure conditions?

6. What kind of solute (solid, liquid, or gas) has a solubility that is least affected by the conditions under which the solution is made?

7. If a solid solute is not dissolving well in water, what can you do to increase its solubility?

8. If a chemist makes a solution in a beaker using water and a solute that dissolves endothermically in water, would you expect the beaker to feel hot or cold?

9. What is the difference between molality and molarity?

10. If you wanted to protect water from freezing, which compound would accomplish this best:

$NaNO_3$, $Mg(NO_3)_2$, or $Al(NO_3)_3$?

PRACTICE PROBLEMS FOR MODULE #11

1. Hot tubs generally contain water at a temperature of 38 °C. In order to keep the water clean, several solid solutes are dissolved in the water. Many hot tub manufacturers warn that if the water gets too cold, it will get cloudy and filled with a fine solid. Why?

2. If you have an open glass of soda pop sitting on the counter, it gets "flatter" than the same open glass of soda pop that sits in the refrigerator. Why?

3. The following reaction is performed in a lab:

$$3Na_2SO_4\ (aq)\ +\ 2Al(NO_3)_3\ (aq) \rightarrow Al_2(SO_4)_3\ (s)\ +\ 6NaNO_3\ (aq)$$

If 345 mL of 1.25 M aluminum nitrate is added to an excess of sodium sulfate, how many grams of aluminum sulfate will be produced?

4. Aqueous sodium bromide and silver nitrate will react to form solid silver bromide (a compound used in photography) and sodium nitrate:

$$NaBr\ (aq)\ +\ AgNO_3\ (aq) \rightarrow AgBr\ (s)\ +\ NaNO_3\ (aq)$$

If a chemist needs to produce 100.0 grams of silver bromide, how many liters of 2.5 M silver nitrate must be added to an excess of sodium bromide?

5. One popular test for lead (Pb^{2+}) contamination in drinking water is the "chloride test." This test involves adding NaCl to water. Any Pb^{2+} in the water will immediately react with the Cl^-, forming a solid precipitate:

$$Pb^{2+}\ (aq)\ +\ 2NaCl\ (aq) \rightarrow PbCl_2\ (s)\ +\ 2Na^+\ (aq)$$

Measuring the mass of this precipitate can tell you what concentration of Pb^{2+} is in the water. If an excess of NaCl is added to 1.6 liters of water and 25.0 grams of $PbCl_2$ are produced, what was the molarity of Pb^{2+} in the water?

6. If a solution is made by mixing 50.0 moles of KOH to 3.4 kilograms of water, what is the solution's molality?

7. If a solution is made by mixing 35.0 g of $CaBr_2$ with 657 grams of water, what is the solution's molality?

8. If you want to lower water's freezing point 15 °C by adding NaCl, what must the molality of the salt solution be (K_f for water is 1.86 °C/m)?

9. What is the freezing point of a solution made by mixing 35.0 grams of NH_3 with 350.0 grams of water (K_f for water is 1.86 °C/m)?

10. What is the boiling point of a solution made by mixing 100.0 g $ZnCl_2$ with 750.0 grams of water (K_b for water is 0.512 °C/m)?

MODULE #12: The Gas Phase

Introduction

Since the last module concentrated on chemistry as it happens in the liquid phase, it is only natural that this module should emphasize the gas phase. Gases are an important part of God's creation. We learned in Module #4 that the mixture of gases that makes up earth's atmosphere is perfectly designed to support life. In some of the experiments you have performed, gases were produced as part of the chemical reaction. In addition, we have learned that Gay-Lussac's Law allows us to relate the volumes of gases that appear within a chemical equation. Thus, we have already touched on the subject of gases quite a bit.

In this module, we will learn a few more things about how the physical characteristics of gases change in response to changes in the environment. In addition, we will learn a way to relate the number of moles of a gas to such things as the volume that the gas occupies, the pressure it exerts, and the temperature of its surroundings. This particular technique will become very useful in helping us perform stoichiometric calculations on chemical equations that contain gases.

The Definition of Pressure

In order to understand the gas phase, we must first understand the concept of **pressure**.

Pressure - The force per unit area exerted on an object

Mathematically, we would say:

$$P = \frac{F}{A} \tag{12.1}$$

where P is the pressure, F is the force exerted on an object, and A is the area over which this force is applied.

Now, although Equation (12.1) is a very important equation, it is not one that you have to memorize, because chemists don't really use it. When you take physics, however, this equation will become much more important for you to know. For right now, the only reason I present this equation is to illustrate the standard unit for pressure. The unit for force is called the "Newton," and your experience in geometry should indicate to you that the unit for area is m^2. Pressure, then, has the units of Newtons/m^2. This unit is renamed, however, to the **Pascal** (abbreviated **Pa**) in order to honor Blaise Pascal (Pas kal'), a French mathematician/scientist/philosopher. Pascal was a brilliant man who revolutionized the study of probability as well as the sciences of chemistry and physics. In addition, he was a celebrated philosopher who provided several convincing arguments for the validity of the Christian faith.

The important thing that you have to realize now is that a gas exerts pressure on any object that comes into contact with it. How does a gas exert pressure? In Module #4, we learned about the kinetic theory of matter. This theory states that the molecules (or atoms) that make up a substance are in constant motion. In solids, the molecules (or atoms) vibrate back and forth. In the liquid phase, molecules (or atoms) move around their container, staying relatively close to one another. In the gas phase, the molecules (or atoms) move around their container quickly, staying far apart from each other.

Since gas molecules move around quickly, they tend to collide with anything that gets into their way. Thus, if an object comes into contact with a gas, the molecules (or atoms) that make up the gas will begin striking the object, exerting a force. The average force per unit area that these collisions exert on the object is called the gas pressure.

Now since our atmosphere is composed of a mixture of gases, the atmosphere must exert pressure on any object with which it comes into contact. Thus, every person, animal, and object on earth is subject to a pressure exerted on it by the atmosphere. This "atmospheric pressure" is rather constant. It varies a bit, depending on the weather conditions, but on average, it is equal to 101.3 kiloPascals (kPa). Since atmospheric pressure is reasonably constant, we actually use this value as an alternative unit for measuring pressure. We say that a gas exerts 1.000 **atmosphere** (abbreviated **atm**) if it exerts 101.3 kPa of pressure. Thus, we have this new unit for pressure: the atm. Mathematically, we would say:

$$\mathbf{1.000\ atm = 101.3\ kPa} \qquad (12.2)$$

Chemists use both of these units to measure pressure, so you will have to be comfortable with both of them.

Just to make things a bit more confusing, chemists often use two other units for pressure: **millimeters of mercury** (abbreviated **mmHg**) and **torr**. If you take physics, you will learn more about these units. For right now, however, all you need to know is that they are alternative units for pressure and they are both equal. Their relationship to atmospheres is given below:

$$\mathbf{760.0\ torr = 1.000\ atm} \qquad (12.3)$$

$$\mathbf{760.0\ mmHg = 1.000\ atm} \qquad (12.4)$$

In the end, there are four units that chemists use for pressure: Pa, atm, torr, and mmHg. You need to know that all of these are pressure units, and you also need to know how they relate to atmospheres. In other words, you need to know the three relationships presented in Equations (12.2) - (12.4).

Boyle's Law

Now that you know the definition of pressure and the various units used to measure it, you can begin to see how the pressure of a gas is affected by the conditions under which it is stored. In the mid-1600s, an Irish scientist named Robert Boyle did several experiments to determine how the pressure of a sample of gas relates to its volume. He determined that as long as the temperature of the gas stayed the same throughout the experiment, the product of a gas' volume and its pressure is always the same. In mathematical terms,

$$PV = \text{constant} \qquad (12.5)$$

This relationship is called **Boyle's Law**. It tells us that since the product of pressure and volume must stay constant, when the volume of a gas is increased, the pressure will decrease. Conversely, if the volume of a gas is decreased, its pressure will increase. Of course, it works the other way as well. If the pressure of a gas increased, its volume will decrease in response. If the pressure is decreased, the volume of the gas will increase. This is illustrated in Figure 12.1:

FIGURE 12.1
Boyle's Law

Illustration by Megan Whitaker

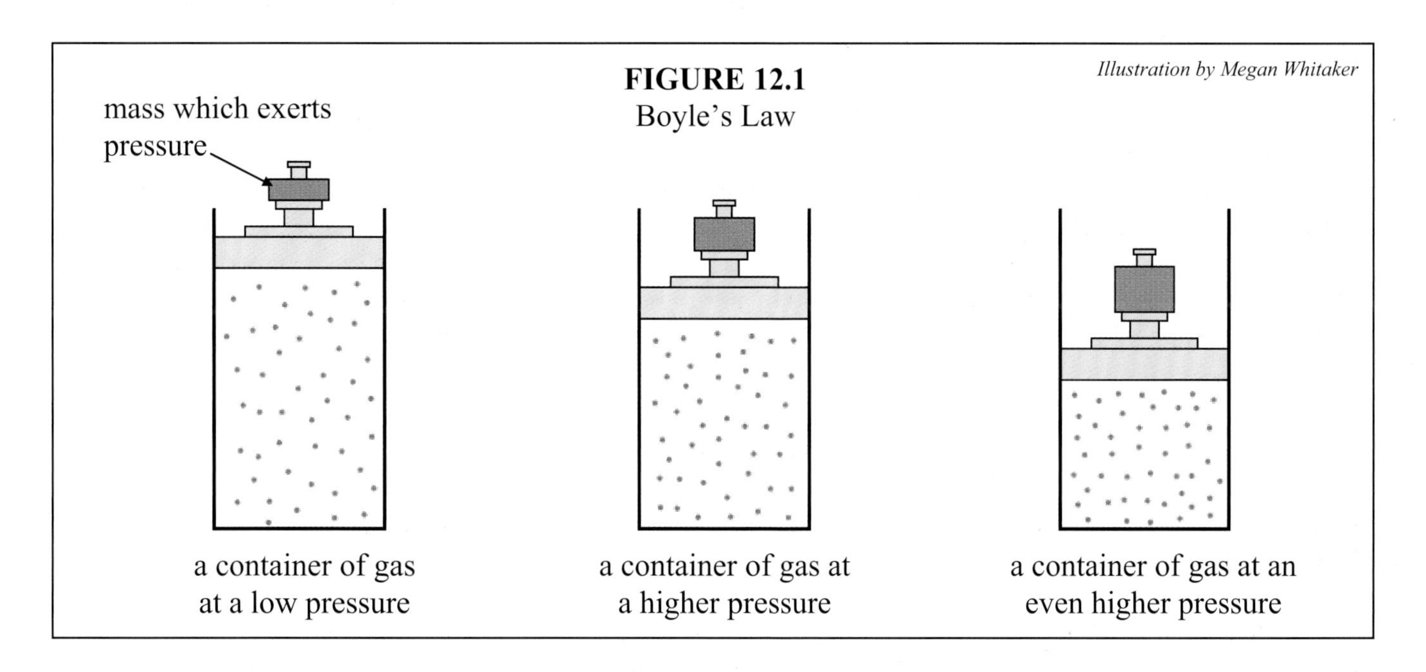

In this figure, a gas is contained in a piston with a top that is free to move up and down. In the drawing on the far left, a small mass is sitting on top of the piston. The weight of the mass pushes down on the top of the piston, exerting a pressure on the gas. In the middle drawing, the only thing that has changed is the size of the mass. The amount of gas, temperature, size of the piston, etc., are all the same as they were in the drawing on the left. However, since the mass in the middle drawing is larger, it exerts a larger pressure on the gas. As a result, the volume of the gas decreases. In the drawing on the right, the mass has been increased even more. This results in an even larger pressure, which further reduces the volume of the gas. That's Boyle's Law.

If we think about the kinetic theory of matter and the definition of pressure, this law should make sense. After all, if I increase the volume that a gas occupies, the molecules (or atoms) of the gas can spread out. Thus, they will make fewer collisions with any object that comes into contact with the gas. Based on our definition of pressure, then, fewer collisions will result in lower pressure. On the other hand, if I decrease the volume that a gas can occupy, the molecules (or atoms) of the gas cannot spread out. Since they have less space to move around in, they will collide with objects in the gas a lot more often. Based on our definition of pressure, more collisions means higher pressure.

Now, of course, Boyle did not have the benefit of understanding his law based on the kinetic theory of matter, because the kinetic theory of matter wasn't around in the mid-1600s. Instead, he wrote his law based solely on experiment. He studied several gases and found that if he doubled the pressure on a container of gas, the volume of the gas within the container decreased by a factor of 2. After hundreds and hundreds of such experiments on several different gases, he assumed that his law was correct, even though he didn't have a good understanding as to why.

There is an alternative way of expressing Boyle's Law. Since Equation (12.5) tells us that the product of a gas' volume and its pressure is always the same, we could also say that the product of a gas' volume and its pressure under one set of conditions must be equal to the product of the gas' volume and its pressure under any other set of conditions. Mathematically, we could say:

$$P_1V_1 = P_2V_2 \tag{12.6}$$

where "P_1" and "V_1" are the pressure and volume of a gas under one set of conditions, and "P_2" and "V_2" are the pressure and volume of the same gas under another set of conditions. Now it is important for you to understand that Equations (12.5) and (12.6) both say the same thing. They are both ways of mathematically representing Boyle's Law; they just make the statement in two different ways.

Now, although Boyle's Law is important to know, Equations (12.5) and (12.6) need not be memorized. It turns out that Boyle's Law can be combined with the law you are going to learn next (Charles's Law). When the two laws are combined, the mathematical equation that results is the important one to remember. Thus, you need only understand Boyle's Law from a conceptual point of view. We will put the mathematics with the concepts a little later on.

If you purchased the MicroChem kit discussed in the introduction, you can perform Experiment #8 in that kit to get more experience with Boyle's Law.

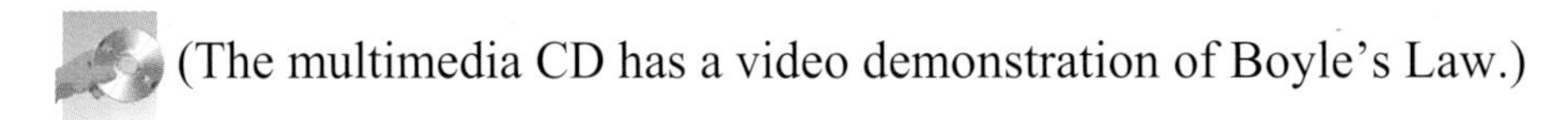

(The multimedia CD has a video demonstration of Boyle's Law.)

Before I leave this section, I want to spend a moment discussing the man behind this famous law. Robert Boyle is considered by many to be the founder of modern chemistry, because he took the unscientific field of alchemy and turned it into a scientific endeavor. His scientific accomplishments go well beyond the law that we are studying here. For example, he demonstrated the existence of a vacuum, which scientific thought in his day considered impossible. He also invented an ingenious vacuum pump that was used widely by other scientists. He was the first to define the term "element" properly, he invented a primitive kind of litmus paper, he demonstrated that water expands when it freezes, and he provided strong evidence for the existence of atoms. Clearly, his contributions to science were significant.

Robert Boyle was also a devout Christian. In fact, his Christian faith provided the motivation for changing alchemy into a scientific discipline. Boyle considered science to be a means of glorifying God, and he was eager to uncover the "secrets" of God's creation. Thus, rather than trying to turn common metals such as lead into precious metals such as gold (which was alchemy's aim), he wanted to learn as much as he could about how God had designed the world. Boyle's attitude towards science and faith is best summed up in a quote from his final address to the Royal Society, an organization of scientists in England. In that address, he told the assembled scientists, "Remember to give glory to the One who authored nature." (Dan Graves, *Scientists of Faith*, Kregel Resources 1996, p. 63) Sadly, many scientists today ignore Boyle's admonition, and science has suffered as a result. Boyle wanted to make sure that the important aspects of his work continued after his death, so his will established a lecture series designed to convince people of the reality of the Christian faith.

Charles's Law

In the late 1700s, a French scientist named Jacques Charles began building on the work of Robert Boyle by investigating the effect of temperature on a gas' volume. He found that when a gas' temperature was raised under conditions of constant pressure, the volume occupied by the gas increased as well. This is shown in the figure below:

FIGURE 12.2
Charles's Law

Illustration by Megan Whitaker

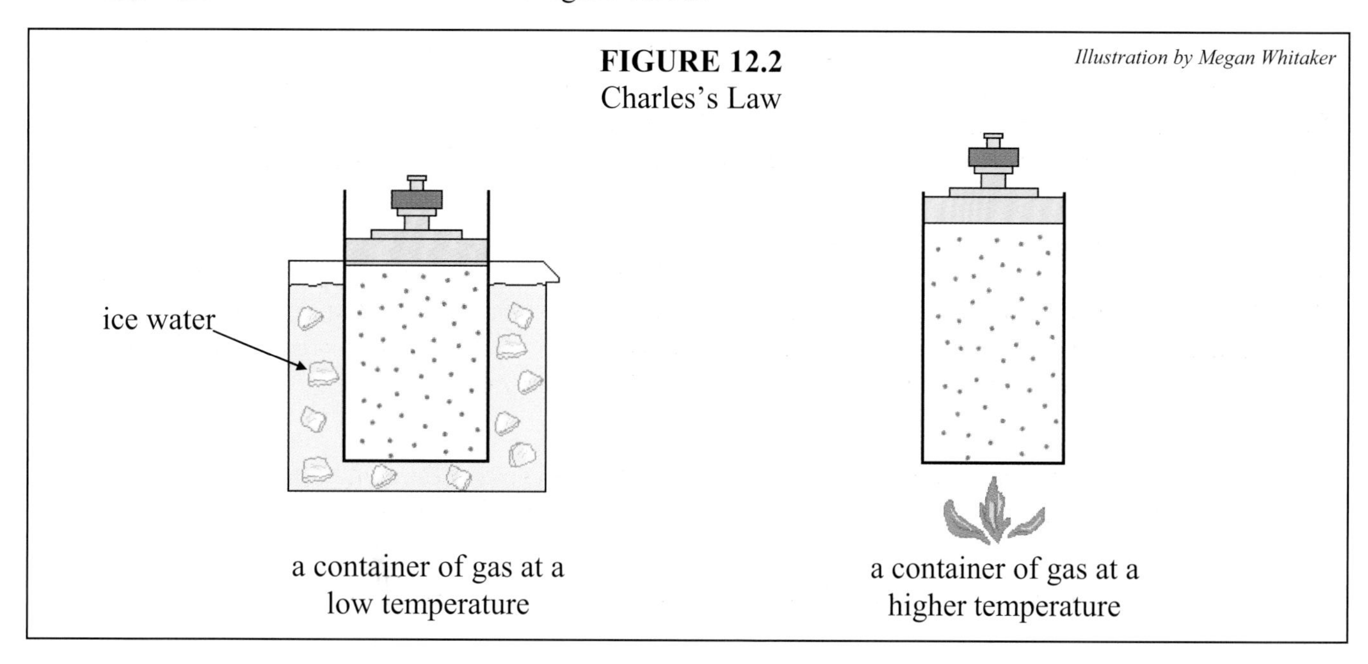

In the figure, a container of gas at a certain pressure is placed in ice water so that its temperature is low. The container is then put over a flame to increase its temperature. During the entire time, the pressure is constant, because the size of the mass on the top of the piston is the same. Notice that when the temperature is warmer, the gas occupies a greater volume. If you perform this kind of experiment with different gases and different temperatures, you can produce a graph that looks like this:

FIGURE 12.3
The Effect of Temperature on the Volume of a Gas at Constant Pressure

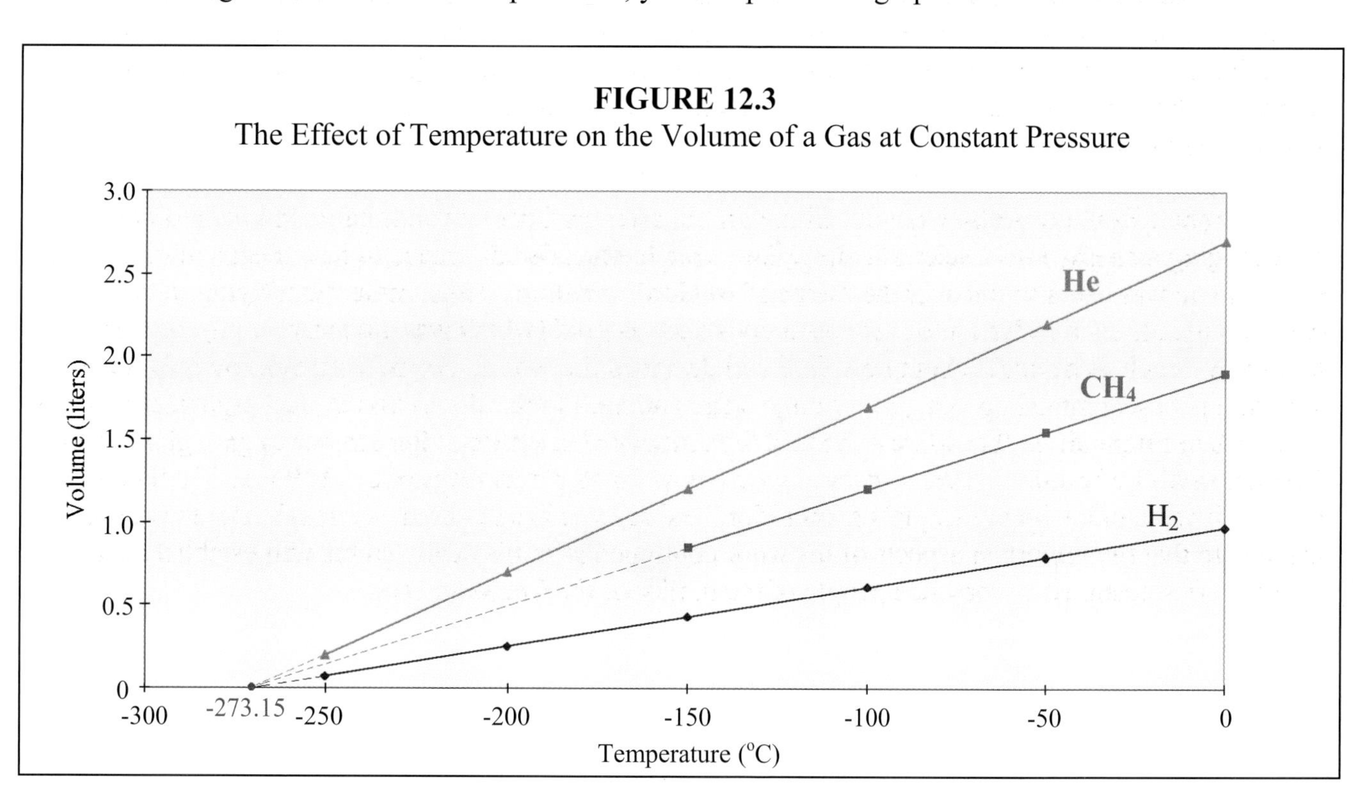

If you study the figure, you should see two important aspects of the data. First, the volume of a gas seems to depend linearly on its temperature. If the temperature of a gas is doubled, its volume doubles. Conversely, if the temperature is halved, the volume of the gas decreases by a factor of 2 as well. This fact is called **Charles's Law**:

Charles's Law - At constant pressure, the temperature and volume of a gas are linearly proportional.

This law can be mathematically expressed as follows:

$$\frac{V}{T} = \text{constant} \qquad (12.7)$$

In this equation, "V" is the volume of the gas and "T" is its temperature. You should be able to see how this equation expresses Charles's Law. If the quotient of volume and temperature must remain constant, then when temperature is increased, volume must increase proportionally to keep the quotient the same.

In terms of the kinetic theory, you should see why Charles's Law makes sense. If the temperature of a gas increases, the gas molecules (or atoms) have more energy. Thus, they want to move around a lot more quickly. However, Charles specifically constructed his experiments so that the pressure exerted by the gas could not change. Thus, the only way a gas can move around more quickly without making more collisions with other objects (and hence raising the pressure) is to spread out. If the gas molecules spread out and occupy a larger volume, they can move around more quickly without colliding with other objects more frequently. This will keep the gas pressure from rising.

Charles's Law can also be expressed in terms of an equation similar to Equation (12.6). Since volume divided by temperature must always remain the same, the volume divided by the temperature under one set of conditions should be the same as the volume divided by the temperature under another set of conditions:

$$\frac{V_1}{T_1} = \frac{V_2}{T_2} \qquad (12.8)$$

Remember that Equations (12.7) and (12.8) really say the same thing; they just express it in different ways. Also, as was the case with Boyle's Law, these two equations needn't be memorized. Instead, the combined gas law that you will learn in the next section will be the important equation to remember.

Before we move on to the combined gas law, however, we must look once more at Figure 12.3. After all, I said that there were two important aspects to the data in that figure, but so far I have only discussed one. Notice, if you will, that each gas has a different line on the graph. The line that goes through helium's data is higher on the graph than the one that goes through methane's (CH_4) data. In addition, the line that goes through methane's data is higher on the graph than the one that goes through hydrogen's data. Notice, however, that if we follow the lines backward far enough (see the dashed part of each line), they all converge to the same temperature when the volume of the gas reaches zero. That temperature is -273.15 ^{o}C.

It turns out that no matter what gas you put on that graph, the line that goes through that gas' data points will always hit -273.15 °C when the volume of the gas is zero. Now if you think about that fact for a minute, you will come to the remarkable conclusion that no matter how hard you try, you can never cool something to a lower temperature than -273.15 °C. After all, a gas (which is made up of matter) can never reach a volume of zero. If that were the case, the gas would no longer be matter because, by definition, all matter takes up some space. Thus, since you can never get to zero volume, you can never get to the place where all of the lines in Figure 12.3 converge. Thus, you can never get to -273.15 °C.

This might sound a little odd to you. After all, in order to get a substance's temperature a bit lower, all you need to do is cool it a little more, right? Well, that's right up to a certain point. However, once the temperature of a substance gets near -273.15 °C, it gets harder and harder to cool. In the end, you can never quite get there. For centuries, scientists have tried to cool things to a temperature of -273.15 °C or lower, but they have never succeeded.

You should recognize the number "273.15." In Module #2, you learned that to convert temperature from Celsius to Kelvin, you had to add 273.15 to the Celsius temperature. You also learned that the Kelvin temperature was called an **absolute temperature scale** because you could never actually reach 0.00 Kelvin. Now you know why that is the case. Since you can never reach -273.15 °C, and since the Kelvin temperature scale is determined by adding 273.15 to all Celsius temperatures, you can never reach (or go below) 0.00 Kelvin. As of May 2001, the lowest temperature ever reached was 0.0000001 K.

If you purchased the MicroChem kit discussed in the introduction, you can perform Experiment #9 in that kit to get more experience with Charles's Law and absolute zero.

Before we leave this section, I need to digress a bit. Notice that in Figure 12.3, the portions of the lines closest to -273.15 °C are dashed. The reason that the lines are dashed in that region of the graph is to indicate that no data has been collected for the temperatures on that portion of the line. In other words, for helium and hydrogen, there are no data points for temperatures lower than -250 °C. For methane (CH_4), there are no data points for temperatures lower than -150 °C. It turns out that at those low temperatures, it is nearly impossible to get an accurate measurement of volume for the gases in the graph. As a result, the experiments that give us the data presented in Figure 12.3 could not be performed at temperatures significantly lower than those listed.

If there are no data points in that region of the graph, why do we draw in a dashed line? Well, the three sets of data follow a very steady trend from 0 °C all the way down to -250 °C. Since the data seem to follow the lines so well for such a big temperature range, *there is no reason to expect that they will stop following that trend.* As a result, even though there are no data points below -250 °C, it is probably safe to assume that the data follows the same trend, and thus we continue to draw the line without any data points to guide it. This process is called **extrapolation** (ik strap' uh lay' shun).

Extrapolation - Following an established trend in the data even though there is no data available for that region

Whether you realize it or not, you use the process of extrapolation all of the time.

For example, suppose a person lies to you a lot, and suppose further that you catch him in several of those lies. If that is the case, you have a lot of data telling you that this person is a liar. The more lies you catch him in, the more data you have telling you this. As a result, you will eventually get to the point where you no longer believe what he is telling you. Therefore, when he tells you something, you may not have any data to indicate that this particular statement is a lie. However, following the trend of all the other data you have collected, you will just assume it is a lie. That's an extrapolation.

Extrapolation is a good and necessary part of science, but you must be careful when you do it. Think about the example I just gave you: Suppose, unbeknownst to you, the person who you think is a liar has actually turned over a new leaf and is starting to tell the truth. Based on your extrapolation, you would have assumed he was lying and therefore would have been wrong. Thus, extrapolations can lead to bad conclusions, if you are not careful. In general, an extrapolation is good as long as the amount of extrapolation you are doing is small compared to the data that you have. For example, in Figure 12.3, we have data that spans a temperature range of 0 °C to -250 °C. That's a range of 250 degrees. In the end, we had to extrapolate an additional 23.15 degrees to get all of the lines to converge. Well, compared to 250 degrees, an additional 23.15 degrees is pretty small; thus, we can say that our extrapolation is probably okay.

Some scientists, however, take extrapolation to an extreme and get into all sorts of trouble. This is how the theory of evolution got started. Charles Darwin, an English naturalist who lived in the mid-to-late-1800s, noticed that living organisms had the ability to change and adapt to their environment. While studying a set of islands called the Galapagos Islands, he noticed that each island had finches (a type of bird), but the finches were all different.

On islands where the birds could find food by sticking their beaks into crevices in the trees and logs, the finches had long, narrow beaks. On islands where woods were scarce, however, the finches had short, fat, hard beaks that allowed them to burrow for food. He imagined that at one time, both of these types of finches were the same. When the finches began living on separate islands, however, their species began to adapt to the different food sources, and after many, many generations, they developed different kinds of beaks that were appropriate for the different food supply on each island.

Darwin was absolutely right on this point. Today, scientists have shown quite conclusively that species do have the ability to adapt and change in response to their environment. For example, in 1977, there was a major drought on Daphne, one of the Galapagos Islands. Researchers had been measuring the beak sizes of finches on that island for some time, and they continued to measure beak sizes long after. They found that the very next generation of finches on the island had beaks that were (on average) about 5% larger than the generation of finches that existed prior to the drought. Since the drought caused a shortage of seeds on the island, the finches with larger beaks were better able to crack open the few, tough seeds left on the island. Thus, the size of the finches' beaks varied in response to the drought.

In 1983, there were strong rains on the same island. This resulted in an abundance of seeds for the finches on the island. Sure enough, scientists who were measuring beak sizes noticed that (on average) the next generation of finches had smaller beak sizes. Once again, then, the finches adapted to a change in their environment. Since seeds were plentiful, a large beak provided no specific advantage for survival. Thus, the finch beaks began to decrease in size again. These two instances really showed that Darwin's idea was right. The population of finches could indeed adapt to changes in their surroundings from generation to generation.

Even though Darwin was right when he said that species have the ability to change and adapt to their environment, he was dead wrong when he tried to extrapolate his data. He said that since species have the ability to change, they should be able to change into a different species. In other words, if a population of finches can, through several generations, slowly develop different beaks, why can't they also develop different wings, heads, bodies, and feet so that they change into eagles? Why can't finches, after many years of such change, develop into a completely new species? Darwin thought that this could, indeed, happen. This idea became the foundation for the theory of evolution, and it is, unfortunately, still believed by many scientists today.

The problem is that Darwin made a big mistake in extrapolation. He took small changes that he observed in animals and extrapolated them into huge changes. Darwin noticed that the finches on the Galapagos Islands had changed their beaks, their feather color, and (to some extent) their body sizes to adapt to the environment of each different island. Those kinds of changes, however, are *very small* compared to the kinds of changes necessary to turn a finch into a completely different species of bird! Thus, Darwin took a *small amount of data and tried to make a huge extrapolation with it!* This is precisely the opposite of what a careful scientist would do. A careful scientist will only make small extrapolations based on large sets of data. Darwin was not careful and, as a result, has caused all sorts of problems for both the scientific community and the population as a whole.

The Combined Gas Law

Now that you have seen both Boyle's Law and Charles's Law, it is time to combine the two. Remember, the experiments that Boyle performed involved changing the volume of a gas and measuring its pressure or vice versa. During all of those experiments, however, the temperature was held constant. Likewise, when Charles performed his experiments, he changed the temperature of a gas and measured its volume or vice versa. During those experiments, however, the pressure exerted by the gas was held constant.

Even though the experiments performed by Boyle and Charles were quite different, we can actually combine their results. The product of a gas' pressure and volume, divided by the gas' temperature, remains the same despite the conditions under which the gas is stored. Mathematically, we could say:

$$\frac{PV}{T} = \text{constant} \tag{12.9}$$

As we did with both Boyle's Law and Charles's Law, we can restate this mathematical equation as follows:

$$\frac{P_1V_1}{T_1} = \frac{P_2V_2}{T_2} \tag{12.10}$$

In this equation, "P_1", "V_1", and "T_1" represent the pressure, volume, and temperature of a gas under one set of conditions, while "P_2", "V_2", and "T_2" represent the pressure, volume, and temperature of the gas under any other set of conditions.

Looking back on Equations (12.6) and (12.8), it should be rather apparent how Equation (12.10) combines both Boyle's Law and Charles's Law. As I said previously, this particular equation is important. It is an extremely useful equation that allows you to predict how the characteristics of a gas change when the conditions to which it is exposed change. Since this equation is so powerful, it is imperative that you memorize it and become proficient in its use.

As you will soon see, Equation (12.10) is very simple to use, but there is one very important fact that you must remember:

When using any equation in this module, you must always use the Kelvin temperature scale.

This is a very important fact. If you notice, Equations (12.7) through (12.10) all contain a temperature term in the denominator of a fraction. What do the rules of mathematics tell us about variables that appear in the denominator of a fraction? Those variables can never equal zero, because the fraction would become undefined at that point. The Kelvin temperature scale, as we have previously discussed, is defined so that you can never reach a value of 0.00 Kelvin. As a result, this is the *only* temperature scale that can be used in the equations that are presented in this module.

Even though it is very important to make sure that you keep the temperatures in this equation in units of Kelvin, there are no constraints on the units you use for pressure and volume. As long as both pressures (P_1 and P_2) have the same units, it doesn't matter whether those units are atm, torr, Pascal, or mmHg. In the same way, as long as both volumes (V_1 and V_2) have the same units, it doesn't matter whether those units are L, mL, cm^3, or m^3. Let's see how this works in a few example problems:

EXAMPLE 12.1

A weather balloon is filled with helium before it is launched. On the ground, the gas has a temperature of 25 °C and a pressure of 1.0 atm. The volume of the balloon is 4.5 m^3. As the balloon rises, however, the pressure and temperature change to 0.30 atm and 15 °C. What is the new volume of the balloon?

This problem gives us the pressure, temperature, and volume of a gas under one set of conditions. The conditions are then changed, and the new pressure and temperature are given. Using Equation (12.10), we can use this information to determine the new volume:

$$\frac{P_1V_1}{T_1} = \frac{P_2V_2}{T_2}$$

In this problem, P_1 = 1.0 atm, V_1 = 4.5 m^3, and T_1 = 25 °C, whereas P_2 = 0.30 atm and T_2 = 15 °C. Since we are solving for the new volume, V_2 is the unknown variable in the equation. Before we can use Equation (12.10), however, we must first convert the temperatures into Kelvin:

$$T_1 = 25 + 273.15 = 298 \text{ K}$$

$$T_2 = 15 + 273.15 = 288 \text{ K}$$

Remember, since we are adding in these equations, we do not count significant figures. Instead, we look at decimal places. Since the gas temperatures have their last significant figure in the ones place, the answer must be reported to the ones place as well.

Now that we have the proper units for temperature, we can use algebra to rearrange Equation (12.10) and solve for V_2:

$$V_2 = \frac{P_1 V_1 T_2}{T_1 P_2}$$

Sticking in the values for the known variables:

$$V_2 = \frac{1.0 \cancel{\text{atm}} \cdot 4.5 \text{ m}^3 \cdot 288 \cancel{\text{K}}}{298 \cancel{\text{K}} \cdot 0.30 \cancel{\text{atm}}} = 14 \text{ m}^3$$

Notice that as long as the pressures have the same unit, they cancel. Thus, we could have used any pressure units, as long as they were the same. In addition, whatever units V_1 has will be the units of V_2. Therefore, the volume of the balloon at a high altitude will be $\underline{14 \text{ m}^3}$.

A gas is stored in a 3.00 liter container at a pressure of 456 torr. If the temperature stays constant, what will the pressure of the gas be if the volume is decreased to 0.75 liters?

In this problem, you are only given V_1 (3.00 liters), P_1 (456 torr), and V_2 (0.75 liters). You might think, therefore, that you have not been given enough information to solve Equation (12.10). However, if we look at the equation, we should see something:

$$\frac{P_1 V_1}{T_1} = \frac{P_2 V_2}{T_2}$$

The problem tells us that the temperature remains constant, which means that $T_1 = T_2$. What does algebra tell us about equal variables that appear in the same place on both sides of the equation? It tells us that they cancel each other out. Thus, since T_1 is equal to T_2, they cancel out:

$$\frac{P_1 V_1}{\cancel{T_1}} = \frac{P_2 V_2}{\cancel{T_2}}$$

$$P_1 V_1 = P_2 V_2$$

This second equation might look familiar. It is the equation that describes Boyle's Law. Thus, the combined gas law reduces to Boyle's Law under conditions of constant temperature. This is typically the way we use the combined gas law. If a problem tells us that one of the variables stays constant, that variable will cancel on both sides of the equation.

Now that we have a useful equation, we can solve for P_2, plug in our numbers, and calculate our answer:

$$P_2 = \frac{P_1 V_1}{V_2} = \frac{456\ \text{torr} \cdot 3.00\ \cancel{\text{L}}}{0.75\ \cancel{\text{L}}} = 1.8 \times 10^3\ \text{torr}$$

Thus, when the volume is decreased under conditions of constant temperature, the pressure increases to $\underline{1.8 \times 10^3\ \text{torr}}$.

ON YOUR OWN

12.1 A gas is stored in a metal cylinder with a volume that cannot change. If the gas is packed into the cylinder at a temperature of 0.00 °C and under a pressure of 3,415 kPa, what will the pressure inside the cylinder be when it is stored at room temperature (25.0 °C)?

12.2 At room temperature (25 °C) and normal atmospheric pressure (1.00 atm), the average person's lungs have a total capacity of about 5.5 liters. If a person takes a deep breath (filling his or her lungs to capacity) and then dives deep into a lake where the temperature is 19 °C and the pressure is 915 torr, what will be the new volume of the air in his or her lungs?

12.3 If you fill a balloon with air to a volume of 2.31 liters at room temperature (25 °C), what volume will the balloon have if it is stuck in the freezer (-5 °C)? Assume that the pressure stays constant.

Ideal Gases

Now that you have some experience working with the combined gas law, it is important to learn *when you can* use this law and *when you can't*. It turns out that this very useful law cannot be applied to all gases under all conditions. In order for this law to apply, the gas in question must behave in an "ideal" manner. Such a gas is called an **ideal gas** and has the following properties:

1. **The molecules (or atoms) that make up the gas are very small compared to the total volume available to the gas.**

2. **The molecules (or atoms) that make up the gas must be so far apart from one another that there is no attraction or repulsion between them.**

3. **The collisions that occur between the gas molecules (or atoms) must be elastic. In other words, when the molecules (or atoms) collide, no energy can be lost in the collision. Also, no energy can be lost when the molecules (or atoms) of the gas collide with the walls of the container.**

Thus, in order to use the combined gas law (or any equations presented in this module), you will have to know that the gas you are working with is ideal. How will you know that? It's actually rather simple. Gases tend to behave ideally at high temperatures and low pressures. If you think about it, that should make sense.

After all, for a gas to be ideal, its molecules (or atoms) must be very small compared to the total volume of the gas, and they must be far apart from each other. Thus, the larger total volume that the gas occupies, the more ideal its behavior will be. According to Boyle's Law, when do gases occupy a large volume? At low pressures. In the same way, Charles's Law tells us that gases tend to occupy large volumes at high temperatures. Thus, gases tend to behave ideally at high temperatures and low pressures.

You might be asking yourself what I mean by "high" temperatures and "low" pressures. Is 10 °C a high temperature? What about 0.9 atmospheres? Is that a low pressure? In order to help us determine whether or not a given pressure or temperature is high or low, we define something called **standard temperature and pressure**, which we abbreviate as **STP**.

Standard Temperature and Pressure (STP) - A temperature of 273 K and a pressure of 1.00 atm

This is a very important definition because we use it as a reference to determine whether or not a gas will behave in an ideal manner. If the pressure is significantly larger than standard pressure, we say that the pressure is too high to guarantee that a gas will behave in an ideal manner. In the same way, if the temperature is significantly lower than standard temperature, we say that the temperature is too low to ensure ideal behavior. In other words:

A gas with a temperature that is close to (or larger than) 273 K and a pressure that is near (or lower than) 1.00 atm will behave in an ideal fashion. Under those conditions, we can use all of the equations we will develop in this module.

It is important that you be able to recognize whether or not a gas will behave ideally based on its pressure and temperature.

Dalton's Law of Partial Pressures

Now that we know when a gas behaves ideally, we can learn a few more laws that relate to ideal gases. The next one we will learn about is **Dalton's Law of Partial Pressures**:

Dalton's Law of Partial Pressures - When two or more ideal gases are mixed together, the total pressure of the mixture is equal to the sum of the pressures of each individual gas.

We could restate Dalton's Law mathematically as follows:

$$P_T = P_1 + P_2 + P_3 + ... \qquad (12.11)$$

where P_T is the total pressure of the gas, and P_1, P_2, etc. are the pressures due to each individual gas. These individual pressures are often referred to as the **partial pressures** of each gas.

Dalton's Law might look very simple to you, but it actually tells us a great deal. For example, suppose I have two steel containers that each contain a gas at a pressure of 5 atms. Let's say that container #1 holds oxygen gas at a pressure of 5 atms while container #2 holds nitrogen gas at the same pressure. Dalton's Law tells us that if I were to transfer all of the nitrogen from container #2 into container #1, the resulting pressure in container #1 would be 10 atms. Sounds simple, doesn't it?

Consider another situation. Suppose that container #1 and container #2 each hold 5 atms of oxygen gas. What is the resulting pressure when I transfer all of the gas from container #2 into container #1? It is still 10 atms. Whether container #2 holds oxygen, nitrogen, or any other ideal gas at a pressure of 5 atms, the resulting pressure when the contents of the second container are transferred to the first container will remain 10 atms. This tells us something important:

The pressure of an ideal gas does not depend on the identity of the gas.
It depends only on the quantity of that gas!

We will use this fact in a moment to restate Dalton's Law in a completely different way. For right now, however, we need to learn an important application of Dalton's Law: dealing with **vapor pressure**.

Vapor Pressure

If you were to fill a glass with water and set it out on a table, what would eventually happen? Obviously, all of the water would eventually evaporate and the glass would end up empty. Think for a moment about *why* this occurs. Since the water evaporates, we know that it turns from its liquid phase into its gas phase. How does it do that, however? After all, the water isn't *boiling***.** If the water isn't boiling, why is it changing from liquid to gas?

Remember, according to the kinetic theory of matter, the molecules in a glass of liquid water are all moving around. Some of those molecules move around pretty quickly, while others move rather slowly. As the molecules move around, some of them will encounter the surface of the water. If they are moving fast enough, they can hit the surface of the water with enough energy to actually escape from the liquid. When they do that, they turn from the liquid phase into the gas phase. Thus, evaporation occurs because molecules of a liquid are sometimes able to escape from the surface of that liquid, provided they have sufficient energy to do so.

Since this is happening continuously, there are water molecules that continuously hit the surface of the water and escape. Thus, at any one time, there is always some vapor that sits on the top of a sample of liquid. That vapor is composed of the molecules that have escaped the liquid by crashing through the surface and turning into a gas. Well, every gas exerts pressure; thus, the vapor that sits above a sample of liquid exerts a certain amount of pressure:

Vapor pressure - The pressure exerted by the vapor which sits on top of any liquid

In other words, whenever you have a container of liquid, there will be gas floating on top of it. The pressure that this gas exerts on its surroundings is called "vapor pressure."

Now, as you might expect, the vapor pressure of any liquid depends on the chemical nature of that liquid. For example, gasoline has a very large vapor pressure, because it evaporates rather easily. On the other hand, water has a reasonably low vapor pressure, because it does not evaporate as easily. What you may not know, however, is that the vapor pressure of any given liquid also depends on temperature. For example, the table below lists the vapor pressure of water at several different temperatures:

TABLE 12.1

The Vapor Pressure of Water

Temperature (°C)	*Vapor Pressure (torr)*	*Temperature (°C)*	*Vapor Pressure (torr)*
20	**17.5**	**29**	**30.0**
21	**18.7**	**30**	**31.8**
22	**19.8**	**40**	**55.3**
23	**21.1**	**50**	**92.5**
24	**22.4**	**60**	**149.4**
25	**23.8**	**70**	**233.7**
26	**25.2**	**80**	**355.1**
27	**26.7**	**90**	**525.8**
28	**28.3**	**100**	**760.0**

Notice the general trend in this table. As the temperature increases, so does the vapor pressure. Based on the kinetic theory of matter, this makes sense. As the temperature increases, the molecules in a sample of water increase in energy. Since each molecule has more energy, any molecule that reaches the surface of the liquid has an increased chance of escaping the liquid and turning into a gas. As a result, there is more gas floating on top of the liquid, increasing the vapor pressure. This is a general rule:

The vapor pressure of any liquid increases with increasing temperature.

Thus, to determine a liquid's vapor pressure, we must know both the identity of the liquid and its temperature.

Look at the vapor pressure of water at 100 °C. It is equal to 760.0 torr. Does that number mean anything to you? According to Equation (12.3), 760.0 torr is the same as 1.000 atm, which is normal atmospheric pressure. A temperature of 100 °C should also mean something to you. It is the boiling point of water. It isn't a coincidence that the vapor pressure of water is 1.000 atm at its boiling point. That is, in fact, the true definition of boiling point:

Boiling point - The temperature at which the vapor pressure of a liquid is equal to normal atmospheric pressure

Have you ever heard that water boils at a lower temperature at higher altitudes? This definition tells you why. At high altitudes, atmospheric pressure is significantly lower than 760 torr. Thus, the temperature to which water must be heated for its vapor pressure to equal atmospheric pressure is lower. As a result, the boiling point of water is lower.

(The multimedia CD has a video demonstration of boiling water at room temperature to illustrate the definition of boiling point.)

Now that you understand vapor pressure, we can finally relate it to Dalton's Law. When chemists make a gas, that gas is often collected over a liquid. For example, in Experiment 6.1, you mixed baking soda and vinegar. The reaction produced gaseous carbon dioxide, which you collected in a balloon. Think for a minute about what the balloon contained. Did it contain pure carbon dioxide? No, it didn't. You see, the carbon dioxide you collected was in the presence of liquid water. Since liquid water was present, its vapor was also present. Thus, the gas inside the balloon was a mixture of carbon dioxide *and* water vapor.

How much of the mixture was carbon dioxide and how much was water vapor? That's where Dalton's Law can help you. After all, Dalton's Law tells you that the total pressure of the gas mixture is just the sum of the partial pressure of carbon dioxide and the partial pressure of water. Well, if we know the total pressure, we can look up the partial pressure of the water vapor by simply looking at Table 12.1. Then we can use algebra to determine the partial pressure of carbon dioxide. This tells us how much carbon dioxide was in the balloon.

If you don't quite understand what I'm saying, don't worry. There is an example problem below that will allow you to see exactly how all of this works. Before we go through that example, however, I need to introduce a bit of terminology. It turns out that chemists often produce gas in chemical reactions that contain water. The gas bubbles out of the water, and then the chemist collects it in some container, like a balloon. Since this happens so often, chemists have a phrase for it. When a chemist makes gas in the presence of water, we call it either "collecting gas over water" or "collecting gas in the presence of water." Thus, whenever you see one of those phrases in a problem, you must realize that the gas collected is a mixture of the gas produced and water vapor. To find out how much of the gas was produced, you must use Dalton's Law to account for the vapor pressure of water. The following example shows how that is done:

EXAMPLE 12.2

A chemist produces hydrogen gas in an experiment and collects that gas in the presence of water. If the total pressure of gas collected was 1.115 atm and the temperature was 26 °C, what was the pressure of hydrogen gas collected?

This problem tells you that a chemist collects hydrogen gas in the presence of water. Thus, the gas collected is actually a mixture of hydrogen gas and water vapor. Dalton's Law tells us that the total pressure of that mixture is simply the sum of the individual pressures of water vapor and hydrogen:

$$P_T = P_{\text{hydrogen}} + P_{\text{water vapor}}$$

The problem tells us what the total pressure is, and, since we are also given the temperature, we can use Table 12.1 to determine the vapor pressure of the water. According to the table, water at a temperature of 26 °C has a vapor pressure of 25.2 torr. Now, of course, our pressure units do not match. The total pressure is in atm while the pressure of the water vapor is in torr. I will fix that by converting torr to atm:

$$\frac{25.2\,\cancel{\text{torr}}}{1} \times \frac{1\text{ atm}}{760\,\cancel{\text{torr}}} = 0.0332\text{ atm}$$

Now that we have consistent units, we can use Dalton's Law:

$$1.115\text{ atm} = P_{\text{hydrogen}} + 0.0332\text{ atm}$$

Using algebra, we can then solve for $P_{hydrogen}$:

$$P_{hydrogen} = 1.115 \text{ atm} - 0.0332 \text{ atm} = 1.082 \text{ atm}$$

Thus, the chemical reaction produced 1.082 atm of hydrogen gas.

Make sure you understand this by performing the following problem:

ON YOUR OWN

12.4 A chemist collects carbon dioxide over water. The pressure of gas in the collection container was 961 torr. If the temperature was 22 °C, what pressure of carbon dioxide was collected? (Refer to Table 12.1.)

An Alternative Statement of Dalton's Law

The way we have used Dalton's Law in the previous examples and "On Your Own" problems is not the only way you can use this important law. It turns out that there is an alternative way of expressing it. Before we can learn this, however, we must learn to use yet another concentration unit called the **mole fraction**. Mole fraction is always used in the context of a mixture. Mathematically, the mole fraction of any component in a mixture (abbreviated as "X") can be defined in the following manner:

$$X = \frac{\text{\# of moles of component}}{\text{total \# of moles in the mixture}} \tag{12.12}$$

Before we learn what mole fraction tells us, let's make sure we understand its definition by studying the next example and performing the "On Your Own" problem that follows.

EXAMPLE 12.3

A mixture of gases contains 3.0 moles of oxygen gas, 5.0 moles of nitrogen gas, and 1.0 mole of neon gas. What is the mole fraction of each component in the mixture?

According to Equation (12.12), we calculate the mole fraction of any component in a mixture by taking the number of moles of that component and dividing by the total number of moles in the mixture. According to the problem:

$$\text{Total number of moles} = 3.0 \text{ moles} + 5.0 \text{ moles} + 1.0 \text{ mole} = 9.0 \text{ moles}$$

Now that we have the total number of moles in the mixture, we can simply use Equation (12.12):

$$X_{O_2} = \frac{3.0 \text{ moles}}{9.0 \text{ moles}} = \underline{0.33}$$

$$X_{N_2} = \frac{5.0 \text{ moles}}{9.0 \text{ moles}} = \underline{0.56}$$

$$X_{Ne} = \frac{1.0 \text{ mole}}{9.0 \text{ moles}} = \underline{0.11}$$

Notice a couple of things about mole fraction. First, since the unit "mole" cancels in the equation, mole fraction has no units. If you remember, we call this a **dimensionless quantity**. Second, notice that the three mole fractions add up to 1.00. That should always be the case. If you add up the mole fractions of each component in a mixture, the sum must equal 1. Because of significant figures, they might add to 0.99 or 1.01, but in general, they should add to 1.

ON YOUR OWN

12.5 A tank of compressed air contains 672.0 grams of oxygen, 2,184.0 grams of nitrogen, and 40.0 grams of argon. What is the mole fraction of each component in this mixture?

Now that you have learned to use the equation for mole fraction, it is time to learn what it means. The mole fraction tells us what fraction of the total number of molecules (or atoms) in the mixture is represented by the component of interest. For example, suppose you have a gas mixture that contains carbon dioxide and water vapor. If the mole fraction of carbon dioxide is 0.10, it tells you that one-tenth (0.10) of the molecules in the mixture are carbon dioxide molecules. Thus, for every 100 molecules in the mixture, 10 will be carbon dioxide molecules and the rest will be water molecules.

The reason we need to understand mole fraction is that you can use this concept to restate Dalton's Law. Remember, Dalton's Law tells us that the pressure of a gas does not depend on the identity of the gas. It depends only on the quantity. Thus, if we have a mixture of gases, the partial pressure of each gas should depend only on the amount of that gas in the mixture. Mathematically, we would say:

$$P_1 = X_1 P_T \qquad (12.13)$$

This equation tells us that the partial pressure of any gas in a mixture depends on the mole fraction of that gas and the total pressure of the gas mixture.

Think about what this means: If I have a mixture of nitrogen and oxygen, and I know the total pressure that the mixture exerts on its container, I can calculate the partial pressure of each gas by taking the gas' mole fraction and multiplying it by the total pressure. If this seems a little confusing right now, don't worry. A couple of example problems should flesh it all out:

EXAMPLE 12.4

A gas mixture contains 40.0 grams of carbon monoxide and 10.0 grams of nitrogen dioxide. If the total pressure of this gas mixture is 1.2 atms, what is the partial pressure of each individual gas?

Equation (12.13) tells us that if we know the total pressure of the gas mixture and the mole fraction of each component, we can calculate the partial pressure of each component. We were given the total pressure of the gas mixture, and we were also given enough information to determine the mole fraction of each component of the mixture. To determine the mole fractions, we must first convert from grams to moles:

$$\frac{40.0 \text{ \sout{g CO}}}{1} \times \frac{1 \text{ mole CO}}{28.0 \text{ \sout{g CO}}} = 1.43 \text{ moles CO}$$

$$\frac{10.0 \text{ \sout{g NO}}_2}{1} \times \frac{1 \text{ mole NO}_2}{46.0 \text{ \sout{g NO}}_2} = 0.217 \text{ moles NO}_2$$

We can now use this information to calculate the mole fractions of each gas:

$$\text{Total moles} = 1.43 \text{ moles} + 0.217 \text{ moles} = 1.65 \text{ moles}$$

$$X_{CO} = \frac{1.43 \text{ \sout{moles}}}{1.65 \text{ \sout{moles}}} = 0.867$$

$$X_{NO_2} = \frac{0.217 \text{ \sout{moles}}}{1.65 \text{ \sout{moles}}} = 0.132$$

Notice that the mole fractions do not quite add up to 1. Rather, they add up to 0.999. This is just due to significant figures, so don't worry about it. Now that we have the mole fractions, we can use Equation (12.13) to get the partial pressures of each gas:

$$P_{CO} = 0.867 \cdot 1.2 \text{ atms} = 1.0 \text{ atm}$$

$$P_{NO2} = 0.132 \cdot 1.2 \text{ atms} = 0.16 \text{ atm}$$

This mixture, then, is made up of 1.0 atm of CO and 0.16 atm of NO_2.

A gas mixture contains 456 torr of oxygen, 112 torr of CO_2, and 501 torr of nitrogen. What is the mole fraction of each component in the mixture?

At first, you might think that this problem gives you the wrong information. After all, to calculate mole fraction, don't you need either moles or grams? Not when you have Equation (12.13).

Since this equation relates pressure and mole fraction, you can calculate mole fraction as long as you are given the pressure.

According to the equation, to calculate the mole fraction of any component in the mixture, all I have to know is the total pressure of the mixture and the partial pressure of the component. Since we were given the partial pressure of each component, we can use Equation (12.11) to calculate the total pressure:

$$P_T = P_{oxygen} + P_{carbon\ dioxide} + P_{nitrogen}$$

$$P_T = 456\ torr + 112\ torr + 501\ torr = 1069\ torr$$

Now that we have both the total pressure of the mixture and the partial pressure of each component, we can use Equation (12.13) to calculate the mole fraction of each component:

$$P_1 = X_1 \cdot P_T$$

$$X_1 = \frac{P_1}{P_T}$$

Plugging in all three sets of numbers:

$$X_{oxygen} = \frac{456\ \cancel{torr}}{1069\ \cancel{torr}} = \underline{0.427}$$

$$X_{carbon\ dioxide} = \frac{112\ \cancel{torr}}{1069\ \cancel{torr}} = \underline{0.105}$$

$$X_{nitrogen} = \frac{501\ \cancel{torr}}{1069\ \cancel{torr}} = \underline{0.469}$$

I hope that you are beginning to see how powerful Dalton's Law can be. You can use it to determine the pressures of individual components in a mixture or, more importantly, you can use the partial pressure of a gas to determine its concentration in terms of mole fraction. Make sure you understand this by performing the following problems:

ON YOUR OWN

12.6 A sample of artificially made air contains 78 grams of nitrogen and 21 grams of oxygen. If the pressure of the air sample is 760 torr, what is the partial pressure of each gas?

12.7 A chemist produces hydrogen gas in a chemical reaction and collects it over water at 23 °C. If the pressure of the gas inside the collection vessel is 781 torr, what is the mole fraction of the hydrogen gas in the vessel?

The Ideal Gas Law

The last thing we need to learn about gases is probably the most important. As we have seen in other modules, the most important thing we can learn about a substance is the number of moles of that substance. Usually, we can get this information by knowing the identity of the substance and its mass. For gases, however, the mass is a very difficult thing to measure. Thus, it would be nice to be able to relate the number of moles of a gas to things that are easy to measure. The **ideal gas law** does this. The ideal gas law is:

$$PV = nRT \tag{12.14}$$

In this equation, "P" is the pressure of the gas, "V" is its volume, "T" is its temperature, and "n" is the number of moles of that gas. "R" refers to a special physical constant. It is called the **ideal gas constant** and is a relatively important quantity in the study of chemistry.

The ideal gas constant can have several values, depending on what units are used in Equation (12.14). For our purposes, we will learn only one value:

$$\mathbf{R = 0.0821\ \frac{L \cdot atm}{mole \cdot K}}$$

As with all centered, boldfaced facts, I expect you to memorize this. Before you do that, however, make sure you know what it means. Just like Planck's constant (h) that you learned in Module #7, the ideal gas constant never changes. However, the value of the constant is different if you use different units.

For example, suppose you use m^3 as your volume unit. After all, if you take length times width times height, you get volume. If your length, width, and height are measured in meters, you will get a volume unit of m^3. Also, suppose you measure pressure in Pascals. When you do that, you will use a different value for the ideal gas constant, because you are using different units. For those units, the value of the ideal gas constant is $8.315\ \frac{Pa \cdot m^3}{mole \cdot K}$, which can also be expressed as $8.315\ \frac{J}{mole \cdot K}$. In this course, you will usually work in liters and atmospheres. That's why I asked you to memorize the value of R in those units.

Looking at the ideal gas equation, you can see that it relates four pieces of information concerning a gas. It relates a gas' volume, pressure, temperature, and number of moles. If I am given three of those four pieces of information, I can use Equation (12.14) to determine the fourth. Notice how this works in the following example:

EXAMPLE 12.5

A gas is in a 12.2-liter container at a pressure of 13.5 atm and a temperature of 25 °C. How many moles of gas are present?

This problem gives us P, V, and T, and wants us to calculate n. To do this, we start with Equation (12.14) and rearrange it using algebra in order to solve for n:

$$PV = nRT$$

$$n = \frac{PV}{RT}$$

Now we can't just go plugging in our numbers willy-nilly. We first have to examine our units. First, notice that volume is in liters and pressure is in atms. That tells us we will use $R = 0.0821 \frac{L \cdot atm}{mole \cdot K}$. However, since R has temperature units of K (and since we already know that all temperatures in this module must be converted to K), we need to turn 25 °C into 298 K. Now we can plug in our numbers:

$$n = \frac{13.5 \cancel{atm} \cdot 12.2 \cancel{L}}{0.0821 \frac{\cancel{L} \cdot \cancel{atm}}{mole \cdot \cancel{K}} \cdot 298 \cancel{K}} = 6.73 \text{ moles}$$

Thus, this container contains 6.73 moles of gas.

So you see that using the ideal gas law is actually rather simple, as long as you have your units correct. Make sure you understand how this works by doing the following problem:

ON YOUR OWN

12.8 How many liters does 1.00 mole of an ideal gas occupy at STP?

Using the Ideal Gas Law in Stoichiometry

By now you have probably guessed that I can take virtually any subject in chemistry and relate it back to stoichiometry. Well, I can do it with the ideal gas equation too! Remember, stoichiometry is done in moles. Thus, if I can figure out a way to determine the number of moles of gas I have, I can do stoichiometry on that gas. Perform the following experiment to see how this is done.

EXPERIMENT 12.1

Using the Ideal Gas Equation to Determine the Amount of Acid in Vinegar

Supplies:

- Mass scale
- Plastic 2-liter bottle

Note: A sample set of calculations is available in the solutions and tests guide. It is with the solutions to the practice problems.

- Round balloon with an 8-inch diameter
- Vinegar
- Baking soda
- Seamstress' tape measure (A piece of string and a ruler will work as well.)
- Thermometer
- Weather report that contains the atmospheric (sometimes called barometric) pressure for the day (You can get this information at www.weather.com. If you can't find a weather report or access the internet, assume that the atmospheric pressure is 1.00 atm.)

1. Clean out the 2-liter bottle and fill it with 200.0 grams of vinegar (a quantitative measurement). The best way to do this is to measure the mass of your bottle with the scale and then slowly add vinegar while the bottle is still on the scale. You have added 200.0 grams of vinegar when the mass reaches 200.0 more grams than the original mass of the bottle.
2. Fill the balloon with at least 30 grams of baking soda. The baking soda is the excess reactant here, so more than 30 grams is fine.
3. Once you have filled the bottle and the balloon, attach the balloon to the bottle as you did in Experiment 6.1. Once again, be sure that you have an airtight seal between the bottle and the balloon.
4. Hold the balloon up so that the baking soda mixes with the vinegar.
5. Gently shake the balloon so that all of the baking soda falls into the vinegar.
6. Once the bubbling subsides, gently swirl the contents of the bottle to make sure that the baking soda has fully mixed with the vinegar.
7. Now you need to measure the radius of the balloon. To accomplish this, you need to measure the distance around the widest part of the balloon. You can use a seamstress' tape measure to do this. If you don't have such a tape measure, take a piece of string and wrap it once around the widest part of the balloon. Carefully cut the string so that it is just the length necessary to wrap around the balloon, then use a ruler to measure the length of the string.
8. If your ruler or tape measure is marked in inches, convert to centimeters by multiplying your measurement by 2.54. Once you have the measurement in centimeters, divide it by 10. This will give you the circumference of the balloon in decimeters.
9. Take the circumference of the balloon in decimeters and divide it by 2π. Since the circumference of a sphere is $2\pi r$, dividing the circumference of the balloon by 2π gives us the balloon's radius.
10. Take the radius you just calculated and plug it into the volume equation for a sphere:

$$V = \frac{4}{3}\pi r^3$$

11. Your answer is the volume of the balloon in dm^3, which (believe it or not) is the same as liters. Thus, by measuring the circumference of the balloon, you were able to calculate the volume in liters. This represents the volume of the gas produced in the reaction.
12. As we learned in Experiment 6.1, the reaction that occurs when baking soda and vinegar are mixed is:

$$\underset{\text{acetic acid in vinegar}}{C_2H_4O_2\,(l)} + \underset{\text{baking soda}}{NaHCO_3\,(s)} \rightarrow H_2O\,(l) + CO_2\,(g) + \underset{\text{sodium acetate}}{NaC_2H_3O_2\,(aq)}$$

13. The calculation that you just performed gives you the volume of gas produced in this reaction. Because the gas was collected over water, it will be a mixture of water vapor and carbon dioxide.

14. Look at the weather report to determine the atmospheric pressure for today. If you can't find it, 1.00 atm is a good assumption. It is likely that the atmospheric pressure will be listed with a unit of inches. That represents inches of mercury. If you divide the pressure in inches by 29.92, you will get the pressure in atms.
15. The atmospheric pressure corresponds to the pressure of the gas inside your balloon, because the atmosphere is pushing against the balloon. The problem is, there is not only CO_2 in the balloon, but there is also water vapor. Thus, to determine the pressure of the CO_2 in the balloon, you must subtract the vapor pressure of water. To do this, measure the temperature of the room and determine the vapor pressure of water for that temperature by looking at Table 12.1. Convert that pressure to atm by dividing by 760.0 and subtracting it from the atmospheric pressure. The answer represents the pressure of the CO_2 formed in the experiment.
16. Take your measured temperature and convert it to K. This will give you the temperature of the CO_2 gas formed in the experiment.
17. Now look what you have. You know the volume of CO_2 gas (in liters), the pressure of the gas (in atm), and the temperature of the gas (in K). You can use that information and the ideal gas law (with $R = 0.0821 \frac{\text{l} \cdot \text{atm}}{\text{mole} \cdot \text{K}}$) to determine the number of moles of gas produced. Once you have that, you can use it in stoichiometry to determine how much of the limiting reactant (acetic acid) you had.
18. According to the equation, for every 1 mole of CO_2 produced, 1 mole of acetic acid should have reacted. Thus, the number of moles you just calculated is equal to the number of moles of acetic acid you started with.
19. The molecular mass of acetic acid is 60.0 amu, so take the number of moles you just calculated and multiply it by 60.0. This gives you the number of grams of acetic acid you started with.
20. Clean up your mess.

Don't get so lost in the math that you fail to realize what you did. By measuring the pressure, volume, and temperature of a gas, you were able to calculate the number of moles. That allowed you to perform stoichiometric calculations to learn about the other substances in the reaction you just studied. If you aren't quite sure exactly why you did all of the math that I told you to do, don't worry about that, either. Real-life experiments are always more difficult than the problems that I expect you to do.

Now that you've had a little practical experience with using the ideal gas law in stoichiometry, I will present a couple of example problems that have the same level of difficulty as the problems that I expect you to be able to do. After you have studied those examples, do the "On Your Own" problems that follow.

EXAMPLE 12.6

One way that hydrogen gas can be produced commercially is to react a strong acid with a metal:

$$2HCl\ (aq) + Zn\ (s) \rightarrow H_2\ (g) + ZnCl_2\ (aq)$$

If 512 grams of HCl are reacted with excess Zn at T = 24 °C and P = 0.95 atm, what volume of hydrogen gas will be produced?

In this stoichiometry problem, we are given the amount of limiting reactant and asked to calculate how much product will be made. This is nothing new; it just uses the things you have already learned in a different way. We first convert the amount of limiting reactant to moles:

$$\frac{512 \text{ g HCl}}{1} \times \frac{1 \text{ mole HCl}}{36.5 \text{ g HCl}} = 14.0 \text{ moles HCl}$$

We can then use stoichiometry to determine the number of moles of H_2 produced:

$$\frac{14.0 \text{ moles HCl}}{1} \times \frac{1 \text{ mole } H_2}{2 \text{ moles HCl}} = 7.00 \text{ moles } H_2$$

Now we get to the point where we apply the things we learned in this module. Instead of asking how many moles or how many grams of H_2, the problem asks how many liters. For that, we need to use the ideal gas law. Looking at our units, we will need to convert temperature into Kelvin so that the units for "R" will work out. Once we do that, we can then use the ideal gas law:

$$PV = nRT$$

$$V = \frac{nRT}{P} = \frac{7.00 \text{ moles} \cdot 0.0821 \frac{\text{L} \cdot \text{atm}}{\text{mole} \cdot \text{K}} \cdot 297 \text{ K}}{0.95 \text{ atm}} = 180 \text{ L}$$

The reaction, then, will produce <u>180 liters</u> of hydrogen.

To make magnesium hydroxide, a major component in some stomach antacids, hot water vapor is passed over solid magnesium:

$$2H_2O\ (g) + Mg(s) \rightarrow Mg(OH)_2\ (s) + H_2\ (g)$$

If 4.50 liters of water vapor at 200.0 °C and 4.94 atm are passed over excess magnesium, how many grams of magnesium hydroxide will be produced?

In this problem, we are once again given the amount of limiting reactant and asked to determine how much product is made. The problem here is that the amount of limiting reactant is not given in grams or moles. Instead it is given in P, V, and T. Thus, we must use the ideal gas law to determine the number of moles of limiting reactant. Looking at the units, we must convert the temperature to Kelvin, and then we can use the ideal gas law to solve for the number of moles:

$$n = \frac{PV}{RT} = \frac{4.94 \text{ atm} \cdot 4.50 \text{ L}}{0.0821 \frac{\text{L} \cdot \text{atm}}{\text{mole} \cdot \text{K}} \cdot 473.2 \text{ K}} = 0.572 \text{ moles}$$

Now that we have the moles of limiting reactant, this becomes a stoichiometry problem:

$$\frac{0.572\ \cancel{\text{moles } H_2O}}{1} \times \frac{1\ \text{mole } Mg(OH)_2}{2\ \cancel{\text{moles } H_2O}} = 0.286\ \text{moles } Mg(OH)_2$$

$$\frac{0.286\ \cancel{\text{moles } Mg(OH)_2}}{1} \times \frac{58.3\ \text{g } Mg(OH)_2}{1\ \cancel{\text{mole } Mg(OH)_2}} = 16.7\ \text{g } Mg(OH)_2$$

Thus, 16.7 g of magnesium hydroxide will be produced.

ON YOUR OWN

12.9 In an automobile engine, gasoline (an important component of which is octane, C_8H_{18}) is burned to make gaseous carbon dioxide and water. A gallon of octane has a mass of about 3.2×10^3 grams. If a gallon of octane is burned in excess oxygen at P = 1.00 atm and T = 180.0 °C, how many liters of CO_2 are produced? The unbalanced chemical equation is:

$$C_8H_{18}\ (l) + O_2\ (g) \rightarrow CO_2\ (g) + H_2O\ (g)$$

12.10 Methanol (commonly called wood alcohol) can be produced as follows:

$$CO\ (g) + 2H_2\ (g) \rightarrow CH_4O\ (l)$$

If 500.0 liters of hydrogen gas are reacted with excess CO at STP, how many grams of CH_4O can be produced?

Now that you are done studying ideal gases, you might wonder how chemists deal with non-ideal gases. After all, we already said that gases behave ideally only when the pressure is near or below standard pressure and the temperature is near or above standard temperature. What happens if you are at a pressure and temperature where gases do not perform ideally? It turns out that nonideal gases are very difficult to deal with, so you will not have to worry about them in this course. However, since STP is very near room temperature and pressure, it is safe to assume that all gases under conditions of room temperature and pressure are ideal.

ANSWERS TO THE "ON YOUR OWN" PROBLEMS

12.1 This problem asks you to predict how a gas will change when you change some of the conditions under which it is stored. This means that you need to use the combined gas law (Equation 12.10).

$$\frac{P_1V_1}{T_1} = \frac{P_2V_2}{T_2}$$

According to this problem, $P_1 = 3{,}415$ kPa, $T_1 = 0.00\ ^{\circ}C$, and $T_2 = 25\ ^{\circ}C$. Also, the problem states that the cylinder's volume cannot change; thus, the volumes cancel out:

$$\frac{P_1\cancel{V_1}}{T_1} = \frac{P_2\cancel{V_2}}{T_2}$$

We can now rearrange the equation to solve for the new pressure:

$$\frac{P_1T_2}{T_1} = P_2$$

Before we put in the numbers and determine the new pressure, we first have to convert the temperatures to Kelvin, because all temperatures in this equation must be in Kelvin:

$$T_1 = 0.00 + 273.15 = 273.15\ K$$

$$T_2 = 25.0 + 273.15 = 298.2\ K$$

Now we can put in the numbers and determine the new pressure:

$$\frac{3{,}415\ kPa \cdot 298.2\ \cancel{K}}{273.15\ \cancel{K}} = 3.728 \times 10^3\ kPa$$

The new pressure, then, is <u>3.728×10^3 kPa</u>.

12.2 This is obviously another combined gas law problem, with $P_1 = 1.00$ atm, $V_1 = 5.5$ liters, $T_1 = 25\ ^{\circ}C$, $P_2 = 915$ torr, and $T_2 = 20\ ^{\circ}C$. The problem asks us to determine the new volume, so we have to rearrange Equation (12.10) to solve for V_2:

$$\frac{P_1V_1T_2}{T_1P_2} = V_2$$

Before we can plug in the numbers, however, we need to convert the temperatures to Kelvin:

$$T_1 = 25 + 273.15 = 298\ K$$

$$T_2 = 19 + 273.15 = 292\ K$$

Additionally, we need to make the pressure units the same. We can do this by converting torr into atm or vice versa. I will choose to do the latter:

$$P_1 = \frac{1.00 \cancel{\text{atm}}}{1} \times \frac{760.0 \text{ torr}}{1.000 \cancel{\text{atm}}} = 7.60 \times 10^2 \text{ torr}$$

Now we can plug in the numbers:

$$\frac{7.60 \times 10^2 \cancel{\text{torr}} \cdot 5.5 \text{ liters} \cdot 292 \cancel{\text{K}}}{298 \cancel{\text{K}} \cdot 915 \cancel{\text{torr}}} = V_2$$

$$4.5 \text{ liters} = V_2$$

So when the person dives deep, the volume of his or her lungs decreases to 4.5 liters.

12.3 In this combined gas law problem, we are asked to calculate the volume of a balloon under conditions of constant pressure. Thus, we need to rearrange the equation to solve for V_2, realizing that the pressures cancel because they are the same:

$$\frac{\cancel{P_1} V_1}{T_1} = \frac{\cancel{P_2} V_2}{T_2}$$

$$\frac{V_1 T_2}{T_1} = V_2$$

To be able to calculate the new volume, we first need to get the temperatures into Kelvin:

$$T_1 = 25 + 273.15 = 298 \text{ K}$$

$$T_2 = -5 + 273.15 = 268 \text{ K}$$

Now we can plug in our numbers:

$$\frac{2.31 \text{ L} \cdot 268 \cancel{\text{K}}}{298 \cancel{\text{K}}} = 2.08 \text{ L}$$

Thus, the balloon shrinks to a volume of 2.08 liters.

12.4 This problem requires you to recognize that when a gas is collected over water, the gas is contaminated with water vapor. Thus, the 961 torr of gas is the total pressure of carbon dioxide *plus* water vapor. To determine the partial pressure of water vapor, we need only look at Table 12.1. At 22 °C, the vapor pressure of water is 19.8 torr. Thus, Dalton's Law becomes:

$$P_T = P_{\text{carbon dioxide}} + P_{\text{water vapor}}$$

$$961 \text{ torr} = P_{\text{carbon dioxide}} + 19.8 \text{ torr}$$

$$P_{\text{carbon dioxide}} = 961 \text{ torr} - 19.8 \text{ torr} = 941 \text{ torr}$$

Thus, only 941 torr of carbon dioxide was collected.

12.5 Mole fraction is defined as the number of *moles* of component divided by the total number of *moles*. Right now, the problem gives us *grams*, not moles. Thus, we must first convert from grams to moles:

$$\frac{672.0 \text{ g } O_2}{1} \times \frac{1 \text{ mole } O_2}{32.0 \text{ g } O_2} = 21.0 \text{ moles } O_2$$

$$\frac{2{,}184.0 \text{ g } N_2}{1} \times \frac{1 \text{ mole } N_2}{28.0 \text{ g } N_2} = 78.0 \text{ moles } N_2$$

$$\frac{40.0 \text{ g Ar}}{1} \times \frac{1 \text{ mole Ar}}{39.9 \text{ g Ar}} = 1.00 \text{ moles Ar}$$

Realize that we had to use N_2 and O_2 for nitrogen and oxygen because they are both homonuclear diatomics. Now that we have the number of moles of each component, we can calculate the total number of moles in the mixture:

$$\text{Total number of moles} = 21.0 \text{ moles} + 78.0 \text{ moles} + 1.00 \text{ moles} = 100.0 \text{ moles}$$

Plugging that into Equation (12.12):

$$X_{O_2} = \frac{21.0 \text{ moles}}{100.0 \text{ moles}} = \underline{0.210}$$

$$X_{N_2} = \frac{78.0 \text{ moles}}{100.0 \text{ moles}} = \underline{0.780}$$

$$X_{Ar} = \frac{1.00 \text{ moles}}{100.0 \text{ moles}} = \underline{0.0100}$$

The mole fractions are dimensionless and they all add up to 1.

12.6 This problem gives us the total pressure and the amount of each gas, and asks us to determine the partial pressures of those gases. This is clearly a Dalton's Law problem using Equation (12.13). We must therefore use the amounts of each gas to determine the mole fraction of each component:

$$\frac{78 \text{ g } N_2}{1} \times \frac{1 \text{ mole } N_2}{28.0 \text{ g } N_2} = 2.8 \text{ moles } N_2$$

$$\frac{21 \text{ g } O_2}{1} \times \frac{1 \text{ mole } O_2}{32 \text{ g } O_2} = 0.66 \text{ moles } O_2$$

Total moles = 2.8 moles + 0.66 moles = 3.5 moles

$$X_{N_2} = \frac{2.8 \text{ moles}}{3.5 \text{ moles}} = 0.80$$

$$X_{O_2} = \frac{0.66 \text{ moles}}{3.5 \text{ moles}} = 0.19$$

The mole fractions do not add up to exactly 1 because of significant figures. Now we can finally use Equation (12.13):

$$P_1 = X_1 \cdot P_T$$

$$P_{N_2} = 0.80 \cdot 760 \text{ torr} = \underline{6.1 \times 10^2 \text{ torr}}$$

$$P_{O_2} = 0.19 \cdot 760 \text{ torr} = \underline{1.4 \times 10^2 \text{ torr}}$$

12.7 This problem combines what we learned about vapor pressure and what we learned about using Dalton's Law to calculate mole fraction. We are told that the total pressure of the gas once it is collected is 781 torr. However, since it was collected over water, we should recognize that it is contaminated with water vapor. Thus, we really have a mixture of water vapor and hydrogen gas. According to Table 12.1, water vapor has a pressure of 21.1 torr at 23 °C. We can use this information and Equation (12.11) to calculate the partial pressure of hydrogen in this mixture:

$$P_T = P_{\text{hydrogen}} + P_{\text{water vapor}}$$

$$781 \text{ torr} = P_{\text{hydrogen}} + 21.1 \text{ torr}$$

$$P_{\text{hydrogen}} = 781 \text{ torr} - 21.1 \text{ torr} = 7.60 \text{ x } 10^2 \text{ torr}$$

Now that we have the total pressure and the partial pressure of hydrogen, we can figure out the mole fraction of hydrogen:

$$P_{hydrogen} = X_{hydrogen} \cdot P_T$$

$$X_{hydrogen} = \frac{P_{hydrogen}}{P_T} = \frac{7.60 \times 10^2 \ \cancel{torr}}{781 \ \cancel{torr}} = \underline{0.973}$$

12.8 In this problem, we are given pressure and temperature (STP means 1.00 atm and 273 K) and the number of moles. We are then asked to calculate V. We can do this by rearranging the ideal gas law:

$$PV = nRT$$

$$V = \frac{nRT}{P}$$

Once again, we have to examine our units. In this case, they work because temperature is in Kelvin:

$$V = \frac{1.00 \ \cancel{moles} \cdot 0.0821 \frac{L \cdot \cancel{atm}}{\cancel{mole} \cdot \cancel{K}} \cdot 273 \ \cancel{K}}{1.00 \ \cancel{atm}} = 22.4 \text{ L}$$

At STP, then, one mole of an ideal gas occupies <u>22.4 liters</u>.

12.9 In this stoichiometry problem, we are given the amount of limiting reactant and asked to calculate how much product will be made. First, we must balance the equation:

$$2C_8H_{18} (l) + 25O_2 (g) \rightarrow 16CO_2 (g) + 18H_2O (g)$$

Now we can convert the amount of limiting reactant to moles:

$$\frac{3.2 \times 10^3 \ \cancel{g \ octane}}{1} \times \frac{1 \text{ mole octane}}{114.2 \ \cancel{g \ octane}} = 28 \text{ moles octane}$$

We can then use stoichiometry to determine the number of moles of CO_2 produced:

$$\frac{28 \ \cancel{moles \ octane}}{1} \times \frac{16 \text{ moles } CO_2}{2 \ \cancel{moles \ octane}} = 2.2 \times 10^2 \text{ moles } CO_2$$

Now we need to use the ideal gas law. Looking at our units, we will need to convert temperature into Kelvin before we use the ideal gas law:

$$PV = nRT$$

$$V = \frac{nRT}{P} = \frac{2.2 \times 10^2 \ \cancel{moles} \cdot 0.0821 \frac{L \cdot \cancel{atm}}{\cancel{mole} \cdot \cancel{K}} \cdot 453.2 \ \cancel{K}}{1.00 \ \cancel{atm}} = \underline{8.2 \times 10^3 \text{ L}}$$

12.10 In this problem, we are once again given the amount of limiting reactant and asked to determine how much product is made. The problem here is that the amount of limiting reactant is not given in grams or moles. Instead it is given in P, V, and T. Thus, we must use the ideal gas law to determine the number of moles of limiting reactant. Since STP means P = 1.00 atm and T = 273 K, the ideal gas law becomes:

$$n = \frac{PV}{RT} = \frac{1.00\ \cancel{\text{atm}} \cdot 500.0\ \cancel{\text{L}}}{0.0821\ \frac{\cancel{\text{L}} \cdot \cancel{\text{atm}}}{\text{mole} \cdot \cancel{\text{K}}} \cdot 273\ \cancel{\text{K}}} = 22.3\ \text{moles}$$

Now that we have the moles of limiting reactant, this becomes a stoichiometry problem:

$$\frac{22.3\ \cancel{\text{moles H}_2}}{1} \times \frac{1\ \text{mole CH}_4\text{O}}{2\ \cancel{\text{moles H}_2}} = 11.2\ \text{moles CH}_4\text{O}$$

$$\frac{11.2\ \cancel{\text{moles CH}_4\text{O}}}{1} \times \frac{32.0\ \text{g CH}_4\text{O}}{1\ \cancel{\text{mole CH}_4\text{O}}} = 358\ \text{g CH}_4\text{O}$$

Thus, 358 g are produced.

REVIEW QUESTIONS FOR MODULE #12

1. Define pressure and name three units used to measure it.

2. Write Boyle's Law in your own words.

3. Which of the laws you studied in this module leads to the Kelvin temperature scale?

4. What general rule must a careful scientist follow when extrapolating data?

5. What are the properties of an ideal gas?

6. Under what conditions do gases behave ideally?

7. What is standard temperature and pressure?

8. Two steel containers of equal volume contain two different gases at the same temperature. The first container holds 1.3 moles of N_2 gas and the second contains 1.3 moles of SO_3 gas. Compare the pressures of the two containers.

9. A liquid's temperature is lowered from 50 °C to 10 °C. Did the the liquid's vapor pressure increase, decrease, or stay the same during that time?

10. The mole fraction of nitrogen gas in compressed air is 0.78. If you have 1,000 molecules of compressed air, how many will be nitrogen molecules?

PRACTICE PROBLEMS FOR MODULE #12

1. A balloon is filled with hydrogen at a temperature of 20 °C and a pressure of 755 mmHg. If the balloon's original volume was 1.05 liters, what will its new volume be at a higher altitude, where the pressure is only 625 mmHg? Assume the temperature stays the same.

2. A bacterial culture isolated from sewage produced 46.1 mL of methane gas at 25 °C and 780 torr. What would the volume of the methane be at STP?

3. A steel container with a volume that cannot change is filled with 123.0 atms of compressed air at 20.0 °C. It sits out in the sun all day, and its temperature rises to 55.0 °C. What is the new pressure?

4. In an experiment similar to Experiment 6.1, a chemist collects carbon dioxide over water at 27 °C. If the gas in the collection vessel has a pressure of 787 torr, what is the pressure of carbon dioxide in the vessel? (Use Table 12.1.)

5. As part of an environmental experiment, the mixture of gases released from a smokestack is collected in a balloon. If the mixture is analyzed and found to contain 15.0 g of SO_2 gas, 12.1 g of NO gas, and 2.5 g of SO_3 gas, what is the mole fraction of each component in the mixture?

6. If the total pressure of the gas mixture in the problem above was 1.1 atm, what was the partial pressure of each gas?

7. A steel container is filled with 5.00 atms of nitrogen, 2.00 atms of oxygen, and 0.50 atms of argon. What is the total pressure in the vessel and the mole fraction of each gas?

8. What is the volume of 16.1 grams of chlorine gas at 1.00 atm and 25 °C?

9. Silver tarnishes in the presence of oxygen and hydrogen sulfide according to the following equation:

$$4Ag\ (s) + 2H_2S\ (g) + O_2\ (g) \rightarrow \underset{\text{tarnish}}{2Ag_2S\ (s)} + 2H_2O\ (l)$$

Usually, H_2S is the limiting reactant. If a silver spoon is placed in a container that holds 56.7 mL of H_2S at 1.23 atm and 22 °C, how many grams of silver on the spoon will tarnish?

10. One common rocket engine uses the following reaction:

$$7H_2O_2\ (aq) + N_2H_4\ (g) \rightarrow 2HNO_3\ (g) + 8H_2O\ (g)$$

If the rocket engine contains 1500.0 grams of H_2O_2 and an excess of N_2H_4, what volume of water vapor will be produced when the rocket fires up to a temperature of 540 °C and a pressure of 1.2 atm?

MODULE #13: Thermodynamics

Introduction

In Module #11, we learned about endothermic and exothermic processes. Specifically, I told you that when solutes dissolve exothermically, they release energy, causing the solution to heat up. Similarly, when solutes dissolve endothermically, they absorb energy from their surroundings, causing the solution to cool down. In this module, we will examine exothermic and endothermic chemical reactions.

Remember that when a solute dissolves in a solvent, no *chemical* change is occurring. Way back in Module #4 we learned that when things dissolve, they undergo a *physical* change. As you might expect, however, chemical changes can also be accompanied by either a release or absorption of energy. In fact, almost all chemical changes either release or absorb energy. Thus, the definitions of "exothermic" and "endothermic" can also be applied to chemical reactions. In this module, we will learn how this is done.

Enthalpy

Up to this point, our discussions of chemical reactions have focused on the substances that participate in those reactions. It is now time to turn our attention to the energy change that accompanies the reaction. As I mentioned above, chemical reactions can be exothermic or endothermic. If a reaction is exothermic, energy is released in the process. This means that we can think of energy as a *product* in the chemical reaction. After all, if energy is released during the reaction, it is as if the reaction produced energy. In the same way, if a chemical reaction is endothermic, we can think of energy as a reactant in the chemical equation.

Now remember, the First Law of Thermodynamics tells us that energy cannot be created or destroyed. Thus, energy is never really produced by a chemical reaction, and neither is it really consumed. Nevertheless, for the purpose of our discussion, we will think of energy and chemical reactions in the following way:

When a reaction is exothermic, energy will be thought of as a product in the reaction. If the reaction is endothermic, energy will be considered a reactant.

Once again, make sure you understand that this is just a convenient way of thinking about the energy changes that accompany chemical reactions. It is not really correct, because reactants in a chemical reaction get consumed and products get produced. The First Law of Thermodynamics tells us that energy cannot be consumed or produced. It can only change forms.

If chemical reactions don't really produce or consume energy, where does the energy come from? When I perform a chemical reaction that is exothermic, the entire reaction will get hot. Since energy is required to raise the temperature of any object, I know that energy had to have come from somewhere. If the reaction didn't really produce the energy, how did it get there? The answer to this requires you to remember something that we learned way back in Module #2.

Remember when we first discussed energy? I told you back then that there are two types of energy: potential and kinetic. Potential energy is defined as energy that is stored, while kinetic energy is defined as energy in motion. It turns out that in chemistry, we can think of potential energy as the energy that is stored up in the chemical bonds of a substance. Kinetic energy, on the other hand, can be thought of as the heat associated with a chemical reaction. Thus, when energy is absorbed or released in a chemical reaction, it is considered kinetic energy. The substances that are a part of the chemical reaction, however, contain potential energy.

If we think about things in this way, we can finally see where the energy associated with a chemical reaction comes from. If a chemical reaction is exothermic, some of the potential energy that is stored up in the reactants is converted into kinetic energy. This kinetic energy then heats up the things surrounding the reaction, making the reaction hot. On the other hand, if a chemical reaction is endothermic, some of the kinetic energy in the surroundings is absorbed and converted into the potential energy stored in the products of the reaction. Since this situation causes the things surrounding the chemical reaction to lose kinetic energy, they feel cold.

This, then, is the real explanation for where the energy in a chemical reaction comes from. Potential energy in the chemical bonds of the substances involved can be converted into kinetic energy, or vice versa. Now that we know the real explanation, however, we will still use the convention that exothermic reactions have energy as a product, and endothermic reactions have energy as a reactant. This convention, even though it is not physically accurate, is just too convenient to ignore.

If this discussion has all been a little bit too theoretical for you, let's bring it down to practical terms by giving you a couple of examples. If you react HCl (which you should recognize as an acid) with NaOH (which you should recognize as a base), the following chemical reaction will occur:

$$HCl + NaOH \rightarrow H_2O + NaCl \qquad (13.1)$$

Remember, based on our discussion of acid/base chemistry in Module #10, you should have been able to predict the products of this reaction, given only the reactants. If you were to actually perform this reaction, you would feel the reaction vessel get hot, because this reaction is exothermic. Given the fact that we now know this is an exothermic reaction, I could rewrite Equation (13.1) as follows:

$$HCl + NaOH \rightarrow H_2O + NaCl + \text{energy} \qquad (13.2)$$

This is what I mean when I say that we will consider energy to be a product in an exothermic reaction.

In the same way, if I mix NH_4SCN and $Ba(OH)_2(H_2O)_8$ together, the following reaction will occur:

$$2NH_4SCN + Ba(OH)_2(H_2O)_8 \rightarrow 2NH_4OH + Ba(SCN)_2 + 8H_2O \qquad (13.3)$$

When you perform this reaction, you will feel the reaction vessel get cold, indicating that this is an endothermic reaction. Since we know it is endothermic, I could re-write the equation in the following way:

$$2NH_4SCN + Ba(OH)_2(H_2O)_8 + \text{energy} \rightarrow 2NH_4OH + Ba(SCN)_2 + 8H_2O \qquad (13.4)$$

This is what I mean when I say that we will consider energy as a reactant in endothermic reactions.

Now that we understand how to think about the energy changes that accompany chemical reactions, it is time to learn some more terminology. First, since every substance in nature contains potential energy, we have a name for it. The energy stored in a substance is called the **enthalpy** (en' thuhl pee) of that substance. For some odd reason, chemists use "H" to symbolize enthalpy. In a chemical reaction, substances are both consumed and produced. As a result, enthalpy changes in a chemical reaction. When this happens, energy is either released or absorbed. We call the energy associated with a chemical reaction (the same energy that appears in Equations [13.2] and [13.4]) the **change in enthalpy**.

<u>Change in Enthalpy (ΔH)</u> - The energy change that accompanies a chemical reaction

Since enthalpy is abbreviated with an "H," the change in enthalpy is symbolized by "ΔH."

It turns out that chemists (even this one) often get very sloppy with their terminology. Thus, even though ΔH represents the change in enthalpy, we often refer to it as just enthalpy. Thus, we might call the ΔH of a particular reaction the "enthalpy of the reaction," when it really should be referred to as the "change in enthalpy" of that reaction. You will therefore have to be comfortable with ΔH being referred to as both "change in enthalpy" and "enthalpy." The latter term is, in fact, wrong, because enthalpy refers to the energy associated with a *particular* substance, not a group of substances that undergo change. Nevertheless, this sloppy terminology exists in chemistry, so you must get used to it.

Now remember, ΔH represents the energy change associated with a chemical reaction. As a result, it will have an energy unit attached to it. Remember from Module #2 that the two major energy units are Joules and calories. These units are related to each other as follows:

$$1 \text{ calorie} = 4.184 \text{ Joules} \quad (13.5)$$

Since most chemical reactions involve quite a bit of energy, the units you will probably see attached to ΔH are either kiloJoules (kJ) or kilocalories (kcal).

It turns out that kJ and kcals are not the only units you will see attached to ΔH. As you might expect, the energy released or absorbed by a chemical reaction depends on the *amount* of reactants that are consumed. After all, if I burn 1.00 kg of wood, I will produce more heat than if I burn 1.00 g of wood. As a result, the value of ΔH really depends on the amount of reactants undergoing the chemical change under consideration. In order to account for this, chemists sometimes define the ΔH of a reaction on a mole-by-mole basis. Thus, you will sometimes see ΔH with the units kJ/mole or kcals/mole. These units just tell you that the energy change represented by ΔH occurs when one mole of reactant gets consumed in the reaction. Thus, a ΔH of 50 kJ/mole simply means that after 1 mole of reactant undergoes the chemical change, there is an energy change of 50 kJ.

You not only have to recognize the units that are attached to ΔH, but you also need to understand that ΔH can be either positive *or* negative. By convention, we say that:

ΔH is positive for endothermic reactions, and ΔH is negative for exothermic reactions.

Thus, if you are told that a reaction has a ΔH of -123 kJ/mole, you know that the reaction is exothermic (ΔH is negative), and you can think of the reaction as producing 123 kJ of energy for every mole of reactant consumed. On the other hand, a reaction with a ΔH of 56 kJ/mole is endothermic and thus has 56 kJ as a reactant. Let's see how we use all of this terminology in an example problem:

EXAMPLE 13.1

When CH_4 undergoes complete combustion, the change in enthalpy of the reaction is -803.1 kJ. Write a balanced chemical equation representing this process. Include energy in your equation.

This problem first requires us to remember what a complete combustion reaction is. If you recall, complete combustion was defined as reacting a substance with oxygen to produce carbon dioxide and water. Thus, the unbalanced chemical equation is:

$$CH_4 + O_2 \rightarrow CO_2 + H_2O$$

Balancing the equation gives us:

$$CH_4 + 2O_2 \rightarrow CO_2 + 2H_2O$$

Now, according to the example, ΔH is negative, implying that the reaction is exothermic. Thus, energy is a product:

$$CH_4 + 2O_2 \rightarrow CO_2 + 2H_2O + \underline{803.1\text{ kJ}}$$

Solve the following problem, which also provides you with some review because it requires you to predict certain chemical equations. If you don't remember how to predict formation, decomposition, and combustion equations, refer to Module #5. If you can't figure out the acid/base reaction, refer to Module #10.

ON YOUR OWN

13.1 Write balanced chemical equations for each of the following processes, being sure to include energy as either a reactant or a product:

a. The formation of H_2SO_4, which has a ΔH of -814 kJ.
b. The decomposition of K_3PO_4, which has a ΔH of 2,030 kJ.
c. The reaction between H_2CO_3 and KOH, which releases 21 kcals.

Determining ΔH for a Chemical Reaction by Experiment

Now that you understand what the enthalpy of a chemical reaction is and how it relates to the chemical equation, the next logical step is to figure out how we can predict it. After all, in both the example problem and the "On Your Own" problem from the last section, I told you the ΔH for several different chemical reactions. How did I figure them out? Well, of course, I looked in a few books. But how did the authors of those books figure the ΔH's out?

It turns out that there are three ways to determine the ΔH of a chemical reaction. The first method is experimental. Way back in Module #2 you did an experiment using a technique we called calorimetry. This technique can be applied to chemical reactions as a method of determining ΔH. Remember what calorimetry does. In a calorimetry experiment, we measure the temperature change of water inside a calorimeter. We use that temperature change to learn about the energy associated with the process that we studied. To refresh your memory on calorimetry, and to learn how ΔH can be experimentally determined, perform the following experiment:

EXPERIMENT 13.1

Determining the ΔH of a Chemical Reaction

Note: A sample set of calculations is available in the solutions and tests guide. It is with the solutions to the practice problems.

Supplies:

- Two Styrofoam coffee cups
- Thermometer
- Vinegar
- Mass scale
- Measuring tablespoon and ½ teaspoon
- Lye (This is commonly sold in supermarkets with the drain cleaners. A popular brand is Red Devil® Lye. If you cannot find lye, any *powdered* drain cleaner ought to work.)
- Safety goggles

1. In order to perform this experiment, you must be able to measure out a relatively small mass of lye. The problem is, your mass scale is not very good at measuring small masses. Thus, what we need to do is figure out how many grams are in a teaspoon of lye. Then we can use measuring spoons as a quick way of measuring out a small mass of lye. To do this, measure the mass of 10 tablespoons of lye.
2. Remembering that each tablespoon represents 3 teaspoons, realize that the mass you just measured is the mass of 30 teaspoons of lye. Now take that mass and divide it by 30. The result is the mass of 1 teaspoon of lye. You should recognize this process as another example of calibration. You have just calibrated your measuring spoons to tell you the mass of lye.
3. In a moment, you will add ½ teaspoon of lye to some vinegar. Calculate what mass that is by taking the mass of 1 teaspoon of lye and dividing by 2. This will be the mass of lye that you will use in the experiment.
4. Now we can begin the real experiment. Nest the two coffee cups as you did in Experiment 2.2 to make a calorimeter.
5. Pour 100.0 mL (½ cup) of vinegar into your calorimeter.
6. Place your thermometer in the vinegar and let it sit for three minutes. After that time, record the temperature of the vinegar. This will be the initial temperature in the experiment.
7. Now add ½ teaspoon of lye to the vinegar, and begin stirring the mixture with your thermometer.
8. As time goes on, the following reaction will take place:

$$\underset{\text{lye}}{NaOH\ (s)} + \underset{\text{acetic acid in vinegar}}{C_2H_4O_2\ (aq)} \rightarrow H_2O\ (l) + NaC_2H_3O_2\ (aq)$$

9. This reaction is exothermic; thus, you should see the temperature begin to rise. Read the temperature every 30 seconds. Once the temperature falls or stays constant for two consecutive readings, you can end the experiment.
10. Now you are ready to do your calculations. Remember, calorimetry experiments use the change in temperature to determine the heat transferred. The equation we use is:

$$q = mc\Delta T$$

In this equation, "q" is the amount of heat transferred, "m" is the mass of the sample being heated up, "c" is the specific heat capacity of the thing being heated up, and "ΔT" is the change in temperature. Now, if you remember your calorimetry experiment from Module #2, you will remember that we can sometimes ignore the calorimeter. We will do that here. Thus, all we need to consider are the contents of the calorimeter.
11. The specific heat (c) of vinegar is $4.1 \frac{J}{g \cdot {}^{o}C}$.
12. The mass of the vinegar in the calorimeter will be its volume (100.0 mL) times its density (which equals 0.99 g/mL). In addition to this mass, however, you must add the mass of ½ teaspoon of lye, because that was also in the calorimeter. In the end, then, the mass of the calorimeter contents is the mass of the vinegar plus the mass of the lye added.
13. Subtract the initial temperature in the experiment from the final temperature to get ΔT.
14. Now you can calculate the heat absorbed by the contents of the calorimeter. Take the specific heat given in step #11 times the mass you calculated in step #12, and then multiply it by the ΔT you calculated in step #13. This gives you the heat absorbed by the calorimeter's contents.
15. Since the heat absorbed by the calorimeter's contents came from the chemical reaction, you have just calculated the ΔH of the reaction!
16. We aren't quite finished, however. The ΔH we calculated was the amount of energy released when the mass of lye that *we used in the experiment* reacted with vinegar. It would be more useful if we could determine the ΔH *per mole* of NaOH that reacts. To do that, all we have to do is take the ΔH we just calculated and divide it by the number of moles of NaOH that we used.
17. Take the mass of lye and divide it by the molar mass of NaOH. That will give you the number of moles of NaOH that you used in the reaction.
18. Divide the ΔH that you calculated in step #14 by the number of moles you calculated in step #17. The result is the ΔH of the reaction in J/mole.
19. Clean up your mess.

One reason I wanted you to perform Experiment 13.1 was to give you a review of calorimetry. The other reason was to get you to appreciate how ΔH's have been measured over the years. We have, quite literally, hundreds of thousands of ΔH's compiled in today's chemistry literature. This means that chemists have done hundreds of thousands of experiments just like the one you performed in order to provide us with this information. You should now appreciate how much work this is! Finally, performing the experiment should have illustrated to you how useful the technique of calorimetry is for the pursuit of chemical knowledge. Now that we are done with all that, however, it is time for you to learn some easier methods for determining the ΔH of a chemical reaction.

Determining the ΔH of a Chemical Reaction Using Bond Energies

When a chemical reaction occurs, the valence electrons of the atoms that make up the reactants must rearrange themselves into the configurations necessary to make the products. This process requires a lot of chemical bond breaking and chemical bond making. For example, let's go back to the burning of methane:

$$CH_4 + 2O_2 \rightarrow CO_2 + 2H_2O$$

If we were to look at the Lewis structures of the molecules in this equation, they would look like this:

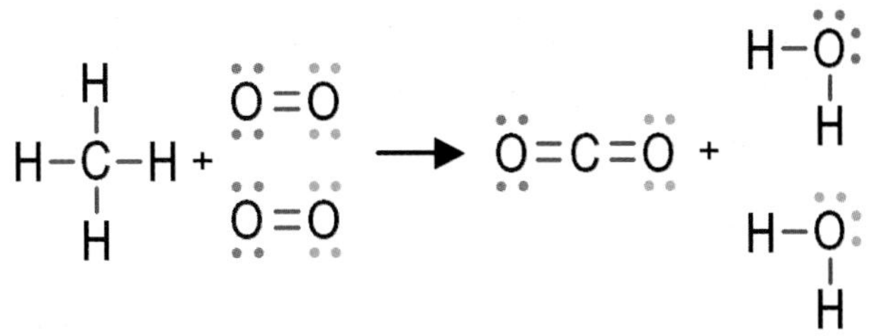

Looking at the equation in this form, we can see that in order for the reactants to turn into the products, several bonds have to be broken and several others have to be formed.

Look, for example, at the CH_4 molecule. It contains a carbon atom bonded to four hydrogen atoms by four separate single bonds. In order to form the products, those four bonds must be broken. That way, the hydrogens will be free to bond with the oxygen atoms to make water, while the carbon atom will be free to bond to the other oxygen atoms to make carbon dioxide. For this to happen, of course, the double bonds that hold the two oxygen atoms together in each oxygen molecule must also be broken. Thus, to make this chemical reaction work, all the bonds on the reactants side of the equation must be broken.

Once the bonds are broken, the job is only halfway done. Now all the bonds on the products side must be formed. This means that two double bonds between the carbon atom and the oxygen atoms in carbon dioxide must be formed. In addition, four single bonds between the hydrogen atoms and the other oxygen atoms must be formed in order to make two water molecules. After all this happens, the chemical reaction is finished.

Think for a moment about the energy required to make all this happen. First, it takes energy to break bonds. This should make sense to you. After all, if you want to break anything, you have to expend some energy. Thus, as the bonds on the reactants side of the equation are being broken, energy is being absorbed by the reactants. What may not make as much sense to you (but is nevertheless true), is that when bonds are formed on the products side of the equation, energy is released. Thus, each time a bond forms on the products side of the equation, the product being formed releases energy. If we look at the balance between the energy absorbed and the energy released, we will see why some chemical reactions are endothermic and others are exothermic.

Suppose, for example, a chemical reaction involves reactants with very strong bonds and products with very weak bonds. Since the reactant bonds are strong, it takes a lot of energy to break them. Thus, the reactants tend to absorb a lot of energy. On the products side of the reaction, however, the bonds are weak. This means that when they are formed, the products release only a small amount of energy. In this situation, then, the reactants had to absorb lots of energy, and the products only released a little. On balance, the chemical reaction absorbed more energy than it released; thus, there was a net effect of absorbing energy. This reaction, then, would be considered endothermic.

On the other hand, suppose you had a chemical reaction in which the reactants contained weak bonds and the products contained strong bonds. In this situation, the reactants would not need to absorb much energy, because their bonds would be easy to break. On the other hand, when strong bonds are formed, a lot of energy is released. Therefore, this chemical reaction would release more energy than it absorbed. This makes the reaction, on balance, an exothermic reaction.

One thing that you should see from this discussion is that whether a reaction is exothermic or endothermic, *energy is both absorbed and released during the course of the reaction.* The difference between exothermic and endothermic reactions depends on *how much* energy is absorbed and *how much* energy is released. In exothermic reactions, more energy is released than absorbed. In endothermic reactions, more energy is absorbed than released.

Of course, for any of this discussion to be of practical use, there has to be some way of determining whether a set of chemical bonds is weak or strong. If we want to predict whether or not a chemical reaction is exothermic or endothermic, we need to know how strong all the bonds in the reaction are. How is that done? Well, believe it or not, chemists can actually measure the strength of a chemical bond, which is typically called the **bond energy**. The process by which bond energies are measured is a bit beyond the scope of this course, but the bond energies themselves are easily understood and applied to this course. Table 13.1 summarizes several chemical bond energies:

TABLE 13.1
Bond Energies

Chemical Bond	Bond Energy (kJ/mole)	Chemical Bond	Bond Energy (kJ/mole)
C-H	411	H-H	432
C-Cl	327	H-O	459
C-C	346	H-Cl	428
C-N	305	O=O	494
C-O	358	N≡N	942
C=C	602	N-H	386
C=O	799	Cl-Cl	240
C≡C	835	Br-Br	190

Notice the unit used to measure bond energy. It is the same as the unit used to measure ΔH, kJ/mole. Realize, of course, that the unit could also be kcals/mole or, for that matter, any energy unit divided by moles. What does this unit mean when applied to the measurement of bond strength? It tells us how much energy is required to break one mole of those bonds. Alternatively, it tells us how much energy will be released when one mole of those bonds are formed.

For example, suppose you had one mole of Cl_2 molecules. You could write the Lewis structure of Cl_2 and find out that the two Cl atoms that make up this molecule are held together with one Cl to Cl single bond. According to our table, the energy of that bond is 240 kJ/mole. This means two things. First, it means that if I wanted to destroy one mole of Cl_2 molecules, I would have to expend

240 kJ of energy to break all their bonds. Likewise, this number also tells me that if I formed one mole of Cl_2 molecules by letting 2 moles of Cl atoms join together, 240 kJ of energy would be released in the process. Thus, the bond energy tells us not only how much energy is required to break it, but it also tells us how much energy is released when the bond is formed.

Before we see how these bond energies can be applied to learn about the energetics of a chemical reaction, I want to point out that you needn't memorize anything on Table 13.1. No chemist I know has these things memorized. A good chemist, however, knows exactly where to look them up. That's what I will expect of you. Throughout the rest of this module, several more tables will be presented. You will not be required to know any of these tables, but you will be required to look at them in order to solve problems. All the tables you will need to solve the practice problems and the test are presented in Appendix A near the end of the text. Feel free to use that appendix whenever you solve any problems, including the ones on the test.

Now that we have this table of bond energies, how do we use it to analyze the energetics of a chemical reaction? It's actually relatively simple. As I mentioned before, to determine whether or not a chemical equation is exothermic or endothermic, all we have to do is figure out how much energy is absorbed to break the reactants' bonds and how much energy is released when the products' bonds form. The difference between the two will be the ΔH of the reaction. Mathematically:

$$\Delta H = \text{Energy required to break bonds} - \text{Energy released when bonds form} \qquad (13.6)$$

The table of bond energies gives us all the information we need to use Equation (13.6), as long as we know what bonds are breaking and what bonds are forming. To do this, of course, we need to be able to look at a chemical equation in terms of its Lewis structures. That's the only way you will know what kinds of bonds are broken and what kinds of bonds are formed. Thus, in order to solve the problems in this module, you will need to be able to do Lewis structures. If you're a bit shaky on those, review Module #8.

Before I show you a couple of examples of this kind of problem, I need to point out one last thing. Look at Equation (13.6) for a moment. Suppose the energy required to break bonds is greater than the energy released when bonds are formed. In this case, the chemical reaction will be endothermic. What sign will ΔH have? It will be positive. On the other hand, an exothermic reaction has more energy released when the products' bonds are formed than energy required to break the reactants' bonds. When this happens, Equation (13.6) tells us that ΔH will be negative. Thus, Equation (13.6) is consistent with our previous definition of ΔH. If we calculate a ΔH from Equation (13.6) and it ends up positive, this tells us that the reaction is endothermic. A negative ΔH, however, indicates an exothermic reaction. Let's see how all of this works in a couple of example problems:

EXAMPLE 13.2

To make ammonia, nitrogen and hydrogen are reacted at high temperatures according to this equation:

$$N_2(g) + 3H_2(g) \rightarrow 2NH_3(g)$$

What is the ΔH for this reaction? Is it endothermic or exothermic?

This problem requires the use of Equation (13.6). In order to use this equation, however, we need to know what bonds are being broken and what bonds are being formed. Thus, we need to rewrite the chemical equation in terms of Lewis structures:

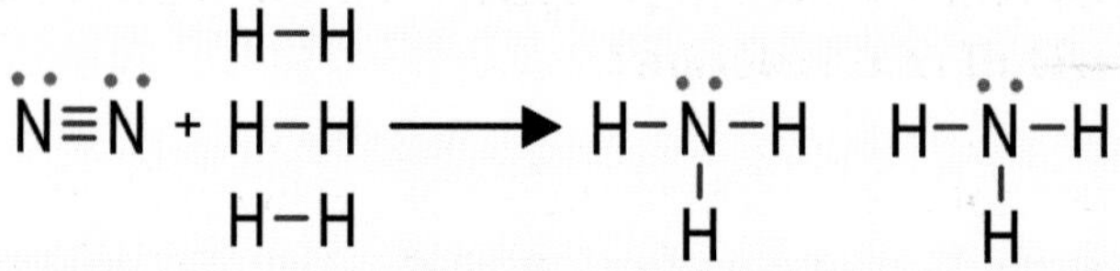

$$:N \equiv N: \;+\; \begin{matrix} H-H \\ H-H \\ H-H \end{matrix} \;\longrightarrow\; \begin{matrix} \ddot{N} \\ H-\,|\,-H \\ H \end{matrix} \quad \begin{matrix} \ddot{N} \\ H-\,|\,-H \\ H \end{matrix}$$

Make sure you understand each of these Lewis structures. You have to be able to draw Lewis structures to solve this kind of problem. Also, realize why I drew three Lewis structures of H_2 and two Lewis structures of NH_3. I did that because the chemical equation tells me that I have three H_2 molecules and two NH_3 molecules, so I had to draw that many Lewis structures to be able to see all the bonds involved in this chemical reaction.

Now that I can see all the bonds involved in the chemical equation, I can look in Table 13.1 and get the energy of each bond. The first bond I see is a nitrogen-to-nitrogen triple bond. According to Table 13.1, this bond has an energy of 942 kJ/mole. The other bonds involved in the reaction are a H-H bond (energy is 432 kJ/mole) and a N-H bond (energy is 386 kJ/mole). Now that I have the bonds involved in the chemical reaction and their energies, I am ready to use Equation (13.6).

Remember, though, in order to calculate the total amount of energy required to break all the reactant bonds, I need to count *how many* of each bond I need to break. According to my Lewis structure equation, there is only one N≡N bond, so the energy required to break that bond is (1 mole) x (942 kJ/mole). The equation shows me three H-H bonds, however. Since each individual H-H bond has an energy of 432 kJ/mole, the total amount of energy needed to break these bonds is (3 moles) x (432 kJ/mole). Finally, when I look at the products side, I see a total of six N-H bonds. Each of those bonds has an energy of 386 kJ/mole, so the products release (6 moles) x (386 kJ/mole) of energy when they form. Thus, Equation (13.6) looks like this:

$$\Delta H = (1\ \cancel{\text{mole}}) \times (942\ \frac{\text{kJ}}{\cancel{\text{mole}}}) + (3\ \cancel{\text{moles}}) \times (432\ \frac{\text{kJ}}{\cancel{\text{mole}}}) - (6\ \cancel{\text{moles}}) \times (386\ \frac{\text{kJ}}{\cancel{\text{mole}}}) = -78\ \text{kJ}$$

Thus, the change in enthalpy for this reaction is <u>-78 kJ</u> and, since ΔH is negative, the reaction is <u>exothermic</u>.

I want you to notice a couple of things about the way I solved this problem. First, let's deal with the units. Remember, the stoichiometric coefficients in a chemical equation refer to the number of moles. In the equation, 1 mole of N_2 reacts with 3 moles of H_2 to make 2 moles of NH_3. Thus, the moles cancel, leaving just "kJ" as the unit for ΔH. This is typically the way we report the change in enthalpy for a reaction. We simply list the energy produced when the reactants are combined in exactly the mole amounts given by the stoichiometric coefficients.

Now let's deal with the significant figures. Remember, the stoichiometric coefficients in a chemical equation are exact. Thus, what we are really doing in Equation (13.6) is adding and subtracting. We are adding one 942 with three 432's and then subtracting six 386's. As a result, we must use the rule of addition and subtraction when determining the significant figures here. Since all of the bond energies have their last significant figure in the ones place, our answer must be reported to the ones place. That's why the answer is -78 kJ.

The complete combustion of methyl alcohol (CH_3OH) is used to power some experimental automobiles. What is the ΔH of that reaction?

In this problem, we are not even given a chemical equation, so our first job is to get an equation. According to the definition of complete combustion, the unbalanced chemical equation is:

$$CH_3OH + O_2 \rightarrow CO_2 + H_2O$$

Balancing the equation gives us:

$$2CH_3OH + 3O_2 \rightarrow 2CO_2 + 4H_2O$$

Next, we must get the equation into Lewis structure form. Before we do that, however, look at the way methyl alcohol's chemical formula has been written. It has been written in this way to help you make the Lewis structure. Normally, I have referred to methyl alcohol as CH_4O. Often, however, chemists will write a chemical formula in such a way as to illustrate the Lewis structure. That's what I have done here. The three H's are grouped together next to the C to make it clear that those three H's are bonded to that C. The last H is written after the O to make it clear that it is bonded to the O. This should help you in writing the Lewis structure:

```
                                          H-O:
                                            |
    H                                       H
    |            O=O                      H-O:
  H-C-O-H                                   |
    |                       O=C=O           H
    H          + O=O  -->                +
    H                       O=C=O         H-O:
    |                                       |
  H-C-O-H        O=O                        H
    |                                     H-O:
    H                                       |
                                            H
```

Now that we have the equation in Lewis structure form, we can count up all of the bonds, look up the bond energies in Table 13.1, and apply Equation (13.6). On the reactants side of the equation, there are six C-H bonds (each worth 411 kJ/mole), two C-O bonds (each worth 358 kJ/mole), two O-H bonds (each worth 459 kJ/mole) , and three O=O bonds (each worth 494 kJ/mole). On the products side of the equation, there are four C=O bonds (which will release 799 kJ/mole each) and eight O-H bonds (which will release 459 kJ/mole each). In the end, then, Equation (13.6) looks like this:

$$\Delta H = (6 \cancel{\text{moles}}) \times (411 \frac{\text{kJ}}{\cancel{\text{mole}}}) + (2 \cancel{\text{moles}}) \times (358 \frac{\text{kJ}}{\cancel{\text{mole}}}) + (2 \cancel{\text{moles}}) \times (459 \frac{\text{kJ}}{\cancel{\text{mole}}}) +$$

$$(3 \cancel{\text{moles}}) \times (494 \frac{\text{kJ}}{\cancel{\text{mole}}}) - (4 \cancel{\text{moles}}) \times (799 \frac{\text{kJ}}{\cancel{\text{mole}}}) - (8 \cancel{\text{moles}}) \times (459 \frac{\text{kJ}}{\cancel{\text{mole}}})$$

$$\Delta H = -1{,}286 \text{ kJ}$$

In the end, then, $\underline{\Delta H = -1{,}286 \text{ kJ}}$ for this reaction, which makes it exothermic. Of course, you should have already realized that this was an exothermic reaction, because combustion is always exothermic.

Notice a couple of things in these two problems. First of all, I refer to the oxygen-to-hydrogen bonds as O-H bonds, but I use the bond energy in Table 13.1 for the H-O bonds. This is because the bond energy is independent of the order of the atoms. The H-O bond energy is the same as the O-H bond energy. Second, notice that I subtract each of the terms from the products side. It is important to remember to do that. Each term from the reactants side must be added to the equation, while each term from the products side must be subtracted.

Now that you have seen how we use Table 13.1 and Equation (13.6) to calculate ΔH, make sure you understand the process by completing the following "On Your Own" problems. Remember, one place you can really mess up is in making your Lewis structures. Be very careful that the Lewis structures you are using are correct. Otherwise, your answer will definitely be wrong!

ON YOUR OWN

13.2 Hydrochloric acid (HCl) is widely used as a cleaning agent in both home and industrial use. To make HCl, chemists react hydrogen gas and chlorine gas at high temperatures. What is the ΔH of this process? Is it exothermic or endothermic?

13.3 In an acetylene torch, C_2H_2 is burned to make the hot, blue flame that cuts or welds metal. Assuming that the combustion is complete, what is the ΔH of this reaction?

Hess's Law

Before I begin this section, I think it is necessary to warn you that the discussion which follows gets a little deep. If you feel lost while you are reading, don't worry too much about it. Read the section through and pick up as much as you can. Before I get to the example problems, I will summarize what you need to know from this section, which is significantly less than I am going to cover. In fact, what you need to know is rather simple, even though the discussion that leads up to it is not! However, for a complete discussion of the topic, I think it is necessary to provide you with all the following information, and I will let your ability in chemistry determine how much of the information you actually absorb.

So far, we have discussed one experimental method of determining the ΔH of a reaction and one non-experimental method. Remember, however, that I said there were three ways to measure the ΔH of a reaction. The third method makes use of **Hess's Law**, named after Germain Henri Hess, a famous Russian chemist whose experimental work led to its development:

Hess's Law - Enthalpy is a state function and is therefore independent of path.

What this law means, much less how it applies to calculating the ΔH of a chemical reaction, is probably a bit of a mystery to you at this point. That's okay, though. Once I have explained a few things to you, not only Hess's Law but also how it applies to calculating ΔH should become a little more clear.

First of all, a **state function** is any quantity that depends solely on the final destination, not on the way you get to that destination. For example, King's Island is an amusement park in Cincinnati, Ohio. People from all over the midwestern United States come to this park for a fun-filled day of rides, shows, and other attractions. Two people can leave from the same city and take completely different routes to the park. One of them may take a route that is made up mostly of highways. Most likely, he will have to travel a little bit out of the way in order to stay on the highway but, since highway travel is faster, he might end up saving time that way. On the other hand, if you want to take the straightest route to King's Island, you will have to travel the back roads. Thus, one person might travel the highway route and the other might travel the back roads. In the end, however, they will both get to King's Island and have the same amount of fun. In this case, fun could be described as a state function. Regardless of the route that each person took, they both got to the same location (King's Island) and therefore had the same amount of fun once they arrived.

In this scenario, fun might be a state function, but many other things will not be state functions. For example, the number of miles traveled will not be a state function. The person who took the straightest route to King's Island will have traveled fewer miles than the person who took the highway route. Time will also not be a state function. Most likely, one of the routes to King's Island will take longer than the other. Thus, some of the quantities associated with a trip to King's Island depend on the path taken and therefore are not state functions, while other quantities are independent of the path taken and therefore are state functions.

What Hess's Law tells us is that enthalpy is also a state function. In other words, the amount of energy contained in a substance is independent of how it is made. For example, each of the chemical equations below represents a way of making water:

$$2H_2 + O_2 \rightarrow 2H_2O \tag{13.7}$$

$$2H_2O_2 \rightarrow 2H_2O + O_2 \tag{13.8}$$

Suppose I made two moles of water using Equation (13.7) and measured the enthalpy of that water. If I used Equation (13.8) instead and isolated just the water, the enthalpy of that water would be exactly the same as the enthalpy of the water I made using Equation (13.7). Both chemical reactions represent two different paths that I could use for making water. Regardless of which path I used, however, the enthalpy of the water I created would be exactly the same. That's what Hess's Law tells us.

How does that help us calculate the ΔH of a chemical reaction? Well, remember what ΔH represents. It represents the *change in enthalpy* that occurs within a chemical reaction. Thus, if we somehow knew the enthalpy of each substance in a chemical equation, we could simply take the difference of the enthalpies of the products and the enthalpies of the reactants to get the change in enthalpy of the reaction, which would be ΔH. Hess's Law tells us that the enthalpy of a given substance is *independent* of the way it was made. Therefore, if we have some standard way of measuring the enthalpy of a bunch of substances, we could use that information to calculate the ΔH of any reaction that contained those substances, regardless of how those substances were made in preparation for that particular chemical reaction.

It turns out that it is very hard to actually measure the enthalpy of an individual substance. The only thing that chemists can measure is the change in enthalpy that occurs in a chemical reaction. Thus, you might think that this whole discussion was in vain, because we can't actually measure the enthalpy of an individual substance. Luckily enough for us, however, we can get around this problem.

To get around this problem, we will define something called the **enthalpy of formation** (abbreviated as ΔH_f):

Enthalpy of formation (ΔH_f) - The ΔH of a formation reaction

Remember that a formation reaction is defined as a reaction that forms a single substance from its elemental components. Thus, the enthalpy of formation of any given substance is just the ΔH of the formation reaction that creates that substance. This, of course, is something that can be measured experimentally. All you have to do is run the formation reaction in a calorimetry experiment (like Experiment 13.1), and you will have the enthalpy of formation.

How does this help? Well, it turns out that we can define the enthalpy of any element in its elemental form to be zero. Thus, since all the reactants of a formation reaction are elements in their elemental form, the total enthalpy of the reactants in a formation reaction is zero. In addition, since there is only one product in a formation reaction (the substance of interest), the ΔH of the formation equation is, in essence, the enthalpy of that substance. Thus, having the ΔH_f of a substance is equivalent to having the enthalpy of the substance. Therefore, if we had a table of the ΔH_f's of several substances, we could calculate the ΔH of any reaction that uses those substances!

Before I give you that table, I have to introduce a point of terminology. It turns out that the ΔH of a chemical reaction can change depending on the conditions under which the reaction occurs. Thus, when we report ΔH's, it is important to specify the conditions under which the reaction has been run. Well, most chemistry is done at room temperature (25 ^{o}C) and normal atmospheric pressure (1.00 atm). Thus, we call these **standard conditions**. Note that this is *not* standard temperature and pressure (STP). STP is defined as 0.00 ^{o}C and 1.00 atm. Instead, a temperature of 25 ^{o}C and a pressure of 1.00 atm are considered the *standard conditions* under which most chemistry is done.

In this course, all the ΔH's you will calculate will refer to standard conditions. Thus, in the table below, you will find the ΔH_f's for several molecules under standard conditions. To denote the fact that they are under standard conditions, we put a "^{o}" after the symbol, and we call them **standard enthalpies of formation**. Thus, a symbol of "ΔH_f^{o}" tells you that the enthalpy of formation was measured at 25 ^{o}C and 1.00 atm.

TABLE 13.2
Standard Enthalpies of Formation in kJ/mole

Substance	ΔH_f^o	Substance	ΔH_f^o	Substance	ΔH_f^o	Substance	ΔH_f^o
H_2O (g)	-242	NO_2 (g)	33.2	$Ca(OH)_2$ (s)	-987	C_2H_6O (l)	-278
H_2O (l)	-286	NH_3 (g)	-45.9	$CaCO_3$ (s)	-1207	C_2H_4 (g)	52.5
SO_3 (g)	-396	CO_2 (g)	-394	C_2H_6 (g)	-84.7	C_4H_{10} (g)	-126
NO (g)	90.3	CaO (s)	-635	CH_3OH (l)	-239	H_2SO_4 (l)	-814

Notice a couple of things about these standard enthalpies of formation. First of all, they can be either positive or negative. This should make sense to you. After all, they are just the ΔH's of formation reactions under standard conditions. Thus, some formation reactions are exothermic while others are endothermic. Also, notice that the enthalpy of formation depends on the state in which the substance finds itself. For example, the enthalpy of formation of liquid water is different from the enthalpy of formation of gaseous water. Once again, this should make sense once you realize that it takes energy to turn water from a liquid to a gas. Thus, the heat released by a reaction that makes gaseous water should be less than the heat released by a reaction that makes liquid water, because the first reaction will have to expend some of its extra energy turning water from a liquid into a gas.

Now that we have a table that gives us the enthalpies of formation for several substances, we can define the standard change in enthalpy (ΔH^o) of any reaction that contains those substances as:

$$\Delta H^o = \Sigma\, \Delta H_f^o \text{ (products)} - \Sigma\, \Delta H_f^o \text{ (reactants)} \qquad (13.9)$$

If you aren't familiar with it from algebra, the "Σ" symbol stands for "summation." What Equation (13.9) tells us, then, is that the ΔH^o of any chemical reaction is simply the sum of the ΔH_f^o for the products minus the sum of the ΔH_f^o for the reactants. It is important to make sure that you use products minus reactants, because that will keep the sign of ΔH^o consistent with our previous definitions. That way, a negative ΔH^o will still indicate an exothermic reaction and a positive ΔH^o will still indicate an endothermic reaction. Equation (13.9) is often referred to as Hess's Law, because it is a direct consequence of that law.

Please note that since we are using standard enthalpies of formation in Equation (13.9), the change in enthalpy that we calculate using that equation will also refer to standard conditions. That's why there is a "o" after the ΔH in Equation (13.9). When we use Equation (13.9) and Table 13.2, then, the change in enthalpy that we determine will only be valid under standard conditions. Since most of the chemistry you do takes place at room temperature and normal atmospheric pressure, however, this is really not much of a limitation.

Now that you've made it through this rather deep discussion of Hess's Law, it is time to summarize the things that you really need to understand. As I said at the beginning of this section, don't worry if you didn't understand everything I just discussed. You really need to know only a few things from the reading. First, you need to know the definition of Hess's Law and what a state function is. You also need to know the definition of enthalpy of formation and Equation (13.9). Finally, you just have to remember that:

The standard enthalpy of formation of an element in its elemental form is zero.

Thus, chemicals like Ca (s), Na (s), K (s), and Fe (s) all have $\Delta H_f^o = 0$. In addition, the homonuclear diatomics like H_2 (g), N_2 (g), O_2 (g), and Cl_2 (g) have $\Delta H_f^o = 0$. Let's see how all of this pans out in a couple of example problems.

EXAMPLE 13.3

What is the ΔH° of the following reaction?

$$2NO\ (g) + O_2\ (g) \rightarrow 2NO_2\ (g)$$

Is it an exothermic or an endothermic reaction?

Looking at Table 13.2, we have the ΔH_f^o for all of the substances in this equation except O_2 (g). However, we know that any element in its elemental form has $\Delta H_f^o = 0$. Since O_2 (g) is the elemental form of oxygen (because oxygen is a homonuclear diatomic and a gas at room temperature), we know that its $\Delta H_f^o = 0$. Thus, now we know the ΔH_f^o of all substances in the equation. Remember, when looking for the ΔH_f^o of a substance, you must match both the substance *and* the phase. If, for example, this chemical reaction contained NO (l), we could not solve the problem because our table does not list NO (l); it only lists NO (g). In addition, if the reaction used O_2 (l) instead of O_2 (g), we could not solve the problem, because we are working under standard conditions. Under standard conditions, oxygen is a gas, not a liquid. Thus, if the equation used O_2 (l), we could not assume that its ΔH_f^o is zero. However, since we do have all of the standard enthalpies of formation that we need, we can simply apply Equation (13.9):

$$\Delta H^o = (2\ \cancel{\text{moles}}) \times (33.2\ \frac{\text{kJ}}{\cancel{\text{mole}}}) - (2\ \cancel{\text{moles}}) \times (90.3\ \frac{\text{kJ}}{\cancel{\text{mole}}}) - (1\ \cancel{\text{mole}}) \times (0\ \frac{\text{kJ}}{\cancel{\text{mole}}})$$

$$\Delta H^o = -114.2\ \text{kJ}$$

Notice that I had to multiply each ΔH_f^o by the stoichiometric coefficient of the substance in the chemical equation. Also notice that I added all of the terms from the products side and subtracted all of the terms from the reactants side. This is the proper way to apply Equation (13.9) to a chemical reaction. Thus, the ΔH° for this reaction is <u>-114.2 kJ</u>, and it is therefore <u>exothermic</u>.

Some gasoline companies put a small amount of liquid ethyl alcohol (C_2H_6O) into their gasoline. What is the ΔH° for the complete combustion of this chemical?

We could solve this problem using bond energies as was explained in the previous section. However, to illustrate the use of Hess's Law, we will solve it using the concepts discussed in this section. First, we write the balanced chemical equation:

$$C_2H_6O\ (l) + 3O_2\ (g) \rightarrow 2CO_2\ (g) + 3H_2O\ (g)$$

Notice that I added phase symbols in this equation. I must have them in order to use Table 13.2. In Module #5, I told you to assume that the water and carbon dioxide produced in combustion reactions were in the gas phase, so there is nothing new here.

Looking in Table 13.2, I can find ΔH_f^o's for all of the substances in their proper phase except O_2 (g). However, since it is an element in its elemental form, the ΔH_f^o is zero. Thus, I can now apply Equation (13.9):

$$\Delta H^\circ = (2 \cancel{\text{moles}}) \times (-394 \frac{\text{kJ}}{\cancel{\text{mole}}}) + (3 \cancel{\text{moles}}) \times (-242 \frac{\text{kJ}}{\cancel{\text{mole}}}) - (1 \text{ mole}) \times (-278 \frac{\text{kJ}}{\cancel{\text{mole}}}) - (3 \cancel{\text{moles}}) \times (0 \frac{\text{kJ}}{\cancel{\text{mole}}})$$

$$\Delta H^\circ = -1{,}236 \text{ kJ}$$

The ΔH^o for this reaction is -1,236 kJ.

So you see that even though the discussion of Hess's Law was very complicated, the application of it is very simple. If you understand the examples you have just read and can perform the "On Your Own" problems that follow, you have understood this section well enough to move on.

ON YOUR OWN

13.4 When you use a pocket cigarette lighter (hopefully not to light cigarettes!), you are burning gaseous butane (C_4H_{10}). Assuming it is complete combustion, what is the ΔH^o for this reaction? Do not use bond energies to solve this problem. Use Hess's Law.

13.5 In order to make the shells that house and protect them, many shellfish take solid lime (CaO) from the ocean floor and react it with gaseous carbon dioxide that has been exhaled by marine organisms. The resulting calcium carbonate is the major component of their shell. What is the ΔH^o for this reaction?

Before we leave this section, I need to answer a question that you should be asking yourself. Since you now have two non-experimental methods of determining ΔH, which should you use? In one of the example problems from this section and one of the "On Your Own" problems, I specifically told you to use Hess's Law instead of bond energies. The reason I did that is because using Hess's Law is a bit more accurate, since it also takes into account the phase of the substances involved. Thus, if you have the ΔH_f^o of each substance in the equation with the proper phase, you should always use Hess's Law. This is good, because Hess's Law is less work than determining ΔH from bond energies!

The only time you should use bond energies is if you don't have the ΔH_f^o's necessary to solve the problem, or if the problem doesn't tell you the phases of the substances in the reaction. After all, if you don't know the phase, you can't use Table 13.2! When you use bond energies to calculate ΔH, it is important to realize that your answer is only an approximation. The only way to accurately calculate the ΔH of a reaction is by taking the phase of each substance into account. You can only do that using Hess's Law.

You should use Table 13.2 when you solve all the problems in this module, including the test problems. Thus, you will also find a copy of Table 13.2 in Appendix A at the end of the text. Make sure you use that appendix when you are working on problems from this module!

Applying Enthalpy to Stoichiometry

Now that we know three different ways to determine the ΔH of a chemical reaction, we are ready to apply this newly learned concept. Where better to apply it than to the wonderful world of stoichiometry? As I have said many times before, stoichiometry is the backbone of chemistry; thus, I try to relate nearly everything that we study (in one way or another) to this terribly important subject.

Remember, once we have determined both the value and the sign of ΔH, we can include it in the equation, as you did in "On Your Own" problem 13.1. Well, once something becomes part of a chemical equation, it can be used in stoichiometry. Study the next two examples to see how this is done, and then perform the "On Your Own" problems that follow to cement this new skill into your head.

EXAMPLE 13.4

We determined previously that the ΔH^{o} for the combustion of ethyl alcohol is -1,236 kJ. How much energy can be produced by burning 150.0 grams of ethyl alcohol in excess oxygen?

From our last example problem, we determined that the balanced chemical equation for the combustion of ethyl alcohol is:

$$C_2H_6O\ (l) + 3O_2\ (g) \rightarrow 2CO_2\ (g) + 3H_2O\ (g)$$

Once we determined the ΔH^{o} of the reaction, we could have simply put it in the chemical equation. Since ΔH^{o} turned out to be negative, energy can be considered a product in the reaction. Thus:

$$C_2H_6O\ (l) + 3O_2\ (g) \rightarrow 2CO_2\ (g) + 3H_2O\ (g) + 1{,}236\text{ kJ}$$

We can start doing stoichiometry as soon as we convert the amount of ethyl alcohol into moles:

$$\frac{150.0\ \cancel{\text{g } C_2H_6O}}{1} \times \frac{1\text{ mole } C_2H_6O}{46.1\ \cancel{\text{g } C_2H_6O}} = 3.25\text{ moles } C_2H_6O$$

Now, according to our equation, 1 mole of C_2H_6O will give us 1,236 kJ of energy. Therefore, our stoichiometric relationship is:

$$1\text{ mole } C_2H_6O = 1{,}236\text{ kJ}$$

Using that relationship gives us:

$$\frac{3.25\ \cancel{\text{moles } C_2H_6O}}{1} \times \frac{1{,}236\text{ kJ}}{1\ \cancel{\text{mole } C_2H_6O}} = \underline{4{,}020\text{ kJ}}$$

One way of producing butane (C_4H_{10}) for cigarette lighters is to force ethane (C_2H_6) to react with itself:

$$2C_2H_6\ (g) \rightarrow C_4H_{10}\ (g) + H_2\ (g)$$

If you have 1.23 kg of ethane and would like to turn it all into butane, how many kJ of energy would be necessary?

Since this question asks about the energy associated with a chemical reaction, we must determine ΔH° in order to solve the problem. This is easy to do, however, since we have all of the information related to this problem available in Table 13.2. Thus, according to Hess's Law,

$$\Delta H^{\circ} = (1\ \cancel{\text{mole}}) \times (-126\ \frac{\text{kJ}}{\cancel{\text{mole}}}) + (1\ \cancel{\text{mole}}) \times (0\ \frac{\text{kJ}}{\cancel{\text{mole}}}) - (2\ \cancel{\text{mole}}) \times (-84.7\ \frac{\text{kJ}}{\cancel{\text{mole}}}) = 43\ \text{kJ}$$

Since ΔH° is positive, energy can be thought of as a reactant in this problem:

$$2C_2H_6\ (g) + 43\ \text{kJ} \rightarrow C_4H_{10}\ (g) + H_2\ (g)$$

Now we can use the energy in stoichiometry. First, we must convert the amount of ethane into moles:

$$\frac{1.23 \times 10^3\ \cancel{\text{g } C_2H_6}}{1} \times \frac{1\ \text{mole } C_2H_6}{30.1\ \cancel{\text{g } C_2H_6}} = 40.9\ \text{moles } C_2H_6$$

Then we use the chemical equation:

$$2\ \text{moles } C_2H_6 = 43\ \text{kJ}$$

We can use that relationship, then, to convert from moles of ethane to kJ of energy:

$$\frac{40.9\ \cancel{\text{moles } C_2H_6}}{1} \times \frac{43\ \text{kJ}}{2\ \cancel{\text{moles } C_2H_6}} = \underline{880\ \text{kJ}}$$

ON YOUR OWN

13.6 Ethane gas (C_2H_6) has also been considered as a fuel for automobiles because, like octane, it releases a lot of energy when it is burned. How many kJ of energy are released when 250.0 grams of ethane undergo complete combustion?

13.7 When mined, coal has a small amount of sulfur in it. When that coal is burned in a power plant, the sulfur burns right along with it, producing SO_3 gas. When the SO_3 gas reacts with the water in the air, it produces sulfuric acid according to the following equation:

$$H_2O\ (g) + SO_3\ (g) \rightarrow H_2SO_4\ (l)$$

The sulfuric acid produced in this manner falls to the earth as acid rain. How much energy is produced when 50.0 grams of H_2SO_4 are produced in this way?

Energy Diagrams

Have you ever heard the phrase, "A picture is worth a thousand words"? Chemists often agree with this statement; thus, we have developed a way to draw a "picture" of the energy associated with a chemical reaction. This "picture," called an **energy diagram**, tells a chemist about a chemical reaction without using any numbers or words. For example, suppose I was studying the following chemical reaction:

$$C_2H_6O\ (l) + H_2\ (g) \rightarrow C_2H_6\ (g) + H_2O\ (g) \qquad (13.10)$$

I could use Hess's Law to calculate the ΔH^o of this reaction and communicate this result to other chemists. The sign of ΔH^o would tell the chemists whether the reaction was exothermic or endothermic, giving them an idea of the energetics of the reaction.

Instead of calculating ΔH^o, however, I could communicate the energetics of this reaction by simply drawing a picture. The picture would look something like this:

FIGURE 13.1
An Energy Diagram for the Reaction Given by Equation (13.10)

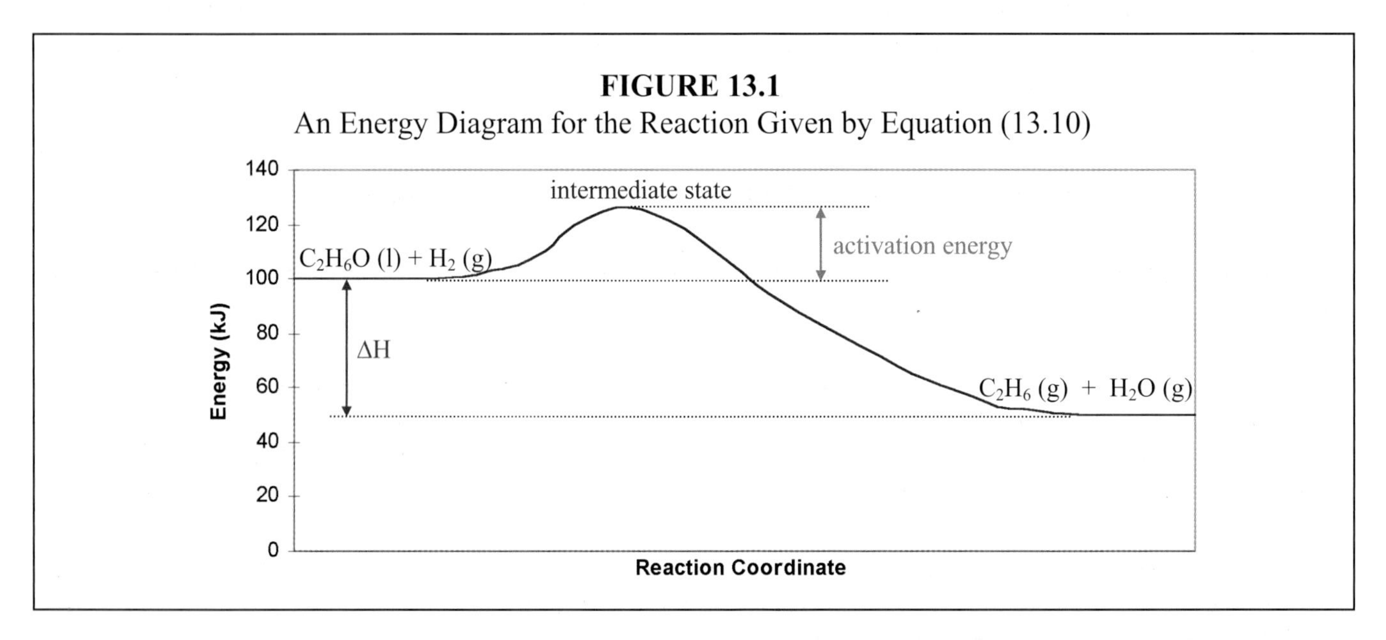

Notice first of all that this picture is actually a graph. On the vertical (y) axis of the graph, the energy of the substances in the chemical reaction is plotted. On the horizontal (x) axis of the graph, we plot the **reaction coordinate**, which basically tells us how close the reaction is to completion. For example, when the reaction coordinate is zero, we say that the reaction hasn't started yet; therefore, only the reactants are present. That's why we have only the reactants listed on the left side of the graph. As x gets larger, however, the reaction starts. The reactants come together and interact. At the top of the hump, which represents the midpoint of the reaction, we say that some **intermediate state** (sometimes called the **activated complex**) exists. By the time x reaches the far right side of the axis, however, the reaction has completed, and we have only products. That's why only the products are listed on the right side of the graph. Thus, when you look at the horizontal part of this graph, just remember that the far left side of the graph represents the reaction before it begins, while the far right side represents the end of the reaction. The points in between represent the reaction as it progresses.

Now let's look at the vertical (y) axis, which represents the energy of the substances in the chemical reaction. For this particular reaction, notice that the reactants (the left side of the graph) are much higher on the vertical axis than the products (the right side of the graph). This tells us that the reactants have more energy than the products. What does this tell you about the ΔH of the reaction? If the reactants have more energy than the products, then once the reaction completes, the excess energy from the reactants has to go somewhere. Where does it go? It gets released into the surroundings, usually in the form of heat. Thus, this energy diagram tells us that the reaction is exothermic.

The energy diagram tells us a bit more than that, however. If we look at the difference between the vertical (y) position of the products and the vertical (y) position of the reactants, it actually tells us how much energy was released into the surroundings. That's why I have labeled the vertical difference between the reactants and the products as "ΔH." Thus, if I can read the graph well enough, I can actually calculate the value of ΔH from it.

Notice, also, that this graph gives us more information than just the ΔH of the reaction. In between the left and right sides of the graph, there is a big hump. This hump represents the energy of the intermediate state of the reaction. Notice that the energy of the intermediate state of the reaction is higher than the energy of the reactants and the products. This is always the case. The intermediate state of the reaction has more energy than either the reactants or the products. This tells us that in order to reach the intermediate state of the reaction, some extra energy has to be put into the reaction. This energy, which is necessary to start the reaction, is called **activation energy**.

Activation energy - The energy necessary to start a chemical reaction

All reactions, even exothermic ones, have a certain amount of activation energy that is necessary to get the reaction started.

We are, in fact, very lucky that all chemical reactions require some amount of activation energy in order to start. After all, in order to burn paper, the only two things we need are paper and oxygen. Well, the paper you are looking at right now is in the presence of oxygen, but it is not bursting into flames. The reason the paper isn't burning right now is that the chemical reaction between paper and oxygen requires some activation energy. If I were to provide that activation energy (by lighting a match and sticking it under the paper, for example), the reaction could begin and the paper would burst into flames. The activation energy is represented by the difference between the energy of the reactants and the energy of the intermediate state, as is labeled in the graph.

The explanation for why all chemical reactions have an activation energy is really quite simple. If you think about it, in order for any chemical reaction to proceed, the reactants must get close enough to each other to start rearranging their electrons. After all, as I have said many times before, a chemical reaction is just a rearrangement of the valence electrons in the molecules that make up the reactants. The problem is, under normal conditions, molecules don't like to get close enough to do this, because the electrons that must be rearranged repel each other. Thus, in order for the molecules in a chemical reaction to get close enough to rearrange their electrons, they must be given enough energy to overcome the repulsion that their electrons feel from each other. This energy is what we call activation energy. Each reaction has its own, unique activation energy, since each set of reactants has a different repulsive force between their electrons.

In the end, then, you see that energy diagrams can be very informative. Realize, of course, that the energy diagram you just studied is not the only type of energy diagram that exists. Consider, for example, the following one:

FIGURE 13.2
An Energy Diagram for an Endothermic Reaction

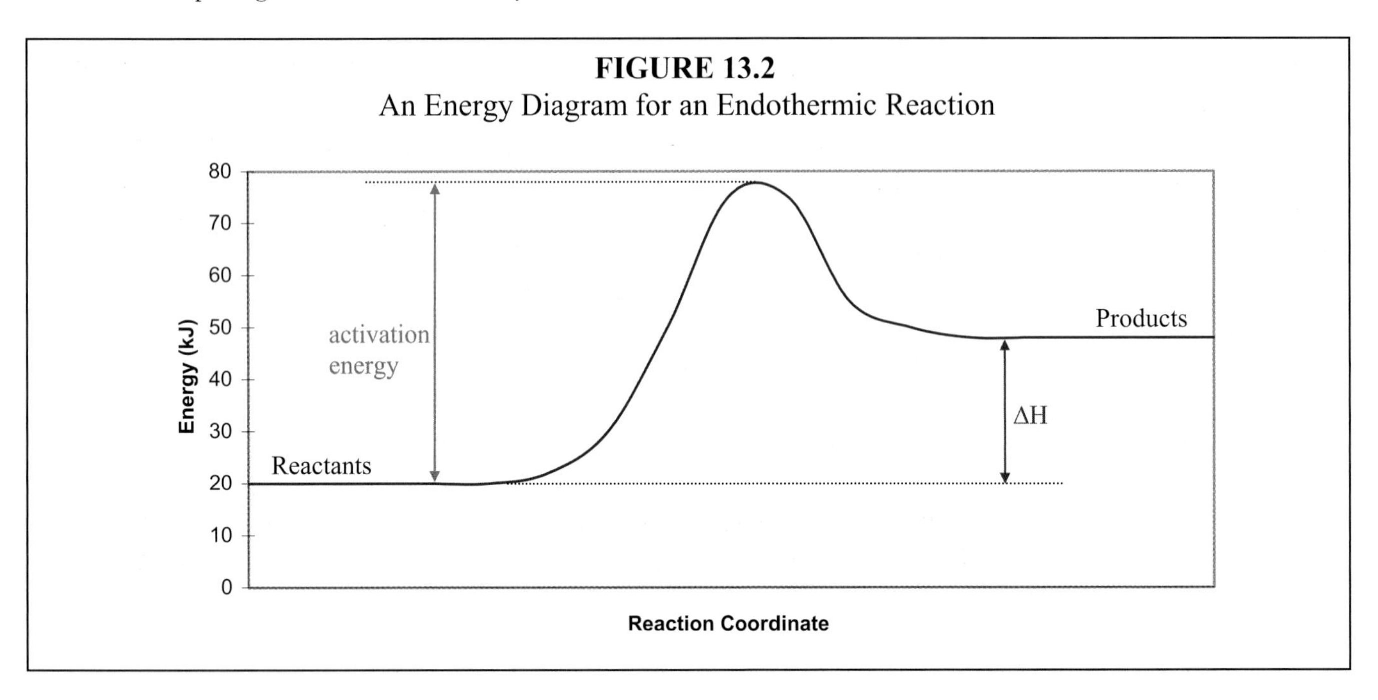

This diagram looks similar to the one presented earlier, but there is one important difference. In this case, the energy of the reactants is *lower* than the energy of the products. What does this tell you? It tells you that the reactants have to absorb energy in order to reach the energy of the products. In other words, this is an endothermic reaction. Once again, the activation energy and ΔH can be read directly from the graph, as is indicated.

Study the next example and do the "On Your Own" problem that follows in order to make sure you understand how to interpret these energy diagrams.

EXAMPLE 13.5

Consider the following energy diagrams:

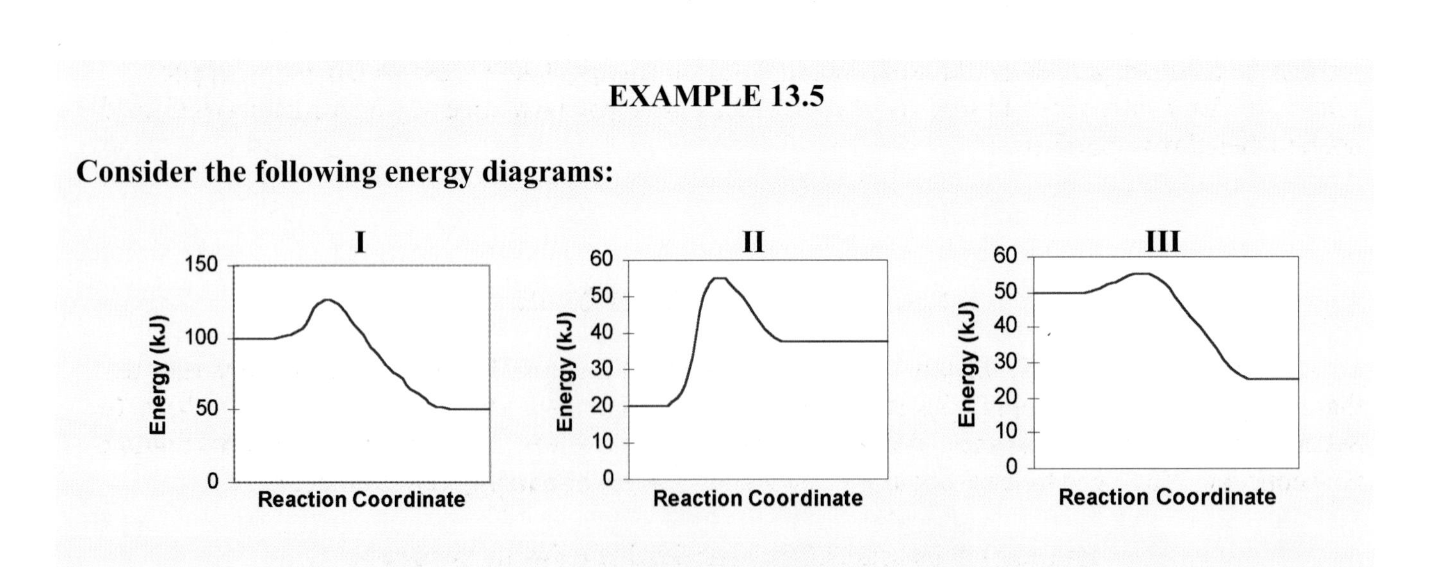

a. Which diagram represents the most exothermic reaction? What is its ΔH?

To solve this problem, we are looking for the diagrams that represent exothermic reactions. Any diagram in which the reactants (the left side of the diagram) have a higher vertical position than the products (the right side of the diagram) would represent an exothermic reaction. Thus, both I and III represent exothermic reactions. The most exothermic is the one that has the largest difference between the vertical position of the reactants and products. Thus, diagram I represents the most exothermic reaction. To get the ΔH of the reaction, we simply must read the difference in energy between the products and the reactants. The products (on the right side of the graph) look to be at an energy (y-axis position) of 50 kJ, while the reactants (on the left side of the graph) are at a position of 100 kJ. To calculate ΔH, we take the energy of the products and subtract the energy of the reactants. Thus, $\Delta H = 50 \text{ kJ} - 100 \text{ kJ} = -50 \text{ kJ}$. Notice that by taking the energy of the products minus the energy of the reactants, I have the proper sign for ΔH. Thus, $\Delta H = -50 \text{ kJ}$.

b. Which diagram represents the reaction that is easiest to get started?

The reaction that is easiest to start has the lowest activation energy. Thus, we are looking for the reaction that has the lowest difference in energy between the reactants and the intermediate state. This would clearly be diagram III.

c. Which diagram represents an endothermic reaction? What is its ΔH?

An endothermic reaction is one in which the products have a higher energy than the reactants. This is the case in diagram II. Once again, to get ΔH, we take the energy of the products (which looks to be about 40 kJ) minus the energy of the reactants (which looks to be about 20 kJ). Thus, $\Delta H = 20 \text{ kJ}$.

ON YOUR OWN

13.8 Suppose you perform two experiments. In the first, you run a reaction that we will call "reaction A." This reaction is very hard to start, but once it gets started, it gets very, very hot. In the second experiment, you run "reaction B." This reaction is incredibly easy to start, but it gets only slightly warmer than before you began the experiment. Draw an energy diagram for each reaction, based solely on what I have just told you. (Note that since I gave you no numbers, your vertical axis should not be labeled with numbers.)

The Second Law of Thermodynamics

Way back in Module #2, you learned about the First Law of Thermodynamics, which tells us that energy cannot be created or destroyed; it can only change form. Now it's time for you to learn the Second Law of Thermodynamics. Before you can do that, however, you need to be acquainted with a new concept. You need to learn about the fascinating subject of **entropy** (en' truh pee):

Entropy - A measure of the disorder that exists in any system

Since entropy is a measurement, it must have units. These units are always an energy unit divided by both moles and Kelvin. Thus, $\frac{\text{Joules}}{\text{mole}\cdot\text{K}}$, $\frac{\text{kcals}}{\text{mole}\cdot\text{K}}$, and $\frac{\text{kJ}}{\text{mole}\cdot\text{K}}$ are all acceptable units for measuring entropy. For some strange reason, entropy is abbreviated with a capital "S."

Now it might sound odd to you that chemists can measure disorder, but it turns out that disorder is one of the most fundamental things we can learn about a system. For example, suppose you have a priceless vase sitting on a shelf somewhere. Suppose further that something happens that causes the vase to fall onto the floor and shatter into a million pieces. Which situation, do you think, represents more disorder? Clearly, when the vase is shattered, it is more disordered than when it was sitting on the shelf in one piece. Thus, we would say that while the vase is sitting on the shelf, it is in a state of low entropy, because it is not very disordered. However, when the vase shatters, it moves into a state of high entropy, because now it is very disordered.

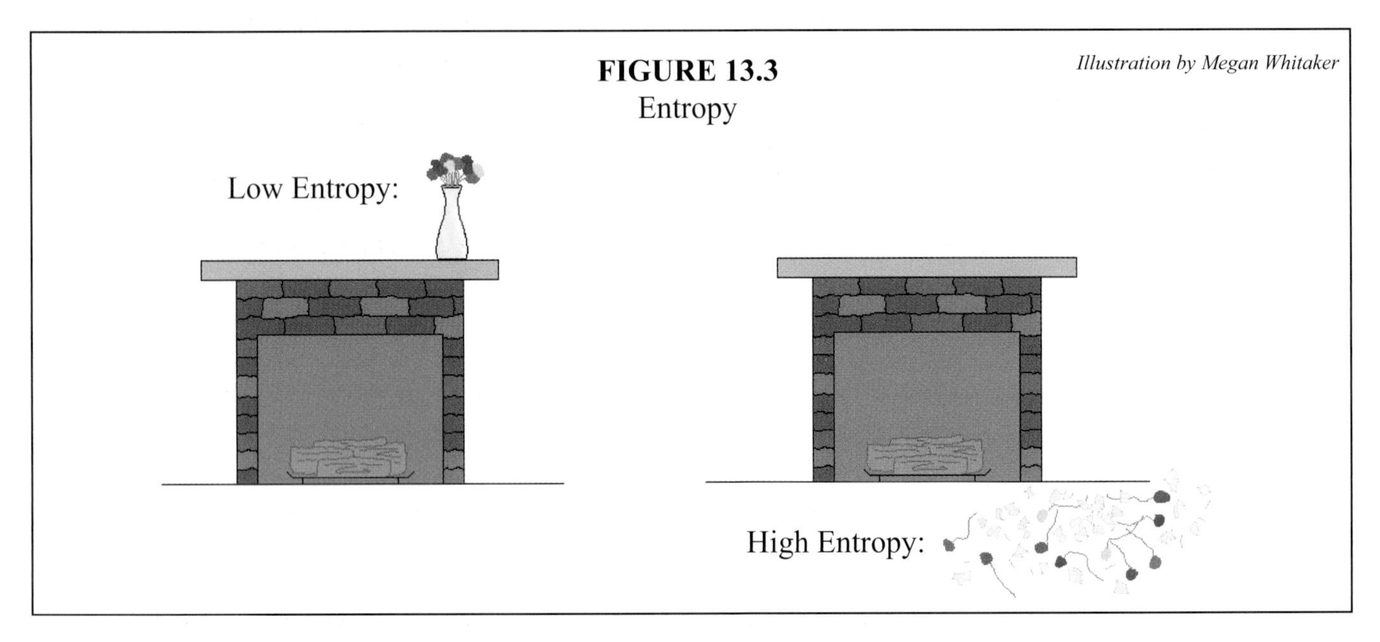

FIGURE 13.3
Entropy

Although you can imagine a vase falling from a shelf and shattering into a million pieces, can you ever imagine the reverse process happening? Can you imagine a situation where a shattered vase spontaneously rearranges itself into a whole vase again? Of course not! Even though you know that this will never happen, do you know why? The reason that this cannot happen is because of the **Second Law of Thermodynamics**:

<u>The Second Law of Thermodynamics</u> - The entropy of the universe must always either increase or remain the same. It can never decrease.

Think for a moment about how this law would apply to our vase. If the shattered vase were to spontaneously rearrange itself into a whole vase again, it would be going from a high entropy state to a low entropy state. This would mean that the entropy of the universe would decrease, because the entropy of the vase decreased. This would violate the second law. That's why broken things do not spontaneously fix themselves. It would be a violation of the Second Law of Thermodynamics.

Now do you see why the concept of entropy is so important to a scientist? To a large extent, the Second Law of Thermodynamics determines what types of events are possible in nature and what types of events are not possible. Thus, this law is just as important and just as fundamental to scientists as is the First Law of Thermodynamics. The problem is, however, that this law is much harder to understand than the First Law of Thermodynamics. As a result, there is a lot of confusion concerning what the Second Law of Thermodynamics prohibits and what it does not prohibit. At the end of this section, I will spend a considerable amount of time trying to explain to you exactly how the Second Law of Thermodynamics should be interpreted. For right now, however, you need to have a firm grounding in the concept of entropy, so let's start with that.

As I said before, entropy is a measure of the disorder that exists in a system. As a chemist, you should be very comfortable in determining whether a given state represents high entropy or low entropy. In the case of the vase, it was rather simple to determine which state was high entropy and which was low entropy. However, there are more subtle differences in entropy of which you must be aware.

First and foremost, the entropy of a substance changes when the substance changes phase. In the solid phase, the molecules (or atoms) that make up a substance are held in place and are only allowed to vibrate. Since the molecules cannot move around very much, they are not very disordered. As a result, the solid state of matter has the lowest amount of entropy. In the liquid phase, however, molecules are free to move around. Their movement is somewhat restricted, however, because they must stay relatively close to their neighbors. As a result, the entropy of the liquid state is larger than that of the solid state. The gaseous state, however, is the phase with the highest entropy because the molecules (or atoms) that make up a substance move freely and quickly. You need to remember the relative entropy of the phases of matter:

The solid state has the least amount of entropy associated with it. The entropy of the liquid state is higher, and the entropy of the gaseous state is the highest of all.

Although this fact is very important, there are other things to consider besides the phase in which a substance finds itself.

In general, whenever a substance's temperature increases, its entropy increases. This is because as temperature increases, the motion of the molecules (or atoms) in a substance increases as well. The faster molecules (or atoms) begin moving, the harder it is to keep track of them. As a result, the entropy increases. This is another important fact:

The entropy of a system increases with increasing temperature.

Thus, if I have two identical samples of matter and raise one's temperature, the sample with the higher temperature is more disordered.

Finally, the more matter you have in a system, the more disordered it is. This should make sense to you based on your own experience. Consider your own room. If you only had a few things to store in your room, it would probably be very neat and tidy. However, the more things you have in your room, the harder it is to keep your room neat. The same problem exists in nature.

The entropy of a system increases as the matter it contains increases.

Thus, a sample of gas that contains 5 moles has more entropy than a sample of the same gas that contains only 3 moles.

Now that you know some general rules about entropy, it is time that you learned how a chemist applies these concepts. First of all, since it is very difficult to measure the entropy of a system, chemists are usually more interested in the change in entropy that a system undergoes. This should sound a little familiar to you. We already learned that enthalpy was very hard to measure, whereas the change in enthalpy of a system is easier to measure. The same thing holds true for entropy as well. Thus, chemists are usually more interested in ΔS than they are in S. In fact, we can even restate the Second Law of Thermodynamics in terms of ΔS:

$$\Delta S_{universe} \geq 0 \quad (13.11)$$

You should be able to see how Equation (13.11) restates the Second Law of Thermodynamics. If the entropy of the universe must always either increase or stay the same, then the change in entropy of the universe must be either zero or positive.

What chemists concentrate on, of course, is trying to determine the ΔS of a chemical reaction. There are two ways that chemists typically do this. The first way we will discuss is a qualitative method that doesn't allow us to calculate a *number* for the ΔS of a chemical reaction. Rather, this method allows us to determine the *sign* of ΔS for a chemical reaction. The sign of ΔS is an important thing to know, because it relates to the Second Law of Thermodynamics, as illustrated by Equation (13.11).

To determine the sign of ΔS for a chemical reaction, all we have to do is look at the number of molecules (or atoms) on each side of the equation as well as the phase of the substances involved. We can then use some of the facts we learned above. In general, the side of the reaction that contains the most molecules (or atoms) in the gas phase has the highest entropy. If no gases are present, then the side that has the most molecules (or atoms) in the liquid or aqueous phase has the highest entropy. Finally, if all substances in the chemical equation are solids, then the side with the largest number of molecules has the highest entropy. If none of these situations is true, then you simply cannot determine the sign of ΔS this way. If this all sounds a bit confusing, don't worry. An example problem should clear everything up.

EXAMPLE 13.6

For each reaction below, determine the sign of ΔS:

a. $Zn\ (s) + HCl\ (aq) \rightarrow ZnCl_2\ (aq) + H_2\ (g)$

The reactants side of the equation has no molecules in the gas phase, while the products side of the reaction has one molecule in the gas phase. Thus, the products side has a higher entropy. This means that when the reaction occurs, the substances go from a lower entropy state to a higher entropy state. Thus, ΔS is positive.

b. N_2 (g) + $3H_2$ (g) → $2NH_3$ (g)

The reactants side of this equation contains four molecules in the gas phase (one nitrogen molecule and three hydrogen molecules). The products side contains only two molecules in the gas phase (two ammonia molecules). Thus, the reactants side of this equation has the highest entropy. This means that as the reaction occurs, the chemicals go from a higher entropy to a lower entropy. Thus, ΔS is negative. Now you might be tempted to say that since ΔS is negative for this reaction, the reaction can never proceed. After all, the Second Law of Thermodynamics says that $\Delta S_{universe}$ must always be greater than or equal to zero. It turns out that using the Second Law of Thermodynamics is a bit more complicated than that, however. We will soon see that a reaction can proceed, even if its ΔS is negative. Some very important conditions must be met first, however.

c. $2NH_4NO_3$ (s) + $Ba(OH)_2(H_2O)_8$ (s) → $Ba(NO_3)_2$ (aq) + $2NH_4OH$ (aq) + $8H_2O$ (l)

This equation has no molecules or atoms in the gas phase on either side of the chemical equation. Thus, we must examine the number of molecules or atoms in the liquid or aqueous phase. Based on that, the products side clearly has the highest entropy. This means that ΔS is positive.

Make sure you can determine the sign of ΔS for a chemical reaction by solving the following problem:

ON YOUR OWN

13.9 Determine the sign of ΔS for the following reactions:

a. $C_2H_4O_2$ (aq) + $NaHCO_3$ (s) → $NaC_2H_3O_2$ (aq) + H_2O (l) + CO_2 (g)
b. $2H_2O_2$ (g) → $2H_2O$ (g) + O_2 (g)
c. $2AgNO_3$ (aq) + $Mg(OH)_2$ (aq) → $2AgOH$ (s) + $Mg(NO_3)_2$ (aq)

Now if you need a little more information than just the sign of ΔS, there is a way to determine a number for the change in entropy of a chemical reaction. It turns out that even though it is very difficult to measure the entropy of an individual substance, it can be done. Table 13.3 tabulates the absolute entropies of the same substances contained in Table 13.2. This, of course, makes determining ΔS very simple. You can use an equation very similar to Equation (13.9) to calculate the change in entropy between the products and the reactants, which is the definition of ΔS:

$$\Delta S^o = \Sigma\ S^o\ (\text{products}) - \Sigma\ S^o\ (\text{reactants}) \qquad (13.12)$$

Thus, determining the change in entropy of a chemical reaction is as simple as using Hess's Law, provided you have the proper table. There is one difference between calculating ΔS and ΔH, however. When calculating ΔH, you follow the rule that the ΔH_f of an element is zero. When calculating ΔS, however, the S of an element is *not* zero. This is very important:

All substances, even elements, have an absolute entropy.

TABLE 13.3

Standard Absolute Entropies in $\frac{J}{mole \cdot K}$

Substance	S°	Substance	S°	Substance	S°	Substance	S°
H_2O (g)	189	NO_2 (g)	240	$Ca(OH)_2$ (s)	76.1	C_2H_6O (l)	161
H_2O (l)	69.9	NH_3 (g)	193	$CaCO_3$ (s)	92.9	C_2H_4 (g)	219
SO_3 (g)	257	CO_2 (g)	214	C_2H_6 (g)	230	C_4H_{10} (g)	260
NO (g)	211	CaO (s)	38.2	CH_3OH (l)	127	H_2SO_4 (l)	157

Notice a few things about this table. The entropies listed here are all positive. This is because the table does not contain values of ΔS; it contains values of S. As a result, they all must be positive, because *all* substances have some level of disorder and thus some absolute entropy. Second, notice that, compared to the values of ΔH tabulated in Table 13.2, these numbers are quite small. Table 13.2 contained numbers with the unit of kJ/mole, whereas these numbers are in $\frac{J}{mole \cdot K}$. This means that these numbers are about 1,000 times smaller than the ones tabulated in Table 13.2.

Since the use of Table 13.3 and Equation 13.12 is exactly the same as that of Hess' Law, I will not give you an example of how to solve a problem using them. Instead, just use the same ideas you learned before to solve the problem that follows:

ON YOUR OWN

13.10 In Problem 13.5, you determined the ΔH° of the following reaction:

$$CaO\ (s) + CO_2\ (g) \rightarrow CaCO_3\ (s)$$

Use what you just learned to calculate ΔS° for this same reaction.

The Proper Application of the Second Law of Thermodynamics

As I said before, even though the Second Law of Thermodynamics is a very, very, important law in science, it is also a very misunderstood and misused law. For example, you might think that the Second Law of Thermodynamics makes evolution impossible. After all, the whole theory of evolution is predicated on the belief that simple (disordered) life forms can evolve into complex (ordered) life forms. The Second Law of Thermodynamics states that disorder must always stay the same or increase. Evolution requires just the opposite, so the Second Law of Thermodynamics says that evolution cannot happen, right?

Although this argument sounds quite nice, it is, nevertheless, wrong. There are spontaneous processes that occur every day that result in a decrease of disorder. For example, what happens to a bucket of water when it is left outside on a cold, winter day? It freezes. Well, as we learned earlier, the solid state of matter has lower entropy than the liquid state. Thus, when water freezes, it goes from a more disordered state to a less disordered state. In other words, for this process, ΔS is negative.

This, on the surface, seems to violate the Second Law, which states that $\Delta S_{universe}$ must always be greater than or equal to zero.

Well, it turns out that this process only *seems* to contradict the Second Law of Thermodynamics. It seems that way because we have only considered *part* of the process. You see, when water freezes, it affects the molecules that surround it. Think about it this way. When water freezes, its molecules slow down and don't move as much as they did in the liquid state. This means that they lose energy. Where does that energy go? It is actually released into the surroundings. As a result, when water freezes, its surroundings actually heat up a bit! This may sound odd to you, but it is nevertheless true. The very process of freezing requires a release of energy. When a system releases energy, its surroundings heat up. Thus, when water freezes, the things surrounding the water heat up.

What happens to a substance's entropy when it heats up? The entropy increases. Thus, when the water freezes, the entropy of the things surrounding the water actually increases. This increase in entropy of the surroundings offsets the decrease in entropy of the water. As a result, the *total* ΔS of the universe (water + its surroundings) does not change. This is in agreement with the Second Law of Thermodynamics, because it says that ΔS can be positive, *or zero.*

Thus, when you look at a process in terms of the Second Law of Thermodynamics, you *must* look at everything that is affected. This includes the system that you are interested in, *plus everything else that is affected by it*. Let's imagine one more example. Suppose you come home one afternoon and realize that your house is a mess. You then decide to clean it up. When you start to clean your house, you begin to transform it from a disordered (high entropy) state into a relatively ordered (low entropy) state. At first glance, then, cleaning the house seems to violate the Second Law of Thermodynamics. What are we missing?

Well, as you clean up the house, you begin to move around. This agitates the molecules that make up the surrounding air, causing them to move faster. This increases their entropy. Plus, as you begin to work, your body starts to burn calories to supply you with the energy you need. This heats up your body, increasing its entropy. If you were able to measure the amount of increased entropy in the air and your body, you would find out that it is larger than the decrease in entropy caused by the fact that you cleaned house. In the end, then, the *total* entropy of the universe increased, in agreement with the Second Law.

How does this apply to evolution? Well, if an advanced species (which is highly ordered) did evolve from a simple species (which is less ordered), then that organism would have, indeed, experienced a decrease in entropy. To make this consistent with the Second Law of Thermodynamics, then, the organism's surroundings would have to become more disordered so that the decrease in entropy of the organism could be offset. This is the only way that evolution could be consistent with the Second Law of Thermodynamics.

Thus, suppose a single-celled creature were to evolve into a cell that would cooperate with other cells. Evolutionists will tell you that this can happen if that cell's DNA were to undergo a mutation that would make the DNA a little more advanced. The problem is that the very instant that the DNA mutated and became more ordered, there would have to be a corresponding increase in entropy in the organism's surroundings so that the *total* entropy of the universe would still increase or at least stay the same. No scientist on earth can come up with a single idea as to *how* that can happen.

As far as anyone can tell, mutating DNA does not significantly increase the entropy of the surroundings.

So you see that the Second Law of Thermodynamics doesn't really disprove evolution, but it does attach some pretty strong conditions to it. The Second Law says that *if* evolution were to occur, then *each step of that evolution* must be accompanied by an *immediate* increase in the entropy of the surroundings. If anyone ever comes up with a mechanism by which this might happen, evolution could at least be made *consistent* with the Second Law. Currently, however, no one can come up with even a vague notion as to how this might happen. As a result, the current theories of evolution are not consistent with the Second Law. This does not mean that the Second Law *disproves* these theories, however. It just means that no one has come up with the proper mechanism. In my opinion, however, such a mechanism does not exist.

Gibbs Free Energy

As we have seen, then, if a process is to occur, that process must result in no change to the entropy of the universe or an increase in the entropy of the universe. In order to determine whether or not this happens, you must consider not only the entropy of the process, but also the entropy of any surroundings that this process affects. This might sound like a daunting task. However, as far as chemical reactions go, it is rather easy!

You see, we can already determine the ΔS of the chemical reaction as we learned earlier; thus, we are halfway there. The only other thing we have to do in order to determine whether or not a reaction will really happen is to determine the effect that it has on the entropy of its surroundings. It turns out that this is relatively simple as well!

You see, the main way that a chemical reaction can alter the entropy of its surroundings is through its enthalpy. If a chemical reaction has a negative ΔH, it releases energy into its surroundings. As a result, the entropy of the surroundings increases. Alternatively, if a chemical reaction has a positive ΔH, it absorbs energy from its surroundings, decreasing their entropy. Thus, to determine whether or not a chemical reaction is consistent with the Second Law of Thermodynamics, we need only look at its ΔS and ΔH. The value of ΔS will tell us about the change in entropy of the reaction itself, and the ΔH will allow us to learn about the change in entropy of the reaction's surroundings.

There is a very simple formula that does all of this for us. It relates ΔS and ΔH into something called the **Gibbs Free Energy**, which we label as ΔG:

$$\Delta G = \Delta H - T \cdot \Delta S \qquad (13.13)$$

In this equation, T is the temperature of the surroundings (in Kelvin). The Gibbs Free Energy actually can tell us whether or not a reaction is consistent with the Second Law. If it is, we say that the reaction is **spontaneous**, because we know that it can proceed. If it is not consistent with the Second Law, we call it **not spontaneous**, because the reaction cannot occur.

It turns out that as long as the ΔG for any reaction is negative, the reaction is consistent with the Second Law. If, however, the ΔG is positive, the reaction is not consistent with the Second Law. This important fact must be remembered:

When $\Delta G < 0$, then the reaction is consistent with the Second Law and is thus spontaneous. If $\Delta G > 0$, then the reaction cannot proceed, because it violates the Second Law.

It turns out that this statement is a bit of an approximation. You see, there are ways that a reaction can affect its surroundings other than through ΔH. Since the equation for ΔG does not consider any other factors, it is not an exact way of determining the spontaneity of a reaction. However, it is such a good approximation that most chemists are content to use it, so we definitely will. If you take advanced chemistry, you will learn a more precise way of relating ΔG to chemical reactions.

Let's look at Equation (13.13) and see what it tells us. If a reaction is to be spontaneous, then ΔG has to be negative. Looking at Equation (13.13), how can this happen? Well, if ΔH is negative and ΔS is positive, the first term in Equation (13.13) will be negative and the second term ($T \cdot \Delta S$) will be positive. If a negative term has a positive term subtracted from it, then the result will *always* be negative. As a result, whenever a reaction is exothermic (has a negative ΔH) and also has a positive ΔS, it will *always* be a spontaneous reaction. On the other hand, if ΔH is positive and ΔS is negative, ΔG will *always* be positive. Thus, reactions that are endothermic and have a negative ΔS will *never* be spontaneous.

This, then, is one qualitative way of determining whether or not a reaction is spontaneous. If we know whether the reaction is exothermic or endothermic, we can look at the equation and determine the sign of its ΔS. If the reaction is endothermic and ΔS is negative, the reaction will never occur. If, however, the reaction is exothermic and ΔS is positive, the reaction will always be spontaneous.

Now, of course, the two possibilities I just discussed are not the only possibilities for a chemical reaction. Suppose the ΔH of a chemical reaction is negative and the ΔS is also negative. In this case, the first term in Equation (13.13) will be negative, but the $T \cdot \Delta S$ term will also be negative. Thus, you will have one negative number subtracted from another negative number. Sometimes, the result will be positive (indicating a non-spontaneous reaction), and sometimes the result will be negative (indicating a spontaneous reaction). In those cases, you will have to actually calculate a value of ΔG to determine whether or not the reaction is spontaneous.

Of course, calculating a value for ΔG isn't too hard, since we already know how to calculate ΔH and ΔS. Thus, one way to calculate ΔG is to calculate ΔH and ΔS for a reaction and plug the result into Equation (13.13). There is, however, another way to calculate ΔG. It turns out that ΔG is also a state function. Thus, we can use Hess's Law to calculate ΔG^o, as long as we have a table of ΔG_f^o's.

TABLE 13.4

Standard Gibbs Free Energies of Formation in $\frac{kJ}{mole}$

Substance	ΔG_f^o	Substance	ΔG_f^o	Substance	ΔG_f^o	Substance	ΔG_f^o
CH_4 (g)	-50.8	NO_2 (g)	240	$CaSO_4$ (s)	-1320	C_2H_6O (l)	-175
C_6H_6 (l)	125	H_2O (l)	-237	$CaCO_3$ (s)	-1129	$CaCl_2$ (s)	-750
CO (g)	-137	H_2O (g)	-229	H_2CO_3 (aq)	-623	HCl (aq)	-131
CO_2 (g)	-394	CaO (s)	38.2	CH_3OH (l)	-166	$Ca(OH)_2$ (s)	-897

The numbers in this table, then, will allow us to use the following equation as a means of determining ΔG:

$$\Delta G^o = \Sigma\, \Delta G_f^o\,(\text{products}) - \Sigma\, \Delta G_f^o\,(\text{reactants}) \qquad (13.14)$$

Once again, the ΔG_f^o of any element in its elemental form is zero, as is its ΔH_f^o. Now, since we have two different methods for calculating ΔG, does it matter which we use? Yes, it does. You see, the values given for ΔG_f^o in Table 13.4 are all for a specific temperature: 298 K. Thus, if you are working with a reaction at 298 K, you can use Equation (13.14) and Table 13.4 to calculate ΔG^o. Anytime you are working on a reaction at a temperature other than 298 K, you need to use Equation (13.13). Let's see how this works in a few example problems:

EXAMPLE 13.7

In Example 13.3, we studied the following chemical reaction:

$$\mathbf{2NO\,(g) + O_2\,(g) \rightarrow 2NO_2\,(g)}$$

and found that its ΔH^o was -114.2 kJ. If its ΔS^o is -147 $\frac{\text{J}}{\text{mole}\cdot\text{K}}$, what is its ΔG at 532 ^oC (805 K)? Is it a spontaneous reaction at that temperature?

In this problem, we are working with a reaction at a temperature other than 298 K. As a result, we must use Equation (13.13). This isn't hard, though, because we already know ΔS^o and ΔH^o. The only thing we have to notice is that their units are not, at this point, consistent. The value for ΔS^o has Joules in it while the value for ΔH^o has kJ in it. We must therefore change one set of units. I will choose to convert ΔS^o:

$$\frac{-147\,\cancel{\text{J}}}{\text{mole}\cdot\text{K}} \times \frac{1\text{ kJ}}{1{,}000\,\cancel{\text{J}}} = -0.147\,\frac{\text{kJ}}{\text{mole}\cdot\text{K}}$$

Now I can use Equation (13.13):

$$\Delta G = -114.2\,\frac{\text{kJ}}{\text{mole}} - (805\,\cancel{\text{K}})\cdot(-0.147\,\frac{\text{kJ}}{\text{mole}\cdot\cancel{\text{K}}}) = 4\,\frac{\text{kJ}}{\text{mole}}$$

Notice that since ΔS has the Kelvin temperature unit in it, the temperature I use in the equation must also be in Kelvin. Since this ΔG is positive, we can say that this reaction is <u>not spontaneous</u>. Thus, the decrease in entropy of the reaction is not offset by the increase in entropy of the surroundings caused by the release of heat.

Is there any temperature at which the above reaction does become spontaneous?

Since ΔG must be negative for the reaction to be spontaneous, this question really asks for the temperature at which ΔG falls below zero. Since $\Delta G = \Delta H - T\cdot\Delta S$, that means we are looking for the following situation:

$$\Delta H - T\cdot\Delta S < 0$$

$$-114.2\ \frac{\text{kJ}}{\text{mole}} - \text{T}\cdot -0.147 \frac{\text{kJ}}{\text{mole}\cdot\text{K}} < 0$$

We can solve inequalities just like regular algebraic equations, as long as we remember that if we multiply or divide both sides by a negative, we must switch the direction of the inequality sign:

$$\text{T}\cdot 0.147 \frac{\text{kJ}}{\text{mole}\cdot\text{K}} < 114.2\ \frac{\text{kJ}}{\text{mole}}$$

$$\text{T} < \frac{114.2\ \frac{\cancel{\text{kJ}}}{\cancel{\text{mole}}}}{0.147\ \frac{\cancel{\text{kJ}}}{\cancel{\text{mole}}\cdot\text{K}}}$$

$$\text{T} < 777\ \text{K}$$

Thus, for any temperature under 777 K, this reaction is spontaneous.

What is the ΔG for the following reaction at 298 K? Is it spontaneous?

$$\textbf{Ca (s) + 2HCl (aq)} \rightarrow \textbf{CaCl}_2\textbf{ (s) + H}_2\textbf{ (g)}$$

This problem asks us for the ΔG of a reaction at 298 K. Thus, we can use Table 13.4 and Equation (13.14) to answer it. Recognizing that the ΔG_f^o's for both Ca (s) and H_2 (g) are zero, Equation (13.14) becomes:

$$\Delta G^o = (1\ \cancel{\text{mole}}) \times (-750 \frac{\text{kJ}}{\cancel{\text{mole}}}) - (2\ \cancel{\text{moles}}) \times (-131\ \frac{\text{kJ}}{\cancel{\text{mole}}}) = \underline{-490\ \text{kJ}}$$

This reaction, then, is spontaneous at 298 K.

As I have cautioned you in the past, do not get so wrapped up in solving problems that you forget the chemical significance of what you are doing. Remember, ΔG is so important because it takes inventory of all of the entropy of both the chemical reaction and its surroundings. By doing that, we can use it to determine whether or not any chemical equation is consistent with the Second Law and therefore whether or not it will occur naturally. Finish this module with the following problems:

ON YOUR OWN

13.11 The ΔH^o of a certain chemical reaction is 10.0 kJ whereas its ΔS^o is 123 J/K. At what temperatures is this reaction spontaneous?

13.12 Is the following synthesis of ethyl alcohol a spontaneous reaction at 298 K?

$$2CH_4\ (g) + O_2\ (g) \rightarrow C_2H_6O\ (l) + H_2O\ (l)$$

ANSWERS TO THE "ON YOUR OWN" PROBLEMS

13.1 a. A formation reaction involves using the elements in the compound as reactants and the compound itself as the only product. The unbalanced equation, then, is:

$$H_2 + S + O_2 \rightarrow H_2SO_4$$

Remember, both hydrogen and oxygen are homonuclear diatomics; that's why they each get subscripts of "2." Now we balance the equation and, since ΔH is negative, add the energy as a product:

$$\underline{H_2 + S + 2O_2 \rightarrow H_2SO_4 + 814\ \text{kJ}}$$

b. When a compound decomposes, we write a chemical equation in which the compound is the sole reactant and its constituent elements are the products. Thus, the unbalanced chemical reaction is:

$$K_3PO_4 \rightarrow K + P + O_2$$

Once again, oxygen is a homonuclear diatomic, so it gets a subscript of "2." Now we balance the equation and, since ΔH is positive, add energy as a reactant:

$$\underline{K_3PO_4 + 2030\ \text{kJ} \rightarrow 3K + P + 2O_2}$$

c. You should recognize H_2CO_3 as an acid and KOH as a base. Thus, this is an acid/base reaction that produces salt and water:

$$H_2CO_3 + KOH \rightarrow H_2O + K_2CO_3$$

Now we just balance the equation and, since the reaction releases energy, add that energy as a product:

$$\underline{H_2CO_3 + 2KOH \rightarrow 2H_2O + K_2CO_3 + 21\ \text{kcals}}$$

13.2 To solve virtually any problem involving chemical reactions, we first must find the balanced chemical equation. Realizing that hydrogen and chlorine are both homonuclear diatomics, we can say that the unbalanced equation is:

$$H_2 + Cl_2 \rightarrow HCl$$

This equation is rather simple to balance:

$$H_2 + Cl_2 \rightarrow 2HCl$$

Now we just need to convert it to Lewis structure form:

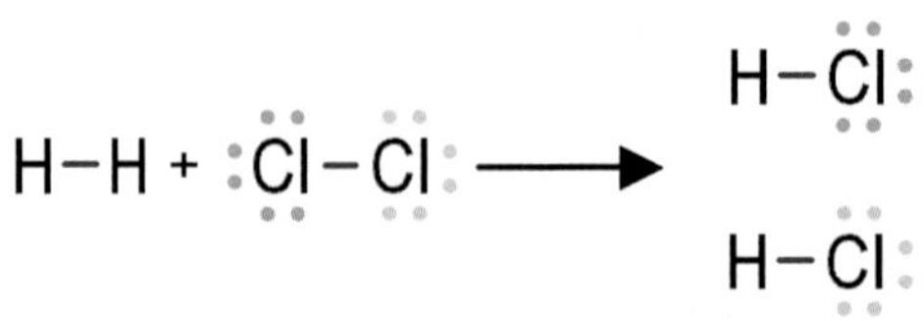

H–H + :C̈l–C̈l: ⟶ H–C̈l: H–C̈l:

In this reaction, then, one H-H bond and one Cl-Cl bond are broken while two H-Cl bonds are formed. According to Table 13.1, the bond energies for these bonds are 432 kJ/mole, 240 kJ/mole, and 428 kJ/mole, respectively. Thus, we can calculate ΔH as:

$$\Delta H = (1\ \cancel{\text{mole}}) \times (432\ \frac{\text{kJ}}{\cancel{\text{mole}}}) + (1\ \cancel{\text{mole}}) \times (240\ \frac{\text{kJ}}{\cancel{\text{mole}}}) - (2\ \cancel{\text{moles}}) \times (428\ \frac{\text{kJ}}{\cancel{\text{mole}}}) = -180\ \text{kJ}$$

The change in enthalpy for this reaction, then, is -180 kJ, indicating that it is an exothermic reaction. Note that I could only report my answer to the tens place in this problem, because the bond energy of the Cl-Cl bond is reported to the tens place.

13.3 The reaction in question is a combustion reaction, and by definition, the unbalanced equation would be:

$$C_2H_2 + O_2 \rightarrow CO_2 + H_2O$$

This balances to:

$$2C_2H_2 + 5O_2 \rightarrow 4CO_2 + 2H_2O$$

In order to use bond energies to calculate ΔH, however, we need to see the Lewis structures of the chemicals involved:

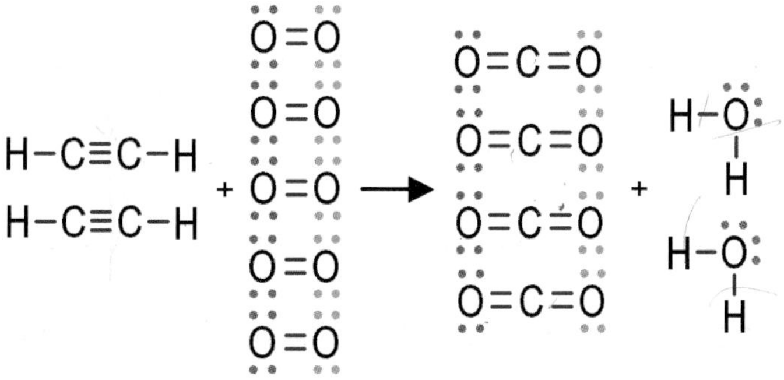

H−C≡C−H
H−C≡C−H
+ O=O, O=O, O=O, O=O, O=O →
O=C=O, O=C=O, O=C=O, O=C=O
+ H−O−H, H−O−H

Now we can see exactly what bonds need to be broken and what ones need to be formed to make this reaction work. We need to break two C≡C bonds (each of which has an energy of 835 kJ/mole), four C-H bonds (each of which has an energy of 411 kJ/mole), and five O=O bonds (each of which has an energy of 494 kJ/mole). In the process, eight C=O bonds (worth 799 kJ/mole each) and four H-O bonds (worth 459 kJ/moles each) are formed. Equation (13.6), then, looks like this:

$$\Delta H = (2\ \cancel{\text{moles}}) \times (835\frac{\text{kJ}}{\cancel{\text{mole}}}) + (4\ \cancel{\text{moles}}) \times (411\ \frac{\text{kJ}}{\cancel{\text{mole}}}) + (5\ \cancel{\text{moles}}) \times (494\ \frac{\text{kJ}}{\cancel{\text{mole}}}) -$$
$$(8\ \cancel{\text{moles}}) \times (799\ \frac{\text{kJ}}{\cancel{\text{mole}}}) - (4\ \cancel{\text{moles}}) \times (459\ \frac{\text{kJ}}{\cancel{\text{mole}}})$$

$$\Delta H = -2{,}444\ \text{kJ}$$

The ΔH for this reaction, then, is -2,444 kJ, and it is exothermic.

13.4 Since the mean guy who wrote this problem says that we have to use Hess's Law to solve it, we will use Hess's Law. First, however, we need a balanced equation:

$$2C_4H_{10}\ (g)\ +\ 13O_2\ (g)\ \rightarrow\ 8CO_2\ (g)\ +\ 10H_2O\ (g)$$

Now that we have a balanced chemical equation, we must check to see if Table 13.2 contains all of the substances in their proper phases. Realizing that we don't need the ΔH_f^o for O_2 (g) since it is the elemental form of oxygen, we should see that Table 13.2 contains everything that we need. Now we just have to apply Equation (13.9):

$$\Delta H^\circ = (8\ \cancel{\text{moles}}) \times (-394 \frac{\text{kJ}}{\cancel{\text{mole}}}) + (10\ \cancel{\text{moles}}) \times (-242 \frac{\text{kJ}}{\cancel{\text{mole}}}) - (2\ \cancel{\text{moles}}) \times (-126 \frac{\text{kJ}}{\cancel{\text{mole}}}) - (13\ \cancel{\text{moles}}) \times (0 \frac{\text{kJ}}{\cancel{\text{mole}}})$$

$$\Delta H^\circ = \underline{-5.320 \times 10^3\ \text{kJ}}$$

13.5 To solve this problem, we need to first determine the chemical equation. According to the problem, CaO (s) and CO_2 (g) are the reactants, because the problem says that they are reacted together. The product is an ionic compound containing the polyatomic ion CO_3^{2-}. Based on our rules for determining the formula of an ionic compound from its name (Modules #8 and #9), you should have determined that the product was $CaCO_3$ (s). Thus, our chemical equation is:

$$CaO\ (s)\ +\ CO_2\ (g)\ \rightarrow\ CaCO_3\ (s)$$

Table 13.2 contains all of these substances in their proper phase, so we just need to apply Equation (13.9):

$$\Delta H^\circ = (1\ \cancel{\text{mole}}) \times (-1207 \frac{\text{kJ}}{\cancel{\text{mole}}}) - (1\ \cancel{\text{mole}}) \times (-635 \frac{\text{kJ}}{\cancel{\text{mole}}}) - (1\ \cancel{\text{mole}}) \times (-394 \frac{\text{kJ}}{\cancel{\text{mole}}}) = \underline{-178\ \text{kJ}}$$

From this ΔH^o, you see that the reaction is exothermic. This is one way that shellfish can keep themselves warm if they get too cold. They just produce more shell, and the exothermic reaction that makes the shell warms the shellfish!

13.6 Since this problem asks about energy in relationship to a chemical reaction, the first thing we must do is calculate the ΔH of the reaction. To do that, however, we need a balanced chemical equation for the combustion of ethane:

$$2C_2H_6\ (g)\ +\ 7O_2\ (g)\ \rightarrow\ 4CO_2\ (g)\ +\ 6H_2O\ (g)$$

Since all of the substances in this equation that are not elements appear in Table 13.2, we can use Hess's Law to determine the ΔH^o:

$$\Delta H^{\circ} = (4 \cancel{\text{moles}}) \times (-394 \frac{\text{kJ}}{\cancel{\text{mole}}}) + (6 \cancel{\text{moles}}) \times (-242 \frac{\text{kJ}}{\cancel{\text{mole}}}) - (2 \cancel{\text{moles}}) \times (-84.7 \frac{\text{kJ}}{\cancel{\text{mole}}}) = -2{,}859 \text{ kJ}$$

This means that energy is a product in the reaction:

$$2C_2H_6\ (g) + 7O_2\ (g) \rightarrow 4CO_2\ (g) + 6H_2O\ (g) + 2{,}859 \text{ kJ}$$

Now we are finally ready to do stoichiometry. To do this, we first convert to moles of ethane:

$$\frac{250.0 \cancel{\text{g } C_2H_6}}{1} \times \frac{1 \text{ mole } C_2H_6}{30.1 \cancel{\text{g } C_2H_6}} = 8.31 \text{ moles } C_2H_6$$

Then we realize that the chemical equation tells us that 2 moles of ethane give us 2,859 kJ of energy:

$$\frac{8.31 \cancel{\text{moles } C_2H_6}}{1} \times \frac{2{,}859 \text{ kJ}}{2 \cancel{\text{moles } C_2H_6}} = 1.19 \times 10^4 \text{ kJ}$$

Thus, burning 250.0 g of ethane provides $\underline{1.19 \times 10^4 \text{ kJ}}$ of energy.

13.7 To solve this problem, we first need to calculate ΔH. This can be done with Hess's Law, since we already have the equation, and everything in the equation is in Table 13.2:

$$\Delta H^{\circ} = (1 \cancel{\text{mole}}) \times (-814 \frac{\text{kJ}}{\cancel{\text{mole}}}) - (1 \cancel{\text{mole}}) \times (-242 \frac{\text{kJ}}{\cancel{\text{mole}}}) - (1 \cancel{\text{mole}}) \times (-396 \frac{\text{kJ}}{\cancel{\text{mole}}}) = -176 \text{ kJ}$$

This means that energy is a product in the reaction:

$$H_2O\ (g) + SO_3\ (g) \rightarrow H_2SO_4\ (l) + 176 \text{ kJ}$$

Now we can do stoichiometry:

$$\frac{50.0 \cancel{\text{g } H_2SO_4}}{1} \times \frac{1 \text{ mole } H_2SO_4}{98.1 \cancel{\text{g } H_2SO_4}} = 0.510 \text{ moles } H_2SO_4$$

$$\frac{0.510 \cancel{\text{moles } H_2SO_4}}{1} \times \frac{176 \text{ kJ}}{1 \cancel{\text{mole } H_2SO_4}} = 89.8 \text{ kJ}$$

Thus, when 50.0 grams of H_2SO_4 are made in this way, <u>89.8 kJ</u> of energy are produced.

13.8 For reaction A, we know two things. It is very hard to start. This tells us that the difference in the energy of the reactants and the intermediate state is large. In other words, our energy diagram needs to contain a large hump. Secondly, we know that the reaction is highly exothermic. This means that the reactants must have an energy that is much larger than the products. Putting that information into an energy diagram:

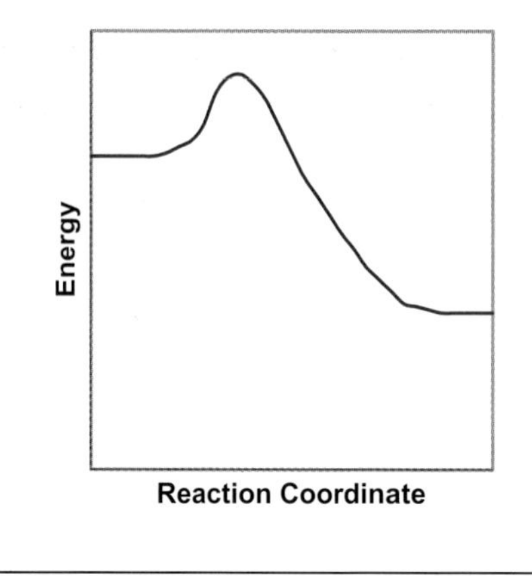

For reaction B, we know that it was easy to start. This means that the difference in energy between the reactants and the intermediate state is small. In other words, this energy diagram has a small hump. Secondly, although the reaction is exothermic, it doesn't release much energy. Thus, the reactants do have a higher energy than the products, but not by much:

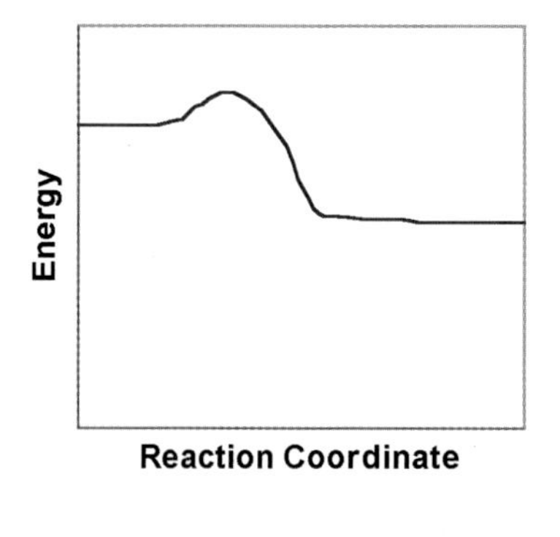

13.9 a. This reaction has one gas molecule on the products side of the equation and none on the reactants side. This means that the products have more entropy than the reactants, so <u>ΔS is positive</u>.

b. In this reaction, there are two molecules of gas on the reactants side and three on the products side. Thus, the products have higher entropy than the reactants, and therefore <u>ΔS is positive</u>.

c. This reaction has no molecules or atoms in the gas phase, thus we must look at the liquid (or aqueous) phase. There are three molecules in the aqueous phase on the reactants side and only one molecule in the aqueous phase on the products side. As a result, <u>ΔS is negative</u>.

13.10 Table 13.3 has all of the information we need, so we just set this up like a Hess's Law problem:

$$\Delta S^{o} = (1\ \cancel{\text{mole}}) \times (92.9 \frac{\text{J}}{\cancel{\text{mole}} \cdot \text{K}}) - (1\ \cancel{\text{mole}}) \times (38.2 \frac{\text{J}}{\cancel{\text{mole}} \cdot \text{K}}) - (1\ \cancel{\text{mole}}) \times (214 \frac{\text{J}}{\cancel{\text{mole}} \cdot \text{K}}) = \underline{-159 \frac{\text{J}}{\text{K}}}$$

13.11 This problem gives us ΔH and ΔS and asks us at what temperature the reaction is spontaneous. In other words, we need to see for what temperatures ΔG is negative. Thus:

$$\Delta H - T\Delta S < 0$$

In order to use this equation, though, we need to get our units consistent:

$$\frac{123 \cancel{J}}{K} \times \frac{1 \text{ kJ}}{1{,}000 \cancel{J}} = 0.123 \frac{\text{kJ}}{K}$$

Now we can use the equation:

$$10.0 \text{ kJ} - T \cdot 0.123 \frac{\text{kJ}}{K} < 0$$

We solve an inequality just like a regular algebraic equation, except when we multiply or divide both sides by a negative value. If that happens, we must switch the direction of the inequality sign:

$$- T \cdot 0.123 \frac{\text{kJ}}{K} < -10.0 \text{ kJ}$$

$$T > \frac{-10.0 \cancel{\text{kJ}}}{-0.123 \frac{\cancel{\text{kJ}}}{K}}$$

$$\underline{T > 81.3 \text{ K}}$$

13.12 This question asks you to determine the spontaneity of a chemical reaction. The only way you can do this is to calculate ΔG. Since this reaction is supposed to run at 298 K, we can use Table 13.4 and Equation (13.14):

$$\Delta G^\circ = (1 \cancel{\text{mole}}) \times (-237 \frac{\text{kJ}}{\cancel{\text{mole}}}) + (1 \cancel{\text{mole}}) \times (-175 \frac{\text{kJ}}{\cancel{\text{mole}}}) - (2 \cancel{\text{moles}}) \times (-50.8 \frac{\text{kJ}}{\cancel{\text{mole}}}) = \underline{-3.10 \times 10^2 \text{ kJ}}$$

This means that the reaction is spontaneous at 298 K .

NOTE: All Lewis structure illustrations in this module by Megan Whitaker

REVIEW QUESTIONS FOR MODULE #13

1. In a chemical reaction, where is the potential energy? Where is the kinetic energy?

2. If a chemical reaction has a positive ΔH, will the beaker that contains the reaction feel hot or cold once the reaction is finished?

3. Why is Hess's Law a more exact way of determining ΔH than the technique that uses bond energies?

4. What is a state function? Give two examples.

5. Which of the following substances will have a ΔH_f^o of zero?

$$NaOH\ (aq),\ Na^+\ (aq),\ O_2\ (g),\ O\ (g),\ Cl_2\ (g),\ H_2\ (l)$$

6. Which of the following diagrams indicates an endothermic reaction? What is the ΔH of that reaction?

I

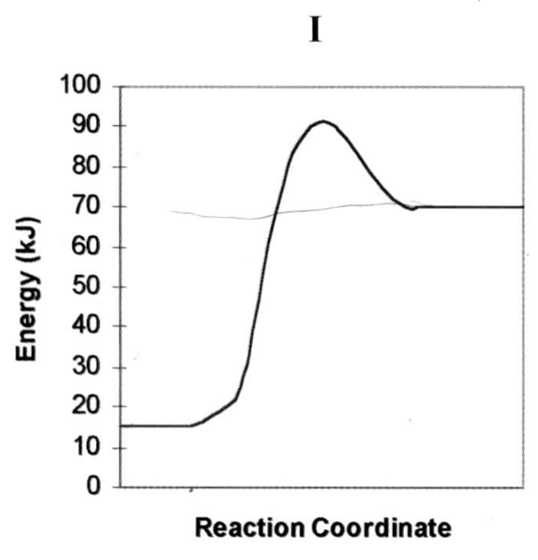

II

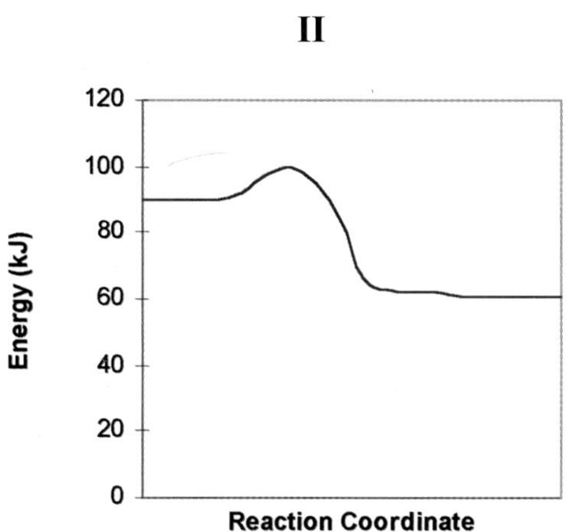

7. Draw an energy diagram for a reaction that has a large activation energy and a ΔH of zero.

8. If you have two 50.0 kg blocks of copper, and one is at a temperature of 50 °C while the other is at a temperature of 25 °C, which has more entropy?

9. When you study, your brain neurons store information. This causes the brain to become more ordered. Why doesn't that fact contradict the Second Law of Thermodynamics?

10. If a chemical reaction is exothermic but has a negative ΔS, what could you do to the temperature in order to make it possible to run the reaction?

PRACTICE PROBLEMS FOR MODULE #13

(You will need to use Tables 13.1 - 13.4)

1. The "chlorine smell" of pool water is not, in fact, due to chlorine. When chlorine hits water, some of it reacts with the water in the following way:

$$2Cl_2 + 2H_2O \rightarrow 4HCl + O_2$$

The smell you get from a chlorinated pool is often the smell of the HCl produced in this reaction. What is the ΔH of this reaction?

2. Benzene (C_6H_6) is a toxic liquid with a ΔH_f^o of 49.00 kJ/mole. The best method for disposing of benzene is to burn it in the presence of excess oxygen. What is the ΔH^o for this reaction? (You should know all of the phases in this problem, so use Hess's Law).

3. When coal is burned in a power plant, it usually contains a small amount of sulfur contamination. The sulfur burns along with the coal to make SO_2. This sulfur dioxide then begins to travel up the smokestack, where it reacts with oxygen as follows:

$$2SO_2\ (g) + O_2\ (g) \rightarrow 2SO_3\ (g)$$

If the ΔH^o of this reaction is -198 kJ, what is the ΔH_f^o of SO_2 (g)?

4. Using the ΔH you got in #2, calculate the number of kJ of energy released when 500.0 grams of benzene are burned in excess oxygen.

5. Suppose you synthesized ammonia with the following reaction:

$$2NO_2\ (g) + 7H_2\ (g) \rightarrow 2NH_3\ (g) + 4H_2O\ (g)$$

How much energy will be released if 35.0 g of NH_3 are formed?

6. What is the sign of ΔS for the following reactions?

a. $NaHCO_3\ (aq) + HCl\ (aq) \rightarrow NaCl\ (aq) + H_2O\ (l) + CO_2\ (g)$
b. $HCl\ (g) + NH_4OH\ (g) \rightarrow NH_4Cl\ (s) + H_2O\ (l)$
c. $SO_2\ (g) + 3F_2\ (g) \rightarrow SF_6\ (g) + O_2\ (g)$

7. What is the ΔS^o of the following reaction?

$$H_2O\ (l) + CaO\ (s) \rightarrow Ca(OH)_2\ (s)$$

8. If the ΔH of a certain reaction is -1023 kJ/mole and the ΔS is -324 $\frac{\text{Joules}}{\text{mole} \cdot \text{K}}$, what is the temperature range for which this reaction is spontaneous?

9. Determine the activation energy and ΔH for the following reaction:

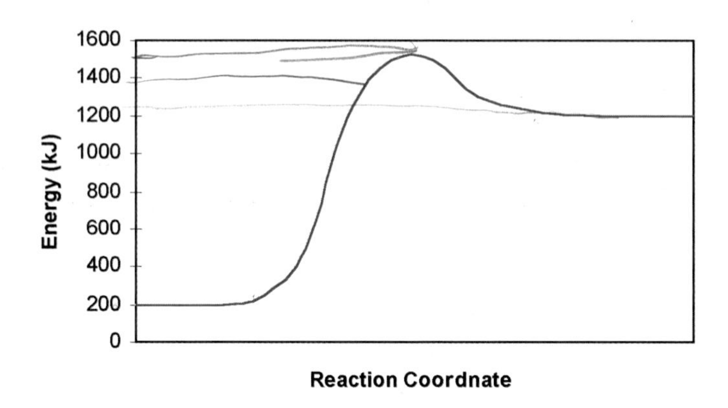

Is this reaction endothermic or exothermic?

10. Is the following reaction spontaneous at 298 K?

$$2C_6H_6\ (l)\ +\ 12H_2\ (g)\ +\ 3O_2\ (g)\ \rightarrow\ 6C_2H_6O\ (l)$$

Cartoon by Speartoons, Inc.

MODULE #14: Kinetics

Introduction

In the previous module, we spent a great deal of time learning how we could determine whether or not a reaction was spontaneous. We learned that a reaction had to meet certain enthalpy and entropy requirements before it could really occur. Since enthalpy and entropy are both concepts in the field of thermodynamics, chemists often call these requirements "thermodynamic criteria." Thus, we could say that a reaction will not occur spontaneously unless it meets certain thermodynamic criteria.

When studying a chemical reaction, however, there is another thing we must consider besides just the thermodynamic criteria involved: We must understand how quickly the reaction proceeds. Consider, for example, the decomposition of hydrogen peroxide:

$$2H_2O_2\ (aq) \rightarrow 2H_2O\ (l) + O_2\ (g)$$

The ΔG^o for this reaction is -190 kJ/mole, indicating that it is a spontaneous reaction at room temperature and normal atmospheric pressure. Despite this fact, if you watched a sample of aqueous hydrogen peroxide all day, you would never see bubbles of oxygen being produced, as you would expect from the above reaction. In fact, bottles of hydrogen peroxide solution can be left on the shelf for years and they will still have some hydrogen peroxide solution in them once they are opened. The solution will not be as concentrated as it was when it was purchased, but there will still be some hydrogen peroxide in the bottle.

If the decomposition of hydrogen peroxide is spontaneous (as our thermodynamic criteria say it is), why can't you watch it decompose? Why does it sit on the shelf without decaying away completely? The answer is really quite simple. Since the reaction is spontaneous, it does, in fact occur. The problem is, however, that it occurs *very, very slowly*. Thus, even though this reaction is thermodynamically spontaneous, it is not necessarily a good way of producing oxygen, since the reaction proceeds so slowly.

In studying a chemical reaction, then, we need to consider two very important things. First, we need to look at the thermodynamic criteria. After all, if the reaction will not occur, there is no need to continue to study it. That would get rather dull. Once we have examined the reaction using our thermodynamic criteria, however, we must also look at how fast the reaction can go. Even if the reaction is thermodynamically spontaneous, if it doesn't proceed quickly enough, there is no use running it.

Reaction Kinetics

When we study how fast a reaction proceeds, we call that a study of the **reaction kinetics** (kih net' iks). In kinetics, the first important definition we come to is the **rate** of the chemical reaction:

Reaction rate - The rate of change of a product in a chemical reaction

In mathematical terms, we could say:

$$R = \frac{\Delta[\text{product}]}{\Delta t} = -\frac{\Delta[\text{reactant}]}{\Delta t} \quad (14.1)$$

At first, this equation might look quite mysterious, but it isn't really that bad. First of all, the "R" stands for rate. Secondly, chemists often abbreviate concentration with square brackets. Thus, "[product]" just means the concentration of a product in the reaction, while "[reactant]" just refers to the concentration of a reactant in the equation. As always, the "Δ" means "change in," and "t" refers to time.

Thus, Equation (14.1) says that if you measure how much the concentration of a reaction product changes and divide it by the time it took to make that change, you will get the rate of the chemical reaction. Think about the units that will result from such a calculation. If I take concentration (usually expressed in molarity) and divide by time (usually expressed in seconds), I will get a unit of molarity/second, which is usually abbreviated as M/s. This is the unit for the rate of a chemical reaction.

Notice that you can also calculate the rate of a chemical reaction by using the change in concentration of one of the reactants in the equation. If you do that, however, you need to take the negative of that change. This should make sense. In a chemical reaction, products are made; therefore, throughout the course of a chemical reaction, the concentration of the products increases. As a result, the change in concentration of a product is positive. Since reactants are used up in a chemical reaction, however, their concentration decreases. The change in concentration of the reactants, therefore, is negative. That's why the negative sign is present in Equation (14.1). Since the change in the concentration of a reactant is negative, we need to take the negative of that change to ensure that the rate of a chemical reaction is always positive.

Factors That Affect the Kinetics of a Chemical Reaction

Since we now know that chemical reactions do not occur instantaneously but instead take a certain amount of time depending on the reaction rate, it is important to determine what factors affect the rate of a chemical reaction. For example, if a reaction is slow, is there any way to speed it up? Alternatively, suppose a reaction happens too rapidly. Is there a way to slow it down? Perform the following experiment to figure that out:

EXPERIMENT 14.1

Factors That Affect Chemical Reaction Rates

Supplies:

- Liquid toilet bowl cleaner (The Works® was used when this experiment was tested, but any liquid that contains hydrochloric acid or hydrogen chloride should work. Stay away from the thickened toilet bowl cleaners that cling, however.)
- Antacid tablets (TUMS® works best. In principle, any antacid tablet that has calcium carbonate as its main ingredient should work.)
- Sharp, nonserrated knife
- Spoon

- Stirring rod (Something thin enough to stir the contents of the test tubes.)
- Four test tubes (Small glass containers will work.)
- Flame heater (An oven burner will work.)
- Large beaker (A short, fat glass will work.)
- Small beaker (A short, fat glass will work. If you are not using test tubes, make sure that one of your glass containers fits into this.)
- Rubber gloves
- Safety goggles

1. Fill your small beaker ¾ of the way full with tap water and start heating it. The water needs to come to a boil.
2. While you are waiting for the water to boil, put on the gloves.
3. Fill three of the test tubes halfway with toilet bowl cleaner. We will call these tube #1, tube #2, and tube #3.
4. Put the three tubes in the large beaker (not the one being heated) so that they stand up. The large beaker will act as a test tube holder.
5. To the last test tube (#4), add only enough toilet bowl cleaner to fill it 1/8 of the way full. Use tap water to dilute the toilet bowl cleaner until the test tube is ½ full, like the others.
6. Stir the resulting diluted solution.
7. Put that tube in the large beaker as well.
8. Take an antacid tablet and use the knife to cut the tablet into four equal pieces.
9. Take one of those pieces and grind it into a fine powder with a spoon. When you are done grinding, fill the spoon with the powder you just created.
10. Once the water in the beaker is boiling, remove the beaker from the flame so that the boiling stops.
11. Put test tube #3 into the hot water, so that the toilet bowl cleaner in it gets hot.
12. Place a piece of antacid tablet into test tubes #1, #3, and #4.
13. Take the powdered antacid in the spoon and pour it into tube #2.
14. Observe what goes on in each test tube.
15. The antacid should be bubbling, indicating the production of a gas. Use your stirring rod to stir the contents of each test tube several times over the course of five minutes.
16. After the five minutes are up, observe what is left in each of the test tubes.
17. Clean up your mess.

Let's evaluate what happened in each test tube in the experiment. In test tube #1, there will probably be a small amount of antacid tablet left. In test tube #2, however, all of the powder should be gone. In test tube #4, there should still be antacid tablet present. As a matter of fact, there should be a lot more antacid tablet left in test tube #4 than there was in test tube #1. Finally, there should be little or no antacid tablet left in test tube #3. What explains these results?

The reaction that took place in this experiment comes from the hydrochloric acid in the toilet bowl cleaner reacting with the calcium carbonate in the antacid tablet:

$$\underset{\text{acid in the toilet bowl cleaner}}{2HCl\ (aq)} + \underset{\text{calcium carbonate in the antacid tablet}}{CaCO_3\ (s)} \rightarrow H_2O\ (l) + CaCl_2\ (aq) + CO_2\ (g)$$

The carbon dioxide gas formed in this reaction was responsible for the bubbles that you saw.

By observing the amount of antacid tablet left after a certain amount of time, you are essentially learning about the rate of the chemical reaction. Since test tube #4 still had a lot of antacid left after five minutes, the reaction in test tube #4 proceeded rather slowly. On the other hand, since there was little or no antacid left in test tubes #3 and #2, the reaction proceeded rather quickly in those test tubes. Finally, since there was still antacid left in test tube #1, the reaction proceeded slowly in that test tube. However, since there was less antacid left in test tube #1 than there was in test tube #4, the reaction in test tube #1 was faster than the one in test tube #4.

Now that we see the relative rates of the chemical reactions, we can draw some general conclusions from this experiment. First of all, the only difference between the reaction in test tube #1 and the one in test tube #3 was the temperature. Test tube #3 (with a reaction that was faster than that of test tube #1) was hotter, indicating that chemical reactions have a faster rate at higher temperatures. In addition, the only difference between the reaction in test tube #1 and test tube #4 was the concentration of the toilet bowl cleaner. Test tube #4, remember, had diluted toilet bowl cleaner in it. Since the reaction in test tube #1 was faster than the one in test tube #4, we can conclude that dilute reactants lead to lower reaction rates. Finally, the only difference between test tube #1 and test tube #2 was that the antacid in test tube #2 was powdered. Since that led to a reaction with a higher rate, we can conclude that powdered solids react more quickly than nonpowdered solids. We will see *why* these factors influence reaction rates in a moment. For right now, just understand how this experiment leads us to these conclusions.

Before leaving this experiment, please notice one thing. Throughout the previous paragraph, I compared every reaction to the one that occurred in test tube #1. That's because the reaction in test tube #1 was performed under "normal" conditions (regular concentration of toilet bowl cleaner, room temperature, uncrushed antacid tablet). In the other test tubes, I changed only one thing in comparison to test tube #1. This is a commonly used experimental technique called a **controlled experiment**. In such an experiment, one test is done under standard conditions, and all other tests are compared to those standard conditions. The test done under standard conditions is referred to as the **control**. Ideally, only one thing should change between the control and any other test. This allows you to make better conclusions. Thus, in this experiment, test tube #1 was our control, and test tubes #2-4 each had only one condition different from the control. This allowed us to make firm conclusions regarding how temperature, reactant concentration, and the state of a solid affect the rate of a chemical reaction.

Now that the experiment has shown us how certain factors affect the rate of a chemical reaction, it is time to learn *why* they do. In Module #7, you learned that the valence electrons in an atom are responsible for its chemical behavior. In Module #8, you then learned that the way molecules are formed is through the transfer or the sharing of those valence electrons. Ionic compounds are formed when valence electrons are transferred, while covalent compounds are formed when valence electrons are shared.

Based on these principles, it should be easy to see that for a chemical reaction to proceed, the reactants must be able to get close enough to each other so that their electrons can either be transferred or rearranged. In order to do this, the reactants must essentially collide into one another. That's the only way that they can get close enough to do what must be done.

In order to react, chemicals must collide with one another so that their electrons can either be transferred or rearranged.

This is the guiding principle through which we can understand why certain factors influence the rate of a chemical reaction.

Consider, for example, the effect of temperature. We learned through our experiment that when temperature increases, the rate of a chemical reaction increases. If we think about this result in terms of the principle we just learned, the reason for it should become obvious. After all, if the reactants must collide together in order for the chemical reaction to proceed, then the faster each one is moving, the more likely it is for a collision to occur. Well, based on the kinetic theory of matter, when temperature increases, the motion of the molecules or atoms in a substance increases. Thus:

Higher temperature increases chemical reaction rate because the reactant molecules (or atoms) move faster, increasing the chance for a collision.

This is why increasing temperature increases the rate of a chemical reaction. It is a fact worth remembering.

Next, consider the effect of concentration. In the experiment, we learned that as the concentration of a reactant decreases, the chemical reaction rate decreases as well. Once again, this can be understood completely in terms of collisions between reactant molecules. Remember, the concentration of a substance tells us how many molecules (or atoms) of that substance exist within a certain volume. When concentration decreases, there are fewer molecules (or atoms) in the same amount of volume. In other words, the reaction vessel is less crowded. Since the reaction vessel is less crowded, the chance for collisions between reactant molecules (or atoms) decreases. Thus:

Decreasing concentration decreases the chemical reaction rate because there are fewer reactant molecules (or atoms) to collide with one another.

Of course, if we want to slow a reaction down even more, we can decrease *all* reactant concentrations rather than just the concentration of one reactant. After all, the fewer reactants there are in a certain volume, the less crowded it will be, and the lower the reaction rate.

Finally, we need to see why crushing the antacid tablet increased the rate of our chemical reaction. When a solid object gets crushed, the molecules get spread out over a larger area. Chemists generally call this an increase in the **surface area** of a reactant. Thus, chemists usually say that increasing the surface area of a reactant will increase the rate of the reaction. Once again, we can understand this effect in terms of collisions. When you crush a solid, it can be more evenly distributed through the reaction vessel. As a result, the chance for collision between this reactant molecule (or atom) and another reactant molecule (or atom) goes up. After all, a collision is more likely when the molecules (or atoms) involved mingle closely together. When a solid exists as a chunk, its molecules (or atoms) cannot easily mingle with the other substances in the reactant vessel. Crush the solid, however, and its molecules (or atoms) can travel throughout the vessel, making them much more likely to collide with something else.

Increasing the surface area of a reactant increases reaction rate because the molecules (or atoms) of the reactant can more easily mingle with the molecules (or atoms) of the other reactants.

(The multimedia CD has a video demonstration of the effect of surface area on the rate of a chemical reaction.)

In the end, then, we can see that the major factors affecting the rate of a chemical reaction can all be understood in terms of the collisions that must take place between reactants in a chemical reaction. It is important for you to know these three factors, how they affect rate, and why. Many times a chemist must make decisions about how to change the rate of a chemical reaction. These facts help a chemist to make such decisions. See if you understand these principles by solving the following problems:

ON YOUR OWN

14.1 Two containers are filled with the same mass of both nitrogen and hydrogen gas. If the first container is significantly larger than the second, in which container will the reaction be fastest?

14.2 The muscles in an animal's body are run by chemical reactions. When these reactions occur, they contract the muscle, making it move. A biologist is studying a lizard and notices that when the lizard is in a cold place, it moves very slowly, but when it is in a warm place, it moves very quickly. Explain this observation in terms of chemical kinetics.

The Rate Equation

Now that we know the factors that affect the rate of a chemical reaction, it is time to make our study of chemical kinetics a little more detailed. It turns out that chemists can actually write mathematical equations to determine the rate of a chemical reaction. These mathematical equations are based on the concentration of the reactants in the chemical equation. This, of course, should make sense to you, since we just learned that the concentration of the reactants in a chemical reaction has profound influence on the rate of the chemical reaction.

Consider, for example, the following "generic" chemical equation:

$$aA + bB \rightarrow cC + dD$$

This expression represents a general chemical equation that has reactants A and B that react to form the products C and D. The little letters (a, b, c, and d) represent the stoichiometric coefficients that exist for each of the substances (A, B, C, and D) in the chemical equation.

For this generalized chemical reaction, we can develop a mathematical equation that we call the **rate equation**:

$$R = k[A]^x[B]^y \tag{14.2}$$

In this equation, "R" represents the rate of the chemical reaction, "[A]" represents the concentration of reactant A, "[B]" stands for the concentration of reactant B, and "k" is called the **rate constant** for the chemical reaction. The letters "x" and "y" represent exponents for the concentrations of the reactants.

Now if all of this seems rather confusing, don't worry. We first have to pick the rate equation apart bit by bit to understand each of its components. First, let's start with the two concentration terms. If we look at Equation (14.2), what happens to the reaction rate (R) when the concentration of either reactant ([A] or [B]) increases? Based on the equation, you would expect R to increase. Thus, this equation accurately predicts a fact that we have already learned: When reactant concentration increases, the rate of the chemical reaction increases.

Of course, chemists are detail-oriented people. They not only want to know that reaction rate increases when reactant concentration increases, but they also want to know *how much* it increases. For example, if I double the concentration of a reactant in a chemical equation, will the reaction rate double? Will it triple? Will it quadruple? In fact, all of those things are possible. For some chemical reactions, doubling the concentration of a reactant doubles the reaction rate. For others, however, doubling the concentration of a reactant quadruples the reaction rate. For other reactions, doubling the concentration of a reactant will increase the reaction rate by a factor of 8! This is where x and y come in. These exponents allow you to determine exactly how much reaction rate changes when reactant concentration changes. These exponents are called the **order** of the chemical reaction with respect to the reactant they affect. Thus, "x" is the order of the chemical reaction with respect to reactant A while "y" is the order of the reaction with respect to reactant B. If you add the two together (x + y), you get the **overall order of the chemical reaction**.

Finally, the "k" term is a number that is unique for each chemical reaction. Its value is totally independent of the concentration of any reactant, and that is why it is called the rate constant. The important thing to remember is that there is only one way to determine x, y, and k: by experiment. In order to figure out the rate equation for any chemical reaction, you must do some carefully controlled experiments. You will learn how to evaluate those experiments in this course, and if you purchased the Microchem kit discussed in the introduction, you can perform a similar experiment yourself.

Before we jump into a study of these experiments, however, let's clear up any remaining confusion by applying what we have learned to an example problem:

EXAMPLE 14.1

Write the rate equation for the following chemical reaction:

$$N_2 + 3H_2 \rightarrow 2NH_3$$

According to Equation (14.2) you get the rate equation by just taking the concentration of each reactant and raising it to the power of "x" or "y":

$$\underline{R = k[N_2]^x[H_2]^y}$$

Now, of course, that wasn't very hard, but it also wasn't very useful. This is, indeed, the rate equation for the reaction written above, but it tells us virtually nothing. In order to make the rate equation useful, we need to figure out what x, y, and k are. That's where the experiments come into play.

Using Experiments to Determine the Details of the Rate Equation

As we learned in the section above, the rate equation is really pretty useless unless we can figure out the values of k, x, and y. Once we have those, however, the rate equation becomes very powerful, because it allows us to predict the rate of the chemical reaction regardless of the concentrations of either reactant. Thus, figuring out these three terms is a pretty important task. Well, it turns out that there is a standard experimental technique you can perform on any chemical reaction that allows you to calculate the values of k, x, and y.

To perform such an experiment, the first thing you need to do is run the reaction of interest and experimentally measure its rate. That's the tricky part. Generally, in order to measure the rate of a chemical reaction, you need some pretty fancy equipment. As a result, there are very few reactions you can study this way in a high school chemistry course. Even though we cannot *perform* very many experiments like this, we can still learn how to analyze the data from such experiments so that we can determine the values of k, x, and y in the rate equation. Study the example that follows. It will show you how these experiments are constructed, and it will show you how to determine k, x, and y from the resulting data.

EXAMPLE 14.2

A chemist runs an experiment designed to determine the rate equation for the following reaction:

$$2ICl + H_2 \rightarrow I_2 + 2HCl$$

To do this, she runs the reaction three times, varying the concentration of the reactants:

Trial	Initial Concentration of ICl (M)	Initial Concentration of H_2 (M)	Instantaneous Reaction Rate (M/s)
1	0.25	0.25	0.0102
2	0.25	0.50	0.0204
3	0.50	0.50	0.0408

Use this information to determine the rate equation for this reaction.

Before we begin to actually solve the problem, let's look at the table and see what it tells us. The columns of the table are labeled to tell you what values they contain. The first column just tells you which trial the data come from. The chemist ran the reaction a total of three times. Each time she ran it, she called it a "trial" run of the reaction. The next two columns contain the initial concentrations of the two reactants. In other words, these are the concentrations that the chemist began the experiment with. The "(M)" just tells you that these concentrations are expressed in molarity. The last column of the table tells you the instantaneous rate the chemist measured when she mixed the reactants together. If you don't understand why the term "instantaneous" is used, don't worry. I will explain that later in the module. For right now, just realize that this column represents the experimentally measured rate of the chemical reaction. The "(M/s)" just tells us that the units are molarity per second, which, as you should recall, is the standard unit for chemical reaction rate.

Now that we know what the data table is all about, let's start to solve the problem. First of all, let's write down the rate equation to the best of our abilities. First of all, since ICl and H_2 are the reactants, we know that the basic rate equation should look something like this:

$$R = k[ICl]^x[H_2]^y$$

As always, however, we have no idea what k, x, and y are, so as it stands, this rate equation is pretty useless. However, we can figure out these terms in the equation if we look at the data from the experiment.

In the first row of data in the table, we see that the chemist started out with equal concentrations of both reactants. She then measured the reaction rate to be 0.0102 M/s. The next row of data tells us that the second time she ran the experiment, she doubled the concentration of H_2, but she left the concentration of ICl alone. It turns out that these concentrations were chosen very carefully. If the concentration of ICl stayed the same between these two trials, what does that tell you about any rate changes that you see between the two trials? Well, since the concentration of ICl was not changed, any difference between trial 1 and trial 2 must be due solely to the influence of the concentration of H_2. Well, between trial 1 and trial 2, the concentration of H_2 doubled. What happened to the rate? It doubled as well. This information is all we need to calculate the value for y.

Think about it. If the concentration of H_2 doubled and the rate also doubled, what does that tell you about the value for y? It tells you that $y = 1$. After all, if doubling the concentration of H_2 doubles the rate of the reaction, then rate and H_2 concentration are linearly proportional to each other. Thus, $y = 1$. Do you get the idea here? In order to determine the exponent associated with a reactant's concentration, all we have to do is hold everything else constant and see what happens to the rate when the concentration of only that reactant changes. This will tell us what the exponent must be.

Now that we have determined y, we can work on figuring out the value for x. In order to do this, we must look at two trials that keep the concentration of H_2 the same and vary only the concentration of ICl. Which trials do that? In trials 2 and 3, the concentration of H_2 stays the same (0.50 M), but the concentration of ICl doubles. Thus, any change in rate between trials 2 and 3 is due solely to the influence of ICl. What happened? The concentration of ICl doubled and the rate doubled as well. This tells us that $x = 1$. Thus, our rate equation has now become:

$$R = k[ICl][H_2]$$

We are almost done now. All we have to do is figure out a value for k. How do we do that? Well, we just wrote down the rate equation that now contains only one unknown: k. We also have three trials of data that tell us the values of every other variable in the equation. Thus, to calculate the value of k, all we have to do is choose *any one* of the three trials and put that data into the equation. The only unknown in the equation will be k, and we can then solve for it. As I said, I can choose any one of the three trials, so I will just use the first one. In that trial, $[ICl] = 0.25$ M, $[H_2] = 0.25$ M, and $R = 0.0102$ M/s. Putting those numbers in the rate equation gives us:

$$R = k[ICl][H_2]$$

$$0.0102\ \frac{M}{s} = k \cdot (0.25\,M) \cdot (0.25\,M)$$

Solving for k gives us:

$$k = \frac{0.0102\ \frac{\not{M}}{s}}{(0.25\ \not{M}) \cdot (0.25\ M)} = 0.16\ \frac{1}{M \cdot s}$$

Since k is constant regardless of the concentration of the reactants, this k is the same for all three trials, and for any other trials that the chemist might run later on at the same temperature. Thus, if we had used the data in either trial 2 or trial 3 to calculate k, we would have gotten the exact same answer. Notice the units on k. When we took the rate in M/s and divided by the first concentration, the molarity unit canceled. However, when we divided by the second concentration there was nothing to cancel that molarity unit. Also, nothing canceled the seconds unit. That's why k has such a strange unit.

In the end, then, the final rate equation for this reaction is:

$$R = (0.16\ \frac{1}{M \cdot s})\ [ICl][H_2]$$

Now that we have the final rate equation, we can predict the rate of this chemical reaction for any concentration of reactants. Thus, if the chemist wished to run a fourth trial with [ICl] = 1.0 M and $[H_2]$ = 0.75 M, we could determine the rate of the reaction under these conditions by just plugging those numbers into our rate equation:

$$R = (0.16\ \frac{1}{\not{M} \cdot s}) \cdot (1.0\ \not{M}) \cdot (0.75\ M) = 0.12\ \frac{M}{s}$$

That's why the rate equation is so important. Once we have figured it out, it can be used to determine how fast the reaction goes no matter what the concentration of the reactants.

If this long and complicated example seemed a little confusing, don't worry about it. This is an important subject, so I will do two more examples in a moment to help you become more confident in solving problems like the one you just read. Before we do that, however, I think it is helpful to sit back and review exactly how problems like these are approached and why they give us the information that they do.

The first thing that we must do in analyzing an experiment such as the one given in Example 14.2 is to look at the data that results from each trial. We need to find two trials in which the concentration of one reactant didn't change, but the concentration of the other did change. We can then say that any difference in the rate between the two trials is due solely to the reactant whose concentration did change. Then, by comparing the change in the reaction rate to the change in concentration of the reactant, we can determine the value of the exponent that is attached to the concentration of that reactant. We can then use the same procedure to determine the value of the exponent that is attached to the other reactant. Finally, after the exponents are determined, we can determine the value for the rate constant by using the data from any one trial, plugging it into the rate equation, and solving for k. Let's see how this all pans out in two more example problems:

EXAMPLE 14.3

A chemist studies the following reaction:

$$2NO\ (g) + Cl_2\ (g) \rightarrow 2NOCl\ (g)$$

In order to determine the rate equation, he runs the reaction three different times and comes up with the following data:

Trial	Initial Concentration of NO (M)	Initial Concentration of Cl_2 (M)	Instantaneous Reaction Rate (M/s)
1	0.50	0.50	0.019
2	1.00	0.50	0.076
3	1.00	1.00	0.152

What is the rate equation for this reaction?

Once again, we can start solving this problem by first writing down as much of the rate equation as we already know:

$$R = k[NO]^x[Cl_2]^y$$

Now we need to figure out the values of x and y. To do this, we need to find two trials in which the concentration of one reactant does not change but the concentration of the other reactant does change. In trials 1 and 2, the concentration of Cl_2 remains the same, but the concentration of NO doubles. The data from these two trials will allow us to calculate the value for x. The concentration of NO doubled from trial 1 to trial 2, but the rate went up by a factor of 4. What does x need to be in order to explain this data? The NO concentration increased by a factor of 2, but the rate increased by a factor of 4. What do I need to do to turn a factor of 2 into a factor of 4? I need to square it. Thus, x = 2. That way, when the concentration of NO doubles, the rate will go up by a factor of 4.

Now that we have determined x, we need to determine y. To do this, we need to find two trials in which the concentration of NO stayed the same, but the concentration of Cl_2 changed. Trials 2 and 3 fit this bill. In these trials, the concentration of NO stayed the same, but the concentration of Cl_2 doubled. This will give us the information we need to determine y. The concentration of Cl_2 doubled, and the rate doubled as well. This means y = 1, since rate and concentration of Cl_2 are directly proportional to one another. Thus, our rate equation becomes:

$$R = k\,[NO]^2[Cl_2]$$

Notice, then, that the exponents in a rate equation do not have to be the same. This particular rate equation tells us that in order to increase the rate of the chemical reaction, it is more effective to increase the concentration of NO than it is to increase the concentration of Cl_2. After all, the rate depends on the square of the concentration of NO, and it only depends on the concentration of Cl_2 to the first power. Thus, any increase in the concentration of NO will increase the rate much more than the same increase in the concentration of Cl_2.

Now the only thing left to do is to determine the value for k. To do this, we choose any trial and plug the data from that trial into our rate equation. The only unknown will be k, so we can solve for it:

$$R = k[NO]^2[Cl_2]$$

$$0.019\ \frac{M}{s} = k \cdot (0.50\ M)^2 \cdot (0.50\ M)$$

$$k = \frac{0.019\ \frac{\cancel{M}}{s}}{(0.25\ M^2) \cdot (0.50\ \cancel{M})} = 0.15\ \frac{1}{M^2 \cdot s}$$

Notice the units on the rate constant this time. Since the rate equation has a square in it, I had to square the 0.50 M concentration of NO. When I squared that concentration, I got 0.25 M^2. This resulted in a M^2 term that could not cancel, making the units on the rate constant look strange indeed. Thus, we come to a very important point. The units for the rate constant depend on the rate equation. As a result, you cannot know the units for the rate constant until you have determined the exponents in the rate equation. This is a very important point and must be remembered:

The units for the rate constant vary. They depend on the exponents of the rate equation.

Now that we have the value for the rate constant, we can write down the final rate equation:

$$\underline{R = (0.15\ \frac{1}{M^2 \cdot s}) \cdot [NO]^2[Cl_2]}$$

In order to determine the rate equation for the following reaction:

$$\mathbf{C_4H_9Br + OH^- \rightarrow C_4H_9OH + Br^-}$$

a chemist performs a rate study and gets the following results:

Trial	Initial Concentration of C_4H_9Br (M)	Initial Concentration of OH^- (M)	Instantaneous Reaction Rate (M/s)
1	0.10	0.10	0.0010
2	0.20	0.10	0.0020
3	0.10	0.20	0.0010

What is the rate equation for this reaction?

Right now, we know the rate equation will take this form:

$$R = k[C_4H_9Br]^x[OH^-]^y$$

To determine x, we need to find two trials in which the concentration of OH^- didn't change but the concentration of C_4H_9Br did. Trials 1 and 2 fit the bill. In these trials, the concentration of C_4H_9Br doubled, and the reaction rate doubled. This tells us that x =1. To determine y, we look for two trials in which the concentration of C_4H_9Br didn't change but the concentration of OH^- did. Those would be trials 1 and 3. In those trials, the concentration of OH^- doubled, but the rate stayed the same! What does this tell us? It tells us that the reaction rate is *independent* of the concentration of OH^-. What value does y need to be in order for the rate equation to be independent of the concentration of OH^-? It needs to equal zero, because *any* number raised to the zero power is 1. Thus, the rate equation is:

$$R = k[C_4H_9Br][OH^-]^0$$

Since any number raised to the zero power is equal to 1, we usually just drop the $[OH^-]$ term, since it always equals 1. Thus, the rate equation can also be expressed as

$$R = k[C_4H_9Br]$$

Now we just need to solve for k by using any of the three trials:

$$R = k[C_4H_9Br]$$

$$0.0010 \frac{M}{s} = k \cdot (0.10 \text{ M})$$

$$k = \frac{0.0010 \frac{\cancel{M}}{s}}{(0.10 \cancel{M})} = 0.010 \frac{1}{s}$$

Thus, the final rate equation is:

$$\underline{R = (0.010 \frac{1}{s}) \cdot [C_4H_9Br]}$$

In the end, then, we see that one of our earlier statements was not really correct. Initially, we said that increasing the concentration of the reactants in a chemical equation will increase the rate of the reaction. Now we see that this is not necessarily true for all reactants in a chemical equation. In this particular reaction, you can increase the concentration of OH^- all you want, and you will never, ever, affect the rate of the reaction. We must keep this in mind:

Sometimes, the rate of a chemical reaction will be unaffected by the concentration of one of its reactants. In this case, the exponent of the rate equation is zero for that reactant.

If you purchased the MicroChem kit discussed in the introduction, you can perform Experiment #15 in that kit. It shows you a similar experimental method you can use to determine the rate equation.

Now that you have seen some examples of how to determine the rate equation for a chemical reaction, it is time to cement your newfound skills with a few problems:

ON YOUR OWN

14.3 A chemist does a rate analysis on the following chemical reaction:

$$H_2\ (g) + 2NO\ (g) \rightarrow N_2O\ (g) + H_2O\ (l)$$

He comes up with the following data:

Trial	Initial Concentration of H_2 (M)	Initial Concentration of NO (M)	Instantaneous Reaction Rate (M/s)
1	0.35	0.30	0.002835
2	0.35	0.60	0.01134
3	0.70	0.60	0.02268

What is the rate equation for this reaction?

14.4 Ozone can be destroyed by certain pollutants, one being the nitrogen dioxide that is expelled from the exhaust of cars and planes. The reaction that governs this process is:

$$NO_2\ (g) + O_3\ (g) \rightarrow NO_3\ (g) + O_2\ (g)$$

A chemist does the following experiment to determine how fast this ozone destruction occurs:

Trial	Initial Concentration of NO_2 (M)	Initial Concentration of O_3 (M)	Instantaneous Reaction Rate (M/s)
1	0.000050	0.000010	0.044
2	0.000025	0.000020	0.022
3	0.000050	0.000020	0.044

What is the rate equation for this reaction?

Rate Orders

Now that you know how to use experimental data to determine rate equations, we need to review some terminology. Remember that the exponents in a rate equation are referred to as the "order" of the chemical reaction. Thus, a chemist would look at the following rate equation:

$$R = k[HNO_3][Cu(OH)_2]^2$$

and say that this reaction is first order with respect to HNO_3 (because the exponent for HNO_3 is 1) and second order with respect to $Cu(OH)_2$ (because the exponent for $Cu(OH)_2$ is 2). In addition, a chemist would say that this rate equation has an overall order of 3, because the sum of the exponents is equal to 3. Thus, chemists often communicate the rate equation in terms of reaction order. You need to be familiar with this kind of communication, so study the following example and perform the "On Your Own" problems that follow in order to make sure you understand it.

EXAMPLE 14.4

A chemist does a rate analysis of the following reaction:

$$\mathbf{HCl + NaOH \rightarrow H_2O + NaCl}$$

The experiment reveals that this reaction is first order with respect to each of its reactants. What is the rate equation and the overall order of the reaction?

If the reaction is first order with respect to each reactant, then the exponent attached to each reactant is 1. The rate equation, then, is:

$$\underline{R = k[HCl][NaOH]}$$

Since we have no data to calculate k, we just have to leave it as an unknown in the equation. The overall reaction order is just the sum of all exponents in the rate equation. Since there are two exponents and each of them is 1, the overall order of the reaction is $1 + 1 = \underline{2}$.

ON YOUR OWN

14.5 The following reaction:

$$2PO + Br_2 \rightarrow 2POBr$$

is determined by experiment to be second order with respect to PO and first order with respect to Br_2. What is the rate equation? What is the overall order of the reaction?

14.6 The following reaction:

$$C_7H_{15}Cl + NaF \rightarrow C_7H_{15}F + NaCl$$

is determined to be first order with respect to $C_7H_{15}Cl$ and zero order with respect to NaF. What is the rate equation and what is the overall order of the reaction?

You might have already noticed that in some of the rate equations we have developed so far in this module, the orders of the rate equation are often the same as the stoichiometric coefficients in the chemical equation. For example, in the first problem of Example 14.3, the stoichiometric coefficient for each reactant in the chemical equation was the same as the exponent attached to that coefficient in the rate equation. This was also the case in "On Your Own" problems 14.3 and 14.5. It is often tempting to simply take the stoichiometric coefficients in a chemical equation and use them as exponents in the rate equation, but that is not correct! You must always remember:

The reaction rate orders for a chemical reaction must be determined experimentally; they are NOT equal to the stoichiometric coefficients in the chemical equation.

It turns out that if you take a second year of high school chemistry, or when you take chemistry in college, you might learn a way to determine the reaction rate orders from the stoichiometric coefficients in a chemical equation. Unfortunately, such a topic is beyond the scope of this particular course, because it is quite complicated.

Using Rate Equations

Now that you have a thorough understanding of rate equations, we need to spend some time using them to learn something more about how chemical reactions proceed. Consider, for example, the following chemical reaction:

$$H_2\,(g) + I_2\,(g) \rightarrow 2HI\,(g) \qquad (14.3)$$

Experimentally, we can determine that the rate equation is:

$$R = (0.012\,\frac{1}{M^2 s})\cdot[H_2][I_2]^2$$

So, if we decided that we wanted to run this reaction, we could calculate the rate of the chemical reaction for any set of reactant concentrations.

Suppose, then, that we wanted to run this reaction with initial reactant concentrations of $[H_2] = 0.50$ M and $[I_2] = 0.50$ M. We could calculate the rate of this reaction to be:

$$R = (0.012\,\frac{1}{\cancel{M^2} s})\cdot(0.50\text{ M})\cdot(0.50\,\cancel{M})^2 = 0.0015\,\frac{M}{s}$$

But what does this rate tell us? Well, according to the definition of reaction rate, this number tells us that every second, the concentration of the reactants will decrease by 0.0015 M. But wait a minute. If the reactant concentration decreases, what will happen to the rate? It will go down! Thus, when the reaction is halfway finished, half of the reactants will be gone and only half will be left. Thus, the concentrations of the reactants will only be 0.25 M each. At that point, the rate will be:

$$R = (0.012\,\frac{1}{\cancel{M^2} s})\cdot(0.25\text{ M})\cdot(0.25\,\cancel{M})^2 = 0.00019\,\frac{M}{s}$$

This is almost a factor of 10 lower than the initial rate of the chemical reaction!

So we see that as soon as a reaction begins, the concentrations of the reactants decrease, which causes the reaction rate to decrease as well. Thus, the rate that we calculate from the rate equation is not the reaction rate throughout the entire reaction. In fact, the reaction rate will go down immediately, because the reactant concentrations keep going down. As a result, we must be sure to make this distinction when we use the rate equation. Thus, we call the rate calculated from rate equation an **instantaneous rate**. It tells us that at the instant the reactants have the concentration we used, the rate will be what we calculate. In the next instant, however, the concentration of the reactants will be lower, reducing the rate from what we had just calculated.

To really bring this fact home, examine Figure 14.1:

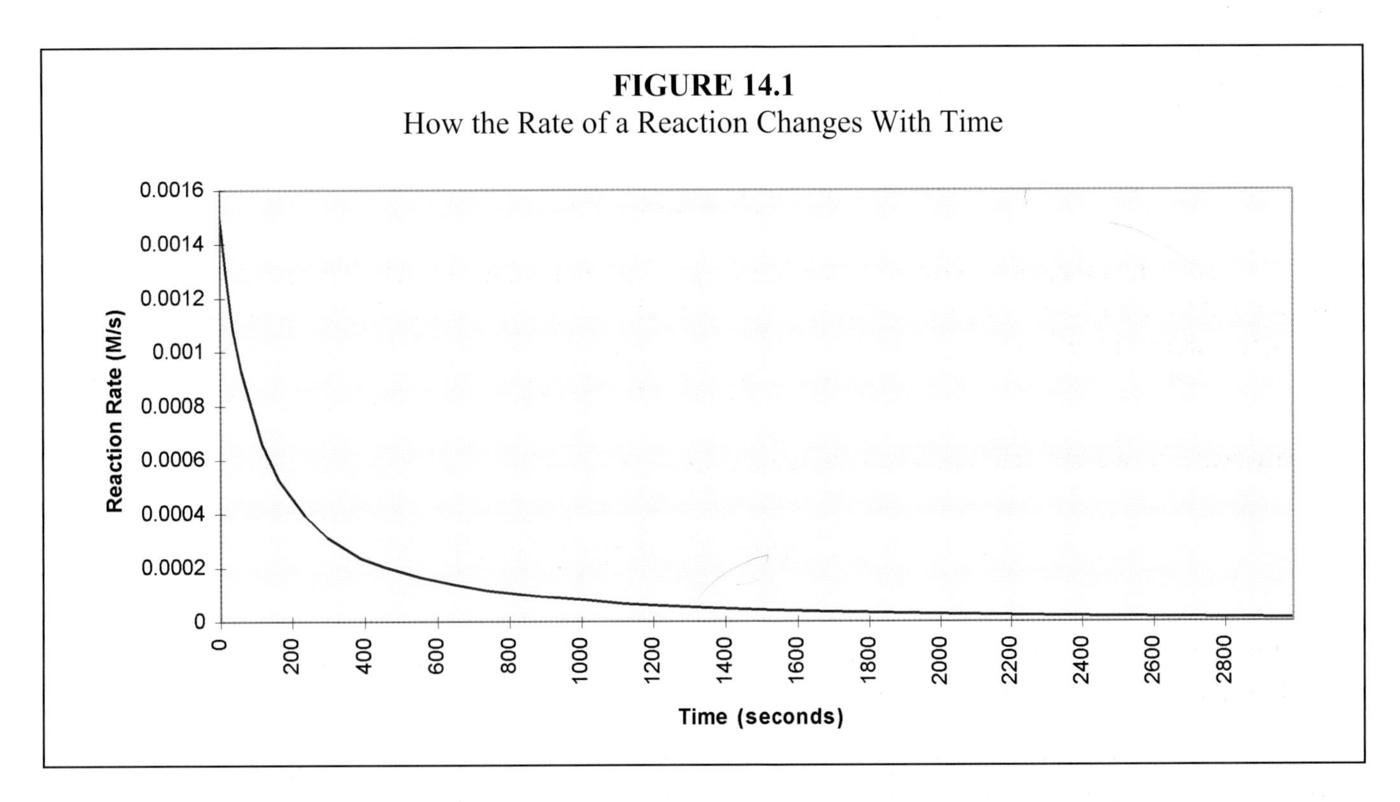

FIGURE 14.1
How the Rate of a Reaction Changes With Time

In this figure, the rate of the reaction we have been discussing is plotted versus the time we allow the reaction to run. Notice that when we first start the reaction (at time = 0), the rate is what we calculated initially: 0.0015 M/s. However, right away, the reaction rate lowers. This is because as soon as the reaction is allowed to proceed, the concentration of the reactants lowers, reducing the rate of the reaction. After a few thousand seconds, the rate of the reaction has lowered so much that is it very close to zero.

This graph tells us something very important about how chemical reactions proceed. When a chemical reaction begins, its rate is the largest that it will ever be. In the first few minutes of the reaction, the rate slows down enormously. Notice that for this reaction, in only 10 minutes (600 seconds), the rate of the reaction dropped from 0.0015 M/s to about 0.00015 M/s, a factor of 10 less than the original rate. After those first few minutes, the rate becomes very slow and does not change much. Notice that for this reaction, in the next 10 minutes, the reaction rate dropped from about 0.00015 M/s to about 0.00006 M/s, which represents only a 60% change. In general, then, reactions occur with a flurry of early activity, and then the reaction rate drops off so quickly that the reaction slows down to a crawl.

In Figure 14.2, the concentrations of the reactants in this chemical reaction are plotted for the same time period:

FIGURE 14.2
How the Concentration of a Reactant Changes With Time

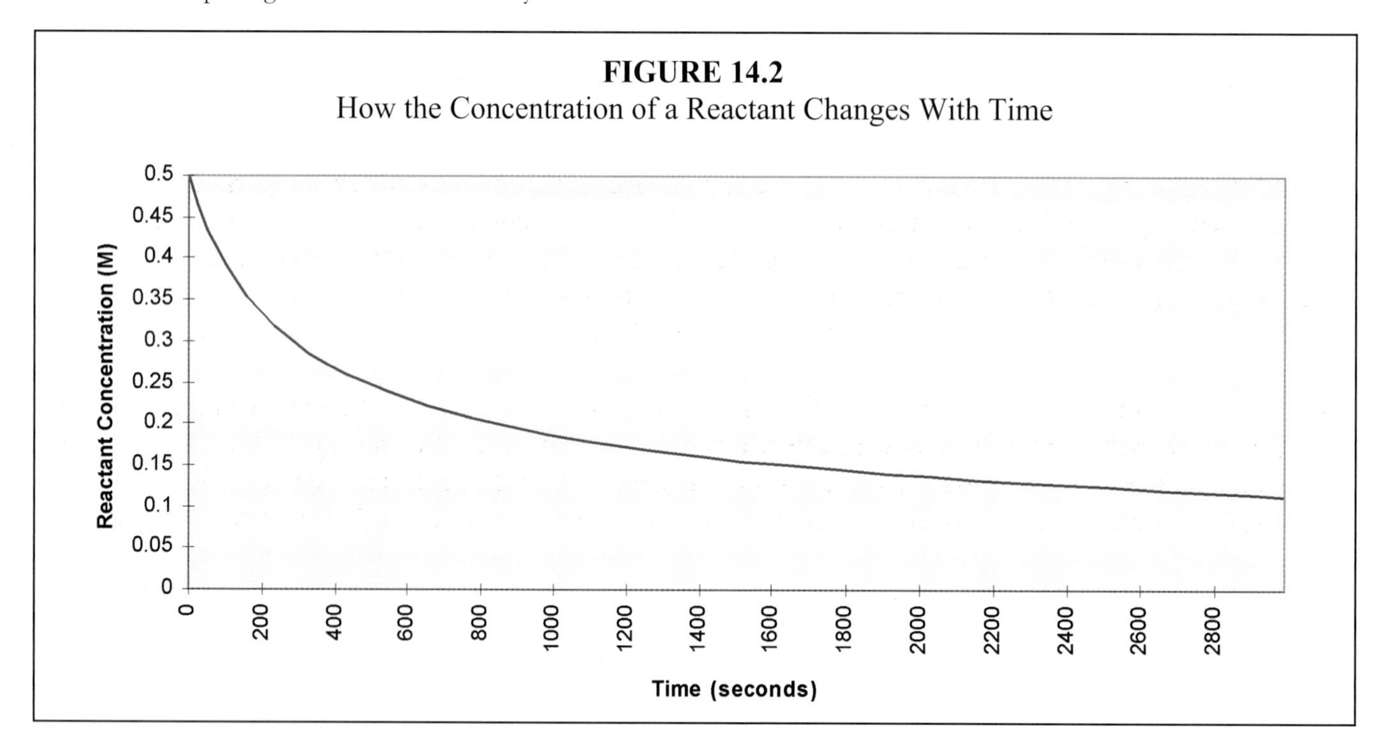

Notice that the concentration of the reactant first drops very quickly. This is a reflection of the fact that at first, the reaction is proceeding quickly. However, as time goes on, the reactant concentration starts changing quite slowly, because the reaction rate has dropped so much that the reaction is proceeding at a crawl.

The most important thing to notice about this figure, however, is that it looks like the reactant concentration will never really reach zero. If you extrapolate the curve in Figure 14.2 out to longer times, it will just get flatter and flatter. It will never actually reach zero! This tells you that the reactants will never really go completely away in a chemical reaction. Thus,

Chemical reactions never actually finish. They just keep getting slower and slower until they proceed imperceptibly slowly. Thus, for any chemical reaction, no matter how long you wait, there will always be some reactants left.

This probably goes against your preconceived notions, but it is nevertheless true.

In all of our stoichiometry problems, we assume that the limiting reactant runs out during the course of a chemical reaction. Does this discussion mean that all of our stoichiometry rules are wrong? No, not really. For the vast majority of chemical reactions, even though the reaction never really finishes, it comes so close to finishing that there is almost none of the limiting reactant left. Thus, for purposes of stoichiometry, etc., we can go ahead and assume that the limiting reactant in a chemical reaction is always used up. Nevertheless, you must remember that this is only an approximation. In reality, when the concentration of the reactants gets very low, the reaction begins to proceed so slowly that it never truly finishes.

Temperature Dependence in the Rate Equation

There is one final thing that we must consider when discussing the rate equation. We learned near the beginning of this module that the rate of a chemical reaction depends on temperature. The higher the temperature, the higher the reaction rate. The rate equation, however, has no term that involves temperature. The rate equation accurately portrays the fact that as the concentration of reactants goes up, the rate of the reaction goes up. However, there seems to be no temperature dependence in the rate equation. Does this mean that our rate equation is wrong? No, of course not.

It turns out that temperature is in the rate equation; it is just hidden so that you cannot really see it. The temperature dependence of reaction rate is taken care of by the rate constant. You see, despite its name, the rate constant is not totally constant. It changes with temperature. Figure 14.3 shows this change by plotting the rate constant for the reaction depicted in Equation (14.3) versus temperature.

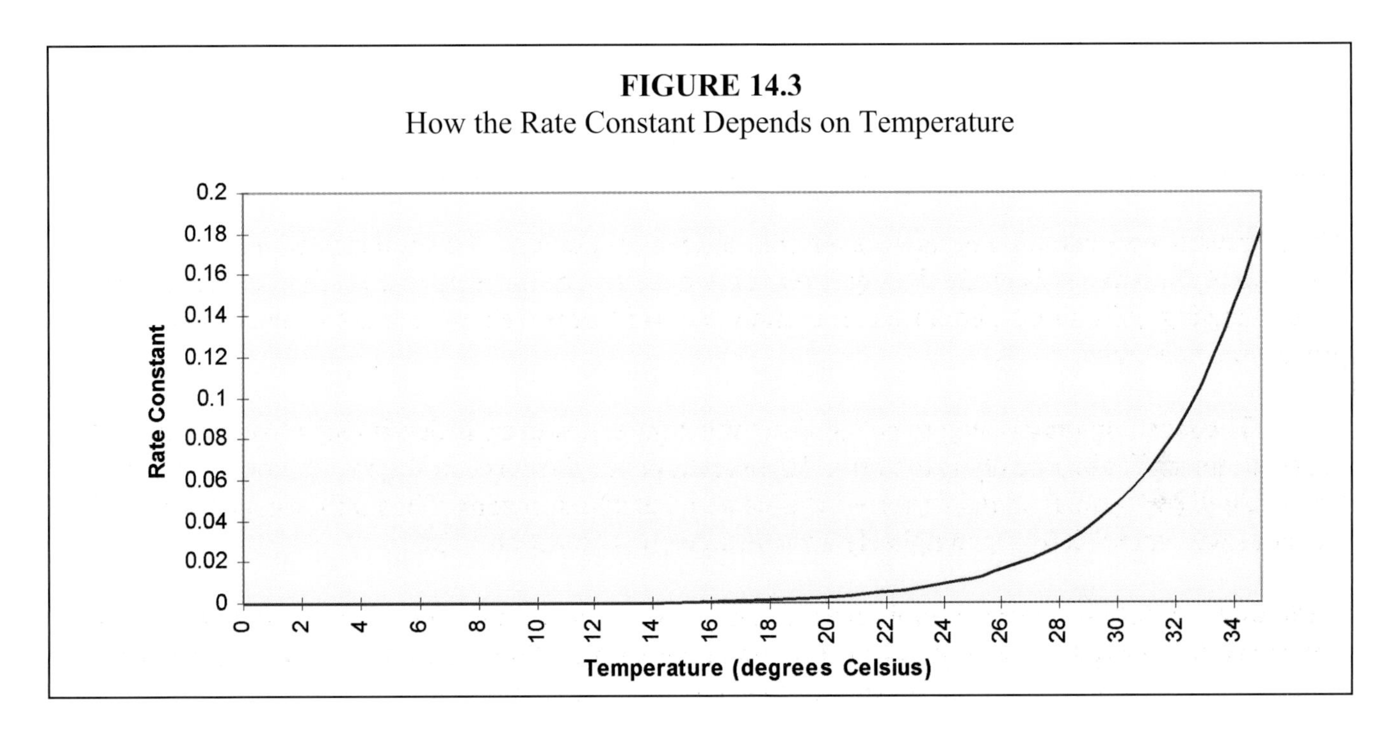

Notice from the figure that for temperatures below 14 °C, the rate constant for this reaction is very near zero. This basically means that at these temperatures, the reaction proceeds rather slowly. After all, since the rate constant multiplies the concentration terms in the rate equation, if the rate constant is near zero, then the rate itself is rather small. Notice also that after 14 °C, the rate constant shoots up quickly, indicating that the rate increases dramatically with temperature, as we learned earlier in the module. The fact that the rate constant increases with increasing temperature is an important fact to remember:

The temperature dependence of reaction rate is determined by the variation of the rate constant with temperature. Since the rate constant increases dramatically with increasing temperature, the reaction rate does as well.

There is actually an equation that relates the rate constant to the temperature at which the reaction is run. It is called the "Arrhenius (uh ree' nee us) equation," but the use of that equation is a bit beyond the scope of this course. Instead, I will teach you a general rule of thumb that is applicable for many chemical reactions at the conditions under which you will typically work:

The rate of many chemical reactions doubles for every 10 °C increase in temperature.

Now please understand that this is just a general "rule of thumb" that is not applicable in all cases. Nevertheless, it is a good estimate for standard conditions and at least gives you some idea of the temperature dependence of reaction rate. Study the next example and do the "On Your Own" problems that follow to make sure you remember this rule and can use it.

EXAMPLE 14.5

A chemist performs the following reaction at a temperature of 25 °C:

$$2NO(g) + Cl_2(g) \rightarrow 2NOCl(g)$$

In an earlier example, we determined that its rate equation was as follows:

$$R = k \cdot [NO]^2[Cl_2]$$

The rate constant for a temperature of 25 °C is 0.15 $\frac{1}{M^2 \cdot s}$. What will the instantaneous rate of the reaction be at 25 °C if he runs the reaction with both reactants at a concentration of 0.15 M? What will the rate be if he runs the reaction at 45 °C instead of 25 °C?

To answer the first question in this problem, we merely have to plug all of our numbers into the rate equation. After all, the problem gave us the rate equation, the concentrations of the reactants, and the rate constant. That's all we need to find rate:

$$R = k \cdot [NO]^2[Cl_2]$$

$$R = (0.15 \frac{1}{\cancel{M^2} \cdot s}) \cdot (0.15 \cancel{M})^2 (0.15\ M) = \underline{0.00051 \frac{M}{s}}$$

We can't simply use the rate equation to answer the second question, however, because the rate constant given is *only* for 25 °C, not for 45 °C. However, we have this rule of thumb that for every 10 °C increase in the temperature, the rate of a chemical reaction doubles. Since 45 °C is 20 °C above 25 °C, the reaction rate doubles (for the first 10 °C increase in temperature) and then doubles again (for the second 10 °C increase in temperature). In the end, then, the reaction rate doubled, and then doubled again. We can calculate what the new rate will be by simply multiplying the old rate by 2, and then multiplying by 2 again:

$$\text{At 45 °C: } R = 0.00051 \frac{M}{s} \times 2 \times 2 = \underline{0.0020 \frac{M}{s}}$$

To use our rule of thumb, then, you can always find the reaction rate at a new temperature by multiplying the old reaction rate by 2 for each 10 °C increase in the temperature. Alternatively, if the temperature is lowered, simply divide the old rate by 2 for every 10 °C that the temperature is lowered.

ON YOUR OWN

14.7 A chemist runs a chemical reaction at 100 °C and determines the instantaneous rate to be 0.12 M/s. She decides to speed up the reaction by running it at 150 °C. What is the new reaction rate?

14.8 In "On Your Own" problem 14.5, you determined that the rate equation for the following reaction:

$$2PO + Br_2 \rightarrow 2POBr$$

is:

$$R = k[PO]^2[Br_2]$$

If a chemist runs this reaction at 25 °C with reactant concentrations of 1.5 M each and then runs the reaction again at 5 °C with the same reactant concentrations, what will the reaction rates be in both cases? The rate constant for this reaction at 25 °C is $1.12 \frac{1}{M^2 \cdot s}$.

If you purchased the MicroChem kit discussed in the introduction, you can perform Experiment #16 in that kit, which demonstrates that our general rule of thumb doesn't always work!

Before we leave this section, you might be asking yourself why the rate constant is called a constant when, in fact, it changes depending on the temperature. Well, even though the rate constant varies with temperature, it does not vary with the concentration of reactants. Thus, no matter what the concentration of the reactants, the value of the rate constant does not change. Since most reactions are always carried out at the same temperature and the only thing that varies is the reactant concentrations, it is convenient to think about "k" as a constant. Just remember it this way: The rate constant is, indeed, constant, as long as you are always working at the same temperature.

Catalysts and Reaction Rate

The last thing that we need to consider when it comes to reaction kinetics is the topic of **catalysts** (kat' uh lists). We have already learned that there are three things we can do to increase the rate of a chemical reaction. We can increase the temperature at which the reaction is being run, we can increase the concentration of one or more of the reactants, and we can increase the surface area of any solids that are in the reaction. There is, however, one more thing that we can do. We can add a catalyst. Catalysts can dramatically increase the rate of a chemical reaction. Perform the following experiment to see how this can happen:

EXPERIMENT 14.2
The Effect of a Catalyst on the Decomposition of Hydrogen Peroxide

Supplies:

- 100 mL beaker (A small glass will do.)
- Watch glass (A small saucer that covers the beaker will do.)
- Graduated cylinder (A 1/8 measuring cup will work.)
- ¼ measuring teaspoon
- Hydrogen peroxide (available at supermarkets and drug stores)
- Baker's yeast (Any kind of bread yeast, even bread machine yeast, will do.)
- Safety goggles

1. Pour 25 mL (1/8 cup) of hydrogen peroxide into your beaker.
2. Observe the clear liquid for a moment. Not much happening, is there? In fact, there is something happening. As I mentioned early in this module, the hydrogen peroxide is decomposing into water and oxygen according to the following equation:

   $$2H_2O_2\ (aq) \rightarrow 2H_2O\ (l) + O_2\ (g)$$

   The problem is, this reaction proceeds incredibly slowly. Thus, even though the hydrogen peroxide is decomposing, it is doing so at such an imperceptibly slow rate that you cannot see it happening. We will change this.
3. Measure out ¼ of a teaspoon of yeast and pour it on top of the hydrogen peroxide. You should start to see some bubbles forming.
4. Quickly cover the beaker with the watch glass.
5. Grasp the beaker with your hand and place one finger on the center of the watch glass to hold it in place. Then gently swirl the yeast into the hydrogen peroxide. You should start to see many bubbles forming. In fact, bubbles should form so quickly that they rise up the beaker and smash into the watch glass. That's why the watch glass is there. It stops the bubbles from overflowing out of the beaker.
6. What happened here? The yeast acted as a catalyst and sped up the otherwise incredibly slow decomposition of hydrogen peroxide. What's neat about a catalyst is that although the catalyst speeds the reaction up, *it doesn't actually get used up in the reaction.* It's hard for you to see, because the yeast has spread out in the liquid, but there is still ¼ of a teaspoon of yeast in the solution. Not a single milligram of yeast was used up to make this reaction go fast! You can confirm this if you want by adding another 25 mL of hydrogen peroxide. You will see that the bubbles still form. You can add another 25 mL, and the bubbles will still form. In fact, the yeast that you put in the beaker will decompose as much hydrogen peroxide as you want to add, because the yeast never gets used up.
7. Clean up your mess.

The behavior demonstrated by this experiment is important to remember:

Catalysts speed up the reaction rate without actually getting used up in the chemical reaction.

Now you might be asking yourself how in the world a catalyst can speed up a chemical reaction without being used up in the process. In general, a catalyst speeds up a chemical reaction by reducing the activation energy of that reaction. It turns out that the lower the activation energy of a chemical reaction, the faster the reaction can proceed. This is an important point:

Decreasing the activation energy of a reaction increases its rate.

If we think about this in terms of an energy diagram, it should make some sense. Look at Figure 14.4:

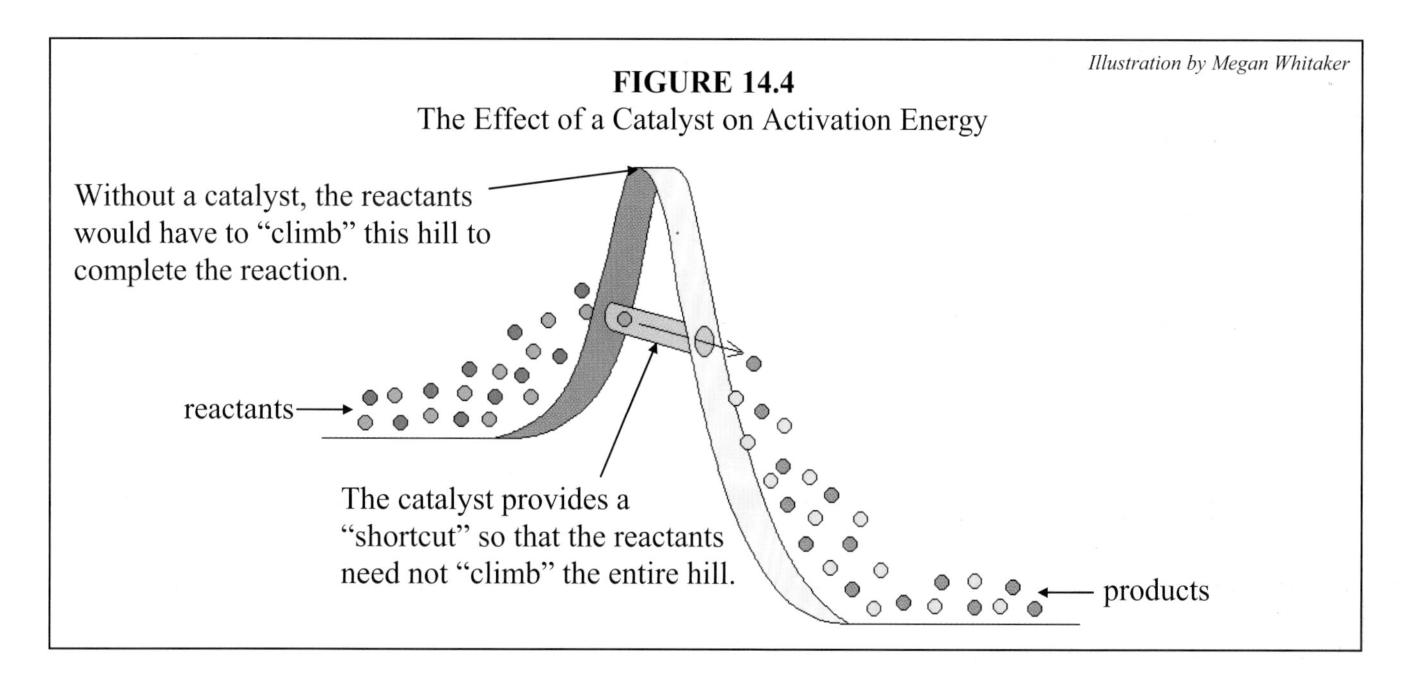
Illustration by Megan Whitaker
Without a catalyst, the reactants would have to "climb" this hill to complete the reaction.
reactants
The catalyst provides a "shortcut" so that the reactants need not "climb" the entire hill.
products

FIGURE 14.4
The Effect of a Catalyst on Activation Energy

The figure shows a stylized version of an energy diagram. The reactants (the dark green and light green balls) have a certain energy, and the products (the orange and yellow balls) have a lower energy. For the reactants to form products, they would normally have to "climb" the hump in between, which represents the activation energy of the reaction. A catalyst provides a "shortcut" for the reactants, so that they need not climb all of the way up the hill. This speeds up the reaction.

Catalysts speed up reaction rates by lowering the activation energy of the reaction.

For a given reaction, then, there can be more than one energy diagram. If the reaction is done without a catalyst, the hump between the reactants and products will be one height. If the reaction is done with a catalyst, the reactants and products will be in the same place on the diagram, but the hump between them will be much smaller.

Now, of course, you might be asking yourself *how* does a catalyst reduce the activation energy of a reaction? Well, there are, in fact, two ways that catalysts do their job. As a result, there are two types of catalysts: **heterogeneous catalysts** and **homogeneous catalysts**.

Heterogeneous catalysts - Catalysts that are in a different phase than the reactants

Homogeneous catalysts - Catalysts that have the same phase as at least one of the reactants

I will give examples of each type of catalyst and how they end up reducing the activation energy of the reaction.

In Module #5, I mentioned that catalytic converters were put in automobiles in the late 1970s in order to reduce the amount of carbon monoxide (a poison) in automobile exhaust. Well, a catalytic converter is an example of a heterogeneous catalyst. You see, carbon monoxide is naturally converted into carbon dioxide according to the equation:

$$2CO\ (g) + O_2\ (g) \rightarrow 2CO_2\ (g)$$

Although this reaction occurs spontaneously, it is very slow. Thus, when carbon monoxide is produced in automobile exhaust, this reaction must be sped up so that the carbon monoxide has been converted into carbon dioxide before it leaves the exhaust pipe. The catalytic converter accomplishes this task.

A catalytic converter is a ceramic structure coated with a metal catalyst, which is usually platinum, rhodium, and/or palladium. The exhaust flows through this structure, exposing the exhaust to the catalyst as well as oxygen. The catalyst works by attracting carbon monoxide molecules and oxygen molecules to its surface. As the molecules get close to the catalyst's surface, they also get close to each other. In fact, they get so close to each other that they begin to exchange electrons and the reaction starts. Thus, by attracting carbon monoxide molecules and oxygen molecules to its surface, the metal causes these molecules to get much closer than they would normally get to each other, forcing the reaction to proceed. Figure 14.5 illustrates this process:

FIGURE 14.5
A Heterogeneous Catalyst

Illustration by Megan Whitaker

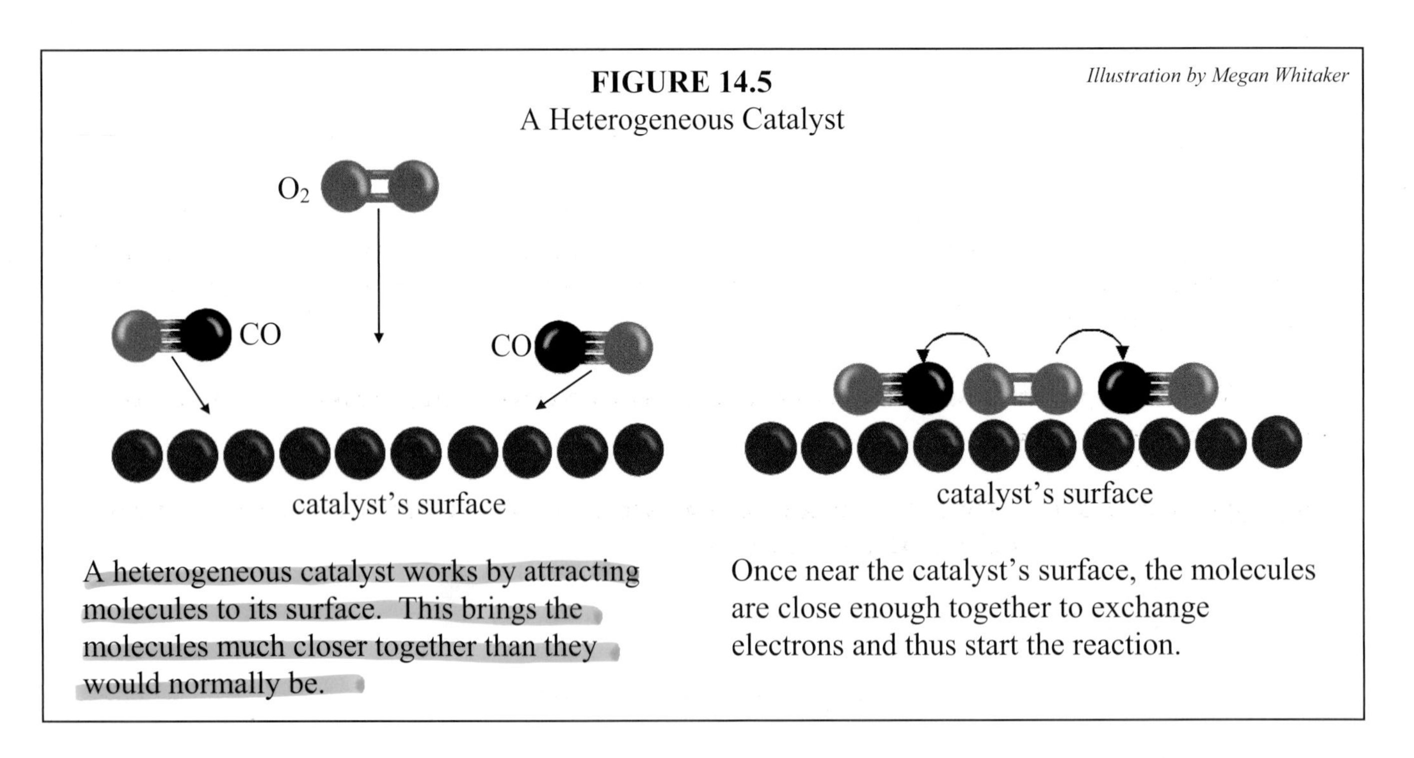

A heterogeneous catalyst works by attracting molecules to its surface. This brings the molecules much closer together than they would normally be.

Once near the catalyst's surface, the molecules are close enough together to exchange electrons and thus start the reaction.

Another way to understand this process is to think about it in terms of the energy necessary to get molecules close together. Since the electrons in a molecule repel the electrons in another molecule, molecules do not like to get too close to one another. However, since the catalyst in a catalytic converter attracts the molecules to its surface, the catalyst is, in fact, pulling the molecules close together. In the end, this reduces the amount of energy that the molecules themselves need to exert in order to get close enough to react together.

In contrast to heterogeneous catalysts, homogeneous catalysts work in an entirely different way. Instead of forcing the reactants to get closer together, a homogeneous catalyst actually participates in the chemical reaction, entirely changing the way in which the chemicals interact. For example, the oxidation of sulfur dioxide is a spontaneous reaction:

$$2SO_2\ (g)\ +\ O_2\ (g)\ \rightarrow\ 2SO_3\ (g)$$

Although this reaction is spontaneous, it is also very slow. Nitrogen monoxide gas, however, is known to speed this reaction up immensely, without ever being used up in the reaction. Thus, NO (g) is a catalyst for the reaction. Since it is in the same phase as the reactants, it must be a homogeneous catalyst.

Here's how it works. Without NO, there must be a collision between sulfur dioxide molecules and an oxygen molecule in order for the reaction to go. In the presence of NO, however, the reaction completely changes. First, the NO reacts with the oxygen to make NO_2:

$$2NO\ (g)\ +\ O_2\ (g) \rightarrow\ 2NO_2\ (g)$$

It turns out that NO and O_2 react much faster than SO_2 and O_2. Thus, this reaction is very fast. After that reaction is complete, the NO_2 molecules then react with the SO_2 molecules:

$$2NO_2\ (g)\ +\ 2SO_2\ (g)\ \rightarrow\ 2SO_3\ (g)\ +\ 2NO\ (g)$$

This ends up making the SO_3 gas that was supposed to be made in the original equation.

Do you see what has happened here? The NO gas reacts in the first step of this process, and in the next step, it is produced again. Thus, even though it participates in the reaction, it never gets used up because each time two NO's are destroyed in the first step, two more NO's are made in the second step. As a result, the number of NO's present never really changes.

When a single chemical reaction can be written as a series of other chemical reactions, we call the series a **reaction mechanism**.

Reaction mechanism - A series of chemical equations that tells you the step-by-step process by which a chemical reaction occurs

Thus, we could say that the reaction:

$$2SO_2\ (g)\ +\ O_2\ (g)\ \rightarrow\ 2SO_3\ (g)$$

will proceed by the following mechanism when catalyzed with NO gas:

Step 1: $2NO\ (g)\ +\ O_2\ (g) \rightarrow\ 2NO_2\ (g)$

Step 2: $2NO_2\ (g)\ +\ 2SO_2\ (g)\ \rightarrow\ 2SO_3\ (g)\ +\ 2NO\ (g)$

(The multimedia CD has a video demonstration of how the reaction mechanism affects the rate of a chemical reaction.)

Now if you don't exactly understand all of this, don't get worried. The only thing you need to understand is the difference between homogeneous and heterogeneous catalysts. They both reduce the activation energy of a chemical reaction, but they go about it in different ways. Homogeneous catalysts change the steps that the reaction takes. Heterogeneous catalysts, on the other hand, force the reactants in a chemical equation close together by attracting them to a surface. You also need to know the definitions of heterogeneous and homogeneous catalysts, and you need to be able to look at a reaction mechanism and see which substance in that mechanism is a catalyst. The following problems will test your knowledge on these points:

ON YOUR OWN

14.9 To catalyze the following reaction:

$$C_2H_4\ (g) + H_2\ (g) \rightarrow C_2H_6\ (g)$$

solid platinum is often used. Is the platinum a heterogeneous or a homogeneous catalyst?

14.10 Two energy diagrams for the same reaction are presented below. Which one represents the reaction with a catalyst?

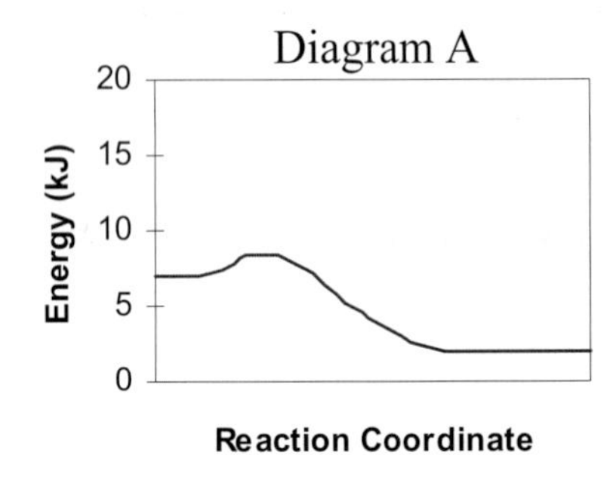

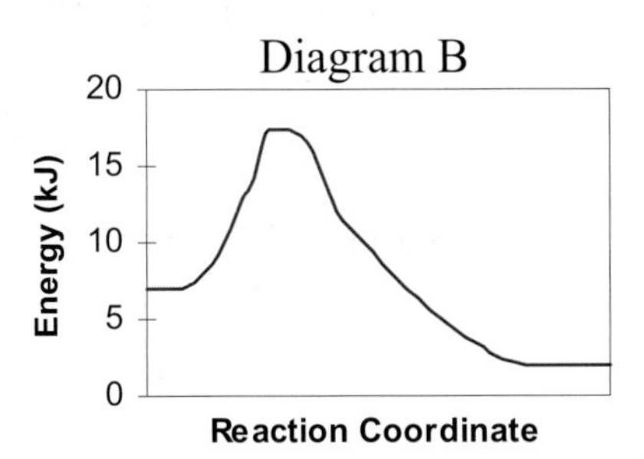

14.11 Identify the catalyst in the following reaction mechanism:

Step 1: $2H_2O_2 + 2I^- \rightarrow 2H_2O + 2IO^-$

Step 2: $IO^- + IO^- \rightarrow O_2 + 2I^-$

ANSWERS TO THE "ON YOUR OWN" PROBLEMS

14.1 In the smaller of the two vessels, the volume is lowest. Since concentration is number of moles divided by volume, the gases in the smaller container will have much larger concentrations. This means that the reaction in the smaller vessel will proceed faster, because its reactants have larger concentrations.

14.2 The rate of a chemical reaction depends on the temperature. When the lizard is cold, the chemical reactions that run its muscles proceed very slowly. As a result, the muscles move slowly. When the lizard is warm, the chemical reactions that run its muscles have a faster rate, so the muscles can move faster. Note that the external temperature does not affect the motion of mammals (such as humans, dogs, and cats) because mammals are warm-blooded. Thus, our body temperatures, and hence the temperatures at which the chemical reactions in our bodies run, are mostly independent of the outside temperature.

14.3 The rate equation will take on the form:

$$R = k[H_2]^x[NO]^y$$

To determine x and y, we look at trials where the concentration of one reactant stayed the same and the concentration of the other reactant changed. In trials 1 and 2, the concentration of H_2 remained the same but the concentration of NO doubled. When that happened, the rate increased by a factor of 4. This means that $y = 2$. In the same way, between trials 2 and 3, the NO concentration remained constant, but the H_2 concentration doubled. When that happened, the rate doubled. This means $x = 1$. The rate equation, then, looks like:

$$R = k[H_2][NO]^2$$

To solve for k, we can use the data from any trial and plug it into our rate equation. We can then solve for k:

$$R = k[H_2][NO]^2$$

$$0.002835\,\frac{M}{s} = k\cdot(0.35\text{ M})\cdot(0.30\text{ M})^2$$

$$k = \frac{0.002835\,\frac{\cancel{M}}{s}}{(0.090\text{ M}^2)\cdot(0.35\,\cancel{M})} = 0.090\,\frac{1}{M^2\cdot s}$$

The overall rate equation, then is

$$R = (0.090\,\frac{1}{M^2\cdot s})\cdot[H_2][NO]^2$$

14.4 The rate equation will take on the form:

$$R = k[NO_2]^x[O_3]^y$$

To determine x and y, we look at trials where the concentration of one reactant stayed the same and the concentration of the other reactant changed. In trials 1 and 3, the concentration of NO_2 remained the same but the concentration of O_3 doubled. When that happened, the rate stayed the same. This means that $y = 0$, because the rate is not affected by the change in concentration of O_3. In the same way, between trials 2 and 3, the O_3 concentration remained constant, but the NO_2 concentration doubled. When that happened, the rate doubled. This means $x = 1$. The rate equation, then, looks like:

$$R = k[NO_2][O_3]^0\text{, which is the same as } R = k[NO_2]$$

To solve for k, we can use the data from any trial and plug it into our rate equation. We can then solve for k:

$$R = k[NO_2]$$

$$0.044\ \frac{M}{s} = k \cdot (0.000050\ M)$$

$$k = \frac{0.044\ \frac{\not{M}}{s}}{(0.000050\ \not{M})} = 8.8 \times 10^2\ \frac{1}{s}$$

The overall rate equation, then, is:

$$\underline{R = (8.8 \times 10^2\ \frac{1}{s}) \cdot [NO_2]}$$

14.5 If the reaction is second order with respect to PO, then the exponent attached to PO in the rate equation is a 2. First order in Br_2 means that its exponent is 1. Thus, the rate equation is:

$$\underline{R = k[PO]^2[Br_2]}$$

The overall order is the sum of the exponents, or $2 + 1 = \underline{3}$.

14.6 If the reaction is first order with respect to $C_7H_{15}Cl$, then its exponent is 1. The exponent for NaF, however, is 0, because the reaction is zero order with respect to NaF. Thus, the rate equation is:

$$R = k[C_7H_{15}Cl][NaF]^0$$

Usually, we drop exponents of zero from the rate equation, so the real answer is:

$$\underline{R = k[C_7H_{15}Cl]}$$

This simply means that the concentration of NaF has no effect on the rate of the reaction. The overall reaction order is the sum of the exponents, which is just $\underline{1}$.

14.7 In order to calculate the rate of a chemical reaction at an elevated temperature, you just multiply the old rate by 2 for every 10 °C increase in temperature. Thus, when the chemist went from 100 °C to 150 °C, he increased the temperature by five 10-degree units. Therefore, we must multiply the old rate by five 2's in order to get the new rate:

$$R = 0.12\ \frac{M}{s} \times 2 \times 2 \times 2 \times 2 \times 2 = \underline{3.8\ \frac{M}{s}}$$

14.8 To get the rate of the reaction at 25 °C, we just need to plug things into the rate equation:

$$R = k \cdot [PO]^2[Br_2]$$

$$R = (1.12 \frac{1}{\cancel{M^2} \cdot s}) \cdot (1.5\ \cancel{M})^2 (1.5\ M) = \underline{3.8\ \frac{M}{s}}$$

To get the rate of the reaction at a lower temperature, we just divide the old rate by 2 for every 10 °C decrease. The chemist dropped the temperature by two 10-degree increments, thus:

$$R = 3.8\ \frac{M}{s} \div 2 \div 2 = \underline{0.95\ \frac{M}{s}}$$

14.9 The platinum must be a <u>heterogeneous catalyst</u> because it has a different phase than the reactants.

14.10 <u>Diagram A</u> has the lowest activation energy, so it must represent the catalyzed reaction.

14.11 A catalyst is not used up in the reaction. Thus, the substance that is both used in the first step of the mechanism and then produced in the last step of the mechanism is the catalyst. <u>I^-</u> is used in step 1 and produced in step 2. This makes it the catalyst for the reaction.

REVIEW QUESTIONS FOR MODULE #14

1. If two molecules "want" to react together to form new chemicals, what is the first thing that must happen?

2. A chemist runs a reaction at a low temperature and then runs the same reaction again at a high temperature. Which reaction runs faster?

3. Explain in your own words why the rate of a chemical reaction increases with increasing reactant concentration.

4. A chemistry book lists the rate constants for two different reactions. The first rate constant is $0.12 \frac{1}{s}$ and the second is $34.5 \frac{1}{M \cdot s}$. Even though the rate equations for these reactions are not given, an astute student says that the overall order of these two chemical reactions must be different from each other. In addition, the student says that the second reaction is faster than the first if the concentrations of the reactants are roughly the same. Why can the student make these statements?

5. The order of a chemical reaction with respect to one of its reactants is zero. If you double the concentration of that reactant, what happens to the rate?

6. A student is studying a chemistry book and reads that for the following reaction:

$$CH_3Cl + NaF \rightarrow CH_3F + NaCl$$

the rate equation is:

$$R = k[CH_3Cl][NaF]$$

The student looks at the rate equation and says that since temperature does not appear in the rate equation, the rate of this reaction is obviously not affected by temperature. Is the student correct? Why or why not?

7. What do we call a chemical that increases the rate of a chemical reaction without getting used up in the process?

8. A chemist studies two different reactions and finds that the activation energy for reaction #1 is significantly higher than that of reaction #2. Assuming that the reactant concentrations are roughly the same in the reactions, which reaction is the fastest?

9. What is the main difference between a heterogeneous catalyst and a homogeneous catalyst?

10. What is the purpose of a catalytic converter in an automobile?

PRACTICE PROBLEMS FOR MODULE #14

1. Milk sours because of a chemical reaction that takes place between some of the chemicals that make up the milk. Why does a carton of milk left out on a countertop sour sooner than a carton of milk left in the refrigerator?

2. A chemist does a reaction rate analysis on the following reaction:

$$2NO\ (g)\ +\ O_2\ (g)\ \rightarrow\ 2NO_2\ (g)$$

She collects the following data:

Trial	Initial Concentration of NO (M)	Initial Concentration of O_2 (M)	Instantaneous Reaction Rate (M/s)
1	0.0125	0.0253	0.0281
2	0.0250	0.0253	0.112
3	0.0125	0.0506	0.0562

What is the rate equation for this reaction?

3. A chemist does a reaction rate analysis of the following reaction:

$$C_4Cl_9OH\ +\ F^-\ \rightarrow\ C_4Cl_9F\ +\ OH^-$$

He collects the following data:

Trial	Initial Concentration of C_4Cl_9OH (M)	Initial Concentration of F^- (M)	Instantaneous Reaction Rate (M/s)
1	0.25	0.25	0.0202
2	0.25	0.50	0.0202
3	0.50	0.50	0.0404

What is the rate equation for this reaction?

4. The following reaction:

$$Br_2\ (g)\ +\ I_2\ (g)\ \rightarrow\ 2IBr\ \ (g)$$

is determined by experiment to be second order with respect to I_2 and first order with respect to Br_2. Its rate constant is $1.1\text{x}10^{-3}\ \frac{1}{M^2 \cdot s}$. What is the overall order of the reaction? If the reaction were run with both reactants at a concentration of 0.5 M, what would the instantaneous rate be?

5. A chemist runs a reaction with a rate of 0.0167 M/s at 100 °C. If the reaction is run at 50 °C, what will its rate be?

6. A chemist runs a chemical reaction at 25 °C and decides that it proceeds far too slowly. As a result, he decides that the reaction rate must be increased by a factor of 16. At what temperature should the chemist run the reaction to achieve this goal?

7. Which of the following graphs accurately depicts the dependence of the rate constant on temperature?

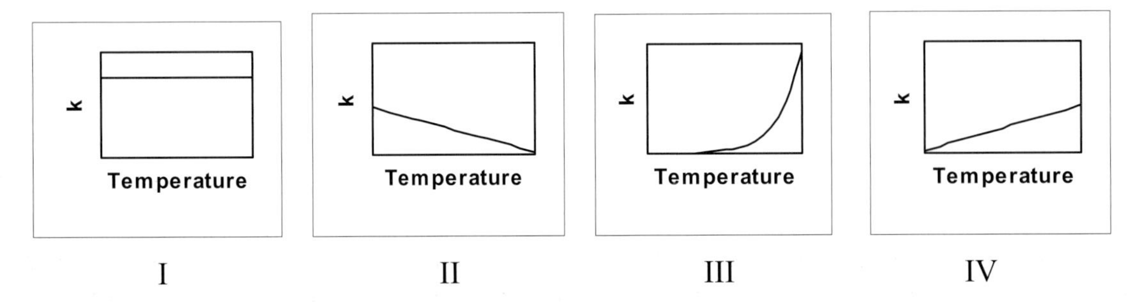

8. A chemist wants to use a Cu (s) catalyst on the following reaction:

$$2H_2 (g) + O_2 (g) \rightarrow 2H_2O (g)$$

Would the copper be considered a heterogeneous catalyst or a homogeneous catalyst?

9. Energy diagrams are presented below for the same reaction. One diagram is for the reaction without a catalyst, one is for the same reaction with a catalyst that increases the reaction rate by a factor of 3, and one is for the same reaction with a catalyst that increases the reaction rate by a factor of 10. Identify the diagram for each case. Note: "reaction" is abbreviated as "rxn" in the diagrams below.

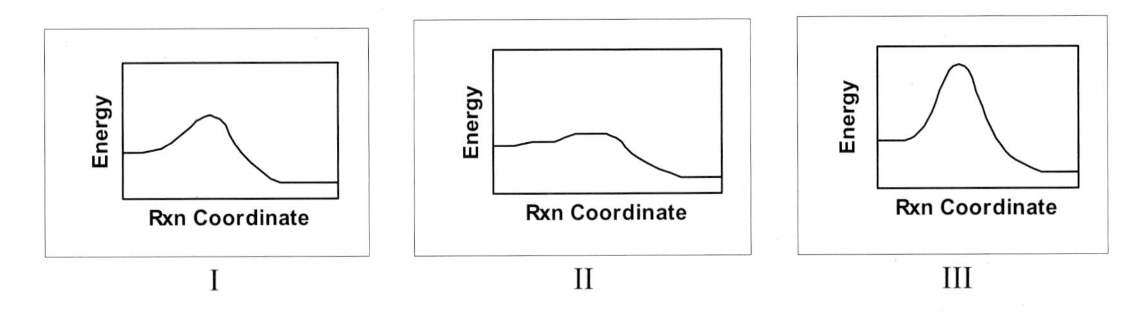

10. In the reaction mechanism below, indicate which substance is acting like a catalyst.

Step 1. $2Cl (g) + 2O_3 (g) \rightarrow 2O_2 (g) + 2OCl (g)$

Step 2. $2OCl (g) \rightarrow O_2 (g) + 2Cl (g)$

Is this a heterogeneous or a homogeneous catalyst?

MODULE #15: Chemical Equilibrium

Introduction

Welcome to the revolution! What I mean to say is that this module will revolutionize the way you think about chemical reactions. You see, up until this point, we have talked about chemical reactions in only one way: reactants forming products. In other words, if I were to present you with a chemical equation like this one:

$$NO_2 \text{ (g)} + CO \text{ (g)} \rightarrow NO \text{ (g)} + CO_2 \text{ (g)}$$

you would say that the NO_2 and CO react to form NO and CO_2. This is the way you've been taught.

It turns out, however, that I could look at the equation another way. Instead of assuming that the NO_2 and CO are reactants, I could assume that they are products. In other words, I could write the equation "backwards":

$$NO_2 \text{ (g)} + CO \text{ (g)} \leftarrow NO \text{ (g)} + CO_2 \text{ (g)}$$

Since the arrow is now pointing the other way, this equation indicates that the NO and CO_2 react to form NO_2 and CO. Which chemical equation do you think is correct? Well, it turns out that *both* are correct.

If I were to start out with pure NO_2 and pure CO and mix them together, at first, I would see them react to form NO and CO_2. As time went on, however, some of the NO and CO_2 that had been formed would actually collide into each other and turn back into NO_2 and CO! As I watched the reaction for a while, there would eventually come a time where for every NO_2 and CO that react to form NO and CO_2, an NO and a CO_2 would react to form NO_2 and CO. Thus, as I watched the reaction, I would see it going *both ways*. This kind of situation is called **chemical equilibrium** (ee' kwuh lib' ree uhm), and makes up the bulk of what we will cover in this module.

The Definition of Chemical Equilibrium

Since the reaction we have been discussing can actually operate both "forwards" and "backwards," we often use the following notation in the chemical equation:

$$NO_2 \text{ (g)} + CO \text{ (g)} \leftrightharpoons NO \text{ (g)} + CO_2 \text{ (g)} \qquad (15.1)$$

The double arrow indicates that this particular reaction can operate in either direction. If NO_2 reacts with CO to make NO and CO_2, we call that the **forward reaction**. If, however, NO reacts with CO_2 to make NO_2 and CO, we call that the **reverse reaction**.

Now if this chemical reaction is going forwards and backwards all of the time, what happens? Does the chemical reaction ever stop? Does it just keep going back and forth and back and forth? Well, the answers to these questions lie in a discussion of the *rate* of both chemical reactions. Suppose I started off with a container of NO_2 and another container of CO. When I mixed these two substances together, they would begin to react according to Equation (15.1). Now, since there would not be any NO or CO_2 initially, the reverse reaction would not go. However, since there would be plenty of NO_2

and CO, the forward reaction would start to go. Since the concentrations of these reactants would be high, the rate of the forward reaction would be high. As time went on, however, the rate of the forward reaction would decrease, because the concentration of the reactants would decrease. Of course, this forward reaction would produce NO and CO_2, allowing the reverse reaction to start running as well.

At first, the rate of the reverse reaction would be low, because not much NO and CO_2 would have been made; therefore, the concentration of these reactants would be very low. As time went on, however, the forward reaction would keep making more NO and CO_2, increasing their concentration. This would keep increasing the rate of the reverse reaction. At the same time, since the forward reaction would be using up NO_2 and CO, their concentrations would go down. This would decrease the rate of the forward reaction.

Do you see what's happening here? When the reaction begins, the rate of the forward reaction is high and the rate of the reverse reaction is zero. As time goes on, however, the rate of the forward reaction *decreases* and the rate of the reverse reaction *increases*. Eventually, the *two rates will equal each other*. At that point, every time one forward reaction occurs, a reverse reaction will occur as well. In other words, every time an NO_2 molecule and a CO molecule are used up to make a NO molecule and an CO_2 molecule, the reverse reaction will also happen. As a result, despite the fact that both the forward and reverse reactions are going like crazy, the average amount of each substance *will not change*. This situation is referred to as **chemical equilibrium**.

Chemical equilibrium - The point at which both the forward and reverse reactions in a chemical equation have equal reaction rates: When this occurs, the amounts of each substance in the chemical reaction will not change, despite the fact that both reactions still proceed.

Now if all of this sounds a little confusing so far, don't worry. This is such an important concept that I want to explain it again using Figure 15.1. Even if you think you fully grasped the discussion the first time, carefully study Figure 15.1 and the explanation that follows:

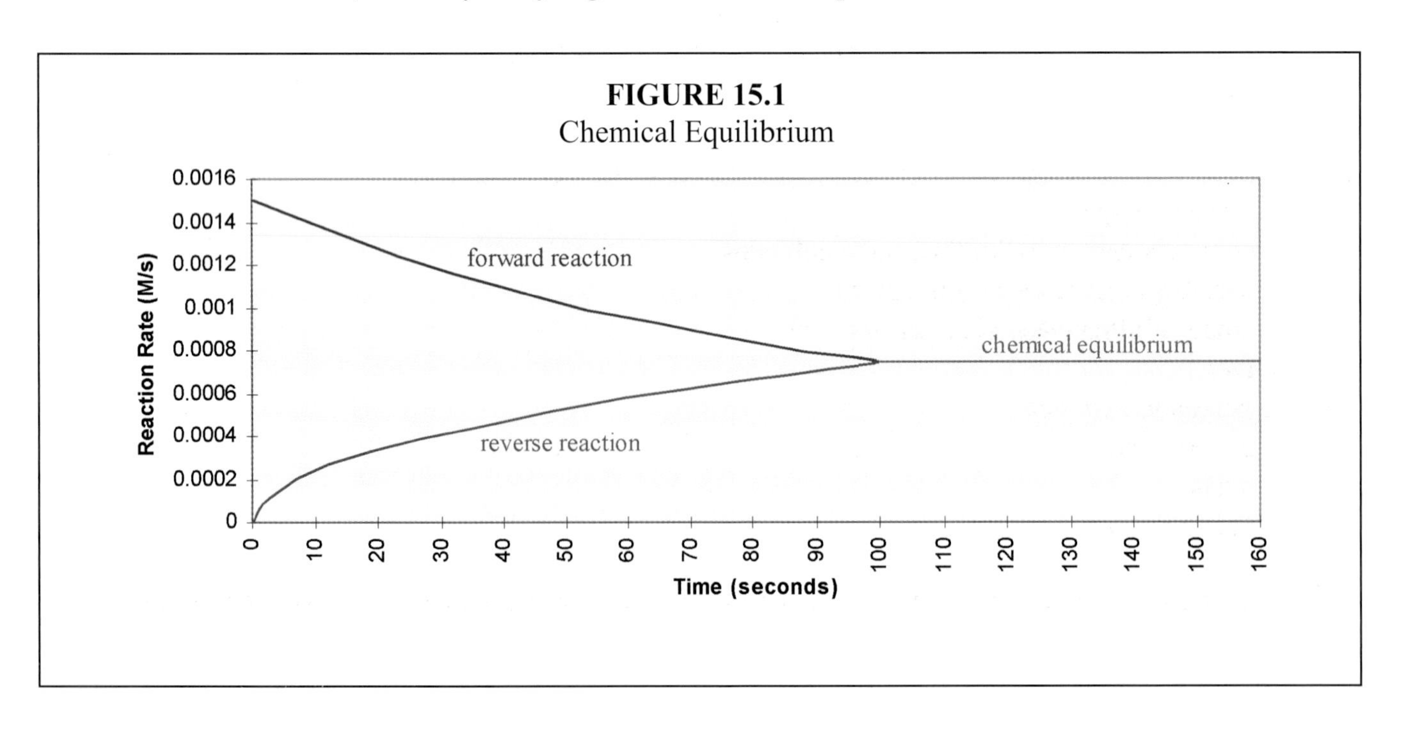

In the figure, the rates of the forward and reverse reactions of Equation (15.1) are plotted versus time. Notice that when the reaction starts, the rate of the forward reaction is high and the rate of the reverse reaction is zero. This is because (as I stated earlier) we started with lots of NO_2 and CO and no CO_2 or NO. Once the forward reaction got started, however, its rate went down. At the same time, since NO and CO_2 were being formed by the forward reaction, the rate of the reverse reaction began to increase. In about 100 seconds (according to the graph), the forward reaction rate had decreased and the reverse reaction rate had increased enough so that they became equal. At that point, every time a forward reaction occurred, a reverse reaction did as well. As a result, the concentrations of the substances didn't change, and therefore the rate didn't change. No matter how long you watched this chemical reaction after the first 100 seconds, the rates of the forward and reverse reactions would stay equal; thus, the concentrations of all the substances in the chemical reaction would never change. This is what we refer to as chemical equilibrium.

So chemical equilibrium is a very interesting state. If you were to look only at the concentrations of the substances in the chemical equation, you would notice that their concentrations did not change. As a result, you might think that nothing is really happening in the reaction. However, if you were able to study the individual molecules involved, you would see that they would be reacting together like mad. Despite the fact that there is a lot of reacting going on, the amount of each substance does not change because every time a forward reaction occurs, a reverse reaction also occurs.

Now it's very important to realize that, as illustrated in Figure 15.1, chemical reactions do not *start* at equilibrium. When a chemical reaction begins, the rate of the forward and reverse reactions will not be equal. Thus, it takes *time* for a reaction to achieve chemical equilibrium. According to Figure 15.1, the reaction that we have been studying took about 100 seconds to reach equilibrium. Once there, however, the reaction will stay at equilibrium forever, unless it is disturbed by some outside intervention. Perform the following experiment to get an idea of how reactions reach and maintain equilibrium:

EXPERIMENT 15.1

A Demonstration of Equilibrium

Supplies:

- Three plastic two-liter bottles (Plastic milk cartons will work as well.)
- Four small cups
- Two bowls that are taller than the small cups
- A serrated knife (like a steak knife)
- A small, Phillips-head screwdriver
- Water
- A person to help you
- A kitchen counter
- Towels
- Safety goggles

NOTE: This experiment will get the counter wet. Use the towels to keep the mess to a minimum.

1. Use the knife to cut the tops off two of the plastic bottles. **Be careful - do not cut yourself in the process!** You want to remove each top completely, so that the bottles have large openings at their tops. The openings need to be large enough to easily dump the contents of a cup into the bottle without spilling anything (see the picture below).
2. Use the screwdriver to make a small hole near the bottom of the two bottles. **Be careful - don't poke yourself in the process!** On one of the bottles, make the hole about two inches from the bottom of the bottle. We will call this "bottle #1." On the other bottle, make the hole about an inch from the bottom of the bottle. We will call this "bottle #2."
3. Put the two bowls on the counter upside down.
4. Stand each of the bottles on a bowl with the hole pointing toward you.
5. Place a small cup under each hole, and place the other cups right next to those cups. The setup should look something like this:

6. Fill the remaining two-liter bottle (the one you haven't cut up) with water.
7. Have your helper stand in front of bottle #2.
8. You need to stand in front of bottle #1.
9. Have your helper hold bottle #1 as you use the bottle you filled in step #6 to pour water into it. Fill bottle #1 with as much water as possible without risking spillage. As soon as the water level reaches the hole, water will start to come out of the hole. Use your free hand to hold the cup and catch the water as it comes out. As the water level rises, the water will shoot out farther from the hole, so you will need to follow the water stream to keep catching the water.
10. Once bottle #1 is as full as you can get it, put down the bottle you have been using to pour water. Grab the other cup and quickly pull the cup that is getting filled with water out from under the stream of water and replace it with the other cup.
11. Tell your helper to let go of bottle #1.
12. Pour the water in the cup that is full into bottle #2.
13. Once the cup that is currently being filled with water gets pretty full, pull it from under the stream of water and replace it with the cup in your other hand.
14. Once again, take the full cup and pour its contents into bottle #2.

15. Eventually, the water level in bottle #2 will rise above its hole. At that point, water will start coming out of the hole. Have your helper catch the water with one of the cups while he holds the other cup in his free hand. He will lose some water at first, because the water will just trickle out of the hole. That's fine; just have him do the best that he can.
16. Once both you and your helper are catching water with your cups, start counting to 20. Each time you reach 20, you should switch the cups so that the cup that has been filling is out from under the water, and the cup in your other hand is filling with water. Then, dump the contents of the full cup into the other person's bottle. You should always dump water into bottle #2, and your helper should always dump water into bottle #1. Once you have each dumped your water, count to 20 again and repeat the process.
17. Keep doing this for several minutes. Once you both have the rhythm down, start looking at the water level in the bottles. After a while, the average water level in each bottle should not change significantly. Despite the fact that water is leaving each bottle through the hole and entering each bottle through the top, the average water level in each bottle should stay constant.
18. Clean up your mess.

The experiment you just did provides an excellent illustration of equilibrium. Bottle #1 represents the reactants in a chemical equilibrium, while bottle #2 represents the products. At first, there was lots of water in bottle #1 (lots of reactants) and no water in bottle #2 (no products). However, bottle #1 began losing water (the forward reaction began making products). When this happened, bottle #2 began to fill with water (products were forming). Eventually, the water level in bottle #2 rose to the point where it began to lose water (the reverse reaction began running). As a result, some of the contents of bottle #2 started going back into bottle #1.

Since the holes were at different heights in the bottles, and since the water levels in the bottles were different, the bottles lost water at different rates. This represents the fact that the forward and reverse reactions in an equilibrium have different rates. Eventually, the rate of water loss in bottle #1 (the rate of the forward reaction) decreased, and the rate of water loss in bottle #2 (the rate of the reverse reaction) increased to the point where the two rates were equal. At that point, despite the fact that water was being continually transferred between the bottles, the average amount of water in the bottles did not change. This is just like chemical equilibrium, where the forward and reverse reactions are both running, but the average amounts of the products and reactants do not change.

 (The multimedia CD has a video demonstration of chemical equilibrium.)

The Equilibrium Constant

Now that we know a little bit about *how* a chemical reaction reaches equilibrium, it is time to learn about *where* a chemical reaction ends up once it has reached equilibrium. For example, when the reaction represented by Equation (15.1) reached equilibrium, how many NO_2 and CO molecules were left? Before we got to this module, we would have assumed that by the end of the reaction, there would be no CO and NO_2 left, because they all would have reacted. Now that we know about chemical equilibrium, however, we know that the reaction reaches some point where the forward reaction is running at the same rate as the reverse reaction; thus, NO_2 and CO always remain. The question is, how much remains? If I were to start out with 1 mole of each, for example, how much would be left at equilibrium?

In order to determine this, chemists have developed the **equilibrium constant.** This constant, abbreviated with a capital "K," tells us all we need to know about how much reactant and how much product exist at equilibrium. For the generic chemical equilibrium reaction:

$$aA + bB \leftrightarrows cC + dD$$

we define the equilibrium constant as:

$$K = \frac{[C]^c_{eq}[D]^d_{eq}}{[A]^a_{eq}[B]^b_{eq}} \qquad (15.2)$$

As always, the square brackets simply stand for concentration. The "eq" subscript on each concentration term reminds us that in order to calculate the equilibrium constant, you must always use the concentrations of the substances once equilibrium is reached. Before we actually learn what the equilibrium constant is and what it means, let's make sure we understand the equation by studying a couple of examples and solving some problems:

EXAMPLE 15.1

The following reaction:

$$N_2 (g) + O_2 (g) \leftrightarrows 2NO (g)$$

reaches equilibrium when the concentration of N_2 is 0.354 M, the concentration of O_2 is 0.124 M, and the concentration of NO is 4.51 M. What is the equilibrium constant for this reaction?

According to Equation (15.2), we can calculate the equilibrium constant for any reaction by taking the equilibrium concentrations of the products (raised to their stoichiometric coefficients) and dividing them by the equilibrium concentrations of the reactants (raised to their stoichiometric coefficients). For this particular reaction, the equilibrium constant is given by:

$$K = \frac{[NO]^2_{eq}}{[N_2]_{eq}[O_2]_{eq}}$$

Since the concentrations given are the concentrations at equilibrium, we can simply plug them into the formula:

$$K = \frac{(4.51\,\cancel{M})^2}{(0.354\,\cancel{M})(0.124\,\cancel{M})} = 463$$

Notice that the units of M^2 cancel out, leaving the equilibrium constant with no units. Just like the rate constant, the units of the equilibrium constant can be different for different reactions, so you must always be careful to keep track of the units while you are calculating K.

A chemist is studying the following reaction:

$$4NH_3\,(g) + 7O_2\,(g) \leftrightharpoons 4NO_2\,(g) + 6H_2O\,(g)$$

If the chemist starts with 1.2 M NH_3, 3.4 M O_2, and no NO_2 or H_2O, he finds that the reaction reaches equilibrium when the concentration of NH_3 reaches 0.80 M and the concentration of O_2 reaches 2.7 M. At that point the concentration of NO_2 is 0.40 M, and the concentration of H_2O is 0.60 M. What is the equilibrium constant for this reaction?

According to Equation (15.2), the equilibrium constant for this particular reaction can be calculated with the following formula:

$$K = \frac{[NO_2]_{eq}^4 [H_2O]_{eq}^6}{[NH_3]_{eq}^4 [O_2]_{eq}^7}$$

Now remember, only *the concentrations at equilibrium* can be used to calculate K. Thus, the first two concentrations are irrelevant in the problem, because those are the concentrations that the chemist *started with.* We must use the concentrations at equilibrium, which are given in the next part of the problem. Plugging those numbers into the equation gives us:

$$K = \frac{(0.40\text{ M})^4 (0.60\text{ M})^6}{(0.80\text{ M})^4 (2.7\text{ M})^7} = \underline{2.8 \times 10^{-6}\ \frac{1}{\text{M}}}$$

Notice that this time one of the molarity units in the denominator did not cancel, leaving the equilibrium constant with a unit of 1/M.

ON YOUR OWN

15.1 What is the value of the equilibrium constant for the following reaction?

$$N_2\,(g) + 3H_2\,(g) \leftrightharpoons 2NH_3\,(g)$$

The equilibrium concentration of N_2 is 0.121 M, the equilibrium concentration of H_2 is 0.234 M, and the equilibrium concentration of NH_3 is 1.14 M.

15.2 A chemist is studying the following reaction:

$$2SO_3\,(g) \leftrightharpoons 2SO_2\,(g) + O_2\,(g)$$

If the chemist starts with 1.30 M SO_3, and no SO_2 or O_2, he finds that the reaction reaches equilibrium when the concentration of SO_3 reaches 0.20 M. At that point the concentration of SO_2 is 1.10 M, and the concentration of O_2 is 0.55 M. What is the equilibrium constant for this reaction?

Now, of course, it doesn't do us much good to calculate the equilibrium constant unless we know what it means and how we can use it. So that's what we have to learn now. The first thing that you need to know about the equilibrium constant is that it never changes, as long as the temperature doesn't change. In other words, once you calculate the equilibrium constant for a particular reaction, you never have to calculate it again at that temperature. No matter what the initial concentrations of the substances in the chemical equation are, they will always change so that, at equilibrium, the equilibrium constant is always the same.

The more important thing to remember about the equilibrium constant is that it can be used as a gauge to tell you the relative concentrations of reactants and products when the reaction reaches equilibrium. In other words, the equilibrium constant can tell you *where* the reaction will end up at equilibrium. For example, suppose the equilibrium constant for a chemical reaction is very large. This tells you that at equilibrium, there will be a lot of products and only a few reactants. After all, look at Equation (15.2). The equilibrium constant is calculated with the concentration of the products in the numerator of the fraction and the concentration of the reactants in the denominator. The only way such a calculation will result in a large number is if the number in the numerator is larger than the number in the denominator. That means the concentration of products will be greater than the concentration of the reactants.

On the other hand, if the equilibrium constant is small, it tells you that at equilibrium, there will be a lot of reactants and only a few products. Once again, think about Equation (15.2). The only way that the equation can work out to a small number is when the number in the numerator is smaller than the number in the denominator. This will happen when the concentration of reactants is greater than the concentration of the products. Finally, when the equilibrium constant is close to 1, then there will be approximately equal amounts of reactants and products when the reaction reaches equilibrium.

When an equilibrium reaction has more products than reactants at equilibrium, we say that it is "weighted toward the products side of the equation." On the other hand, when there are more reactants than products at equilibrium, we say that the reaction is "weighted toward the reactants side of the equation." Finally, when there are roughly the same amounts of products and reactants at equilibrium, we say that the equilibrium is "balanced between reactants and products." Therefore, the principles we just learned can be restated as follows:

- **When K is large, the equilibrium is weighted toward the products side of the equation.**
- **When K is small, the equilibrium is weighted toward the reactants side of the equation.**
- **When K is near unity, the equilibrium is balanced between reactants and products.**

These are important principles to remember.

ON YOUR OWN

15.3 In example 15.1, the equilibrium constants of two equilibrium reactions were calculated. For each reaction, tell whether the equilibrium is weighted toward reactants, weighted toward products, or balanced between reactants and products.

At this point, it's time to let you in on a little secret. It turns out that *all* chemical reactions are, in fact, equilibrium reactions. Thus, to be perfectly correct, *every* equation that you have ever seen in this course should be written with a double arrow rather than a single arrow. Does this mean that everything we've learned up to this point is wrong? No, of course not. As we just learned, when the equilibrium constant for a reaction is large, the reaction is weighted toward the products side of the equation. Thus, when equilibrium is achieved, there are a lot of products and only a few reactants. Well, if the equilibrium constant for a chemical reaction is *very* large, then that equilibrium is *heavily* weighted toward the products side of the equation. At some point, the equilibrium constant becomes so large that once the reaction has hit equilibrium, there are mostly products and virtually no reactants present. At that point, chemists just say that the amount of reactants present for such a reaction is so small that we can just ignore them. In the same way, if the equilibrium constant is very, very small, we can say that the reverse reaction is really what happens, because only reactants exist at equilibrium.

Thus, when the equilibrium constant for a reaction is very large, we replace the double arrow with a single arrow, indicating that, as far as we are concerned, the reaction completely runs out of reactants. When the equilibrium constant is very small, we write the reaction in reverse. From now on, then, when you see a reaction that contains a single arrow, remember that the reaction is, in fact, an equilibrium, but the equilibrium is so heavily weighted toward the products side that we just assume that there are no reactants present when the reaction finishes.

ON YOUR OWN

15.4 Three chemical equilibria (the plural of equilibrium) are written below. Next to each equation, the value for the equilibrium constant is given. If any of these equations can be written with a single arrow, then do so, making sure that the arrow is pointing in the proper direction:

a. $2NO_2\ (g) + Cl_2\ (g) \leftrightarrows 2NOCl\ (g) + O_2\ (g)$ $\qquad K= 4.6 \times 10^4$

b. $2Cl_2\ (g) + 2H_2O\ (g) \leftrightarrows 4HCl\ (g) + O_2\ (g)$ $\qquad K = 3.2 \times 10^{-14}\ M$

c. $CH_4\ (g) + Cl_2\ (g) \leftrightarrows CH_3Cl\ (g) + HCl\ (g)$ $\qquad K = 1.2$

A Few More Details Concerning the Equilibrium Constant

Now that you've had a chance to become a bit familiar with the concept of the equilibrium constant and how it is calculated, it is time to throw you a curveball. Equation (15.2) tells us that we calculate K by taking the products of the equation (raised to their stoichiometric coefficients) and dividing them by the reactants of the equation (raised to their stoichiometric coefficients). It turns out, however, that there are a few things we ignore in the equation for K, despite the fact that they might appear in the chemical equation.

First, any solids that happen to be in the chemical equation are not used when calculating K. Thus, for the following reaction:

$$2NaOH\ (aq) + MgCl_2\ (aq) \leftrightarrows Mg(OH)_2\ (s) + 2NaCl\ (aq)$$

we would calculate the equilibrium constant with the following equation:

$$K = \frac{[NaCl]_{eq}^2}{[MgCl_2]_{eq}[NaOH]_{eq}^2}$$

Notice that, despite the fact that $Mg(OH)_2$ is a product, it is not included in the equation for the equilibrium constant. This is because $Mg(OH)_2$ is a solid, and no solids are ever included in the equation for the equilibrium constant. You must remember this:

When a solid appears in a chemical equilibrium, it is not included in Equation (15.2).

Now you might be wondering why solids are not included in the equilibrium constant. After all, if it is a part of the chemical equation, shouldn't it be included? Well, remember that in the solid state, molecules are very close to one another. The distance between them is governed by the properties of the particular molecule involved and, for a given temperature, that distance cannot change. As a result, *the concentration of a solid is constant*. Think about it this way: At a given temperature the *density* of a solid is always the same. Thus, the number of grams per mL is constant. Well, I could take the density of a solid and convert grams to moles by looking at the periodic chart. I could also convert mL to L. What would I have then? I would have *moles per liter*, which is the *concentration*. Thus, the *concentration* of a solid is constant for any given temperature. This means that while the *mass* of a solid might increase or decrease through the course of a chemical reaction, its *concentration* (moles per liter) remains the same.

Along the same lines, when calculating the equilibrium constant, we can ignore liquids that appear in the chemical reaction. The rationale is the same. Since the density (and thus the concentration) of a pure liquid is constant for a given temperature, liquids should not be included in Equation (15.2). Please note that this only applies to *liquids*, not aqueous substances. For example, in the following reaction:

$$2H_2O_2\ (aq) \leftrightarrows 2H_2O\ (l) + O_2\ (g)$$

the equilibrium constant would be calculated as follows:

$$K = \frac{[O_2]_{eq}}{[H_2O_2]_{eq}^2}$$

Even though the reaction contains liquid water, we ignore it as a part of the equation for the equilibrium constant. However, we *do not* ignore the aqueous H_2O_2. This, then, is another important principle:

When a liquid appears in a chemical equation, we do not include it in Equation (15.2). This applies *only* to the liquid phase; it does not apply to the aqueous phase.

Study the following example and then solve the "On Your Own" problems so that you are sure you know how to apply these new rules.

EXAMPLE 15.2

What is the equation used to calculate the equilibrium constant in the following reaction?

$$NH_4Cl\ (s) \leftrightharpoons NH_3\ (g) + HCl\ (g)$$

To solve this problem, we just use Equation (15.2), recognizing that we ignore the solid in the equation :

$$K = [NH_3][HCl]$$

What is the expression for the equilibrium constant for this reaction?

$$CH_3CO_2H\ (aq) + C_2H_5OH\ (aq) \leftrightharpoons CH_3CO_2C_2H_5\ (aq) + H_2O\ (l)$$

To solve this problem, we once again use Equation (15.2), remembering that we ignore the liquid in the equation:

$$K = \frac{[CH_3CO_2C_2H_5]}{[CH_3CO_2H][C_2H_5OH]}$$

ON YOUR OWN

15.5 Give the expression for the equilibrium constant of this reaction:

$$2Li\ (s) + 2H_2O\ (l) \leftrightharpoons 2LiOH\ (aq) + H_2\ (g)$$

15.6 What is the equation for the equilibrium constant for the following reaction?

$$Ba(OH)_2H_{16}O_8\ (s) + 2NH_4SCN\ (s) \leftrightharpoons 2NH_4OH\ (aq) + Ba(SCN)_2\ (aq) + 8H_2O\ (l)$$

Using the Equilibrium Constant to Predict the Progress of a Reaction

Besides telling you the concentration of reactants and products at equilibrium, the equilibrium constant can also be used to predict whether or not a chemical reaction has actually reached equilibrium yet. In order to do this, you must first know the value for the equilibrium constant for whatever temperature at which you are running the reaction. Next, you can measure the concentrations of each substance in the chemical reaction at any given time. You can then plug those concentrations into the equation for the equilibrium constant. If the reaction has reached equilibrium, by definition, the answer you get should equal the value of the equilibrium constant. If it does not, you know that the equation is not at equilibrium yet.

This exercise will tell you more than just that. It will actually allow you to predict what the reaction will do next in order to reach equilibrium. What I mean is this: If a reaction has not yet reached equilibrium, it must either make more products (which will lower the rate of the forward reaction) or make more reactants (which will lower the rate of the reverse reaction). When a chemical reaction does something like this, we say that the reaction "shifts." The equilibrium constant can be used to predict the direction that a chemical reaction will shift if it is not already at equilibrium. Study the following examples to see how that is done:

EXAMPLE 15.3

A chemist is studying the following reaction:

$$N_2\,(g) + O_2\,(g) \leftrightarrows 2NO\,(g)$$

She knows that the equilibrium constant for this reaction at this temperature is 0.050. Suppose she knows that the concentrations of each substance are as follows: $[N_2]$ = 1.5 M, $[O_2]$ = 1.3 M, and [NO] = 0.020 M. Determine whether or not the reaction has reached equilibrium yet. If it has not, determine the direction it will shift in order to reach equilibrium.

First, we have to see whether or not the reaction has reached equilibrium. To do this, we just plug the concentration values into Equation (15.2). If the result is equal to 0.050, then we know that the reaction is at equilibrium. If not, then the reaction has not reached equilibrium yet.

$$\frac{[NO]_{eq}^2}{[N_2]_{eq}[O_2]_{eq}} = \frac{(0.020\,\cancel{M})^2}{(1.5\,\cancel{M})(1.3\,\cancel{M})} = 0.00021$$

According to the problem, the correct value of the equilibrium constant is 0.050. Since this is not the answer we got, we are left to conclude that the concentrations we used were not, in fact, equilibrium concentrations. This means that the reaction is not at equilibrium.

To determine which way the reaction will shift in order to reach equilibrium, we just need to compare our calculation to the actual value of K. Our calculation is smaller than the correct value. This means that, in order to reach equilibrium, the concentrations need to change in such a way as to increase the value of the equation we used. How can that happen? Well, in the equation, the concentration of NO is in the numerator of the fraction. Thus, if the concentration of NO is increased, the value we calculate from the equation will increase, bringing it closer to the true value for K. How can we get the concentration of NO to increase? The reaction just needs to shift toward the products side of the equation. This will both increase the concentration of NO (because more NO will be made) and at the same time decrease the concentrations of N_2 and O_2 (because they will be used up). Both of these things will result in a larger value calculated from Equation (15.2), bringing the reaction closer to equilibrium. Thus, the reaction needs to make more products. In chemical terminology, we would say that the reaction must shift toward the products.

The same chemist studied the same reaction under different concentrations. Suppose this time the chemist measures the following concentrations: $[N_2] = 0.100$ M, $[O_2] = 0.200$ M, $[NO] = 1.00$ M. Is the reaction at equilibrium under these conditions? If not, how will the reaction need to shift in order to reach equilibrium?

Once again, we have to see whether or not the reaction has reached equilibrium. To do this, we just plug the concentration values into Equation (15.2). If the result is equal to 0.050, we know that the reaction is at equilibrium. If not, the reaction has not reached equilibrium yet.

$$\frac{[NO]_{eq}^2}{[N_2]_{eq}[O_2]_{eq}} = \frac{(1.00\ \cancel{M})^2}{(0.100\ \cancel{M})(0.200\ \cancel{M})} = 50.0$$

According to the problem, the correct value of the equilibrium constant is 0.050. Since this is not the answer we got, we are left to conclude that the concentrations we used were not, in fact, equilibrium concentrations. Thus, the reaction is not at equilibrium.

To determine which way the reaction will shift in order to reach equilibrium, we just need to compare our calculation with the value of K. Our calculation is larger than the correct value. This means that, in order to reach equilibrium, the concentrations need to change in such a way as to decrease the value of the equation we used. How can that happen? Well, in the equation, the concentration of NO is in the numerator of the fraction. Thus, if the concentration of NO is decreased, the value we calculate from the equation will decrease, bringing it closer to the value for K. How can we get the concentration of NO to decrease? The reaction just needs to make more reactants. This will both decrease the concentration of NO (because NO will be used up) and at the same time increase the concentrations of N_2 and O_2 (because they will be made). Both of these things will result in a smaller value calculated from Equation (15.2), bringing the reaction closer to equilibrium. Thus, the reaction must shift toward the reactants.

So you see that all we need to do to predict which way a reaction will shift to reach equilibrium is take the concentrations at the moment and plug them into Equation (15.2). If the number we get is equal to the equilibrium constant, then the reaction is at equilibrium and will not change. If the number we get is less than the equilibrium constant, the reaction needs to shift toward the products. If, however, the number we calculate is greater than K, the reaction must shift toward reactants in order to reach equilibrium. Make sure that you grasp this concept by solving the following problem:

ON YOUR OWN

15.7 A chemist is studying the following reaction:

$$2NH_3\ (g) \leftrightharpoons N_2\ (g) + 3H_2\ (g)$$

The reaction has an equilibrium constant of 17 M^2. The chemist measures the concentrations of each substance to be $[NH_3] = 0.40$ M, $[N_2] = 1.5$ M, $[H_2] = 1.3$ M. Is the reaction at equilibrium? If not, which way will it shift in order to attain equilibrium?

Le Chatelier's Principle

Earlier in this module, I said that a reaction that has attained equilibrium will stay at equilibrium forever, unless it is disturbed by some outside influence. Well, now it's time to learn how a chemical equilibrium behaves when such a thing happens. For example, suppose I have the following equilibrium:

$$H_2\,(g) + I_2\,(g) \leftrightharpoons 2HI\,(g)$$

If I start with just H_2 and I_2, the forward reaction will occur at a fast rate, and the reverse reaction won't happen at all. As soon as some HI is formed, however, the reverse reaction will start to occur as well. As time goes on, the reverse reaction will get faster while the forward reaction gets slower and, eventually, equilibrium will be achieved.

Let's suppose now that equilibrium has been achieved. What would happen if I added some H_2 into the system? Before the H_2 was added, the reaction was at equilibrium. This means that the rate of the forward reaction was equal to the rate of the reverse reaction. If suddenly there were extra H_2 added, the rate of the forward reaction would go up, because the concentration of H_2 would have increased. This would throw the reaction out of equilibrium. Since the forward reaction would suddenly be faster than the reverse reaction, more HI would be made. This would start to slow down the forward reaction (because H_2 and I_2 would be used up), and it would start to increase the rate of the reverse reaction (because more HI would be made). Eventually, the reaction rates would equal each other again, and a new equilibrium would be established. This process is illustrated in Figure 15.2:

FIGURE 15.2
The Effect of Adding More Reactant to an Equilibrium

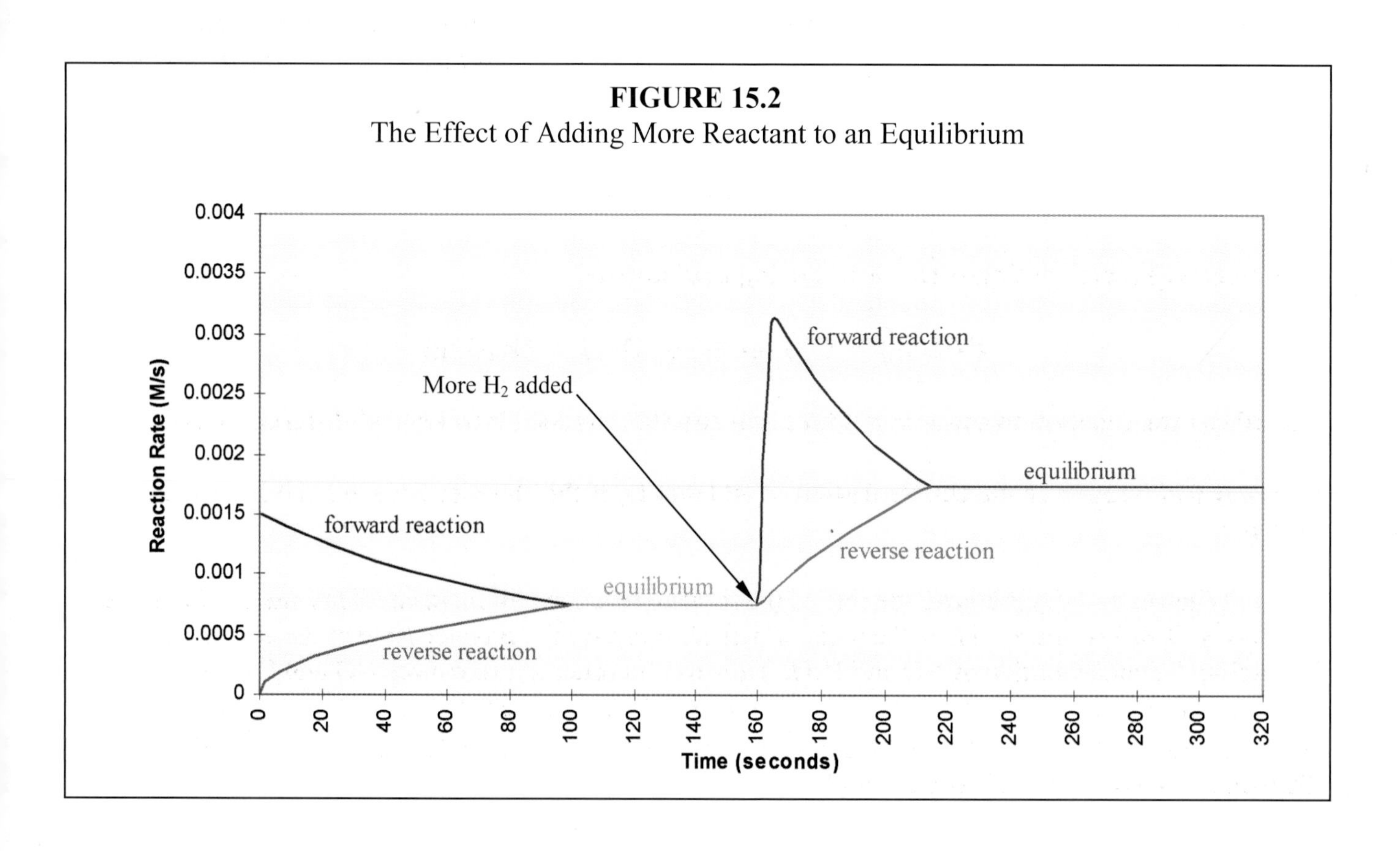

When more H_2 was added to the system, the equilibrium was upset. However, in order to restore equilibrium, the reaction simply shifted in such a way as to make the forward and reverse reaction rates equal again. This is a general principle, called **Le Chatelier's** (luh shat' tee ayes) **principle:**

Le Chatelier's Principle - When a stress (such as a change in concentration, pressure, or temperature) is applied to an equilibrium, the reaction will shift in a way that relieves the stress and restores equilibrium.

Le Chatelier's principle allows us to predict how an equilibrium responds to stress and is therefore a useful principle to apply.

Let's look back to our original example again. We already determined that if our equilibrium is stressed by adding more H_2 to the system, the equilibrium will relieve the stress by making more HI. This will lower the rate of the forward reaction and raise the rate of the reverse reaction, eventually bringing the system back to equilibrium. Thus, we know that at the new equilibrium point, the concentration of HI will be larger than it was at the first equilibrium point. In other words, the equilibrium will shift to the products side of the equation in order to relieve the stress.

What about the concentration of I_2? At the new equilibrium, will the concentration of I_2 be greater, lower, or the same as the original equilibrium? Well, when the rate of the forward reaction increases, more HI will be made. The only way that the reaction can do this is to use up I_2. So, in the end, the concentration of I_2 will be less at the new equilibrium, because when the reaction shifted to restore equilibrium, I_2 was used up. This is the kind of reasoning you must be able to do when it comes to Le Chatelier's principle. Study the following example to make sure you can reason this way:

EXAMPLE 15.4

A chemist is experimenting with the following equilibrium:

$$4NH_3\ (g)\ +\ 5O_2\ (g)\ \leftrightharpoons\ 4NO\ (g)\ +\ 6H_2O\ (g)$$

She allows the reaction to come to equilibrium and then stresses it in several different ways.

a. What will happen to the concentration of NO and O_2 if the chemist adds extra NH_3 to the system?

When extra NH_3 is added, the rate of the forward reaction will increase. This means that more NO and H_2O will be made. Thus, the reaction will shift toward the products side of the equation, increasing the concentration of NO and H_2O. However, in order to make more NO and H_2O, the reaction will have to use up some O_2. That will reduce the concentration of O_2. Thus, the concentration of O_2 will decrease (because it will be used up), and the concentration of NO will increase (because more of it will be made).

b. What will happen to the concentration of H_2O and NH_3 if the chemist removes O_2 from the system?

When O_2 is removed from the system, the forward reaction slows down, making the reverse reaction suddenly faster than the forward one. Thus, the equilibrium will begin making extra NH_3 and O_2. In other words, the equilibrium will shift toward the reactants side. This will increase the concentration of NH_3. In order to make extra NH_3 and O_2, however, the reaction must use up NO and H_2O, which will cause their concentrations to decrease. When the system reaches its new equilibrium, the concentration of H_2O will be lower (because it will be used up), and the concentration of NH_3 will be greater (because more of it will be made).

c. What will happen to the concentration of H_2O and NH_3 if the chemist adds extra NO?

When the chemist adds extra NO, the reverse reaction speeds up. This results in more NH_3 and O_2 being made, shifting the equilibrium to the reactants side. In order to do that, however, the reaction must use up NO and H_2O. Thus, when the new equilibrium is achieved, the concentration of H_2O will be lower (because it will be used up), and the concentration of NH_3 will be greater (because it will be produced).

 (The multimedia CD has a video demonstration of Le Chatelier's principle.)

Before we go on, I want to revisit something I mentioned earlier. When discussing equilibria, I told you that all chemical reactions are, in fact, equilibrium reactions. However, if the equilibrium constant is large, the equilibrium will be so heavily weighted toward the products that we can ignore the few reactants that exist. In that case, we can assume that the reactants react completely to form products. Consider, however, the following reaction:

$$CaCO_3\ (s) \leftrightharpoons CaO\ (s) + CO_2\ (g)$$

The equilibrium constant for this reaction at 25 °C is 0.86 M. Thus, according to the analysis we did before, we cannot consider this a complete reaction. In fact, this reaction is weighted toward the reactants side of the equation, since $K < 1$.

However, Le Chatelier's principle provides us a way to *force* this reaction to go to completion so that no $CaCO_3$ remains. How is that possible? Well, consider what would happen if the CO_2 were simply allowed to escape the reaction vessel while the reaction took place. What would that do to the equilibrium? If that were to happen, the concentration of CO_2 would decrease. This would stress the equilibrium. As a result, the equilibrium would shift to the products side of the equation, making more CaO and CO_2 and using up $CaCO_3$. If the CO_2 were continually allowed to escape, the equilibrium would continually shift to the products side of the equation until no more $CaCO_3$ remained. In this case, then, despite the fact that the equilibrium constant for this reaction is less than 1, the reaction would go until all of the reactant is used up. In fact, the only way you could get this reaction to reach equilibrium is to run it in an enclosed vessel so that the carbon dioxide could not escape. Under those conditions, then, the concentration of CO_2 would not be continually decreasing, so the reaction could reach equilibrium.

Le Chatelier's principle, then, tells us how to force an equilibrium to react so that the limiting reactant is always used up. All you have to do is continually remove a product. If you do that, the equilibrium will continually be stressed and will continually shift to the products side of the equation to relieve the stress. As long as you continually remove one or more products, this will happen until the limiting reactant is completely used up.

This concept is used quite often in the commercial production of chemicals. For example, ammonia is made commercially via its formation reaction:

$$N_2\ (g)\ +\ 3H_2\ (g)\ \leftrightarrows\ 2NH_3\ (g)$$

This reaction is run at high temperatures (usually around 400-600 °C) to increase the rate of the reaction, but at those temperatures, the equilibrium constant is not very large (less than 10 1/M). As a result, the equilibrium has more products than reactants, but it certainly cannot be treated as a reaction that uses up all of its reactants. At equilibrium, there are still significant quantities of nitrogen and hydrogen gas present.

Well, when you are in the business of making ammonia, you want to produce all of the ammonia that you can from the reactants that you have. Thus, in order to force this reaction to go to completion, the ammonia is continually removed from the reaction. As a result, the equilibrium is continually stressed so that it continually shifts to the products side of the equation. As a result, all of the nitrogen and hydrogen gas ends up reacting to form ammonia.

So you see that although all chemical reactions are, in fact, equilibria, you can stress an equilibrium so that it shifts to the products side of the equation. If you continue to do this throughout the course of the reaction, the reaction will never be able to reach equilibrium, and it will eventually use up all of its limiting reactant, regardless of the value of the equilibrium constant.

ON YOUR OWN

15.8 Assume that the following reaction has achieved equilibrium.

$$CO_2\ (g)\ +\ H_2\ (g)\ \leftrightarrows\ CO\ (g)\ +\ H_2O\ (g)$$

a. What will happen to the concentration of CO and H_2 if extra H_2O is added to the system?

b. What will happen to the concentration of CO_2 and H_2O if CO is removed from the system?

c. What will happen to the concentration of CO and H_2 if extra CO_2 is added to the system?

Pressure and Le Chatelier's Principle

Believe it or not, we are not finished with Le Chatelier's principle yet. This is a pretty important topic in the world of equilibrium, so we are going to dwell on it for another moment or two. In the definition of Le Chatelier's principle, I mentioned that stress can come in three different forms. I can stress an equilibrium by changing concentration, but I can also do it by changing either the pressure or the temperature. In this section, I want to explore what happens when you stress an equilibrium by changing the pressure.

Before we can do that, however, I have to remind you of something from an earlier part of the module. Remember that when we use Equation (15.2), we are supposed to ignore solids and liquids. Well, the same thing is true with Le Chatelier's principle. If I change the amount of a solid or liquid in a chemical reaction, the equilibrium is not stressed. As a result, nothing happens. This is an important fact to remember:

Le Chatelier's principle ignores solids and liquids as a source of stress to the equilibrium.

As I mentioned before, the aqueous phase is not the same as the liquid phase. Thus, we cannot ignore aqueous substances in the reaction. We can only ignore substances in the solid or liquid phase.

Now that that's out of the way, we can learn how changes in pressure stress a chemical equilibrium. When I increase pressure, what happens to the volume of any gases present? According to the gas laws that we learned in Module #12, increasing the pressure of a gas will lower its volume. Well, what happens to concentration when volume is lowered? According to the definition of concentration, volume is in the denominator of the fraction. Thus, when volume is lowered, concentration goes up. That's the effect of increasing the pressure on a chemical reaction. The concentration of the gases in the reaction increases.

What happens to the concentration of solids, liquids, or aqueous substances when the pressure increases? Nothing, really. Only gases are affected by a change in pressure. As a result, if an equilibrium contains no gases, a change in pressure does not cause it any stress. If it does contain gases, however, a change in pressure does produce a stress. In fact, whichever side of the equation has more gas molecules is more affected by a change in pressure. Consider, for example, the following equilibrium:

$$CaCO_3\ (s) \leftrightarrows CaO\ (s) + CO_2\ (g)$$

In this equation, only the products side of the equation has a gas in it. As a result, an increase in pressure will only increase the concentration of CO_2. When that happens, the reverse reaction will speed up, and the equilibrium will shift toward the reactants. If, on the other hand, the pressure decreases, that will decrease the concentration of CO_2, causing the reverse reaction to slow down, making the forward reaction faster than the reverse. Thus, the equilibrium will shift to the products side of the reaction.

In the end, then, the equilibrium will tend to shift based on the side of the equation that has more total gas molecules. If pressure is increased, the reaction will shift away from the side that contains more gas molecules. On the other hand, if the pressure is reduced, the reaction will shift toward the side with the most gas molecules on it. If both sides have the same number of gas molecules on them, then both sides of the reaction are affected equally and, as a result, the reaction will not shift. These are important rules to remember:

When an equilibrium is subjected to an increase in pressure, it will shift away from the side with the largest number of gas molecules. If pressure decreases, the equilibrium will shift toward the side with the largest number of gas molecules. Finally, if there are no gases in the equation or if the number of gas molecules are the same on both sides, nothing will happen.

Make sure you can apply these rules by studying the example below and solving the "On Your Own" problems that follow:

EXAMPLE 15.5

A chemist studies the following reaction, which is already at equilibrium:

$$NH_4Cl\ (s) \leftrightharpoons NH_3\ (g) + HCl\ (g)$$

a. Which way will the reaction shift if pressure is increased?

There are two molecules of gas on the products side and none on the reactants. An increase in pressure causes a shift away from the side with the largest number of gas molecules. Thus, the reaction will shift toward the reactants side of the equation.

b. What will happen to the amount of NH_4Cl present if pressure is decreased?

A decrease in pressure causes a shift toward the side with the largest number of gas molecules. Since the products side has the largest number of gas molecules, the reaction will shift that way. When this happens, NH_4Cl will be used up, so the amount of NH_4Cl present will decrease. Note that the *concentration* of the NH_4Cl did not change, because the concentration of a solid is constant. However, the *amount* did change.

c. Which way will the reaction shift if more NH_4Cl is added?

We were just taught that when solids (or liquids) are added or taken away from an equilibrium, it does not stress the system. Thus, the reaction will not shift. Remember, the *amount* of a solid (or liquid) can change as a result of the shift in an equilibrium, but the *concentration* of a solid (or liquid) does not change. Thus, changing the amount of a solid (or liquid) in a chemical equilibrium does not change the concentration and thus cannot stress the equilibrium.

ON YOUR OWN

15.9 Assume that the following reaction has already attained equilibrium:

$$2SO_2\ (g) + O_2\ (g) \leftrightharpoons 2SO_3\ (g)$$

a. Which way will the reaction shift if the pressure is increased?
b. Which way will the reaction shift if the pressure is decreased?

Temperature and Le Chatelier's Principle

Once again, we are still not quite done with Le Chatelier's principle. We still have to figure out how a change in temperature affects an equilibrium. This requires you to remember a fact we learned in Module #13. We learned that when a reaction is exothermic, its ΔH is negative, and energy can be considered a product of the reaction. Endothermic reactions, on the other hand, have positive ΔH's and are considered to have energy as a reactant. With these facts in mind, let's see how temperature stresses an equilibrium.

Consider the following reaction:

$$2H_2O\ (g) \leftrightharpoons 2H_2\ (g) + O_2\ (g) \qquad (\Delta H^o = 242\ kJ)$$

Because $\Delta H^o > 0$, we can think of energy as a reactant in the equation. Thus, we can think of the equation this way:

$$2H_2O\ (g) + 242\ kJ \leftrightharpoons 2H_2\ (g) + O_2\ (g)$$

Now, when we raise the temperature of the system, what are we really doing? We are adding energy. For this reaction, then, when we raise the temperature, we are really adding extra reactant. What did we learn about equilibria that have extra reactants added? The forward reaction increases in rate, making more products and less reactant. Thus, when the temperature of this reaction is raised, more H_2 and O_2 are made while H_2O is used up.

In the same way, if we were to lower the temperature of this reaction, it would be like taking a reactant away. When that happens, the rate of the forward reaction slows down, making the reverse reaction faster than the forward reaction. This causes the reaction to shift to the reactants side of the equation, making more H_2O and using up H_2 and O_2. The general trend, then, is:

When temperature is raised, an equilibrium will shift away from the side of the equation that contains energy. When temperature is lowered, the reaction will shift toward the side that contains energy.

Perform the following experiment to get more experience with Le Chatelier's principle and temperature changes:

EXPERIMENT 15.2

Temperature and Le Chatelier's Principle

Supplies:

- Two test tubes
- Two eyedroppers
- Two beakers
- Two small cups
- Clear ammonia solution (This is sold with the cleaning supplies in most supermarkets. It must be clear. A colored solution will mess up the colors you are supposed to see.)

- White vinegar (It must be white. Colored vinegar will mess up the colors you are supposed to see.)
- Flame or stove
- Pot for boiling water on the stove
- Water
- Ice (preferably crushed)
- A few leaves of red cabbage (It must be red cabbage, not regular cabbage.)
- Safety goggles

1. Put about 200 mL of water into the pot and add the leaves of red cabbage.
2. Put the pot on the stove and allow the water to boil for about three minutes.
3. While you are waiting for the water to boil, pour a small amount of ammonia solution into one of the cups, and then mark the cup so that you know it contains ammonia.
4. Pour a small amount of the white vinegar into the other cup and mark the cup so that you know it contains vinegar.
5. Fill the larger beaker with a mixture of ice and water. The mixture should have a lot of ice and only a little water, because the ice will melt throughout the experiment.
6. Fill the smaller beaker about ¾ of the way with water.
7. Once the water in the pot has been boiling for three minutes, remove it from the stove.
8. Place the smaller beaker (the one with just water on it) on the stove or on a heating flame so that the water will begin to boil.
9. Use an eyedropper to put 20 drops of cabbage water into each of the test tubes. **Be careful - the cabbage water is hot!!**
10. Rinse out the eyedropper.
11. Use that same eyedropper to add 15 drops of vinegar to each test tube.
12. Swirl the test tubes to mix the cabbage water and vinegar. The solution in each test tube should be pink.
13. Use the other eyedropper to add ammonia to one of the test tubes. Add the ammonia drop by drop. Swirl the test tube after each drop is added. As soon as the solution turns any shade of green, stop adding ammonia.
14. Repeat step #13 for the other test tube. When you are done, each solution should be approximately the same shade of green.
15. Place both test tubes in the solution of ice water that is in the big beaker.
16. Allow them to sit in the ice water for at least one minute.
17. Pull them both out of the ice water and compare the color of the solutions in the test tubes. They should be the same.
18. Put one of the test tubes back into the ice water.
19. Once the water in the smaller beaker is boiling, take the other test tube and place it in the boiling water. **Once again, be careful**. As long as you hold the test tube by the very top, you will not get burned.
20. Allow the test tube to sit in the boiling water for 30 seconds, and then pull it out.
21. Pull the other test tube out of the ice water and compare the color of the solutions. The solution that was in the boiling water should be noticeably bluer than the one in the ice water. If it is not, try putting it back into the boiling water for another 30 seconds.
22. Now put both test tubes back into the ice water.
23. After both test tubes have been in the ice water for a few minutes, pull the test tubes out again.
24. Compare the color of the solutions. They should be the same again.

25. If you like, repeat steps 18-24 again. Each time, the solution that gets hot should become more blue than the test tube in the ice water and each time, the solutions should be roughly the same color once they have been in the ice water for a while. Do not allow the test tube to sit in the hot water for more than 30 seconds at a time, however, because you do not want to drive the ammonia out of solution.
26. Clean up your mess.

What happened in the experiment? When you mixed the ammonia and vinegar in the test tubes, you set up the following equilibrium:

$$NH_3\ (aq)\ +\ C_2H_4O_2\ (aq) \leftrightharpoons NH_4^+\ (aq)\ +\ C_2H_3O_2^-\ (aq)$$

This equilibrium is heavily weighted toward the products, but nevertheless, it is an equilibrium, so some of the reactants remain. The cabbage water you added was an indicator, telling you how much acid or base was present. When you added just vinegar (the source of $C_2H_4O_2$), the cabbage water turned pink to indicate the presence of an acid. When you added the ammonia to the solution, it eventually turned a shade of green, indicating the fact that the acid was neutralized by the base.

It turns out that this reaction is exothermic. Thus, energy is a product in the reaction. When you put one test tube into ice water, you lowered its temperature. This, in effect, removed a product from the reaction. As a result, the reaction shifted toward the products in order to relieve the stress. That caused the cabbage water to turn a bit more green. When you put the other test tube in the boiling water, you raised its temperature. This, in effect, added a product to the equation, causing the equilibrium to shift toward the reactants. Thus, the amounts of acid and base present in the test tube changed, and the color of the cabbage water changed accordingly.

This experiment, then, is an excellent illustration of how Le Chatelier's principle works with temperature changes. The cabbage water provided an indication of the relative amounts of reactants and products. A blue color indicated more reactants, while a green color indicated more products. When the temperature of the reaction was lowered, the equilibrium shifted to the products, because energy (a product) was being removed from the equilibrium. This caused the solution to turn green. When the temperature of the reaction was increased, the reaction shifted toward the reactants, because energy (a product) was being added. As a result, the solution turned blue. Make sure you can do this kind of reasoning by studying the example below and solving the problems that follow:

EXAMPLE 15.6

For the following equation:

$$H_2\ (g)\ +\ F_2\ (g) \leftrightharpoons 2HF\ (g)$$

ΔH^o = -273 kJ. If the reaction has already reached equilibrium:

a. What will happen to the concentrations of each substance when the temperature is raised?

When the temperature is raised, equilibria shift away from the side that contains energy. Since the ΔH^o for this reaction is less than zero, it is exothermic, making energy a product. Thus, the reaction will shift toward the reactants. Therefore, the HF concentration will decrease while the H_2 and F_2 concentrations will increase.

b. What will happen to the concentrations of each substance if the temperature is lowered?

When the temperature is lowered, equilibria shift toward the side that contains energy. Since the ΔH for this reaction is less than zero, it is exothermic, making energy a product. Thus, the reaction will shift toward the products. Therefore, the HF concentration will increase while the H_2 and F_2 concentrations will decrease.

ON YOUR OWN

15.10 Consider the following reaction that has already attained equilibrium:

$$CaCO_3\ (s) \leftrightharpoons CaO\ (s) + CO_2\ (g)$$

The $\Delta H = 132$ kJ.

a. What will happen to the amount of $CaCO_3$ if the temperature is raised?

b. What will happen to the concentration of CO_2 if the temperature is lowered?

Acid/Base Equilibria

Now that we've learned the basics of chemical equilibrium, it is time to examine a specific application of this interesting phenomenon. In Module #10, I introduced you to the subject of acid/base chemistry. I couldn't teach you everything there was to know about acids and bases, however, because in order to understand all of it, you need to understand chemical equilibrium. With the basic material of this module under your belt, however, you can now learn about acids and bases in a bit more depth.

Remember that we defined an acid as any molecule that donates an H^+. Similarly, a base was defined as anything that accepts an H^+. Thus, the molecule HCl (hydrochloric acid) acts as an acid in the following reaction:

$$HCl\ (aq) + OH^-\ (aq) \rightarrow H_2O\ (l) + Cl^-\ (aq) \qquad (15.3)$$

because it donates an H^+ to the OH^-. Similarly, the hydroxide ion acts as a base because it accepts an H^+. It turns out, however, that this reaction is actually a two-step process. In the first step, the H^+ ion must first pull away from the HCl molecule as follows:

Step 1: $$HCl\ (aq) \rightarrow H^+\ (aq) + Cl^-\ (aq) \qquad (15.4)$$

Only then can the H^+ go on and be donated to the OH^- ion:

Step 2: $$H^+\ (aq) + OH^-\ (aq) \rightarrow H_2O\ (l) \qquad (15.5)$$

From what you learned in the previous module, you should recognize that Equations (15.4) and (15.5) represent the reaction mechanism of Equation (15.3).

So you see that if an acid is going to donate an H^+ ion, it must first go through a reaction like Equation (15.4) where the H^+ actually separates from the molecule. Without this step, the acid cannot do its job. As a point of terminology, reactions like Equation (15.4) are called **acid ionization reactions**.

Acid ionization reaction - The reaction in which an H^+ separates from an acid molecule so that it can be donated in another reaction

Since this step is the most critical in allowing an acid to do its job, we need to look at it in some detail and see how well this reaction works for different acids.

For example, the acid we have been using is HCl. This is considered a **strong acid**, which means its acid ionization reaction works very well. When a solution of HCl is made, virtually all of its molecules break apart into H^+ and Cl^-. As a result, we never really bother to write the reaction as an equilibrium, because this particular reaction has such a huge equilibrium constant that we can assume there are no reactants left once the reaction finishes.

Other acids, however, are not nearly as good at ionization. Acetic acid ($C_2H_4O_2$) is an example of such an acid. This acid's ionization reaction is represented by the following equation:

$$C_2H_4O_2\,(aq) \leftrightarrows H^+\,(aq) + CH_3CO_2^-\,(aq) \qquad (15.6)$$

Notice that this equation is written as an equilibrium. That's because the H^+ does not separate well from the acetic acid molecule. As a result, this reaction reaches equilibrium after only a few products are made. In other words, this is an equilibrium that is heavily weighted toward the reactants side of the equation.

Now remember, the first acid we looked at (HCl) was so good at separating its H^+ from the rest of the molecule that its ionization reaction wasn't even written as an equilibrium. As a result, we call HCl a strong acid. However, the acid we are considering now (acetic acid) does not ionize very well. As a result, the equilibrium constant for its ionization is only 1.8×10^{-5} M. Since that equilibrium constant is small, we know that this reaction produces few products. When this reaction comes to equilibrium, then, it has mostly reactants present with only a few products present. Since acetic acid's ionization reaction does not produce many H^+ ions, we can say that acetic acid is a **weak acid**.

So now we see that some acids do very well at giving up H^+ ions, and we call them strong acids. There are many acids like acetic acid, however, that do not do very well at forming H^+ ions. We call these weak acids. It's important to be able to tell the difference between weak and strong acids. Luckily, that's not very difficult. All we have to do is look at the value of the equilibrium constant for the acid's ionization reaction. The larger the equilibrium constant, the stronger the acid. As a rule of thumb, an acid is considered weak if the equilibrium constant is less than one. As soon as the equilibrium constant gets larger than that, we consider the acid to be a strong acid.

Since the equilibrium constant of an acid's ionization reaction is so important in determining the strength of an acid, we give it a special name. We call it the **acid ionization constant**, which is usually abbreviated as $\mathbf{K_a}$.

Acid ionization constant - The equilibrium constant for an acid's ionization reaction

Since the acid ionization constant is simply the equilibrium constant for an acid's ionization reaction, we can say that the equation for the ionization constant for acetic acid looks like this:

$$K_a = \frac{[H^+][CH_3CO_2^-]}{[C_2H_4O_2]} \quad (15.7)$$

So, if we are given the value of this ionization constant, or if we are given the equilibrium concentrations with which to calculate its value, we can determine the strength of an acid.

If the ionization constant of an acid is large, the acid is considered a strong acid. If the ionization constant is not large, we call it a weak acid. In general, the larger the ionization constant, the stronger the acid.

Not surprisingly, just as there can be strong and weak acids, there can also be strong and weak bases. However, since a base's job is quite different from an acid's job, the reactions that tell us whether a base is strong or weak are quite different. In order to determine the strength of a base, we have to look at how well it *accepts an H^+ from water.* Thus, for the base NH_3 (ammonia), the reaction that determines its strength as a base is:

$$NH_3\,(aq) + H_2O\,(l) \leftrightarrows NH_4^+\,(aq) + OH^-\,(aq) \quad (15.8)$$

The base ionization constant, then, is given by the equilibrium constant for this reaction:

$$K_b = \frac{[NH_4^+][OH^-]}{[NH_3]} \quad (15.9)$$

Once again, the larger the K_b, the stronger the base.

EXAMPLE 15.7

Nitrous acid (HNO_2) has an ionization constant equal to 4.5×10^{-4} M. Give the equation for this ionization constant. Is this acid stronger or weaker than acetic acid ($K_a = 1.8 \times 10^{-5}$ M)?

The ionization constant is the equilibrium constant for the acid ionization reaction. In order to determine the ionization reaction, you simply take the acid in its aqueous phase and remove an H^+. When you remove an H^+ from HNO_2, you are left with NO_2^-. If that doesn't make sense to you, review Module #10, where we first learned to reason this way. In the end, then, the aqueous acid is the reactant, and the H^+ and NO_2^- (both in aqueous phase) will be the products:

$$HNO_2\,(aq) \leftrightarrows H^+\,(aq) + NO_2^-\,(aq)$$

The equilibrium constant for this reaction is the ionization constant, K_a:

$$K_a = \frac{[H^+][NO_2^-]}{[HNO_2]}$$

That is the equation for the ionization constant. Since the value given for this ionization constant is slightly larger than acetic acid, this is a slightly stronger acid than acetic acid.

Methyl amine, CH_5N, is a base with a K_b of 4.4×10^{-5}. Give the equation for its ionization constant. Would this be considered a strong or weak base?

The base ionization reaction involves having the base accept an H^+ ion from water. If CH_5N accepts an H^+, it becomes CH_6N^+. When water gives up that H^+, it becomes OH^-. All of this takes place in water, so the water phase is liquid and everything else is aqueous:

$$CH_5N\ (aq) + H_2O\ (l) \leftrightharpoons CH_6N^+\ (aq) + OH^-\ (aq)$$

The ionization constant, then, is just the equilibrium constant of this reaction. Remember that we must ignore water here since it is in its liquid phase:

$$K_b = \frac{[OH^-][CH_6N^+]}{[CH_5N]}$$

Since the value given for this ionization constant is quite low, this would be considered a weak base.

ON YOUR OWN

15.11 What is the equation for the acid ionization constant of HF?

15.12 What is the equation for the base ionization constant of PH_3?

The pH Scale

Since the strength of an acid is really dependent on how many H^+ ions it can make in an aqueous solution, chemists have developed a special scale to help us determine concentration of H^+ in an aqueous solution. This scale is called the **pH scale**. The letters pH stand for "**potential hydrogen**."

The pH scale is much like a scale of 1 to 10 that judges use to rate gymnasts, divers, skaters, and the like. This scale, however, ranges from 0 to 14. When a solution has a pH of zero, we consider it a very strongly acidic solution. Thus, it has a lot of H^+ ions. When a solution has a pH of 14, it is strongly basic. When a solution is purely neutral (neither acid nor base) it has a pH of 7. Generally speaking:

Solutions with pH 0 - 1.9 are considered strongly acidic; solutions with pH 2 - 6.9 are weakly acidic; solutions with pH 7.1 - 12 are weakly basic; and solutions with pH 12.1 - 14 are strongly basic. A pH of 7 indicates a neutral solution.

In other words, the lower the pH, the more acidic a solution is, whereas the higher the pH, the less acidic (and therefore more basic) the solution is.

If you purchased the MicroChem kit discussed in the introduction, you can perform Experiment #11 to get more experience with the pH scale and pH indicators.

Now it is important for you to realize that pH is a measure of the *overall* acidity of a solution. For example, the pH of a 1.0 M solution of acetic acid is about 2.3. As I mentioned before, acetic acid is a weak acid. However, the pH of a 0.0050 M solution of HCl is *also* 2.3, despite the fact that HCl is a strong acid. Why does the solution made with a weak acid have the same pH as a solution made with a strong acid? Look at the concentrations. The weak acid has a concentration of 1.0 M, while the strong acid has a concentration that is 200 times smaller. That's what's nice about pH. It is a measure of the acidity of a *solution*, which takes both the strength of the acid *and* the concentration of the acid into account. Remember, the chemistry of any substance depends on its concentration. A weak acid can still make a strongly acidic solution, if its concentration is large enough.

The other nice thing about the pH scale is that it is very flexible. For every one unit change of pH, there is a tenfold change in the acidity of the solution. In other words, a solution with a pH of 3 is 10 times more acidic than a solution with a pH of 4, and it is 100 times more acidic than a solution with a pH of 5.

You will find the pH scale on a lot of products, especially things that come into contact with hair or skin. It turns out that your body prefers everything it comes into contact with to be neither acidic nor basic. Thus, the best thing for your skin, hair, etc. is something that has a pH of 7. We often call something with a pH of 7 "pH balanced." You probably hear or read that in advertisements a lot. It simply means that the product is neither acidic nor basic. It is purely neutral.

ON YOUR OWN

15.13 Equal concentration aqueous solutions are made with the following four chemicals. The pH of each solution is written next to the chemical formula:

$$C_6H_6OH \ (pH = 5.1), \ KOH \ (pH = 13.5), \ C_2H_7N \ (pH = 8.7), \text{ and } HBr \ (pH = 1.0)$$

Identify the acids and the bases. In addition, indicate their relative strengths.

Acid Rain

Now that we are back onto the subject of acids and bases, I think it is time to bring up an environmentally oriented subject again. You have probably heard something about our society's problem with acid rain. In order to truly understand this problem, however, you need to understand a bit about where acid rain comes from.

The first thing to learn is that rain is naturally acidic. When most people hear that, they are usually shocked. Isn't acid dangerous? Not necessarily. As you learned in Module #10, many of the foods you eat and drink are acidic. The acids that are in those foods provide a pleasant sour taste. The key is that to be safe, acids have to be weak. The acid in normal rain is carbonic acid, the same acid you drink in soda pop. The carbonic acid gets into the rain because the atmosphere has carbon dioxide in it. When that carbon dioxide mixes with the water in a raindrop, the following reaction occurs:

$$CO_2\ (g)\ +\ H_2O\ (l)\ \leftrightharpoons\ H_2CO_3\ (aq)$$

The pH of the resulting solution is usually around 5.3.

Thus, natural rain is indeed slightly acidic. In a way, then, all rain is acid rain. The problem is, however, that pollutants in the air can mix with the water in a raindrop just like CO_2 can. For example, coal that is burned in power plants is often contaminated with sulfur. The sulfur burns right along with the coal, making sulfur trioxide. When this mixes with the water in a raindrop, it makes sulfuric acid:

$$SO_3\ (g)\ +\ H_2O\ (l)\ \rightarrow\ H_2SO_4\ (aq)$$

This puts more acid in the raindrop, making it more acidic.

In addition, a major source of pollutants on this planet is the automobile. When an automobile burns gasoline, it also burns some of the nitrogen in the surrounding air. This makes nitrogen dioxide. Like carbon dioxide and sulfur trioxide, this can react with the water in a raindrop to form acid as well:

$$2NO_2\ (g)\ +\ H_2O\ (l)\ \rightarrow\ HNO_3\ (aq)\ +\ HNO_2\ (aq)$$

Once again, this adds more acid to the raindrop, making it more acidic.

In the end, then, with all of this extra acid in the rain, the pH of the rain goes down. There have been instances when the pH of rain was measured to be 3.2. Now remember, every 1 pH unit represents a tenfold change in acidity. Thus, this pH represents a raindrop that is more than 100 times more acidic than it should be! At such pH's, rain can be hazardous to living organisms.

Thus, acid rain is a problem today. However, many people do not realize that acid rain is *less* of a problem today than it was 30 years ago. This is because the air we are breathing today is *significantly cleaner* than the air we were breathing 30 years ago. Although this little fact may sound surprising in the light of the environmental hysteria that's out there today, it is nevertheless true. Government regulations over the past 30 years have succeeded in cutting the total amount of pollutants in the atmosphere significantly. This means there is less sulfur trioxide and nitrogen dioxide in the air, which means acid rain is less of a problem. Of course, that doesn't mean we should ignore the problem of acid rain; it just means that the problem isn't as bad as it used to be!

ANSWERS TO THE "ON YOUR OWN" PROBLEMS

15.1 For this particular reaction, the equilibrium constant is given by:

$$K = \frac{[NH_3]_{eq}^2}{[N_2]_{eq}[H_2]_{eq}^3}$$

Since the concentrations given are, indeed, the equilibrium concentrations, we can simply plug them into our equation:

$$K = \frac{(1.14\text{ M})^2}{(0.121\text{ M})(0.234\text{ M})^3} = 838\ \frac{1}{\text{M}^2}$$

15.2 According to Equation (15.2), the equilibrium constant for this particular reaction can be calculated with the following formula:

$$K = \frac{[SO_2]_{eq}^2[O_2]_{eq}}{[SO_3]_{eq}^2}$$

Now remember, only *the concentrations at equilibrium* can be used to calculate K. Thus, the first concentration is irrelevant in the problem, because it is the concentration that the chemist *started with.* We must use the concentrations at equilibrium, which are given in the next part of the problem:

$$K = \frac{(1.10\ \cancel{\text{M}})^2(0.55\text{ M})}{(0.20\ \cancel{\text{M}})^2} = \underline{17\text{ M}}$$

15.3 In the first reaction of Example 15.1, the equilibrium constant was calculated to be 463. Since this is a number much larger than 1, we would have to say that this reaction is weighted toward the products side of the equation.

In the second reaction, the equilibrium constant was calculated to be 2.8×10^{-6} 1/M. Thus, since K is a small number, the reaction is weighted toward the reactants side of the equation.

15.4 a. If the equilibrium constant of a reaction is very large, the double arrow can be replaced by a single arrow indicating that the reaction leads to all products and no reactants. Since this reaction has a very large equilibrium constant, it qualifies:

$$\underline{2NO_2\,(g) + Cl_2\,(g) \rightarrow 2NOCl\,(g) + O_2\,(g)}$$

b. This reaction has a very small equilibrium constant. That means there are virtually no products. Thus, the double arrow in this reaction can be replaced by a single arrow pointing toward the reactants:

$$2Cl_2\,(g) + 2H_2O\,(g) \leftarrow 4HCl\,(g) + O_2\,(g)$$

Since this looks awkward, we usually turn the reaction around so that it is written in the direction that it actually goes:

$$\underline{4HCl\,(g) + O_2\,(g) \rightarrow 2Cl_2\,(g) + 2H_2O\,(g)}$$

c. In this case, the equilibrium constant is close to 1. This means that the equilibrium is actually well balanced; thus, we cannot remove the double arrow in this case.

15.5 To solve this problem, we simply use Equation (15.2), recognizing that we ignore Li since it is a solid. We also ignore H_2O because it is in its liquid state. Thus, the equilibrium constant is given by the following equation:

$$K = [LiOH]^2[H_2]$$

15.6 In this equation, there are two solids to ignore. We can also ignore the water, since it is in its liquid state. Thus, Equation (15.2) becomes:

$$K = [NH_4OH]^2[Ba(SCN)_2]$$

15.7 We first have to see whether or not the reaction has reached equilibrium. To do this, we just plug the concentration values into Equation (15.2). If the result is equal to 17 M^2, then we know that the reaction is at equilibrium. If not, then the reaction has not reached equilibrium yet.

$$\frac{[N_2]_{eq}[H_2]_{eq}^3}{[NH_3]_{eq}^2} = \frac{(1.5 \text{ M})(1.3 \text{ M})^3}{(0.400 \text{ M})^2} = 21 \text{ M}^2$$

According to the problem, the correct value of the equilibrium constant is 17 M^2. Since this is not the answer we got, we are left to conclude that the concentrations we used were not, in fact, equilibrium concentrations. This means, then, that the reaction is not at equilibrium. To determine which way the reaction will shift in order to reach equilibrium, we just need to compare our calculation to the value of K. Our calculation is larger than the correct value. This means, then, that in order to reach equilibrium, the concentrations need to change in such a way as to decrease the value of the equation we used. Thus, the reaction must shift toward the reactants.

15.8 a. If extra H_2O is added to the system, the rate of the reverse reaction will increase. This will make more CO_2 and H_2. In order to do this, however, the reaction will use up CO and H_2O. Thus, at the new equilibrium, the concentration of CO will be lower (because it will be used up), and the concentration of H_2 will be higher (because it will be made).

b. If CO is removed from the system, the rate of the reverse reaction will decrease, making the forward reaction suddenly faster than the reverse. This will make more CO and H_2O. In order to do this, however, the reaction will use up CO_2 and H_2. Thus, at the new equilibrium, the concentration of CO_2 will be lower (because it will be used up), and the concentration of H_2O will be higher (because it will be made).

c. If extra CO_2 is added to the system, the rate of the forward reaction will increase. This will make more CO and H_2O. In order to do this, however, the reaction will use up CO_2 and H_2. Thus, at the new equilibrium, the concentration of H_2 will be lower (because it will be used up), and the concentration of CO will be higher (because it will be made).

15.9 a. When the pressure is increased, an equilibrium will shift away from the side with the most gas molecules. There are 3 gas molecules on the reactants side (2 + 1), and only 2 on the products side. Thus, the reaction will shift away from the reactants side of the equation, which is toward the products side of the equation.

b. When pressure is lowered, an equilibrium will shift toward the side with the most gas molecules on it. Thus, the reaction will shift toward the reactants side of the equation.

15.10 a. The ΔH tells us that this equation is endothermic, making energy a reactant. Thus, when temperature is raised, the equilibrium will shift as if extra reactants had been added. This means that the equilibrium will shift toward the products side of the equation. Since that means $CaCO_3$ will be used up, the amount of $CaCO_3$ present will decrease.

b. When temperature is lowered, energy is taken away. Since energy is a reactant in this reaction, lowering the temperature is like removing a reactant. Thus, the equilibrium will shift toward reactants. This will cause CO_2 to be used up, and the concentration of CO_2 will go down.

15.11 The ionization constant is simply the equilibrium constant for the acid ionization reaction. In order to determine the ionization reaction, you simply take the acid in its aqueous phase and remove an H^+. When you remove an H^+ from HF, you are left with F^-. In the end, then, the aqueous acid is the reactant, and the H^+ and F^- (both in aqueous phase) will be the products:

$$HF\ (aq) \leftrightharpoons H^+\ (aq) + F^-\ (aq)$$

The equilibrium constant for this reaction is the ionization constant, K_a:

$$K_a = \frac{[H^+][F^-]}{[HF]}$$

15.12 The base ionization reaction involves having the base accept an H^+ ion from water. If PH_3 accepts an H^+, it becomes PH_4^+. When water gives up that H^+, it becomes OH^-. All of this takes place in water, so the water phase is liquid and everything else is aqueous:

$$PH_3\ (aq) + H_2O\ (l) \leftrightharpoons PH_4^+\ (aq) + OH^-\ (aq)$$

The ionization constant, then, is just the equilibrium constant of this reaction:

$$K_b = \frac{[OH^-][PH_4^+]}{[PH_3]}$$

15.13 When pH is 7, the solution is neutral. Less than 7 means acid, and the lower the pH, the more acidic. More than 7 means base, and the higher the pH, the more basic. Thus, the HBr solution is strongly acidic; the C_6H_6OH solution is weakly acidic; the C_2H_7N solution is weakly basic; and the KOH solution is strongly basic.

REVIEW QUESTIONS FOR MODULE #15

1. What is the definition of chemical equilibrium?

2. Does a chemical reaction actually stop once equilibrium is reached?

3. A chemist is studying two chemical equilibria. The first one has an equilibrium constant equal to 1,213 M, while the second has an equilibrium constant equal to 0.344 M^2. Which reaction will make more products?

4. After taking a cursory glance through a chemistry book, a student says that sometimes a chemical reaction is written with a single arrow and sometimes it is written with a double arrow. The student says that this means some chemical reactions are equilibrium reactions and some are not. What is wrong with the student's statement?

5. Why do we ignore solids in the equilibrium constant and when using Le Chatelier's principle?

6. Why are acid ionization reactions important in chemistry?

7. What are the limits of the pH scale?

8. Three solutions have the following pH levels:

Solution A: pH = 11
Solution B: pH = 8
Solution C: pH = 2

Which is (are) the acidic solution(s)?

9. Three acid solutions of equal concentration have the following pH levels:

Solution A: pH = 4
Solution B: pH = 6
Solution C: pH = 1

Which solution is made with the acid that has the *lowest* ionization constant?

10. Where does acid rain come from?

PRACTICE PROBLEMS FOR MODULE #15

1. For the following reaction:

$$2NH_3\ (g) \leftrightharpoons N_2\ (g) + 3H_2\ (g)$$

The equilibrium concentrations are: $[NH_3]$ = 0.24 M, $[N_2]$ = 0.45 M, and $[H_2]$ = 0.63 M. What is the value of the equilibrium constant for this reaction?

2. A chemist is studying the following equilibrium:

$$2Pb(NO_3)_2\ (s) \leftrightharpoons 2PbO\ (s) + 4NO_2\ (g) + O_2\ (g)$$

He starts out with 10 g of $Pb(NO_3)_2$ and, at equilibrium, has 2.02 g of PbO. The concentrations of NO_2 and O_2 at equilibrium are 0.18 M and 0.045 M, respectively. What is the value of the equilibrium constant?

3. Three chemical equilibria are written below, along with their equilibrium constants. If any of these equations can be written with a single arrow, do so.

a. $2NOCl\ (g) + O_2\ (g) \leftrightharpoons 2NO_2\ (g) + Cl_2\ (g)$ $K = 2.2 \times 10^{-5}$

b. $2SO_3\ (g) \leftrightharpoons 2SO_2\ (g) + O_2\ (g)$ $K = 1.6\ M$

c. $2NO_2 \leftrightharpoons N_2O_4$ $K = 9.3 \times 10^{4}\ 1/M$

4. A chemist is studying the following reaction:

$$H_2CO_3\ (aq) \leftrightharpoons CO_2\ (g) + H_2O\ (l) \qquad K = 2.3 \times 10^{4}$$

She measures the concentrations of H_2CO_3 and CO_2 when there is 1.23×10^{3} kg of water present. If their concentrations are 2.3 M and 3.5 M, respectively, is the reaction at equilibrium? If not, which way must the reaction shift to attain equilibrium?

5. The following reaction:

0,1156 0,0178 0,0784

$$2SO_2\ (g) + O_2\ (g) \leftrightharpoons 2SO_3\ (g)$$

has an equilibrium constant equal to 4.4 1/M. If the following concentrations are present: $[SO_2]$ = 0.340 M, $[O_2]$ = 0.154 M, $[SO_3]$ = 0.280 M, is the reaction at equilibrium? If not, which way must it shift to reach equilibrium?

6. Consider the following reaction that has reached equilibrium:

$$CaCO_3 (s) + H_2O (l) \leftrightharpoons H_2CO_3 (aq) + CaO (aq)$$

a. What will happen to the concentration of H_2CO_3 if extra $CaCO_3$ is added?

b. What will happen to the concentration of CaO if more H_2O is added?

c. What will happen to the amount of CaO if H_2CO_3 is removed?

7. Consider the following reaction that has reached equilibrium:

$$H_2 (g) + F_2 (g) \leftrightharpoons 2HF (g) \qquad \Delta H = 541 \text{ kJ}$$

a. What will happen to the concentration of H_2 if the temperature is raised?

b. What will happen to the concentration of HF if the temperature is lowered?

8. Consider the following reaction that has reached equilibrium:

$$N_2 (g) + 3H_2 (g) \leftrightharpoons 2NH_3 (g) \qquad \Delta H = -92.2 \text{ kJ}$$

a. What will happen to the concentration of H_2 if the temperature is raised?

b. What will happen to the concentration of NH_3 if the pressure is raised?

c. What will happen to the concentration of N_2 if the pressure is lowered?

9. What is the equation for the acid ionization constant of HCN?

10. What is the equation for the base ionization constant of SO_4^{2-}?

MODULE #16: Reduction/Oxidation Reactions

Introduction

Have you ever wondered how a battery works? You stick a D-cell into a flashlight, turn on the switch, and the light comes on. Now we all know that in order for the light to work, it must have electricity. How is the battery supplying this electricity? Does a battery *store* electricity and then just release it when the switch is turned on? If a battery does store electricity, how does it get there in the first place? What is electricity, anyway? These questions will be answered in this module. Of course, in order to be able to answer these questions, we need to cover a lot of background material, so we'll begin with that.

Oxidation Numbers

In order to understand the processes that govern batteries, we first need to learn about **oxidation numbers**. All along we've talked about two fundamentally different types of compounds: ionic and covalent. The big difference between the two is that ionic compounds are made when electrons are given up by one atom and taken by another, while covalent compounds are made when atoms share their electrons. Oxidation numbers provide a way to bridge the gap between these two types of compounds.

As we learned in Module #9, even though atoms share electrons in a covalent molecule, some atoms in the molecule have a bigger share of those electrons than others. We learned that the more electronegative an atom is, the bigger the share of electrons it receives. In other words, when two atoms share electrons, the more electronegative of the two actually ends up with the bigger portion of this electron pair. Thus, in many covalent molecules, electrons are not shared equally. As a result, there are fractional charges that exist in these molecules, and we call them "polar covalent" molecules. This is all review, but it's important that you see where we're going with this.

Since the more electronegative atoms get a larger portion of the electrons they share, it might be instructive to see what charges *would* develop on the atoms of a molecule, *if* the more electronegative atoms got *all* of the electrons they share. When we do this, we are calculating the **oxidation number** of the atoms.

Oxidation number - The charge that an atom in a molecule would develop if the most electronegative atoms in the molecule took the shared electrons from the less electronegative atoms

For example, the HCl molecule is made up of a hydrogen atom and a chlorine atom sharing one pair of electrons, as shown in the following Lewis structure:

H:C̤̈l:

Even though these atoms share a pair of electrons, the Cl is much more electronegative than the H. As a result, it gets more than its fair share of that electron pair. When we calculate oxidation numbers, we are asking what charges would develop if the Cl just took that electron pair and stopped sharing it. If that happened, the Cl would suddenly have eight electrons, and H would have none. Since Cl usually has seven electrons, the fact that it would now have eight electrons indicates that it would have a -1 charge. In the same way, a hydrogen atom usually has one electron. If the hydrogen atom were left with no electrons, it would have a +1 charge. We say, then, that the oxidation number of hydrogen in this molecule is +1 while the oxidation number of Cl is -1.

Do you see now why I say that oxidation numbers bridge the gap between ionic and covalent compounds? By calculating oxidation numbers, you are actually calculating *possible* charges on the atoms in a molecule *assuming* that the electrons were not shared but were transferred from one atom to another. Thus, oxidation numbers treat all compounds as ionic, attempting to assign charges to all atoms that make up the compound. Now remember, oxidation numbers *aren't real*. In our example above, H *does not* have a charge of +1 and Cl *does not* have a charge of -1. Oxidation numbers are a way of saying "what if?" They don't have any significance in the real world.

If oxidation numbers don't have any real physical or chemical meaning, why are we bothering to learn them? Well, it turns out that oxidation numbers are an excellent "bookkeeping" technique that will allow us to determine where electrons travel during a chemical reaction. As we will see later, knowing where electrons go during a chemical reaction is a major key to understanding the reactions we will study in this module. Even though oxidation numbers have no real physical or chemical significance to them, they are a powerful *tool* that we can use in analyzing chemical equations. Thus, we need to learn how to calculate and use them.

The first rule to learn in calculating oxidation numbers is that in any compound, the sum of all oxidation numbers (times the number of atoms that have that oxidation number) must equal the electrical charge on the compound. For example, we determined the oxidation numbers for H and Cl in the HCl molecule. The H has an oxidation number of +1, and the Cl has an oxidation number of -1. They add to zero, which is the overall charge of the molecule.

The sum of all oxidation numbers in a molecule must equal the charge of that molecule.

This simple rule can mean all the difference in calculating the oxidation numbers of atoms.

EXAMPLE 16.1

What is the oxidation number of iron in Fe?

The sum of all oxidation numbers must add up to the total charge of the substance. Since Fe has no charge, and since there is only one atom here, the oxidation number must be <u>0</u>.

What is the oxidation number of iron in Fe^{3+}?

Once again, the sum of all oxidation numbers must equal the charge. Since the charge is +3, and since there is only one atom in the substance, the oxidation number of Fe must be <u>+3</u>.

Suppose you know that the oxidation number of Mg in $MgBr_2$ is +2. What is the oxidation number of Br in that same compound?

The total charge of $MgBr_2$ is zero. This means that all oxidation numbers must add up to zero. Since the oxidation number of Mg is +2, this means that the sum of all Br's oxidation numbers must be -2. Since there are two Br's, each must have an oxidation number of -1 to make this happen.

Suppose you know that the oxidation number of H is +1 and O is -2 in the compound $HBrO_3$. What is the Br's oxidation number?

Since H has an oxidation number of +1 and there are three O's, each with an oxidation number of -2, the sum of the oxidation numbers so far is $1 + 3 \times (-2) = -5$. In order for all of the oxidation numbers to add up to the total charge (which is 0), the Br must have an oxidation number of +5.

This example brings up a good point. Notice that Br had an oxidation number of -1 in $MgBr_2$ and an oxidation number of +5 in $HBrO_3$. The same atom can have different oxidation numbers depending on the molecule it is in. Remember, oxidation number tells us the charge that Br would have if the more electronegative atoms got all of the shared electrons. In $MgBr_2$, bromine is the most electronegative atom. Thus, it gets all of the electrons, making it negative. In $HBrO_3$, however, the oxygen is more electronegative. Thus, Br loses its shared electrons and becomes positive.

Try the following problem to make sure you understand this part of calculating oxidation numbers:

ON YOUR OWN

16.1 What is the oxidation number of nitrogen in each of the following substances?

a. N
b. N^{3-}
c. NO_3^- (In this compound, O's oxidation number is -2.)

(The multimedia CD has a video demonstration about the same atom having different oxidation numbers in different compounds.)

Determining Oxidation Numbers

Of course, you need to be able to calculate the oxidation numbers of *all* atoms in a molecule. I really shouldn't have to tell you any of them. I just had you do "On Your Own" problem 16.1 as a warmup to the real thing, which is often a bit difficult. You see, the best way to calculate the oxidation numbers of all atoms in a molecule is to draw a Lewis structure of the molecule and then give the shared electrons to the most electronegative atoms. Then you can count the electrons on each atom and determine the charge they would have under those conditions. This is what I did in the previous section to show you the oxidation numbers of H and Cl in HCl.

The problems with figuring out oxidation numbers this way are pretty enormous. First of all, many of the compounds that participate in the reactions we are going to study have extremely difficult Lewis structures. As a result, first-time chemists often get very frustrated figuring them out. Also, even though you learned in Module #8 how to tell which atoms are more electronegative than others, there are many times when the rules you learned are broken. As a result, it is often very difficult to determine which atoms get the shared electrons. In the end, then, the Lewis structure method cannot really be taught to chemistry students at either the high school or college level.

Are we stuck then? Not really. You see, there are some rules we can develop that allow us to learn the oxidation numbers of certain atoms. Once you learn those rules, you can usually use those oxidation numbers to help you calculate the oxidation numbers of other atoms. For example, the first four rules are:

1. When a substance has only one type of atom in it (F_2, O_3, Ca^{2+}, or Mg for example) the oxidation number for that atom is equal to the charge of the substance divided by the number of atoms present.

2. Group 1A metals (Li, Na, K, Rb, Cs, and Fr) always have oxidation numbers of +1 in molecules that contain more than one type of atom.

3. Group 2A metals (Be, Mg, Ca, Sr, Ba and Ra) always have oxidation numbers of +2 in molecules that contain more than one type of atom.

4. Fluorine always has a -1 oxidation number in molecules that contain more than one type of atom.

What's nice about these four rules is that they are *always true*. The problem is, after these four rules, things get a little fuzzy, because there are a few exceptions to the next rules. In general, though, the next rules *usually* apply:

5. When grouped with just one other atom that happens to be a metal, H has an oxidation number of -1. In *all other cases* in which it is grouped with other atoms, H has an oxidation number of +1.

6. Oxygen has an oxidation number of -2 in molecules that contain more than one type of atom.

In fact, there are exceptions to both rule #5 and rule #6. However, you will be happy to know that for this course, we will ignore those exceptions. Thus, as far as you are concerned, rules 1-6 always are true. After that, though, things get really fuzzy. If you are really stuck and none of these rules apply, you can follow this general (and often not true) guideline:

7. If all else fails, assume that the atom's oxidation number is the same as what it would take on in an ionic compound. The atoms that are most likely to follow this rule are in groups 3A, 6A, and 7A.

This means that you might be able to assume that chlorine has a -1 oxidation number because it is in group 7A, and those atoms take on a -1 charge in ionic compounds. Similarly, it might be a good guess that S has a -2 oxidation number because all group 6A atoms take a -2 charge in ionic compounds. This general principle, however, has so many exceptions that it should only be used as a last resort!

In the end, we have six rules that you can assume will always work, and one general principle that has many exceptions but is a good guess when all else fails. How do we use these rules to determine oxidation numbers? In general, we end up trying to determine the oxidation numbers of all atoms using rules #1-6. If that doesn't work, before we resort to rule #7, we see if rules #1-6 apply to all but one of the atoms in the molecule. If they do, then we can determine the oxidation number of the last atom the same way we did in the previous section, by using the condition that all oxidation numbers have to add up to the charge on the molecule. Thus, you use rule #7 *only* when rules #1-6 do not apply to either all or all but one of the atoms in the molecule. This may sound really complex right now, but some practice will make you a virtual master at this!

EXAMPLE 16.2

What is the oxidation number of each atom in the following molecules?

a. H_2

According to rule #1, H's oxidation number is 0 divided by 2, which is 0.

b. $Ca(OH)_2$

In this compound, we can apply our rules to every atom. Ca is a group 2A metal, so its oxidation number is +2 (rule #3). Rule #6 tells us that oxygen has an oxidation number of -2. Finally, since H is not grouped with just one other atom, its oxidation number is +1 (rule #5). As a check, we need to make sure that these numbers add up to the total charge (0) of the molecule. There are one Ca, two O's, and two H's in the molecule. Adding the oxidation numbers:

$$(+2) + 2 \text{ x } (-2) + 2 \text{ x } (+1) = 0.$$

Thus, the oxidation numbers are: Ca: +2, O: -2, H: +1.

c. F_2CO

Rule #4 tells us that F is always -1. Rule #6 says that oxygen is -2. We have no rule for C, but that's okay. Since we know all but one of the atoms, we can figure out C by making sure that the oxidation numbers add up to the overall charge of the molecule (0). There are two F's and one O, so the total oxidation number so far is 2 x (-1) + (-2) = -4. To make all of the numbers add up to 0, the C must have an oxidation number of +4. Thus, the oxidation numbers are: F: -1, C: +4, O: -2.

d. $NaSO_4^-$

Rule #2 tells us that Na's oxidation number is +1, since it is a group 1A metal. Rule #6 says that O must have a -2 oxidation number. Once again, we have no rule for S, but we can figure out its oxidation number because the sum of all oxidation numbers must equal the charge of the molecule, which is -1. Since Na is +1 and O is -2, we have (+1) + 4 x (-2) = -7. In order for the sum of oxidation numbers to be -1, S must have an oxidation number of +6. Thus, the oxidation numbers are Na: +1, S: +6, O: -2.

e. C_2H_4

According to rule #5, when H is grouped with just one other atom that happens to be a metal, its oxidation number is -1; otherwise it is +1. In this molecule, H is grouped with just one other atom, but that atom is not a metal. Thus, H's oxidation number is +1. We have no rules for C, but we know that the sum of oxidation numbers must equal the overall charge (0). Since H is +1, we have a +4 sum so far. Thus, in order for them to add to zero, the two C's must account for a sum of -4. Thus, each individual C must have a -2 oxidation number. In the end, then, the oxidation numbers are C: -2, H: +1.

f. NCl_3

This is the first time that rules #1-6 do not apply. As a result, we must fall back on our general principle (#7). According to this principle, groups 3A, 6A, and 7A are the most likely to follow the rule. This molecule contains a group 7A atom, so we can apply the principle to it. This means that Cl will have an oxidation number of -1. If you don't understand how I got this, you need to review Module #8, where we learned how to predict the charge of an atom in an ionic compound. Since Cl has an oxidation number of -1, and since there are three of them, N must have an oxidation number of +3 in order for the oxidation numbers to add up to the total charge (0). Thus, the oxidation numbers are: N: +3, Cl: -1.

g. $MgCoBr_4$

In this compound, we know that Mg is +2 because it is a group 2A metal (rule #3). For the other two, rules 1-6 do not apply. Thus, we are stuck with rule #7. We will apply this to Br for two reasons. First, the rule states that group 7A is a group that likely follows the rule, and Br is in group 7A. Also, Co is not in an "A" group, and our rules for determining charge apply only to groups 1A - 8A. Thus, Br is the only atom we can apply this rule to. Since Br is in group 7A, it develops a -1 charge when in ionic compounds. That leaves us to determine Co's oxidation number by making use of the fact that all oxidation numbers add up to the overall charge (0). Thus, the oxidation numbers are Mg: +2, Co: +2, Br: -1.

Try this yourself to make sure you can calculate oxidation numbers:

ON YOUR OWN

16.2 What are the oxidation numbers of all atoms in the following compounds?

a. $LiNH_2$ b. N_2H_2 c. $Ca(NO_2)_2$ d. CO_2 e. BF_4^- f. PO_4^{3-} g. ClNO h. S_8

Before you go on in this module, make sure that you understand how to calculate the oxidation numbers of all atoms in a molecule. The rest of this module depends on you knowing this important skill. If you are still having trouble at this point, please find another chemistry book and see how it explains this important topic. You must be able to do this in order to go on!

Oxidation and Reduction

Now that you have learned how to calculate oxidation numbers, we can go on to learn about reduction/oxidation reactions, which are often called **redox reactions**. The first thing we must do to accomplish this is to learn the definitions of these terms:

Oxidation - The process by which an atom loses electrons

Reduction - The process by which an atom gains electrons

It's so important for you to remember these two definitions that I will give you a mnemonic to help you. Leo is a name often used to refer to a lion. When lions are angry (when they study chemistry, for example) they growl. Thus, we could say:

LEO says GER

"LEO" stands for "Lose Electrons Oxidation" and "GER" stands for "Gain Electrons Reduction." Use this mnemonic to help you keep these definitions straight.

In order to understand redox reactions, you must be able to recognize when an atom is undergoing oxidation and when it is undergoing reduction. This is why oxidation numbers are so important. Let's say an atom has an oxidation number of +5. If it undergoes a chemical reaction, it will end up in a different molecule; thus, its oxidation number might change. Let's suppose it does. Let's suppose that after a chemical reaction, its oxidation number changed to +3. What happened? Well, to go from +5 to +3, the number of electrons owned by this atom changed. Since electrons are negative, the atom must have *gained* two electrons. That's the only way to change the oxidation number from +5 to +3. Since the atom gained electrons (GER), we say that it underwent reduction. As a point of terminology, when an atom undergoes reduction, we often say that it has been **reduced**. Similarly, when an atom undergoes oxidation, we say that it has been **oxidized**.

Thus, in order to determine whether an atom is oxidized or reduced, we must look at how its oxidation number changes. We then determine whether negatives had to be added to or subtracted from the original oxidation number in order to get the new one. If negatives must be added, electrons were gained, and therefore the atom was reduced. If negatives must be subtracted, the atom lost electrons and therefore was oxidized. Study the example and perform the "On Your Own" problems that follow to cement these ideas in your head.

EXAMPLE 16.3

A sulfur atom goes from an oxidation number of -2 to one of +6. Has it been oxidized or reduced? How many electrons participated in this change?

Remember, to determine oxidation or reduction, we must think in terms of electrons. Thus, we have to ask ourselves whether we must add negative numbers to or subtract negative numbers from the original oxidation number to get the final one. In order to go from -2 to +6, negative numbers would have to be *subtracted*. This means that the atom *lost* electrons, therefore it was oxidized.

The second question simply asks how many electrons it took to make the oxidation happen. Well, in order to go from -2 to +6, I must subtract a -8, because:

$$-2 - (-8) = +6$$

Thus, the sulfur must have lost eight electrons.

A carbon atom goes from a -2 oxidation number to a -4 oxidation number. Was it oxidized or reduced? How many electrons participated in the change?

In terms of electrons, there is only one way to go from -2 to -4. You must gain negatives. Thus, you must gain electrons. Therefore, the atom was reduced. To see how many electrons participated, we just figure out how many negatives must be gained to go from -2 to -4:

$$-2 + (-2) = -4$$

Thus, two electrons participated in the change.

c. An atom goes from a +5 oxidation number to +3. Was it oxidized or reduced? How many electrons participated in the change?

In order to go from +5 to +3, we must add two negatives. Since negatives were added, the atom was reduced. Since it took two negatives, two electrons participated in the change.

d. An atom goes from a -6 oxidation number to -1. Was it oxidized or reduced? How many electrons participated in the change?

In order to go from -6 to -1, five negatives must be subtracted. Since negatives must be subtracted, the atom lost electrons. Therefore, it was oxidized. Since five negatives must be subtracted, five electrons participated in the change.

ON YOUR OWN

16.3 An atom changes its oxidation number from +5 to +1. Was it oxidized or reduced? How many electrons participated in this change?

16.4 An atom goes from an oxidation number of -6 to -3. Was it oxidized or reduced? How many electrons participated in this change?

Recognizing Reduction/Oxidation Reactions

We are now finally ready to actually study redox reactions. We call a chemical reaction a redox reaction if, during the process of the chemical reaction, an atom's oxidation number changes. For example, in the following reaction:

$$Mg\ (s) + 2HCl\ (aq) \rightarrow H_2\ (g) + MgCl_2\ (aq)$$

The oxidation numbers of both Mg and H changed. On the reactants side of the equation, Mg has an oxidation number of 0, H is +1, and Cl is -1. On the products side of the equation, however, H has an oxidation number of 0. In addition, Mg has an oxidation number of +2, while Cl still has an oxidation number of -1. Let me rewrite this reaction, putting the oxidation number above each atom so that you can see this clearly:

$$\overset{0}{Mg}\ (s) + 2\overset{+1}{H}\overset{-1}{Cl}\ (aq) \rightarrow \overset{0}{H_2}\ (g) + \overset{+2}{Mg}\overset{-1}{Cl_2}\ (aq)$$

So you see that Mg went from an oxidation number of 0 to +2, while H went from an oxidation number of +1 to 0. We could therefore say that Mg was oxidized while H was reduced. This makes the reaction a redox reaction. Notice that the oxidation number of chlorine did not change. This just means that Cl was neither oxidized or reduced. Thus, not every atom in a redox reaction needs to change oxidation numbers. If *any* atoms change oxidation numbers in a reaction, it is a redox reaction.

Compare that reaction to the following one:

$$NaOH\ (aq) + HCl\ (aq) \rightarrow H_2O\ (l) + NaCl\ (aq)$$

In this reaction, no atom changed oxidation number. On the reactants side, the oxidation numbers are: Na: +1, O: -2, H: +1 (in both molecules), and Cl: -1. On the products side, the oxidation numbers are still: Na: +1, O: -2, H: +1, and Cl: -1.

$$\overset{+1}{Na}\overset{-2}{O}\overset{+1}{H}\ (aq) + \overset{+1}{H}\overset{-1}{Cl}\ (aq) \rightarrow \overset{+1}{H_2}\overset{-2}{O}\ (l) + \overset{+1}{Na}\overset{-1}{Cl}\ (aq)$$

Thus, in this reaction, no atom changed its oxidation number, so this is not a redox reaction. Make sure that you can identify redox reactions by studying the example and solving the "On Your Own" problem that follows.

EXAMPLE 16.4

For each of the following, determine whether or not it is a redox reaction. If it is a redox reaction, identify the atoms whose oxidation numbers changed and tell whether they were oxidized or reduced.

a. $Ca (s) + Cl_2 (g) \rightarrow CaCl_2 (s)$

On the reactants side, the oxidation numbers are as follows: Ca: 0, Cl: 0. On the products side, the oxidation numbers are: Ca: +2, Cl: -1. The oxidation numbers did change, so this is a redox reaction. Since Ca went from 0 to +2, negatives were lost. This means that Ca was oxidized. On the other hand, Cl went from 0 to -1, indicating that negatives were gained. This means that Cl was reduced.

b. $Ag^{+} (aq) + OH^{-} (aq) \rightarrow AgOH (s)$

On the reactants side, the oxidation numbers are: Ag: +1, O: -2, H: +1. On the products side, Ag: +1, O: -2, H: +1. Thus, this is not a redox reaction.

c. $3IF_5 (aq) + 2Fe (s) \rightarrow 2FeF_3 (aq) + 3IF_3 (aq)$

On the reactants side, the oxidation numbers are: F: -1, I: +5, Fe: 0. On the products side, the oxidation numbers are Fe: +3, F: -1 (in both molecules), I: +3. Thus, this is a redox reaction, because Fe and I changed oxidation numbers. Since Fe went from 0 to +3, it lost three electrons. This means that Fe was oxidized. On the other hand, I went from +5 to +3, indicating that it gained two electrons. Thus, I was reduced.

If you purchased the MicroChem kit discussed in the introduction, you can perform Experiment #6 in that kit. It will give you some experience working with redox reactions.

ON YOUR OWN

16.5 For each of the following, determine whether or not it is a redox reaction. If it is a redox reaction, identify the atoms whose oxidation numbers changed and tell whether they were oxidized or reduced.

a. $Cu^{2+} (aq) + Zn (s) \rightarrow Zn^{2+} (aq) + Cu (s)$

b. $2H_2O \rightarrow 2H_2 + O_2$

c. $NaCl + AgF \rightarrow AgCl + NaF$

d. $Ca (s) + 2HNO_3 (aq) \rightarrow Ca(NO_3)_2 (aq) + H_2 (g)$

An Important Characteristic of Reduction/Oxidation Reactions

Hopefully you noticed something while studying the example and solving the "On Your Own" problems. In each one of the problems, if the reaction was a redox reaction, there were always two atoms that changed their oxidation numbers. In addition, one of those atoms was always oxidized and the other was always reduced. This is, in fact, the way it has to happen, and that's why we call them reduction/oxidation reactions:

When a redox reaction occurs, there will always be one atom that is oxidized and one atom that is reduced.

Thus, just like acids can only work if there is a base (and we therefore call those reactions acid/base reactions), reduction can only occur if oxidation occurs as well (and we therefore call those reactions reduction/oxidation reactions).

Now if you think about this, it should make sense. After all, an oxidation number tells us what charge an atom would take if it got full control of the electrons for which it already owns the largest share. Thus, if the oxidation number of an atom changes, it has, in effect, lost or gained electrons. Well, since electrons cannot simply be created out of thin air, if one atom gains electrons, another atom has to lose them. Thus, for an atom to be reduced (gain electrons), another atom must be oxidized (lose electrons).

The fact that atoms gain and lose electrons in a redox reaction forms the basis of how a battery works. Therefore, to understand batteries, you need to understand this point. Whereas many chemical equations involve simply the *rearrangement* of electrons, redox reactions involve the *transfer* of electrons. In a redox reaction, one atom gains electrons and another loses. In the next section, you will see how, under the proper circumstances, this ends up making a battery. Before you learn that, however, do the following experiment that utilizes a reduction/oxidation reaction.

EXPERIMENT 16.1
Invisible Writing

Supplies:
- Large bowl
- Blank notebook paper without lines
- Scissors
- Cotton swab or small paintbrush
- Lemon juice (This must be *real* lemon juice, either fresh or from concentrate.)
- Iodine solution (This is available in large drugstores. The pharmacy department within your supermarket will probably not have this and neither will a small corner drugstore. A major chain will have it, however. Please note that *iodide* will not work. Iodide, as you should know, is I^-, while iodine is I_2.)
- Water
- Measuring cups
- Medicine dropper
- Stirring rod (A **plastic** spoon or knife will work as well. Do not use any flatware around the iodine solution, as iodine will rust metal.)
- Safety goggles

1. Take the notebook paper and cut it into a square that will fit in the bottom of the bowl.
2. Wet your cotton swab in the lemon juice and begin using it to write your first name on the paper. The lemon juice will appear clear as you write with it, but you will be able to see what you are writing because the cotton swab will leave a wet trail. Make sure the trail left by the cotton swab is wet enough so that you can easily see what you are writing. You will probably have to dip your

cotton swab in the lemon juice several times. Once you have finished writing your name in this way, allow the paper to dry completely, so that your name has essentially vanished.

3. While you are waiting for the paper to dry, fill the bowl with 1 cup of water.
4. Add to that water 40-50 drops of iodine solution. Mix the water and iodine together thoroughly with the stirring rod.
5. Once the paper is dry, examine it. The name that you wrote on the paper should be pretty much invisible.
6. Put the paper (face up) into the iodine/water solution. Be sure to fully submerge the paper into the solution.
7. Observe what happens. You should see the paper slowly start to turn blue. This is because all paper has starch in it. Starch and iodine react to form a very complex molecule that happens to be blue in color. As you wait, however, you should see that the name you wrote with lemon juice will not turn blue. As a result, the paper will be blue except where you wrote your name. Your name, therefore, will be visible.
8. Clean up your mess.

What happened in the experiment? The blue iodine/starch compound could form only where there was no lemon juice. This is because in order to form this compound, the iodine must have an oxidation number of 0. In an iodine solution, the I_2 molecule is present, which has iodine with an oxidation number of 0. However, when citric acid (an acid found in lemon juice) reacts with iodine, it *oxidizes* iodine to its +5 oxidation number. As a result, wherever the citric acid is, there is no iodine with a 0 oxidation number. Thus, the blue iodine/starch compound could not form wherever the citric acid was, leaving your name white.

How Batteries Work

Now that we have taken the time to learn how to analyze redox reactions, we are finally able to see how a battery works. Remember from the last section that redox reactions occur because electrons are transferred from one chemical compound to another. For example, consider the following simple equation:

$$Cu^{2+}\ (aq)\ +\ Zn\ (s)\ \rightarrow\ Zn^{2+}\ (aq)\ +\ Cu\ (s)$$

This is clearly a redox reaction because the reactants side has Cu with an oxidation number of +2 and Zn with an oxidation number of 0. On the products side, however, the Cu now has a 0 oxidation number while Zn has the +2 oxidation number. Thus, in this reaction, Cu was reduced and Zn was oxidized.

Now the reason that this happens is that Zn gives two electrons to the copper. Thus, electrons are transferred between zinc and copper. So, if I were to put solid zinc into a solution of Cu^{2+} ions, I would slowly see the solid zinc disappear, because it would begin to turn into aqueous Zn^{2+}. Remember, aqueous means "dissolved in water," so once the zinc turns into Zn^{2+}, it dissolves into the water. To an observer, then, it would look like the zinc was disappearing. At the same time, I would start to see solid copper begin to appear. Before the reaction, the copper was dissolved as Cu^{2+}; however, when the reaction begins, it forms solid copper, which would settle down at the bottom of the container. Thus, if I *mixed* these two chemicals together, I could observe this reaction.

Rather than mixing the two chemicals, however, what if I just put them in two separate beakers and placed the beakers next to each other, like this:

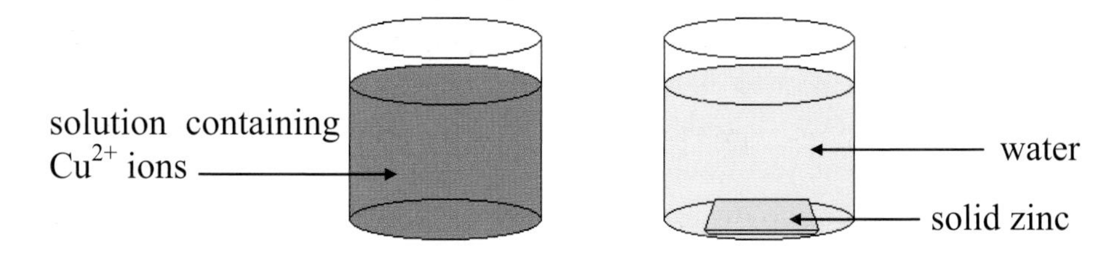

Would anything happen? Of course not! In order for the reaction to occur, the reactants would have to get close enough together to exchange electrons. Since they can't even touch each other, there's no way to get the electrons transferred. Or is there?

Suppose we were to put the two chemicals in contact, but not directly. Suppose instead of mixing the chemicals together, we connected them with wires as illustrated in Figure 16.1:

FIGURE 16.1
A Simple Battery

Illustration by Megan Whitaker

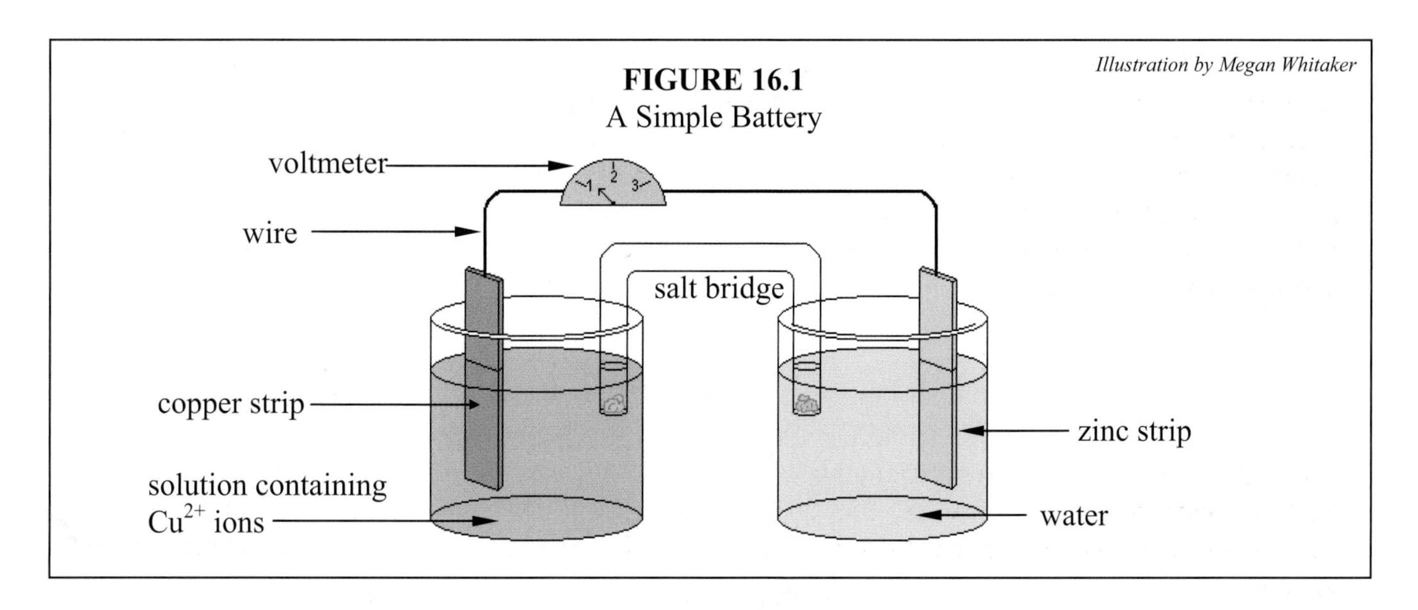

Believe it or not, under these circumstances, *the reaction would proceed!* You see, in order to get this reaction to work, all you need to do is get electrons from the zinc to the Cu^{2+}. The wire allows this to happen, because it conducts electrons. Thus, electrons will flow out of the zinc electrode, through the wire, into the copper strip, and into the Cu^{2+} solution. When they leave the zinc, the zinc becomes Zn^{2+}. As the electrons arrive in the Cu^{2+} solution, the Cu^{2+} turns into Cu. Thus, the reaction occurs, despite the fact that the reactants are in different containers! If you watched this setup for a while, you would notice that the zinc strip would slowly disappear as it turned into Zn^{2+} and dissolved into solution. You would also notice the copper strip get thicker, as Cu^{2+} turned into copper right on the copper strip.

Electricity, of course, is just electrons traveling through a wire. Since this chemical reaction causes electrons to run through a wire, it is "making" electricity. Thus, this setup can be used to run an electrical device. For example, Figure 16.1 shows a voltmeter connected to the wire. A voltmeter measures the potential difference between two points. For example, if you hook a voltmeter up to the ends of a D-cell battery, the voltmeter will read about 1.5 volts. This indicates the voltage of the battery. In Figure 16.1, notice that the voltmeter is reading just slightly more than one volt. Thus, this chemical reaction is supplying almost the same voltage as a D-cell battery.

Now, of course, things are a bit more complicated than this. For example, notice the thing labeled "salt bridge" in the figure. In order for this battery to work, the salt bridge must be there, but its job is a very difficult one to explain. Think about it this way. When the first few electrons leave the solid zinc, it turns into Zn^{2+}. As a result, the entire solution in the beaker containing zinc develops a positive charge. As the solution in the beaker becomes more and more positive, electrons will be less and less likely to leave it, because that positive charge will attract the electrons, keeping them from leaving. In the same way, as electrons reach the solution containing Cu^{2+}, that solution will start becoming negative. This will repel electrons traveling down the wire, making them less likely to get there. In the end, then, after a few electrons were transferred in this way, the battery would stop working, because electrons could no longer travel across the wire.

The salt bridge keeps this from happening. Remember, "salt" is just another term for "ionic compound." Thus, a salt bridge is just made up of an aqueous ionic compound. When ionic compounds are aqueous, their ions break apart and are able to move independently. Thus, when the beaker that holds the zinc begins to become positively charged, the negative ions in the salt bridge travel to that solution. This cancels out the positive charge that is developing and, as a result, more electrons are free to travel down the wire. In the same way, as the solution that contains the Cu^{2+} becomes more negative, the positive ions in the salt bridge are attracted to that beaker. This cancels out the negative charge that is developing, making sure that electrons are not hampered as they move towards the Cu^{2+}. If you don't quite understand all of this, don't worry. Just keep in mind that a salt bridge is something that contains an aqueous ionic compound and in order for the battery to work, the salt bridge must be there to balance out the charges that otherwise would develop in the beakers.

Look again at Figure 16.1. Make sure you understand what's going on here. The electrons leave the zinc. We know this happens because, according to the chemical equation presented earlier, zinc's oxidation number is changing from 0 to +2. This means that the zinc is losing electrons. We know that the electrons are traveling to the Cu^{2+} solution because the equation tells us that the copper's oxidation number is changing from +2 to 0. This means that the copper is being reduced, which tells us that it is gaining electrons. This is why we know that electrons flow from the zinc to the copper.

Now remember, this is a very simple picture of a battery. If you look at a battery, you will see that somewhere, the ends are labeled as positive (+) and negative (-). In the battery pictured in Figure 16.1, which beaker would be the positive side and which would be the negative side? Well, electrons are flowing away from the zinc. Since electrons are negative, they are repelled by other things that are negative. This means that the container of zinc can be thought of as negatively charged. In the same way, since electrons travel towards the container of Cu^{2+}, it must be positive, because electrons travel towards positive charges.

Before we can continue, we need to nail down some terminology. Chemists call the negative side of the battery the **anode** (an' ohd). This is because it is a source of negative ions, which are usually called **anions** (an' eye uns). The positive side of the battery is called the **cathode** (kath' ohd), because it is the place where you can find positive charges, which are usually called **cations** (kat' eye uns). Also, the term "battery" is a bit too specific for chemists, because there are many different types of chemical reactions like this. Thus, chemists call this kind of setup a **Galvanic cell**, after Italian physiologist Luigi Galvani, who first discovered that electricity can flow between two metals. He actually discovered this while dissecting a frog. He noticed that when his steel scalpel touched a brass hook, the frog's leg would twitch. He realized that this meant he was somehow making electricity. Of course, now we know that a redox reaction occurred between the two metals, causing electricity to flow through the frog's nerves, activating a response in the frog's muscles. Finally, the metal strips are usually called **electrodes** (ee lek' trohds), which means that they form the electrical contact between the wire and the contents of the beakers.

In the end, then, we can label our picture as is done in Figure 16.2:

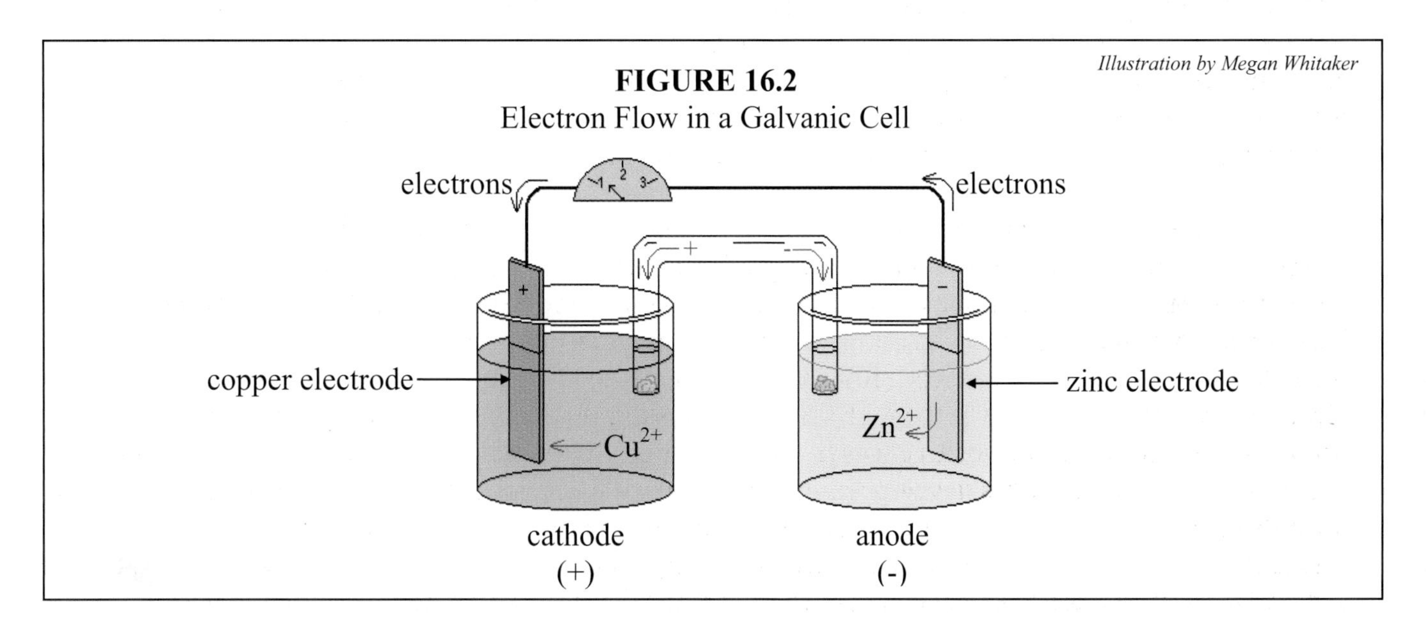

FIGURE 16.2
Electron Flow in a Galvanic Cell

From your own experience, you know that batteries eventually stop working. Now, hopefully, you can see why. A battery produces electricity because electrons flow from one reactant in the battery's chemical reaction to another reactant. Thus, as long as the reaction can proceed, the battery will produce electricity. However, once one of the reactants is completely used up, the chemical reaction stops. This stops the flow of the electrons and causes the battery to go dead. Thus, a battery does not "store" electricity. It stores chemicals. The electricity a battery produces is made through the reaction of those chemicals. When the chemicals get used up, the battery stops working.

What about rechargeable batteries? When a battery uses up its chemicals, it is dead. How does a battery recharger work? Well, the details of this are a bit too complicated to get into here, but I can provide you with an abbreviated explanation. A battery charger produces an electrical force that actually causes the battery's chemical reaction to reverse. Thus, when a battery is dead, it has all products and no reactants. A battery recharger will force those products to react with each other and reform the reactants. Once that happens, the battery is ready to go again. This doesn't work for all batteries, however. That's why many batteries are not rechargeable.

Now, of course, a real battery is a little more complicated than the simple picture we have discussed so far but, nevertheless, all of the principles that exist in this simple battery are the same as those that exist in a real one. In a moment, I will show you some examples of real batteries and how they work. First, however, I want you to get comfortable with analyzing a Galvanic cell. Given the chemical equation, I would like you to be able to draw a figure similar to Figure 16.2, labeling the anode and cathode and identifying the electron flow. This will take a bit of practice, so study Example 16.5 and then solve the "On Your Own" problems that follow.

EXAMPLE 16.5

A Galvanic cell runs on the following chemical reaction:

$$2Ag^{+}\,(aq) + Mg\,(s) \rightarrow 2Ag\,(s) + Mg^{2+}\,(aq)$$

Draw a diagram of the Galvanic cell, indicating the electron flow, labeling the anode and cathode, and indicating the positive and negative sides of the Galvanic cell.

The first thing that we have to do is look at the reactants and see which is being oxidized and which is being reduced. Since Ag^{+} turns into Ag, it must be gaining electrons, so it is reduced. This means that the container that holds the Ag^{+} will have the electrons flowing into it. The Mg, on the other hand, loses electrons, because it goes from an oxidation number of 0 to an oxidation number of +2. This means that the Mg is oxidized, and the container that holds it will have electrons leaving it. Since the electrons are flowing from the Mg to the Ag^{+}, the container that holds Mg is negative (electrons are repelled by negative charge) and is therefore the anode. The container that holds the Ag^{+} is positive (electrons are attracted to the positive charge), and is therefore the cathode. In the equation, Ag^{+} is aqueous. This means that the Ag^{+} is dissolved, so the solution that holds it will be clear. The Mg, on the other hand, is solid, so there should be a magnesium electrode in its beaker. That will be the source of the solid magnesium. As always, there must be a salt bridge as well. In the end, then, the picture should look something like this:

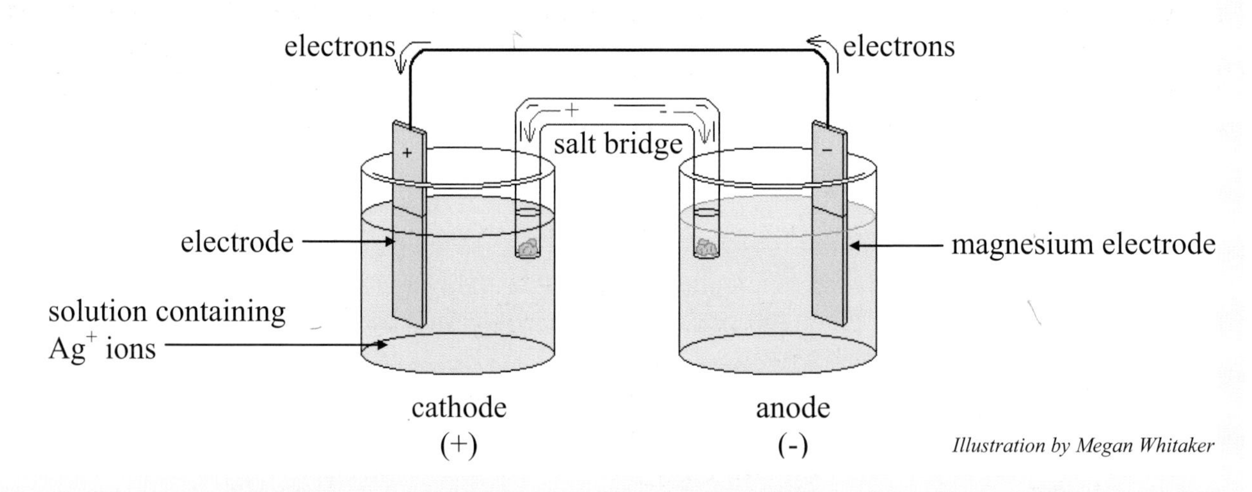

Illustration by Megan Whitaker

Notice that I did not label the electrode that is in the Ag^{+} solution. That's because it does not need to be any specific metal. It just needs to conduct electricity. The electrode on the right must be magnesium, because solid magnesium is a reactant in the equation. Thus, solid magnesium must be present. Since the only other reactant needed is Ag^{+}, and since Ag^{+} is in the solution, there is no need for the electrode in that solution to be silver. It can be any metal. Usually, we do use the metal version of the ion in that ion's beaker, but we need not. Since silver is kind of expensive, a chemist would probably not use silver as the electrode in this case. Thus, when you are drawing Galvanic cells, you need not describe what an electrode is made of unless the electrode *must* be made of something. In this problem, for example, you would be required to label the magnesium electrode on the right, because that electrode must be made of magnesium. You would not be required to label the metal that makes up the electrode on the left, however, since any metal would do.

Finally, notice that I also did not include the voltmeter in the diagram. After all, the voltmeter is not a necessary part of the Galvanic cell. Electrons will flow through the wire with or without the voltmeter being there. I just used it in Figures 16.1 and 16.2 to emphasize that electricity will flow through the wires. It would be hard for you to draw the voltmeter, because you would need to determine the voltage, which is different for every Galvanic cell. It turns out that there is a way to do that, but I don't want to teach it here. If you take advanced chemistry later on, you will learn how to calculate Galvanic cell voltages in that course.

(The multimedia CD has a video demonstration about a reaction very similar to the one you just analyzed.)

A Galvanic cell runs on the following chemical reaction:

$$I_2 \text{ (aq)} + Fe(s) \rightarrow 2I^- \text{ (aq)} + Fe^{2+}$$

Draw a diagram of the Galvanic cell, indicating the electron flow, labeling the anode and cathode, and indicating the positive and negative sides of the Galvanic cell.

In this reaction, iodine is going from an oxidation number of 0 to an oxidation number of -1. This indicates that it is gaining electrons. Thus, the solution holding aqueous iodine will have electrons flowing into it. This makes that container positive (electrons are attracted to a positive charge), and it will be the cathode. The Fe is going from an oxidation number of 0 to an oxidation number of +2. This means it loses electrons. Since it is losing electrons, the electrons are flowing away from the container holding the solid Fe. This makes that container the negative side of the Galvanic cell, and it is therefore the anode. The picture, then, looks like this:

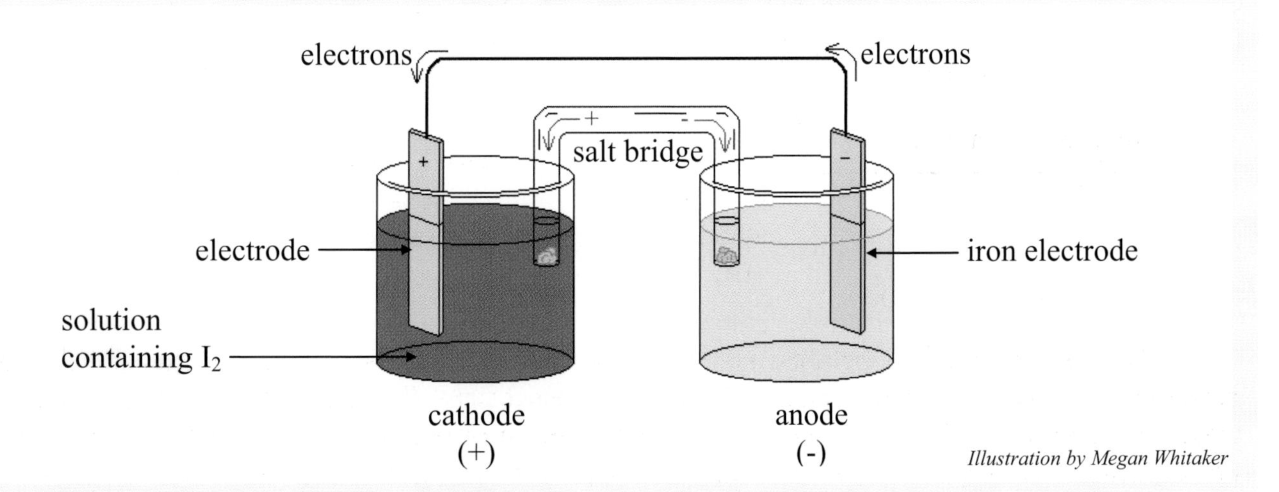

Illustration by Megan Whitaker

Notice once again that the metal that makes up the electrode on the left is not mentioned. That's because it could be any metal, since the reactant on that side of the Galvanic cell is I_2, which is in solution. The iron electrode is labeled, however, because that electrode must be iron, since iron is a reactant in the equation. Also, notice the color of the solution that contains I_2. I used that color because it is roughly the correct color for such a solution. I would not expect you to know that, however, so you need not use colors in your diagrams.

If you purchased the MicroChem kit discussed in the introduction, you can perform Experiment #17, which will give you some experience with making simple Galvanic cells.

ON YOUR OWN

16.6 A Galvanic cell runs on the following chemical equation:

$$2H^{+}\ (aq)\ +\ Mn\ (s)\ \rightarrow\ H_2\ (g)\ +\ Mn^{2+}\ (aq)$$

Draw a diagram for this Galvanic cell, labeling the electron flow, the anode and cathode, and the positive and negative sides of the Galvanic cell.

16.7 A Galvanic cell runs on the following chemical equation:

$$3Cl_2\ (aq)\ +\ 2Al\ (s)\ \rightarrow\ 2Al^{3+}\ (aq) +\ 6Cl^{-}\ (aq)$$

Draw a diagram for this Galvanic cell, labeling the electron flow, the anode and cathode, and the positive and negative sides of the Galvanic cell.

<u>Real Batteries</u>

As I said before, batteries in real life are a bit more complicated than those we have discussed here. This is primarily for two reasons. First, the reactions that run batteries are a bit more complex than those that we have studied so far. Also, since we try to make batteries as small as possible, their architecture is significantly more complicated. Nevertheless, batteries are such an important part of our everyday life that we need to at least have a rudimentary understanding of why they work.

Let's start our discussion with the battery used in your automobile. This battery is called a "**lead-acid battery**," and it runs on the following reaction:

$$PbO_2\ (s)\ +\ Pb\ (s)\ +\ 2H_2SO_4\ (aq)\ \rightarrow\ 2PbSO_4\ (s)\ +\ 2H_2O\ (l) \qquad (16.1)$$

Notice that this reaction is much more complicated than the ones we have discussed so far. Thus, we need to analyze it carefully. Let's first see what is being oxidized and what is being reduced. Here are the oxidation numbers for the atoms in the reaction:

$$\overset{+4\ -2}{PbO_2}\ (s)\ +\ \overset{0}{Pb}\ (s)\ +\ \overset{+1+6-2}{2H_2SO_4}\ (aq)\ \rightarrow\ \overset{+2+6-2}{2PbSO_4}\ (s)\ +\ \overset{+1\ -2}{2H_2O}\ (l)$$

Before I get into the analysis of what happens in this reaction, note that while you could have come up with all of the oxidation numbers on the reactants side of the equation, it would have been hard for you to come up with the oxidation numbers in $PbSO_4$. In order to get the right oxidation numbers for that compound, you would have had to recognize that it is an ionic compound made with the sulfate ion (SO_4^{2-}) and a Pb^{2+} ion. Using those ions, you could have come up with the proper oxidation numbers.

Don't worry about that, however. I would never expect you to analyze an equation that is this difficult.

Now let's look at what happened to the oxidation numbers in this reaction. First, notice that the oxidation numbers of H, O, and S did not change. Thus, they are not being oxidized or reduced. What atoms are being oxidized and reduced? The *lead* atoms are. Notice that the Pb in PbO_2 has a +4 oxidation number. On the products side of the equation, the only lead atom is in $PbSO_4$, and in that compound, Pb has an oxidation number of +2. Thus, the Pb in PbO_2 went from an oxidation number of +4 to one of +2 in $PbSO_4$. Thus, the Pb in PbO_2 gained electrons and was therefore reduced. On the other hand, the Pb in Pb (s) has an oxidation number of 0, and the only lead atom on the products side of the equation (the Pb in $PbSO_4$) has an oxidation number of +2. Thus, the Pb (s) went from an oxidation number of 0 to an oxidation number of +2. It therefore lost electrons and was oxidized.

Do you think it is strange that two different lead atoms started out with two different oxidation numbers on the reactants side of the equation and ended up with the same oxidation number on the products side of the equation? You shouldn't. After all, one lead atom (the one in PbO_2) was reduced and another lead atom (the one in solid Pb) was oxidized. Thus, one atom was oxidized and the other was reduced. They just ended up making two lead atoms that happened to have the same oxidation number!

I want to make one more point about this equation before I discuss the battery itself. Notice that H_2SO_4 is a reactant. That's why this kind of battery is called a "lead-acid battery." It uses two forms of lead [solid lead and lead(IV) oxide] as well as an acid (sulfuric acid). This acid is always present in an automobile battery, and that's why you must use it with caution. If you were to tip an automobile battery over and spill its contents on yourself, the acid would begin eating away at your clothes and skin. Thus, you must always handle automobile batteries with extreme care.

Now let's look at the way a lead-acid battery is put together. Figure 16.3 contains a simplified illustration of a lead-acid battery:

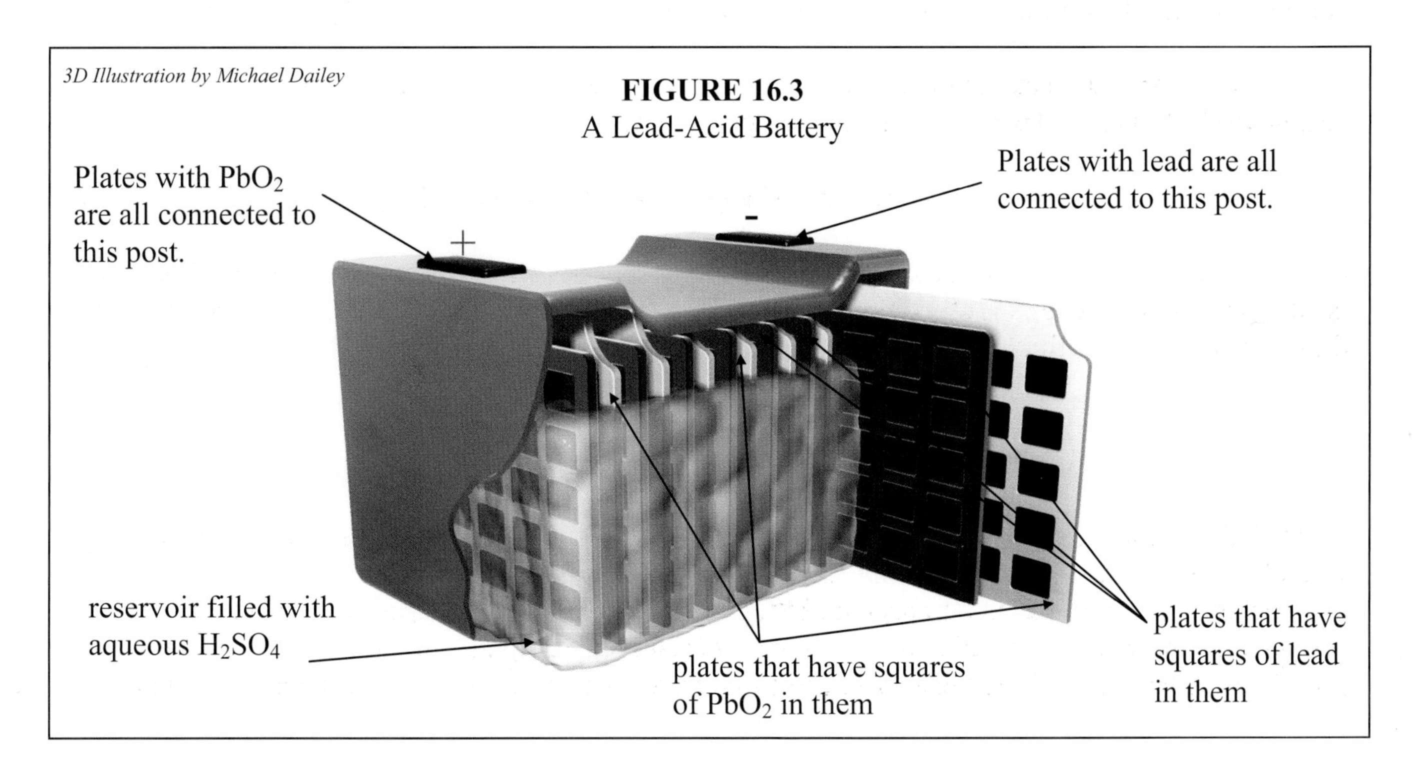

FIGURE 16.3
A Lead-Acid Battery

Notice from the figure that there are two different kinds of plates in a lead-acid battery: ones that contain squares of lead and others that contain squares of lead(IV) oxide. Those are two of the reactants in Equation (16.1). The third reactant, sulfuric acid, is in the reservoir, so the plates are constantly bathed in it. The plates containing solid lead are connected to the negative post of the battery. That's because solid lead is being oxidized in the reaction. Thus, it is losing electrons. This means that electrons flow away from the solid lead, and therefore the solid lead makes up the anode of the battery. This means (of course) that the plates containing lead(IV) oxide make up the cathode, and they are all connected to the positive post of the battery. This should make sense, since the lead(IV) oxide is being reduced. Thus, electrons flow to the lead(IV) oxide, making it the positively charged cathode.

Notice one more thing from the figure: There are several plates containing lead, and there are several plates containing lead(IV) oxide. Why are there several plates? Well, it turns out that the reaction represented by Equation (16.1) produces only about 2 volts. Most automobile batteries run at higher voltages. Thus, in order to get a higher voltage, the individual plates that make up the reaction are hooked together. If done in the right way, this allows the voltages to add. Thus, seven plates of lead hooked up to the negative side of the battery (the anode), and seven plates of lead(IV) oxide hooked up to the positive side of the battery (the cathode) will give the battery about 14 volts instead of just 2 volts.

As you probably know, a lead-acid battery is rechargeable. If you hook up a battery charger to a lead-acid battery, it will force the reaction to run in reverse, converting products to reactants so that the battery can be used again. In automobiles, once the battery is used to start the engine, it can be recharged by the alternator or generator in the engine.

Before I leave the discussion of the lead-acid battery, you might be wondering where the salt bridge is. Remember, a battery must have a salt bridge to balance out the charges that develop during the reaction. Well, remember that the plates are bathed in aqueous sulfuric acid. What happens when sulfuric acid is added to water? The following reaction occurs:

$$H_2SO_4\ (aq) + H_2O\ (l) \rightarrow H_3O^+\ (aq) + HSO_4^-$$

Notice that this reaction produces two ions: H_3O^+ and HSO_4^-. Those ions do the job of the salt bridge in a lead-acid battery. They move in response to the charges that develop during the chemical reaction, making sure that the anode and cathode do not build up excess charge that would keep the electrons from flowing.

Let's continue our discussion of real batteries by briefly discussing a typical flashlight battery. One major difference between a flashlight battery and a lead-acid battery is that a flashlight battery does not have any aqueous solutions in it. Since aqueous solutions take up far too much room, we could never make a reasonably sized flashlight battery that way. Because there are no aqueous solutions in this kind of battery, it is called a **dry cell**.

There are many, many different kinds of dry cells, so chemists must further distinguish between them. The dry cell that we use as a flashlight battery is called an **alkaline cell**, because the salt bridge it uses is made from KOH, a base. The alkaline cell uses solid zinc as the anode and MnO_2 as the cathode. As the battery begins to produce electricity, the MnO_2 turns into Mn_2O_3, and the zinc turns into Zn^{2+}. When this happens, it actually breaks down the inner construction of the battery. This is why a typical alkaline battery, unlike a lead-acid battery, cannot be recharged. The inner construction of the battery actually degrades during the reaction. Thus, even if you hooked a standard alkaline battery up to a battery recharger, it would not recharge, because the inner construction of the battery would be unable to reform.

Corrosion

Lest you think that all redox reactions are useful reactions, you should realize that corrosion, one of the most destructive chemical processes in the world, is also a redox reaction. For example, when iron is exposed to water and oxygen, the following reaction occurs:

$$2Fe\ (s) + 2H_2O\ (l) + O_2\ (g) \rightarrow 2Fe(OH)_2\ (s)$$

In this reaction, the iron goes from an oxidation number of 0 to +2, while the oxygen goes from an oxidation number of 0 to -2. Thus, iron is oxidized and oxygen is reduced, making this a redox reaction. The $Fe(OH)_2$ is further oxidized by the oxygen in the air to form Fe_2O_3, which is typically what we call "rust." This process turns strong, useful iron into brittle, useless rust. It has been estimated that replacing rusted iron and trying to prevent or slow down the corrosion process costs the United States' economy as much as 276 *billion* dollars every year! Thus, although redox reactions are very useful in making batteries, they also can be harmful.

A Few Final Words

Believe it or not, once you do the problems and take the test at the end of this module, you will be finished with your first year of chemistry. If you sit back and think about all that you've learned, you should be rather impressed with yourself. After all, you can do quite a bit more chemistry now than you could when you started the course.

If you were able to maintain an "A" or a "B" while working through these modules, you obviously have a talent for chemistry and should consider a career in it. If you earned a "C," you should not be discouraged from a career in chemistry or some other natural science, because this course is designed so that a student getting a "C" or better will, in the end, be very well prepared for college chemistry.

Regardless of your grade, however, you should have gained a lot from this course. You should be able to appreciate how *difficult* natural sciences like chemistry are. You should also have more appreciation for the amazingly intricate place that we call "earth." Along with that, you should have a much deeper respect for the wisdom and power of the One who created it.

Regardless of what you intend to do with the remainder of your academic career, I would encourage you to continue exploring creation, and continue learning more about the marvelous Creator who gave it to us!

ANSWERS TO THE "ON YOUR OWN" PROBLEMS

16.1 a. Since the overall charge is 0 and there is only one atom, the oxidation number is 0.

b. Since the overall charge is -3 and there is only one atom, the oxidation number is -3.

c. The overall charge is -1, so all oxidation numbers must add up to that. Since O is -2 and there are three O atoms, the total so far is -6. In order for the sum of oxidation numbers to be -1, N must be +5.

16.2 a. Rule #2 says Li has an oxidation number of +1. Rule #5 tells us that in this molecule, H has an oxidation number of +1. We have no rule for N, but we can figure it out because all oxidation numbers need to add up to the total charge (0). The oxidation numbers from Li and H sum up to +3. Thus, N must be -3. The oxidation numbers, then, are: Li: +1, N: -3, H: +1.

b. Rule #5 tells us that, since N is not a metal, H's oxidation number is +1. Thus, to make all oxidation numbers add up to the total charge (0), N must be -1. The oxidation numbers, then, are N: -1, H: +1.

c. Rule #3 says that Ca is +2. Rule #6 tells us that O is -2. There is one Ca and four O's, so the total oxidation number so far is (+2) + 4 x (-2) = -6. There are two N's, and they must combine to make +6 in order for the oxidation numbers to sum up to the total charge (0). Thus, each N must have a +3. The answers, then, are: Ca: +2, N: +3, O: -2.

d. Rule #6 tells us that O has a -2 oxidation number. Since the total charge is zero, C must have a +4 oxidation number. C: +4, O: -2

e. Rule #4 tells us that F is always -1. In order for the oxidation numbers to add up to the total charge (-1), that means B must be +3. B: +3, F: -1

f. Rule #6 tells us that O is -2. Since the oxidation numbers must add up to the total charge (-3), P must be +5. P: +5, O: -2

g. Rule #6 tells us that O is -2. We have no ironclad rules for N or Cl, so now we must resort to general principle #7. This works best on groups 3A, 6A, and 7A. Since Cl is in 7A, we will use it on Cl. Thus, Cl has an oxidation number of -1. In order to get all oxidation numbers to add up to the overall charge (0), N must be +3. Cl: -1, N: +3, O: -2

h. Rule #1 says that when a molecule is made up of only one type of atom, the oxidation number is the charge (0) divided by the number of atoms (8). S: 0

16.3 To go from +5 to +1, four negatives must be added. This means that electrons were gained, which means the atom was reduced. Since four negatives were added, four electrons participated.

16.4 To go from -6 to -3, three negatives must be subtracted. This means that electrons were lost, which means the atom was oxidized. Since three negatives were subtracted, three electrons participated.

16.5 a. On the reactants side of the equation, the oxidation numbers are: Cu: +2, Zn: 0. On the products side, Cu: 0, Zn: +2. This means that both Cu and Zn changed oxidation numbers, telling us that this is a redox reaction. Since Cu went from +2 to 0, it gained electrons; therefore, Cu was reduced. Since Zn went from 0 to +2, it lost electrons; therefore, Zn was oxidized.

b. On the reactants side of the equation, the oxidation numbers are: H: +1, O: -2. On the products side, H: 0, O: 0. This means that both H and O changed oxidation numbers, telling us that this is a redox reaction. Since H went from +1 to 0, it gained electrons; therefore, H was reduced. Since O went from -2 to 0, it lost electrons; therefore, O was oxidized.

c. On the reactants side of the equation, the oxidation numbers are: Na: +1, Cl: -1, Ag: +1, F: -1. On the products side, Na: +1, Cl: -1, Ag: +1, F: -1. This means that no atom changed oxidation number, telling us that this is not a redox reaction.

d. On the reactants side of the equation, the oxidation numbers are: Ca: 0, H: +1, N: +5, O: -2. On the products side, Ca: +2, H: 0, N: +5, O: -2. This means that both H and Ca changed oxidation numbers, telling us that this is a redox reaction. Since H went from +1 to 0, it gained electrons; therefore, H was reduced. Since Ca went from 0 to +2, it lost electrons; therefore, Ca was oxidized.

16.6 In this reaction, H^+ is going from an oxidation number of +1 to an oxidation number of 0. This indicates that it is gaining electrons. Thus, the solution holding aqueous H^+ will have electrons flowing into it. Since electrons are attracted to a positive charge, this must be the positive side and therefore the cathode. The Mn is going from an oxidation number of 0 to an oxidation number of +2. This means it loses electrons. Since it is losing electrons, the electrons are flowing away from the container holding the solid Mn. This makes that container the negative side of the battery, and it is therefore the anode. Since solid Mn is a reactant, the electrode on the anode must be Mn. Since the other reactant is aqueous H^+, there is no need to identify the electrode on the cathode. As always, there must be a salt bridge as well. The picture, then, looks like this:

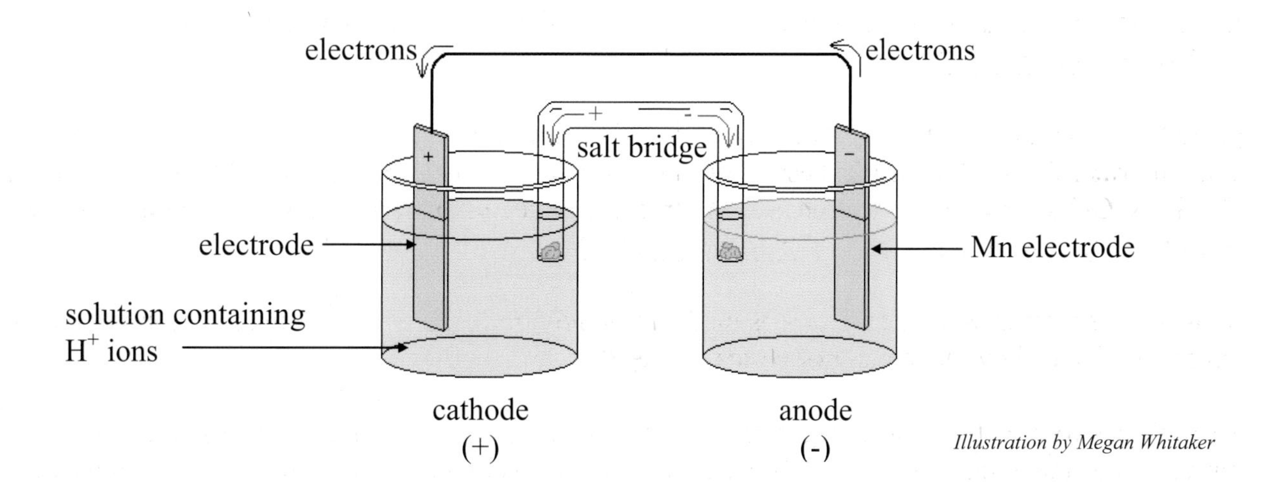

Illustration by Megan Whitaker

16.7 The first thing that we have to do is look at the reactants and see which is being oxidized and which is being reduced. Since chlorine goes from an oxidation number of 0 to an oxidation number of -1, it must be gaining electrons, so it is reduced. This means that the container that holds the chlorine will have the electrons flowing into it, so it must be positive. The Al, on the other hand, loses electrons, because it goes from an oxidation number of 0 to an oxidation number of +3. This means that the Al is oxidized, and the container that holds it will have electrons leaving it; thus, it must be negative. Since the electrons are flowing from the Al to the Cl_2, the container that holds Al is negative and is therefore the anode, while the container that holds the Cl_2 is positive and is thus the cathode. In the equation, Cl_2 is aqueous. This means that the chlorine is dissolved in water. The Al, on the other hand, is solid, so it must be used as the electrode in its container. As always, there must be a salt bridge as well. In the end, then, the picture should look something like this:

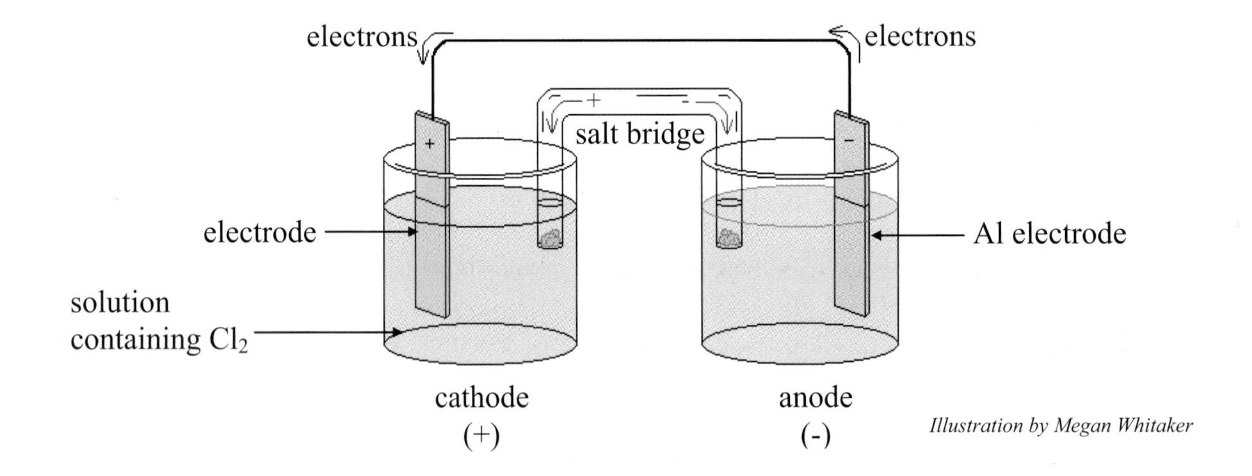

Illustration by Megan Whitaker

REVIEW QUESTIONS FOR MODULE #16

1. Define oxidation number.

2. An atom has an oxidation number of -3 in one molecule, but it has an oxidation number of +3 in a second molecule. What can we say about the electronegativities of the other atoms that make up the two molecules?

3. What must the oxidation numbers of all atoms in a molecule add up to?

4. Define oxidation and reduction.

5. Would there ever be a case in which oxidation could occur without reduction?

6. Is the anode of a battery positive or negative? What about the cathode?

7. When figuring out oxidation numbers, what rule do you use only as a last resort? Why is it only a last resort?

8. What are the three reactants needed for a lead-acid battery?

9. Why are some batteries rechargeable and others not?

10. Name two differences between an alkaline dry cell and a lead-acid battery.

PRACTICE PROBLEMS FOR MODULE #16

1. Give the oxidation numbers of all atoms in the following substances:

a. Fe b. Fe^{2+} c. Cl_2 d. O_3 e. S^{2-}

2. Give the oxidation numbers of all atoms in the following compounds:

a. MnO_2 b. H_2SO_4 c. CO_3^{2-} d. $MgCl_2$ e. KNO_3 f. SF_6 g. $IrCl_6^{3-}$ h. VS

3. An atom changes its oxidation number from +3 to -1. Is it oxidized or reduced? How many electrons did it take to do this?

4. An atom changes its oxidation number from 0 to +2. Is it oxidized or reduced? How many electrons did it take to do this?

5. An atom changes its oxidation number from -3 to 0. Is it oxidized or reduced? How many electrons did it take to do this?

6. An atom changes its oxidation number from -1 to -3. Is it oxidized or reduced? How many electrons did it take to do this?

7. Determine whether or not each of the following is a redox reaction. If it is, determine which atom is being reduced and which is being oxidized.

a. $NiO\ (s) + Cd\ (s) \rightarrow CdO\ (s) + Ni\ (s)$
b. $H_2SO_4\ (aq) + 2NH_3\ (aq) \rightarrow 2NH_4^+\ (aq) + SO_4^{2-}\ (aq)$
c. $2VO_3^-\ (aq) + Zn\ (s) + 8H^+\ (aq) \rightarrow 2VO^{2+}\ (aq) + Zn^{2+}\ (aq) + 4H_2O\ (l)$
d. $Mg(NO_3)_2\ (aq) + 2NaOH\ (aq) \rightarrow Mg(OH)_2\ (s) + 2NaNO_3\ (aq)$
e. $2Na\ (s) + Cl_2\ (g) \rightarrow 2NaCl\ (s)$

8. A Galvanic cell runs on the following reaction:

$$Co\ (s) + Cu^{2+}\ (aq) \rightarrow Co^{2+}\ (aq) + Cu\ (s)$$

Draw a diagram for this Galvanic cell, labeling the electron flow, the anode and cathode, and the positive and negative sides of the Galvanic cell.

9. A Galvanic cell runs on the following reaction:

$$Al\ (s) + Fe^{3+}\ (aq) \rightarrow Al^{3+}\ (aq) + Fe\ (s)$$

Draw a diagram for this Galvanic cell, labeling the electron flow, the anode and cathode, and the positive and negative sides of the Galvanic cell.

Glossary

The numbers in parentheses indicate the page number where the term was first discussed.

Absolute temperature scale - The Kelvin temperature scale: It is not possible to reach or go below 0 Kelvin. (46)

Accuracy - An indication of how close a measurement is to the true value (19)

Acid - A molecule that donates H^+ ions (322)

Acid ionization constant - The equilibrium constant for an acid's ionization reaction (514)

Acid ionization reaction - The reaction in which an H^+ separates from an acid molecule so that it can be donated in another reaction (514)

Activated complex - Another name for the intermediate state of a chemical reaction (436)

Activation energy - The energy necessary to start a chemical reaction (437)

Alkaline dry cell - A dry cell made with a potassium hydroxide salt bridge (544)

Amphiprotic compounds - Compounds that can act as either an acid or a base, depending on the situation (324)

Amphoteric compounds - Another name for amphiprotic compounds (324)

Amplitude - A measure of the height of the crests or the depths of the troughs on a wave (214)

Anion - A negative ion (538)

Anode - The electrode at which oxidation occurs (538)

Atom - The smallest chemical unit of matter (84)

Atomic mass - The number on the periodic chart that is below an element's symbol. It represents the average mass of the atom's isotopes in atomic mass units. (142)

Atomic mass units - The mass unit used on the periodic chart: 1.00 amu = 1.66×10^{-24} g (143)

Atomic number - The number on the periodic chart that is above an element's symbol. It represents the number of protons in the nucleus of the atom. (142)

Atomic radius - The average radius of an atom (265)

Avogadro's number - The number of molecules or atoms in a mole: 6.02×10^{23} (146)

Base - A molecule that accepts H^+ ions (322)

Bent - A molecular shape in which a central atom is bonded to two other atoms and the bonds are 105 degrees apart (293)

Boiling - The process by which a substance changes from its liquid phase to its gas phase (109)

Boiling point - The temperature at which the vapor pressure of a liquid is equal to normal atmospheric pressure (397)

Boiling-point elevation - When a solute is dissolved in a solvent, the boiling point of the solution will be higher than that of the pure solvent (374)

Bond energy - The energy required to break a bond (424)

Boyle's Law - At constant temperature, the pressure and volume of a gas are inversely related. (385)

Calibration - The process of using certain physical measurements to define the scale of a measuring device (42)

Calorie (cal) - The amount of heat necessary to warm one gram of water one degree Celsius (48)

Calorimeter - An experimental device that measures the heat released or absorbed during a chemical or physical change (54)

Calorimetry - An experimental process that measures the heat released or absorbed during a chemical or physical change (53)

Catalyst - A substance that alters the rate of a chemical reaction without changing concentration throughout the course of the reaction (479)

Catalytic converter - A device on an automobile that converts gaseous carbon monoxide produced by the engine into gaseous carbon dioxide (141)

Cathode - The electrode at which reduction occurs (538)

Cathode ray tube - Another name for a Crooke's Tube (202)

Cation - A positive ion (538)

Celsius - A temperature scale defined so that water freezes at 0° and boils at 100° (41)

Change in Enthalpy (ΔH) - The energy change that accompanies a chemical reaction (419)

Charles's Law - At constant pressure, the temperature and volume of a gas are linearly proportional. (388)

Chemical change - A change that affects the type of molecules or atoms in a substance (106)

Chemical equation - A representation of a chemical reaction (115)

Chemical equilibrium - The point at which both the forward and reverse reactions in a chemical equation have equal reaction rates: When this occurs, the amounts of each substance in the chemical reaction will not change, despite the fact that both reactions still proceed. (492)

Chemical formula - A notation that indicates the number of type of each element in a compound (86)

Chemical reaction - A process by which one or more substances change into one or more different substances (115)

Chemical symbol - An abbreviation for an element (87)

Complete combustion reaction - A reaction in which O_2 is added to a compound containing carbon and hydrogen, producing CO_2 and H_2O (138)

Compound - A substances that can be decomposed into elements by chemical means (78)

Condensing - The process by which a substance changes from its gas phase to its liquid phase (109)

Cones - The cells on the eye's retina that detect different energies of light. These cells are responsible for our ability to see colors. (223)

Continuous theory of matter - The idea that substances are composed of long, unbroken blobs of matter (69)

Covalent bond - A shared pair of valence electrons that holds atoms together in covalent compounds (265)

Covalent compound - A compound formed by atoms that share electrons (87)

Crookes tube - An experimental apparatus developed by William Crookes. It consists of a glass tube filled with a small amount of gas. Electrodes in the tube allow for the passage of electricity through the gas. (201)

Dalton's Law of Partial Pressures - When two or more ideal gases are mixed together, the total pressure of the mixture is equal to the sum of the pressures of each individual gas. (395)

Decomposition - The process by which a substance is broken down into its constituent elements (74)

Decomposition reaction - A reaction that changes a compound into its constituent elements (134)

Density - An object's mass divided by the volume that the object occupies (28)

Derived unit - A unit formed by the multiplication and/or division of other units (14)

Dilution - Adding water to a solution in order to decrease the concentration (337)

Dimensionless quantity - A quantity with no units (81)

Diprotic acid- An acid that can donate two H^+ ions (329)

Discontinuous theory of matter - The idea that substances are composed of tiny, individual particles like grains of sand (70)

Double bond - A total of four electrons shared between atoms (270)

Dry cell - A battery made with no aqueous solutions (544)

Electromagnetic radiation - Another term for light, including all wavelengths, both visible and not visible (221)

Electromagnetic spectrum - The total range of wavelengths of light that come from the sun (219)

Electron - One of the three particles that make up the atom. It is negatively charged and orbits the nucleus of the atom (205)

Electron configuration - A notation that lists the number of electrons that occupy each orbital in an atom (234)

Electron-dot diagram - Another name for a Lewis structure (250)

Electronegativity - A measure of how strongly an atom attracts extra electrons to itself (261)

Element - Any substance that cannot be decomposed into less massive substances (74)

Empirical formula - A chemical formula that tells you a simple, whole-number ratio for the atoms in a molecule (186)

Endothermic process - A process that absorbs heat (364)

Endpoint of the titration - The point in a titration where the acid and base have completely reacted with one another. It is usually accompanied by an indicator's change in color (342)

Energy - The ability to do work (37)

Energy diagram - A diagram that shows the energy of the reactants, intermediate state, and products in a chemical reaction (436)

Enthalpy of formation (ΔH_f) - The ΔH of a formation reaction (430)

Entropy - A measure of the disorder that exists in any system (439)

Equilibrium constant - A value that describes the concentrations at which a reaction comes to equilibrium. The equilibrium constant varies with temperature (496)

Excess reactant - The reactant or reactants that are left over at the end of a chemical reaction (172)

Exothermic process - A process that releases heat (364)

Extrapolation - Following an established trend in the data even though there is no data available for that region (389)

Factor-label method - A method of conversion using the multiplication of fractions (9)

Fahrenheit - A temperature scale defined so that water freezes at 32° and boils at 212° (41)

Food calorie (Cal) - 1,000 chemistry calories (cal) (48)

Formation reaction - A reaction that starts with two or more elements and produces one compound (137)

Freezing - The process by which a substance changes from its liquid phase to its solid phase (109)

Freezing-point depression - When a solute is dissolved in a solvent, the freezing point of the solution will be lower than that of the pure solvent (370)

Frequency - The number of wave crests (or troughs) that pass a given point each second (217)

Galvanic cell - An electricity-producing cell that runs on a redox reaction (538)

Gay-Lussac's Law - The stoichiometric coefficients in a chemical equation relate the volumes of gases in the equation as well as the number of moles of substances in the equation. (174)

Graduated cylinder - A device used for measuring the volume of liquids (18)

Gram - The metric unit of mass (4)

Ground state - The lowest possible energy state for a given substance (232)

Heat - Energy that is transferred as a consequence of temperature differences (38)

Hess's Law - Enthalpy is a state function and is therefore independent of path. (429)

Heterogeneous catalysts - Catalysts that are in a different phase than the reactants (481)

Heterogeneous mixture - A mixture with a composition that is different depending on what part of the sample you are observing (103)

Homogeneous catalysts - Catalysts that have the same phase as at least one of the reactants (481)

Homogeneous mixture - A mixture with a composition that is always the same no matter what part of the sample you are observing (103)

Homonuclear diatomic - A molecule composed of two identical atoms (115)

Hypothesis - An educated guess that attempts to explain observations (39)

Ideal gas - A gas whose molecules (or atoms) are very small compared to the total volume of the gas, make elastic collisions with one another as well as the walls of the container, and have no attraction between one another (394)

Incomplete combustion reaction - A reaction in which O_2 is added to a compound containing carbon and hydrogen, producing either CO or C, along with H_2O (140)

Indicator - A substance that turns one color in the presence of acids and another color in the presence of bases (320)

Insoluble - Unable to dissolve (355)

Intermediate state - A complex formed in the process of a chemical reaction when the reactants and products are interacting (436)

Ion - An atom that has gained or lost electrons and thus has become electrically charged (253)

Ionic compound - A compound formed by ions (87)

Ionization - The process by which an atom turns into an ion by gaining or losing electrons (258)

Ionization potential - The amount of energy needed in order to take an electron away from an atom (258)

Isotopes - Atoms with the same number of protons but different numbers of neutrons (206)

Isotopic enrichment - The process by which the abundance of one isotope in an element is increased. This is typically used in order to make the fuel for nuclear bombs. (208)

Joule - The metric unit for energy (40)

Kelvin - The absolute temperature scale: It is not possible to reach or go below 0 Kelvin. (41)

Kinetic energy - Energy that is in motion (40)

Kinetic Theory of Matter - The theory that the atoms or molecules which make up a substance are in constant motion, and the higher the temperature, the greater their speed (111)

Le Chatelier's Principle - When a stress (such as a change in concentration, pressure, or temperature) is applied to an equilibrium, the reaction will shift in a way that relieves the stress and restores equilibrium. (505)

Lead-acid battery - A battery that runs on the oxidation of lead and the reduction of lead(IV) oxide in the presence of sulfuric acid (542)

Lewis structure - A schematic representation of the valence electrons in an atom or molecule (249)

Limiting reactant - The reactant that runs out first in a chemical reaction. It determines the amount of products made. (167)

Liter - A metric unit of volume (5)

Mass - A measure of the amount of matter in an object (4)

Mass number - The total number of neutrons and protons in an atom (207)

Matter - Anything that has mass and takes up space (1)

Melting - The process by which a substance changes from its solid phase to its liquid phase (109)

Melting point - The temperature at which a substance changes from its solid phase to its liquid phase (112)

Meniscus - The curved surface of a liquid, typically in a glass container (18)

Metal - An element that tends to give up its electrons. Metals are found on the left side of the jagged line on the Periodic Table of Elements, with the exception of hydrogen. (77)

Meter - The metric unit of length (5)

Mixture - A substance that contains different compounds and/or elements (99)

Model - A constructed image of something that we cannot see with our eyes (209)

Molality (m) - The number of moles of solute per kilogram of solvent (367)

Molar mass - The mass of one mole of a given compound (187)

Molarity (M) - A concentration unit that tells how many moles of a substance are in a liter of solution. It is determined by taking the number of moles of a substance and dividing by the number of liters of solution. (334)

Mole - A group of atoms or molecules that number 6.02×10^{23} (145)

Mole fraction - The number of moles of a given component in a mixture divided by the total number of moles in the mixture (399)

Molecular formula - A chemical formula that provides the number of each type of atom in a molecule (186)

Molecular mass - The mass of a single molecule (144)

Molecule - More than one atom bound together to form a compound (85)

Neutralization reaction - A reaction between an acid and base that neutralizes both, typically forming salt and water (331)

Neutron - One of the three particles that make up the atom. It is electrically neutral and is in the nucleus of the atom (205)

Newton - The metric unit of force (5)

Nonmetal - An element that tends to take electrons from other elements. Non-metals are found on the right side of the jagged line on the Periodic Table of Elements. Hydrogen is also a non-metal, even though it is on the left side of the jagged line. (77)

Nucleus - The center of the atom that contains the neutrons and protons (212)

Octet rule - Most atoms strive to attain eight valence electrons. (249)

Orbital - A specific shape that confines the position of an electron relative to the nucleus (228)

Oxidation - The process by which an atom loses electrons (531)

Oxidation number - The charge that an atom in a molecule would develop if the most electronegative atoms in the molecule took the shared electrons from the less electronegative atoms (525)

Partial pressure - The pressure exerted by a single gas in a mixture of gases (395)

Particle/Wave Duality Theory - The theory that light sometimes behaves as a particle and sometimes behaves as a wave (213)

Periodic property - A characteristic of atoms that varies regularly across the periodic chart (259)

pH scale - A scale that measured the acidity of a solution. The lower the pH, the more acidic the solution. A pH of 7 indicates a neutral solution (516)

Phase - One of three states of matter: solid, liquid, or gas (109)

Phase change - The process by which a substance changes from one phase (solid, liquid, or gas) to another phase (solid, liquid, or gas) (109)

Photon - A “particle” of light (213)

Physical change - A change in which the atoms or molecules in a substance stay the same (106)

Physical constant - A measurable quantity in nature that does not change (216)

Planck’s constant - The physical constant that relates the energy of light to its frequency: 6.63×10^{-34} J/Hz (221)

Planetary model - Another name for the Rutherford model (211)

Plum pudding model - A model that said the atom is made of a positive gel (the pudding) with negative particles (the plums) suspended in the gel (209)

Polar bond - A covalent bond in which the electrons are shared unequally between the atoms involved (299)

Polar covalent molecule - A covalent molecule that has fractional charges on some or all of its atoms (303)

Polyatomic ions - Ions which are formed when a group of atoms gains or loses electrons (285)

Polyprotic acid - An acid that can donate more than one H^+ ion (329)

Potential energy - Energy that is stored (40)

Precipitation - The process by which a solid solute leaves a solution and turns back into its solid phase (360)

Precision - An indication of the scale on the measuring device that was used (19)

Pressure - The force per unit area exerted on an object (383)

Products - The substances found on the right side of a chemical equation (116)

Proton - One of the three particles that make up the atom. It is positively charged and is in nucleus of the atom (205)

Proton acceptor - Another name for a base: Bases accept H^+ ions, which are protons. (322)

Proton donor - Another name for an acid: Acids donate H^+ ions, which are protons. (322)

Pure substance - A substance that contains only one element or compound (99)

Purely covalent bond - A covalent bond in which the electrons are shared equally between the atoms involved (298)

Purely covalent molecule - A covalent molecule that has no fractional charges on any of its atoms (303)

Pyramidal - A molecular shape in which a central atom is surrounded by 3 bonds which form a pyramid. The bonds are 107 degrees apart. (292)

Quantum assumption - The assumption that a physical quantity (such as energy) cannot have any value, but is restricted to have only discrete values (226)

Quantum mechanical model - The modern-day model of the atom in which electrons whirl around the nucleus in various paths called "orbitals." (228)

Reactants - The substances found on the left side of a chemical equation (116)

Reaction mechanism - A series of chemical equations that tells you the step-by-step process by which a chemical reaction occurs (483)

Reaction rate - The rate of change of a product in a chemical reaction (459)

Reduction - The process by which an atom gains electrons (531)

Rods - The cells on the eye's retina that detect low levels of light (223)

Rutherford model - A model that said the atom is made of a dense, postively-charged nucleus with electrons orbiting the nucleus in circles (211)

Saturated solution - A solution in which the maximum amount of solute has been dissolved (359)

Scientific law - A description of the natural world that has been confirmed by an enormous amount of data (39)

Significant figure - A digit in a measurement that is either non-zero, a zero that is between two significant figures, or a zero at the end of the number and to the right of the decimal (21)

Slug - The English unit of mass (5)

Solubility - The maximum amount of solute that can dissolve in a given amount of solvent (358)

Solute - The substance being dissolved in order to make a solution (353)

Solution - The result of one or more solutes being dissolved in a solvent (353)

Solvent - The substance in which the solute is being dissolved in order to make a solution (353)

Specific heat - The amount of heat necessary to raise the temperature of 1 gram of a substance by 1 degree Celsius (49)

Spectrometer - A device that analyzes light emitted or absorbed by a substance (225)

Spectroscopy - The process by which individual wavelengths of light emitted by a substance are analyzed. This process can be used to identify the elements in a substance. (225)

Standard enthalpy of formation - The ΔH of a formation reaction at $T = 298$ K and $P = 1.00$ atm (430)

Standard Temperature and Pressure (STP) - A temperature of 273 K and a pressure of 1.00 atm (395)

State function - Any quantity that depends solely on the final destination, not on the way you get to that destination (429)

Stoichiometric coefficients - The numbers that appear to the left of the chemical formulas in a chemical equation. They represent the number of moles of each substance. (173)

Stoichiometry - The process by which the amount of one substance in a chemical reaction is related to the amount of another substance in a chemical reaction (156)

Sublimation - The process by which a solid turns directly into a gas, without going through the liquid phase (131)

Tetrahedron - A molecular shape in which a central atom is surrounded by 4 bonds, each of which is 109 degrees apart from the others (291)

The First Law of Thermodynamics - Energy cannot be created or destroyed. It can only change form. (40)

The Law of Definite Proportions - The proportion of elements in any compound is always the same. (78)

The Law of Mass Conservation - Matter cannot be created or destroyed; it can only change forms. (70)

The Law of Multiple Proportions - If two elements combine to form different compounds, the ratio of masses of the *second* element that react with a fixed mass of the *first* element will be a simple, whole-number ratio. (82)

The Second Law of Thermodynamics - The entropy of the universe must always either increase or remain the same. It can never decrease. (440)

Theory - A hypothesis that has been confirmed by experimental data (39)

Titration - The process of slowly reacting a base of unknown concentration with an acid of known concentration (or vice versa) until just enough acid has been added to react with all of the base. This process determines the concentration of the unknown base (or acid). (341)

Transition metal - An element that rests in the "d-orbital" block of the Periodic Table of Elements (257)

Trigonal - A molecular shape in which a central atom is surrounded by 3 bonds, each of which is 120 degrees apart from the others (292)

Triple bond - A total of six electrons shared between atoms (270)

Triprotic acid- An acid that can donate three H^+ ions (329)

Valence electrons - The electrons that exist farthest from an atom's nucleus. They are generally the electrons with the highest energy level number. (248)

Vapor pressure - The pressure exerted by the vapor which sits on top of any liquid (396)

Visible spectrum - The range of light wavelengths that are visible to the human eye (215)

VSEPR theory - Valence Shell Electron Pair Repulsion theory, which states that molecules will attain whatever shape keeps the valence electrons of the central atom as far apart from one another as possible (291)

Wavelength - The distance between the crests (or troughs) of a wave (214)

Work - The force applied to an object times the distance that the object travels parallel to that force (37)

APPENDIX A
TABLES & FORMULAE

TABLE 1.1
Physical Quantities and Their Base Units

Physical Quantity	Base Metric Unit	Base English Unit
Mass	gram (g)	slug (sl)
Distance	meter (m)	foot (ft)
Volume	liter (L)	gallon (gal)
Time	second (s)	second (s)

TABLE 1.2
Common Prefixes Used in the Metric System

PREFIX	NUMERICAL MEANING
micro (μ)	0.000001
milli (m)	0.001
centi (c)	0.01
deci (d)	0.1
deca (D)	10
hecta (H)	100
kilo (k)	1,000
Mega (M)	1,000,000

TABLE 1.3
Relationships Between English and Metric Units

Measurement	English/Metric Relationship
Distance	1 inch = 2.54 cm
Mass	1 slug = 14.59 kg
Volume	1 gallon = 3.78 L

TABLE 2.1
Specific Heats of Common Substances

Substance	Specific heat $\left(\frac{J}{g \cdot {}^\circ C}\right)$
Copper	0.3851
Iron	0.4521
Glass	0.8372
Aluminum	0.9000

TABLE 3.1
Prefixes for Naming Covalent Compounds

Prefix	Meaning	Prefix	Meaning
mono	one	**hexa**	six
di	two	**hepta**	seven
tri	three	**octa**	eight
tetra	four	**nona**	nine
penta	five	**deca**	ten

TABLE 9.1
Important Polyatomic Ions

Ion Name	Formula	Ion Name	Formula
ammonium (uh moh' nee uhm)	NH_4^+	cyanide (sigh' uh nide)	CN^-
hydroxide (hye drox' ide)	OH^-	carbonate (kar' bun ate)	CO_3^{2-}
chlorate (klor' ate)	ClO_3^-	chromate (krohm' ate)	CrO_4^{2-}
chlorite (klor' ite)	ClO_2^-	dichromate (dye krohm' ate)	$Cr_2O_7^{2-}$
nitrate (nye' trate)	NO_3^-	sulfate (suhl' fate)	SO_4^{2-}
nitrite (nye' trite)	NO_2^-	sulfite (suhl' fite)	SO_3^{2-}
acetate (as' uh tate)	$C_2H_3O_2^-$	phosphate (fahs' fate)	PO_4^{3-}

The Visible Spectrum (1 nm = 10^{-9} m)

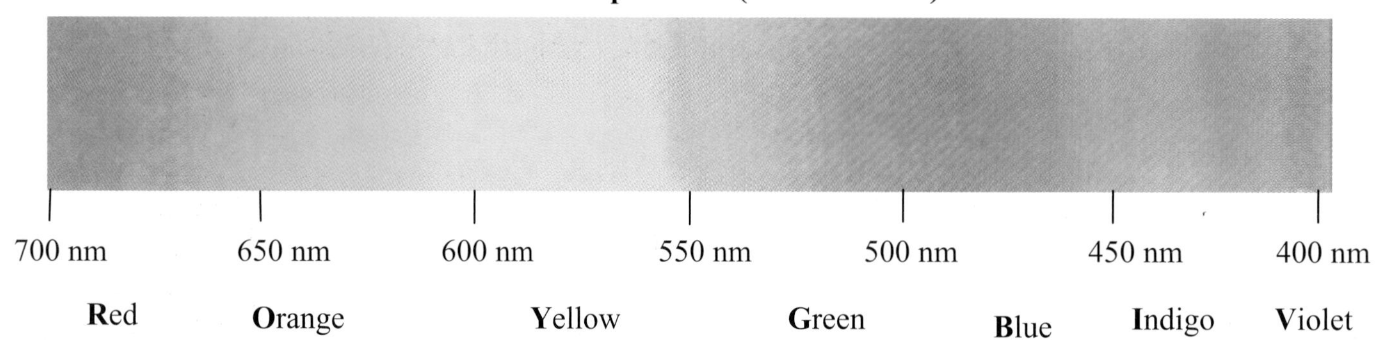

The Electromagnetic Spectrum

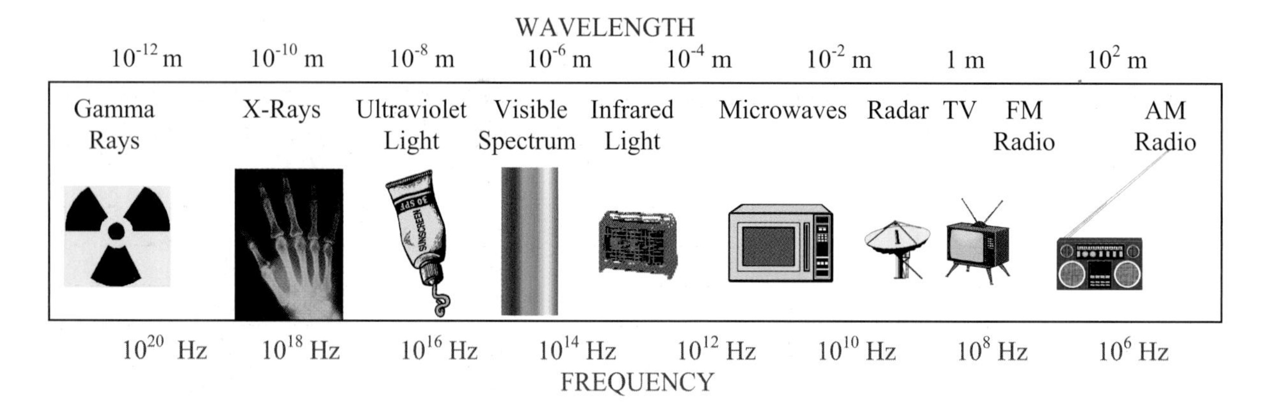

TABLE 10.1
Some Common Acids

Formula	Name	Formula	Name
HCl	hydrochloric (hy' droh klor' ik) acid	H_2SO_4	sulfuric (sul fyur' ik) acid
HBr	hydrobromic (hy' droh brohm' ik) acid	H_3PO_4	phosphoric (fos for' ik) acid
HF	hydrofluoric (hy' droh flor' ik) acid	H_2CO_3	carbonic (kar bon' ik) acid
HNO_3	nitric (nye' trik) acid	$C_2H_4O_2$	acetic (uh see' tik) acid

TABLE 11.1
A Summary of How Solutes Dissolve in Liquid Solvents

SOLUTE PHASE	REQUIREMENTS FOR DISSOLVING
SOLID	The solvent molecules must be attracted to the solute molecules so strongly that the solvent molecules can get between the solute molecules (or ions) and pull them far apart from each other.
LIQUID	The solvent molecules need only be attracted to the solute molecules a little, because the solvent does not need to separate the solute molecules very much. The solvent merely needs to get between the solute molecules.
GAS	The solvent molecules must be attracted to the solute molecules enough to pull the solute molecules closer to one another.

TABLE 12.1
The Vapor Pressure of Water

Temperature (°C)	Vapor Pressure (torr)	Temperature (°C)	Vapor Pressure (torr)
20	17.5	29	30.0
21	18.7	30	31.8
22	19.8	40	55.3
23	21.1	50	92.5
24	22.4	60	149.4
25	23.8	70	233.7
26	25.2	80	355.1
27	26.7	90	525.8
28	28.3	100	760.0

TABLE 13.1
Bond Energies

Chemical Bond	Bond Energy (kJ/mole)	Chemical Bond	Bond Energy (kJ/mole)
C-H	411	H-H	432
C-Cl	327	H-O	459
C-C	346	H-Cl	428
C-N	305	O=O	494
C-O	358	N≡N	942
C=C	602	N-H	386
C=O	799	Cl-Cl	240
C≡C	835	Br-Br	190

TABLE 13.2
Standard Enthalpies of Formation (in kJ/mole)

Substance	ΔH_f^o	Substance	ΔH_f^o	Substance	ΔH_f^o	Substance	ΔH_f^o
H_2O (g)	-242	NO_2 (g)	33.2	$Ca(OH)_2$ (s)	-987	C_2H_6O (l)	-278
H_2O (l)	-286	NH_3 (g)	-45.9	$CaCO_3$ (s)	-1207	C_2H_4 (g)	52.5
SO_3 (g)	-396	CO_2 (g)	-394	C_2H_6 (g)	-84.7	C_4H_{10} (g)	-126
NO (g)	90.3	CaO (s)	-635	CH_3OH (l)	-239	H_2SO_4 (l)	-814

TABLE 13.3

Standard Absolute Entropies (in $\frac{J}{mole \cdot K}$)

Substance	S^o	Substance	S^o	Substance	S^o	Substance	S^o
H_2O (g)	189	NO_2 (g)	240	$Ca(OH)_2$ (s)	76.1	C_2H_6O (l)	161
H_2O (l)	69.9	NH_3 (g)	193	$CaCO_3$ (s)	92.9	C_2H_4 (g)	219
SO_3 (g)	257	CO_2 (g)	214	C_2H_6 (g)	230	C_4H_{10} (g)	260
NO (g)	211	CaO (s)	38.2	CH_3OH (l)	127	H_2SO_4 (l)	157

TABLE 13.4

Standard Gibbs Free Energies of Formation (in $\frac{kJ}{mole}$)

Substance	ΔG_f^o	Substance	ΔG_f^o	Substance	ΔG_f^o	Substance	ΔG_f^o
CH_4 (g)	-50.8	NO_2 (g)	240	$CaSO_4$ (s)	-1320	C_2H_6O (l)	-175
C_6H_6 (l)	125	H_2O (l)	-237	$CaCO_3$ (s)	-1129	$CaCl_2$ (s)	-750
CO (g)	-137	H_2O (g)	-229	H_2CO_3 (aq)	-623	HCl (aq)	-131
CO_2 (g)	-394	CaO (s)	38.2	CH_3OH (l)	-166	$Ca(OH)_2$ (s)	-897

SCIENTIFIC LAWS

Boyle's Law:

$$PV = \text{constant, or, } P_1V_1 = P_2V_2$$

Charles' Law:

$$\frac{V}{T} = \text{constant , or, } \frac{V_1}{T_1} = \frac{V_2}{T_2}$$

Combined Gas Law:

$$\frac{PV}{T} = \text{constant , or, } \frac{P_1V_1}{T_1} = \frac{P_2V_2}{T_2}$$

Dalton's Law of Partial Pressures:

$$P_T = P_1 + P_2 + P_3 + \ldots \text{, or, } P_1 = X_1P_T$$

Ideal Gas Law:

$$PV = nRT\text{, where } R = 0.0821 \frac{L \cdot atm}{mole \cdot K}$$

FORMULAE

Density (ρ)	$\rho = \frac{m}{V}$
Temperature Conversions	$^\circ C = \frac{5}{9}(^\circ F - 32)$ $K = {}^oC + 273.15$
Heat (q)	$q = m \cdot c \cdot \Delta T$
Frequency (f)	$f = \frac{c}{\lambda}$
Energy (E)	$E = h \cdot f$
Molarity (M)	$M = \frac{\#\text{moles}}{\#\text{ liters}}$
Dilution Equation	$M_1V_1 = M_2V_2$
Molality (m)	$m = \frac{\#\text{ moles solute}}{\#\text{ kg solvent}}$
Freezing Point Depression	$\Delta T = -i \cdot K_f \cdot m$
Boiling Point Elevation	$\Delta T = i \cdot K_b \cdot m$
Pressure (P)	$P = \frac{F}{A}$
Mole Fraction (X)	$X = \frac{\#\text{of moles of component}}{\text{total number of moles in the mixture}}$
Change in Enthalpy (ΔH)	$\Delta H = \Sigma \Delta H_f(\text{products}) - \Sigma \Delta H_f(\text{reactants})$
Change in Entropy (ΔS)	$\Delta S = \Sigma S(\text{products}) - \Sigma S(\text{reactants})$
Gibbs Free Energy (ΔG)	$\Delta G = \Delta H - T\Delta S$, or $\Delta G = \Sigma \Delta G_f(\text{products}) - \Sigma \Delta G_f(\text{reactants})$
Reaction Rate	$R = \frac{\Delta[\text{product}]}{\Delta t} = -\frac{\Delta[\text{reactant}]}{\Delta t}$
Rate Equation	$R = k[A]^x[B]^y$
Equilibrium Constant (K)	$K = \frac{[C]_{eq}^{\ c}[D]_{eq}^{\ d}}{[A]_{eq}^{\ a}[B]_{eq}^{\ b}}$

LIST OF ELEMENT NAMES AND SYMBOLS

The list is compiled in the order that they appear on the chart

Name	Symbol
Hydrogen	H
Helium	He
Lithium	Li
Beryllium	Be
Boron	B
Carbon	C
Nitrogen	N
Oxygen	O
Fluorine	F
Neon	Ne
Sodium	Na
Magnesium	Mg
Aluminum	Al
Silicon	Si
Phosphorus	P
Sulfur	S
Chlorine	Cl
Argon	Ar
Potassium	K
Calcium	Ca
Scandium	Sc
Titanium	Ti
Vanadium	V
Chromium	Cr
Manganese	Mn
Iron	Fe
Cobalt	Co
Nickel	Ni
Copper	Cu
Zinc	Zn
Gallium	Ga
Germanium	Ge
Arsenic	As
Selenium	Se
Bromine	Br
Krypton	Kr
Rubidium	Rb
Strontium	Sr
Yttrium	Y
Zirconium	Zr
Niobium	Nb
Molybdenum	Mo
Technetium	Tc
Ruthenium	Ru
Rhodium	Rh
Palladium	Pd
Silver	Ag
Cadmium	Cd
Indium	In
Tin	Sn
Antimony	Sb
Tellurium	Te
Iodine	I
Xenon	Xe
Cesium	Cs
Barium	Ba
Lanthanum	La
Cerium	Ce
Praseodymium	Pr
Neodymium	Nd
Promethium	Pm
Samarium	Sm
Europium	Eu
Gadolinium	Gd
Terbium	Tb
Dysprosium	Dy
Holmium	Ho
Erbium	Er
Thulium	Tm
Ytterbium	Yb
Lutetium	Lu
Hafnium	Hf
Tantalum	Ta
Tungsten	W
Rhenium	Re
Osmium	Os
Iridium	Ir
Platinum	Pt
Gold	Au
Mercury	Hg
Thallium	Tl
Lead	Pb
Bismuth	Bi
Polonium	Po
Astatine	At
Radon	Rn
Francium	Fr
Radium	Ra
Actinium	Ac
Thorium	Th
Protactinium	Pa
Uranium	U
Neptunium	Np
Plutonium	Pu
Americium	Am
Curium	Cm
Berkelium	Bk
Californium	Cf
Einsteinium	Es
Fermium	Fm
Mendelevium	Md
Nobelium	No
Lawrencium	Lr
Rutherfordium	Rf
Dubnium	Db
Seaborgium	Sg
Bohrium	Bh
Hassium	Hs
Meitnerium	Mt
Ununnilium	Uun
Unununium	Uuu
Ununbium	Uub

APPENDIX B
EXTRA PRACTICE PROBLEMS

EXTRA PRACTICE PROBLEMS FOR MODULE #1

1. Two students measure the mass of an object. The teacher knows that the mass is 4.5 g. The first student uses one scale and reports a mass of 4.4 g. The second student uses a different scale and reports a mass of 4.210 g. Which student is more precise? Which is more accurate?

2. How many cg are in 13.1 g?

3. If an object has a volume of 45.1 mL, what is its volume in liters?

4. If a road is 13.1 miles long, how many inches long is it? (1 foot = 12 inches, 1 mile = 5.280×10^3 feet).

5. The volume of a box is given by the equation:

$$V = \text{length} \cdot \text{width} \cdot \text{height}$$

If a box measures 1.2 m by 3.4 m by 0.50 m, what is its volume in liters?

6. Convert the number 0.00341 to scientific notation.

7. Convert the number 1.18×10^7 into decimal notation.

8. How many significant figures are in the number 4500?

9. How many significant figures are in the number 0.0405?

10. How many significant figures are in the number 0.04500?

11. If the mass of a liquid is 23.13 grams and its volume is 35.0 mL, what is its density?

12. The density of a metal is 2.12 grams per mL. What is the volume of 45 g of the metal?

13. What is the mass of 0.67 liters of gold? The density of gold is 19.3 g/mL.

EXTRA PRACTICE PROBLEMS FOR MODULE #2

1. Convert 34.5 °F into °C.

2. Convert 250.0 °F into K.

3. A liquid has a specific heat of 2.4 $\frac{J}{g \cdot {}^{o}C}$. A 75-gram sample of the liquid is initially at 25.0 °C. If 1625 J of energy are removed from the liquid, what is the new temperature?

4. A total of 506 J of heat are added to 50.0 g of water initially at 15.0 °C. What is the final temperature of the water?

5. How much heat is necessary to raise the temperature of a 15.0 g metal from 15.0 °C to 25.0 °C? ($c_{metal} = 1.91 \frac{J}{g \cdot {}^{o}C}$)

6. A 500.0-g mass of unknown metal at 100.0 °C is dropped in a calorimeter initially at 25.0 °C. There are 120.0 g of water in the calorimeter. The final temperature of the water, calorimeter, and metal is 37.0 °C. What is the c of the metal? You can ignore the calorimeter.

7. A 250.0-g mass of metal at 112.4 °C is dropped in a calorimeter (c = 1.40$\frac{J}{g \cdot {}^{o}C}$, m = 5.0 g) containing 150.0 g of water. The water and calorimeter are initially at 25.0 °C. At the end of the experiment, the temperature of the water, calorimeter, and metal is 37.0 °C. What is the c of the metal?

8. A 50.0 g chunk of metal (c = 0.506 $\frac{J}{g \cdot {}^{o}C}$) is heated to 60.0 °C. It is then dropped into a calorimeter (c = 1.40$\frac{J}{g \cdot {}^{o}C}$, m = 5.0 g) that contains 100.0 g of an unknown liquid. The temperature of the liquid raises from 15.0 °C to 20.0 °C. What is the c of the liquid?

9. A 150.0 g piece of metal (c = 0.560 $\frac{J}{g \cdot {}^{o}C}$) is dropped into a calorimeter which contains 40.0 g of water. The metal was initially at 55.0 °C and the calorimeter was initially at 25.0 °C. The mass of the calorimeter is 5.0 g. If the temperature of the calorimeter increased to 29.0 °C, what is the c of the calorimeter?

10. A 75.0-g sample of copper (c = 0.3851 $\frac{J}{g \cdot {}^{o}C}$) is initially at 100.0 °C. It is dropped into a calorimeter (c = 1.40$\frac{J}{g \cdot {}^{o}C}$, m = 5.0 g) that is initially at 25.0 °C. The final temperature of the experiment is 27.5 °C. What mass of water is in the calorimeter?

EXTRA PRACTICE PROBLEMS FOR MODULE #3

1. How can you experimentally determine whether a compound is ionic or covalent?

2. How can you determine from the periodic chart whether a compound is ionic or covalent?

3. Which of the following molecules are covalent?

   a. SO_2 b. $MgCl_2$ c. NaCl d. C_2H_4O

4. Which of the following molecules are ionic?

   a. NO_2 b. $RbNO_3$ c. PH_3 d. Na_2SO_4

5. In making calcium chloride, 100.0 g of calcium plus 100.0 g of chlorine makes 156.5 g of product along with some left over calcium. How much chlorine and calcium should be added together in order to make 1.000 kg of calcium chloride without any left overs?

6. A chemist adds 50.0 g of nitrogen to 50.0 grams of hydrogen. This ends up making 60.7 g of ammonia and some leftover hydrogen. What masses of nitrogen and hydrogen should be reacted in order to make 100.0 g of ammonia with no leftovers?

7. If you react 12.0 grams of carbon with 4.0 grams of hydrogen, you will make methane with no leftovers. If a chemist reacts 50.0 grams of carbon with 50.0 g of hydrogen, how many grams of which element will be left over after the methane is made?

8. In making aluminum sulfide, 54.0 g of aluminum and 100.0 grams of sulfur make 150.3 g of aluminum sulfide and leftover sulfur. How much aluminum and how much sulfur should be used to make 100.0 g of aluminum sulfide with no leftovers?

9. Give the chemical formula for each of the following compounds:

   a. diphosphorus hexaoxide b. dinitrogen tetrahydride c. nitrogen monoxide

10. Give the names of the following compounds:

    a. SO_2 b. $CaCl_2$ c. OCl_2 d. Na_2S

EXTRA PRACTICE PROBLEMS FOR MODULE #4

1. If a liquid goes through a phase change and all you know is that the molecules sped up and moved farther apart, what phase did the liquid turn in to?

2. Classify the following as an element, compound, homogeneous mixture, or heterogeneous mixture:

   a. A bowl of milk and cereal
   b. A bottle of H_2SO_4
   c. A copper wire
   d. Kool-Aid dissolved in water

3. Classify the following as physical or chemical changes:

   a. A window is broken.
   b. Eggs, flour, sugar, milk, and yeast are baked into bread.
   c. Water boils.
   d. Gasoline is burned in an automobile engine.

4. What is the chemical formula of chlorine gas?

5. Balance the following equation:

$$HCl\ (aq) + Ca\ (s) \rightarrow CaCl_2\ (aq) + H_2\ (g)$$

6. Balance the following equation:

$$HBr\ (aq) + Na\ (s) \rightarrow NaBr\ (s) + H_2\ (g)$$

7. Balance the following equation:

$$AgCl\ (aq) + Zn\ (s) \rightarrow ZnCl_2\ (aq) + Ag\ (s)$$

8. Balance the following equation:

$$MgCl_2\ (aq) + H_2S\ (aq) \rightarrow MgS\ (s) + HCl\ (aq)$$

9. Balance the following equation:

$$C_5H_{12} + O_2 \rightarrow CO_2 + H_2O$$

10. Balance the following equation:

$$CO_2\ (g) + H_2O\ (l) \rightarrow C_{18}H_{32}O_{16}\ (s) + O_2\ (g)$$

EXTRA PRACTICE PROBLEMS FOR MODULE #5

1. Balance the following equation:

$$FeCl_3 + Na_2CO_3 \rightarrow Fe_2C_3O_9 + NaCl$$

2. Balance the following equation:

$$NH_4Cl + BaO_2H_2 \rightarrow BaCl_2 + NH_3 + H_2O$$

3. Balance the following equation:

$$P_4O_{10} + H_2O \rightarrow H_3PO_4$$

4. Balance the following equation:

$$CoS_2 + O_2 \rightarrow Co_2O_3 + SO_2$$

5. What is the balanced chemical equation for the formation of C_2H_4O?

6. In the formation of C_2H_4O, how many moles of C are required to make 15 moles of C_2H_4O?

7. What is the balanced chemical equation for the combustion of $C_{16}H_{34}$?

8. In the combustion of $C_{16}H_{24}$, how many moles of CO_2 will be made when 13.2 moles of $C_{16}H_{24}$ are burned?

9. What is the mass of a K_2CO_3 molecule? (1.00 amu = 1.66×10^{-24} g)

10. How many moles of H_2CO_3 are in a 125 gram sample of the compound?

11. What is the mass of a $CaCl_2$ sample if it contains 0.172 moles of the compound?

12. A sample of KCl has a mass of 50.0 g. How many moles is that?

13. How many grams are contained in a 4.51 mole sample of CO_2?

EXTRA PRACTICE PROBLEMS FOR MODULE #6

1. Nitric acid is manufactured with the following reaction:

$$3NO_2 + H_2O \rightarrow 2HNO_3 + NO$$

How many grams of nitrogen dioxide are required to produce 5.89×10^3 kg HNO_3 in excess water?

2. Tungsten metal, W, is produced by the following reaction:

$$WO_3 + 3H_2 \rightarrow W + 3H_2O$$

How many grams of tungsten can be made from 1.0×10^6 g of hydrogen in excess WO_3?

3. How many grams of WO_3 would be used in the above reaction under the conditions specified in #2?

4. Carbon tetrachloride is made according to the following reaction:

$$CS_2 + 3Cl_2 \rightarrow CCl_4 + S_2Cl_2$$

How much CS_2 is needed to make 16 kg of CCl_4?

5. To make a phosphorus fertilizer ($CaH_4P_2O_8$), agricultural companies use the following reaction:

$$Ca_3P_2O_8 + 2H_2SO_4 + 4H_2O \rightarrow CaH_4P_2O_8 + 2CaH_4SO_6$$

If 2.50×10^3 grams of $Ca_3P_2O_8$ are reacted with excess H_2SO_4 and H_2O, how many grams of fertilizer can be made?

6. In the following reaction:

$$Mg\ (s) + 2H_2O\ (l) \rightarrow Mg(OH)_2\ (s) + H_2\ (g)$$

What mass of Mg must you use to make 5.6×10^4 g of $Mg(OH)_2$?

7. Which of the following are empirical formulas?

a. SO_3 b. $K_2S_2O_3$ c. $C_6H_{12}O_6$ d. P_2H_4

8. Decomposition of 100.0 g of a dark green powder yields 39.7 g K, 27.9 g Mn, and an unknown amount of oxygen. What is the empirical formula?

9. The mass of the molecule above is 197.1 amu. What is the molecular formula?

10. A compound has an empirical formula of CH_2O. If its molecular mass is 60.0 amu, what is its molecular formula?

EXTRA PRACTICE PROBLEMS FOR MODULE #7

($1 \text{ nm} = 10^{-9}$ m, $c = 3.0 \times 10^{8}$ m/sec, $h = 6.63 \times 10^{-34}$ J/Hz)

1. Give the number of protons, electrons, and neutrons in the following atoms:

   a. ^{23}Na b. ^{40}Ar c. ^{55}Fe d. ^{238}U

2. If you have an yellow light bulb and a blue one, which emits waves with the largest wavelength? Which emits light of higher frequency? Which emits the higher energy light?

3. What is the frequency of light whose wavelength is 425 nm?

4. What is the wavelength of light whose frequency is 6.15×10^{14} Hz?

5. What is the energy of light that has a frequency of 7.05×10^{14} Hz?

6. What is the energy of light that has a wavelength of 1.95×10^{-7} m?

7. What is the wavelength of light that has an energy of 1.50×10^{-18} J?

8. Give the full electron configurations of the following atoms:

   a. Sc b. Cl c. Ca

9. Give the abbreviated electron configurations for the following atoms:

   a. P b. Mo c. Ba

10. What is wrong with the following electron configurations?

    a. $1s^2 2s^6 2p^2$

    b. $1s^2 2s^2 2p^6 3s^2 3p^6 4s^2 3d^8 4p^6 5s^1$

    c. $1s^2 2s^2 2p^6 3s^2 3p^6 4s^2 3d^{10} 4p^6 4d^{10} 5s^2$

EXTRA PRACTICE PROBLEMS FOR MODULE #8

1. Draw the Lewis structures for the following atoms:

   a. Sr b. Ge c. S

2. Give the chemical formulas for the following molecules:

   a. aluminum oxide b. calcium sulfide c. chromium(III) chloride d. potassium nitride

3. Order the following atoms in terms of increasing ionization potential: Sr, Ra, Mg.

4. Which atom has the greatest desire for extra electrons: Ge, Ca, or Br?

5. Order the following in terms of increasing atomic radius: P, Mg, Al, Ar.

6. What is the Lewis structure for NF_3?

7. What is the Lewis structure for CCl_2S?

8. What is the Lewis structure for SiS_2?

9. What is the Lewis structure for CS?

10. What is the Lewis structure for SBr_2?

11. Which of the molecules in Problems #6-10 would be the hardest to break, based on their Lewis structures?

EXTRA PRACTICE PROBLEMS FOR MODULE #9

1. Give the chemical formulas for the following compounds:

a. sodium sulfate   b. magnesium nitrate   c. calcium carbonate   d. aluminum chromate

2. Name the following compounds:

a. $(NH_4)_2S$ b. KNO_3 c. $Mg_3(PO_4)_2$ d. $AlPO_4$

3. Determine the shape of a CS molecule. Give its bond angle and draw a picture of it.

4. Determine the shape of a SiI_4 molecule. Give its bond angle and draw a picture of it.

5. Determine the shape of a NF_3 molecule. Give its bond angle and draw a picture of it.

6. Determine the shape of a CCl_2S molecule. Give its bond angle and draw a picture of it.

7. Determine the shape of a SF_2 molecule. Give its bond angle and draw a picture of it.

8. Identify each of the molecules in Problems #3-7 as polar covalent or purely covalent.

9. Classify each of the following molecules as ionic, polar covalent, or purely covalent:

a. $SiFCl_3$ b. SiS_2 c. AlF_3 d. N_2 e. CCl_4 f. PF_3

10. Which of the substances in problem #9 would you expect to dissolve in water?

EXTRA PRACTICE PROBLEMS FOR MODULE #10

1. Identify the acid and base in the following reaction:

$$H_2CO_3 + H_2O \rightarrow H_3O^+ + HCO_3^-$$

2. Identify the acid and base in the following reaction:

$$KOH + C_2H_6O \rightarrow KC_2H_5O + H_2O$$

3. Identify the acid and base in the following reaction:

$$2HCl + Mg(OH)_2 \rightarrow 2H_2O + MgCl_2$$

4. What is the reaction between HI and $Al(OH)_3$?

5. What is the reaction between H_3PO_4 and NaOH?

6. What is the reaction between H_3PO_4 and the covalent base CH_5N?

7. Give the concentration (in M) of each of the following solutions:

   a. 1.11 moles of HCl are dissolved in enough water to make 1.5 liters of solution.
   b. 23.1 grams of NaOH are dissolved in enough water to make 500.0 mL of solution.
   c. 14.5 grams of H_2CO_3 are dissolved in enough water to make 200.0 mL of solution.

8. A chemist has a stock solution that is 6.78 M HCl. He wants to make 500.0 mL of 1.15 M HCl. What should he do?

9. You need to make 50.0 mL of a 4.5 M solution of KCl. All you can find is a 5.11 M solution of KCl. What do you do?

10. A chemist needs to determine the concentration of an unlabeled bottle of HBr. She titrates 15.0 mL of it with 1.14 M NaOH. It takes 45.0 mL of NaOH to reach the endpoint. What is the concentration of the acid?

11. 50.0 mL of an unknown concentration of $Mg(OH)_2$ is titrated with 1.00 M HCl. The endpoint is reached when 3.51 mL of acid are added. What is the concentration of the $Mg(OH)_2$?

EXTRA PRACTICE PROBLEMS FOR MODULE #11

(K_f for water is 1.86 $\frac{°C}{m}$, K_b for water is 0.512 $\frac{°C}{m}$)

1. You are having trouble dissolving a gas into a liquid. What should you do to increase the solubility of the gas?

2. The following reaction is performed in a lab:

$$3K_2CO_3\ (aq)\ +\ 2Al(NO_3)_3\ (aq) \rightarrow Al_2(CO_3)_3\ (s)\ +\ 6KNO_3\ (aq)$$

If 191 mL of 1.25 M aluminum nitrate is added to an excess of potassium carbonate, how many grams of aluminum carbonate will be produced?

3. Potassium cyanide can be made in the following way:

$$KOH\ (aq)\ +\ HCN\ (aq) \rightarrow H_2O\ (l)\ +\ KCN\ (aq)$$

How many mL of a 1.51 M solution of HCN would be required to make 50.0 grams of potassium cyanide? Assume that there is excess KOH.

4. When copper is mixed with nitric acid, the following reaction occurs:

$$3Cu\ (s)\ +\ 8HNO_3\ (aq) \rightarrow 3Cu(NO_3)_2\ (aq)\ +\ 2NO\ (g)\ +\ 4H_2O\ (l)$$

How many mL of a 3.50 M solution of HNO_3 would be required to completely use up 50.0 grams of solid copper in this reaction?

5. What is the molality of a solution in which 50.0 g of $Mg(NO_3)_2$ are dissolved in 500.0 grams of water?

6. How many grams of $CaCl_2$ would you have to add to 125 g of water to make a solution with a concentration of 2.0 m?

7. If you wanted to protect water from freezing, which compound would accomplish this best, $Ca_3(PO_4)_2$, $Al(NO_3)_3$, or NaCl? Assume that the molality of the solution is the same in each case.

8. What is the freezing point of a solution that is made by dissolving 10.0 grams of KF in 100.0 grams of water?

9. If you want to lower water's freezing point 5.00 °C by adding $CaCl_2$, what must be the molality of the salt solution?

10. What is the boiling point of a solution made by mixing 100.0 g $(NH_4)_2S$ with 750.0 grams of water?

EXTRA PRACTICE PROBLEMS FOR MODULE #12

1. A balloon is filled with air so that its volume is 5.6 L at 25.0 °C. It is then lowered into a vat of liquid nitrogen so that it reaches a temperature of -196.0 °C. Assuming that the pressure stays constant, what is the new volume?

2. A balloon is filled with helium at a temperature of 25.0 °C and a pressure of 755 mmHg so that its volume is 6.65 L. It then rises to an altitude where the temperature is 15.0 °C and the pressure is 625 mmHg. What is its volume at that altitude?

3. A chemist collects 56.7 mL of nitrogen gas at 25 °C and 740 torr. What would the volume of the nitrogen be at STP?

4. A mixture of nitrogen and oxygen gas are stored at a temperature of 25.0 °C and a pressure of 1.0 atm. If the partial pressure of nitrogen in the mixture is 0.7 atm, what is the partial pressure of oxygen in the mixture?

5. A mixture of gases has a pressure of 1.5 atm. The mixture is analyzed and found to contain 0.200 g of H_2, 4.00 g of CO_2, and 1.45 g of N_2. What is the partial pressure of each gas in the mixture?

6. What is the volume of 12.1 grams of nitrogen gas at STP?

7. What is the pressure of 5.0 grams of nitrogen gas when the temperature is 21.0 °C and the gas occupies 1.00 L of space?

8. A chemist performs the following reaction:

$$CS_2\ (s)\ +\ 3O_2\ (g)\ \rightarrow\ CO_2\ (g)\ +\ 2SO_2\ (g)$$

If the chemist starts with 110.0 grams of CS_2 and an excess of O_2, what volume of sulfur dioxide will be produced at a temperature of 341 °C and a pressure of 2.1 atm?

9. It is difficult to make solid titanium, but since it is used in the manufacture of airplane engines and frames, it must be produced. When mined, titanium ore is mostly TiO_2. To make solid titanium, the titanium ore must first be converted to titanium tetrachloride via the following reaction:

$$3TiO_2\ (s)\ +\ 4C\ (s)\ +\ 6Cl_2\ (g)\ \rightarrow\ 3TiCl_4\ (g)\ +\ 2CO_2\ (g)\ +\ 2CO\ (g)$$

How many liters of Cl_2 gas at STP would be required to completely convert 100.0 grams of TiO_2 into $TiCl_4$?

10. Heating table salt and liquid sulfuric acid produces gaseous HCl, a strong acid:

$$NaCl\ (s)\ +\ H_2SO_4\ (l)\ \rightarrow\ NaHSO_4\ (s)\ +\ HCl\ (g)$$

How many liters of HCl (g) will be produced if 100.0 g of H_2SO_4 are heated in excess NaCl? Assume the pressure is 1.00 atm and the temperature is 350.0 °C.

EXTRA PRACTICE PROBLEMS FOR MODULE #13

1. Which of the following substances will have a ΔH_f of zero?

$$KOH\ (aq),\ Cl^-\ (aq),\ CO_2\ (g),\ F\ (g),\ F_2\ (g),\ O_2\ (l),\ O_2\ (g)$$

2. Which of the following diagrams indicates an endothermic reaction? What is the ΔH of that reaction?

I

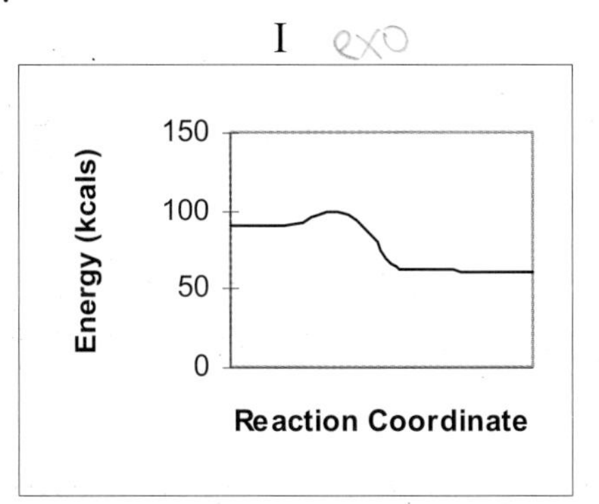

II

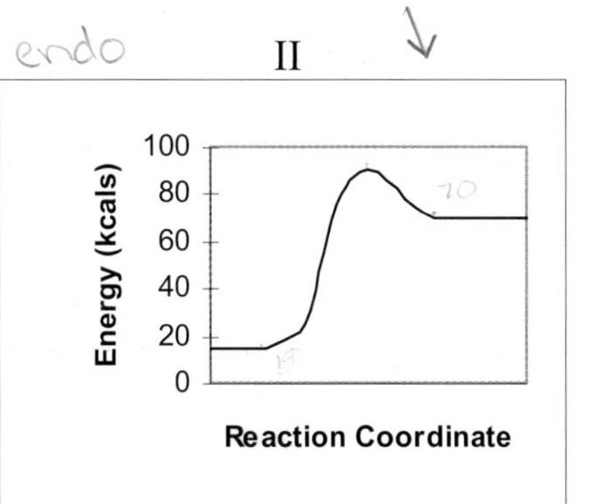

3. For the diagrams above, which represents the reaction with the lowest activation energy?

4. Is it possible for a reaction to occur if the reaction results in a decrease in the entropy of the chemicals involved in the reaction?

5. What is the sign of ΔS for the following reactions?

a. $NaCl\ (aq) + H_2O\ (l) + CO_2\ (g) \rightarrow NaHCO_3\ (aq) + HCl\ (aq)$
b. $NaCl\ (s) + H_2SO_4\ (l) \rightarrow NaHSO_4\ (s) + HCl\ (g)$
c. $4NH_3\ (g) + 5O_2\ (g) \rightarrow 4NO\ (g) + 6H_2O\ (g)$
d. $H_2SO_4\ (aq) + 2NH_3\ (aq) \rightarrow (NH_4)_2SO_4\ (aq)$

6. What is the ΔH of the following reaction? (use bond energies)

$$CO_2 + 4HCl \rightarrow CCl_4 + 2H_2O$$

7. Use Hess' Law to determine the ΔH^o of:

$$4NH_3\ (g) + 5O_2\ (g) \rightarrow 4NO\ (g) + 6H_2O\ (g)$$

8. If the standard absolute entropy of O_2 (g) is 205.0 $\frac{\text{Joules}}{\text{mole} \cdot \text{K}}$, what is the ΔS^o of the reaction in Problem #7?

9. If the ΔH^o of a certain reaction is -623 kJ/mole and the ΔS^o is -114 $\frac{\text{Joules}}{\text{mole} \cdot \text{K}}$, what is the temperature range for which this reaction is spontaneous?

EXTRA PRACTICE PROBLEMS FOR MODULE #14

1. A chemistry book lists the rate constant for a reaction as 34.5 $\frac{1}{\text{M} \cdot \text{s}}$. What is the overall order of the reaction?

2. The order of a chemical reaction with respect to one of its reactants is three. If you double the concentration of that reactant, what happens to the rate?

3. A chemist does a reaction rate analysis on the following reaction:

$$2NO\ (g) + Cl_2\ (g) \rightarrow 2NOCl\ (g)$$

She collects the following data:

Trial	Initial Concentration of NO (M)	Initial Concentration of Cl_2 (M)	Instantaneous Reaction Rate (M/s)
1	0.050	0.050	0.113
2	0.100	0.050	0.452
3	0.100	0.100	0.904

What is the rate equation for this reaction?

4. A chemist does a reaction rate analysis on the following reaction:

$$C_3H_6Br_2 + 3I^- \rightarrow C_3H_6 + 2Br^- + I_3^-$$

He collects the following data:

Trial	Initial Concentration of $C_3H_6Br_2$ (M)	Initial Concentration of I^- (M)	Instantaneous Reaction Rate (M/s)
1	0.100	0.100	0.234
2	0.200	0.100	0.468
3	0.100	0.200	0.468

What is the rate equation for this reaction?

5. A chemist runs a chemical reaction at 25 °C and decides that it proceeds far too slowly. As a result, he decides that the reaction rate must be increased by a factor of 8. At what temperature should the chemist run the reaction to achieve this goal?

6. In the reaction mechanism below, indicate what substance is acting like a catalyst.

Step 1. $2N_2O\ (g) + Cl_2\ (g) \rightarrow 2N_2\ (g) + 2ClO\ (g)$
Step 2. $2ClO\ (g) \rightarrow Cl_2\ (g) + O_2\ (g)$

Is this a heterogeneous or a homogeneous catalyst?

7. What does a catalyst do to the size of the hill between reactants and products in an energy diagram?

EXTRA PRACTICE PROBLEMS FOR MODULE #15

1. Three solutions have the following pH levels:

Solution A: pH = 3 Solution B: pH = 6 Solution C: pH = 12

Which is (are) the acidic solution(s)?

2. Three acid solutions of equal concentration have the following pH levels:

Solution A: pH = 4 Solution B: pH = 6 Solution C: pH = 1

Which solution is made with the acid that has the *largest* ionization constant?

3. A chemist is studying the following equilibrium:

$$NH_4^+ (aq) + CO_3^{2-} (aq) \leftrightharpoons NH_3 (aq) + HCO_3^- (aq)$$

He starts out with NH_4^+ at a concentration of 2.0 M and a CO_3^{2-} concentration of 2.0 M. The concentrations of NH_4^+, CO_3^{2-}, NH_3, and HCO_3^- at equilibrium are 0.80 M, 0.80 M, 1.2 M, and 1.2 M, respectively. What is the value of the equilibrium constant?

4. The following reaction:

$$2SO_2 (g) + O_2 (g) \leftrightharpoons 2SO_3 (g)$$

has an equilibrium constant equal to 4.4 1/M. If the following concentrations are present: $[SO_2]$ = 0.280 M, $[O_2]$ = 0.294 M, $[SO_3]$ = 0.560 M, is the reaction at equilibrium? If not, which way must it shift to reach equilibrium?

5. Consider the following reaction that has reached equilibrium:

$$2NH_3 (g) \leftrightharpoons N_2 (g) + 3H_2 (g) \qquad \Delta H = 92.2 \text{ kJ}$$

a. What will happen to the concentration of H_2 if the temperature is raised?
b. What will happen to the concentration of NH_3 if the pressure is raised?
c. What will happen to the concentration of N_2 if the concentration of NH_3 is lowered?

6. Consider the following reaction that has reached equilibrium:

$$2CO (g) \leftrightharpoons CO_2 (g) + C (s) \qquad \Delta H = -172.5 \text{ kJ}$$

a. What will happen to the concentration of CO if the temperature is raised?
b. What will happen to the concentration of CO if the pressure is raised?
c. What will happen to the concentration of CO if more carbon is added?
d. What will happen to the amount of carbon if more CO is added?

7. What is the equation for the acid ionization constant of NH_4^+?

EXTRA PRACTICE PROBLEMS FOR MODULE #16

1. Give the oxidation numbers of all atoms in the following substances:

a. Mg b. K^+ c. P_4 d. N_2

2. Give the oxidation numbers of all atoms in the following compounds:

a. $KMnO_4$ b. $NaCHO_2$ c. $Zn(OH)_2$ d. SCl_2

3. An atom changes its oxidation number from +2 to -1. Is it oxidized or reduced? How many electrons did it take to do this?

4. An atom changes its oxidation number from 0 to +4. Is it oxidized or reduced? How many electrons did it take to do this?

5. Determine whether or not each of the following reactions is a redox reaction.

a. $H_2CO_3\ (aq) + 2NH_3\ (aq) \rightarrow 2NH_4^+\ (aq) + CO_3^{2-}\ (aq)$
b. $2Mg\ (s) + O_2\ (g) + 2H_2O\ (l) \rightarrow 2Mg^{2+}\ (aq) + 4OH^-\ (aq)$
c. $Zn(NO_3)_2\ (aq) + 2NaOH\ (aq) \rightarrow Zn(OH)_2\ (s) + 2NaNO_3\ (aq)$
d. $2H_2O\ (l) + 2Cl^-\ (aq) \rightarrow H_2\ (g) + Cl_2\ (g) + 2OH^-\ (aq)$

6. For the redox reactions in problem #5, which substance is oxidized? Which is reduced?

7. For the redox reactions in problem #6, how many electrons are transferred in each reaction?

8. A Galvanic cell runs on the following reaction:

$$3Mg\ (s) + 2Al^{3+}\ (aq) \rightarrow 3Mg^{2+}\ (aq) + 2Al\ (s)$$

Draw a diagram for this Galvanic cell, labeling the electron flow, the anode and cathode, and the positive and negative sides of the Galvanic cell.

9. A Galvanic cell runs on the following reaction:

$$F_2\ (aq) + Sn\ (s) \rightarrow 2F^-\ (aq) + Sn^{2+}\ (aq)$$

Draw a diagram for this Galvanic cell, labeling the electron flow, the anode and cathode, and the positive and negative sides of the Galvanic cell.

10. A Galvanic cell runs on the following reaction:

$$2Cr\ (s) + 3Cu^{2+}\ (aq) \rightarrow 2Cr^{3+}\ (aq) + 3Cu\ (s)$$

Draw a diagram for this Galvanic cell, labeling the electron flow, the anode and cathode, and the positive and negative sides of the Galvanic cell.

APPENDIX C
A COMPLETE LIST OF LAB SUPPLIES

Items in boldface, blue type are found in the laboratory equipment set that is sold with the course. The other materials are available at supermarkets, hardware stores, or drug stores.

Module #1

- **Safety goggles**
- A meterstick (A yardstick will work as well; a 12-inch ruler is not long enough.)
- Two 8-inch or larger balloons
- Two pieces of string long enough to tie the balloons to the meterstick
- Tape
- A tall glass
- A paper towel
- A sink full of water
- Book (not oversized)
- Metric/English ruler or rulers
- Water
- Vegetable oil
- **Graduated cylinder** or measuring cups
- Maple syrup (Natural syrup does not work as well as something like Mrs. Butterworth's®.)
- **Mass scale**

Module #2

- **Safety goggles**
- **Thermometer**
- **250 mL beaker** (A small glass container will do, but make sure it is safe for boiling water.)
- Water (preferably distilled water, which can be purchased at any supermarket)
- Ice (preferably crushed)
- **Alcohol burner** or something like a hot plate or stove
- **Mass scale**
- Two Styrofoam cups
- A chunk of metal that has mass of at least 30 grams (a lead sinker or a very large steel nut, for example)
- Boiling water (either in a pot or a beaker)
- Kitchen tongs

Module #3

- **Safety goggles**
- **100-mL beaker** (A juice glass can be used instead.)
- **250-mL beaker** (A juice glass can be used instead.)
- **Watch glass** (A small saucer can be used instead. It must cover the mouth of the 100-mL beaker or juice glass listed above.)
- Teaspoon
- Lye (This is commonly sold in supermarkets with the drain cleaners. A popular brand is Red Devil® Lye. If you cannot find lye, any *powdered* drain cleaner ought to work.)
- White vinegar
- Several leaves of red (often called "purple") cabbage
- Water
- Small pot for boiling water
- Measuring cup
- Stove
- **Mass scale**
- **Stirring rod** (A spoon will work.)
- Rubber cleaning gloves
- Distilled water (available at grocery stores - half a gallon is plenty)
- Baking soda
- Sugar
- 2 pieces of wire (preferably insulated), each of which is at least 15 cm long
- Scissors or wire cutters to strip insulation from wire (if it is insulated)
- Tape (preferably black electrical tape)
- 9-volt battery (***DO NOT*** use an electrical outlet in place of the battery. The energy contained in a wall socket can easily kill you!)

Module #4

- **Safety goggles**
- **Two beakers - 100 mL and 250 mL** (or two glasses - one large and one small)
- Sand (Kitty litter is an acceptable substitute, but don't use the kind that clumps.)
- Table salt
- **Funnel**
- Water
- **Filter paper** (You can cut a circle out of the bottom of a coffee-maker filter.)
- Stirring rod (or small spoon)
- Teaspoon
- Heat source such as a stove or **alcohol burner**
- **Plastic graduated cylinder** (A rain gauge will work as well. A measuring cup may work, but if you have a frost-free freezer, the opening of whatever you use should be very small.)
- Egg in its shell
- Toilet bowl cleaner (The list of ingredients must include hydrochloric acid or hydrogen chloride. The Works® was the brand used when this experiment was tested. You can use vinegar if you cannot find a proper toilet bowl cleaner, but the experiment will have to sit overnight in order for it to work. However, if you *do* use vinegar, the egg will be *much* more interesting to observe!)

- Tall glass
- Spoon
- Rubber cleaning gloves
- Rectangular metal can with a lid (Turpentine and paint thinner are usually sold in such cans. The can must be empty and thoroughly rinsed out.)
- Hot pads
- Two glass canning jars or peanut butter jars, both the same size
- Food coloring (any color)
- A pan and stove to boil water

Module #5

- **Safety goggles**
- **Eyedropper**
- **Stirring rod** (or spoon)
- Water
- Large glass (at least 16 ounces)
- **Graduated cylinder** (Measuring cups and measuring spoons can be used, but they will be less precise.)
- Dishwashing liquid (It must be the kind used for washing dishes by hand, NOT the kind used in automatic dishwashers. Preferably, the brand should be Joy® or Sunlight®.)
- Large bowl (It should have a diameter larger than 10 inches but smaller than 12 inches. If you don't have a bowl in that size range, then use one larger than 12 inches; it will just make it a little harder to use the ruler.)
- Pepper
- Ruler

Module #6

- **Safety goggles**
- Baking soda
- Vinegar
- String or tape measure
- **Graduated cylinder** (or measuring cups and spoons)
- Ruler
- Round balloon
- Plastic 2-liter bottle (or other large bottle)
- **Mass scale**
- **Funnel** or butter knife (Read the experiment to see what is meant.)

Module #7

- **Safety goggles**
- Comb
- Aluminum foil
- Cellophane (Scotch®) tape
- Two plain white sheets of paper (There cannot be lines on them.)
- A bright red marker (A crayon will also work, but a marker is better.)

Module #8

There are no experiments in this module.

Module #9

- **Safety goggles**
- Glass of water
- Vegetable oil
- Styrofoam or paper cup
- Comb
- Pen
- **Two test tubes** (Thin glasses will work, but they must be transparent)
- Table salt

Module #10

- **Safety goggles**
- **Red litmus paper**
- **Blue litmus paper**
- Apple
- Orange juice or soda pop
- Toilet bowl cleaner (Both The Works® and Lime-Away® have been tested, but any toilet bowl cleaner designed to combat lime should work.)
- Bar soap (make sure it doesn't say "pH balanced")
- All-purpose cleaner (Windex® and 409® have been tested, but any spray cleaner not specifically designed for toilets should work.)
- Powdered drain unclogger (like Dran-O® or Red Devil® Lye) or scouring powder (like Comet®).
- **4 test tubes** (small cups will work)
- **Watch glass** (a small saucer will work)
- **Stirring rod**
- Rubber gloves (Gloves are recommended whenever you use powdered drain uncloggers and toilet bowl cleaners, because these chemicals are caustic.)
- **Eyedropper**
- **Mass scale**
- Distilled water (available at any grocery store)
- White sheet of paper (no lines)

- **Stirring rod** (or a small spoon)
- A few leaves of red cabbage (it must be red cabbage, not regular cabbage)
- **2 beakers** (If you don't have beakers, one should be a short, fat glass that is transparent, and the other can be a small pot to boil water in.)
- **Alcohol burner** or stove for heating
- **Graduated cylinder** (Measuring cups and spoons will work, but the experiment will be much harder.)
- Clear ammonia solution (This is sold with the cleaning supplies in most supermarkets. It must be clear. A colored solution will mess up the endpoint.)
- Clear vinegar (Once again, colored vinegar will mess up the endpoint.)

Module #11

- **Safety goggles**
- **250 mL beaker** (A short, fat glass or canning jar might work, but **be careful**. Glasses tend to crack when subjected to the temperature extremes of this experiment. If you don't have a beaker, you may want to just read this experiment, unless you aren't afraid of losing a glass or two.)
- **100 mL beaker** (Another short, fat glass or canning jar might work.)
- **Graduated cylinder** (A ¼ measuring cup will work.)
- **Alcohol burner** (A stove will work.)
- **Stirring rod** (A spoon will work.)
- **Thermometer**
- **Mass scale**
- **Filter paper** (You can cut circles out of the bottom of a coffee filter to make your own filter paper.)
- **Funnel**
- **Two beakers** (Two saucepans will work.)
- **Test tube** (A tall, thin glass will work, but it must fit easily in the saucepans.)
- Ice
- Cold soda pop (Pepsi®, Coke®, Sprite®, etc. It must be carbonated.)
- Lye (This is commonly sold in supermarkets with the drain cleaners. A popular brand is Red Devil® Lye. If you cannot find lye, any *powdered* drain cleaner ought to work.)
- Rubber gloves
- Water
- Sink
- Tablespoon

Module #12

- **Safety goggles**
- **Mass scale**
- Plastic 2-liter bottle
- Round balloon with an 8-inch diameter
- Vinegar
- Baking soda
- Seamstress' tape measure (A piece of string and a ruler will work as well.)
- **Thermometer**
- Weather report that contains the atmospheric (sometimes called barometric) pressure for the day (If you don't have this, don't worry. You can assume that the atmospheric pressure is 1.00 atm.)

Module #13

- **Safety goggles**
- Two Styrofoam coffee cups
- **Thermometer**
- Vinegar
- **Mass scale**
- Measuring tablespoon and ½ teaspoon
- Lye (This is commonly sold in supermarkets with the drain cleaners. A popular brand is Red Devil® Lye. If you cannot find lye, any *powdered* drain cleaner ought to work.)

Module #14

- **Safety goggles**
- Liquid toilet bowl cleaner (The Works® was used when this experiment was tested, but any liquid that contains hydrochloric acid or hydrogen chloride should work. Stay away from the thickened toilet bowl cleaners that cling, however.)
- Antacid tablets (TUMS® works best. In principle, any antacid tablet that has calcium carbonate as its main ingredient should work.)
- Sharp, nonserrated knife
- Spoon
- **Stirring rod** (Something thin enough to stir the contents of the test tubes.)
- **Four test tubes** (Small glass containers will work.)
- **Alcohol burner** (An oven burner will work.)
- **Large beaker** (A short, fat glass will work.)
- **Small beaker** (A short, fat glass will work. If you are not using test tubes, make sure that one of your glass containers fits into this.)
- Rubber gloves
- **Watch glass** (A small saucer that covers the beaker will do.)
- **Graduated cylinder** (A 1/8 measuring cup will work.)
- ¼ measuring teaspoon
- Hydrogen peroxide (available at supermarkets and drug stores)
- Baker's yeast (Any kind of bread yeast, even bread machine yeast, will do.)

Module #15

- **Safety goggles**
- Three plastic two-liter bottles (Plastic milk cartons will work as well.)
- Four small cups
- Two bowls that are taller than the small cups
- Serrated knife (like a steak knife)
- Small, Phillips-head screwdriver
- Water
- Person to help you
- Kitchen counter
- Towels
- **Two test tubes**
- **Two eyedroppers**
- **Two beakers**
- Two small cups
- Clear ammonia solution (This is sold with the cleaning supplies in most supermarkets. It must be clear. A colored solution will mess up the colors you are supposed to see.)
- White vinegar (It must be white. Colored vinegar will mess up the colors you are supposed to see.)
- **Alcohol burner** or stove
- Pot for boiling water on the stove
- Ice (preferably crushed)
- A few leaves of red cabbage (It must be red cabbage, not regular cabbage.)

Module #16

- **Safety goggles**
- Large bowl
- Blank notebook paper without lines
- Scissors
- Q-Tip or small paintbrush
- Lemon juice (This must be *real* lemon juice, either fresh or from concentrate.)
- Iodine solution (This is available in large drugstores. The pharmacy department within your supermarket will probably not have this and neither will a small corner drugstore. A major chain will have it, however. Please note that *iodide* will not work. Iodide, as you should know, is I^-, while iodine is I_2.)
- Water
- Measuring cups
- **Medicine dropper**
- **Stirring rod** (A **plastic** spoon or knife will work as well. Do not use any flatware around the iodine solution, as iodine will rust metal.)

INDEX

-A-

-B-

-C-

-D-

-E-

-F-

-G-

-H-

-I-

-J-

-K-

-L-

-M-

-N-

-O-

-P-

-Q-

-R-

-S-

-T-

-U-

-V-

-W-

-Y,Z-

Conversion Factors

1.00 inch = 2.54 cm

1.000 slug = 14.59 kg

1.00 gallon = 3.78 L

1.00 amu = 1.66×10^{-24} g

1.000 atm = 101.3 kPa

1.000 calorie = 4.184 Joules

1 food calorie (Cal) = 1,000 calories (cal)

760.0 torr = 1.000 atm

760.0 mm Hg = 1.000 atm